CHROMATOGRAPHY

SOLE DISTRIBUTORS FOR THE UNITED STATES OF NORTH AMERICA:

D. Van Nostrand Company, Inc.
120 Alexander Street, Princeton, N.J. (Principal office)
257 Fourth Avenue, New York 10, N.Y.

SOLE DISTRIBUTORS FOR CANADA:

D. Van Nostrand Company (Canada), Ltd.
25 Hollinger Road, Toronto 16, Canada

SOLE DISTRIBUTORS FOR THE BRITISH COMMONWEALTH
EXCLUDING CANADA:

D. Van Nostrand Company, Ltd.
358 Kensington High Street, London, W. 14, England

CHROMATOGRAPHY

A REVIEW OF PRINCIPLES AND APPLICATIONS

by

EDGAR LEDERER

Professor of Biochemistry, Sorbonne; Directeur de Recherches,
Institut de Biologie Physico-Chimique, Paris

and

MICHAEL LEDERER

Maître de Recherches, Institut du Radium, Paris

SECOND, COMPLETELY REVISED AND ENLARGED,
EDITION

ELSEVIER PUBLISHING COMPANY

AMSTERDAM   LONDON   NEW YORK   PRINCETON

1957

First English Edition 1953
First Reprint 1954
Second Reprint 1955
Second English Edition 1957
First Reprint 1958
Second Reprint 1961

PART OF THE FIRST ENGLISH EDITION HAS BEEN
TRANSLATED FROM THE ORIGINAL FRENCH TEXT BY

A. T. JAMES

NATIONAL INSTITUTE FOR MEDICAL RESEARCH,
MILL HILL, LONDON

Library of Congress Catalog Card Number 56-13146

REPRINTED IN THE NETHERLANDS BY
KRIPS REPRINT COMPANY

Preface to the second edition

The unexpected success of the first edition and the ever increasing development of chromatographic methods prompted us to prepare a second edition of this book. It was planned to bring the first edition published in 1953 up to date and where possible to eliminate work which had become obsolete.

To achieve this we have added material to the proofs as recently as September 1956. The new edition contains 50 % more pages, illustrations and R_F tables than the first and twice the number of references.

The most spectacular progress in the last few years has been witnessed in the field of gas-liquid chromatography and it is felt that the relevant chapters of this edition will prove unsatisfactory when this book comes out of press. The authors have therefore added an appendix of references published since this book went to press; these should help the reader to find his way to the most recent developments in this field.

Since the publication of the first edition the following books on chromatography have been published:

Hais and Macek, *Papirova chromatografie (1269)*; Linskens, *Papierchromatographie in der Botanik (2031)*; Pollard and McOmie, *Chromatographic methods of inorganic analysis (2539)*; Schoen, *Analisi cromatografica su carta (2871)*; Smith, *Inorganic chromatography (3038)*; Williams, *The elements of chromatography (3587)*.

Our thanks are due to the following authors for their permission to reproduce figures: W. C. Bauman, J. Beukenkamp, G. L. Brown, W. E. Cohn, A. Deutsch, K. V. Giri, Y. Hashimoto, A. T. James, D. H. James, F. Kaiser, K. A. Kraus, T. K. Lakshmanan, L. S. Lerman, S. Lieberman, W. Matthias, G. R. Noggle, H. W. Patton, E. A. Peterson, E. F. Phares, C. G. S. Phillips, H. K. Prins, J. H. Purnell, W. Rieman, K. Savard, G. T. Seaborg, H. A. Sober, H. Svensson, M. Watson, V. Zbinovsky, L. P. Zill, and to the editors of *Analytical Chemistry, Analyst, Archives of Biochemistry and Biophysics, Acta Chemica Scandinavica, Chemische Berichte, Journal of the American Chemical Society, Journal of Biological Chemistry, Journal of Physical Chemistry, Nature*, and *Naturwissenschaften*.

Paris, February 1957

E. LEDERER
M. LEDERER

Preface to the first edition

Chromatography is an analytical method for the purification and separation of organic and inorganic substances; it is particularly useful for the fractionation of complex mixtures, the isolation of unstable substances and the separation of closely related compounds (isomers, homologues, etc.).

The importance of chromatography in modern chemistry can best be illustrated by the following quotation from P. Karrer's Congress lecture to the International Congress of Pure and Applied Chemistry in London in 1947: "... no other discovery has exerted as great an influence and widened the field of investigation of the organic chemist as much as Tswett's chromatographic adsorption analysis. Research in the field of vitamins, hormones, carotenoids and numerous other natural compounds could never have progressed so rapidly and achieved such great results if it had not been for this new method, which has also disclosed the enormous variety of closely related compounds in nature."

Since then the rapid development of ion exchange and paper chromatography and their application to inorganic compounds has extended still further the usefulness of chromatography.

In speaking of chromatography we intend to use the term in its widest sense. Gordon, Martin and Synge (*1172*) in 1944 defined chromatography as "the technical procedure of analysis by percolation of fluid through a body of comminuted or porous rigid material, irrespective of the nature of the physico-chemical processes that may lead to the separation of substances in the apparatus".

More recently, Martin (*2151*) wrote: "The essence of the chromatogram is the uniform percolation of a fluid through a column of more or less finely divided substance, which selectively retards, by whatever means, certain components of the fluid". Strain* has given a similar definition: "Chromatography is an analytical technique for resolution of solutes, in which separation is made by differential migration in a porous medium, and migration is caused by flow of solvent."

These definitions are very general and cover very different physico-chemical phenomena, such as adsorption, ion exchange and partition between two solvents, which are encountered in separations on columns or on paper.

* Quoted from *Chem. Eng. News*, 1952, *30*, 1372.

In the following pages we distinguish between three types of chromatography: adsorption chromatography, ion exchange chromatography and partition chromatography; it should however be kept in mind that this distinction is sometimes more or less arbitrary and that quite often it is not certain which of the above phenomena is operating in a given separation.

This book represents a review of the chromatographic methods developed in the last 10 to 12 years. It would be quite impossible to list all papers in which chromatography has been employed, since in certain fields (carotenoïds, steroids, etc.) no preparative work is done without the use of chromatography. We have endeavoured to refer to papers which contribute either to the development of new methods or to the application of chromatography to new groups of substances.

Chromatography differs from standard laboratory methods of separation, e.g. fractional distillation, in which more or less the same operations and apparatus are employed, in that a wide variety of techniques, adsorbents and solvents can be used.

Anyone wishing to apply chromatography to a certain category of chemical substances should be able to find in this book information on existing methods in a given field. For more detailed information the original papers should be consulted.

In view of the fact that a book by A. J. P. Martin and A. T. James on the Theory of Chromatography will be published by the Elsevier Publishing Company, we thought it unnecessary to encroach on this field; no detailed discussion of this subject is therefore included in this book.

The present work is a combined and revised edition of two monographs previously published in French*. The enormous impact of chromatography on organic and biochemistry can be seen from the fact that bringing Volume I, published in 1949, up to date involved the inclusion of a further 700 references and of at least 100 pages of text.

References up to July 1952 have been included in the text; at the end of the book the reader will find references to papers published later.

The newcomer to chromatography will find the following books very useful as an introduction to chromatographic technique:

H. H. Strain, *Chromatographic Adsorption Analysis* (1942); L. Zechmeister and L. von Cholnoky, *Principles and Practice of Chromatography* (1943); T. I. Williams: *An Introduction to Chromatography* (1946).

L. Zechmeister's *Recent Progress in Chromatography* contains 3000 references to papers published between 1940 and 1947.

Three recent monographs are devoted solely to paper chromatography: F. Cramer: *Papierchromatographie* (2nd edition 1953); J. Balston and B. E. Talbot: *A Guide to Filter Paper and Cellulose Powder Chromatography* (1952);

---

* *Progrès récents de la Chromatographie.* Première partie: *Chimie Organique et Biologique*, by E. Lederer, 146 pp., Hermann et Cie., Paris, 1949. Deuxième partie: *Chimie Minérale*, by M. Lederer, 131 pp., Hermann et Cie., Paris, 1952.

R. J. Block, R. LeStrange and G. Zweig: *Paper Chromatography*; *A Laboratory Manual* (1952).

H. G. Cassidy's book *Adsorption and Chromatography* (1951) gives a detailed account of modern knowledge of adsorption phenomena.

Among the great number of shorter review articles on chromatography the following may be mentioned: Bosch (*322*), Moore and Stein (*2269*), Strain and Murphy (*3146*).

Acknowledgements

The authors express their grateful acknowledgements to the following for their kind permission to reproduce illustrations: A. W. Adamson, S. Berlingozzi, E. M. Bickoff, P. Boulanger, J. C. Boursnell, G. E. Boyd, I. Brattsten, I. E. Bush, M. Calvin, C. E. Carter, H. G. Cassidy, I. L. Chaikoff, S. Claesson, S. P. Datta, R. Dedonder, C. E. Dent, M. S. Dunn, T. D. Fontaine, C. Fromageot, C. S. Hanes, D. H. Harris, F. J. R. Hird, R. T. Holman, J. K. N. Jones, M. Jutisz, P. Karrer, E. Kawerau, B. H. Ketelle, J. G. Kirchner, Roberta Ma, M. Macheboeuf, A. J. P. Martin, C. S. Marvel, A. E. Mirsky, H. K. Mitchell, S. Moore, L. S. Myers, S. M. Partridge, A. Polson, T. Reichstein, A. Rolovick, E. V. Rowsell, R. W. Schayer, H. Schmid, C. W. Shoppee, F. H. Spedding, F. H. Stodola and the Western Regional Laboratory, H. Svensson, A. Tiselius, E. R. Tompkins, S. Udenfriend, C. S. Watson, T. Wieland, F. P. Winteringham, W. Q. Wolfson, G. R. Wyatt and G. Ziéglé, and the editors of *Analytical Chemistry*, *Analyst*, *Angewandte Chemie*, *Annals of the N. Y. Academy of Science*, *Arkiv Kemi Min. Geol.*, *Biochemical Journal*, *Biochimica et Biophysica Acta*, *Bull. soc. chim. France*, *Experientia*, *Transactions of the Faraday Society*, *Helvetica Chimica Acta*, *Journal of the American Chemical Society*, *Journal of Biological Chemistry*, *Journal of the Chemical Society*, *Nature* and *Science*.

The authors are particularly thankful to Dr. A. T. James for translation of some parts of the French original text, and to Dr. Ulli Eisner for her help in correcting the proofs.

Paris, May 1953   E. LEDERER
                                                        M. LEDERER

Contents

DIVISION II — ION EXCHANGE CHROMATOGRAPHY

DIVISION IV — CHROMATOGRAPHY OF ORGANIC SUBSTANCES

A brief history of chromatography

Accounts of the early history of chromatography have recently been given by Weil and Williams (*3485, 3486*), Farradane (*937*) and Zechmeister (*3677*). While it seems certain that different authors have observed separations of substances by filtration through columns of finely divided adsorbents, it is evident that the Russian botanist Tswett (*3314*) was the first to be aware of the great possibilities of chromatography. He described in detail the separation of pigments and colourless substances by filtration through columns, followed by development with pure solvents. The edition by Richter and Krasnosselskaja* of selected papers of Tswett will be welcomed by those who are interested in this subject.

The first paper of Tswett, published in 1903, contains a study of more than 100 adsorbents used in conjunction with several different solvents and a comparison of the efficiency of column and batch adsorption. The chromatography of a leaf extract on a column of inulin is described as follows:

"The adsorption phenomena observed during filtration through the powder are particularly interesting. The liquid emerging from the lower end of the funnel is at first colourless, then yellow (carotene) whilst at the top of the column of inulin a green ring is formed below which there soon appears a yellow band. When pure ligroin is filtered through the column, both bands start spreading and move down the column". (Translated from p. 21 of the aforementioned selected papers of Tswett.)

The following account was given by Tswett of the purification of lecithin:

"A few cm³ of the solution (from egg yolk) were filtered through a column of inulin. The filtrate was at first colourless, then yellow, and contained much fatty material. Filtration of ligroin through the column was continued until no transparent spot was obtained when a drop of the filtrate was allowed to evaporate on tissue paper. A mixture of ligroin and alcohol was then passed through the column. The filtrate, which was pale yellow, afforded on evaporation a wax-like solid having the characteristic properties of the lecithin complex." (Translated from p. 25 of the selected papers of Tswett.)

Further descriptions of chromatographic techniques are contained in a paper published by Tswett in 1906 and in his book *The chromophylls in the plant and animal world* (1910) (*3314*).

Tswett died prematurely in 1920 (for a biographical sketch, see Dhéré,

---

* M. S. Tswett: *Chromatographic Adsorption Analysis*. Selected papers edited by A. A. Richter and T. A. Krasnosselskaja. Academy of Sciences of the USSR Press, 1946.

768, and Zechmeister, *3675**) and his method was used only very rarely during the following years.

It was only in 1931, when Kuhn and Lederer (*1821*) separated the carotenes and xanthophylls on a preparative scale on columns of alumina and calcium carbonate, that the possibilities of Tswett's method were fully realised. Rapid development of the new field by workers such as Brockmann, Karrer, Winterstein, Zechmeister, *etc.*, soon enabled the organic chemist to apply this technique to the separation and isolation of a wide variety of compounds.

In 1938, Reichstein (*3089*) introduced the *liquid, or flowing chromatogram*, thus extending the applicability of the method to colourless substances.

During the years 1940–1943, Tiselius (*3252–3257*) worked out the techniques of *frontal analysis* and *displacement development*.

Partition chromatography on silica gel, introduced by Martin and Synge (*2156*) in 1941, further extended the scope of chromatography to a great range of biologically important substances. *Paper chromatography* was first described by Consden, Gordon and Martin (*618*) in 1944 and has become a major tool for biochemical analysis and research.

The development of *gas-liquid chromatography*, by James and Martin (*1537*) in 1952, has opened a new field in analytical chemistry which promises to have a wide application, both in research and in industry.

Prominent contributions to the *theory of chromatography* were made since 1940 by Wilson (*3591*), De Vault (*759*), Martin and Synge (*2156*), Glueckauf (*592, 1140–1146*) and Weiss (*2399, 3495*).

In 1947, a series of articles released by the American Atomic Energy Commission, notably by Boyd (*351–353*), Spedding (*3062,3069, 3070*), and Tompkins (*3282, 3284, 3285*) described the use of *ion exchange chromatography* for the separation of fission products and rare earth mixtures. The use of paper chromatography in inorganic chemistry was introduced in 1948 by M. Lederer (*1914*), and Linstead and collaborators (*76*). Chromatography has thus become an important technique in inorganic chemistry as well.

---

* See also: *Michael Tswett's erste chromatographische Schrift* (37 pp.) by G. Hesse and H. Weil (M. Woelm, Eschwege, Germany, 1954).

DIVISION I

ADSORPTION CHROMATOGRAPHY

CHAPTER 1

General chromatographic techniques

1. INTRODUCTION

In its classical form, as described by Tswett (*3314*), chromatography consists in separating substances by filtering their solution through a column of a finely powdered adsorbent, filled into a glass tube, and then washing ("developing") the column with a solvent. Tswett and the early workers after 1931 were mostly concerned with pigments and thus observed the formation of coloured zones on the column of adsorbent, each zone corresponding to a pure pigment. The developing of the column with the pure solvent was clearly an important step in the fractionation process, as the zones were seen to separate more and more from one another during the development. As soon as separation was judged to be sufficient, development was stopped and the zones containing the adsorbed pigments were separated mechanically with a spatula, either with or without extrusion of the whole adsorbent column from its tube. Each portion of adsorbent was then treated separately with an appropriated solvent for elution of the adsorbed pigment.

This simple and very effective procedure is still used extensively in work with pigments. When working with colourless substances, the column can be cut up arbitrarily in several portions, each being eluted separately.

2. LIQUID OR FLOWING CHROMATOGRAM

The above mentioned technique has been superseded for the separation of colourless substances by the method introduced by Reichstein and his school (*2656, 2657, 3089*) known as the "liquid" or "flowing" chromatogram. The method consists in washing the column successively with a series of solvents of stronger and stronger eluting power (e.g. light petroleum, benzene, ether, acetone, alcohol or appropriate mixtures of these solvents), each filtrate being collected separately. For details of this method which is very largely applied, see under Elution, p. 38. An important modification, "gradient elution" is discussed on p. 41.

3. FRONTAL AND DISPLACEMENT ANALYSIS

Tiselius, Claesson and their collaborators (*574–583, 3252–3260, 3262*) invented a modification of adsorption chromatography which they called "frontal analysis". A solution of the substances to be studied is forced continuously through a column and their concentrations on leaving the column are measured by the use of an optical system registering changes in refractive index based on the "Schlieren" method due to Toepler.

The measurement of concentration gradients by optical methods is practised also in sedimentation studies (ultracentrifuge), diffusion and electrophoresis. Numerous optical instruments are used in addition to the aforementioned "Schlieren" method and these are adequately treated in most accounts of electrophoretic methods. Svensson (*3163, 3164, 3167*), in describing a new interferometric method has also surveyed the available techniques. Other instruments have been described by Holman (*1430*) and by Hellström and Borgiel (*1360*).

The degree of adsorption of different substances is expressed as the "*specific retention volume*" (or retardation volume), that is the volume of liquid passing through the column per g of adsorbent before the substance in question leaves the column. Table 99 p. 289 shows the retention volumes of amino acids and peptides (in ml/g of adsorbent). These figures allow the characterisation of the affinity of a variety of substances for a given adsorbent.

Claesson (*577*) showed that in an homologous series the degree of adsorption can be correlated with the boiling point, always rising with increasing boiling point. Hall and Tiselius (*1278*) have confirmed this for the case of isomeric compounds with similar functional groups. Thus the boiling points of 2-methyl-2-butanol, 3-methyl-1-butanol and 1-pentanol are respectively 101.8°, 130.5° and 138°. The retention volumes observed were 76, 97 and 113 ml. Table 1 shows the same phenomenon for isomeric butyric and valeric acids. This correlation can be very useful for choosing the proper "carrier" in "carrier displacement" (see p. 8).

TABLE 1.   Correlation of boiling point and retention volume (Hall and Tiselius, *1278*)

Substance	B.p. °	Retention volume
2-Methylpropionic acid	154.4	9.0
n-Butyric acid	163.5	18.5
3-Methylbutyric acid	176.7	20.1
n-Valeric acid	187.0	27.5
4-Methylvaleric acid	207.7	42.0

Fig. 1 shows the frontal analysis curve for a mixture of four substances *A*, *B*, *C* and *D*. The point at which each substance leaves the column is indicated by a step, the number of steps indicates the number of substances in the mixture. This "frontal analysis" does not, however, allow the separation of the substances.

In order to obtain a separation it is necessary to wash the column with pure solvent. Fig. 2 shows the case of *elution development* of a mixture *E*, *F*, *G*

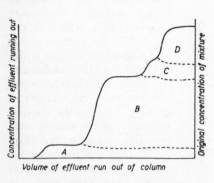

Fig. 1. Frontal analysis of a mixture $A + B + C + D$, according to Tiselius (from *2157*)

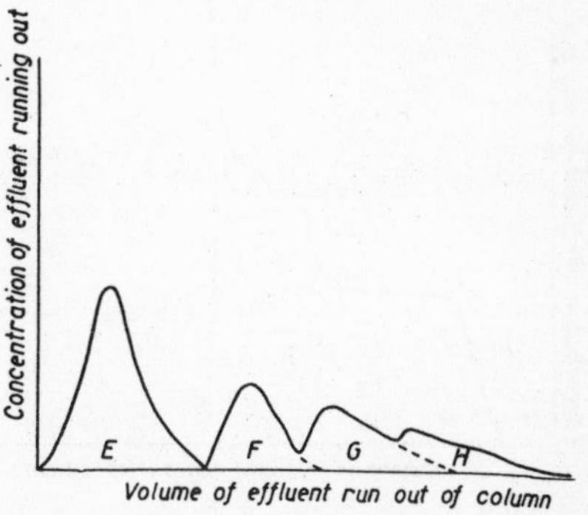

Fig. 2. Elution development of a mixture $E + F + G + H$, according to Tiselius (from *2157*)

and *H*. It can be seen that substance *E* leaves the column free from the others and is immediately followed by *F*. Complete separation of *F*, *G* and *H* is however impossible as *G* begins to leave the column before *F* is completely eluted, the same phenomenon being repeated in an exaggerated form for *G* and *H*. This difficulty, found very often in chromatography, is due to the fact that many substances are adsorbed more strongly from dilute solution; quantitative elution thus becomes more and more difficult, as the substances "tail" through the column.

Tiselius (*3257*) has since shown that it is advantageous to develop the chromatogram with a solution of a substance more strongly adsorbed than the substances to be separated. This method is called *displacement development*. The band of the substance the least adsorbed leaves the column, followed immediately by the next, one displacing the other, the last being displaced by the substance chosen for the displacement (e.g. phenol for peptides, ephedrine for polysaccharides, Fig. 3). The advantage of this method is that the back of the zone remains sharp and does not tail through the column (being continuously displaced by the following zone). Difficulties do arise, however, as is shown by the work of Holman and Hagdahl (*1435*) on the separation of fatty acids by displacement. The disadvantage of the method is that the zones leave the column immediately behind one another without being separated by a zone of pure solvent. This difficulty has been overcome by the use of "carriers" (see p. 8).

The phenomenon of displacement of one band by another is often observed during the chromatographic separation of a mixture. The band moves through the column as a sharp zone whereas the same substance when run by itself on a similar column would show a diffuse tailing band on elution.

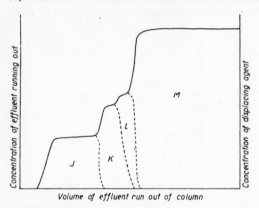

Fig. 3. Displacement development of a mixture $J + K + L$ with displacing agent M, according to Tiselius (from *2157*)

Tiselius has shown that in the course of displacement development the substances rapidly attain a stationary concentration which is expressed in the elution curve as a step whose length is dependent on the quantity of the substance. This stationary concentration (the height of the step) enables the characterisation of the eluted substance to be carried out (Fig. 3). Tiselius has reviewed his methods in several papers (*3255, 3258, 3259, 3261*).

Drake (*812*) has studied some factors governing frontal sharpness in adsorption analysis; various methods of packing and various shapes of columns affect the sharpness of the fronts. The superiority of long, narrow columns and of low pressure heads was shown.

The scaling up of a separation previously carried out on a milligram scale is often difficult as the zones show many irregularities of shape. The front of a zone instead of being horizontal can assume the shape shown in Fig. 4a. Claesson (*580*) has invented a means of correcting this fault. It consists in joining two or three columns of decreasing diameter to-

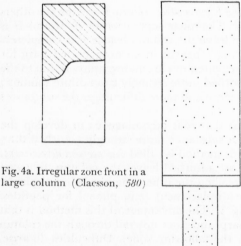

Fig. 4a. Irregular zone front in a large column (Claesson, *580*)

Fig. 4b. A double sectioned column (Claesson, *580*)

gether. A column of large diameter possesses at its base a compartment containing no adsorbent and is mounted on a second smaller column (Fig. 4b).

When the hump in the front of the zone enters the compartment, mixing occurs with the pure solvent already there and the diluted solution enters the second column. As in general the more dilute part of a zone is held back more strongly than the concentrated part of the zone, it is rapidly caught up by the region of higher concentration leaving the large column. In this way a uniform horizontal front is formed in the second column. Fig. 5 shows the result obtained with and without Claesson's modification.

Claesson recommends the use of a series of columns each of which is one-fifth of the size of the one preceding (e.g. from top to bottom: 40 × 100 mm,

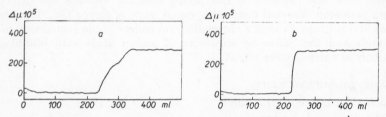

Fig. 5. Frontal analysis of 2 % sucrose in water. Filter a) 20,000 π, b) 20,000 π + 5,000 π
(Claesson, *580*)

20 × 50 mm, and 10 × 50 mm). The same effect is not obtained either by placing an empty compartment between two columns of the same diameter, or by using a conical column.

Hagdahl (*1254*) has also shown that with frontal analysis and displacement development a series of filters of diminishing size connected by narrow tubes greatly improves the sharpness of separation, the irregularities of the front in one section being more or less corrected in the succeeding section. Such composite columns are available from LKB-Produkter, Stockholm 12 (Hagdahl, *1255*).

A column system made of perspex is described by Hall (*1277*).

Some important applications of the Tiselius techniques may be cited here:

Frontal analysis of ethers and organic sulphides on silica and alumina is described by Hurd *et al.* (*1488*). Wetterholm (*3519*) has studied the frontal analysis of glycols and glycerols, Holman (*1432*) has examined in detail the displacement analysis of lipid soluble substances, such as hydrocarbons, alkyl iodides, bromides, chlorides, mercaptans, nitriles, esters, and Holman and Williams (*1438*) have described the displacement analysis of un-saturated acids on charcoal (with aqueous ethanol as solvent). Li, Tiselius *et al.* (*2003*) examined ACTH peptides on Carboraffin Supra + Hyflo Supercel (1 : 9) by displacing with 0.4% Zephiran chloride. More recently, Porath and Li (*2562*) have described elution and displacement analysis of insulin and adrenocorticotropic peptides on pre-treated charcoal (Darco G-60) (see p. 36). For frontal analysis and displacement analysis of acids, sugars and gases, see also Claesson (*577*).

4. CARRIER DISPLACEMENT

Tiselius and Hagdahl (*3269*) have invented an improvement on displacement development called *carrier displacement*. In ordinary displacement development the zones are in close contact, thus effective quantitative separations are rendered difficult. By interposing a number of substances of intermediate adsorption affinities (carriers) between the substances the zones can be separated successfully from each other.

In a preliminary communication (*3269*) the separation of a number of amino acids and peptides by interposing aliphatic alcohols was described using activated charcoal containing equal parts of Super Cel. The amino acids and peptides in such a chromatogram collect at the boundaries of the alcohols and the following sequences of amino acids with homologous alcohols as carriers were reported:

Valine: *tert*.butanol/water.

Leucine: *sec*.butanol/*tert*.butanol.

Methionine: *sec*.butanol/*tert*.butanol.

Leucyl-glycyl-glycine: *iso*amyl/*n*-butyl alcohol.

Phenylalanine: *n*-amyl/*iso*amyl alcohol.

Glycyl-tryptophan: benzyl alcohol/*n*-amyl alcohol.

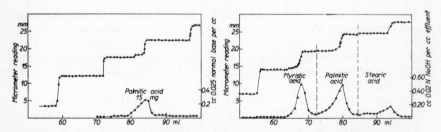

Fig. 6. Carrier displacement separation of myristic, palmitic and stearic acids in a carrier system of methyl esters of these acids; filter column capacity 40 cc, solvent 95% ethanol, carriers 50 mg methyl laurate, 80 mg methyl myristate, 120 mg methyl palmitate, displacer 1.0% methyl stearate. Acid quantities indicated on curves (Holman, *1431*)

Very successful separations of the homologous fatty acids were also obtained with methyl esters as carriers (*1431*). (See Fig. 6 and the chapter on fatty acids.) Similar work on the C_{18} fatty acids was carried out by Kurtz (*1839*), who terms the process "amplified chromatography". The carrier displacement of peptides, using normal alcohols with 8 to 10 carbon atoms as carriers has been described by Porath (*2558*) and Li *et al.* (*2000*).

5. CONTINUOUS CHROMATOGRAPHY

An extensive report on the development of a continuous chromatographic apparatus was made by Svensson *et al.* (*3165, 3166*) giving references to relevant preliminary work on this question.

The **principle** of the apparatus (Fig. 7) described by these authors is

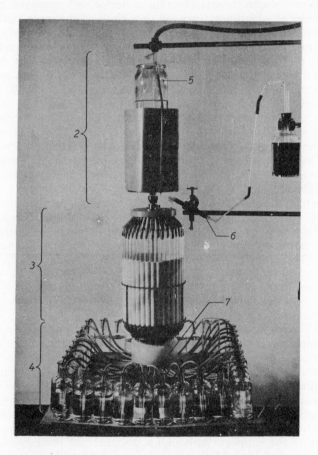

Fig. 7. Experimental apparatus for continuous chromatographic separation. 1 Wooden frame; 2 Feeding arragement; 3 Rotating chromatographic column; 4 Fraction collector. (Svensson *et al., 3166*)

best understood by considering two concentric cylinders with plane bottom, the annular space between the cylinders being filled with the sorbing agent, and the bottom being equipped with a number of equidistant outlet tubes. The mixture to be separated is continuously fed from one fixed point above the upper surface of the sorbent bed (Fig. 7, No. 6), while an eluting agent is fed to the entire remaining part of the same surface (Fig. 7, No. 5). Both solutions are fed slowly enough so that each drop has time to drain into the column before the next drop falls at the same spot; yet the rate of flow has to be fast enough to keep the upper surface constantly wet. Below the cylindrical column there is a fraction collector with many compartments, the inlet holes of which are arranged in a circle of the same radius as that of the column (Fig. 7, No. 7). The column is slowly rotated, while the feeding arrangement and the fraction collector are kept stationary. Every solute will then acquire the same tangential speed, which is determined by the angular velocity of the column. Each solute will have a specific rate of vertical migration, however, which depends upon its strength of sorption. Consequently, each solute will acquire its own flow spiral in the rotating column. Components without any sorption will form the steepest spiral, while for a sorbed component, the steepness of the spiral will be smaller the stronger the sorption of the component. Since the sample feeding arrangement and the fraction collector are fixed, each component will continuously run out into one or in a few adjacent compartments of the collector.

One model in which the sorbent container is sectioned, i.e. consists of a series of separate columns, is shown in Fig. 7. The above authors report satisfactory performance during one month's uninterrupted separation.

Solms has constructed a continuous apparatus using essentially the same principles as above and employing a paper curtain (*3045*). See also the patent by Olsen (*2412*).

6. ELECTROCHROMATOGRAPHY

Strain (*3135*) first described a combination of chromatography and electrophoresis using two electrodes inserted in an alumina column. By this method dyestuffs could be separated by the application of 175–200 v. Schoofs and Lecoq (*1897, 2874*) used essentially the same method for the separation of alkaloids and of inorganic ions.

There exists a considerable literature on electrophoresis inside packed columns, gels and filter paper strips and in most cases such a non-homogeneous electrophoresis is considerably influenced by adsorption effects on the surface of the support. Although a discussion of these methods is not within the scope of this book the main developments will be mentioned here.

(a) *Electrophoresis inside packed columns*

Columns packed with glass wool, asbestos fibre (*488*) or glass powder (*1261*) have been employed for proteins and their hydrolysates. Haglund and

Tiselius (*1261*) obtained promising results by removing the solution contained in the column after electrophoresis in small fractions using an automatic fraction collector. (See also *3168*.)

(b) *Electrophoresis inside gels*

Silica gel was first used by Martin *et al.* (*2157*) for the electrophoretic separation of peptides. The separation of numerous protein mixtures even with high molecular weights was achieved by Gordon *et al.* (*1167*) inside slabs of agar jelly. By suitable techniques the proteins could be isolated. Nucleotides (*1173*), porphyrins (*2439*), polyphenols (*2489*) and inorganic ions (*1942*) were separated inside agar; the transparency of the medium permits the use of spectrophotometers for the quantitative estimations and diffusion measurements *in situ*.

(c) *Electrophoresis inside filter paper*

The separation of amino acids and peptides on a paper strip moistened with a buffer was first described by Wieland and Fischer (*3560*) employing a simple plastic frame and allowing the ends of the paper strip to dip into beakers holding the electrodes. (See also Biserte, *254*.)

The most important application of this method is in the analysis of serum proteins. Various arrangements of the paper and methods for cooling the paper have been proposed by Durrum (*837*), Cremer and Tiselius (*653*), Turba and Enenkel (*3321*), Macheboeuf *et al.* (*2084*). Paper electrophoresis can be employed almost as widely as paper chromatography and for example separations of sugars (as their borate complexes) (Consden and Stanier, *626*), dyestuffs (Evans and Walls, *927*), purines and pyrimidines (Wieland and Bauer, *3557*), indole derivatives (von Denffer *et al.*, *735, 736*), inorganic ions (McDonald *et al.*, *2073*, Kraus and Smith, *1792*, Lederer and Ward, *1944, 1945*, Strain, *3146, 3147*), have been described, which in some cases offer advantages as to speed and scope over paper chromatography.

(d) *Continuous electrophoretic separations*

Simultaneous development with a solvent and electrophoretic separation was first described by Haugaard and Kroner (*1335*). By the use of a flowing stream of buffer and a central stream of a sample through a porous medium a number of continuous electrophoretic set-ups have been developed. Svensson and Brattsten use a bed of powdered glass (*3167*) in the first apparatus of this kind (see Fig. 8). Others using filter paper sheets or pads were described by Grassmann and Hannig (*1192*), Brattsten and Nilsson (*363*),

Strain *et al.* (*3147*). Monographs on electrophoretic methods have been written by Wunderly (*3633*), McDonald (*2072*) and M. Lederer (*1936*).

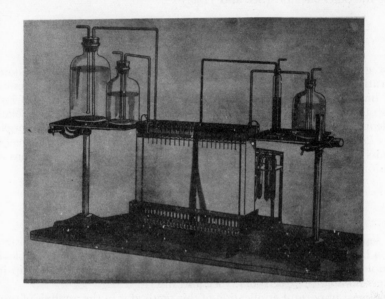

Fig. 8. Apparatus for continuous electrophoresis (Svensson and Brattsten, *3167*)

7. MISCELLANEOUS

(a) *The "chromatostrip"*

Kirchner *et al.* (*1727*) have prepared fluorescent "chromatostrips" by coating a glass strip with a slurry of silicic acid, mixed with starch as binder and $ZnCdS_2$ and Zn silicate as fluorescing agent. The dried strips are spotted with a terpene mixture and a solvent (e.g. 15% ethanol in hexanol) is allowed to ascend the strip in a test tube (see Fig. 60, p. 154). Kirchner *et al.* (*1727*) also described the preparation of non-fluorescent strips with various adsorbents (see also "chromatobar", p. 16). An apparatus for the preparation of chromatostrips was described by Miller and Kirchner (*2223*).

Reitsema (*2669*) has used larger strips, called *"chromatoplates"* for the analysis of essential oils; the use of plates allows the simultaneous running of known substances, with the unknown.

In the senior author's laboratory non-fluorescing chromatostrips or also chromatoplates (of silicic acid containing 5% starch) have proved very useful for the control of column eluates; development of the strips takes only

30 minutes; most substances can be detected on the strips by placing them
in a jar with iodine vapors (Demole, unpublished experiments).

(b) *"Inverted chromatography"*

Lowman (*2052*) has proposed an "inverted chromatography" consisting
in pouring the adsorbent through a column of the solution of substances to
be adsorbed.

CHAPTER 2

Apparatus

1. CHROMATOGRAPHIC COLUMNS

Chromatographic tubes are usually made of glass, drawn out at one end, sometimes fitted with a porous plate at the bottom and carry a tap on the exit tube so that the .filtration can be stopped if desired. Bigger columns usually carry a separating funnel to hold the solvent. The height of the column should be 4 to 10 times the diameter. The reviews of Lederer (*1899*), Strain (*3136*), Williams (*3585*) and Zechmeister and von Cholnoky (*3679*) contain a number of illustrations of columns (Fig. 9).

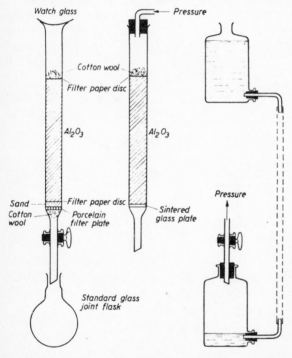

The sizes of tubes listed in Table 2 are recommended by Reichstein and Shoppee (*2658*).

The tube is selected so that, when the appropriate quantity of adsorbent has been introduced, it is about half full.

Georges *et al.* (*1099*) used tapered tubes which facilitated the extrusion of the columns for application of streak reagents.

Fig. 9. Columns and pressure device for chromatography with fractional elution (Reichstein and Shoppee, *2658*)

Gault and Ronez (*1091*) have described tubes consisting of sections with

TABLE 2. Sizes of chromatographic tubes (Reichstein and Shoppee, *2658*)

Amount $Al_2O_3(g)$	Internal diam. (mm)	Usable length (mm)	Bore of tap (mm)
1	8	110	3
2	10	130	3
4	13	160	4
8	16	200	4
15	20	250	4
30	25	300	6
60	32	400	6
125	40	500	6
250	50	600	8
500	65	750	8

interchangeable ground glass joints; the column can thus be opened at different heights.

Tiselius, Claesson *et al.* (*574–583*, *3252–3262*) have described in detail the elaborate apparatus necessary for their "frontal analysis". Tiselius's columns are made in the form of cells or rectangular cuvettes of transparent plastic (Perspex, Lucite, or Plexiglass) or of glass, or metal and have diameters of 20, 10 and

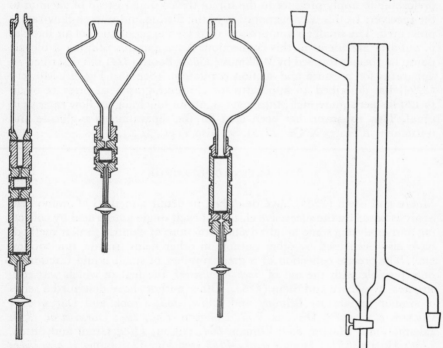

Fig. 10. Sectional drawing, showing some metal filters coupled together by the aid of connection pieces and provided with containers for the eluent (Tiselius *et al.*, *3268*)

Fig. 11. Chromatographic tube with heating mantle

5 mm and heights of 40, 20, 10 and 5 mm (*3255*). Tiselius *et al.* (*3268*) used a demountable apparatus of three columns (Fig. 10) for the separation of

the amino acids into a number of groups. Composite columns have been mentioned on p. 6.

For chromatography at higher temperatures tubes with heating mantles can be used (Fig. 11).

Miller and Kirchner (2220) have described a modification of the usual column technique; they mix the adsorbent (e.g. silicic acid) with plaster of Paris and pour it, after addition of water, into a suitable mould; they thus obtain a "*chromatobar*", a self-supporting column, on which the zones can be located by streaking with a reagent. They have used the chromatobar for separations of terpenes.

2. PRESSURE DEVICES

In working with volatile solvents (light petroleum, ethyl ether, etc.) it is preferable to apply pressure to the top of the column instead of vacuum to the receiver. In this way evaporation of the filtrate and uneven flow-rate is prevented. The small air compressors used for the production of air bubbles in aquaria are useful in this connection. An apparatus utilising a bicycle pump has been proposed by Williams (3585). Booth (314) has described an apparatus for pressure and suction regulation. (See also Fig. 9.) Mowery (2294) has described an apparatus for chromatography at pressure of up to 120 lbs per square inch; thus one can attain much higher flow rates than usual. This apparatus has been used for the separation of D-glucose and D-fructose on Florex XXX (2295). See also Vogt (3426).

3. FRACTION COLLECTORS

Moore and Stein (2265) have described in detail a method of amino acid analysis based on the successive elution of each single amino acid by collecting and analysing some hundred small fractions of eluates; similar methods have also been used by other authors in other fields (sugars, nucleotides, etc.). The precise collection of a great number of small liquid fractions is only possible with the aid of *fraction collectors*, the first of which was constructed by Moore and Stein (2265). Other authors have described a series of similar apparatus (Brimley and Snow, 383; Crook and Datta, 659; Cuckow et al., 672; Desreux, 753; Dodgson et al., 797; Durso et al., 838; Edelman and Martin, 859; Edman, 864; Gilson, 1109; Grant and Stitch, 1188; Harris, 1314; Hough et al., 1458 (see Fig. 12); James et al., 1541; Phillips, 2509; Randall and Martin, 2619). Costly equipment can be avoided, the delivery of equal volumes being obtained by simple mechanical devices (see James et al., 1541; Randall and Martin, 2619).

Recent commercial literature contains many designs of fraction collectors: The main fault found in several commercial models is a turn-table holding insufficient tubes for a satisfactory run. Also one model employs an electrode contact system as mechanism for fractionation. This is certainly unsatisfactory for uncharged solvents as well as for radioactive tracers which might partially deposit on the electrodes.

Of the more recent non-commercial designs we quote Boggs *et al.* (*290*), Brunisholz and Germano (*432*), Delmon (*731*), Dimler *et al.* (*780*), Dustin (*840*), Fraser (*1027*), Grassmann and Deffner (*1190*), Hamilton (*1286*), Henry and Thevenet (*1365*), Hickson and Whistler (*1391*), Lister (*2037*),

Fig. 12. Fraction collector (Hough *et al.*, *1485*)

Mader and Mader (*2100*), Schram and Bigwood (*2876*), Verzele (*3400*), Wingo and Browning (*3596*).

4. INDUSTRIAL APPARATUS

Mair *et al.* (*2107*) have described a 52 foot "laboratory" column. A patent of Meng (*2188*) describes an apparatus consisting of separates zones of adsorbent in which eluting solvent and eluate can be fed into and withdrawn from any selected adsorbent zone independently of the other zones and without any break-up of the adsorption unit. Lynam and Weil (*2065*) suggest that the numerous valves necessary for elution, backwash etc. can be dispensed with, if a movable column is used, which can be shifted for each step.

A centrifuge packed with the adsorbent (called the "*chromatofuge*") was first suggested by Hopf *et al.* (*1444, 1445*). This chromatofuge however is unable to cope with more than pilot scale amounts of adsorbents. Thus the idea was modified by having a stationary disc of adsorbent and centrally feeding it with solution and eluant under a small central pressure and peripheral suction (*2063, 2064*). Numerous advantages are claimed for such radial development. An adsorbent disc 3 feet in diameter and 1 foot deep is claimed to be equivalent to a column 7 feet high and 1 square foot in section. By designing a container, which is sectioned with perforated partitions it is possible to separate zones cleanly. If a number of interchangeable adsorbent containers are used for one chromatographic plant, a semi-continuous scheme may be adopted.

Williams and Hightower (*3580*) describe an industrial installation for the purification of streptomycin on columns of alumina of 80 cm diameter and 4 m high. The industrial chromatography of carotene from alfalfa on columns of charcoal (60 cm diameter and 2 m high) has been described by Shearon and Gee (*2951*, see also *3586*). For a review on industrial chromatography, see Weil (*3484*).

Swinton and Weiss (*3177*) criticise chromatographic development as industrially unfeasible and suggest counter-current adsorption with a-fluidised bed of adsorbent moving against a stream of solution. In further papers the properties of the adsorbents to be used in such systems (*3494, 2232*) are examined, also its application to the purification of a crude penicillin extract (*2231*); a froth flotation process based on the same principle is also discussed (*2248*). Froth chromatography of colloids is also studied by Mokrushin (*2249*).

5. APPARATUS FOR CONTROL OF ELUATES

Clear-cut separation of eluted colourless substances can often be achieved by measurement of physical constants of the eluates. Claesson (*575*) employed an automatic apparatus for the determination of the refractive index of the eluate. Further recording interferometers and refractometers were designed by Holman and Hagdahl (*1437*), Kegeles and Sober (*1671*), Glenn et al. (*1138*), Trenner et al. (*3299*), McCormick (*2069*) and Hellström (*1359*). Jeffrey (*1567*) uses a pH meter for continuous recording of the pH of the effluent; see also Zahn et al. (*3661*). Conductivity recording has been frequently employed, especially in ion exchange chromatography (James et al., *1541*; Partridge and Westall, *2463*; de Verdier and Sjöberg, *763*). Laskowski and Putscher (*1884*) have described a "dielectric indicator" for detecting small changes in the dielectric properties of the eluate. Porter (*2572*) describes a simple device for cutting fractions using a photoelectric cell. Potentiometric control of the eluate is employed in the publications of Kamienski (*1623*) and Waligóra and Bylo (*3453*). For the simultaneous measurement of ultraviolet absorption and radioactivity in eluates see Bradley (*357*).

CHAPTER 3

Adsorbents

1. PARTICLE SIZE

Table 3 shows the particle size of different adsorbents (after Zechmeister and Cholnoky, *3679*). Before using a non-standardized adsorbent it is recommended to rub it through a 150 mesh sieve and collect the part retained by the 200 mesh sieve.

Mowery (*2294*) has described a method for blowing air through adsorbents to remove the finer particles so as to accelerate the flow rate.

TABLE 3.   Average particle size of some adsorbents

Adsorbent	Particle size (μ)
Alumina Merck (standardized Brockmann) . .	7
Acid clay	10
Calcium carbonate precipitated	1.5
Calcium hydroxide	2.5
Calcium sulphate (hydrated)	10.5
Magnesium oxide	1.5
Fuller's earth	3
Floridin	1.5–7
Floridin XXF	1.5–6

The measurement of flow rates has been described by Hesse *et al.* (*1379*) and by Mowery (*2294*).

2. DESCRIPTION AND USE

(a) *Inorganic adsorbents*

Alumina (Al_2O_3) is one of the principal adsorbents used in chromatography. It often contains sodium carbonate and bicarbonate (*2982*) whose presence exerts a marked effect on its adsorptive properties. The alkali of alumina often causes secondary reactions (see p. 61); this can be prevented by washing with dilute acid or with water (followed finally by methanol), then by reactivation at 200° (*925, 2658*). Heating alumina above 500° should be avoided; Krieger (*1802*) has stated that the surface of alumina remains constant up to 528°, then decreases by 15% at 734° and by 40% after heating at 938°. Russell and Cochran (*2756*) have studied the influence of temperature, atmosphere and duration of heating on the surface of various forms of α- and β-aluminium mono- and tri-hydrates and of an amorphous alumina.

Usually 1 g of alumina for chromatography has about 90 square meters surface; an alumina having less than 6 square meters per g is useless as an adsorbent (Jutisz and Teichner, *1614*).

Preparation of alumina for chromatography. We quote in detail a procedure for the preparation and regeneration of alumina recommended by Reichstein and Shoppee (*2658*) (p. 25). Dupont *et al.* (*832*) have described the preparation of alumina by calcination of the hydrate at 500° followed by activation with concentrated HCl. Fuks (*1060*) has also described the preparation and standardization of alumina. The preparation and use of "fibrous" alumina was described by Wislicenus (*3604*). This is a very active but rather costly adsorbent, obtained by treating aluminium amalgam with water.

A patent by Stewart (*3110*) describes the preparation of alumina for chromatography; for older references see Krczil (*1794*). Quite recently, a German firm (Woelm, Eschwege) has begun producing a standardized alkali-free alumina which is obtained by treating pure aluminium with water.

The preparation and properties of amorphous alumina have been described by Imelik *et al.* (*1500*).

Alumina may be treated by stannous chloride to prevent the oxidation of autoxidizable substances (Kofler, *1759*; Stoll, *3128*).

Acid alumina. For adsorption in aqueous media, the properties of alumina can be modified by washing with acid. This activation, used by Kuhn and Wieland (*1824*) for the adsorption of pantothenic acid, transforms the alumina into an ion exchanger (Wieland, *3549*); the adsorbent takes up chloride ions which can be exchanged for mineral or organic anions. This "acid alumina" is used for the adsorption of the dicarboxylic amino acids and acidic peptides (see Chapter 30).

Basic alumina. On heating technical alumina, the alkali carbonate which it contains forms active centres of sodium aluminate; the sodium ions of these centres can be exchanged against inorganic or organic cations (Wieland's "basic column", *3549*); the capacity of this column is however quite low. (See also separations of inorganic ions, Chapter 40).

Bauxite, $Al_2(OH)_4$, has been used by Zechmeister *et al.* (*3694*) for the separation of enzymic hydrolysates of chitin. La Lande (*1871*) recommends it for the refining of sugar.

Aluminium silicate allows the adsorption of sterols (*1793*) and sterol glycosides (*3246*) from oils without the use of solvent.

Magnesia (MgO) often advantageously replaces alumina (*1184, 3144*). As it is often too finely divided to allow a satisfactory filtration, it can be mixed with a filter aid (Celite, Hyflo Supercel, etc., see p. 24). For the preparation of a suitable magnesia, see *1026*. The most active magnesia is obtained by dehydration of the hydroxide.

Magnesium silicate. Magnesol (MgO·2.5 SiO$_2$·H$_2$O) has been used for the separation of sugar acetates (*606*). Magnesium trisilicate (No. 34 of the Philadelphia Quartz Co., Berkeley, Cal.) has been extensively used by Liebermann *et al.* (*2006–2008*) for the separation of steroids; it is less active than alumina. Reichstein *et al.* (*19, 2842*) use magnesium trisilicate (Siegfried, Zofingen) for the chromatography of acetylated glycosides. This adsorbent is better than alumina for the purification of easily saponifiable substances, such as esters, glycerides and lactones (see also p. 61).

The effect of moisture on the chromatographic properties of synthetic hydrated magnesium silicate has been studied by Wolfrom *et al.* (*3618*). Adsorptive capacity increases with decreasing water content.

Calcium hydroxide. The adsorption of the carotenoids on calcium hydroxide has been specially studied by LeRosen (*1971*) as well as by Bickoff (*241*). Karrer and coworkers (*1638–1640, 1642*) have often used it for the separation of carotenoids. Williams (*3585*) recommends, in the case of a difficult elution, that the calcium hydroxide be suspended in water through which CO$_2$ is bubbled; the calcium carbonate formed is a much weaker adsorbent.

A mixture of calcium hydroxide and magnesia allows the separation of the di- and tri-nitrotoluenes (*1276*).

Calcium carbonate is used in the chromatography of xanthophylls (*3673, 3679*), of naphthoquinones (*1823, 1902, 1905*) or other pigments (*1900*). The elution of adsorbed substances can be carried out by dissolving the carbonate in dilute acid.

Stolkowski (*3123*) has shown that the adsorptive capacity of calcium carbonate depends on its crystalline form; vaterite, the unstable crystalline modification, is a much stronger adsorbent than aragonite or calcite. (See also Mathieu (*2171*) for the relationship between adsorption and crystalline structure.)

Dicalcium phosphate (CaHPO$_4$) has been used for the purification of carotene (*2264*).

Tricalcium phosphate (Ca$_3$(PO$_4$)$_2$) is a good adsorbent for enzymes (*3155, 3176*) (see also the chapter on proteins).

Calcium oxalate can be used for the chromatography of anthraquinones and substances related to hypericine (*391, 395*).

Zinc carbonate has been recommended for the chromatography of carotenoids (*1637*) and for coloured derivatives of amino acids (*1641*).

Silica gel* has been used for adsorption of sterols (*380*), of fatty acids and glycerides (*1651, 1652, 1658*), waxes (*2376*), azoated carbohydrates (*2193, 2652*), sugar acetates (*606*) and amino acids (*1054, 2881*). The effect of *high pressures* on catalytic and chromatographic properties of silica gel was studied by Freĭdlin *et al.* (*1031*), the role of *porosity* in silica gel by Neimark *et al.* (*2349*).

The use of silica gel for *partition chromatography* is discussed on p. 107. For the preparation of silica gels, see *1054, 1168, 1315, 1512, 3303, 522* (see also p. 108).

The relation between water content and adsorptive strength of silicic acid-celite mixtures has been studied by Trueblood and Malmberg (*3307*) and by Kay and Trueblood (*1665*).

Greater adsorptive strength was associated with lower content of "free water". When all the "structural water" (i.e. water removable only by ignition) is driven off, the adsorptive power disappears almost completely.

The effect of particle size on the efficacy of silicic acid for the separation of 2,4-dinitrophenylhydrazones of lower aldehydes has been studied by Malmberg (*2112*); a procedure for obtaining a satisfactory adsorbent from commercially available silicic acid is described.

Fuller's earth. The natural clays called fuller's earth are chiefly hydrous magnesium aluminosilicates and are widely used in the petroleum industry for the decoloration of oil. The name fuller's earth is derived from their early use in fulling, the operation of removing grease from woolen goods. For details on the properties and uses of these clays, see Mantell (*2125*).

Various brands of fuller's earth (some called *Filtrol* or *Filtrol-Neutrol* †) have been used for the chromatography of basic hydrosoluble substances (amino acids: *198, 3319*; pteridines: Forrest and Mitchell, *1001*).

Bentonites are also hydrous magnesium aluminosilicates of the montmorillonite group and can be activated by treatment with acid (whereas fuller's earth is not activatable). *Superfiltrol* † is such an activated bentonite. Bentonite has been used for chromatographic separations of vitamin D from vitamin A and sterols (*930*) and for separation of 2,4-dinitrophenyl-hydrazones of aldehydes and ketones (*3535*).

An acid calcium silicate (silene EF) has been found useful for the chromatography of carbohydrates and polyalcohols (*1099*).

Water soluble salts have also found use as adsorbents.

Brockmann (*387*) has separated azobenzene derivatives on anhydrous copper sulphate; substances which could not be separated on alumina were easily separated on $CuSO_4$. Other water-soluble salts, such as anhydrous zinc, manganese, aluminium and magnesium sulphates can also be used to separate azobenzene derivatives. Aluminium sulphate was also used by Brockmann for separations of hydroxyanthraquinones. Elution of strongly

---

* The terms *silica gel, silica* or *silicic acid* refer to hydrated silica precipitates the composition of which varies somewhat according to the precipitation method used.

† Filtrol Corp., Los Angeles 14, Calif. (U.S.A.).

adsorbed compounds is easily achieved by dissolution of the adsorbent in
water. Vitamin A has been adsorbed on sodium carbonate *(632)*.

(b) *Organic adsorbents*

Charcoal has been extensively used by Tiselius for frontal analysis of sugars,
amino acids and other substances *(3252, 3253, 3259, 3268)*. Charcoal has the
specific property of adsorbing strongly aromatic substances (e.g. amino
acids) *(1054, 2880)*; this can be explained by the fact that the –C–C–
spacings in graphite are of the same order as those in benzene. Cassidy
(527, 532) and Claesson *(574)* have used charcoal for adsorption of fatty
acids (see also *900*). Benzene elutes from charcoal a "petroleum-like"
material (Cassidy, *528*, Cason and Gillies, *524*). For details of charcoal as
adsorbent see Mantell *(2125)*, Cassidy *(527, 532)* and Weiss *(3490–3492)*.
Emmett *(898)* has reviewed in detail the adsorptive properties of charcoal
and the methods used for pore size measurements.

Weiss *(3490–3492)* has described activated carbon whose adsorptive properties were modified by depositing on it a film of a non-electrolyte or a fatty
acid. The coating of the surface of an adsorbent by a film of some other substance may be a useful way of preparing adsorbents with special and uniform properties (see p. 36).

Powdered *sucrose* is the best adsorbent for the separation of chlorophyll
(2123, 3143–3145, 3679); it can also separate xanthophylls (see Chapter 34).

Lenoir *(1958)* has suggested the adsorption of colouring matters on a
column packed with discs of textile fabric.

(c) *Mixed adsorbents*

Carlton and Bradbury *(510)* have studied the use of *mixed adsorbents* in
chromatography. Their data show that either the mixture behaves as one
of the two adsorbents in the pure form, the second acting simply as diluent,
or, adsorption is shared between the two adsorbents, in which case adsorption varies almost linearly with percentage composition of the mixture.
An exception was observed with boron oxide mixed with silicic acid, or
Florisil, or alumina, where certain amines were more strongly adsorbed
than by either of the pure adsorbents.

(d) *Classification of adsorbents*

Strain *(3136)* has classified adsorbents on the basis of their adsorptive
capacity; this list (Table 4) is reproduced in the form given by Williams
(3585).

This order, however, must not be considered as applying in every
case; it does not take account of the special affinities and incom-
patibilities which often occur or the possibility of increasing or decreasing

TABLE 4. Adsorbents classified on the basis of their adsorption capacity
          (after Strain, *3136*)

	Weak	Medium	Strong
Increasing activity	Sucrose	Calcium carbonate	Activated magnesium silicate
	Starch	Calcium phosphate	Activated alumina
	Inulin	Magnesium phosphate	Activated charcoal
	Talc	Magnesia	Activated magnesium oxide
	Sodium carbonate	Precipitated calcium hydroxide	Fuller's earth

the activity of various adsorbents (especially alumina) by appropriate
treatment.

Hesse *et al.* (*1379*) have studied the saturation with dyestuffs of different
adsorbents, under standardized conditions and found the following order of
activity of adsorbents:

activated charcoal > silica gel > franconite > floridin > acid alumina >
basic alumina > Cr_2O_3 > ZnS > sugar charcoal > Al_2O_3 (Merck) >
CaF_2 > CaO.

Mantell (*2125*) in his book on adsorption gives useful information on the
principal adsorbents and their industrial uses.

Deitz (*726*) has published a list of adsorbents produced in the U.S.A.
with an indication of the size of their surface.

For the electronic structure of some adsorbents see Meunier and Vinet
(*2199*, p. 55–57).

(e) *Filter aids*

The *filter aids*, generally diatomaceous earths (kieselguhr), are usually mixed
with those finely divided adsorbents which when used by themselves prevent
sufficiently high rates of flow. The preparations often used are Hyflo Supercel and Celite (*2698, 3569*). These powders serve sometimes as adsorbents
(*2688, 3139, 3570*). The adsorbent can also be mixed with paper pulp in
order to render the column more permeable (*3433*). Martin (*2150*) recommends kieselguhr as perfectly inert support for the stationary phase in
partition chromatography.

3. THE PREPARATION OF SOME ADSORBENTS

Neutral alumina. Commercial alumina (Alumina Prolabo for Chromatography, Merck
alumina Brockmann, Neuhausen alumina, or alumina of the Aluminum Ore Co. Ohio, etc.)
is covered with distilled water and a slight excess of 0.5 *N* HCl is added with stirring. After
standing for an hour the liquid is decanted and the alumina repeatedly washed with distilled
water until the washings are free from chloride ions. It is then dried at 100–150° for at
least 12 hours.

Preparation of Al_2O_3 after Reichstein and Shoppee (2658). "Activated and partly standardized Al_2O_3 is available commercially. Equivalent and very active preparations are obtained by heating technical pure $Al(OH)_3$ for about three hours with stirring at 380°–400°; such preparations always contain free alkali (or sodium carbonate) which, however, is usually not deleterious, and in general give the best separations. In certain cases these preparations are too active and lead to condensation of ketones or aldehydes, elimination of alcoholic hydroxyl groups, etc., but by homogeneous addition of moisture their activity can be reduced and standardized (e.g., by the use of selected dyestuffs under defined conditions)."

"For sensitive substances (ketones, lactones, readily hydrolizable esters, etc.) neutralized aluminium oxide is used. Aluminium oxide, prepared in the laboratory by activation at 380°–400° or obtained commercially, is boiled repeatedly with distilled water until the extract is neutral; the filtered material is then washed with methanol and reactivated at 160°–200° (internal temperature) at 10 mm pressure. Preparations reactivated at 200° are too active for many purposes and a temperature of 180° is generally sufficient. The product still contains traces of alkali, which are not harmful and is rather less effective than the alkali containing oxide in regard to separation."

"Neutralization is more rapidly achieved by neutralization of the first aqueous suspension with dilute nitric acid (use of hydrochloric, sulphuric or acetic acid is rather more dangerous) followed by boiling with distilled water, treatment with methanol and reactivation as above. The product contains nitrate ions, which are not always harmless."

Regeneration. "Used aluminium oxide is repeatedly extracted, first with boiling methanol and then with boiling water (with addition of some sodium hydroxide if necessary) and subsequently treated as described above. Slight discolouration of the regenerated material is of no importance."

Acid alumina (after Wieland, 3549). Commercial alumina is mixed with 3 or 4 volumes of N HCl, the mixture is well stirred and the suspended particles are decanted a number of times. The alumina is washed with water until the wash water is feebly acid to litmus paper. Drying is carried out at 100°.

Fromageot *et al.* (1054) prepare acid alumina in the following way: 250 g of commercial alumina is treated with 1,250 ml of $2N$ HCl at the boiling point whilst stirring for 20 minutes. The alumina is poured into a column (diameter, 7 cm) and is treated with 1,250 ml of boiling $2N$ HCl. It is then transferred to an Erlenmeyer flask and washed 4 or 5 times with 1 litre quantities of water, each time the finer particles being decanted. It is then oven dried at 100°.

Charcoal (after Schramm and Primosigh, 2880). The charcoal (Carbo activatus granulatus Schering) previously ground and sieved, is boiled for some minutes with 5 to 10 parts of 20% acetic acid in order to remove nitrogen-containing impurities, then centrifuged hot and washed with hot water. It is then suspended in water and treated for a few minutes with 50 mg of KCN per 100 g of charcoal. After centrifuging, it is washed many times with hot water. Fromageot *et al.* (1054) treat the charcoal (Activite 50X P from la Compagnie Activite, 66 rue d'Auteuil, Paris) with 2 mg of ephedrine per 0.5 g of charcoal (instead of KCN) and carry out the adsorption in solutions saturated with H_2S to prevent any oxidation.

4. THE STANDARDIZATION OF ADSORBENTS

Many methods of standardization of the activity of *alumina* have been proposed.

Brockmann and Schodder (396) measure the activity of alumina by the behaviour on the column of a number of azo-dyes. These dyes, in order of increasing adsorbability, are: azobenzene, p-methoxyazobenzene, benzene-azo-2-naphthol (Sudan yellow), Sudan red (Sudan III), p-aminoazobenzene and p-hydroxyazobenzene. Table 5 shows the behaviour of these dyes on

TABLE 5. Standardization of alumina (Brockmann and Schodder, *396*)

Activity	I	II		III		IV		V
Mixture	1	1	2	2	3	3	4	5
Adsorbed at top of column	methoxy-az.		S. yellow		S. red		amino-az.	hydroxy-az.
Adsorbed at bottom of column	azobenzene	methoxy-az.	methoxy-az.	S.yellow	S.yellow	S.red	S.red	amino-az.
Filtrate	—	azobenzene		methoxy-az.		S. yellow		

(az. = azobenzene S = sudan)

alumina columns more or less deactivated by water (e.g. alumina of activity I adsorbs methoxyazobenzene at the top of the column, and azobenzene at the bottom; with alumina of activity II, azobenzene passes out in the filtrate, methoxyazobenzene is retained at the bottom of the column, etc.). Alumina I is completely activated by calcination, whereas the others are more or less deactivated by a prolonged exposure to damp air. According to Boissonnas (*294*) the addition of 3.3% of water to a completely activated alumina gives an alumina of activity III. See also Boutillon and Prettre (*347*).

Hesse *et al.* (*1379*) have measured the activity of alumina preparations by saturating them with azobenzene. Table 6 shows their figures relating water content of alkali-free alumina with its activity and "azobenzene index".

TABLE 6. Activity of alumina in dependence of water content (Hesse *et al.*, *1379*)

% Water added	Activity (Brockmann & Schodder)	Azobenzene index (10^{-5} mol/g)
0	I	26
3	II	21
6	III	18
10	IV	13
15	V	0

Azobenzene index (Hesse *et al.*, *1379*). 0.5 g of the adsorbent are mixed with 3.00 ml of a 0.1 molar solution of pure azobenzene in pure cyclohexane in a small flask fitted with a ground-in stopper and shaken several times.

After about one hour the supernatant liquid is centrifuged off and a portion poured into a cuvette of a Pulfrich step-photometer and covered immediately. The extinction is measured against pure cyclohexane with a L II Filter and the loss of azobenzene is determined with the aid of a previously prepared calibration curve.

The adsorbed quantity in moles per gram adsorbent (x) is then calculated with the formula

$$x = \frac{a}{m}\,(C_1 - C_2)$$

where a = volume of the dye solution in ml,
m = grams of adsorbent taken,
C_1 initial and C_2 the final concentration of the dye solution in mole/ml.

For the standardization of acid alumina, Hesse *et al.* *(1379)* use the saturation with naphthol orange or Orange GG Cassella; for the standardization of alkaline, cation exchanging alumina, they use methylene blue.

Valentin and Kirchübel *(3359)* determined the activity of alumina by passing a 0.1% light petroleum solution of Sudan red onto the column and measuring the width of the coloured zone, which varies inversely with the activity of the adsorbent.

Müller *(2305)* has developed a very precise method for measurement of the activity of alumina (or any other adsorbent) by measuring the heat evolved by contact between the solvent and the adsorbent. The heat evolved is proportional to the activity of the adsorbent. The activity of the adsorbent is characterised by the heat (Q) evolved in contact with purified light petroleum. Müller has given evidence for the fundamental role played

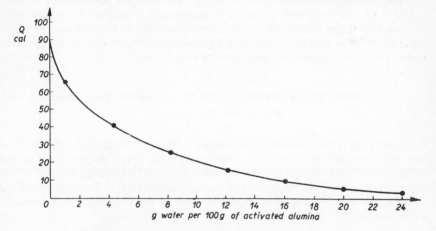

Fig. 13. Heat evolved in relation to water content of alumina (Müller, *2305*)

by water adsorbed by the adsorbent. The addition of 1 % of water to a maximum activated (dried) alumina diminishes markedly its adsorbing capacity (Fig. 13). Müller *(2305)* prepared alumina of any desired degree of activity by the addition of definite amounts of water. (Water is added with a pipette to a known weight of alumina contained in a flask fitted with a ground stopper and shaken for many hours to ensure an even distribution of the water.) For the chemical estimation of the esters of vitamin A, Müller *(2306)* used a column containing three different grades of alumina (see Chapter 35).

For the separation of vitamins A and D and the esters of vitamin A and of carotene, Müller *(2307)* has used columns in 5 parts containing from top to bottom alumina of activities Q 4, 11, 50, 56.5 and 83.5 (the latter representing an alumina completely activated). As an alternative for the calorimetric method Müller recommends the standardization of the alumina with the aid of carotene before carrying out such separations.

Kofler (1758) has proposed a simple and rapid method of measuring the activity of alumina: the colorimetric estimation of the amount of p-hydroxy-azobenzene not adsorbed by a known quantity of alumina (after shaking in a closed system).

Magnesia and carbon can be standardized by known methods (e.g. by the "iodine index", see Mantell, 2125, p. 346).

Wilkie and Jones (3571) have determined the adsorbent strength of magnesia and magnesia-celite mixtures (used for chromatography of vitamin A) in terms of g of Butter yellow No. 4 adsorbed per g of adsorbent.

Brockmann (387) has obtained various grades of bentonite, silica, $CaSO_4$, and $CaCO_3$ by varying the water content. For the standardization of silica gel for adsorption chromatography, see Sen Gupta and Gupta (2944).

For a precise standardization of adsorbents it is preferable to use compounds which are chemically as similar as possible to the substances to be separated. Thus specific influences of adsorbent properties (acidity, alkalinity, pore size, impurities, etc.) can be eliminated.

5. THE CHARACTERIZATION OF ADSORBENTS BY THE METHODS OF LEROSEN

LeRosen (1971) has developed a method of characterising adsorbents. The rate of movement of the solvent is measured, this being proportional to the difference of pressure across the column and inversely proportional to the column length; it is independent of the tube diameter; it becomes constant as soon as the solvent arrives at the bottom of the column. He introduces three factors:

$$S = \frac{\text{length of column containing unit volume of solvent}}{\text{length of tube containing the same volume of solvent.}}$$

V_c = rate of movement of solvent through the column (mm/min)

$$R = \frac{\text{rate of movement of the zone (mm/min)}}{\text{rate of movement of developing solvent}}$$

(this corresponds to the factor R defined by Martin and Synge (2156), see page 104).

The factor S characterizes the degree of packing of the column; it varies for example from 1.97 at the top to 1.79 at the bottom of a column of $Ca(OH)_2$.

V_c varies from column to column (for 14 out of 15 columns from 6.3 to 8.2, for the fifteenth V_c was 10.3).

R expresses the adsorption affinity. Table 7 gives R values for a series of carotenoids.

In measuring R values for various carotenoids, LeRosen was able to calculate the relative positions of the zones obtained on chromatographing

TABLE 7.   Rates of movement of zones relative to that of the developing solvent, on Ca(OH)$_2$ in benzene (LeRosen, *1971*)

	R			R
Capsanthin	0.007	Physaliene		0.590
Zeaxanthin	0.040	δ-Carotene		0.790
Lutein	0.070	Prolycopene		0.885
Lycopene	0.125	β-Carotene		1.000
Cryptoxanthin	0.340			

mixtures. In many cases the relative positions of two substances vary with the adsorbent, for example cryptoxanthin is adsorbed above lycopene on CaCO$_3$ and alumina, but below it on Ca(OH)$_2$. The first observation of this type of behaviour was made for the adsorption of folliculin and equilenin on chalk and alumina (*839*) (see also *3137* and page 49).

The optimum value of V_c, according to LeRosen, lies between 5 and 15 mm/min and corresponds to an average particle diameter of the order of 5 to 15 microns.

LeRosen (*1972*) has applied his method of characterization of adsorbents to a mixture of silicic acid and Celite. S, V_c, R and T_{50} were measured, the latter term representing the time in seconds in which the solvent penetrates to a depth of 50 mm in a dry column 9 mm in diameter and 75 mm long, under the vacuum from a water pump. These terms can be made more precise by the application of suffixes, e.g. $V_{c \text{ benzene}}$, $R_{p\text{-nitraniline}}$, etc. R is preferably measured at the lower front of the zones. For different samples of silicic acid, the value of S is almost constant, whereas those of V_c and T_{50} vary considerably. The best adsorbents are those with values of V_c between 10 and 50 mm/min and of T_{50} between 20 and 100 sec. The values of R differ with differing treatment of the adsorbents; the best chromatograms have been obtained with R values between 0.1 and 0.3; i.e. the speed of movement of the zones should be between one-third and one-tenth of the rate of movement of the solvent. (See also *2253*.)

In studying the system silicic acid-benzene-o-nitraniline, LeRosen (*1973*) found that the rate of displacement of the front of the zones relative to that of the solvent front was independent of the position of the zone and varied with the initial concentration and (to a smaller degree) with the initial volume. (See also *2741*.)

Temperature variation of R was studied by LeRosen and Rivet (*1980*) and found to be negligible for the range of 20° to 35°.

In a later publication, LeRosen et al. (*1977*) relate R to the structure of the compound adsorbed and to the type of bonds which hold the "adsorptive" (a term used by Weil-Malherbe (*3488*) for the substance being chromatographed) to the adsorbent. For the study of the interactions between the adsorptive and the adsorbent the compounds were classified in the following groups:

(1) electron donors,
(2) hydrogen donors (in hydrogen bond formation),
(3) electron acceptors,
(4) odd electron compounds (free radicals),
(5) ionic compounds,
(6) high dipole moment compounds (polar) and
(7) miscellaneous.

Practically all the adsorbents studied by these authors show an adsorption of electron donor substances; the adsorptives acted either as electron donors or as hydrogen donors.

For a quantitative standardization of adsorbent strength and adsorbant specificity towards adsorptives, LeRosen (1976) developed an equation for R in terms of adsorption

$$f = k\,s = \frac{T_a}{T_s} = \frac{1-R}{R}$$

where R = the ratio of rate of movement of the adsorbed compound in the column to the movement of the developing solvent in the column; R_L applies to the front edge of the zone, R_T applies to the rear edge;

f = the proportionality factor in an adsorption isotherm such that the amount of substance adsorbed on the adsorbent in equilibrium with one unit volume of solvent is obtained by multiplying the concentration in solution by f. The value of f may vary with concentration;

k = an equilibrium constant for the adsorption reaction;

s = surface area of the adsorbent in terms of moles per unit of adsorbent as defined for f;

T_s = statistical average time an adsorptive particle spends in solution between adsorptions;

T_a = statistical average time an adsorptive particle spends on the adsorbent during each adsorption.

This expression is similar to those of Martin and Synge (2156) and of Mayer and Tompkins (2176) for partition and ion exchange respectively. The general validity of such rate equations for adsorption chromatography was shown by Trueblood and Malmberg (3306).

To make possible a quantitative treatment in terms of bond energies the following hypothesis was made by LeRosen et al. (1976).

$$f = \left(\frac{1}{M_{sc}}\right)\left[\left(\frac{A_a D_s}{D_d}\right) + \left(\frac{D_a A_s}{A_d}\right) + \left(\frac{D_a^{\mathrm{H}} H_s}{H_a}\right) + \left(\frac{H_a D_s^{\mathrm{H}}}{D_d^{\mathrm{H}}}\right)\right] = \frac{1-R}{R}$$

where M_{sc} = applies to the sum of the molecular weights of all side chains in an adsorptive molecule;

D = donor strength of the substance in respect to the electron pair, D_a refers to the adsorbent, D_d to the developer, D_s to the adsorbed compound. These subscripts are also applied similarly to the terms listed below;

A = acceptor strength of a substance for an electron pair;

D^{H} = donor strength in terms of an electron pair donated to a hydrogen atom in hydrogen bond formation;

H = acceptor action of a hydrogen-bonding hydrogen for an electron pair.

The values for the above terms were taken as follows:

It was assumed that the electron pair of a tertiary nitrogen represented one electron donor unit, an unfilled orbital on a boron-alkyl boron one acceptor unit and the hydrogen of an alcohol one hydrogen-bond hydrogen unit. For the solvents, petroleum ether was taken as unity and the values for other solvents and for the adsorbents obtained form experimental data.

Also the following rules were necessary for the calculation of R values:

1. In aromatic amines the donor strength of the nitrogen was divided by the ratio of the total number of the most important resonance forms to those showing the electron pair on the nitrogen atom.
2. An internally bound hydrogen (in a hydrogen bond) was disregarded; this included the acid hydrogen in case of dimers.
3. The H_a interaction of the adsorbent was not isolated and is included in the A_a value; therefore this term is dropped from adsorption affinity calculations.

The interactions tendencies used by LeRosen et al. (1976) are shown in Table 8.

The validity for the factor $1/M_{sc}$, i.e. the effect of different chain lengths on adsorption, was confirmed in a later paper (Smith and LeRosen, 3018).

Agreement between the calculated results and the observed R values is in most cases excellent.

TABLE 8. Interaction tendencies of substances (LeRosen et al., 1976)

	A	D	D^H	H
Developers				
Petroleum ether	1+	1+	.....	1+
Benzene	25	5.8	.....	4.3
Adsorbents				
Special Filtrol	14,000	1,333	1,300	
Merck reagent silicic acid	4,800	2,570	120	
Florisil	2,000	1,160	260	
Merck heavy powder calcium carbonate . .	33	224	26	
Calcium acid phosphate dihydrate	77	..	42	
Magnesium oxide	47	3,350	190	
Calcium hydroxide	40	11,500	23	
Adsorbed compounds				
Amino N		1.00+		
Alkyl B	1.00+			
Alcohol H				1.00+
Alcohol O		0.17		
Acid or ketone O		0.20		
Nitro group		0.04		
Aromatic ring		0.002		

+ These values are arbitrarily assumed.

An example for the calculation of the R value for ethylaniline on Florisil (a synthetic magnesium silicate, Floridin Co., 200/300 mesh) is given below:

$D_{NH_2} = 1$. Fraction showing electron pair on nitrogen in resonance structures $= \frac{1}{4}$; for ethyl aniline use $D_s = 0.25$.
H_s for hydrogen $= 1$.
$M_{sc} = C_6H_5 + C_2H_5 = 77 + 29 = 106$.
D_d for petroleum ether is (by definition) 1.
H_d for petroleum ether is (by definition) 1.
A_a for florisil was determined as 2000.
D_H for florisil was determined as 260.

$$f = \left(\frac{1}{M_{sc}}\right)\left(\frac{A_a D_s}{D_d} + \frac{D_a^H H_s}{H_d}\right) = \frac{1}{106}\left(\frac{2000 \times 0.25}{1} + \frac{260 \times 1}{1}\right) = 7.15$$

$$R = \frac{1}{f+1} = \frac{1}{7.15+1} = 0.12.$$

The experimental value observed for R is 0.13.

Franc and Latinak (*1016–1018*) discuss the relation between the dipole moment and the R_f value of aromatic compounds and propose a modified LeRosen equation. For further details see the original papers. LeRosen *et al.* (*1974*) also consider the steric factors involved in the chromatography of disubstituted benzenes and find that their contribution to the R values limits the use of the original LeRosen equation.

6. SPECIFIC ADSORBENTS

(a) *Silica gel with "specific holes"*

A promising new approach to the preparation of very specific adsorbents has been tried by Dickey (*773*) following a proposal of Pauling. It consists in preparing silica gel in presence of the molecules (for instance methyl orange) for which specificity is desired. Then the adsorbed substance is extracted and leaves the silica gel with "specific holes" favouring the preferential adsorption of the molecule used in the preparation of the gel. Bernhard (*226*) has recently expanded the experiments of Dickey (*773*). The results obtained are best expressed in terms of "adsorption power" $=$

$$\frac{\text{moles of dye adsorbed/g of gel}}{\text{moles of dye in solution/g of solution}}$$

or of "% excess adsorption" $=$

$$\frac{\text{adsorption power of gel} - \text{adsorption power of control}}{\text{adsorption power of control}}$$

TABLE 9.   Adsorption power of specific silica gel (Bernhard, *226*)

$R_2N—C_6H_4—N=N—C_6H_4—SO_3^-$
I $R = CH_3—$
II $R = CH_3CH_2—$
III $(CH_3)_2N—C_6H_4—N=N—C_6H_4—SO_2NH_2$

	Adsorption power for		
Gel prepared with	*methyl orange*	*ethyl orange*	*III*
0	18	9.2	31
Methyl orange (I)	100	32	144
Ethyl orange (II)	90	74	120
p-Dimethylamino-p′-sulphonamido-azobenzene (III)	106	34	168

TABLE 10.   % Excess adsorption of specific silica gel (Bernhard, *226*)

	% Excess adsorption of		
Gel prepared with	*methyl orange*	*ethyl orange*	*III*
Methyl orange (I)	450	250	370
Ethyl orange (II)	380	700	290
p-Dimethylamino-*p*′-sulphonamido-azobenzene (III)	480	280	450

Tables 9 and 10 show a very definite preferential adsorption by "specific gels"; the specificity for methyl orange and for the sulphonamide III are closely parallel; it is evident that a negative charge at the *p*′-substituent is not a requirement for specificity. Dickey (*733a*) has studied the preparation and properties of "specific" silica gel.

Curti and Colombo (*679*) report an enrichment of 30% *l*-camphor-sulphonic acid by percolation of the racemic compound through a column of silica gel prepared in presence of *d*-camphorsulphonic acid. With mandelic acid an enrichment of only 10% was obtained.

(b) *Urea columns*

Cason *et al.* (*525*) have separated straight chain and branched chain fatty acids on columns of urea; the former are adsorbed, the latter not. This separation is based on the well known property of urea to form crystalline "clathrate" complexes with straight chain substances only; the molecules of the branched chain homologues are too large to fit in the "holes" in the interior of the urea crystals.

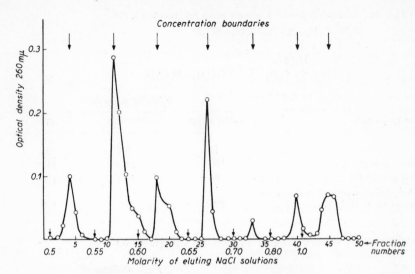

Fig. 14a. Stepwise elution of calf-thymus deoxyribonucleic acid from a column of histone-coated kieselguhr with solutions of sodium chloride (Brown and Watson, *417*)

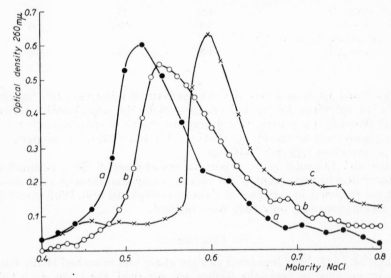

Fig. 14b. Elution diagrams of some deoxyribonucleic acids obtained by eluting with continuously increasing salt solutions (Brown and Watson, *417*)

Curve a — deoxyribonucleic acid from *E.coli*
Curve b — idem from calf thymus
Curve c — idem from human white bood cells

(c) *Histone columns*

Brown and Watson (*417*) have described the chromatographic fractionation of deoxyribonucleic acids on columns of kieselguhr covered with histone (the basic protein which is combined with the nucleic acids in the cells). Fig. 14 shows the elution curves obtained by stepwise elution, or by gradient elution. The specificity of the histone columns is shown by the fact that ribonucleic acids are not retained by them and that deoxyribonucleic acids of different origin give different elution curves (Fig. 14b).

(d) *Separation of enzymes*

Hockenhull and Herbert (*1421*) had already shown that amylase is strongly adsorbed on starch columns; French and Knapp (*1034*) have thus separated amylase from maltase. Lerman (*1967*) has used a similar principle for the purification of tyrosinase; it is known that this enzyme is specifically inhibited by azophenols; Lerman prepared columns of cellulose combined with azophenols (for instance cellulose–$OCH_2C_6H_4N=NC_6H_3(OH)_2$) for the enrichment of tyrosinase. Ghuysen (*1103*) has purified lytic enzymes from culture filtrates of Actinomyces on columns of sand covered with *E. coli* or with staphylococci. These experiments suggest a general method of purification of enzymes by adsorption on their substrates or on specific inhibitors.

(e) *Purification of antibodies*

Lerman (*1968*) has described the purification of antibodies on immunologically specific adsorbents.

An adsorbent specific for antibodies homologous to the simple hapten, *p*-azobenzenearsonate, was prepared from powdered cellulose (Solka-Floc BW 200) by coupling diazotized *p*-(*p*-aminobenzeneazo)-benzenearsonic acid with

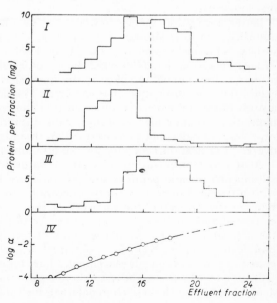

Fig. 15. Gradient elution chromatograms: I. whole serum; II. pooled eluate fractions 13–16 from I; III. pooled eluate fractions 17–24 from I. The arsanilate concentration, *a* (measured in *M*), appearing in each fraction is shown in IV (Lerman, *1968*)

cellulose which had been partially etherified with resorcinol. The arsonate cellulose thus obtained binds about 3 mg of antibody protein per ml of packed moist adsorbent. An analogous cellulose preparation in which *p*-azobenzenesulphonate was substituted for the homologous hapten was found not to adsorb detectable amounts of antibody. Separation of two distinctly different fractions of antibody was achieved by gradient elution with a specific hapten solution (sodium arsanilate) (Fig. 15). When a heterologous hapten (sodium sulphanilate) was used instead of sodium arsanilate, no protein could be eluted.

7. MODIFIED ADSORBENTS

The success of chromatographic separations depends amongst other factors on the uniformity of the adsorbent. Most of these contain a certain amount of impurities which may constitute more or less active centers causing irreversible adsorption or trailing of part of the substances to be separated. This is especially true for charcoal, which has been extensively used for separations of carbohydrates, peptides, antibiotics etc.

Porath and Li (*2562*) explain part of the irreversible adsorption encountered with high molecular weight substances by the simultaneous anchoring of part of the molecules at several points of the surface of the adsorbent. The tendency to irreversible adsorption is greatly reduced by mixing the charcoal with some inert or weak adsorbent, such as diatomaceous earth.

Better separation seems, however, to be obtained by pretreating the adsorbent with a substance ("saturator") which itself is strongly adsorbed. This technique, introduced by Tiselius and collaborators (*3269, 3188*) (see also Weiss, *3490–3492*) blocks especially the strongest adsorption centers. The charcoal modified in this manner is a much weaker adsorbent, exhibiting new adsorption characteristics. Hagdahl *et al.* (*1259*) have made a study of the influence of pretreatment of charcoal for the separation of low molecular compounds. Working with insulin and adrenocorticotropic peptides and *n*-decanol as saturator Porath and Li (*2562*) have studied the degree of saturation necessary for the attainment of optimal separation.

At a low concentration of a strong saturator, such as *n*-decanol, the adrenocorticotropic peptides studied could not be eluted completely, even with a strong displacer (saturated decanol solution). At higher concentrations of the saturator, the recovery of the peptides by displacement was enhanced and finally was quantitative. At an even higher concentration of the saturator, not far from the point of saturation concentration where practically complete recovery is obtained, a part of the peptides was easily eluted, even without displacement. The limit of concentration of the saturator at which elution of the active substance occurs in this manner is determined by the kind of charcoal used, the ratio of charcoal/diatomaceous earth, the concentration of substance, etc. This limit is called the "critical saturation point". Porath and Li point out that for the separation of peptides, chromatography ought to be performed on charcoal saturated as near as possible to the critical saturation point. See also Williams *et al.* (*3583*).

More recently, Porath (*2560*) has described a "step-graded adsorption column", which consists of segments filled with charcoal of different degrees of saturation (for instance from the top to the bottom: charcoal with 25%, 10%, 3%, 1%, 0.3%, 0.1% stearic acid). It is claimed that this arrangement allows chromatography of a mixture consisting of components of widely differing adsorbability with high total yield, in one single chromatogram and that efficient separations are obtained.

CHAPTER 4

Elution

1. PURIFICATION OF SOLVENTS

The purity of solvents is extremely important in chromatography as impurities can influence markedly the course of development (see p. 49). All organic solvents used should be dried and redistilled. Müller (*2305*) has indicated a method for the purification of light petroleum (with concentrated sulphuric acid) which increases markedly the heat evolved in contact with alumina. Dasler and Bauer (*700*) recommend that solvents that might contain peroxides should be purified by passage through an alumina column, the peroxides being quantitatively retained.

When esters are used, great care must be taken to remove the traces of acids and alcohols formed by hydrolysis, as these impurities increase the eluting power. Halogen-containing solvents (chloroform, trichlorethylene, etc.) must be kept over sodium or potassium carbonate in order to remove any free HCl formed.

2. ELUTION

The elution of colouring matters is usually carried out on zones that have been isolated mechanically and so presents no difficulty in principle. With the "liquid chromatogram" (see p. 3) the choice of a convenient series of solvents is more important. For example, the series of solvents used by Steiger and Reichstein (*3089*) for the fractionation of steroids is pentane, benzene-pentane 1 : 4, benzene-pentane 1 : 1, benzene, ether and acetone.

Trappe (*3294*) has established an "eluotropic series of solvents", useful for the fractionation of lipids (see Table 11). Strain (*3136*) has suggested an analogous series. Jacques and Mathieu (*1532*) have stated that the eluting power of solvents is proportional to their dielectric constant ε; they have verified that the eluants function by being themselves adsorbed. In a mixture of two eluants, the solvent with the highest eluting power (having the higher value of ε) is the most strongly adsorbed.

TABLE 11. Eluting power of solvent (in increasing order)

Trappe (3294)	Strain (3136)	Jacques and Mathieu (1532)	ε
Light petroleum	Light petroleum 30–50°	Hexane	1.88
Cyclohexane	Light petroleum 50–70°	Benzene	2.29
CCl_4	Light petroleum 70–100°	Ether	4.47
Trichloroethylene	CCl_4	Chloroform	5.2
Toluene	Cyclohexane	Ethyl acetate	6.11
Benzene	CS_2	Dichlorethane	10.4
Dichloromethane	Anhydrous ether	Butanol-2	15.5
Chloroform	Anhydrous acetone	Acetone	21.5
Ether	Benzene	Ethanol	26
Ethyl acetate	Toluene	Methanol	31.2
Acetone	Esters of organic acids		
n-Propanol	1,2-Dichlorethane		
Ethanol	Alcohols		
Methanol	Water		
	Pyridine		
	Organic acids		
	Mixtures of acids or bases, water, alcohols or pyridine		

The positions of ether and acetone in the series due to Strain do not seem to be justified; these two solvents should be placed after benzene and toluene. Chapter 5 shows that such definite series cannot be established for all substances.

Moseley *et al.* (*2289*) have studied the developing (eluting) power of a series of pure solvents and of two- and three-component mixtures with silicic acid or calcium hydroxide as adsorbent and *o*-nitroaniline as adsorptive.

No correlation of dipole moment of the solvent with adsorption affinity of the adsorptive was found. In the correlation of dielectric constant of the solvent with its developing power some exceptions were encountered; for example, the developing power of acetone was found greater than that of nitrobenzene, whereas acetone and nitrobenzene have dielectric constants of 20.7 and 34.8 respectively. This discrepancy is explained by Moseley *et al.* (*2289*) as the effect of hydrogen bonding: acetone being capable of hydrogen bonding with the

$$-Si \overset{\displaystyle O}{\underset{\displaystyle OH}{\diagup\!\!\!\!\diagdown}}$$

group of silicic acid or the hydroxyls of calcium hydroxide and thus exerting
a displacement effect on the adsorptive, whereas nitrobenzene is much less
capable of hydrogen bonding.

Knight and Groennings (*1747*) have determined the following eluotropic
series on silica gel (increasing order of adsorbability of the solvent):

Heptane
Diisobutylene
Benzene
Isopropyl chloride
Diisopropyl ether
Ethyl ether
Ethyl acetate
sec.-Butyl alcohol
Ethyl alcohol
Water
Acetone
Methanol
Pyruvic acid

TABLE 12.   The eluting power of various solvents in 3 % solution in light petroleum for
β-carotene adsorbed on chalk (Bickoff. *241*)

Eluant	$R_L \times 100$	Eluant	$R_L \times 100$
CCl$_4$	20	Ethyl butyrate	56
n-Amyl ether	22	Butyl acetate	63
Tetrachlorethylene	23	Dioxan	64
Methanol	25	Anethole	65
Toluene	29	Ethyl acetate	67
Benzene	32	Acetone	68
Ethyl ether.	33	Pyridine	72
Chloroform	35	Diacetone alcohol	74
Methylene chloride	38	Cetyl alcohol	77
Ethylene chloride	40	Acetophenone	79
Phenyl ether	42	Octanol	90
p-Cresyl methyl ether	44	Ethanol	93
sym.-Tetrachlorethane	45	Phenyl cellosolve (⁺)	95
Ethyl laurate	55	Butyl cellosolve (⁺⁺)	95
		Methylcarbitol (⁺⁺⁺)	100

(⁺)   C$_6$H$_5$OCH$_2$CH$_2$OH.
(⁺⁺)   CH$_3$CH$_2$CH$_2$CH$_2$OCH$_2$CH$_2$OH.
(⁺⁺⁺) Diethylene glycol monomethyl ether.

Bickoff (*241*) has classified a series of solvents based on their eluting power in 3 % solution in light petroleum for β-carotene adsorbed on chalk (Table 12). It is astonishing to see that methyl alcohol is classified as a weak eluant ($R_L = 25$) whereas ethyl alcohol is described as a strong eluant ($R_L = 93$). This fact has been repeatedly verified.

It is clear that the classification of solvents on the basis of their eluting power has meaning only for a given system under the most precise conditions.

The elution of organic acids adsorbed on alumina often necessitates the use of alcohol or ether containing 5 to 10 % acetic acid or 5 % potassium ethoxide. Appreciable amounts of aluminium acetate or other mineral salts are often found in the filtrates; these may be removed by taking up the residue in ether and washing the solution with water.

The volume of each eluant should be at least double the weight of the adsorbent. In "liquid" of "flowing" chromatography elution should be continued with successive portions of solvent until only traces of substance are eluted.

3. GRADIENT ELUTION

In discussing column chromatography, Tiselius has said that "one of the most important practical problems of chromatography is to eliminate tailing as far as possible". One solution to the problem is "gradient elution", which reduces tailing and thus improves the separation achieved. Tailing of bands is caused by the fact that under usual experimental conditions the adsorption isotherms are not linear, i.e. that the percentage of the substance which is adsorbed depends upon its concentration in solution. Most frequently adsorption increases strongly with dilution of the substance. With the usual form of "flowing chromatogram", where elution is obtained by a succession of different eluants with increasing eluting power, substances which trail on the column can be eluted partially with one solvent and then with the next one (Fig. 16).

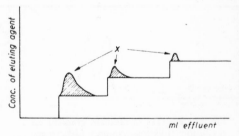

Fig. 16. Schematic representation of a stepwise elution experiment in which one substance occurs three times (Alm, Williams and Tiselius, *42*)

If, however, during elution the composition of the eluting solvent is changed gradually in such a manner that the trailing end of the band comes in contact with a more strongly eluting solvent, trailing can be diminished.

When the equilibrium isotherm is linear, gradient elution may also be advantageous since an increase of the adsorbability at the front of a band and a corresponding decrease at the rear will sharpen the elution curves and change their shape from the Gauss type of curve to one with clear cut ends. Again the "right" gradient must be used otherwise a band may either be cut in two or two components not resolved.

Fig. 17a shows the effect of the steepness of the eluting gradient; resolution of the two substances (dehydroisoandrosterone and androsterone) is improved by slowing down the gradient. With too slow a gradient, however, elution of the substances takes an unduly long time and they are eluted at very low concentrations, the elution curves being broadened out. Lakshmanan and Lieberman (*1867*) point out that "a suitable concentration curve would be one in which the rate of increase of concentration was initially small and increased with time, a requirement satisfied by a concentration curve concave upward".

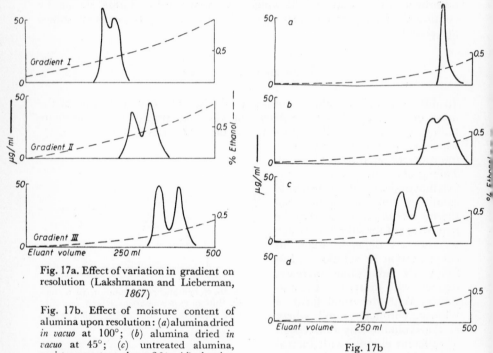

Fig. 17a. Effect of variation in gradient on resolution (Lakshmanan and Lieberman, *1867*)

Fig. 17b. Effect of moisture content of alumina upon resolution: (*a*) alumina dried *in vacuo* at 100°; (*b*) alumina dried *in vacuo* at 45°; (*c*) untreated alumina, moisture content about 3%; (*d*) alumina equilibrated over a saturated solution of NaBr·2H$_2$O, moisture content about 7.5%. Gradient same as III in Fig. 17a (Lakshmanan and Lieberman, *1867*)

Fig. 17b

Fig. 17b shows the importance of using the proper gradient with the proper adsorbent; too weak an adsorbent gives no resolution. In certain cases, too steep a gradient may produce two peaks of one substance.

Devices for producing gradients are described by Alm *et al.* (*42*), Donaldson *et al.* (*802*), Mader (*2099*), Bannister *et al.* (*134*), Bock and Ling (*286*), Lakshmanan and Lieberman (*1867*) and Boman (*305*). See Figs. 18a and b, also Cherkin *et al.* (*563*).

Busch *et al.* (*479*) have applied gradient elution for the separation of organic acids on Dowex-1 columns and Donaldson *et al.* (*802*) separated organic acids on silica with a gradient (see Chapter 22). Tiselius *et al.* (*42, 1259, 3263*) have described the theory of the method in detail and its application to the separation of mixtures of amino acids, sugars and peptides. Alm (*41*) has described the gradient elution of oligosaccharides and Lakshmanan and Lieberman (*1867*) the gradient elution of urinary ketosteroids. For gradient elution with ion exchange resins see also Freiling (*1032*), Stein *et al.* (*3092*) and Moore and Stein (*2270*). Gradient elution in inorganic separations was discussed by Williams (*3582*) and Chalkley and Williams (*542*) using both solvent and temperature gradients applied along the column. Paper chromatography with a gradient was reported by Lederer (*1931*) using the ascending development.

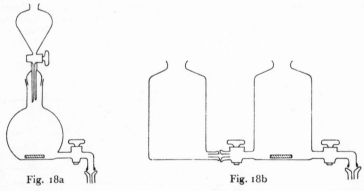

Fig. 18a   Fig. 18b

Fig. 18a. Apparatus for gradient elution, allowing an exponential increase of the eluting force. A flask fitted on a magnetic stirrer and containing the first solvent is surmounted by a separation funnel containing the second solvent. Fig. 18b. Apparatus for gradient elution, allowing a linear increase of the eluting force. Two jars are connected at the bottom by a tube with a stopcock, the first one is mounted on a magnetic stirrer

4. ELUTION CURVES

Recent papers (e.g. Moore and Stein, *2265*) have shown that it is often useful to analyse separate small fractions of the filtrate; in this way changes in the composition of the filtrate may be followed more easily and elution curves may be drawn (quantity of substance eluted as a function of the volume of eluant). These curves enable one to follow the behaviour of the substances to be separated and to judge the degree of separation obtained (see Fig. 17a, 17b). Quantitative results are obtained by measuring the area enclosed by the curves. (See also *3488*.)

5. THE DESCRIPTION OF CHROMATOGRAMS

In the case of chromatography of pigments, it is useful to represent the appearance of a column after development by a drawing indicating the width of the bands and their colour (see Fig. 19). Bickoff (*241*) utilizes an

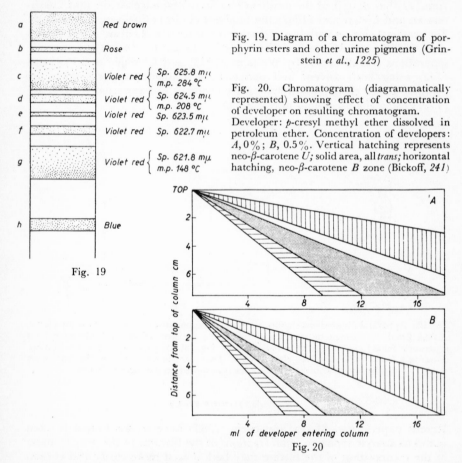

a	Red brown
b	Rose
c	Violet red { Sp. 625.8 mµ / m.p. 284 °C
d	Violet red { Sp. 624.5 mµ / m.p. 208 °C
e	Violet red Sp. 623.5 mµ
f	Violet red Sp. 622.7 mµ
g	Violet red { Sp. 621.8 mµ / m.p. 148 °C
h	Blue

Fig. 19

Fig. 19. Diagram of a chromatogram of porphyrin esters and other urine pigments (Grinstein *et al.*, *1225*)

Fig. 20. Chromatogram (diagrammatically represented) showing effect of concentration of developer on resulting chromatogram.
Developer: *p*-cresyl methyl ether dissolved in petroleum ether. Concentration of developers: *A*, 0%; *B*, 0.5%. Vertical hatching represents neo-β-carotene *U*; solid area, all *trans*; horizontal hatching, neo-β-carotene *B* zone (Bickoff, *241*)

Fig. 20

excellent method of representing the separation of pigments as a function of the nature and quantity of the developing solvent (see Fig. 20).

Elution curves, tables or graphs (*150*) are also very useful.

CHAPTER 5

Chemical constitution and chromatographic behaviour

1. GENERAL CONSIDERATIONS

The relationship between the chemical constitution of a substance and its behaviour on a chromatographic column depend on whether or not adsorption, partition or ion exchange are operating. It is more convenient to study each of these cases separately. It must be emphasized that it is not always easy in any given case to decide which of these phenomena is the most important. A great many cases of partition chromatography probably also involve adsorption (see Chapter 15); ion exchangers can function equally as adsorbents (see Chapter 14) and this makes it more difficult to understand the mechanism of such separations.

The great majority of chromatographic separations carried out in organic media on carbon, alumina, calcium hydroxide, calcium carbonate, magnesia, silica gel, etc. as well as on charcoal in aqueous media, are due to true adsorption. It is necessary to distinguish between non-polar adsorbents (e.g. charcoal) where it is the undissociated molecule which is adsorbed (Kipling, *1723*) and adsorbents such as oxides and salts, where polar forces are acting between the solute and the adsorbent.

Many general rules concerning the position of organic substances on adsorption columns have been produced since the rebirth of Tswett's method in 1931. These were summarized in 1939 by one of the authors (*1899*). They concerned principally the role of double bonds and hydroxyl groups: the more double bonds and hydroxyl groups a molecule contains, the more strongly it is adsorbed. The carotenes arrange themselves on columns of alumina, calcium hydroxide, and magnesia following the number of double bonds; the xanthophylls (carotenols) follow the number of hydroxyl groups. Colour and adsorption are here functions of the same chemical structures, and run side by side. An aldehyde or a ketone is less adsorbed than the corresponding alcohol, esters are in general more adsorbed than hydrocarbons and less than ketones or aldehydes.

A rough classification is shown in the following list, the first members of which are the most strongly adsorbed, the latter members are the least strongly adsorbed:

acids and bases > alcohols, thiols > aldehydes, ketones > halogen containing substances, esters > unsaturated hydrocarbons > saturated hydrocarbons.

A more detailed classification is possible for substances belonging to the same group or possessing the same carbon skeleton.

The effect of functional groups on adsorption affinity has been studied by Brockmann and Volpers (*398*) with a series of *p-substituted stilbenes* and *azobenzenes*. The compounds were adsorbed on alumina from benzene or carbon tetrachloride. These experiments lead to the sequence of functional groups (in order of decreasing adsorption affinity) shown in Table 13.

TABLE 13.   Functional groups in order of decreasing adsorption affinity (Brockmann, *387*)

$R = C_6H_5-CH=CH-C_6H_4-$	$R = C_6H_5-N=N-C_6H_4-$
R—COOH	R—COOH
R—CONH$_2$	
R—OH	R—OH
	R—NH—Ac
	R—O—Ac
R—NH$_2$	R—NH$_2$
R—NH—Ac	R—O—Bz
R—O—Ac	
R—COOCH$_3$	R—COOCH$_3$
R—N(CH$_3$)$_2$ R—O—Bz	R—N(CH$_3$)$_2$
R—NO$_2$	R—NO$_2$
R—OCH$_3$	R—OCH$_3$
R—H	R—H
(Ac = CH$_3$—CO—	Bz = C$_6$H$_5$—CO—)

Compounds separated by a horizontal line could be separated, the others were not fully separable. It can be seen that the same order was found in both series.

Stewart (*3109*) has found the following order of increasing adsorption, for *monosubstituted anthraquinones*: halogeno, nitro, arylamino, alkylamino, amino, acylamino, hydroxy group in the side chain, and hydroxy group attached to the nucleus. Adsorption generally increases with increasing number of substituent groups of the same type. The introduction of further groups of a different type may either increase or decrease the adsorption; thus methyl, halogeno and arylamino groups tend to decrease and amino, methoxy and hydroxy groups to increase the adsorption.

Hydrogen bonding diminishes adsorbability; thus Hoyer (*1465*) has stated that 2-hydroxyanthraquinone is more strongly adsorbed on silica than 1,4,5-trihydroxy- or even 1,4,5,8-tetrahydroxyanthraquinone, the latter substances having their hydroxyl groups in positions where hydrogen bonding occurs.

Of the hydroxymethylanthraquinones shown in Fig. 21, the substances shown at the left side are less adsorbed on silica gel than the isomers at the right side (Hoyer, *1467*). Similar differences were also found for hydroxy-nitroanthraquinones, aminoanthraquinones and mercaptoanthraquinones (Hoyer, *1466*, see also *1464*). Apparent exceptions observed with 1-nitro-2-amino- and 1-nitro-2-hydroxy-anthraquinones could be explained as the result of steric hindrance of the 1-nitro group.

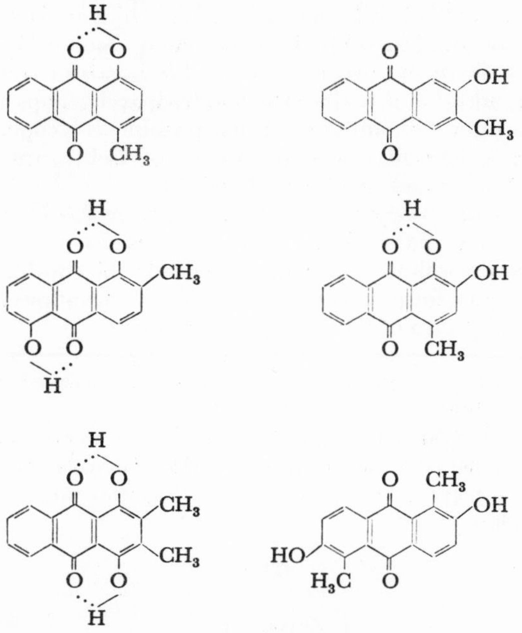

Fig. 21. Hydroxymethylanthraquinones:
those on the left are less adsorbed on silica gel than those on the right (Hoyer, *1467*)

Hydrogen bonding explains also, why 2-nitroresorcinol is less adsorbed (on silica) than *m*- or *p*-nitrophenol, and *o,o'*-dihydroxybenzophenone less adsorbed than *p*-hydroxybenzophenone. Ongley (*2414*) has made similar observations with the hydroxybenzoic acids.

Hydrogen bonding can also have an influence on the eluting power of solvents (see p. 39).

Strain (*3139*) has made an interesting contribution to the problem of the relationship between the nature of the adsorbent and the chemical structure of the substance adsorbed. By studying the relative positions of seven carotenoids on *sucrose* columns he showed that sucrose has a much higher affinity for hydroxy groups than for conjugated double bonds. Cryptoxanthin, with its unique hydroxyl group, is much more strongly adsorbed than lycopene which contains two more double bonds, or than rhodoxanthine which has two more carbonyl groups and one more double bond. Zeaxanthin and lutein, on the other hand, which differ only in the position of one double bond are almost inseparable on a column of sugar. *Magnesia*, on the contrary, exerts an attraction particularly for conjugated double bonds and less so for hydroxyl groups. A difference in position of one double bond allows an excellent separation of α- and β-carotenes and of zeaxanthin and lutein. Lycopene with 13 double bonds (of which 11 are conjugated) is more strongly adsorbed than zeaxanthin with 11 conjugated double bonds and two hydroxyl groups. Here the two isolated double bonds of lycopene exert a greater effect on adsorbability than the two hydroxyl groups of zeaxanthin. (It is true that a rigorous comparison is not possible as lycopene is aliphatic and zeaxanthin is bicyclic.) The properties of Celite are intermediate between those of sugar and magnesia (Table 14).

The differences in adsorption observed by Strain (*3139*) with different solvents, which affect particularly the relative position of rhodoxanthin, are explicable by the influence of polar solvents on the equilibrium between the ketonic and enolic forms of this pigment. These variations are paralleled by changes in the spectra of solutions of the pigments.

TABLE 14.   Position of seven carotenoids on columns of sugar, celite and magnesia (after Strain, *3139*)

Solvent: light petroleum with 4% to 25% of acetone; the most strongly adsorbed pigments are indicated at the top of the columns, the less adsorbed at the bottom. The brackets indicate pigments whose separation is not complete.

Sugar	Celite	Magnesia
{ Zeaxanthin	{ Zeaxanthin	Rhodoxanthin
{ Lutein	{ Lutein	Lycopene
Cryptoxanthin	Rhodoxanthin	Zeaxanthin
Rhodoxanthin	Cryptoxanthin	Lutein
{ Lycopene	{ Lycopene	Cryptoxanthin
{ β-Carotene	{ β-Carotene	β-Carotene
{ α-Carotene	{ α-Carotene	α-Carotene

It is thus understandable that solvents which increase the proportion of groups for which the adsorbent has strong affinity increase also the adsorbability and vice versa (see also the chapter on "Elution").

The reviews of Strain (*3136*), Zechmeister and v. Cholnoky (*3679*), Saenz et al. (*2777–2779*) should be consulted for information on the influence of different substituents (–OH, NH_2, halogens, $-SO_3H$, etc.) on the adsorbability of organic dyestuffs. The separation of stereoisomers is dealt with in Chapter 39.

In certain cases the *form of the surface* of the adsorbent can explain differences in the behaviour of the substances to be adsorbed (see Chapter 19 for the case of paraffins and alkyl-naphthalenes described by Nederbragt and De Jong (*2343*)). Sometimes the *spatial structure* of the molecules to be adsorbed plays an important part in chromatographic behaviour. For example, Kofler (*1759*) has found that methyl groups in the vicinity of the hydroxyl group of vitamin E diminish its adsorption (see Chapter 35).

Charcoal retains specifically those substances possessing an aromatic nucleus (Cassidy, *3433*, Tiselius, *3253*, *3258*, Schramm and Primosigh, *2880*, Fromageot et al., *1054*). This is probably due to the already mentioned fact (p. 23) that the –C–C– spacings in graphite are of the same order as those in benzene.

In the case of strongly polar substances like fatty acids the presence of one or more double bonds does not necessarily increase the adsorption, in fact the contrary is often true. (See the work of Kaufmann, Chapter 22). We have already mentioned the important relation between boiling point and retention volume of members of homologous series found by Claesson (*577*) and extended by Hall and Tiselius (*1278*) to isomeric members of series with the same functional group (see p. 4 and Table 1). An example of the quantitative treatment of the relation between functional groups and adsorption has been given by LeRosen et al. (see pp. 28–32).

2. THE RELATIVE POSITION OF ZONES ON A CHROMATOGRAPHIC COLUMN

The adsorption of an organic substance is a function of its chemical structure, the nature of the adsorbent and of the solvent. Papers published in the years following 1931 allowed the determination of certain rules connecting adsorption and chemical constitution (the number of conjugated double bonds, the number of hydroxyl groups, etc.) and the classification of adsorbents and solvents in series on the basis of their adsorbing and eluting properties. It was supposed that in general the order of the zones of various substances was always the same.

Later on, however, a number of observations have accumulated to show that the order of zones can be varied with the adsorbent (*839*, *1971*, *3137*) or the solvent (*2777*, *2779*, *3577*) used. Finally, Strain (*3138*) gave a large number of examples of the change of zone order with small changes in the nature of the adsorbent, the composition of the solvent mixtures and also the temperature. The examples given by Strain (*3138*) are as follows:

Solvent effects. In light petroleum or benzene, chlorophyll *b* forms a yellow-green band above the yellow band of neoxanthin (xanthophyll $C_{40}H_{56}O_4$) on Celite. After the addition of 25 % of acetone to the light petroleum or benzene, chlorophyll *b* is adsorbed below neoxanthin. Fucoxanthin $(C_{40}H_{56}O_6)$ is adsorbed above violaxanthin $(C_{40}H_{56}O_4)$ on a sugar column with 1,2-dichloroethane as solvent. The order of the zones is inverted if light petroleum containing 0.5 % of methanol or any other aliphatic alcohol is used.

Effect of impurities in the solvents. Traces of alcohol present in the solvents often change zone order. On sugar columns with light petroleum as solvent fucoxanthin is adsorbed below chlorophyll *a*; if 0.5 % of amyl alcohol is added to the solvent, fucoxanthin is adsorbed above chlorophyll *a*. The same effect is observed if 0.1 % cholesterol or ergosterol is dissolved in the light petroleum.

The use of different solvents in developing a chromatogram. On changing solvents in the course of development a change of zone order is often observed. Lutein $(C_{40}H_{56}O_2)$ with dichloroethane as solvent, forms a band below taraxanthin $(C_{40}H_{56}O_4)$ on a column of magnesium oxide - Celite. On washing with light petroleum containing 25 % of acetone, taraxanthin is displaced more rapidly than lutein and moves to a position below the lutein band.

pH effects. It is obvious that pH changes will exert a marked effect on the adsorption of polar substances in aqueous media. One example of inversion of zone order is given by Strain: bromothymol blue is adsorbed below fluorescein on cellulose columns in aqueous ammonia, but in neutral or acid solution bromothymol blue is adsorbed above fluorescein.

The effect of concentration changes. At very high concentrations most substances move more rapidly through the column than at low concentrations. For this reason, an increase in concentration of a strongly adsorbed substance (e.g. fluorescein) can accelerate the rate of movement of the zone to such an extent that it passes the zone of the second substance originally adsorbed further down the column (e.g. bromothymol blue).

Changes in the adsorbent. With dichloroethane as solvent and a mixture of magnesium oxide and Celite as adsorbent, zeaxanthin $(C_{40}H_{56}O_2)$ is adsorbed above fucoxanthin $(C_{40}H_{56}O_6)$; with sugar as the adsorbent zeaxanthin is adsorbed below fucoxanthin. On Celite or magnesium silicate, bromothymol blue in ammoniacal solution is adsorbed above fluorescein; on Celite-magnesium oxide, cellulose or filter paper, bromothymol blue is less adsorbed than fluorescein.

The degree of hydration or the impurities present in the adsorbent can also cause changes in the relative positions of the zones.

*The relation between temperature and adsorption.*On sugar columns, lutein is adsorbed below chlorophyll *a* at 95° in decalin containing 0.5 % propanol, but at 20° under otherwise identical conditions, lutein is adsorbed above chlorophyll *a*.

The adsorption of mixtures. Table 15 (Strain, *3138*) shows six different results obtained by changing solvents and adsorbents with four substances (24 different results are theoretically possible with four components).

Brockmann (*387*) has found different orders of adsorption of the azo-dyes he uses for standardization of alumina (see p. 25) on different adsorbents, such as SiO_2, MgO, $CaSO_4$ and $CuSO_4$. The adsorption sequence of acidic or basic dyes is influenced by the acidity or alkalinity of the adsorbent (Table 16).

The order of these dyestuffs on alumina or $CaSO_4$ columns is also dependent on the solvent used (Brockmann, *387*).

Double zoning. Usually, in chromatographic columns, each substance yields only one zone; Strain (*3140*) has studied the conditions which may lead to the formation of two zones, by one single pigment, a phenomenon which is called "double zoning" and which has also been described by Schroeder (*2883*).

Two zones were obtained by first adsorbing zeaxanthin on magnesia in presence of propylene glycol, a strongly adsorbed substance, and then washing the column with light petroleum containing 25 % of acetone. Adsorption of an acetamide solution on a column of magnesia followed by adsorption of a solution of zeaxanthin also yielded two pigment zones. With weakly adsorbed contaminants, similar phenomena were observed.

When adsorbing xanthophylls from light petroleum on powdered sugar, very narrow bands of pigments of high concentration are formed; if one adds a small concentration of an eluant (0.5 % *n*-propanol) then the pigment is eluted, but as it is only sparingly soluble in the surrounding solvent,

TABLE 15. The zone order of four pigments as a function of the adsorbent and the solvent (Strain, *3138*)

Adsorbent	Sucrose		Magnesia + Celite (1:1)		Sucrose	Celite
	Light petroleum +		Light petroleum +		Light petroleum +	
Solvent	1% ethanol	5% acetone + 0.75 %ethanol	3% ethanol	25% acetone	5% acetone + 0.5% ethanol	
Zone order	fucoxanthin	fucoxanthin	chlorophyll *b*		chlorophyll *b*	
	chlorophyll *b*	chlorophyll *b*	chlorophyll *a*		fucoxanthin	
	zeaxanthin	chlorophyll *a*	fucoxanthin	zeaxanthin	chlorophyll *a*	zeaxanthin
	chlorophyll *a*	zeaxanthin	zeaxanthin	fucoxanthin	zeaxanthin	chlorophyll *a*

TABLE 16.   Adsorption sequence of dyes as influenced by the adsorbent (Brockmann, *387*). Solvent: benzene-light petroleum 1 : 4

Alkaline Al$_2$O$_3$	*HCl treated Al$_2$O$_3$*
Hydroxyazobenzene	Hydroxyazobenzene
Aminoazobenzene	Sudan Red
Sudan Red	Aminoazobenzene
Sudan Yellow	Sudan Yellow
Methoxyazobenzene	Methoxyazobenzene

NaOH treated SiO$_2$	*HCl treated SiO$_2$*
Sudan Red	Sudan Red
Aminoazobenzene	Hydroxyazobenzene
Hydroxyazobenzene	Aminoazobenzene
Sudan Yellow	Sudan Yellow
Methoxyazobenzene	Methoxyazobenzene

the greater part precipitates forming an upper dark band, from which the descending solvent carries away continuously small fractions which will form a lower lighter band; on continuing the development, the upper band of precipitated pigment will dissolve entirely.

By adsorption of two xanthophylls from light petroleum containing a strongly adsorbed impurity, followed by adsorption of more of the pigment solution without impurity, Strain (*3140*) obtained as many as three or four coloured zones. These experiments of Strain show the necessity of working under well-defined conditions and of using solutions of uniform composition.

3. THE IDENTIFICATION OF SUBSTANCES BY CHROMATOGRAPHY

Strain (*3139*) has emphasised the importance of his results for the chromatographic identification of chemical substances. Chromatographic homogeneity can only be proved when a number of adsorbents and solvents have been tried. It is clear that mixtures declared to be inseparable by chromatography could be dealt with by a judicious choice of solvent, adsorbent, concentration and temperature.

This applies equally to the identification of compounds by mixed chromatograms. It is often preferable to carry out this proof of identity with three columns run simultaneously (two columns for each of the separate substances and the third for the mixture) (*1898*). This is necessary because two different pigments are sometimes adsorbed in one apparently homogeneous zone at slightly different heights and with slightly different colours.

CHAPTER 6

The chromatography of colourless substances

With colourless substances a number of possibilities present themselves:

(i) The column can be cut arbitrarily into a number of sections after development and each can be separately eluted.

(ii) The column can be washed successively with a series of solvents of increasing eluting power. This is the most frequently adopted technique since its introduction by Reichstein (*2656, 2657, 3089*); it is usually described as a "liquid chromatogram" although this term is not very appropriate; "flowing chromatogram" is a better term. The series of solvents usually adopted for the chromatography of colourless substances of a lipid nature is given on p. 39.

(iii) Various methods can be used to render the zones visible (see below).

Sometimes colourless substances give rise to visible zones. Waldschmidt-Leitz and Turba (*3450, 3451*) obtained clear zones of peptides on acidic earths and Trappe (*3295*) observed that lipids give rise to almost transparent zones on columns of silica gel. Bendall *et al.* (*182*) obtained visible brownish zones of glycine and creatinine on the synthetic resin Zeo Karb which is almost black.

1. CONVERSION INTO COLOURED DERIVATIVES

(a) *Derivatives of alcohols*

In sugar chemistry, *p*-phenylazobenzoyl chloride, $C_6H_5N = NC_6H_4COCl$ is used to obtain coloured esters (see Chapter 28). The same reagent transforms sterols into coloured esters which form coloured zones on chromatographic columns (*594, 1497*). Bielenberg *et al.* (*243, 244*) coupled phenols with diazotised *p*-nitroanilin, in order to separate chromatographically the derived coloured azo-compounds.

(b) *Derivatives of aldehydes and ketones*

The chromatography of the 2,4-dinitrophenylhydrazones of aldehydes and ketones recommended by Strain (*3136*) has been used by many workers and is reviewed in detail in the chapter on aldehydes and ketones.

(c) *Derivatives of acids*

Silberman and Silberman-Martyncewa (*2983*) describe the chromatographic separation of the bile acids after reaction of their salts with p-(ω-bromomethyl)-azobenzene $C_6H_5-N=N-C_6H_4-CH_2Br$. The esters form orange zones on magnesium carbonate.

(d) *Derivatives of nitrogeous substances*

Karrer *et al.* (*1642*) used p-phenyl-azobenzoyl chloride in order to prepare coloured N-acyl derivatives of amino acids. Their methyl esters are adsorbed on basic zinc carbonate, the coloured zones given by the derivatives of glycine, alanine, leucine, and valine were well separated. The N-(p-phenyl-azobenzene)-sulphonyl derivatives of amino acids, $R-CH(COOH)-NH-SO_2-C_6H_4-N=N-C_6H_5$, have been separated on alumina by Reich. Lemberg *et al.* (*1954, 1955*) separate the aminobenzoic acids after diazotisation and coupling with dimethyl-α-naphthylamine.

Wieland *et al.* (*3546*) and Karrer and Schmid (*1643*) have separated the reineckates of the curare alkaloids.

2. THE ADDITION OF COLOURED INDICATORS

Graff and Skau (*1184*) coloured a column of heavy magnesia with phenol red as a means of following the separation of mixtures of long-chain fatty acids.

Criddle and LeTourneau (*655*) have described a fluorescent-indicator method for the determination of hydrocarbon types in petroleum; a sample containing traces of fluorescent dyes is chromatographed on silica gel, with alcohol as displacing agent. The components of the sample are aligned in the column on the basis of adsorbability: paraffins first, followed by olefins, aromatics and alcohol last. The boundaries between the zones are made visible in ultraviolet light by the fluorescent dyes, and the composition of the sample can be determined by measuring the zones, the lengths of which are proportional to the concentrations of the components of the sample. Ellis and LeTourneau (*886*) have employed a mixture of ordinary and fluorescent dyes as indicators for the determination of hydrocarbon types and total oxygenated solvent content in lacquer thinners. The indicator chromatographic analysis of mixtures of simple organic compounds has also been studied by Knight and Groennings (*1747*).

Trautner and Roberts (*3297*) use dimethylaminoazobenzene as a coloured indicator in the separation of alkaloids on silica gel. The adsorbent is coloured red except in the positions in which the alkaloids are adsorbed.

Organic compounds which react with FeCl$_3$ to give coloured compounds can be made visible by first saturating the alumina with an ether solution of FeCl$_3$ (Valentin and Kirchübel, *3359*). See also Knight and Groennings (*1747*).

3. STREAKING TECHNIQUES

This method, proposed by Zechmeister *et al.* (*3680*), consists in applying a thin streak of a convenient reagent, with a brush, down the length of the column after it has been extruded from the tube. Zechmeister and McNeely (*3683*) used a solution of 1 % KMnO$_4$ in the separation of *cis* and *trans* stilbenes; Wolfrom *et al.* (*247, 248*) used the same solution in the separation of sugar acetates on magnesol; Lew *et al.* (*1994*) used permanganate, pH indicators, and 2,6-dichlorophenol-indophenol with a series of derivatives of sugars. Bell (*177*) showed up the position of methylated sugars on a silica gel column by applying an alcoholic solution of α-naphthol followed by concentrated sulphuric acid (Molisch reaction).

As it is often difficult to extrude the column intact from its tube, Turkevich (*3332*) has suggested carrying out the extrusion by applying air pressure to the bottom of the column. Lew *et al.* (*1995*) have found that the use of aluminium tubes allows an easier extrusion of the columns. Georges *et al.* (*1099*) use tapered glass tubes. LeRosen *et al.* (*1978, 1980*) have tabulated a series of streak reagents which can be used for the detection of a wide variety of substances. Olefins are detected with permanganate, amines with a diazonium reagent or with bromine in carbon tetrachloride, ketones with 2,4-dinitrophenylhydrazine in HCl, etc. Further streak reagents are listed by LeRosen *et al.* (*1979*). See also the chapters on carbohydrates and on proteins.

4. FLUORESCENCE

Examination of the column with ultra-violet radiation suggested by Karrer and also by Winterstein (see *1899, 3136, 3679*) allows the localisation of colourless substances. This technique has been used in the separation of methylglucoses (*2377*), *p*-phenyl-phenacylates of fatty acids (*1728*), alkaloids (*996, 1778*), the isolation of natural polyenes (*3686*), vitamins A and D (*767, 2306*) and the estimation of vitamin E (*1757*) and vitamin K (*1758*). See also Brumberg (*428*).

The preparation of fluorescent derivatives has not been studied systematically from the point of view of chromatography. The fluorescent hydrazine

(9-hydrazino-acridine) prepared by Boxer and Jelinek (*349*) offers certain possibilities for the separation of colourless substances containing carbonyl groups.

Brockmann and Volpers (*397*) have separated colourless substances on alumina columns impregnated with a fluorescent material. The adsorption zones show up in monochromatic ultra-violet light as non-fluorescent bands against a brilliant background. The pentahydroxyflavone morin has been used to make alumina columns fluoresce (300 mg to 500 g of alumina), berberine for silica columns, and morin or diphenylfluorindene sulphonic acid for calcium carbonate or magnesia columns. As examples of separations carried out on alumina impregnated with morin, the following will suffice: the separation of the three synthetic musks (I, II, III) where, contrary to

expectation, the trinitro compound is the least adsorbed; the purification of aromatic aldehydes (separation of vanillin and piperonal); the separation of β-ionone and pseudoionone, and also of phorone and mesityl oxide.

This method is only applicable to those unsaturated compounds which absorb ultraviolet light in the region causing fluorescence; it has been used by White and Dryden (*3536*) for the separation of 3,5-dinitrobenzoates of aliphatic alcohols. Anet *et al.* (*66*) have used morine-coated alumina for the separation of two natural acetylene derivatives.

Sease (*2925*) has described the use of silica gel mixed with 2.5 % of fluorescent zinc sulphide. The wavelengths causing fluorescence lie between 330 and 390 mμ. Sease (*2926*) has carried out the separation of many binary mixtures, e.g. azoxybenzene and nitrobenzene, *p*-nitrobenzoyl bromide and nitrobenzene, cinnamic and salicylic aldehydes, and nitrobenzene and iodoform. The zone order varies with different silica gels.

Brockmann and Volpers (*398*) later used alumina mixed with fluorescent zinc sulphide and have discussed in detail the mechanism of formation of dark zones. Quenching of fluorescence by the adsorbed substance seems to play an important part. For numerous examples of the use of this fluorescent alumina, see the original paper and (*190*). The use of fluorescence in paper chromatography is mentioned in the relevant chapters.

Miller and Kirchner have separated terpenes on fluorescent "chromato-bars" (see p. 16) obtained by incorporating 0.075 % zinc cadmium sulphide and 0.075 % zinc silicate in the adsorbent. The same authors have prepared fluorescent "chromatostrips" (see p. 12) for the separation of terpenes.

The fluorescent indicator method of Criddle and LeTourneau (*655*) is mentioned in Chapter 19.

5. OTHER METHODS

Claesson (581) has suggested a new method of revealing the position of colourless zones on a column by observation of the changes in refractive index of the solution along the length of the column.

The technique used by Tiselius in observing changes in refractive index of the column eluate has already been described (p. 4).

Sease (2927) has described the location of colourless chromatographic zones with an ultraviolet sensitive multiplier phototube. Harvalik (1320) has studied the illumination of columns with infrared radiation and has developed an electronic image converter to convert infrared radiation into visible light so that colourless zones can be detected by a process analogous to the use of ultraviolet fluorescence.

High frequency oscillations were used by Moseley et al. (2252) to detect zones on a chromatogram.

CHAPTER 7

The adsorption chromatography of gases

The early work on gas chromatography deals almost exclusively with low molecular weight hydrocarbons; the reader is referred to Chapter 19 as well as to the literature survey by Janak (*1552* and *1555*). Simultaneously with the development of gas-liquid partition chromatography (see Chapter 17) the advantages of gas-adsorption chromatography were realised. A gas phase eluant permits faster diffusion of the adsorbed substances and hence faster flow rates and in many cases good separations in less than one hour.

Phillips employed first displacement analysis and points out that the adsorption isotherms are not sufficiently linear for satisfactory elution analysis. One of the papers, by James and Phillips (*1542*) deals with the determination of the adsorption isotherms. Fig. 22 shows the type of displacement separation possible on activated charcoal with bromobenzene as displacer.

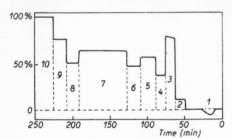

Fig. 22. A typical displacement analysis
(James and Phillips, *1542*)

Components of mixture:

1 water	6 dioxan
2 diethyl ether	7 pyridine
3 chloroform	8 butyl acetate
4 ethyl acetate	9 chlorobenzene
5 thiophen	10 bromobenzene displacer

Janak published a series of valuable technical analytical methods for natural gases (*1553*), the noble gases (*1554*), lower hydrocarbons, halogenated hydrocarbons, N_2O (*1557*), He, Ne and H_2 (*1558*). Using CO_2 as eluant he found good agreement between the shape of the elution curves and the plate theory (see Chapter 15) for a mixture of N_2 and CO. Janak measures:

t_{max}, the time required in seconds for the peak of a band to pass out of a column of given dimensions,

V_{max}, the volume in ml for the elution of the peak,

R_{Fmax}, the ratio of the speed of the peak of a band with the speed of the eluant and

U, the specific adsorptive capacity of the adsorbant for the substance.

These variables depend on the temperature and the adsorbant as shown in Tables 17 and 18.

TABLE 17 R_{Fmax} and U values for active carbon at 20° and 80° (Janak, *1553*)

Gas	t_{max} sec	V_{max} ml	R_{Fmax}	U ($U \times 25.0$ ml)
Temperature: 20° (column length 565 mm, cross section 7.5 mm)				
Hydrogen	8	5.6	0.643	1.25
Nitrogen	17	12	0.3025	2.64
Oxygen	86,87	65	0.307	2.60
Carbon monoxide . . .	94	70.5	0.244	2.82
Methane	44	31	0.117	6.84
Ethane	875	62	0.0059	136.1
Ethylene	435	304	0.0118	67.5
Acetylene	320	224	0.0160	49.8
Cyclopropane	1700			264
Temperature: 80°				
Nitrogen	7	4.9	0.735	1.09
Methane	10	7.0	0.514	1.56
Ethane	55	39	0.0935	8.55
Ethylene	50			
Acetylene	22	16	0.233	3.45
Propane	365	226	0.0160	50.1
Propylene	380	266	0.0135	59.2
Methylacetylene	105	74	0.0490	16.3
n-Butane	440	308	0.0117	68.5
iso-Butane	415	291	0.0124	64.6
Cyclopropane	60	42	0.0857	9.35

The methods of detection employed by Phillips were thermal conductivity measurement (*1542*) and a surface potential detection (*1224*); Janak absorbs the CO_2 used throughout as eluant in an azotometer and thus obtained the volume of separated gas directly in ml. An infra red gas analyser was proposed by Martin and Smart (*2146*) and appears to be more sensitive than other methods suggested for adsorption columns.

In addition to silica gel and charcoal used as adsorbent by the above, Patton *et al.* (*2470*) also employ alumina as adsorbant for the low

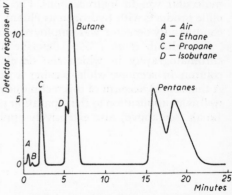

Fig. 23. Elution of straight-run gasoline from alumina by hydrogen, with expanded time scale (Patton *et al.*, *2470*)

TABLE 18. R_{Fmax} and U-values for silica gel B at 20° and 80° (Janak, *1553*)

Gas	t_{max} sec	V_{max} ml	R_{Fmax}	U
Temperature: 20°				
Hydrogen	5·	3.5	0.771	0.78
Nitrogen	7,8	5.3	0.509	1.17
Oxygen	7,8	5.3	0.509	1.17
Carbon monoxide . . .	7,8	5.3	0.509	1.17
Methane	11	7.7	0.351	1.71
Ethane	43	30	0.090	6.69
Ethylene	87	61	0.0443	13.6
Acetylene	240	168	0.0161	37.4
Propane	150	105	0.0267	23.3
Propylene	690	483	0.0056	107.3
Cyclopropane	504	353	0.0077	78.5
n-Butane	600	420	0.0064	93.3
iso-Butane	550	385	0.0070	85.5
Pentane . . . . · .	1350	945	0.0028	210.0
Temperature: 80°				
Methane	5	3.5	0.771	0.78
Ethane	9	6.3	0.428	1.40
Ethylene	11	7.7	0.351	1.71
Acetylene	19	13.3	0.205	2.93
Propane	17	11.9	0.227	2.64
Propylene	22	15.5	0.123	3.45
Cyclopropane	15	10.5	0.257	2.33
Methylacetylene . . .	35	24.5	0.110	5.45
n-Butane	40	28.0	0.0975	6.23
iso-Butane	41	28.7	0.0941	6.38

molecular weight hydrocarbons. Fig. 23 shows a typical separation of gasoline products with hydrogen as eluant. For detection an improved thermoconductivity detector was employed.

Turkeltaub *et al.* (*3696*) describe a "chromathermic" method of gas chromatography in which the development is effected by heating the column in sections while passing a continous stream of gas through it. A theoretical account of this technique is found in a further paper (*3331*) as well as an application to the separation of C_2H_6, C_2H_4, C_3H_8 and C_3H_6 (*28*). Janak (see above) also extensively applies this method.

CHAPTER 8

Secondary reactions caused by the adsorbent

1. ALUMINA

Alumina containing free alkali often gives rise to secondary reactions during the course of adsorption; for example the saponification of glycerides and the autoxidation of fatty acids (*3294*), the deacetylation of acetylated sugars (*372, 3198*), the destruction of vitamin A (*2305*), its transformation into an ether (*2198*), the destruction of vitamin K (*692*), the decomposition of insecticides of the rotenone group (*495, 2384*), the hydration of porphyrins with a vinyl side chain (*967*) and the addition of CH_3OH to the same vinyl group in methanolic solution (*967*). The alkali sensitive alkaloids of *Veratrum viride* are easily inactivated by alkaline alumina (Fried *et al.*, *1043*).

The deleterious action of alumina on some lactones and β-ketoesters is mentioned in Chapter 22.

Prelog and Osgan (*2584*) have observed the transformation of a monocyclic diketone to a bicyclic hydroxyketone on contact with alumina.

Pinckard *et al.* (*2520*) have found that anthrone is oxidized to anthraquinone on some alumina columns.

Schönberg *et al.* (*2872*) explain the deep orange colour formed by addition of activated alumina to a colourless solution of 10-nitroanthrone by isomerisation to 10-nitroanthranol.

Migrations of benzoyl groups have been observed in carbohydrate benzoates after chromatography on weakly alkaline alumina (Schmid and Bolliger (*2860*)

Mycobactin, a mycobacterial growth factor is converted to its crystalline aluminium complex by filtration on alumina columns and elution with alcohol (Francis *et al.*, *1019*).

Halogen derivatives of terpenes, especially tertiary halides are easily dehydrohalogenated on alkaline activated alumina; high yields of the original unsaturated hydrocarbons are obtained from the hydrohalides (Zaoral, *3666*).

Oroshnik *et al.* (*2420*) have observed a prototropic rearrangement of the acetylene $-C\equiv C-CH_2-C=$ to the allene $-C=C=C-C=$ through action of alcoholic alkali, alumina columns or florisil.

Geissman *et al.* (1094) have observed the elimination of one molecule of acetic acid from xanthinin

$$(-C=C-C-CH-CO)$$
$$OCOCH_3$$

to form xanthatin $(-C=C-C=C-CO)$, on columns of alumina. An analogous reaction caused by activated alumina has been reported by Hein (*1356*).

$$
\begin{array}{ccc}
\text{H} & \text{OCOCH}_3 & \\
| & | & \\
R_1-C-C-R_2 & \longrightarrow & R_1-C=\!\!=\!\!C-R_2 \\
| & | & | \quad\quad | \\
\text{CN} & \text{CN} & \text{CN} \quad \text{CN}
\end{array}
$$

The use of acetone as solvent with alumina columns should be avoided as it undergoes condensation reactions with the formation of oily products (*794*).

Many interesting reactions produced by alumina have been reported in the *steroid* field, such as the saponification of steroid benzoates (*487*), the splitting out of a molecule of acetic acid from a sterol acetate with the formation of a double bond (*2200*) (I → II), the debromination of steroids (*2426*) (III → IV) and of lactones (*2761*), the instability of an $\alpha\beta$-unsaturated ketosteroid (*794*). Reichstein and Shoppee (*2658*) report that Δ^5-3-ketones (V) undergo isomerisation to Δ^4-3-ketones (VI) by alumina containing alkali, but not when neutralised Al_2O_3 is employed (cf. Shoppee and Summers, *2970*).

Fieser (*947*) observed a similar migration of the double bond with acid washed alumina. Elks *et al.* (*882*) found that the migration of the double bond of Δ^9-7-keto steroids to Δ^8-7-keto steroids can be avoided by treatment of the alumina with acetic acid and reactivation by heating.

There is often considerable loss with free hydroxy-ketones of the types (VII,

(I) $\xrightarrow{Al_2O_3}$ (II)

(III) $\xrightarrow{Al_2O_3}$ (IV)

(V) → (VI)   (VII) → (VIII)

(IX) → (X)

(XIII) $\xleftarrow{KOH}$ (XI) $\xrightarrow{Al_2O_3}$ (XII)

(XIV) $\xrightarrow{Al_2O_3}$ (XV)

VIII) which cannot be avoided completely by the use of neutralised alumina; it is best to chromatograph the 21-acetates (Reichstein and Shoppee, *2658*).

The same authors report the substitution of Br by Cl on chromatographing the compound (IX) on alumina which had been neutralised by treatment with dilute HCl, then extensively washed with water.

Stavely (*3085*) has described an interesting isomerisation caused by an alumina column: Δ^5-pregnenediol-3,17-one-20 (XI) gives the isomer (XII) having a hexacyclic D ring; an analogous compound (XIII) is produced by the action of KOH on (XI). 17(a)-Hydroxyprogesterone (XIV) undergoes the same isomerisation (to XV). Shoppee and Prins (*2969*) confirmed these observations and noticed that the amount of isomerisation depended on the time of contact with the alumina.

Sarett (*2820*) has observed two interesting reactions of the steroid cyanohydrin (XVI); when passed through an acid-washed alumina column the steroid was dehydrated to give an unsaturated nitrile (XVII) whereas a column of alkaline alumina caused the elimination of a molecule of HCN to regenerate the original ketone (XVIII) in 90 % yield.

(XVIII)   (XVI)   (XVII)

Heusser *et al.* (*1382*) have observed the hydration of a 5,6-oxide (XIX) to a 5,6-diol (XX) on alumina, Liebermann and Fukushima (*2007*) report acyl migrations on chromatographing monoacetates of 3,4-diols on alumina (XXI → XXII) and the replacement of Br by OH in the allylic bromide (XXIII) (Fukushima *et al.*, *1063*). See also Keverling-Buisman *et al.* (*1705*). Sutton and Dutta (*3159*) observed a similar instability of allylic bromides in the aliphatic series (a bromo-oleate being transformed into the corresponding ricinoleate). The same authors state that allyl bromide itself does not react with activated alumina at room temperature, but when they filtered a light petroleum solution of 3-bromocyclohexene through a column of alumina, the adsorbent became warm and turned blue.

(XIX)   (XX)

(XXI)   (XXII)   (XXIII)   (XXIV)

Cremlyn and Shoppee (*654*) have reported that 7β-toluene-*p*-sulphonyl-
oxycholestane (XXV) is partly decomposed by neutralised alumina in
pentane to give cholest-7-ene (XXVI) whilst use of alkaline aluminium
oxide in pentane furnished cholest-7-ene accompanied by cholesten-7α-ol
(XXVII), the formation of which is due to a Walden inversion.

(XXVI)   (XXV)   (XXVII)

Fieser and Stevenson (*949*) have described an isomerisation of a 3-keto-4-
acetate (XXVIII) to a 3-acetyl-4-keto steroid (XXIX)

(XXVIII)   (XXIX)

on alumina; a cyclic acetal is considered to be formed as intermediate
compound.

Acetylated cardiotonic glycosides are best chromatographed not on alumina, but on magnesium silicate to prevent loss of acetyl groups (Aebi,
Schindler and Reichstein, *19, 2842*).

Cholesterol ozonide cannot be eluted from an alumina column (*193*).

Many of these secondary reactions on alumina can be prevented by
removing the free alkali by washing first with acid and then with water, or
simply by washing with water (*925*; see also the section on alumina, p. 19).

Plattner and Pfau (*2526*) have shown that *picrates* of aromatic hydrocarbons are dissociated when passed over alumina, the picric acid being
retained at the top of the column while the hydrocarbons pass out in the
eluate. This is a practical method for the decomposition of these picrates
and is equally applicable to styphnates and trinitrobenzolates (*1907, 1908*).
Picrates of alkaloids, however, are transformed into hydrochlorides if HCl
activated alumina or Wofatit M/HCl are used (Karrer and Schmid, *1643*).
In the same way Wieland *et al.* (*3546*) transformed a β-anthraquinone
sulphonate of a curare alkaloid into the hydrochloride by the use of HCl-
activated alumina. See also p. 473 on preparation of inorganic acids.

2. SILICA AND SILICATES

Arbouzov and Isayeva (72) have reported isomerisation of terpene hydro-carbons by chromatography on silica gel.

Borgström (317) has observed the isomerisation of 2-monoglycerides to 1-monoglycerides during chromatography on silicic acid.

Sterols can undergo isomerisation on activated silica in the presence of halogenated solvents (3295).

Soloway et al. (3047) have observed the isomerisation of the steroid epoxide (XXX) to the methyl ketone (XXXI) by heat or chromatography on silica gel; Leeds et al. (1947) have reported the analogous conversion of (XXXII) to (XXXIII) by heat or silica gel.

(XXX)   (XXXI)   (XXXII)   (XXXIII)

The acid earth adsorbents (Fuller's earth, Jagolite, Superfiltrol) with an incomplete electronic octet absorb carotenoids and vitamin A to give a green or blue coloration. Meunier (2195) has discussed the analogy between this phenomenon and the Carr-Price-reaction (see also Zechmeister and Sandoval, 3689). Meunier and Vinet (2199) have treated in detail the theoretical aspects of the adsorption of mesomeric forms of vitamin A, β-carotene and vitamin D.

Talc can cause acetal formation of aldehydes of the porphyrin series in the presence of methanol (Fischer and Conrad, 967).

3. MnO$_2$

Wald (3447) has described the "chromatographic oxidation" of vitamin A to retinene ($-CH_2OH \longrightarrow -CHO$) on a column of MnO_2. Meunier et al. (2196) have described the oxidative fission of β-carotene and lycopene on MnO_2.

4. CHARCOAL

Charcoal can cause aminolysis of amino acids and also complete oxidation to CO_2, H_2O and NH_3 (Wachtel and Cassidy, 3433). These oxidations can be prevented by the use of small quantities of KCN or water saturated with H_2S (Tiselius, 3256; Schramm and Primosigh, 2880). Some losses are often inevitable in the course of elution of substances adsorbed on charcoal (Tiselius *et al.*, *3268*). See also the chapter on "modified adsorbents".

Picrates of amino acids can be decomposed by adsorption on charcoal; the hydrochlorides are eluted by dilute HCl (Robson and Selim, *2708*). (Picrates of bases, for instance indoles, can also be decomposed by chromatography on $CaSO_4$; Henbest *et al.*, *1363*.)

5. POLARISATION BY ADSORBENTS

Polarisation caused by adsorbents has been reported by Weitz *et al.* (*3498*, *3499*); non-polar colourless substances such as triarylmethanes can be polarised on alumina or silica gel with the formation of coloured zones. By elution with polar solvents the colourless compounds can be recovered unchanged. This phenomenon has been also examined by Buu-Hoi and Cagniant (*490*), concerning triphenylmethane derivatives and by Cruse and Mittag (*668, 669*) for di- and trinitrobenzene. Schönberg *et al.* (*2872*) have described a series of anthrones, nitrotoluenes, spiropyranes and xanthydrols giving deeply coloured adsorbates with activated alumina or silica gel. In the case of the "adsorptiochromism" of hydroxynaphthoquinones described by Green and Dam (*1210*) formation of an alkali salt could be a possible explanation.

6. ION EXCHANGE RESINS

Secondary reactions on ion exchange resins are discussed on page 99.

DIVISION II

ION EXCHANGE CHROMATOGRAPHY

CHAPTER 9

Introduction

The fundamental work on chromatographic separations by the use of ion exchange resins is mainly derived from the studies carried out in connection with the work on atomic energy in the U.S.A.

The separation of the mixture of fission products presented a major problem to analytical chemistry, especially the mixture La, Sm, Eu, Y, Ce, Pr, Nd and Pm, as well as the mixtures Cs and Rb, Sr and Ba, and Zr and Cb in tracer amounts.

It was to solve these problems that synthetic cation exchange resins were employed in combination with the technique of chromatographic elution. The separations of pure rare earth elements were later achieved on a large scale and it is in this field that the chromatographic technique has made its first great contribution to inorganic chemistry. In the course of these separations it was found that rare earths, previously believed to be spectrographically pure contained large amounts of other rare earths. In one case Spedding (*3059*) found, by determining the extinction coefficient of a so-called spectrographically pure standard, the sample obtained by chromatography to analyse to 120 %.

Separations of organic substances are discussed in Chapter 14.

The field of ion exchange was reviewed in the monographs of Nachod (*2332*), Walton (*3471, 3472*), Samuelson (*2799*), Kunin and Myers (*1838*), Martin (*2159*), Cassidy (*530*), Osborn (*2420a*) and Austerweil (*103a*), as well as in articles by Tompkins (*3281*), Kunin (*1832–1835*), Pepper (*2488*), Boyd (*350*), Gapon (*1084*), Applezweig (*71*), Myers (*2328–2330*), Bauman et al. (*163*), D'Ans et al. (*696*), Juda et al. (*1607*) and Davies (*712*). A review of the history of ion exchange was written by Deuel and Hostettler (*754*).

Ion exchange resins are employed in numerous industrial processes such as water softening, sugar refining, metal concentration and catalysis of reactions, in a batch process manner and not in combination with elution from columns. These applications are hence not discussed in this book.

The early cation exchangers were silicates, either natural products such as montmorillonite clay or fuller's earth, or synthetic alumino silicates prepared from aluminium compounds and sodium silicate.

The application of these inorganic exchangers is limited to a narrow pH range as they peptize in alkaline solutions and dissolve in acid. Jagolite, a montmorillonite earth, has been used for adsorption of vitamin E (*2197*), also for carotenoids (*2197*), fuller's earth for vitamin K (*2682*) and for sugars (*1995*). Decalso (a synthetic alumino silicate) for vitamin K (*246*), vitamin B_1 (*612*) and urinary pigments (*884*), permutit (another synthetic alumino silicate) for antibiotics (*1342, 1650*), also for the gonadotropic hormone (*1649*).

Acid-washed alumina has been shown to act as an anion exchanger (*3549*), and alumina containing Na ions as a cation exchanger (see p. 20).

Recently Kraus *et al*. (*1778a*) have used *insoluble zirconium salts* such as tungstates or phosphates for ion exchange separations of the alkali metals.

CHAPTER 10

Synthetic ion exchange resins

1. PROPERTIES

The first synthesis of an ion exchange resin was carried out by Adams and Holmes in 1935 (*14*), who prepared a condensation product of phenol sulphonic acid with formaldehyde. Similar substances were prepared by Liebknecht (*2009*) and Smit (*3013*) by sulphonation of coal.

All these resins possessed reactive OH and COOH in addition to the more important SO_3H exchange groups.

In order to prepare a resin with only one type of reactive group, D'Alelio (*683*) sulphonated a hydrocarbon polymer containing benzene rings (styrene with 10 % divinylbenzene). An analogous resin with basic groups was prepared by reacting the polymer with chloromethyl ether,

$$-CH-CH_2- \quad + \quad CH_3OCH_2Cl \quad \rightarrow \quad -CH-CH_2- \quad + \quad CH_3OH$$
$$CH_2Cl$$

then reacting the chloro groups in the network with tertiary amines,

$$-CH-CH_2- \quad + \quad R_3N \quad \rightarrow \quad -CH-CH_2-$$
$$CH_2Cl \qquad\qquad CH_2\overset{+}{N}R_3 + Cl^-$$

For a review of ion exchange resin synthesis see Craig (*642*).

The functional acidic (or basic) groups of an exchanger will always be occupied by ions of the opposite charge. When holding hydrogen ions a resin is said to be in the "hydrogen form", similarly when holding sodium ions in the "sodium form" etc.

When a polymer containing active groups (e.g. SO_3H) is formed, the ionisation of the respective groups is not changed, thus a *vast sponge-like network* is produced with properties identical with those of the monomer (*162*, *164*). Resins with sulphonic groups or quaternary amine groups are thus highly ionised though very insoluble and react throughout their entirety. Resins with highly ionised groups such as SO_3H and NR_3 are called strong exchange resins and resins with only partially ionised groups such as COOH, OH, NH_2 are called weak exchange resins. The degree of ionisation as well as the similarity to the monomer can be best illustrated by titrating, for example, the hydrogen form of a strong acid resin such as Dowex-50. A titration curve identical to that of a strong acid with a strong base is produced (Fig. 24).

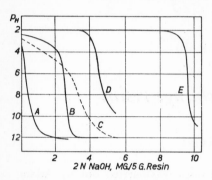

Fig. 24. Titration curves of several cation exchanger functional groups; (A) phenolic OH: (B) methylene sulphonic CH_2SO_3H: (C) carboxyl COOH: (D) and (E) nuclear SO_3H (Tompkins, *3282*)

A similar analogy exists, with weak exchange resins, these giving titration curves typical of weak acids and weak bases. For titration curves see also Topp and Pepper (*3288*).

2. CROSSLINKAGE

Crosslinkage of various degrees may be obtained by copolymerising various quantities of divinylbenzene with the polystyrene. In Dowex resins the amount of crosslinkage is expressed by the percentage of divinylbenzene as no known analytical method for its determination is available.

For example Dowex-50-X4 is prepared from a mixture of styrene and divinylbenzene containing 4 % divinylbenzene.

A commercial pamphlet of the Dow Chemical Co. (*807*) lists the following cross-linkage effects:

Copolymers of styrene containing *low* amounts of divinylbenzene (1–4 %) are characterised as follows:

1. High degree of permeability.
2. Contain a large amount of moisture.
3. Capacities are lower on a wet volume basis.
4. Equilibrium rates are high.
5. Physical stability is reduced.
6. Selectivity for various ions is decreased, but ability to accommodate larger ions is increased.

Copolymers of styrene containing *high* amounts of divinylbenzene (12–16 %) would exhibit characteristics in the opposite direction. An average divinyl-benzene content for the Dowex Fine Mesh Resins series is a resin containing 8 % divinylbenzene and the terms high and low or decreased and increased are all relative to an 8 % crosslinked resin. For implications of degree of crosslinking in the chromatography of organic substances, see the chapters on amino acids and peptides.

3. PARTICLE SIZE

Table 19 gives the conversion of mesh range to particle diameter.

TABLE 19.   Conversion of mesh ranges to particle diameters *(807)*

Mesh Range	Diameter of particles		
	inches	*mm*	*microns*
20–50	0.0331–0.0117	0.84–0.297	840–297
50–100	0.0117–0.0059	0.297–0.149	297–149
100–200	0.0059–0.0029	0.149–0.074	149–74
200–400	0.0029–0.0015	0.074–0.038	74–38
> 400	< 0.0015	< 0.038	< 38

As the particle of an ion exchange resin is decreased to the fine mesh range (50–100 mesh or finer) the following effects are observed (*807*):

1. The time required to reach equilibrium is decreased.
2. The flow rate decreases.
3. The pressure drop across the ion exchange column increases.
4. The bed expansion during a backwash cycle is greater.
5. The efficiency of a given volume of resin increases or the volume of resin required to perform a specific operation is decreased.

4. CHARACTERISTICS OF SOME ION EXCHANGE RESINS

(a) *Properties of resins*

For the selection of a resin for analytical purposes several important features must be known as is evident from theoretical considerations. This information is avaiable in most cases in pamphlets published by the manufacturers of the resins. In Tables 20–23 a number of the resins which are often referred to in the text are tabulated.

TABLE 20.   Duolite ion exchangers (Tompkins, *3281*)

	C_1	C_2	A_2	A_3
Chemical composition	Phenol-formaldehyde type			
Active groups	sulphonic acid		aliphatic amine	
Density (wet)	1.07	1.13	1.07	1.08
Colour	white to pink red		light yellow	—
Capacity in meq./g dried at 80° C	2.75	3.44	9.35	6.65
Capacity in eq./l of wet tamped vol.	0.5	1.2	2.25	2.0
Stability, chemical	unstable to oxidising agents			
Stability, physical	stable at reasonable periods from 77–100° C. but not absolutely stable at these temperatures			

TABLE 21.   Permutit ion exchangers (Tompkins, *3281*)

	Zeo Karb	*Zeo Rex*	*Permutit Q*	*DeAcidite*
Colour	black	amber	yellow	orange
Form	granular	granular	spherical	granular
Density dry g/l	735	925	900	610
General chemical composition. .	sulphonated coal	sulphonated phenolic	sulphonated hydrocarbon	aliphatic amine resin
Exchange groups	COOH & SO_3H	SO_3H	SO_3H	R_3N
Analyses	S = 6%	S = 8.5%	S = 16%	—
Capacity in meq./g.	1.8	3.0	5.0	6.7

Resin	Amberlite IR-120	Amberlite IRC-50	Amberlite IRA-400	Amberlite IRA-401	Amberlite IRA-410	Amberlite IRA-411	Amberlite IR-4B	Amberlite IR-45	Monobed MB-1	Monobed MB-2	Monobed MB-3
Type	Strongly acidic cation exchanger	Weakly acidic cation exchanger	Strongly basic anion exchanger	More porous analog of Amberlite IRA-400	Strongly basic anion exchanger	More porous analog of Amberlite IRA-410	Weakly basic anion exchanger	Weakly basic anion exchanger	Mixture of Amberlite IR-120 and Amberlite IRA-400	Mixture of Amberlite IR-120 and Amberlite IRA-410	Mixture of Amberlite IR-120 and Amberlite IRA-410 with indicator
Active group	nuclear sulfonic acid	carboxylic acid	quaternary amine	quaternary amine	quaternary amine	quaternary amine	polyamine	polyamine	– Same as components –		
Form supplied	Bead Na	Bead H	Bead Cl	Bead Cl	Bead Cl	Bead Cl	Granular OH	Bead OH	Bead H OH	Bead H OH	Bead with indicator H OH
Density (lbs/ft³)	53	43	42	42	44	42	35	42	44	44	44
Moisture content	44–48	45–55	40–50	55–60	35–45	53–63	40–45	37–45	– Same as components –		
Effective size (mm)*	0.45–0.6	0.33–0.5	0.35–0.45	0.40–0.55	0.35–0.45	0.35–0.50	0.4–0.55	0.35–0.50	– Same as components –		
Total exchange capacity kg as CaCO₃ per ft³	41.5	71.0	22.0	17.5	26.0	15.0	65.0	43.0	9.0	11.0	11.0
meq./ml wet resin	1.9	3.5	1.0	0.8	1.2	0.7	3.0	2.0	0.4	0.5	0.5
meq./g dry resin	4.25	10.0	2.5	3.0	3.0	3.0	5.23	5.0	—	—	—
Maximum operating temperature (°F)	250	250	140 (OH form)	140 (OH form)	105 (OH form)	105 (OH form)	105	212	140	105	105
Effective operating pH range	1–14	7–14	0–12	0–12	0–12	0–12	0–7	0–7	0–14	0–14	0–14
Maximum swelling (%)**	5 Na→H	100 H→Na	5 Cl→OH	10 Cl→OH	5 Cl→OH	10 Cl→OH	25 OH→HCl	15 OH→HCl	– Same as components –		

* Sieve opening that will retain 90% of the sample. ** % increase in volume of exchangers when form is changed.

TABLE 23. Dowex ion exchange resins (Information supplied by the manufacturer)

Name	Dowex 50	Dowex 1	Dowex 2	Dowex 3
Type	Strongly acidic cation exchanger	Strongly basic anion exchanger	Strongly basic anion exchanger	Weakly basic anion exchanger
Crosslinkage – Standard % DVB	8%	7.5%	7.5%	Not defined
Approximate density (lbs/ft.^{3}) Standard Material	53	45	45	45
Active group	Nuclear sulfonic acid	Trimethyl benzyl ammonium	Dimethyl ethanol benzyl ammonium	Polyamine
Resin form as shipped	Na (20–50 mesh) H (All other mesh)	Cl	Cl	OH
Form	Spheres	Spheres	Spheres	Spheres
Standard mesh size	20–50	20–50	20–50	20–50
Total dry weight capacity– H$^+$ or Cl$^-$ form	5 meq./g	approximately 3 meq./g	approximately 3 meq./g	approximately 6 meq./g
Total wet volume capacity– H$^+$ or Cl$^-$ form	1.8 meq./ml	approximately 1.1 meq./ml	approximately 1.1 meq./ml	approximately 3 meq./ml
Moisture content– H$^+$ or Cl$^-$ form	55% (H$^+$ form)	45% (Cl$^-$ form)	40% (Cl$^-$ form)	30% (Base form)
Volume change	$\Delta V_H^{Na} = -8\%$	$\Delta V_{Cl}^{OH} = +24\%$	$\Delta V_{Cl}^{OH} = +14\%$	$\Delta V_{Cl}^{OH} = -20$ to -30%

Name	Dowex 50	Dowex 1	Dowex 2	Dowex 3
Selectivity – e.g. K_H^{Na}	$K_H^{Na} = 1.2$	$K_{OH}^{Cl} =$ approximately 15	$K_{OH}^{Cl} =$ approximately 1.5	
Order of selectivity for ions	Ag>Rb>Cs>K>NH$_4$>Na>H>Li	I>NO$_3$>Br>Cl>Acetate>OH>F	I>NO$_3$>Br>Cl>OH>Acetate>F	
Breakage	<15%	<25%	<25%	<25%
Bed expansion	<40% at upflow velocity of 4 gpm/ft²	<35% at upflow velocity of 2 gpm/ft²	<35% at upflow velocity of 2 gpm/ft²	<35% at upflow velocity of 2 gpm/ft²
Pressure drop	approximately 0.5 lb/ft at 5 gpm/ft²	approximately 0.5 lb/ft at 5 gpm/ft²	approximately 0.5 lb/ft at 5 gpm/ft²	approsimately 0.5 lb/ft at 5 gpm/ft²
Special mesh sizes (dry)	50–100, 100–200, 200–400, –400, colloidal	50–100, 100–200, 200–400, –400	50–100, 100–200, 200–400, –400	Not stocked
Special crosslinkages	1% – 16% DVB	1% – 10%	1% – 10%	Not defined
Stability – thermal	Good up to 150° C	OH-form – Fair up to 50° C Cl-form – Good up to 150° C	OH-form – Fair up to 30° C Cl-form – Good up to 150° C	Tentatively limited to 65° C. Upward revision may result with additional factual data
Stability – solvent	Very good	Very good	Very good	Very good
Stability – oxidation	Slow solution in hot 15% HNO$_3$	Slow solution in hot 15% HNO$_3$	Slow solution in hot 15% HNO$_3$	Good
Stability – reduction	Very good	Break down in presence of sulfur containing reducing agents		Unknown

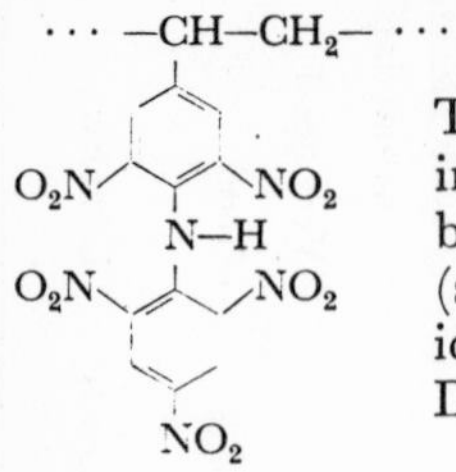

$\cdots$ —CH—CH$_2$— $\cdots$

O$_2$N—⟨ ⟩—NO$_2$
N—H
O$_2$N—⟨ ⟩—NO$_2$

NO$_2$

(b) *Complexing resins*

The preparation of resins containing complexing groups instead of ion exchange groups has been attempted by Skogseid *(3009)* who attached dipicrylamine groups (see formula) for the fixation of potassium; the same ideas were advanced and tried by Mellor *(2184)* and Dwyer *et al.* *(844)*.

(c) *Electron exchange resins*

Electron exchange resins prepared by polymerising vinylhydroquinone alone or mixed with styrene were described by Cassidy *(529, 3351)*. In a later paper by Ezrin and Cassidy *(931)* some of the properties of this electron exchange resin are reviewed. Its titration curve (volt against % oxidised) is almost a straight line from 0.24 V to 0.4 V. Hence it may be considered as an insoluble redox system with an $E_0 = 0.32$ V. No chromatographic applications of this electron exchanger have so far been reported. Several reactions have however been carried out by passing a solution through a column such as the quantitative reduction of dichromate or the oxidation or reduction of iodide-iodine. For such reactions it was also employed in other forms such as dispersed on filter paper or precipitated on diatomaceous earth.

CHAPTER 11

Ion exchange equilibria

When an ion exchange resin in the hydrogen form is immersed in a sodium chloride solution, the hydrogen ions on the resin are replaced by sodium and sodium chloride is converted to hydrochloric acid.

This ion exchange reaction can be considered either as an adsorption system obeying the Langmuir isotherm (*353, 837*) or Freundlich isotherm, or as a reaction obeying the law of mass action (*353*); or as a Donnan equilibrium between the inside of the resin particles and the outside solution (*164, 1217, 1218*). All these theories agree in that the amount of an ion inside the resin particle increases with the concentration of that ion in the solution. However, only the Donnan theory adequately explains the volume changes (swelling) on replacing one ionic species by another. In his application of the Gibbs-Donnan equilibrium to ion exchange resins, Gregor (*1217, 1218*) points out that an equilibrium is reached between the osmotic pressure of the hydrated ion and the back pressure of the elastic net-work of the cross-linked polymer.

Thus the affinity of ions to the resin phase or the "free energy of the exchange reaction" is directly related to the volume change of the resin particles during the reaction. The amount of cross-linking influences the "elasticity" of the resin and hence the equilibrium between ions, as was shown by Reichenberg, Pepper and McCauley (*2654*).

Duncan and Lister (*831*) suggest that secondary adsorption could account for the difficulties in evaluating the activities inside the resin phase.

1. EXCHANGE EQUILIBRIA BETWEEN VARIOUS IONS

(a) *Inorganic cations*

Walton (*3470*) has reviewed the data on the affinity of cations for various exchanges and found that for the Group 1A and 2A metals the affinity increases with atomic weight.

Hydration of the ions decreases the affinity as shown by Jenny (*1573*), and the experiments of Wiegner (*3544, 3545*) indicated that in a range of alcohol-water mixtures the affinities of Cs and Na became more alike as the hydration differences fell.

Extensive investigations were carried out by Boyd, Schubert *et al.* (*353*) on the relative affinities of various cations for cation exchange resins. This work has special importance in the identification of radio-active isotopes of otherwise unknown elements. Table 24 gives the free energy of "salt formation" and the ionic radius of a number of metals studied with Amberlite IR-1.

TABLE 24.   Free energy of "salt formation" and ionic radius of some metals (Boyd *et al.*, *353*)

Ion	Free energy ΔF° (298.1°) cal/mole	Crystal ionic radius
H^+	0	
Li^+	–60	0.78
Na^+	320	0.98
NH_4^+	410	
K^+	530	1.33
Rb^+	615	1.49
Cs^+	860	1.65
Ba^{++}	1680	1.43
Y^{+++}	1830	1.06
La^{+++}	2110	1.22

It was thus shown that the adsorption affinities are determined chiefly by the magnitude of the charge and the radius of the hydrated ions in solution. Russell and Pearce (*2757*) found, contrary to the above, that the rare earth ions of a larger radius are less strongly held than smaller ions, and this observation was also confirmed by Kozak and Walton (*1776*). Ketelle and Boyd (*1702, 1703*) also found that when eluting mixtures of rare earth metals with citrate solutions, the sequence of relative adsorbability is in the order of their ionic radii thus: La, Ce, Pr, Nd, Pm, Sm, Eu, Gd, Tb, Dy, Y, Ho, Er, Tm, Yb, and Lu. Yttrium falls between dysprosium and holmium in this series. Using this rule, Marinsky *et al.* (*2134*) were able to obtain evidence for the existence of element 61 in a mixture of rare earths.

(b) *Inorganic and organic anions*

Kunin and Myers (*1837*) examined the exchange equilibria of anions with Amberlite IR-4B. The relative exchange abilities of the anions studied are: hydroxide > sulphate > chromate > citrate > tartrate > nitrate > arsenate > phosphate > molybdate > acetate = iodide = bromide >

chloride > fluoride. Reactions between the chloride form of the resin with sodium borate desorbed all the chloride without adsorption of borate, due to the extremely small ionisation of boric acid. Organic acids were found to be in the following order: benzoic < oxalic < formic < acetic = citric < salicylic.

2. ELUTION

The *elution* of ions attached to the resin is possible either by displacement with a more strongly adsorbed ion or with highly concentrated solutions of other ions or again the concentration of the adsorbed ion in solution can be decreased by the addition of a complexing agent.

For example in an equimolecular solution of lanthanum and hydrogen ions, the lanthanum is almost completely adsorbed on the resin; however in presence of citrate ions a lanthanum citrate complex is formed and the concentration of free lanthanum ions so reduced as to have practically no lanthanum inside the resin.

Tompkins et al. (3284) first used solutions of citric acid in this manner for the elution of rare earth ions. The pH of such solutions plays an important role as it influences the proportions of H_3Cit, and the ions H_2Cit^-, $HCit^{--}$ and Cit^{---}.

The equilibrium constant of the distribution of an ion between the solution and the resin is called K_d where

$$K_d = \frac{M_s \,/\, \text{mass of resin}}{M_e \,/\, \text{volume of solution}} = \frac{M_s}{M_e} \times \frac{\text{volume of solution}}{\text{mass of resin}}$$

where M_s and M_e are the fractions of the cation M in the resin and liquid phases.

The reaction between the ammonium resin and the citrate complex of the rare earth metals depends on the following chemical equilibria:

$$M^{3+} + 3NH_4R \rightleftarrows MR_3 + 3NH_4^+ \tag{1}$$
$$M^{3+} + nH_xCit^{x-3} \rightleftarrows M(H_xCit)_n{}^{3+n\,(x-3)} \tag{2}$$

where M^{3+} = rare earth ion
  NH_4R = ammonium resin
  MR_3 = the rare earth resin compound
and $M(H_xCit)_n{}^{3+n\,(x-3)}$ = the rare earth citrate complex.

The distribution coefficient K_d varies interdependently with ammonium ion concentration, total rare earth concentration and pH.

By determining K_d for an ion under known conditions it is possible to calculate the dissociation constants of complexes and the activity coefficient of the ions. For details, see Schubert et al. (2891–2896), Vanselow (3377, 3378), and (613, 830, 2177, 2898, 2899), also Division V of this book.

3. THE SEPARATION FACTOR

As a measure of the chromatographic separation possible between two ions the ratio of the two equilibrium constants is employed and called the separation factor a.

$$\frac{K_{d_1}}{K_{d_2}} = \alpha$$

For an equilibrium between a rare earth citrate complex, the rare earth ion and its resin compound (for example $Ce^{+++} + H_2Cit^-$ etc.)

$$K_d = \frac{K_{exchange} \cdot K_{complex}}{K^3_{1\ cit}} \cdot \left(\frac{NH_4R}{NH_4^+}\right)^3 \left(\frac{H^+}{H_3Cit}\right)^3$$

(for derivation see *1702*). In taking ratios between two K_d's the variation due to ammonium ion concentration, H^+ concentration, etc., cancel out. Thus the separation factor is only dependent on the exchange and complexing constants.

$$\alpha = \frac{K^1_{exchange}}{K^2_{exchange}} \cdot \frac{K^1_{complex}}{K^2_{complex}}$$

Improvement of a particular separation can be effected by varying either the resin or the complexing agent. Table 25 gives the effect of various complexing acids on a; the efficiency of various ion exchange resins was compared as shown in Table 26.

TABLE 25. Effect of complexing acids on the separation factor (Tompkins & Mayer, *3285*)

Complexing compound	pH at which K_d of Eu is 21.9	α K_d Pm/K_d Eu
Citric acid	3.05	1.45
Tartaric acid	2.85	1.94
Lactic acid	3.40	1.71
Sulphosalicylic acid	5.4	1.49
Ethyl acetoacetate	7.4	1.33
Oxalic acid	1.7	1.32
Citric acid in 50% EtOH	3.0	1.31
Acetylacetone	7.2	1.91

TABLE 26. Comparison of the efficiencies of several resins for rare earth separations (Eu and Pm) (Tompkins and Mayer, *3285*)

Dowex-50			Dowex-30			Duolite-C			Amberlite IR-1		
pH	$K_d\,Eu$	$\dfrac{K_d\,Pm}{K_d\,Eu}$	pH	$K_d\,Eu$	$\dfrac{K_d\,Pm}{K_d\,Eu}$	pH	$K_d\,Eu$	$\dfrac{K_d\,Pm}{K_d\,Eu}$	pH	$K_d\,Eu$	$\dfrac{K_d\,Pm}{K_d\,Eu}$
2.9	92	1.45	2.61	118	1.56	2.40	230	1.65	2.42	39	1.32
3.08	18.8	1.44	2.82	13	1.57	2.55	85	1.60	2.58	19	1.24
3.25	5.2	1.41				2.80	27	1.62	2.83	8.3	1.30
						3.00	3.1	1.52			

4. DIFFUSION INTO AND INSIDE THE ION EXCHANGE RESIN

According to Boyd *et al.* (*351*) the kinetics of an ion exchange reaction may be divided into five steps:

(i) Diffusion of the A^+ through the solution up to the resin particle.

(ii) Diffusion of A^+ through the resin particle.

(iii) Chemical exchange between A^+ and BR (BR being the resin R with the cation B) at the exchange positions inside the particles.

(iv) Diffusion of the displaced ion B^+ out of the interior of the exchanger.

(v) Diffusion of the displaced ion B^+ through the solution away from the resin particle.

By vigorous mixing (i) and (v) can be made negligible and (iii), the chemical exchange, is either instantaneous or, according to Gregor (*1217, 1218*), non-existent. Thus if by the use of radioactive tracers the rate of exchange in stirred solutions is measured, the diffusion rates inside the particles can be arrived at.

In solutions from 0.1 M up the rate of diffusion through the particle determines the rate of the exchange reaction for any given grain size of resin particles. In very dilute solutions, 0.003 M and less, the limiting rate was the diffusion from the outside to the surface of the resin particle. Diffusion inside the resin particle will depend on the affinity of the ion to the resin; a strongly bonded ion will diffuse much slower through a pore on whose surface active groups are situated, than a less strongly bonded. This retardation is counterbalanced by the faster diffusion of more strongly "adsorbed" ions in the water phase. The rate of absorption of an ion into the resin particle will still be dependent on the equilibrium constant as it will determine the concentration gradient inside the pores of the resin. This was also illustrated by Gapon and Gapon (*1083*) by examination of the cross section of granules of exchange resin dipped into a solution containing two coloured ions. Usually diffusion rates through a resin are 5–10 times slower than in water.

Hale and Reichenberg (*1275*) examined the exchange kinetics of a sulphonated polystyrene resin in which the "chemical process of exchange" was found to be the rate controlling process, the rate of exchange being independent of particle size.

Kressman and Kitchener (*1799*) studied sulphonated phenol formaldehyde resins and their equilibria with NH_4^+ and quaternary substituted ammonium ions. Two mechanisms control the rate: diffusion in the particles and the bounding Nernst film, as was shown by Boyd *et al.* (*351*) above. Also see Baumann and Eichhorn (*164*).

Depending on the comparative size of the resin pores and cations, the energy of activation for diffusion varies from *ca.* 5 kcal/mole for small ions to 8 kcal/mole for large ones (*1799*), for example

$$\left[\underset{CH_3}{\overset{CH_3}{\underset{|}{\overset{|}{\langle \ \rangle - N - CH_2 - \langle \ \rangle}}}} \right]^+$$

A mathematical treatment of the kinetics of ion exchange may be found in Thomas' chapter in "*Ion Exchange*", edited by Nachod (*2332*).

5. NON-EQUILIBRIUM CONDITIONS

Ion exchange can be carried out by pouring a solution of an ion over a column of the resin saturated with another ion, analogous to front analysis.

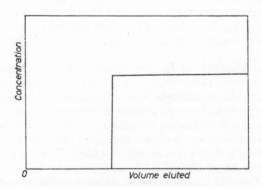

Fig. 25. Equilibrium elution (diagrammatic)

If the rate of flow and the particle size is so as to establish a perfect equilibrium an elution curve as shown in Fig. 25 will be obtained. This has been achieved at very low speeds by Ekedahl *et al.* (*875*). However, it is an ideal condition which is not usually approached in practice. Owing to the rate of

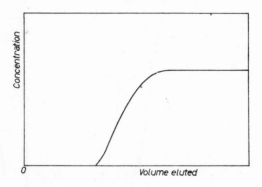

Fig. 26. Non-equilibrium elution (diagrammatic)

flow and the particle size being too large, curves as shown in Fig. 26 are usually obtained.

Boyd *et al.* (*352*) have examined the non-equilibrium conditions from the point of view of incomplete equilibria and without considering vertical

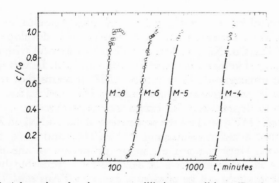

Fig. 27. Adsorption showing non-equilibrium conditions (Boyd *et al.*, *352*)

diffusion which has the same effect. Their experimental results with active tracers on a column saturated with the same but inactive ion are shown in

Figs. 27 and 28 for adsorption and desorption and in Fig. 29 for a band being chromatographically eluted.

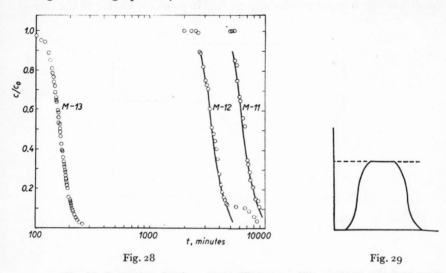

Fig. 28   Fig. 29

Fig. 28. Desorption showing non-equilibrium conditions (Boyd *et al.*, *352*)

Fig. 29. A band eluted, showing at the front and end non-equilibrium conditions (after Boyd *et al.*, *352*)

6. ION EXCHANGE CHROMATOGRAPHY WITH ORGANIC SOLVENTS

Recently a number of investigations deal with the elution of ions with organic solvents or solvent mixtures. Carleson (*506*) developed methods of separation of Zn, Cu, Ni, Co, Mn and Fe on Dowex-50 with such solvents as methyl *n*-propyl ketone containing water and HCl. In this work as in the work on resin impregnated filter paper (*1937*) it seems that the resin plays the part of an inert stationary phase holding HCl and water. The presence of HCl seems to depress any ion exchange properties.

Kember *et al.* (*1679*) studied the separation of Cu and Ni on Zeo Karb 225 and recommend acetone containing 4 % of HCl and 10 % of water as a suitable solvent. Here the ion exchange properties of the resin seem to interfere since in other mixtures of acetone-HCl (reported by the above) sharp separation could not be obtained. See also d'Ans *et al.* (*697*).

Carroll (*513*) who examined the elution sequence of *organic acids* from Dowex-1 with alcohol also concluded that the elution sequence seems to be influenced by the partition coefficients as well as the pK values. For further work with organic solvents and organic compounds see also Kornberg and Pricer (*1769*), Kennedy (*1688*) and Savary and Desnuelle (*2833*).

CHAPTER 12

Theories of chromatographic elution on ion exchange columns

A mathematical theory of chromatographic adsorption was developed by Wilson (3591) and DeVault (759) and later extended by Glueckauf (1142-1146, 592), Weiss and his collaborators (2399) and others. It has not contributed to practical chromatography and has served mainly for the calculation of adsorption isotherms from elution data. The theory of column processes was first developed in analogy to fractional distillation by Martin and Synge (2156) its results and principles are stated in Chapter 15. Using a similar approach Mayer and Tompkins (2176) obtained an equation relating the volume of solvent necessary to elute the peak of a band with the equilibrium constant. This was expressed by them as

$$F_{max} = C \text{ (see also page 106)} \tag{1}$$

where F_{max} = number of V's that have passed through the column, when the peak of a band is eluted.

V = the volume of solution in the column

C = distribution of solute in any plate, i.e. the equilibrium constant corrected for resin volume and solvent volume ratio.

Beukenkamp et al. (237) express this equation as

$$U^X = CV \tag{2}$$

where U^X = the volume that has passed through the column when the concentration of the solute is maximum, i.e. when the peak is eluted. As they prefer to measure the volume which collects till the peak is eluted, which is the volume added to the column + the volume already contained in it they express this as

$$U^X = CV + V$$

where U^X is defined as the volume eluted when the concentration of the solute is maximum.

Equation (1) and (2) merely express mathematically that the least adsorbed substance is first eluted and permits the calculation of the volume required to elute a peak of a band when certain data on the resin and column dimensions are known.

Both Martin and Synge (*2156*) and Mayer and Tompkins (*2176*) proceed then to calculate the number of "theoretical plates", i.e. the length of column required for one equilibrium process between the dissolved and the adsorbed solute. Their work was criticised by Glueckauf (*1147*) who points out that a discontinuous treatment may introduce errors which are obviated by his continuous flow model of a column. He constructed tables giving the purity of the separated products as function of the separation factor (see page 84) and the number of theoretical plates of the column for special conditions when equilibrium is practically obtained, that is with very small particles and low flow rates.

We shall only give the results of a simplified model due to Beukenkamp *et al.* (*237*) which assumes that the elution curves resemble very closely the Gaussian equation. In this case the number of theoretical plates p of a column is

$$p = \left(\frac{2C}{C+1} \right) \left(\frac{U^x}{U_a - U^x} \right) \tag{3}$$

where U_a is the volume required to elute that part of the band where the concentration of the solute equals the peak concentration divided by e (base of the natural system of logarithms). As all terms of this equation may be obtained from an experimental elution graph (provided V is known), this equation permits the calculation of the number of theoretical plates from experimental data.

A similar equation is cited by Schubert (*1180*)

$$p = \frac{2C(C+1)}{W^2} \tag{4}$$

where W is the half width of the elution curve at an ordinate of $1/e$ of the maximum.

Beukenkamp *et al.* (*237*) then proceed also to calculate the relation of C (the equilibrium constant) to the pH and the concentration of the eluant for the general case of a weak tribasic acid H_3A through a column saturated with Cl^-; for details see the original paper.

The column height required for a given separation was calculated by the above authors (*237*) still on the assumption of the elution curve of Gaussian shape. If the separation is "quantitative" that is, that 99.9 % of A is separated from 99.9 % of B then an approximation formula

$$\sqrt{H} = \frac{3.29}{C_2 - C_1} \left(\frac{C_2 + 0.5}{\sqrt{P_2}} + \frac{C_1 + 0.5}{\sqrt{P_1}} \right) \tag{5}$$

may be used, where H = the height of the column, P = the number of plates per centimeter (this of course depends on U_a (see equation (3)) and hence varies for each substance) and $C =$ as before the equilibrium constant. For the use of this equation an experimental elution curve must be available and then it serves to calculate:

(a) if the separation is incomplete what column length must be used for a complete separation or

(b) if the separation is complete, how much smaller a column may be used while maintaining a quantitative separation.

Thus the main use of this expression will be in the calculation of conditions used in routine analyses where such exact information would be valuable. This equation was confirmed for the elution of a mixture of tetrametaphosphate and trimetaphosphate with 0.500 M KCl as shown in Fig. 30. The first column used was 5.8 cm long and 3.8 cm² cross section, as shown in Fig. 30A. From this elution graph a column height of 8.0 cm was calculated and the experimental results with a column 8.3 cm long are shown in Fig. 30B. Fig. 30C shows an elution with a 13.8 cm column. Beukenkamp et al. (237) recommend that the actual column used should be

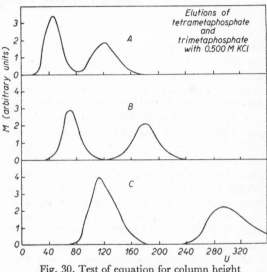

Fig. 30. Test of equation for column height
(Beukenkamp et al., 237)

slightly larger than that calculated by equation (5), in this particular case 8.5 cm.

To summarise: the various theories may be used for calculating adsorption isotherms and equilibrium constants (instability constants of complexes, dissociation constants of weak acids and bases etc.) from elution data. It is also possible to calculate the number of theoretical plates, the amount of contamination of one substance in another on a given part of the elution curve, the minimum length of a column for a given separation and, if certain constants are known of the substance, even the optimum eluant conditions.

Nevertheless for the separation or identification of unknown substances most data are preferably employed on the usual empirical basis and column lengths etc. conveniently judged by trial and error.

CHAPTER 13

Ion exchange papers

Ordinary filter paper contains usually a certain amount of COOH groups the effect of which may not be negligible in chromatography at least in some cases, as has been shown by Schoenfeld and Broda (*2873*), Burma (*458*), Boscott (*325*), Schute (*2907*) and also by Ultée and Hartel (*3344*) who determine the COOH content of paper chromatographically.

It is however desirable to increase the ion exchange capacity of filter paper if it is to be used for ion exchange chromatography.

Carboxyl groups in the paper were formed by Wieland and Berg (*3558*) by oxidation with N_2O_4 after Yackel and Kenyon (*3640*) which produces a paper containing up to 5 % of COOH groups. This paper is then readily soluble in alkali and has to be used only with acid solvents. Wieland and Berg employed this paper for separations of amino acids, inorganic ions (Ag + Pb–Bi–Hg and Sb–As–Sn) as well as for curare alkaloids. This work has been continued by Ströle (*3153*) who studied in detail the variables (pH, buffers, COOH content) for the separation of the basic amino acids. Commercial oxycellulose is used for chromatography of peptides (see Chapter 31). For the preparation of carboxyl, *sulphonic* and *pyridinium* papers see Lautsch *et al.* (*1886*). Sober and Peterson (*3043*) prepare a carboxyl cellulose by treating strongly alkaline cellulose with chloroacetic acid. Proteins were separated on this cellulose (in columns) using phosphate buffers as eluant (see Fig. 91, 92).

Phosphorylated papers were described by various authors (for example Walravens and Chantrenne (*3469*)) and are now available from the manufacturers of Whatman paper in small quantities. Inorganic separations with such a paper were carried out by Kember and Wells (*1682*), for example Fe–Cu–Ni with $2N$ NaCl as solvent. A paper obtained from cotton which had been treated with 2-aminoethylsulphuric acid (also prepared by Whatman) was used for anion exchange separations such as Au–Pt–Pd–Rh with $2N$ HCl as developer.

Jermyn and Thomas (*1587*) prepare a basic paper by oxidation with periodate and condensation with 2-aminoethylhydrazine. Hydrophobic paper which contained free COOH groups was synthesised by Micheel and Albers (*2203*).

A simple way to prepare an ion exchange paper is by dipping paper strips into colloidal suspensions of the respective Dowex resin. These papers have been shown *(1937)* to behave like the corresponding resin column. Thus mixtures of rare earths could be separated with citrate solutions on Dowex-50 paper and mixtures of Cu–Fe–Co with HCl on Dowex-2 paper. Essentially the same work with Zeo-Karb 225 or Amberlite IRA–400 incorporated into paper (home-made from pulp mixed with resin on a Buchner funnel) was reported by Hale *(1273)*.

Some theoretical data for a Dowex resin paper were given by Lederer and Kertes *(1942a)*. The R_M of an ion (for definition of R_M see page 117) varies directly with the pH when an aqueous acid is used as eluant. This relation permits the calculation of the optimum conditions for a separation. Selenite and tellurite ions were studied in detail to confirm this.

CHAPTER 14

The behaviour of organic compounds on ion exchange columns

In this chapter some special features pertaining to the separations of organic substances on ion exchange columns will be discussed.

Such separations have been achieved with numerous groups of ionised substances such as *amino acids*, *purine* and *pyrimidine bases*, *nucleotides*, and borate complexes of *carbohydrates* (see the relevant chapters).

1. ADSORPTION EFFECTS

With amino acids, the equilibria involved in the exchange reaction are those between the unionised, the ionised and the "exchanged" forms. Davies (*711*) discussed the mechanism of this exchange and found that *molecular adsorption* is very large and increases with molecular weight. Similarly, organic acids and bases have been shown to be adsorbed in the unionised form by Davies and Thomas (*714*). (See also p. 97.)

Aromatic substances are much more strongly adsorbed than aliphatic ones. This has been studied in detail by Davies and Thomas (*714*) for acids such as benzoic and phenylacetic as compared to citric and tartaric. The amount adsorbed is independent of particle size except for large molecules.

The affinity of straight chain fatty acids on Duolite A-2 (anion exchanger) was examined by Robinson and Mills (*2703*) and was found to increase depending on the solvent, from acetone → water → shell solvent.

Molecular adsorption of indicators on ion exchange resins was observed by Weiss (*3493*) and by Idler (*1496*). The latter employs adsorbed phenolphthalein as an indicator to detect a band of aspartic acid on the column.

2. ADSORPTION CHROMATOGRAPHY ON ION EXCHANGE RESINS

The possibility of separations depending on adsorption rather than ion exchange on the surface of ion exchange resins have been discussed in a series of papers from the research laboratories of the Dow Chemical Company. One may divide such separations into two types: (i) the separation of organic substances from inorganic salts called *ion exclusion* and (ii) the separation of two or more organic non-ionised substances. As has been shown in ion exchange work, the resins have a very large surface, being completely permeable and thus offer also good possibilities for reversible adsorption.

(a) *Ion exclusion*

Tompkins *et al.* (*3284*) have reported in their work on fission products that a cation exchange resin allows an acid, e.g. HCl or HI to pass through the column without any adsorption. Wheaton and Bauman (*3525*) were able

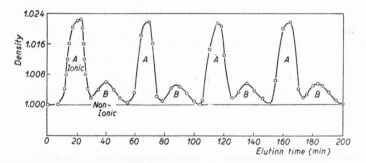

Fig. 31. Ion exclusion: semi-continuous cycles. Flow rate: 0.62 gpm/ft², 1vol. feed: 3 vol. rinse (Wheaton and Bauman, *3525*)

to show that weakly ionised organic acids, such as acetic acid or neutral organic molecules such as polyhydric alcohols are more or less strongly adsorbed and hence if a mixture of HCl and CH₃COOH is passed over a Dowex-50 H⁺ column, HCl is first eluted followed by the organic substance, the separation depending on the adsorption affinity of the organic substance to the resin surface.

With the sodium form of Dowex-50 such separations as NaCl from ethanol are possible; industrial applications for this process are suggested.

The Dow Chemical Company (*808*) recommends for ion exclusion operations Dowex-50 resin with 4–12 % cross-linkage. Pilot plant operations with columns of 35 gallons capacity have been operated and the following separations were achieved in a "cyclic", i.e. semi-continuous process (see Fig. 31): HCl from acetic acid, HCl from dichloracetic acid, trichloracetic from dichloracetic acid, NaCl from ethylene glycol and from higher glycols, NaCl from ethyl alcohol, from formaldehyde, from alkanolamines, from ethylene diamine, from higher amines and ammonium chloride from amino acids. A typical separation is shown in Fig. 32.

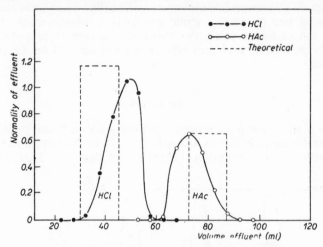

Fig. 32. Ion exclusion: complete separation.
Resin: Dowex 50 × 8% 50–100 mesh. Feed: 15 ml at 1.17 N HCl and 0.66 N acetic acid
(Wheaton and Bauman, *3525*)

(b) *Non-ionic separations with ion exchange resins*

Since it was shown that weakly or unionised substances are retained on ion exchangers it was also evident that they could be separated from each other. Wheaton and Bauman (*3524*) discuss in detail the effect of cross-linking and particle size on the adsorption of numerous substances. Equilibration rates were also investigated in relation to these two factors. To illustrate the possible separations we shall reproduce a table of the K_d values (see p. 83) of organic substances on Dowex resins (Table 27). The few examples so far examined indicate that adsorption chromatography on ion exchange resins has a great future both in the laboratory and in industry.

3. BEHAVIOUR OF LARGE MOLECULES

Detailed studies on the different behaviour of polymers and monomers were made by Deuel *et al.* (*755, 756*). Thus clupein, polygalacturonic acid, polymannuronic acid and polymetaphosphoric acid are not exchanged, being too large to penetrate the pores of the exchanger (Amberlite IR-4B), whilst the corresponding monomers are readily exchanged. The catalytic hydrolysis of esters and disaccharides (with Amberlite IR-120) proceeds with small molecules, e.g. ethyl acetate, maltose, galacturonic methyl ester, but not with their polymers: glycogen, polyvinyl acetate and polygalacturonic methyl ester. On resins with very large pore size, both reaction and exchange with the polymers can take place.

The inability of exchange resins to adsorb large molecules was used by Richardson (*2678*) to purify direct cotton dyes. All inorganic salts may be removed by passage through a cation and an anion exchanger.

Protein molecules, even high molecular peptides, may be unable to penetrate the pores of resins of sufficiently high cross-linkage (see Gilbert and Swallow (*1105*), Thompson (*3236*) and Partridge (*2455*).

Cellulose xanthate (viscose) is not adsorbed on anion exchangers and can thus be freed from contaminating inorganic anions (Samuelson and Gartner, *2803*).

TABLE 27. Distribution constants of organic substances on Dowex resins (Wheaton and Bauman, *3524*) $C^{\circ}{}_{s^{\circ}} = 0.05$

Solute (in aqueous solution)	resin	K_d
Ethylene glycol	Dowex-50-X8 H$^+$	0.67
Sucrose		0.24
D-Glucose		0.22
Glycerol		0.49
Triethylene glycol		0.74
Phenol		3.08
Acetic acid		0.71
Acetone		1.20
Formaldehyde		0.59
Methanol		0.61
Formaldehyde	Dowex-1-X 7.5 Cl$^-$	1.06
Acetone		1.08
Glycerol		1.12
Methanol		0.61
Phenol		17.7
Formaldehyde	Dowex-1-X8 SO$_4{}^{--}$	1.02
Acetone		0.66
Xylose	Dowex-50-X8 Na$^+$	0.45
Glycerol		0.56
Pentaerythritol		0.39
Ethylene glycol		0.63
Diethylene glycol		0.67
Triethylene glycol		0.61
Ethylene diamine		0.57
Diethylene triamine		0.57
Triethylene tetramine		0.64
Tetraethylene pentamine		0.66

4. OCCLUSION

While dealing with the relation of molecular size to pore size, a mention should be made of numerous effects which have been observed with porous zeolites. The term occlusion is usually employed here and differentiation between relatively small molecules was observed. Chabazite for example absorbs all the *n*-heptane from a mixture of heptane and toluene, whilst the toluene does not enter the pores of the mineral at all (see *530*).

5. SPECIFIC ION EXCHANGE RESINS

Suitable modifications of the active groups of synthetic ion exchange resins have been proposed for the preparation of more or less specific or selective resins.

Mercuration of a phenol-formaldehyde polymer produces a resin that selectively removes *mercaptans* from aqueous solutions (Miles *et al.*, *2214*). Reduced glutathione and cysteine are quantitatively retained by the resin and are recovered by elution with 2-mercapto-ethanol. Coenzyme A is also retained and can be eluted with 0,1 M potassium sulfide at pH 7.7. The marked influence of pH on the retention of different compounds by the resin, as well as the differences in the effectiveness of various eluting agents offer interesting possibilities for chromatographic separations of mercaptans.

Grubhofer and Schleith (*1235*) have treated Amberlite XE 64 with quinine and have used the "asymmetric column" thus obtained for the res·lution of the optical antipodes of mandelic acid.

The same authors (*1234*) have diazotized the –NH$_2$ groups of a basic resin and then coupled it with proteins. Enzymes are said to have kept their activity after thus being fixed to the resin.

Isliker (*1522*) has described the purification of *antibodies* by means of antigens linked to ion exchange resins; after fixing the stromata of erythrocytes on anion exchange resins, columns were obtained which adsorbed specifically the iso-agglutinins; similarly, cation exchange resins were transformed into their acid chlorides which were made to react with serum albumin; the thus modified resin fixes specifically the homologous antibody. Viruses can also be fixed on carboxylic resins in the acid chloride form, which are thus made specific for adsorption of the corresponding virus antibodies. Elution of the adsorbed antibodies was possible either by pH changes, or by the action of different carbohydrates having special affinities with the antibodies.

Similarly Manecke and Gillert (*2120*) prepared immunologically specific resins by diazotizing polyaminostyrene and coupling the diazonium salt with antibodies (anti-cows' milk serum); the resins so obtained adsorbed specifically the homologous *antigen*.

6. SECONDARY REACTIONS OF ORGANIC COMPOUNDS ON ION EXCHANGE RESINS

In working with ion exchange resins, it should be·remembered that they are insoluble acids or bases and that a whole series of reactions can be brought about by their action· on organic compounds, under more or less drastic conditions: hydrolysis of esters (Bernhard and Hammet, *227*), hydrolysis or synthesis of glycosides (Mowery, *2296*); hydrolysis of peptide bonds (Paulson *et al.*, *2472*; Dixon, *788*); aldolization, ketolization, crotonization and condensation of aldehydes and ketones (Durr, *836*; Mastagli and Durr, *2167* Austerweil and Pallaud, *104*), hydration of acetylenic bonds (Heilman and Glénat, *1354*), etc.

Carbohydrates are especially susceptible to the action of the strong base resins; Phillips and Pollard (*2513*), Woolf (*3625*) and Hulme (*1478*) report destruction of sugars on Amberlite IRA-400 (OH) or Dowex-2 (OH) with formation of lactic and glycolic acids. Rebenfeld and Pacsu (*2638*), Turton and Pacsu (*3336*) and Sowden (*3054*) have observed isomerisation of D-glucose on Amberlite IRA-400, fructose and mannose being formed to approximately the same extent as from D-glucose in presence of dilute aqueous NaOH.

Buhler *et al.* (*447*) have studied the epimerization and fragmentation of glucose by quaternary ammonium base type anion exchange resins with the aid of glucose-2-^{14}C. Richardson and Hulme (*2677*) have found that even weakly basic anion exchange resins, such as Deacidite G, react in the OH form with sugars to give several acids, including glyceric acid. This difficulty was overcome by using Deacidite G in the acetate form (for the separation of acids from sugars in plant extracts freed from amino acids).

Nucleosides and *nucleotides* can be partly hydrolysed by cation exchange resins; Anderson *et al.* (*54*) report that Dowex-50, even in the ammonium form produces extensive hydrolysis of the glycosidic linkage of deoxyribonucleosides, particularly of adenine deoxyriboside, which remains on the column for the longest period. No such hydrolysis occurred on anion exchange resins (Dowex-2 in the formate form). Lund *et al.* (*2060*) observed partial hydrolysis of ATP to ADP on cation exchange resins, Stadtman and Kornberg (*3081*) decomposition of coenzyme A on Dowex-50 and Kenner *et al.* (*1692*) destruction of uridine nucleotides on anion exchange resins.

Azaserine, containing the labile aliphatic diazo group is destroyed (or irreversibly adsorbed) on Dowex-50 (H$^+$), Zeo Rex (H$^+$), Amberlite IRA-400 (OH$^-$), XE 97 (H$^+$) or XE 98 (OH$^-$) (Fusari *et al.*, *1065*).

DIVISION III

PARTITION CHROMATOGRAPHY

CHAPTER 15

Introduction

General reviews on partition chromatography were made by Martin (*2148, 2149, 2151*), Boulanger and Biserte (*333*), Synge (*3183*).

Monographs on paper chromatography have been published by Balston and Talbot (*131*), Block, LeStrange and Zweig (*282*), Block *et al.* (*281a*), Cramer (*643*), Hais and Macek (*1269*), Linskens (*2031*) and Schoen (*2871*). Numerous review articles have been published on paper chromatography; of the more recent ones we quote Berlingozzi (*215*), Consden (*615*), E. Lederer (*1903, 1904*), M. Lederer (*1922, 1941*), Macheboeuf *et al.* (*2083*), Martin (*2152*), Partridge (*2445*), Pöhm and Wichtl (*2532*), and Wankmüller (*3475*).

When a solution of a substance is shaken with an immiscible solvent, the solute will distribute itself between two phases and when equilibrium is reached, the coefficient $\dfrac{\text{concentration in solvent A}}{\text{concentration in solvent B}}$ is a constant a, where a is termed the *partition coefficient*.

In order to effect a separation of a mixture of amino acids, Martin and Synge (*2155*) utilised the difference of partition coefficients with a battery of solvent-solvent extractors. Although this work with solvent-solvent extractors (counter-current distribution) has been further developed, notably by Craig (*641*), Martin and Synge (*2156*) found that a far more efficient, fractional solvent-solvent extraction is possible by packing columns with silica gel, holding about 50 % water, placing the solution of a mixture on the column, and developing with water-immiscible solvents e.g. chloroform containing small amounts of butanol. The liquid held on the column is termed the stationary phase and the eluant the mobile phase. Other materials are also capable of holding water as a stationary phase and columns of starch, cellulose powder, cotton linters, even asbestos and glass beads will be mentioned in the special sections.

This process has been called *partition chromatography* as distinct from adsorption chromatography.

Consden, Gordon and Martin (*618*) showed that filter paper sheets and strips can also be used as support of a stationary phase in partition chromatography. This technique, called *paper chromatography* — other names proposed: papyrography (*741*) and partography (*2726*) — is probably the most versatile method for analytical work on a micro scale.

Generally water or hydroxylated polar solvents are used as stationary solvents and more or less non-polar solvents as mobile phases.

For lipid soluble substances with very low solubility in hydroxylated solvents (higher fatty acids, weakly polar steroids etc.) it is preferable to hold a lipid phase stationary on such materials as rubber latex (Boldingh, *300*), glass powder (Partridge and Chilton, *2446*), silicone-treated kieselguhr (*1460*) and silicone- or vaselin-treated filter paper (*3602*). The mobile phase in this type of chromatography, called *reversed phase chromatography*, is usually a hydrophilic solvent.

Theory

A theory of partition chromatography, developed by Martin *et al.* (*2156*) considers chromatography analogous to fractional distillation with total reflux. By employing the concept of a theoretical plate, they developed an equation to correlate the rate of movement of a zone with the partition coefficient, viz.

$$R_F = \frac{A_L}{A_L + a A_S} \quad \cdots \cdots \cdots \cdots \cdots \cdots \quad (1)$$

where a = partition coefficient $\dfrac{\text{conc. in water phase}}{\text{conc. in lipoid phase}}$

A_S = the cross sectional area occupied by the stationary phase

A_L = the cross sectional area occupied by the mobile phase

$R_F = \dfrac{\text{distance travelled by the zone}}{\text{distance travelled by the liquid front}}$

The R_F values of amino acids have been shown to agree well with those calculated from the partition coefficients measured by static methods. Also the degree of separation of two substances of known partition coefficients can be calculated for a given length of column or paper strip.

In a later paper Martin (*2150*) developed the theory of partition displacement chromatograms along similar lines to his original theory and arrived at the following equations for the movement of zones in a *buffered column*.

$$R = \frac{T}{M + aS \left(1 + \dfrac{K_a}{H^+} \right)}$$

where R = ratio of rate of movement of zones divided by rate of movement of developing liquid in tube above column

T = cross sectional area of total area in column

M = cross sectional area of mobile phase in column

a = partition coefficient of unionised acid

S = cross sectional area of stationary phase in column

K_a = dissociation constant of acid.

From this relationship the optimum pH range for the separation of two acids, A and B, with partition coefficients α and β is given:

when $\beta/\alpha < 1$ and $K_b / K_a < 1$, a high pH is desirable.

when $\beta/\alpha < K_a / K_b < 1$, a low pH is desirable.

when $K_a / K_b < \beta/\alpha < 1$, a high pH is desirable.

The factors governing enrichment by passing from one phase to another were also considered by the same authors.

Martin and Synge (*2156*) calculated the number of theoretical plates in a silica gel column and found the height equivalent to a theoretical plate to be 0.002 cm. Later estimates are only 0.02 cm (Verzele, *3401*). Thus a column a few cm high has a few hundred theoretical plates, compared to 20 for the best distillation columns, 20 for a column of alumina in Lindner's "radiometric analysis" (*2022*) and a few hundred for ion exchange columns (Nachod, *2332*).

A relationship similar to equation (1) was also developed for ion exchange colums (see p. 89).

A simple and non-mathematical approach can be developed along the kinetic approach first used by Cremer and Muller (*650*). A kinetic picture of the chromatographic process is necessary: the molecules of a given solute are continually moving from the stationary phase to the moving phase and back, and, depending on the partition coefficient, spend a certain average time in each phase. As all molecules do not have the same energy, some will spend more time than the average in the moving phase and some less, thus producing a band with the characteristic concentration curve, similar to the normal distribution curve.

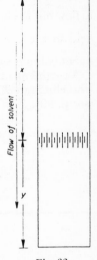

Fig. 33

Now consider the column or paper strip, Fig. 33, and let x be the distance the center of a band of solute has travelled and $x + y$ the distance the solvent has travelled.

Then x is proportional to the solubility of the solute in the moving phase and y proportional to its solubility in the stationary phase.

i.e. $\dfrac{x}{y} \propto \dfrac{1}{\alpha}$ (α is the partition or more generally the equilibrium coefficient)

and $\dfrac{x}{y} = \dfrac{1}{\alpha} k$ where k is a constant accounting for the ratio of the cross section of the moving and stationary phase.

Also $R_F = \dfrac{x}{x+y}$ and $\dfrac{1}{R_F} = \dfrac{x+y}{x} = 1 + \dfrac{y}{x} = 1 + \dfrac{\alpha}{k}$

where k as in the equation of Martin and Synge $= \dfrac{A_L}{A_S}$

$$\text{then } \frac{1}{R_F} = 1 + a\,\frac{A_S}{A_L} = \frac{A_L + a\,A_S}{A_L} \text{ and } R_F = \frac{A_L}{A_L + a\,A_S}$$

an expression identical with that obtained from the theory of fractional distillation.

Also the volume necessary to elute the peak of a band of a solute may be obtained in a similar manner. In Fig. 33, consider x now as the length of a column with the peak of the band just about the flow out. Assume y to be another section of column which is used to collect the effluent.

Then as before, $k\,a = \dfrac{y}{x}$ but y is now proportional to the volume that has to flow out to elute the peak of the band and x proportional to the volume held in the column; thus $\dfrac{\text{peak effluent volume*}}{\text{column volume}} = k\,a$ or the peak effluent volume $= Ka$.

An equivalent expression was obtained by Mayer and Tompkins (*2176*) who eliminate K by expressing the volume in terms of the column contents (see p. 89).

---

* Peak effluent volume $= F_{max}$, see p. 89.

CHAPTER 16

Partition chromatography on columns

1. SILICA GEL

The terms *silica gel*, or *silicic acid* refer to hydrated silica precipitates, the properties of which can vary rather widely according to the precipitation and purification method used.

Silica gel is usually prepared by precipitation of sodium silicate (water-glass) with 10 N HCl, followed by exhaustive washing and drying (Gordon *et al.*, *1168*, also Martin, *2148*). For the preparation of columns it is mixed with 53 % of water and poured into the column as a slurry with the mobile phase. The exact method as given by Gordon *et al.* (*1168*) is as follows:

"Commercial water-glass (140° Tw.-Jos. Crosfield, Ltd., Warrington) is diluted to 3 vol. with distilled water containing a little methyl orange. 10 N HCl is added in a thin stream with vigorous stirring, addition being interrupted at intervals and stirring continued to get efficient mixing. The solution changes first slowly then rapidly to a thick porridge and all but the smallest lumps are broken up by stirring. When the mixture is permanently acid to thymol blue, addition of HCl is stopped and the mixture is kept three hours. It is filtered on a Buchner funnel and washed with distilled water (approx. 2 1/250 g dry gel) without allowing the precipitate to crack. The gel is then suspended in N/5 HCl and aged 2 days at room temperature. It is again filtered and washed in the same way with distilled water (approx. 5 1/250 g dry gel) until the washings are free from methyl orange. Finally, the gel is crumbled and dried at 110° in an air oven. With such a preparation the addition of 53 per cent w/w of indicator solution to the dry gel should be satisfactory with the butanol-chloroform or propanol-cyclohexane mixtures. The dry gel can be stored in a closed vessel for long periods without deterioration."

The gel prepared in this way contains nitrogen. If it is wished to use it for the adsorption of the hexone bases further purification is necessary (*1054*) as follows: 100 g of silica gel is treated for 48 to 55 hours with 500 ml of 20% acetic acid. After filtration on a Buchner funnel the gel is washed with 200 ml of 20% acetic acid and then with boiling water until the filtrate reaches pH 6. Drying is carried out at 110° for 6 hrs. The gel then contains approx. 20 μg of N per g. The complete removal of nitrogenous material is achieved by treating the silica gel in the column in which it is to be used with 200 ml of 0.1 N HCl and then with 200 ml of water.

Columns containing too much water can give rise to a phenomenon similar to water logging on paper chromatograms and a single substance may yield more than one band, as part of the band moves down the column and part is held in a "wet" zone at the top of the column (Lester Smith, *3025*, and Ovenston, *2428*).

The adsorptive powers of silica gel also vary considerably from batch to batch and testing with standard mixtures of known substances for sharpness of the eluted zone is necessary.

The role of water in the chromatographic behaviour of silicic acid was subjected to a study by Kay and Trueblood (*1665*) who concluded that depending on the amount of water held, silicic acid may act as an adsorbent or partition support; usually, both properties contribute to varying degrees to the separation effects obtained.

Tristram (*3303*), Isherwood (*1512*) and Harris and Wick (*1315*) have described the preparation of silica gel for partition chromatography.

In order to render the zones of acidic substances (acetyl amino acids, fatty acids) visible on the column, indicators can be incorporated in the aqueous phase. Methyl orange was first used but was too soluble in the mobile phase. Anthocyanins, also 3,6-disulpho-β-naphthalene-azo-N-phenyl-α-naphthyl-amine (*2005*) have been found more satisfactory.

Commercial *silica gel* can be purified as follows:

"A 50-gram sample of silica gel (80- to 120-mesh) was suspended in 300 ml of concentrated hydrochloric acid and allowed to stand overnight. The yellow supernatant solution was decanted, fresh concentrated hydrochloric acid was added, and the mixture was shaken, and again allowed to stand. This process was repeated until the solution was colorless. The mixture was next filtered with suction on a sintered-glass funnel. The residue on the filter was suspended in water and washed by decantation until free of chloride, filtered as before, and the material was then suspended in 95% ethyl alcohol. This suspension was filtered and washed with 200 ml of 95% ethyl alcohol on the sintered-glass disk. The gel was then washed with 200 ml of absolute ethyl alcohol, suspended in anhydrous ether, filtered, and washed with 500 ml of anhydrous ether. The gel was heated for 24 hours at 100° C and finally dried for 24 hours in a desiccator over phosphorus pentoxide in vacuo."

"Davison Chemical Corp. Grade 70 silica gel, purified and dried in this way, will adsorb its weight of water and still remain dry enough to "gel" with chloroform. (This property is desirable because it allows solutions of organic acids to be taken up with a relatively small amount of gel. The mixture of gel and acids can then be placed on top of the major portion of the column in a compact zone, which permits sharp eluate fractions.)"

"Treatment with hydrochloric acid is necessary to remove inorganic cations which form insoluble salts with organic acids; incomplete washing with hydrochloric acid leads to low recoveries of the acids. The adsorbent properties which are present in some silica gels also are removed in the purification process." (Resnik *et al.*, *2670*).

Silicic acid (Merck or Mallinckrodt) can be purified as follows:

"500 g of reagent grade silicic acid (Merck or Mallinckrodt) are thoroughly mixed with 800 ml of acid ethanol (400 ml of c.p. hydrochloric acid diluted to 2400 ml with 95 per cent ethanol). The mixture is filtered through a Büchner funnel, and the washing procedure is repeated at least three times. The filtrate from the final washing should be colorless. The silicic acid is then washed three times in the same manner with 700 ml portions of 95 per cent ethanol, and finally three times with 800 ml portions of diethyl ether. It is then spread on clean white paper and allowed to dry for 24 to 48 hours. The material is then passed through an 80 mesh sieve and dried in a vacuum desiccator." (Morrison and Stotz, *2284a*).

For separation of fatty acids or penicillins *buffered silica columns* are used (see Chapters 22 and 37). The bands are either made visible with indicators or eluted portions are titrated with alkali. Such gels may be prepared in the following way:

Buffered silica gel (for the purification of penicillin, after Levi and Terjesen, *1988*):
30 g of silica gel prepared as described by Martin and Synge (*1168*) is well mixed in a
mortar with 15 or 20 ml of a solution of potassium phosphate at pH 6.6 (the buffer is
prepared by titrating a solution of 25 parts of KOH in 100 ml of water with an aqueous
solution of 30% of phosphoric acid to pH 6.6 as measured potentiometrically). In this way
a dry powder is obtained which is poured into the column with amyl acetate (the solvent
used for the adsorption of penicillin). 100 g of gel are sufficient for the adsorption of 3.2 g
of the calcium salt of penicillin.

2. KIESELGUHR

The preparation of adequate kieselguhr columns (Martin, *2150*) requires
the tamping down of the slurry in the columns. Owing to the more compact
nature of the silica in the kieselguhr practically no adsorption occurs.

3. STARCH

Starch columns were employed for amino acids by Moore and Stein (*2265*)
(see Chapter 30). The preparation of such columns was described in great
detail and cannot be repeated here (see *3093*). Other uses of starch columns
are mentioned in the special chapters.

4. CELLULOSE

Cellulose columns were used by Hough *et al*. (*1458*) for separating macroquantities of sugars and by Burstall *et al*. (*468*) for macroquantities of
inorganic cations. The latter employs glass tubes treated with dichlorodimethylsilane to prevent creeping of the solvent at the glass surface.

Hough *et al* (*1458*) recommend the following for testing the efficiency of
the column: "The performance of the column may be observed visually by
placing a suitable mixture of the following dyes on the column and noting
the appearance and the separation of the coloured materials as they advance
down the column. It has been found that various dyes move at different
rates on sheet-filter-paper chromatograms. The Rg values of a number of the
more suitable of these coloured materials have been determined on sheet-
paper chromatograms, using the top layer of a mixture of n-butanol (40 %),
ethanol (10 %), water (50 %) as the mobile phase. Rg values determined
were as follows: auramine 1.00, dimethyl yellow 0.95, bromothymol blue
0.83, brilliant cresyl green 0.73, metanil yellow 0.48, cresol red 0.41,
methyl red 0.38, bromocresol green 0.30, bromophenol blue 0.26, and
methyl orange 0.23. If the coloured materials travel through the column in
the form of regular horizontal bands, the column is regarded as satisfactory
for use, the uniformity of packing being thus indicated. These dyes can also

be used as markers, since a dye that moves slightly faster than the fastest-moving component of the sugar mixture, can be incorporated in the mixture to be resolved, and it is then only necessary to examine the eluate after the coloured material has emerged from the column."

Both starch and cellulose columns have essentially the same general behaviour as paper strips; however, in general solvents with high R_F values are used for columns in order to keep the volume necessary for elution within reasonable limits. For displacement analysis on partition columns see Levi (*1987*). Solka-Floc cellulose columns were used for amino acid separations by Carter *et al.* (*521*). *Ascending development* on cellulose columns was employed successfully by Fischer and Behrens (*963*) for the separation of indole derivatives.

Cellulose acetate has been proposed by Boscott (*325*) for separations of phenols and aromatic acids.

5. RUBBER

Rubber columns can be used for "reversed phase" partition chromatograms, i.e. for the separation of lipophilic substances.

Boldingh (*300*) has used a column of moderately vulcanized Hevea rubber saturated with benzene for the separation of higher fatty acids (elution with aqueous methanol).

Nyc *et al.* (*2391*) and Bosch (*323*) have used a powdered vulcanized rubber ("Mealorub", Andresen Corp., Chicago) for the separation of oestrogens.

Partridge and Swain (*2462*) have developed a reversed-phase chromatogram using a commercial chlorinated rubber ("Alloprene", I.C.I. Ltd.; extra high viscosity grade E) for the separation of 2,4-dinitrophenyl-derivatives of amino acids (butanol as stationary, aqueous buffers as mobile phase).

CHAPTER 17

Gas-liquid partition chromatography

Phillips (*2505a*) has recently published a monograph on "Gas chromatography".

The main factor limiting the efficiency of liquid-liquid partition chromatograms is the rate of diffusion in the two liquid phases of the molecules to be separated. Maximum efficiency is obtained by running the columns very slowly. James and Martin (*1537*) introduced the use of a permanent gas as the mobile phase so that the diffusion is much more rapid and high rates of flow can be used even with very long columns. The technique has first been applied to the separation and identification of volatile acids and bases (see pp. 179, 201) by using titration involving an automatic recording burette as the means of detection of the materials in the gas stream; the whole procedure is automatic and rapid. High column efficiencies were obtained, ranging from 750 to 1250 theoretical plates for 4 ft columns, and one 11 ft column had an efficiency of 2000 theoretical plates.

James and Martin (*1539*) have later extended this method to separate "all those substances capable of being distilled even at pressures of only a few millimetres of mercury". The important technical advance in this work was the development of an instrument capable of detecting any eluted substance in the gas stream (Fig. 34). Thermal conductivity measurement was rejected by James and Martin as being too inaccurate and a *gas density meter* was developed (*2152a*).

The instrument measures the difference between the density of the gaseous effluent from the chromatogram and that from a comparison column through which only nitrogen is passed, the two columns being maintainted at the same temperature and with the same nitrogen pressure. The apparatus consists essentially of a series of tubes bored in a copper block (held at the same temperature as the columns) and connected in a manner analogous to a Wheatstone bridge so that a difference of flow rate of the two gas streams lead into the block causes no pressure difference between two points in the system. Two sets of channels are connected by a cross-channel in such a way that any density difference between the two streams causes a pressure difference and hence a flow of gas through the cross channel. This channel contains a flow detector consisting of a small filament (electrically heated) arranged below and equidistant from two connected thermojunctions; any cross flow of gas causes the stream of hot convected gas from the filament to be diverted to one or the other of the thermojunctions, heating one and cooling the other. The thermo-electromotive force produced is fed to a commercial direct current amplifier whose output is led to a recording galvanometer. The galvanometer deflection is linearly related to the density difference of the two gas streams and, unlike the thermo-conductivity method (using a catharometer) is unaffected by changes of gas flow rate. The instrument is highly sensitive, amounts of the level of 1/16 microgram of amyl alcohol/ml of nitrogen being detectable.

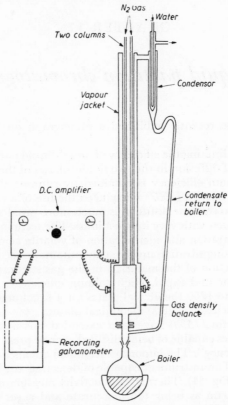

Fig. 34. Schematic lay-out of gas-liquid partition chromatography apparatus, using a gas-density for detection and estimation of the zones (James and Martin, *1539*)

Thus not only titratable acids and bases but also hydrocarbons, alkyl chlorides, alcohols, aldehydes, esters, ethers, and ketones were separated (*1538, 1539*). The separation of *isomers* is much better than with usual chromatographic methods.

There are already numerous analytical applications of this method. Aromatic hydrocarbons, esters and alcohols were separated by Littlewood *et al.* (*2038*). Cropper and Heywood (*662*) employ sodium chloride crystals as support of a stationary phase (silicone grease) in high temperature separations of high molecular fatty alcohols and esters. Fatty acids were also separated by Annison (*68*) and Van de Kamer *et al.* (*3363*).

The theory of gas-liquid chromatography was worked out by James and Martin (*1536*); it differs from the theory of liquid-liquid chromatography in that the mobile phase is compressible and so produces a gradient of gas velocity down the column. Interesting observations were possible by comparing stationary phases which only hold the dissolved molecules by Van der Waals forces (or induced dipoles) and those permitting hydrogen bonding such as in the case of primary and secondary amines with polyether columns. Each homologous series, when relative retention volumes

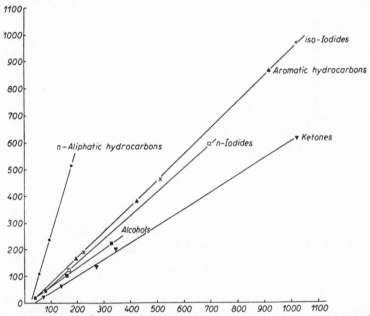

Fig. 35. Graphical representation of relative interaction forces in two types of stationary phase, liquid paraffin (ordinate) and benzyldiphenyl (abscissa) (James and Martin, *1539*)

are plotted with one phase as abscissa and the other as ordinate, gives a straight line whose slope depends on the amount of hydrogen bonding. "In this way the nature of an unknown amine may be closely defined".

James and Martin (*1539*) examined a large number of aliphatic compounds (the series $C_5H_{11}-X$) and stated that the sequence in an aromatic stationary phase (benzyl diphenyl) is (increasing interaction): $H < Cl < Br < I < NH_2 < NO_2 < OH < COCH_3 < CN$ while liquid paraffin or polyether columns gives the sequence $H < OCH_3 < Br,Cl < I < NO_2 < COCH_3 < NH_2 < CN < OH$. Here hydrogen bonding plays a greater part than on the aromatic column.

James and Martin stress the utility of these sequences as well as the regular behaviour of homologous series for the determination of unknown substances.

A typical graphical representation of some of the results is shown in Fig. 35.
The advantages of additions of surface active agents to the stationary
phase were stressed by Purnell and Spencer (*2603*) and are illustrated in

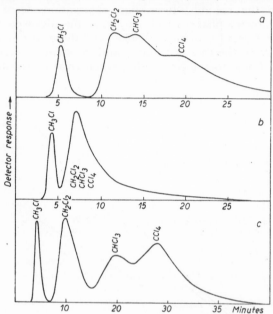

Fig. 36. Elution by nitrogen at 20° of chloromethanes on 38-cm columns.
(a) Kieselguhr (6.3 ml/min); (b) Water-kieselguhr (6.3 ml/min); (c) Water/"Teepol"/
kieselguhr (7.7 ml/min) (Purnell and Spencer, *2603*)

Fig. 36. For further work see also the papers by Ray (*2636*), Harvey and
Chalkley (*1321*) and Pollard and Hardy (*2537*) and the special chapters.

The application of gas-liquid chromatography on a preparative scale has
been reported by Evans and Tatlow (*925a*) and Ambrose and Collerson (*46a*).

The very rapid recent developments of vapour phase chromatography are
described in the Proceedings of the "Vapour Phase Chromatography
Symposium", organised by the Institute of Petroleum (Hydrocarbon
Research Group) in London (May 1956).

Commercial models of gas-liquid chromatography apparatus are available: the "Griffin vapour phase chromatographic apparatus", Griffin and
Tatlock (London); the "Vapor Fractometer", Perkin-Elmer Corporation
(Norwalk, Conn., U.S.A.), the "Kromo-tog", Burrell Corporation (Pittsburgh 19, Pa., U.S.A.), the "Fisher-Gulf Partitioner", Fisher Scientific
(Pittsburgh 19, Pa., U.S.A.), the "Chromacon", Podbielniak Inc. (Chicago
11, Ill., U.S.A.), etc. These apparatuses generally use catharometers for
the detection of eluted substances (as first described by Ray, *2636*).

CHAPTER 18

Paper chromatography

1. MECHANISM

Filter paper in an atmosphere saturated with water vapour absorbs approximately 22 % of water (*1373*). Consden, Gordon and Martin (*618*) considered filter paper as an inert support of an aqueous stationary phase and explained the observed separations as a result of continuous partitions of the substances between the aqueous stationary phase and the water immiscible organic solvent flowing down the paper. This approach was criticized later by numerous authors (*468, 641, 1297, 2265*). Craig (*641*) thought that an equilibrium could not be established sufficiently fast in a system without agitation and hence the separations could not be explained by simple solvent-solvent extraction. Martin (*2151*) however points out that on the basis of known diffusion constants the efficiency of paper chromatograms is of the right order.

A more important objection is as follows: if the separations obtained are due to partition between two solvents, it should be impossible to employ water-miscible solvents, where only one phase exists and hence no partition is possible. However, numerous authors have effected separations of amino acids and other substances with water soluble solvents such as propanol, acetone, ethanol or even with pure water (*190, 1925, 2215, 2265, 2322*). Before deciding on this point it is necessary to examine in detail the exact state of the water-saturated filter paper.

If one exposes dry cellulose to water vapours, a quantity of water, ca. 6 %, is absorbed with a high heat of sorption and high apparent density and a low velocity of diffusion. As more water is absorbed these properties change and become more and more like those of liquid water, especially the diffusion rate. Martin (*2151*) compares the aqueous phase on the paper with a concentrated solution of carbohydrate: "The stationary phase in a cellulose chromatogram should be compared with say a strong solution of glucose, or better of some soluble polysaccharide rather than water saturated with the organic phase. It should therefore cause no surprise that solvents miscible with water can be used. A strong solution of glucose will form two phases with aqueous propanol, the carbohydrate-rich phase containing a relatively higher proportion of water, the other phase a relatively higher of organic solvent."

Hanes and Isherwood (*1297*) adopted a point of view analogous to Martin's; they consider the stationary phase as a water-cellulose complex. A substance in solution will be held more or less strongly in this complex, depending on its hydrophilic properties.

It is thus mainly a matter of definition of terms whether the mechanism of paper chromatography is called an adsorption on, or a partition in the "water cellulose complex".

Horner *et al.* (*1448*) studied the distribution of water between the paper and the solvent refractometrically with various water-solvent mixtures. See also the review of Moore and Stein (*2269*).

True adsorption on paper as well as ion exchange with the free carboxylic groups of the carbohydrate network also exists. This has been studied by Schönfeld and Broda (*2873*) with the aid of radioactive tracers and the adsorption of ions was found to be similar to that of a weak cation exchanger. See also Giles *et al.* (*1107*) and p. 92.

A streaming potential as high as 10 mV per cm was measured by Rutter (*2758*) for pure water flowing through ordinary filter paper, thus indicating adsorption. See also Epshtein (*903*).

Burma and Banerjee (*460*) chromatographed a number of amino acids and sugars with water as solvent and found in most cases no retardation of the spots ($R_F = 1$), showing absence of adsorption.

However numerous dyestuffs and other compounds show considerable adsorption.

"Real partition chromatography"

Tschesche *et al.* (*3308*) calls "real partition chromatography" the development on paper which has been thoroughly moistened with the stationary phase and thus does not absorb its stationary phase from the atmosphere or the developing solvent. This technique is advantageous when the stationary phase is organic and the aqueous phase is mobile as used for the separation of cardiac glycosides. Disappearance of otherwise unavoidable comets is claimed in this technique.

2. CORRELATION OF CHEMICAL CONSTITUTION AND R_F VALUE

As shown, the R_F value depends on the partition coefficient and the relative amounts of the two phases in contact. For a given column or paper it depends solely on the partition coefficient, which is a thermodynamic property of a given molecular species, as specific and distinctive as other phase transition points such as boiling and melting points.

Whilst for the identification by boiling or melting point a pure specimen is required, the R_F value is not influenced by the presence of many impurities. Further, only micro quantities are required for a paper chromatographic identification. To identify a substance one usually measures the R_F value with a number of suitable solvents, as well as the R_F values of reference substances run on the same paper at the same time. It is not possible, however, to identify an unknown substance by R_F value alone. Numerous mixtures of isomers

$$
\begin{array}{cc}
\text{CH}_2\text{OH} & \text{CH}_2\text{NH}_2 \\
| & | \\
\text{for example}\quad \text{CHNH}_2 \quad \text{and} \quad \text{CHOH} \\
| & | \\
\text{CH}_3 & \text{CH}_3
\end{array}
$$

produce only one spot in all solvents examined (*1990*).

The R_M value

A new R value proposed by Bate-Smith and Westall (*158*) is

$$R_M = \log\left(\frac{1}{R_F} - 1\right)$$

It is proportional to the free energy of moving a molecule from one phase to the other. Both Bate-Smith and Westall (*158*) and Martin (*2150*) have shown that the R_M value is made up of additive values representing the groups in the molecule and a constant for the given solvent system, paper etc. For correlations of R_M values of peptides with constituent amino acids see Chapter 31, for phenols Chapter 20. More recently several attempts were made to use R_M values to calculate the R_F values of given compounds (*2655a*) or even to determine structures from a given number of R_F values in different solvents (*2836a*). For this purpose "group constants" must be calculated for the solvent system from R_F values of known compounds as well as the constant for the solvent system (Grundkonstante).

We shall cite one example from Reichl (*2655a*):

Solvent: amyl alcohol-5N HCOOH
Paper: Whatman No. 1
Ascending method

constant for solvent system (Grundkonstante)	0.97
each carbon atom	0.12
each branched chain	0.25
primary OH	—0.73
secondary OH	—0.50
tertiary OH	—0.58
carboxylic group	—0.63
amino group	—1.65
keto group	—0.39

R_F values

	measured	calculated by adding the above constants and converting R_M to R_F
adipic acid	0.75	0.73
citric acid.	0.23	0.23
glycolic acid.	0.43	0.41

In numerous cases, for example peptides, the correlation is not as satisfactory. For determination of structures Schauer and Burlisch (*2836a*) propose the calculation using determinants. This requires the R_F values in sufficient solvent systems and by analysis a knowledge of the number of each group present.

The ΔR_M method

Substances such as acids and bases can be chromatographed, depending on the solvent, in the ionised or non-ionised form. If a solvent pair is chosen in which most other groups than the ionisable ones have low group constants the difference in R_M values will be proportional to the number of the ionisable (or complexed etc.) forms in the molecule. Reichl (*2655b*) bases a method for determining the number of COOH groups on this principle. Using ethyl acetate-acetic acid-water (2:1:1) as the acid solvent and acetone-0.5 $\mathcal{N}$ ammonium acetate as a neutral one he obtains the following results (examples are cited only):

	Solvent 1		Solvent 2		
1 COOH	R_F	$-R_M$	R_F	$-R_M$	ΔR_M
glycine	0.41	−0.17	0.24	−0.49	0.32
lactic acid	0.76	0.50	0.52	0.04	0.46
2 COOH					
aspartic acid	0.39	−0.19	0.06	−1.20	1.01
fumaric acid	0.86	0.78	0.14	−0.79	1.57
oxalic acid	0.40	−0.17	0.04	−1.33	1.16
3 COOH					
aconitic acid	0.82	0.65	0.02	−1.62	2.27
citric acid	0.59	0.16	0.01	−1.92	2.08

Thus it is apparent that the ΔR_M value varies with the number of COOH groups and the estimation of the number of COOH groups may be made. For the determination of free glycol systems in veratrum alkaloids by the same method see Macek and Vejdelek (*2078a*).

Homologous series

Fatty acids and their derivatives show a regular increase in their R_F values with the increase in the number of C atoms. In the partition chromatography of amino acids on paper the relationship between chemical structure and partition coefficient is shown clearly by certain regularities in the position of the spots. Fig. 37 of Polson (*2552*) illustrates this relationship.

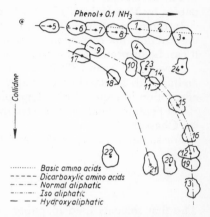

On a two-dimensional chromatogram (e.g. phenol and collidine) the monoamino aliphatic acids (glycine, alanine, α-amino butyric acid, valine and leucine) appear on a curve, the hexone bases occur together, and the dicarboxylic acids are very close to one another (Consden *et al.*, *618*). The position of an unknown spot thus allows, to a limited extent, the drawing of conclusions on the chemical structure of the substance in question.

Fig. 37. Tracing of a two-dimensional chromatogram of a mixture of amino acids. Spots due to acids of related composition fall on smooth curves (Polson, *2552*)

The partition coefficients of members of a homologous series are sufficiently different to allow complete separation. That changes in the constitution of the carbon chain have only small effects on the partition coefficient is shown by the difficulty encountered in the separation of leucine, iso-leucine and nor-leucine, and also of valine and nor-valine.

If the R_M values are plotted, for example, against number of C atoms straight lines are obtained for all homologous series. This makes the calculation of R_F values of unknown members rather accurate as shown in the example below.

Fatty acid hydrazides (Satake and Seki, *2824*), see page 188.

Number of C atoms	R_F measured	R_F calculated from graphical average of a $R_M - CH_2$ line
1	0.11	0.10
2	0.18	0.18
3	0.37	0.33
4	0.54	0.52
5	0.70	0.70
6	0.77	0.83

Inaccuracies near the solvent front occur regularly and are probably caused by changes in the solvent composition in that region which should be usually impoverished in water.

The behaviour of homologous series was discussed further in a series of papers by Van Duin (*3367–3372*). See also Serchi (*2945*).

In those cases where the behaviour of a substance on a partition chromatogram does not correspond to that calculated from its partition coefficient, it can be suspected that *adsorption* is playing some part.

Moore and Stein (*2265*) have tested this in the case of the separation of amino acids on starch (see Chapter 30). For a detailed study of the role of adsorption in paper chromatography see Burma (*458*).

For peptides the R_F values were theoretically calculated by Martin (*2150*), see Chapter 31.

For sugars see Levy (*1993*) and Chapter 28.

3. THE ORGANIC SOLVENT

The first solvents used in paper chromatography were collidine, phenol, butanol and benzyl alcohol. Collidine and phenol have to be purified before use by distillation (for details see Draper and Pollard, *816*); phenol may also be purified by extraction with petroleum ether and washing with water (Mars, *2108*). As pure collidine is a rare chemical in some countries, many workers prefer using a lutidine-collidine mixture.

Water-miscible solvents such as propanol, furfuryl alcohol, acetone, pyridine and tetrahydrofuran are also used.

Bentley and Whitehead (*190*) examined water-miscible solvents for the separation of amino acids. They found that, as the number of carbon atoms of the aliphatic alcohols increased, the R_F value fell. This effect in a series containing the same quantity of water in each solvent indicates that the polarity of the alcohol has an effect on the R_F value. The same observation for inorganic ions was made by Lederer (*1925*). It can also be shown that ΔR_M for a CH_2 group increases regularly with the increase in the number of carbon atoms in the solvent when alcohols are used (M. Lederer, to be published). Thus, when the range of R_F values is suitable, a high molecular alcohol gives better separation for a difference of a CH_2 than a low molecular one.

Bentley and Whitehead (*190*) further showed that when the amount of water is increased in furfuryl alcohol the R_F value of the amino acids rose correspondingly. Lacourt et al. (*1851*) report experiments with solvents, which were not water-saturated and obtained the same relation between polarity and R_F as with water-saturated solvents. Mixtures of organic solvents were examined by Walker and Lederer (*3461*) who found that the R_F of a metal ion in a mixture (1 : 1) of two alcohols is slightly below the average value of the two alcohols.

(a) *Additions of acids to the solvent*

When a molecule can exist in an ionised and an unionised form (weak bases and acids), the R_F of the unionised form is usually higher than that of the ionised molecule. In a water immiscible solvent e.g. butanol-acetic acid-water, an increase in the acid concentrations usually increases the solubility of the solvent for water and the R_F values are generally raised in such a solvent. A number of metallic ions change their R_F values considerably with a change of acid concentration owing to formation of complexes between the metals and the more concentrated acid.

pH variation has a considerable effect on polar organic substances, such as amino acids, and some workers (Landua *et al.*, *1878*, McFarren, *2074*) showed that by suitable buffering of the paper (e.g. by dipping in phosphate buffers and drying) an improvement of the separation can be achieved.

For weak acids and inorganic cations a certain minimum concentration of an acid is necessary to prevent comet formation (see below).

(b) *Comet formation*

When organic acids and bases are chromatographed with neutral solvents, instead of the usual round spots, long trails (comets) are formed. This phenomenon was first observed by Lugg and Overell (*2057*, *2058*) in the chroma-

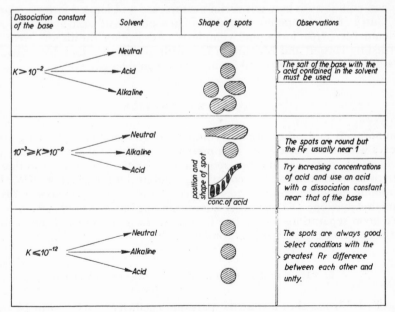

Fig. 38. The shape of spots of weak bases with solvents of various pH (Munier, *2318*)

tography of organic acids and is due to the dissociation of these acids into one or more ionised forms. As each ionised form has a different R_F value such dissociation results in a continuous deposition of material from the spot travelling on the paper.

Addition of an acid to the solvent (swamping acid) can prevent this formation of a comet by inhibiting ionisation.

Munier and Macheboeuf (*2323*) observed the same trailing of spots with alkaloids and other weak bases (K between 10^{-3} and 10^{-10}) (Fig. 38).

Other reactions beside ionisation can also produce a comet. Hanes and Isherwood (*1297*) consider traces of Ca^{++} and Mg^{++} responsible for the trailing and low R_F values of many acids and employ *acid-washed papers* to obviate this effect. For techniques of washing filter paper see Isherwood and Hanes (*1517*). Inorganic ions can either be hydrolysed (e.g. Sb^{+++}) or reduced (e.g. Au^{+++} and Ir^{++++}) to form comets. In the first instance an increased acid concentration and in the second the addition of oxidising agents can prevent comet formation (*1921*).

Adsorption comets occur especially with dyes (Zahn, *3660*).

(c) *Multispots*

One extreme case of comet formation is the formation of two or more discrete spots ("multispots") of one substance. Multispots may be caused by non-equilibrium between various complexed and ionised forms as shown by Curry (*676*) for mono- and dihydrogen phosphate and by Erdem and Erlenmeyer (*910*) for amine complexes. Adsorption multispots were recorded by Hassall and Magnus (*1332*) with monamycin and dyes. See also Waldron-Edward (*3449*) and for inorganic examples page 527.

4. COMPLEX FORMATION

Amino acids and other organic compounds are capable of forming complexes with traces of copper ions present in commercial filter papers, and yield *double spots*, i.e. another spot due to the Cu complex. Addition of 8-hydroxyquinoline or HCN completely inhibits this interference (*618*). For inorganic chromatography solvents containing benzoylacetone (*2541*), dimethylglyoxime (*58*) or other complexing agents are used, which often give good separations where free ions show no R_F differences. For details see the section on inorganic chromatography.

5. THE EFFECT OF TEMPERATURE VARIATION

Wark (*3477*) has shown that the partition coefficient varies with the temperature according to an isochore of the type of Van 't Hoff's isochore. In paper chromatography not only the effect of temperature on the actual partition coefficient but also on the composition of the solvent phases, the hydration of the cellulose, adsorption equilibria and further factors must be considered.

Consden *et al.* (*618*) demonstrated already very early that the main factor influencing R_F value changes with temperature was the water content of the organic phase which in some solvents (e.g. phenol) varies little and in others (e.g. collidine) considerably. Experimental data for sugars and amino acids for 17°, 37° and 57° were reported by Counsell *et al.* (*637*); for water soluble solvents a similar study was carried out by Burma (*455*); for inorganic ions by Tewari (*3223*).

Isherwood and Jermyn (*1518*) and later Alcock and Cannell (*34*) pointed out that a linear relation between the R_M value and the logarithm of the water content of the solvent may be used for those solvents, where the water content is known at various temperatures.

As R_M is a function of log a (the logarithm of the partition coefficient) a direct relationship between R_M and the absolute temperature may be obtained which has been shown (M. Lederer, to be published) to hold well for the usual range of room temperatures namely

$$R_M = \frac{-\Delta H}{2.303 \, R} \cdot \frac{1}{T} + C$$

where ΔH = heat of partition
R = universal gas constant
T = absolute temperature
C = constant

In examining the experimental data of the workers cited above it appears that ΔH is almost the same for a group of similar substances. Thus a change of R_M between two temperatures for one sugar may in most cases be used to predict the the R_F of another sugar at a given temperature, providing the R_F is known at one temperature.

Many authors hence employ thermostatically controlled or insulated chromatographic chambers (*157*).

6. THE PAPER

(a) *Common papers*

The commonly used papers are Whatman No. 1 (specially prepared "for chromatography"), Whatman Nos. 3 and 4, Schleicher and Schüll (SS) 589 and 595, and Munktell (Swedish) Nos. OA and OB.

Kowkabany and Cassidy (*1774*) compared 22 commercial brands of filter paper and classified them by considering the following:

Degree and clarity of separation of amino acid spots
Diffuseness of spots
Formation of a brown front (due to impurities in the paper)

Extent of tails and beards (comets)
Rate of movement of the solvent in descending development.

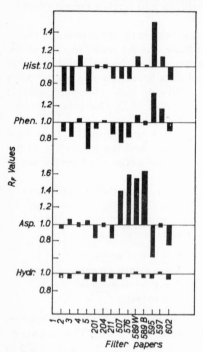

For amino acids, the papers were classified according to the usefulness in the solvent as follows (the best placed first):

Collidine: Whatman No. 3, SS 595, Whatman No. 4, Whatman No. 1.

Phenol: Whatman No. 3, SS 595, Whatman No. 1, Whatman No. 4.

1-Butanol: SS 589 Black Ribbon, SS 595, SS 598, Whatman No. 3.

2-Butanol-formic acid: Whatman No. 3, Whatman No. 1, Whatman No. 2, SS 602.

2-Butanol-ammonia: SS 589 Blue Ribbon, -Red Ribbon, -White Ribbon, Whatman No. 1.

Reeve Angel papers No. 202, 226 and 230 were found unsuitable.

Kowkabany and Cassidy (*1774*) also observed that the R_F values differed considerably from paper to paper, also that aspartic acid is adsorbed to a different degree on various papers. Thus on SS 507, 576, 589 Black, 589 Blue and 589 White Ribbon, aspartic acid gives circular spots; on all other papers an elongated spot.

A comparison of R_F values of amino acids on various papers is shown in Fig. 39. The authors recommend Whatman No. 3 as a generally useful paper with sharply resolved spots; SS 595 also gives excellent separations with generally higher R_F values.

A similar study by Burma (*457*) compares Whatman papers 1, 2, 4, 7, 11, 40, 42, 54 and 3MM for phenol and collidine. Here Whatman No. 1 paper was found most satisfactory followed by 4 and 54.

Fig. 39. R_F values of amino acids on various papers.

R_F values for four amino acids, two each in two different solvents, are shown for fifteen filter papers. In each group the papers were run at the same time in the same cabinet and the same mixture of amino acids was used with each paper. The R_F values are plotted on the ordinate relative to Whatman No. 1, which is taken as the norm.
Hist.: histidine in 1-butanol-ammonia.
Phen.: phenylalanine in 1-butanol-ammonia.
Asp.: aspartic acid in phenol. Hydr.: hydroxy proline in phenol.
On the abscissa are listed the filter papers: *W* refers to White ribbon, *B* to Blue ribbon. The first five are Whatman papers: the next three are Reeve Angel, the rest are Schleicher and Schuell papers (Kowkabany and Cassidy, *1774*)

Of the papers manufactured in France, D'Arches No. 302 has been found satisfactory (*1903*).

It must always be kept in mind that paper is a product of natural origin. Many constituents of living cells can thus be present as impurities and as some cells are not broken during manufacture they may yield such substances as amino acids, peptides or even fatty materials (*481*) on development or washing. It appears impossible to wash paper completely free of certain impurities as each wash may induce a further release of impurities by loosening some fibres. A *glass fibre paper* developed by the Whatman manufacturers may overcome some of these problems.

The usual impurities in filter papers are Ca^{++} and Mg^{++} ions (see comets, p. 121), also Fe^{+++} and Cu^{++} ions (see also complex formation, p. 122); the latter two produce a dark background when H_2S is used as a reagent for inorganic cations. A peptide-like contaminant in filter paper has been noted by Wynn (*3638*).

In addition to the properties examined by Kowkabany and Cassidy (*1774*), uniformity is essential when the paper is used for quantitative determinations by spot area measurements (see Quantitative methods).

Balston and Talbot (*131*) consider the suitability of Whatman papers with special regard to flow rates. The flow rate is faster in the machine direction, i.e. in the direction of the orientation of the fibres, in certain machine-made papers. In other papers an even flow rate in two directions is possible and then produces rounder spots.

(b) *Purified papers*

Purified papers often give faster flow rates due to irreversible swelling on purification. The Whatman papers are shown below in the order of diminishing flow rate:

Fast	No. 15	a thick paper
	No. 4	
Medium	No. 1	
	No. 3 MM	a thick paper
	No. 29	a black paper
Medium Slow	No. 11	a thin paper
	No. 2	
Slow	No. 20	

To prevent tearing of wet paper sheets during manipulations, Whatman "wet strength paper", more dense than the ordinary papers, can be used. In order of diminishing absorbency these are:

		Acid washed	
Very fast	Separa DHC	—	slightly creped surface
	Separa DH	—	
Fast	No. 54	No. 541	
Medium Slow	No. 52	No. 540	
Slow		No. 542	
		No. 544	a little thinner than 542
Very slow	No. 50	—	very smooth surface

Whatman also produces singly and doubly acid washed papers with approximately equal flow rates in the two directions, with slightly grained surface and less dense than the qualitative papers (except No. 120):

	Singly acid washed	Doubly acid washed
Fast	No. 31 good for amino acids	No. 41
	and paper electrophoresis	No. 43
Medium	No. 30	No. 40
Slow	—	No. 120 very thick and dense
		No. 44 slightly thinner than No. 42
Very Slow	No. 32	No. 42

A comprehensive treatise of the physics and chemistry of cellulose fibres was written by Hermans (*1373*).

(c) *Chemically modified paper*

In order to make paper water-repellent for reversed phase chromatography (see also p. 104), the cellulose was acetylated by Burton (*473*), Michael and Schweppe (*2205*), Scott and Golberg (*2922*), Buras and Hobart (*454*), or equipped with butyl-, benzoyl- and phthaloyl groups (*2203*). Further references to acetylated papers are also in the special chapters. Another method for making the paper water repellent is to cover it with rubber, silicones etc. (see the special chapters, also Morin *et al.* (*2278*)). For ion exchange papers see p. 92. *Red-ox papers* were prepared by Ezrin and Cassidy (*931*) and Sansoni (*2818*) but so far no chromatographic use has been found for them.

7. THE MOVEMENT OF THE SOLVENT DURING DEVELOPMENT

Müller and Clegg (*2310–2312*) carried out extensive work on the movement of solvents through filter paper. They confirmed the previous findings of Goppelsroeder in the years 1888–1904 (*1163, 1164*), Jermyn and Isherwood (*1586*) and Karnovsky and Johnson (*1633*), namely that the rate of travel of the solvent depends on the viscosity of the solvent and slows down with time.

Müller and Clegg (*2312*) showed that for short development the height h to which a solvent rises in time t obeys the equation:

$$h^2 = Dt - b$$

where h is in mm, t the time in seconds, b a constant and equivalent to an h_0^2 term and D a constant for a given paper and liquid. This constant D, called the diffusion coefficient, varies with the surface tension, viscosity and density of the solvent and obeys the equation:

$$D = a\, \gamma/\eta\, d + b$$

where a and b are constants depending on the filter paper. Good confirmation of these equations was obtained with water, methanol, ethanol and higher alcohols. Temperature variations of the rate of flow of a solvent could be explained by the temperature variation of the term $\gamma/\eta \ d$. The flow of fluid in the paper was also critically reviewed by Cassidy (*12, 531*).

8. FORMATION OF MORE THAN ONE LIQUID FRONT ("DEMIXION")

When a solvent is allowed to run over a strip of filter paper, the paper dehydrates the solvent to some extent, absorbing ca. 20 % of water of hydration. In a saturated atmosphere this dehydration is made up by some water vapour condensing on the paper and the composition of the solvent remains uniform. It is essential, therefore, to keep the atmosphere saturated with all the constituents of the solvent. This is readily achieved with solvents of which all the constituents are volatile, by placing a lower layer, obtained from the preparation of the solvent, into the developing tank. It is not possible, however, when the solvent (e.g. butanol) contains non-volatile constituents such as hydrochloric or nitric acids. In such solvents, a band poor in hydrochloric acid will precede the main portion of the solvent flowing over the paper and thus two distinct liquid fronts will be formed; the first being mainly that of butanol, the second being butanol with water and hydrochloric acid in the proportions found in the bulk of the solvent. Certain, not yet determined impurities in the paper, give the second front a dark border and leave a dark line which does not fade on drying. The height of the hydrochloric acid ascent depends on the temperature and acid concentration of the solvent. In butanol shaken with 0.3 N HCl, the acid or water front does not travel far at all. In concentrations above 1 N it usually travels 60–75 % of the total distance travelled by the butanol. A good account of demixion was published by Munier and Macheboeuf (*2321*).

9. TECHNIQUE

In the following pages the operations required for a paper chromatographic analysis are discussed.

(a) *Desalting*

Large quantities of HCl or alkalies are usually present in solutions of amino acids and sugars prepared by the hydrolysis of large molecules. Their presence can cause the formation of large spots of high water content (water logging) and can interfere with the reagents used (e.g. $AgNO_3$ reacts with Cl^- instead of the sugar). Thus pyrex tubes should be used for the hydrolysis to keep the electrolyte concentration low (*616*).

In addition to the usual chemical methods for the removal of electrolytes (such as precipitation with $BaCO_3$, etc.), an electrolytic method was developed by Consden *et al.* (*620*) which utilises a circulating mercury cathode and a graphite anode (Fig. 40); this apparatus has been modified by Acher *et al.* (*8*); see also Joseph (*1606*), Astrup *et al.* (*93*), and Katz and Chaikoff (*1646*).

It was observed by Stein and Moore (*3096*) that by this method low results were obtained for arginine, which is converted to ornithine.

Ion exchange resins have been used for desalting the neutral amino acids by Brenner and Frey (*378*) who percolate the solution through a column of Amberlite IR-4B and then through a column of Amberlite IRC-50. This yields an eluate containing all the neutral amino acids, with the anions and cations on their respective exchangers. A column of Nalcite SAR to adsorb the amino acids from samples of saliva (except arginine and lysine, which are lost) is used by Piez *et al.* (*2518*). See also Drèze and DeBoeck (*820*), Drèze *et al.* (*821*).

In a study of the amino acids and polypeptides of biological liquids, Boulanger and Biserte (*336*) achieved the desalting of amino acids and oligopeptides by chromatography on the sulfonated polystyrene resin Permutit-50. This resin retains all cations, all amino acids and some oligopeptides; the adsorbed organic compounds can be eluted by displacement with dilute ammonia; the inorganic cations are not eluted. Some losses of arginine seem inevitable.

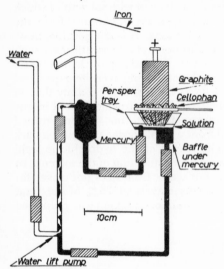

Fig. 40. Diagram of desalting apparatus (Consden *et al.*, *620*)

Boulanger and Biserte (*332*) extract free amino acids from dry plasma with acetone containing 1 % HCl in which the inorganic salts are insoluble. Lipids are removed from this extract with ether.

Sugars can be dissolved from dry residues of hydrolysates with pyridine in which the electrolytes are insoluble (Malpress and Morrison, *2114*). This operation may not be without danger to the sugar molecules.

In numerous instances the removal of electrolytes is not essential. Berry and Cain (*230*) describe the analysis of amino acids in urine without desalting. Ishii and Ando (*1521*) show that the R_F of amino acids is not affected by HCl in phenol or lutidine; there is, however, an effect with butanol as solvent. See also Baliga *et al.* (*127*); a commercial desalter is available from the Research Equipment Corp.*

(b) *Placing the sample on the paper*

Solutions are usually placed on the paper by means of a micro-pipette,

* Research Equipment Corp. (1155 Third Street, Oakland 20, Calif., U.S.A.).

capillary tube or micrometer syringe. The usual volumes employed are 0.002–0.02 ml. The spots should be placed at such distances (2–3 cm) that the chromatograms do not interfere with each other. Numerous authors prefer to place the sample on the paper as a thin line rather than a round spot; see for example Boser (*327*) and Fig. 102.

Urbach (*3353*) developed a special pipette with a platinum wire fitted inside the capillary, which delivers a solution slowly onto the paper and thus permits evaporation of large volumes directly onto the paper as a small spot. A similar technique was employed by Glazko *et al.* (*1136*).

Yanovsky and coll. (*3648*) apply large volumes of solutions with a stationary pipette to a sheet of paper, revolving on a chymograph, which is dried with an infra-red lamp.

Novellie (*2387*) prefers to apply the solution (in large amounts) to the paper by folding the sheet on the line of application and dipping into a trough holding enough solution for several chromatograms. Euler and Eliasson (*922*) first apply the solution to be separated onto a thin strip of paper which is clamped to the sheet used for the chromatographic run with glass rods. Thus relatively large amounts may be concentrated on thin lines. The method was successful with Grycksbo OB paper but not with Whatman No. 1 or No. 4 papers. See also Barker and Perry (*136*) who place the paper on a specially constructed frame for faster drying of large volumes.

(c) *Transferring spots from one sheet to another*

Whenever two-dimensional separations are required, but for reasons of technique the spot is to be run on another paper in the second solvent, it may be excised and clamped to the new sheet (Schlögl and Siegel, *2850*) or eluted directly onto the next paper (Moore and Boylen, *2262*) or even sewn or woven into the new paper (Stoeckli, *3118*; Boggs, *289*). See also Gregory (*1219*).

Reactions may also be performed on the filter paper, preliminary to chromatography; thus enzymic hydrolysis of raffinose and sucrose with invertase can be carried out on the paper before separating the hydrolysis products (Chargaff and Kream, *547*; Williams and Bevenue, *3578*).

(d) *Development*

Usually, solvents such as phenol or butanol are first saturated with water. The water-rich phase so formed is placed into the development chamber in a dish to prevent evaporation of water from the paper or the solvent trough. It is recommended to "equilibrate" the development chamber for 24 hours before commencing development. With very volatile solvents such as ether, continuous recycling of the solvent over the walls of the container may be necessary to maintain a constant atmosphere (see Baker *et al.*, *123*). See also Münz (*2326*).

(i) *Descending development*

Consden *et al.* (*618*) allowed the liquid to run down the paper by gravity, employing a glass trough to hold the solvent and enclosing the whole apparatus in a drain pipe. Many kinds of development chambers are now used (e.g. Fig. 41, 42): Block and Bolling (*281*); Toennies and Kolb (*3277*);

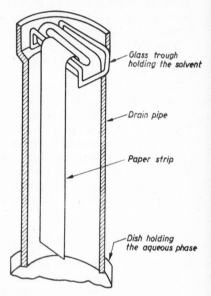

Glass trough holding the solvent

Drain pipe

Paper strip

Dish holding the aqueous phase

Fig. 41. Development chamber: duralumin frame and glass sides (Institut de Biologie Physico-Chimique, Paris)

Fig. 42. Apparatus for uni-dimensional descending development. The lid is not shown (*1922*)

Heyns and Anders (*1384*); Yamaguchi and Howard (*3644*); Alcock and Cannell (*34*); Irrevere and Martin (*1510*); Mitchell (*2241*); Fink *et al.* (*956*).

Winsten (*3597*) developed several strips simultaneously from a petri dish holding the solvent; Miettinen and Virtanen (*2213*) staple a piece of absorbent cotton to the lower end of the filter paper strip for the solvent to run over the paper where it is absorbed by the cotton. Spots of low R_F value may thus be separated. Hird (*1397*) reported a similar technique in which the same problem is solved by merely allowing the solvent to drip off the end of the filter paper (cut in zig-zag at the lower end) for 5 to 6 days. The construction of glass troughs for descending development appears to be difficult: several methods have been described (*95, 618, 2046, 3105*). Troughs made of polythene and supported by a metal frame, or stainless steel troughs may by used in place of the glass troughs. See also Porter (*2572*) and for a trough in sections Wunderly (*3634*).

(ii) *Ascending development*

Williams and Kirby (*3581*) simplified the technique of development by allowing the solvent to run up the filter paper by capillary action instead of running down the paper. A cylinder of the filter paper, with up to ten samples on one sheet, is stood in a dish containing the solvent and the whole enclosed in a battery jar, or earthenware pot (Fig. 43). Usually the same R_F values are obtained by the ascending and descending techniques. In order to decrease the space required by each chromatogram, Ma and Fontaine (*2066*) wind the paper round a specially constructed steel coil; the rolled-up paper is then placed in a measuring cylinder for development (Fig. 44). For other techniques employing paper wound around coils to diminish their size see Alcock and Cannell (*34*) and Simek (*2990*). Stapling and strengthening with strips of adhesive is recommended by Wolfson *et al.* (Fig. 45; *3620*). Other supports for paper strips are described by Singer and Kenner (*2997*) and by Kawerau (Fig. 46; *1663*). A frame capable of holding numerous two-dimensional chromatograms for routine purposes is described by Datta *et al.* (Fig. 47; *701*), also by Brockmann *et al.* (*390*).

Longenecker (*2047*) uses very thin strips of paper, 4–5 mm wide, as well as mercerised cotton, glass wool and thin asbestos paper.

Rockland and Dunn (*2725*) carry out very short chromatograms with small drops using test tubes as containers for the solvent and the paper strip (Fig. 48). This technique was

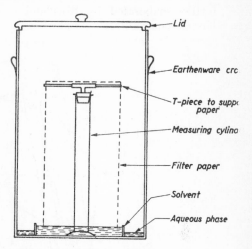

Fig. 43. Apparatus for ascending development (*1922*)

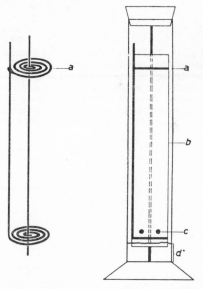

Fig. 44. Apparatus for ascending development (Ma and Fontaine, *2066*) A. paper holder with adjustable coil (a); B. assembled apparatus, showing adjustable coil (a), paper (b), position of test spots (c), and liquid volume (d)

further elaborated by Rockland *et al.* (*2724*), and by Rockland and Underwood (*2729*).

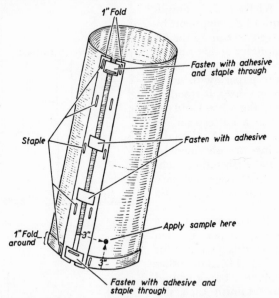

Fig. 45. Technique of folding and fastening large filter paper sheets. When the run is completed, the staples are carefully removed, and the adhesive tape cut. The sheet is hung for drying; if a two-dimensional chromatogram is contemplated, it is reformed into a cylinder and refastened after drying. At this time, excess adhesive from the first run is removed. Spots of material to be fractionated are best applied about 2 in. above the lower border and 2 in. from the vertical border which is to be the lower border when the sheet is run in the second direction (Wolfson *et al.*, *3620*)

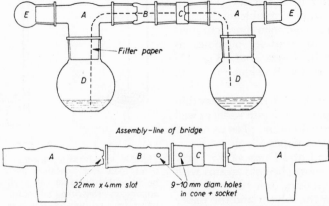

Fig. 46. Bridge unit for filter paper chromatography (Kawerau, *1663*)

For a "continuous" ascending technique see Fischbach and Levine (*959*). Development with both phases simultaneously was investigated by Allouf and Macheboeuf (*39*) who could not find any advantage in this over ordinary development.

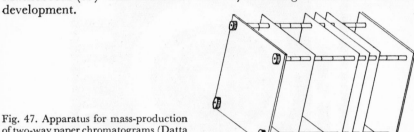

Fig. 47. Apparatus for mass-production of two-way paper chromatograms (Datta *et al.*, *701, 702*)

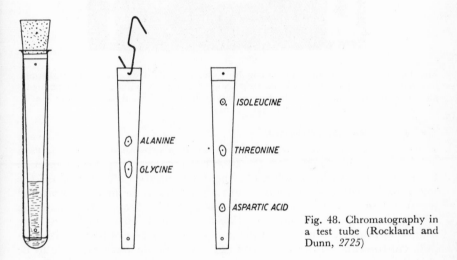

Fig. 48. Chromatography in a test tube (Rockland and Dunn, *2725*)

(iii) *Improvement of unidimensional separations by modifications of the shape of the paper strip*

It was pointed out by numerous authors that when a round spot is placed on a strip of paper the edges are affected in a different way by the developing solvent than the centre since on both sides of the spot only pure solvent is running over the paper. This results in distorted ovoid spots. If it is desired to obtain maximum resolution of close lying spots a shape of paper or spot should be used so that the solvent cannot travel around the substances to be separated but only over them. A very successful shape of paper is shown on the right side of Fig. 49 from the work of Matthias (*2173, 2170*). Similar designs have also been employed by Reindel and Hoppe (*2666*), Ganguli (*1080*) and Schwerdtfeger (*2920*).

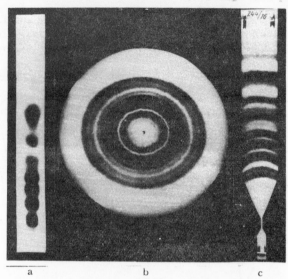

a   b   c

Fig. 49. Amino acids from a protein hydrolysate; a) normal ascending chromatogram. b) circular chromatogram, c) special strip chromatogram; the mixture to be separated is placed at the lower entrance of the narrow part of the strip. Solvent, butanol: acetic acid: water (4:1:1) (Matthias, *2173*)

(iv) *Two-dimensional development*

If one solvent is unable to resolve the mixture of substances to be analysed it is often possible to effect a separation by running one spot on a sheet of filter paper first with one solvent in one direction, then after drying off the first solvent, with a second solvent at right angles to the first. This method was first described by Consden, Gordon and Martin (*618*) for the separation of twenty amino acids (Fig. 50). Two-dimensional separations have been utilised in most cases where a different sequence can be obtained with two different solvents. However all constituents of the first solvent have to be volatile or inert for a successful application of another solvent in the second dimension.

(v) *Multiple development in the same direction*

This technique has been used for separating closely adjacent sugars by Jeanes *et al.* (*1565*), also by Csoban (*671*). Development in the same direction with two different solvents can also be employed (see Burstall and coll., *469*, on the separation of Ta and Nb, also Decker *et al.*, *722*).

(vi) *Radial development*

Lately many papers have appeared describing radial techniques. As introduction to this topic we shall quote Martin's Nobel Prize lecture on the evolution of paper chromatographic techniques: "I was already familiar

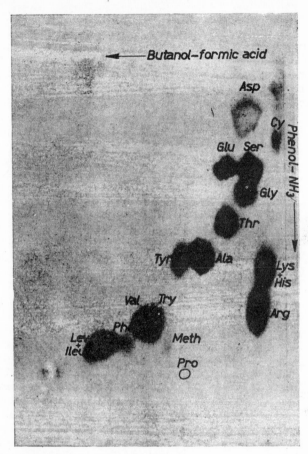

Fig. 50. Two-dimensional chromatogram of amino acids after reaction with ninhydrin. (Hydrolysate of lysozyme; Fromageot and Jutisz, unpublished)

with the use of filter paper chromatograms as used by the dyestuff chemists and adopted at first their technique. A ten centimetre circular paper was cut to a semi-circle with a three centimetre tail about one centimetre wide at the centre. *It was found quicker and more convenient to hang strips of paper* from troughs containing the solvents, and this method was adopted as a routine."

Rutter (*2758*) employed circular discs of filter paper to which the solvent is admitted by cutting a wick 2 mm wide from the periphery to the centre of the disc. The disc, together with a small dish of solvent is housed in a petri dish. Another type of feed for circular development consisting of a cone of filter paper in contact with the centre of the disc chromatogram is employed by Rosebeek (*2735*). See also Zimmermann and Nehring (*3700, 3702*), Berlingozzi and Serchi (*220*), Bersin and Müller (*231*) and Fig. 49. Chromatography on arcs of a circle was employed by Marchal and Mittwer (*2127, 2128*). In a series of papers, Giri and his coworkers (*1111, 1112, 1114, 1124, 1130*) describe not only development techniques for single paper discs but also preparative radial techniques. Saifer and Oreskes (*2781*) study the physical factors in circular chromatography claiming this technique advantageous in a "small laboratory". For further work see Rao (*2246*), Lakshminarayanan (*1868*), Brockmann and Gröne (*388*), Lüderitz and Westphal (*2055*), Ganguli (*1078–1081*), Schwerdtfeger (*2918*), LeStrange and Müller (*1983*) and Chakrabortty and Burma (*540, 541*). A rotating disc for faster development is suggested by Caronna (*512*). For further modifications of radial techniques see also Ambe *et al.* (*46*), Ceriotti (*538*), Carles (*505*) and Erbring and Patt (*909*).

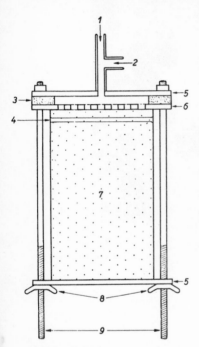

Fig. 51. Diagram of the chromatopile. (1) Connection for a rubber tube for filling the siphon; (2) connection for the siphon tube; (3) rubber gasket; (4) filter paper discs containing the sample; (5) stainless steel plates; (6) perforated stainless steel plates; (7) filter paper disc pile; (8) wing nuts; (9) bolts at four corners of steel plates (Mitchell and Haskins, *2236*)

(vii) *Preparative scale development*

On a preparative scale, a stack of filter paper discs may be clamped together. This is called the *chromatopile* (Fig. 51; *2235, 2236*); after development the discs are separated and the constituents tested for and extracted from the paper. A pack of vertically oriented strips of paper compressed between two steel plates has also been used to separate large quantities of material on paper (the *chromatopack*) (*2571*). Zechmeister (*3678*) discusses the difficulties entailed in the use of stacks of filter paper inside a glass tube. The paper in such a tube must be cut with high precision to fit exactly the diameter of the tube.

The use of thick filter paper sheets is discussed in the various sections (see p. 526) and cellulose columns on p. 521.

The "cellulose sheets" used by Cuendet *et al.* (*673*) are paper sheets $1/8$ inch thick.

Hedén (*1344*) separates 5 mg or more of amino acids by running a row of spots on an ordinary sheet, washing the row towards one end after the first development, and then separating in the second dimension on the same paper. See also Fritz and Bauer (*1050*) and Fischer and Behrens (*962*). Danielson

(*694*) and Hagdahl and Danielson (*1256*) described a new way of rolling paper sheets into a column which now is commercialy avaiable as the "Grycksbo chromatographic filter paper column". It consists of a roll of filter paper closely wound around a central cylindrical rod of inert material and placed in a polyethylene cylinder, which is closed at both ends with polyethylene covers. Liquid inlets and outlets are provided in the covers. Gram quantities of amino acids can be separated on this column.

(e) *Conditions necessary to obtain reproducible R_F values*

Bate-Smith (*157*) has studied the conditions necessary to obtain *reproducible R_F values*. The following recommendations are made: (a) the temperature should be controlled to $\pm 0.5°$; (b) the time of running should be constant; (c) the paper should be equilibrated with the atmosphere in the chamber for 24 hours prior to irrigation with solvent (*1586*); (d) one batch of paper should be used for all determinations; (e) a control substance should be run on every chromatogram. If the R_F differs from the standard by more than ± 0.02, that run should be discarded and fresh solvents used; (f) when solvents are used which undergo reactions, e.g. esterification, they should be allowed to stand for three days at the working temperature prior to use. Both Kowkabany and Cassidy (*1775*) and Zimmermann (*3699*) also point out that the R_F value may vary with the height of the spot above the level of the developing solvent. Zimmermann considers also machine direction of the paper, concentration of substance and the presence of other substances for the reproducibility of R_F values. See also Beck and Ebrey (*171*).

(f) *General techniques for spraying with reagents and detecting spots on the paper*

Solutions of reagents are added to the filter paper with atomisers or painted on with a brush. Numerous designs of atomisers are recorded in the literature and marketed by various firms (for examples see Wingo, *3594*). The general shortcomings of such atomisers is the production of a relatively thin jet which makes uniform spraying over a large chromatogram difficult. Thus when quantitative reactions are desired a mosquito spray gun or atomiser of similar jet size is required.

Gases such as H_2S are passed directly over the paper and solids such as Zn dust are dusted on as a fine powder. If possible, the reactions are carried out so as not to diffuse the spots on the paper. For example, ninhydrin is sprayed on in butanol in which the amino acids are insoluble. Dipping the paper into the reagent is sometimes feasible when a stable insoluble product is formed. See for example Smith (*3031*).

The reagents specific to groups of substances are discussed in detail under the various headings. However there are a few general reagents, which produce spots with practically any substance. Iodine vapour or solution was shown by Brante (*362*) and by Marini-Bettolo and Guarino (*2133*) to yield brown spots or spots paler than the paper with practically any organic compound. This reaction appears to be essentially physical, often a solution of iodine in the spots being formed. After some time the spots disappear and can be shown up again by further addition of I_2 or another reagent.

$KMnO_4$ in dilute H_2SO_4 is a reagent which gives white or light yellow spots on a brown background with most organic compounds (Prochazka, *2597*).

Mitchell (*2243*) uses silver nitrate together with suitable background treatment or light exposure as reagent for reducing agents, halogenated pesticides, halides, phosphates, sulphates, higher fatty acids and some amino acids.

For detection of amphoteric substances, bases or acids, indicators may be sprayed on the paper. By then holding the paper over a volatile acid (CH_3COOH) or base (NH_3), a difference in pH between the background and the spots can be achieved. Tropeolin 00 was used in this manner by Sluyterman and Veenendaal (*3010*), universal indicator by Long *et al.* (*2045*), bromocresol purple by Reid and Lederer (*2662*) and methyl red by Walker (*3459*).

For the detection with ultra-violet rays see the chapter on purines and pyrimidines, also Tennent *et al.* (*3219*), Palladini and Leloir (*2436*). For enhanced light absorption on the paper see Price *et al.* (*2592*).

Differential charring may be employed with suitable solvents for the detection af phosphate esters (Caldwell, *497*). For amperometric detection see De Vries and Van Dalen (*765*) and De Vries (*764*).

(g) *Measurement of R_F values*

R_F values are usually measured with reference to the liquid front. When two liquids fronts are formed, either may be taken but it is important to state to which it refers. Lacourt *et al.* (*1851*) record the band front and band end rather than referring to the center of gravity of a band.

For sugars (*1417*), the R_G value is usually measured where

$$R_G = \frac{\text{Distance travelled by substance}}{\text{Distance travelled by glucose}}$$

Evans and coworkers (*929*) use the R'_F value for phenols, where

$$R'_F = \frac{\text{Distance travelled by spot front}}{\text{Distance travelled by solvent front}}$$

An instrument for rapid R_F measurements was constructed by Phillips (*2508*), consisting of a graduated elastic band from which the R_F values can be read off directly; a transparent scale for the same purpose is described by Rockland and Dunn (*2726*) (the "partogrid"), others by Nettleton and Mefferd (*2357*) and Glazko and Dill (*1135*). Savoia (*2834*) proposes two rulers joined with a hinge which permit direct reading of R_F values.

10. QUANTITATIVE METHODS

(a) *Evaluation of spots*

Generally, quantitative estimation of substances separated on paper can be performed either by examination of the spots on the paper, or by elution of the substances and subsequent determination in the eluate.

Fisher, Parsons and Morrison (*971, 973*) showed that the spot area of round and ovoid spots increases as the logarithm of the spot content. This relationship holds well for a wide range of concentrations. To obtain quantitative results, the spot area of an unknown substance must be compared with spots obtained from the same volume of a known concentration of the same substance developed on the same sheet of paper.

The increase of the spot area with time and with length of development was theoretically treated by Brimley (*382*) and qualitatively confirmed by one of the authors (*1922*).

The uniformity of the paper used is important, and Reid and Lederer (*2662*) found that of a number of Whatman papers only Whatman No. 1 "for chromatography" gave satisfactory results for volatile fatty acids.

The techniques employed to evaluate spots are:

Measurement of spot length: for ovoid regular spots the length of the spot is proportional to the logarithm of the spot content (*973*). Fowler (*1013*) has used this relationship for sugars.

Planimetric measurement: Fisher *et al.* (*973*) obtained an accuracy of $\pm$ 2 %.

Counting squares: The area of the spots can be traced out with a sharp pencil and copied with carbon paper onto graph paper and the number of squares counted. The accuracy of this technique is about $\pm$ 5 % (*2662*).

Weighing excised spots: The spots can be cut out and the piece of paper weighed, giving results equal to the method of counting squares (Overell, private communication).

Approximate analysis by visual comparison: This method was used by Arden *et al.* (*77*) for estimation of UO_2^{++} and gives an accuracy of about 30 %.

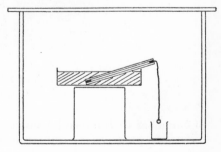

Fig. 52. Washing out of spots. The piece of filter paper containing the required substance is pinched between two pieces of glass resting in a Petri dish. The water in the dish rises between the plates by capillary attraction and is thus supplied to the filter paper. The wash liquor eventually drops off the end of the paper into the little beaker. Evaporation must be prevented by the larger outer vessel (Dent, *740*)

(b) *Combination of paper chromatography with standard quantitative methods*

Numerous methods have been described for the application of the usual quantitative methods to substances chromatographically separated on paper.

For sugars, amino acids etc., the spots are usually cut out, and eluted with water (Fig. 52). It is also possible to extract the spots by merely leaving the paper in contact with water for half an hour, or by the use of a micro-soxhlet extractor. Dumazert and Bozzi-Tichadou (*828*) describe a simple apparatus for the elution of spots; for other methods see the special chapters.

(c) *Use of instruments for measurement of spot intensity*

(i) *Photoelectric cells*

The measurement of spot intensity is required not only in paper chromatographic but also in paper electrophoretic methods as well as in spot colorimetry, diffusion measurements inside filter paper etc.

Fig. 53. Elphor densitometer

The number of commercial instruments available is considerable. Most consist of a photoelectric cell arrangement with direct reading and a device for moving paper strips past a light beam at regular intervals. Direct recording instruments have also been designed. Müller and Clegg (2309, 2310, 2313) carried out extensive work on the instrumental possibilities. See also Fosdick and Blackwell (1008), Block (279, 280), Block and Bolling (281), Bull et al. (449), Crook et al. (660), Eberle (857), Bassir (155), Kutacek and Kolousek (1840), Gorbach (1165), Miettinen and Moisio (2212), De Wael and Cadavieco (766), Röttger (2742).

Of the commercial instruments we shall only mention the recording photodensitometer of Joyce, Loebl and Co. Ltd. (England), the Shandon densitometer, the Elphor photometer (Bender and Hobein, Munich), the transmission densitometer of Baldwin Scientific Instruments (Dartford, Kent), the photometer of Jouan (Paris), the densitometer of the Photovolt Corp., the microphotometer of Dr. B. Lange (Berlin) and the recording photometer of Lerès (Paris).

A typical apparatus is shown in Fig. 53.

Supports for the paper strip to be used in the Beckman-photometer were described by Treiber and Koren (3298) and Ehrmantraut and Weinstock (874). For an instrument measuring reflected radiation see (1238, 3356, 3357).

Hashimoto (1322, 1324) designed an instrument for the measurement of *ultraviolet absorption* of colourless substances on the paper strip. Examples cited are the separation of luteolin, acacetin and rutin (at 2600 Å) and methyl and ethyl xanthates (at 3000 Å). For *high frequency detection* see p. 484. Mori (2274, 2275, 2276) describes a new technique for use with photometry in which the spot is made to travel into a narrow channel formed on the paper. The length of the spot being then measured with greater accuracy than a round spot. It is not quite clear how this could be applied to a large number of spots as in amino acid separations.

Fluorescence was measured by Semand and Fried (2940) and Brown and Marsh (422). For infra-red absorption measurements see Goulden (1183) and Kalkwaif and Frost (1620).

Some authors consider it advisable to decrease the absorption of filter paper in transmission photometry. This problem was mainly dealt with for paper electrophoresis by Grassmann (1193–1195) also by Barrollier (145).

For the *spectrophotometry* of spots on the paper see Campbell and Simpson (499).

(ii) *Photographic methods*

Fisher et al. (973) obtained photographs with sharp spot edges with copy cat (Miles Aircraft), which were measured with a Hilger photomicrometer giving a linear relationship between extinction and the amounts present, over a limited range. Pokrovskii (2600) confirms these results.

Photographic and spectrophotometric methods are widely used for the estimation of purines, pyrimidines and related substances (see Chapter 33).

(d) *Retention analysis*

This method of quantitative estimation was developed by Wieland *et al.* (*3554, 3567*). A strip chromatogram about 6 cm wide is developed in the second dimension, with a reagent for the separated spots. Wherever a spot reacts with the ascending reagent, a V-shaped notch is formed due to the retardation of the flow of reagent in this zone (Fig. 54).

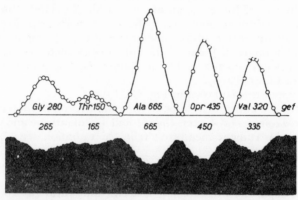

Fig. 54. Retention analysis. Separation of a mixture of glycine, threonine, alanine, hydroxyproline and valine (Wieland and Wirth, *3567*). The upper row of figures represents the values found; theoretical figures are given underneath.

For example, for Cu^{++} ions ascending through spots of amino acids, let a be the adsorption of Cu^{++} in the paper and β the absorption of Cu^{++} by the amino acid; the distance travelled by Cu^{++} in the free paper is then $\dfrac{1}{a} = h$, and the distance travelled in the amino acid spot $\dfrac{1}{(a+\beta)} = \gamma$.

The relationship of the two is

$$\frac{a+\beta}{a} = 1 + \frac{\beta}{a} = \frac{h}{\gamma} \ldots \ldots \ldots (1)$$

The concentration of the amino acid in each point of the V-shaped notch is proportional to β/a, which according to equation (1) is $h/\gamma - 1$.

By plotting $h/\gamma - 1$ against distance on the paper strip, one obtains curves whose area is directly proportional to the concentration of the spot. This method has yielded very good results in many cases, but suffers from the fact that the paper is usually not sufficiently uniform, thus yielding curves with serrated edges. These serrations lower the accuracy as well as reducing the ability to distinguish between closely adjacent bands.

Retention analysis was employed for hydroxy acids (Wieland and Feld,

3559), amino acids (Wieland *et al.*, *3567*) and for glycine, sarcosine and dimethylglycine (Kuhn and Ruelius, *1822*).

11. RADIOACTIVE TRACER TECHNIQUES

Radioactive tracers are now employed for the labelling of substances in chemical and biochemical reactions. .

In combination with chromatography, radioactive tracers can also be used to identify the substances without the use of reagents at very low concentrations. Biochemical processes starting with labelled compounds can be observed and the reaction followed by the gradual disappearance of the activity from one spot and its appearance in another position.

Lissitzky and Michel (*2033*) have reviewed the technique and application of paper chromatography of organic substances containing radioactive isotopes.

A good illustration of the possibilities of radioactive techniques is given by Benson *et al.* (*188*), who employed paper chromatography of radioactive substances for the elucidation of the reactions of photo-synthesis. Green cells containing radioactive phosphorus tracers were exposed to $^{14}CO_2$ and light. The compounds present after various times of exposure were separated by

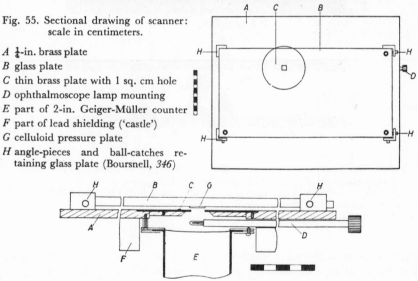

Fig. 55. Sectional drawing of scanner: scale in centimeters.

A $\frac{1}{4}$-in. brass plate
B glass plate
C thin brass plate with 1 sq. cm hole
D ophthalmoscope lamp mounting
E part of 2-in. Geiger-Müller counter
F part of lead shielding ('castle')
G celluloid pressure plate
H angle-pieces and ball-catches retaining glass plate (Boursnell, *346*)

two-dimensional paper chromatography and the radioactivity detected by placing the paper in contact with photographic plates, with and without absorbers, to differentiate between radioactive phosphorus and carbon.

Quantitative results are readily obtained by employing a counting technique rather than photographic plates. A number of simple devices were described by Atkinson (96), Bourne (345), Tomarelli and Florey (3279), Fosdick and Blackwell (1008), Gray et al. (1200), Jones (1596) and Katz and Chaikoff (1646). The apparatus designed by Boursnell (346) is shown in Fig. 55. (The ophthalmoscope is only used for aligning the right portion of the chromatogram.) An accuracy of 1 % was obtained with known amounts of radioactive phosphates. Wieland et al. (3564) determine amino acids by chromatographing their Cu salts containing radioactive Cu.

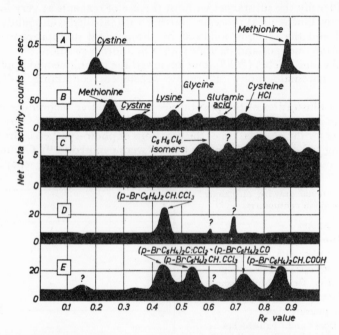

Fig. 56. Experimental radio-chromatograms on paper (Winteringham et al., 3603)
    A chromatogram of protein hydrolysate from wheat grown on ^{35}S
    B strip spotted with known amino acids and exposed to ^{131}I-labelled CH_3I
    C neutron-activated chromatogram of $C_6H_6Cl_6$ isomers
    D neutron-activated chromatogram of a bromine analogue of DDT
    E neutron-activated chromatogram of bromine analogues of DDT derivatives

Winteringham and coll. (3602) extend the use of radioactive techniques by either reacting a developed chromatogram with radioactive reagents, or neutron-activating a developed chromatogram to form radioactive isotopes with the spots on the paper. At the same time a radioactive paper, usually of a lower activity than the spots is formed.

Fig. 56 shows illustrations of some of the results obtained by these methods.

Numerous measurements of small sections of large chromatograms are tedious when applied as a routine technique; Winteringham *et al.* (*3603*) developed a continuous radioactivity scanning device for paper strips. In all

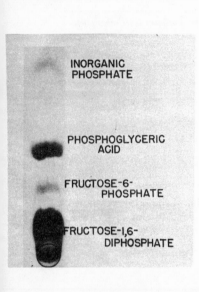

INORGANIC
PHOSPHATE

PHOSPHOGLYCERIC
ACID

FRUCTOSE-6-
PHOSPHATE

FRUCTOSE-1,6-
DIPHOSPHATE

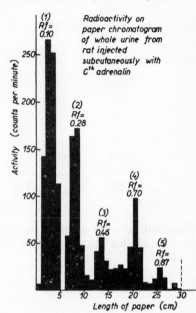

Fig. 57   Fig. 58

Fig. 57. Paper chromatography of ^{32}P-marked hexose phosphates (Benson *et al.*, *188*)
Fig. 58. Radioactivity on paper chromatogram of urine from rat injected subcutaneously with 10 γ dibenamine hydrochloride per g of body wt., followed in 5 minutes by 7 γ β-^{14}C-*dl*-adrenaline per g of body wt. subcutaneously. Developed in butanol 80 parts, glacial acetic acid 20 parts (Schayer, *2837*)

measurements of radioactivity inside filter paper, the effect of self-absorption of the radiation in the paper must be considered, especially as no paper is of uniform thickness. The possible errors incurred with some isotopes for a variation of 10 % in the paper density are shown in Table 28.

Further automatic radioactivity scanning devices were published by numerous authors, for example Rockland *et al.* (*2728*), Williams and Smith (*3584*), Bradley (*357*), Berthet (*234*), Bonet-Maury (*307*), Wingo (*3595*), Lerch and Neukomm (*1966*) and Sternberg (*3102*).

While very useful for the estimation of several similarly labelled compounds, automatic scanning is often unsatisfactory when several different radioactive elements are to be separated since a fixed distance of the counter from the paper may readily detect a high energy radiation but not another with low energy.

Another important application of radioactive tracers is the use of ^{35}S and ^{131}I in the pipsyl derivatives of amino acids (see Chapter 30). Lester Smith (*3022*) used a comparison of radioactive and biological determination

TABLE 28.   Effect of paper density variation on self-absorption (Winteringham *et al.*, *3603*)

Isotope	Proportion of beta particles unabsorbed by paper of mean density 8.785 mg/sq. cm	Variation of self-absorption correction factor corresponding to a variation of ± 10% in paper density
[14]C	0.366	± 7.4%
[35]S	0.431	± 6.5%
[82]Br	0.787	± 2.4%
[131]I	0.828	± 1.9%
[36]Cl	0.843	± 1.7%
[32]P	0.957	± 0.5%

of penicillins for the confirmation of the validity of the latter method. For ^{32}P, see Fig. 57; for ^{35}S, see p. 320; for ^{14}C, see Fig. 58; for ^{131}I, see Fig. 102, p. 407.

DIVISION IV

CHROMATOGRAPHY OF ORGANIC SUBSTANCES

CHAPTER 19

Hydrocarbons

1. CHROMATOGRAPHY OF HYDROCARBONS
IN THE GASEOUS STATE

The recent state of gas chromatography has been reviewed in the book of Phillips (*2505a*).

For the separation of gases by adsorption chromatography, several alterations in technique are required. The eluate cannot be examined by the usual optical or chemical devices; a thermal conductivity cell is usually employed. Constant rate of flow of the eluting gas can be maintained with a compensating flow-meter (Phillips, *2506*). See also p. 58.

Frontal analysis, displacement and elution analysis have been successfully employed for gases as well as desorption by increase of temperature (and/or decrease of pressure), a technique only applicable to gases.

The early literature was reviewed by Claesson (*577*), who also described the separation of hydrocarbons by displacement with esters (ethyl acetate) using a self-recording apparatus. Several brands of activated charcoal were found suitable.

Other simple self-recording instruments are described by Phillips (*2506*) and by Turner (*3333*) for the separation of gas mixtures consisting of light hydrocarbons. Phillips (*2506*) found that the usual principles for good chromatographic separation must also be applied to gases, viz. small particle size, slow flow rate, columns in sections (see Claesson, *577*). Sharp steps could be obtained with hydrocarbons (saturated, unsaturated and aromatic), halogenated hydrocarbons, nitroparaffins, ethers, ketones, alcohols and esters, by displacement with ethyl acetate.

Considerable accuracy in the measurement of step height was obtained, making quantitative analysis feasible.

Hesse and Tschachotin (*1380*) passed a gaseous mixture of benzene and cyclohexane diluted with nitrogen through a tube of silica gel; only the benzene was found to be adsorbed. Benzene and carbon tetrachloride could also be separated in this way.

Cremer and Müller (*650, 651*) employed H_2 as the eluting gas and achieved an accuracy of 1 % with 1–10 mg of the following mixture:

N_2, C_2H_4, C_2H_2, CO_2 and CH_2=CHCl (adsorbent: charcoal or silica). Turkeltaub (*3329, 3330*) has described a method for the analysis of gaseous hydrocarbons on charcoal, eluting with air and separating into three fractions: CH_4, C_2H_6, and the higher hydrocarbons (error 5 %). Desorption by increasing the temperature was considered incomplete by Claesson (*577*) but was found suitable for industrial separation of hydrocarbon mixtures by Small (*3012*).

Kofler (*1760*) has described the separation and purification of organic compounds by means of sublimation through adsorptive substances; a carrier gas is moved through the column under reduced pressure. Other adsorption techniques were patented by Berg (*197*) and by Scheeline (*2839*).

Probably the most efficient and useful method for separation and quanti-

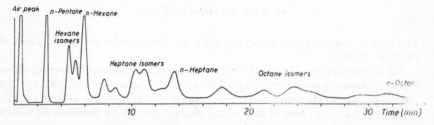

Fig. 59. Separation of aliphatic hydrocarbons from petroleum ether of boiling range 40–120°. Column length, 4 ft.; stationary phase, liquid paraffin; temperature, 78.6°; rate of flow of nitrogen, 28 ml/min; pressure of nitrogen, 36 cm Hg (James and Martin, *1539*)

tative analysis of hydrocarbons in the gaseous state is the *gas-liquid partition chromatography* (for a review of the technique see p. 111). Fig. 59 shows a typical separation of the hydrocarbons of petroleum ether of boiling range 40–120°. Similarly British national petrol was separated into more than fifty components.

Lichtenfels *et al.* (*2004a*) have described separations of complex hydrocarbon mixtures in the C_5 to C_8 range, using hydrogen or helium as carrier gas and dioctyl phthalate or Octoil S (a vacuum pump oil, Consolidated Vacuum Corp., Rochester, N.Y.) as liquid coating of Celite 545.

2. CHROMATOGRAPHY OF HYDROCARBONS
IN THE LIQUID STATE

Hirschler and Amon (*1409*) have described the isolation of pure 2,3,3-trimethyl-butane and -pentane from natural mixtures by adsorption on silica, alumina or charcoal (see also Streiff *et al.*, *3152*). Hibshman (*1389*) has reported the separation of *iso-* and normal paraffins on charcoal columns; only the normal paraffins are adsorbed. A patent of Hirschler (*1406*) describes the selective separation of normal aliphatic hydrocarbons from alicyclic isomers (on silica gel or charcoal); the following mixtures were separated: *n*-heptane and methylcyclohexane; *n*-octane and ethylcyclohexane; *n*-octane and 2,2,4-trimethylpentane. Another patent of Hirschler (*1408*) concerns selective separations of *iso*paraffins on charcoal. A patent of the Standard Oil Development Company (*3083*) describes the separation of *straight*-chain from *branched*-chain paraffins on charcoal or alumina; normal paraffins are selectively adsorbed from mixtures with branched-chain paraffins of the same boiling range, by activated coconut-shell charcoal.

Adsorption analysis of cracked gasolines has become of great industrial importance, as can be judged by the symposium published in *Analytical Chemistry* (1950, *22*, 850–881). Techniques, adsorbent characteristics, and factors affecting separation have been studied in detail by Fink *et al.* (*952, 953*) and Clerk *et al.* (*590*). Displacement techniques have been applied by Dinneen *et al.* (*786*), and Furby (*1064*) has described an adsorption method for the evaluation of petroleum residua and lubricating oil distillates (see also Mair *et al.*, *2107*).

Nederbragt and De Jong (*2343*) have separated paraffins and alkyl-decahydronaphthalenes on floridin, the long chain paraffins being held back more than the saturated naphthalenes. The difference in adsorption can be explained by the existence of pores of diameter 4 × 6.5 Å in the surface of the adsorbent which allow entry only of the paraffins and not of the alkyl-decahydronaphthalenes.

3. ELUTION CHROMATOGRAPHY OF HYDROCARBONS

Mair (*2106*) has obtained good separations of mixtures of aromatic hydrocarbons, paraffins and naphthenes by passage through silica gel. After washing with pentane, the column selectively retains the aromatic hydrocarbons which can be eluted by the addition of methanol. 15 parts of silica gel retain 0.5 to 1 part of aromatic hydrocarbon. For a detailed study of similar separations see also Smit (*3014, 3015*), as well as O'Donnell (*2396*). The latter author uses an apparatus in which the eluate solvent is distilled continuously and conveyed back to the top of the column.

Criddle and LeTourneau (*655*) have developed a method using fluorescent dyes which make the principal hydrocarbon type boundaries (saturated, olefins, aromatics) clearly visible on a column of silica gel Davison. The composition of the sample is determined by measuring the zone length. Ellis and LeTourneau (*886*) have employed a mixture of daylight and fluorescent dyes as indicators for the determination of hydrocarbon types and total oxygenated solvent in lacquer thinners. (See also Conrad, *614*).

Tenney and Sturgis (*3220*) have studied in detail the separability of a great variety of hydrocarbons by elution chromatography; they conclude that on *silica gel* the following separations are possible: mono-olefins from diolefins, mono-olefins from aromatic and partial separation in some cases of acyclic from cyclic olefins. *Alumina* affords a better separation of aromatics, based on the number of aromatic nuclei in the molecule. *Activated charcoal* affords a partial separation of straight chain paraffins from other saturated hydrocarbons.

Silica gel Davison (Code 923) is usually believed to be the best adsorbent for unsaturated hydrocarbons, where isomerisations and polymerisations should be avoided (Saier *et al.*, *2780*).

Saier *et al.* (*2780*) have studied the combination of chromatography on silica gel Davison with infrared absorption techniques for the determination of olefin group types.

A patent of Hirschler and Lipkin (*1410*) describes the isolation of aromatic hydrocarbons from a mixture, on silica gel; a more recent patent of Hirschler (*1407*) describes the separation of different isomeric aromatic hydrocarbons (e.g. ethylbenzene and *o*-xylene) on activated charcoal (see also Rossini, *2739*).

Karr *et al.* (*1634*) have described the fractionation of crude petroleum oils on alumina and bauxite. See also Entel *et al.* (*902*).

Mixtures of mineral oils and vegetable oil can be separated on alumina, the former being eluted with light petroleum, the latter with acetone (Harker, *1310*).

The elution chromatography of the components of lubricating oils has been studied by Irish and Karbum (*1509*).

Hydrocarbons from lubricating oils can be separated on silica gel impregnated with trinitrobenzene. The various aromatic hydrocarbons give yellow to violet zones, due to the formation of trinitrobenzolates (Godlewicz, *1151*).

Neuworth (*2359*) has separated the products of the hydrogenation of coal into three main groups by adsorption on alumina. The first group, the least adsorbed, contained paraffins, naphthalenes and aromatic hydrocarbons of low molecular weight. The second group contained aromatic hydrocarbons with at least three condensed rings and the third group contained sulphur, oxygen and nitrogen compounds.

The chromatographic fractionation of total crude shale oil has been described by Karr *et al.* (*1635*). Smith *et al.* (*3035*) have separated nitrogen compounds from shale oil by adsorption on florisil.

Berenblum (*191*) isolated the carcinogenic hydrocarbon 3,4-benzpyrene from tar, by chromatography on alumina. For the separation of anthracene from tar, see Williams (*3585*); the separation of anthracene oils, Florentin and Heros (*992*); the fractionation of black oils, Lawrence and Barby (*1888*); of coal bitumens, Lahiri and Mikolajewski (*1865*); for the estimation of hydrocarbons in primary tars by "hot" chromatography, Vahrman (*3358*).

The *paper chromatography* of 3,4-benzpyrene and other polycyclic aromatic hydrocarbons has been described by Tarbell *et al.* (*3205*) (see Table 29).

TABLE 29.   Average R_F values of 3,4-benzpyrene, some of its derivatives and some other polycyclic aromatic hydrocarbons using a hexane-dimethylformamide solvent system (Tarbell *et al.*, *3205*)

Compounds	No. of runs (av.)	Av. of R_F values	Colour under ultraviolet light
3,4-Benzpyrene	10	0.54	Blue
Benzpyrene derivatives			
10-Ethyl	4	0.80	Blue
10-Acetyl	4	0.26	Green
α-Hydroxy-10-ethyl	4	0.07	Blue
4′-Hydroxy-1′,2′,3′,4′-tetrahydro . . .	3	0.14	Greenish-blue
4′-Keto-1′,2′,3′,4′-tetrahydro	2	0.37	Greenish-blue
Anthracene	4	0.84	Blue
Naphthacene	3	0.51	Orange
Chrysene	4	0.72	Dark blue
Pyrene	5	0.62	Green
Fluoranthene	4	0.82	Green
12-Methylchlolanthrene	3	0.89	Blue
1,2-Benzanthracene	3	0.66	Blue
1,2,5,6-Dibenzanthracene	3	0.54*	Blue

* Considerable tailing occurs.

4. THE SEPARATION OF TERPENES

Kirchner *et al.* (*1727*) have separated limonene, α-pinene, camphene, *p*-cymene etc., on "chromatostrips", i.e. adsorbent-coated glass strips (see p. 12 and Fig. 60). The best separations were obtained with silicic acid. After development, the location of the spots can be revealed by spraying with suitable reagents. On spraying with fluorescein solution and exposing to bromine vapour, compounds which absorb bromine faster than fluorescein show up as yellow spots on a pink background. Very unreactive compounds can be located with a concentrated sulphuric-nitric acid mixture.

For further developments of this method see Miller and Kirchner (*2221*, *2222*), Ito *et al.* (*1524*).

Bryant (*439*) has described the circular chromatography of terpenes on adsorbent coated glass. The frontal analysis of terpenes has been studied by Varma *et al.* (*3380*).

The preparation of terpeneless essential oils by chromatography on silicic acid has been described by Kirchner and Miller (*1725*).

Terpene oxides, alcohols and ketones can be likewise separated on chromatostrips (*1727*).

The analysis of *essential oils* has been described in several papers by Kirchner and Miller (*1725*, *1726*, *1727*), as well as by Reitsema (*2669*); see also the review of Post (*2575*).

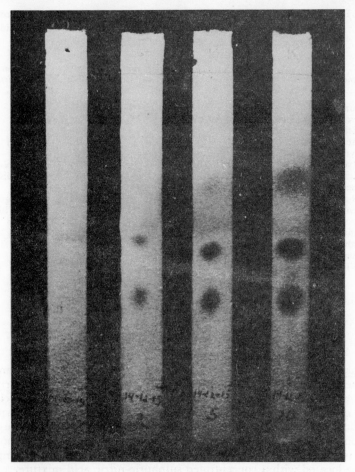

Fig. 60. Separation of *p*-cymene, pulegone and cinnamaldehyde as shown under ultraviolet light on fluorescent strips (Kirchner *et al.*, *1727*)

From top to bottom:

1. *p*-Cymene
2. Pulegone
3. Cinnamaldehyde

From left to right:

1. Blank (spot due to traces of impurity in solvent not removed by distillation)
2. *p*-Cymene, 143γ, pulegone, 3.3γ, cinnamaldehyde, 3.3γ
3. *p*-Cymene, 358γ, pulegone, 8.2γ, cinnamaldehyde, 9.1γ
4. *p*-Cymene, 1.4 mg, pulegone, 32.8γ, cinnamaldehyde, 36.4γ

5. THE ISOLATION OF HYDROCARBONS FROM

UNSAPONIFIABLE MATERIAL

The filtration of light petroleum solutions of unsaponifiable substances through alumina is a simple method of isolating hydrocarbons from extracts of organs (*2586*, *2762*), wool fat (*693*) or ambergris (*1907*). Schuette *et al.* (*2900*) have found chromatography on alumina a better criterion for establishing the purity of a normal C_{24} paraffin isolated from plant material, than transition points. The hydrocarbon squalene ($C_{30}H_{50}$) has been isolated chromatographically from the unsaponifiable matter of an Amanita (*3547*), from the lipids of a Torula (*2655*) and from many vegetable and animal oils. The chromatography of squalene on quilon impregnated *paper* has been described by Dauben *et al.* (*708*). (See also Carotenoids, Chapter 34.)

6. THE SEPARATION OF HYDROCARBON-ESTER MIXTURES

Esters are generally adsorbed more strongly than hydrocarbons. Seidel *et al.* (*2933*) have described the separation of hydrocarbons and esters from lavender essence by filtration through silica gel, only the esters being adsorbed.

CHAPTER 20

Alcohols and phenols

1. ALIPHATIC ALCOHOLS

(a) Column chromatography

Hydroxyl compounds are particularly easy to separate and purify by means of chromatography by virtue of the affinity of the hydroxyl group for a number of adsorbents.

Low molecular weight volatile alcohols can be transformed into coloured derivatives such as p-azobenzene carboxylates (Craw and Sutherland, *648*) or 4-phenylazophenyl urethans (Masuyama, *2168*, *2169*; Davenport and Sutherland, *710*). Good separations of the urethans of methyl and ethyl alcohols can be achieved in this way. The 3,5-dinitrobenzoates of aliphatic alcohols have been separated on columns of silicic acid made fluorescent by means of Rhodamine 6G by White and Dryden (*3536*). Baker and Collis (*125*) have synthesised some carboxylic acids derived from 7-methoxycoumarin, the principal one being 7-methoxycoumarin-3-carboxylic acid, which have the property of forming *fluorescent esters* that can be used in the chromatographic separation of alcohols.

Woolfolk *et al.* (*3626*) have prepared the p-phenyl-azobenzoates of a great number of normal and branched alkanols for identification purposes and have separated some of them by chromatography on alumina.

Neish (*2352*) has separated mg quantities of aliphatic alcohols from methanol to butanol on celite-water columns with CCl_4 and $CHCl_3$ as solvents.

Dal Nogare (*688*) has modified this method using columns of silicic acid or silicic acid-celite.

Van Duin (*3372*) has prepared 4-dialkylamino-3,5-dinitrobenzoates of n-aliphatic alcohols and methyl carbinols up to C_{14} and has separated the coloured esters by partition on columns of silica gel; they are characterized by their retention volume.

Gas-liquid partition chromatography (see p. 111) makes the separation of isomers possible, as is shown in Fig. 61.

(b) *Paper chromatography*

Kariyone and Hashimoto (*1629, 1631*) have described the paper chromatography of simple alcohols in form of their potassium xanthates ROCSSK. The spots can be detected by their dark brown fluorescence in ultra-violet light, or by spraying with Grote's reagent (see Table 30).

TABLE 30. R_F values of potassium xanthates of alcohols (Kariyone and Hashimoto, *1629*)
Solvent: Butanol, 0.5% KOH

Alcohol	R_F	Alcohol	R_F
Methanol	0.23	*iso*Amyl alcohol	0.62
Ethanol	0.35	Benzyl alcohol	0.45
*iso*Propanol	0.44	Octyl alcohol	0.17
*iso*Butanol	0.54	Cyclohexanol	0.04
n-Butanol	0.55	Ethylene-chlorohydrin	0.91

For further work by the same authors see also Kariyone *et al.* (*1631*). Another method worked out by Lederer and Summerfield (unpublished) uses butanol/1.5N NH$_4$OH or *n*-amyl alcohol/1.5N NH$_4$OH as solvent for the xanthates. Acid ammonium molybdate is used as spray reagent yielding blue to purple spots on a white background. Table 31 shows the R_F values with some of the solvents investigated.

Rice *et al.* (*2676*) have described the paper chromatography of 3,5-dinitrobenzoates of simple alcohols (the best solvent being 20% aqueous dioxan). The spots are sprayed with a 0.5% solution of α-naphthylamine to obtain reddish colours which can in some cases be intensified by spraying with 10N KOH. See also Meigh (*2179*).

Momose and Yamada (*2251*) have analyzed qualitatively mixtures of simple alcohols by paper partition of their 3,6-dinitrophthalic half esters. 10 μg of methanol in 1 mg of ethanol could be detected.

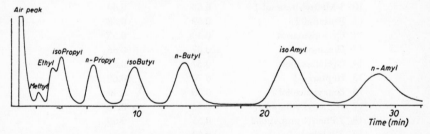

Fig. 61. Separation of alcohols. Column length, 4 ft.; stationary phase, benzyldiphenyl; temperature, 100°; rate of flow of nitrogen, 12 ml/min; pressure of nitrogen, 13 cm Hg (James and Martin, *1539*)

Siegel and Schlögl (*2978*) have used the half esters of 3-nitrophthalic acid or of diphenic acid for similar separations (see Table 32).

TABLE 31. R_F values of alcohol xanthates (M. Lederer and Summerfield, unpublished)

Xanthate	Butanol 50 ml Water 45 ml NH_4OH 5 ml	Amyl alcohol 50 ml Water 45 ml NH_4OH 5 ml	Butanol 25 ml Amyl alcohol 25 ml Water 25 ml NH_4OH 45 ml
Methyl	0.25	0.08	0.10
Ethyl	0.37	0.17	0.20
iso-Propyl	0.50	0.27	0.32
n-Butyl	0.67	0.47	0.52
n-Amyl	0.78	0.58	0.62

TABLE 32. R_F values of alcohol 3-nitrophthalates and diphenates (Siegel and Schlögl, *2978*)

Solvent: isoamyl alcohol – *conc.* ammonia – water (30:15:5)
Reagents: umbelliferone or 4-methylumbelliferone

No. Alcohol	3-Nitrophthalates R_F	Diphenates R_F
1 Methanol	0.15	0.24
2 Ethanol	0.26	0.37
3 Propanol-1	0.42	0.48
4 Propanol-2	0.35	0.47
5 Butanol-1	0.55	0.56
6 2-Methyl-propanol-1	0.53	0.55
7 Butanol-2	0.48	0.54
8 Pentanol-1	0.64	0.61
9 2-Methyl-butanol-1	0.61	0.59
10 3-Methyl-butanol-1	0.63	0.60
11 Pentanol-2	0.59	0.58
12 Cyclopentanol	0.49	0.55
13 Hexanol-1	0.67	0.65
14 Cyclohexanol	0.57	0.62
15 Heptanol-1	0.70	0.69
16 Benzyl alcohol	0.50	0.49
17 2-Phenyl-ethanol-1	0.52	0.56
18 3-Phenyl-propanol-1	0.55	0.60
19 Allylalcohol	0.32	0.41
20 Cinnamic alcohol	0.50	0.53

2. ALIPHATIC AND ALICYCLIC POLYALCOHOLS

(a) Column chromatography

The polyalcohols which show low solubility in organic solvents can be transformed into more lipo-soluble derivatives. Inositol and mannitol in the form of their hexa-acetates can be adsorbed on magnesol, the alcohols being obtained pure after alkaline saponification by passage through an ion exchanger(*247*). Wolfrom *et al.* (*1099, 1995*) have described the chromatography of free polyalcohols. Boissonnas (*294*) has separated the coloured azoates of hexitols on columns of alumina.

Wetterholm (*3519*) has studied frontal analysis of ethylene glycol and glycerol on charcoal.

By chromatographing on alumina the unsaponifiable matter from various vertebrate organs, Prelog, Ruzicka *et al.* (*2580, 2587*) isolated pure batyl alcohol (α-octadecyl glyceryl ether) and chimyl alcohol (α-hexadecyl glyceryl ether), which had previously been found only in fish oils.

Büchi *et al.* (*445*) separated on alumina an addition compound of brein $(C_{30}H_{48}(OH)_2)$ and elemol $(C_{15}H_{25}OH)$ that could not be separated by fractional crystallisation.

Esterification of an alcoholic hydroxyl greatly diminishes its adsorbability; for instance, acetates are much less strongly adsorbed than the free alcohols; separations of mono-, di- and tri-acetates of polyols are thus possible. Polonsky (*2549*) has separated in a single chromatography on alumina, a triterpene triol, its mono-, di- and tri-acetate.

Zager and Doody (*3659*) removed 93–98 % of glycerol from aqueous solutions by filtration on columns of Amberlite IR-400 or Permutit S treated with $Na_2B_4O_7$ solution (formation of acidic borate complexes). Glycols can also be purified by "ion exclusion" (see p. 95).

C_2 to C_4 glycols can be separated by *partition* on silicic acid-celite-water columns (Dal Nogare, *688*).

A method employing silica gel partition columns is described by Neish (*2351*). Glycerol, acetoin and 2,3-butanediol can be separated from each other and from other substances, which react with periodate. The solvents used are ethyl acetate or benzene-butanol.

(b) Paper chromatography

Hough (*1454*) first reported that polyalcohols such as mannitol, sorbitol, glycerol and ethylene glycol can be chromatographed on paper using the usual technique for sugar analysis. By spraying with ammoniacal silver nitrate as little as 1 μg of hexitols and glycerol and 10 μg of ethylene glycol can be detected. Quantitative estimates are carried out with periodate. (See Table 33 for a list of R_F values.)

TABLE 33. R_F values of polyhydric alcohols and some sugar glycosides (Hough, *1454*)

	n-Butanol sat. water	n-Butanol 4 Ethanol 1.1 Water 1.9	n-Butanol 4 Ethanol 1 Water 5	Benzene 1 n-Butanol 5 Pyridine 3 Water 3	n-Butanol 5 Acetic acid 1 Water 2
Trimethylene glycol	0.67	0.62	0.63	0.63	—
Ethylene glycol	0.51	0.54	0.61	0.58	0.58
Glycerol	0.30	0.43	0.37	0.46	0.44
α-Methylgalactoside	0.13	0.31	0.18	0.35	0.31
α-Methylmannoside	—	—	0.31	—	—
β-Methylmaltoside	—	—	0.03	—	—
Sorbitol	0.06	0.21	0.10	0.21	0.17
Dulcitol	0.05	0.21	0.10	0.20	0.18
Mannitol	0.05	0.22	0.10	0.22	0.19
Inositol	0.00	0.10	0.02	0.07	0.05
Sucrose	0.00	0.15	0.03	0.18	0.09

TABLE 34. R_F values of polyhydric alcohols and some hydroxy acids (Buchanan *et al.*, *444*)

Compound	n-Butanol 4 Acetic acid 1 Water 5	n-Butanol 20 'Acetic acid 5 Water 25 Conc. HCl 1	Reaction with lead tetra-acetate	Reaction with periodate
Glucose	0.19	—	+	+
Ethylene glycol	0.64	0.69	+	+
Glycerol	0.48	0.52	+	+
Mannitol	0.21	0.26	+	+
Sorbitol	0.19	0.25	+	+
Inositol	0.09	—	+	+
Lactose	0.04	—	+	+
Maltose	0.05	—	+	+
Raffinose	0.03	—	+	+
Sucrose	0.08	—	+	+
Trehalose	—	0.03	+	+
Glucose-1-phosphate	0.03	—	+	+
α-Methylglucopyranoside	0.34	—	+	+
Gluconic acid	0–0.24 ⎱ 0.41 ⎰	0.18 ⎱ 0.45 ⎰	+	+
DL-Tartaric acid	—	0.48	+	+
meso Tartaric acid	—	0.41	+	+
Erythro-9,10-dihydroxystearic acid	—	0.52–0.94	+	+
Erythro-6,7-dihydroxystearic acid	—	0.91	+	+
Citric acid	—	0.45	—	+
Lactic acid	—	0.68 ⎱ 0.78 ⎰	—	+
Pyruvic acid	—	0.77	—	+
Dihydroxyacetone	—	0.38	+	+

See also Hackman and Trikojus (*1252*); for the separation of ethylene glycol and glycerol on paper, see Viscontini *et al.* (*3423*); for the separation of glycols in food see Bergner and Sperlich (*206*).

Another list of R_F values of polyhydric alcohols was published by Buchanan, Dekker and Long (*444*) (see Table 34). These authors also describe a number of new spraying reagents for glycols and glycosides. A solution of lead tetra-acetate in benzene (1 %) yields white spots on a brown background of PbO_2 with 1,2-glycols.

The paper chromatography of glyceryl ethers (batyl and chimyl alcohol, etc.) has been described by Emmerie (*896*).

The paper chromatography of *inositol* and other cyclitols has been described by Ballou and Anderson (*129*), Serro and Brown (*2947*), Böhm and Richarz (*292*), and Posternak *et al.* (*2576*) (Table 35); the latter authors describe in detail a series of detection methods.

TABLE 35. R_F values of cyclitols and cycloses (Posternak *et al.*, *2576*)

Substances	Butanol 4 Acetic acid 1 Water 5	s-Collidine Water	Phenol Water	Acetone Water (85:15 v/v)
meso-Inositol	0.09	0.13	0.23	0.26
Scyllo-meso-inosose	0.09	0.15	0.20	0.28
Epi-meso-inosose	0.11	0.20	0.30	0.34
Scyllitol	0.09	0.13	0.18	0.26
Epi-inositol	0.11	0.13	0.38	0.28
(—)-Inositol	0.12	0.17	0.20	0.33
Mytilitol	0.11	0.14	—	—
Isomytilitol	0.14	0.17	0.31	0.31
Hydroxy-mytilitol	0.11	0.14	0.18	0.27
Hydroxy-isomytilitol	0.11	0.14	0.18	0.27
Chloro-isomytilitol	0.34	—	—	—
Bromo-isomytilitol	0.36	—	—	—
Iodo-isomytilitol	0.44	—	—	—
Methyl-pentahydroxy-cyclo-hexane	—	0.30	—	—
Methylene-pentahydroxy-cyclo-hexane	0.31	0.35	—	—
Methylene-pentahydroxycyclo-hexane oxide	0.15	0.09	—	—
Inosamine HCl	0.08	0.14	—	—
N-Acetyl-inosamine	0.09	0.14	—	—
Quebrachitol	0.18	—	0.43	0.45
Pinitol	—	—	0.39	0.49
(—)-Viburnitol	0.17	0.27	0.40	0.39
(+)-Quercitol	0.20	0.31	0.45	0.45
Desoxy-scyllitol	0.20	—	—	—
DL-Desoxy-4-meso-inositol	0.20	—	—	—

(Table continued)

(Continued)

Substances	Butanol 4 Acetic acid 1 Water 5	s-Collidine Water	Phenol Water	Acetone Water (85 : 15 v/v)
Cyclohexane-tetrols:				
DL-1,2,3/4	0.37	0.46	0.66	0.54
DL-1,2,4/3	0.37	0.46	0.70	0.54
DL-1,3/2,4	0.36	0.43	—	—
DL-1,2/3,4	0.41	0.57	—	—
1,4/2,3	0.41	0.57	0.60	0.61
Conduritol	0.37	—	0.54	0.61
Cyclohexane-triols:				
DL-1,2/3	0.57	0.77	0.80	0.74
1,3/2	0.57	0.75	0.80	0.74
1,2,3 *cis*	0.57	0.75	0.80	0.77

3. PHENOLS

(a) *Column chromatography*

Gore and Venkataraman (*1179*) have studied the relation between chelation of azophenols and the adsorbability on alumina. Various phenolic compounds resulting from oxidation of lignin have been purified on magnesol columns (Pearl and Dickey, *2482*; Pearl and Beyer, *2480*). Zaprometov (*3667*) has chromatographed tea tannins in ether solution on silica gel.

Separations of simple phenols as coloured azo-derivatives have been described by Bielenberg *et al.* (*243, 244*). Free phenols, with 10 to 13 carbon atoms have been purified on alumina by Lederer and Polonsky (*1909*).

Schneider and Oberkobusch (*2868*) have found that *o*-substituted hydroxybenzenes have less adsorption affinity for alumina than the isomeric *p*- or *m*-substituted compounds.

LeRosen *et al.* (*1974*) have compared the behaviour of *o*- and *p*-substituted phenols on silicic acid, with benzene as solvent.

Benzoates of phenol, resorcinol and phloroglucinol have been separated on morine-coated alumina by Brockmann and Volpers (*397*). The three dihydroxy-phenols could not be separated by this means.

Zahner and Swann (*3662*) have described the partition chromatography of phenol and the isomeric cresols on silica gel.

Sweeney and Bultman (*3173*) have separated creosote oil in nine fractions by chromatography on moist silicic acid, with *iso*-octane as mobile solvent. Franc (*1014, 1015*) has described the determination of *o*-, *m*-, and *p*-cresol and of phenol in commercial tricresol by partition on silica gel. White and Grant (*3534*) have separated *m*- and *p*-cresol on a celite column buffered with 0.5 *M* phosphate (pH 11.5) (cyclohexane as mobile phase).

(b) *Paper chromatography*

Bray and Thorpe (*369*) have published a review on "*Analysis of phenolic compounds of interest in metabolism*" with full details on paper chromatographic methods.

(i) *Free phenols*

Evans *et al.* (*929*) have obtained good separations of phenols using butanol (1 vol.), pyridine (1 vol.), mixed with 2 vols. of a saturated aqueous NaCl solution, as the mobile solvent (Table 36). The spots were detected with diazotized sulphanilic acid. Phenolic acids give elongated spots; this can be overcome as has been shown by Rydel and Macheboeuf (*2768*), by using an acid solvent (butanol-acetic acid) which diminishes the ionisation of the acids. Bate-Smith (*156, 157*) has used butanol-acetic acid or *m*-cresol-acetic acid for separating di- and trihydric phenols and phenolic acids. The spots were visualized by spraying and then heating with ammoniacal silver nitrate, or with $FeCl_3$ (Table 37). Riley (*2684*) has used *n*-amyl alcohol-water or butanol-benzene-water (1:9:10 or 1:19:20) for the separation of various phenols. The spots were visualised by spraying the dried chromato-

TABLE 36. R_F values of phenols (Evans *et al.*, *929*)

Solvent: Butanol 1, pyridine 1, water saturated with NaCl 2.

R_F' is measured, which is the R_F of the leading edge of a spot, instead of the usual centre of the spot

Substance	Colour with diazotised sulphanilic acid	R_F'
Phenol	deep yellow	0.97
p-Hydroxybenzoic acid	deep yellow	0.85
m-Hydroxybenzoic acid	deep yellow	0.74
Salicylic acid	pale yellow	0.65
Catechol	reddish brown	0.96
Protocatechuic acid	reddish brown	0.74
Resorcinol	brown	0.97
β-Resorcylic acid	brown	0.55
Hydroquinone	colourless	0.96
Gentisic acid	colourless	0.51
Pyrogallol	colourless	0.94
Pyrogallol carboxylic acid	colourless	0.71
Gallic acid	colourless	0.43
Hydroxyquinol	brown	0.48
Hydroxyquinol carboxylic acid	brown	0.45
Phloroglucinol	orange	0.96
Tannic acid	colourless	0.99
Guaiacol	orange	0.95
Thymol	yellowish brown	0.95
α-Naphthol	brownish red	0.95
β-Naphthol	orange	0.95

TABLE 37. R_F values of phenols and aromatic acids at 20°. (Bate-Smith, *157*)

Substance	Butanol 4 Acetic acid 1 Water 5	m-Cresol 50 Acetic acid 2 Water 48
Benzoic acid	0.92	0.93
Catechol	0.91	0.74
Cinnamic acid . . . . . :	0.94	0.92
o-Coumaric acid	0.94	0.82
Gallic acid . . . . -	0.68	0.08
m-Hydroxybenzoic acid	0.91	0.72
p-Hydroxybenzoic acid	0.90	0.72
Orcinol	0.91	0.75
Phloroglucinol	0.76	0.16
Phloroglucinol carboxylic acid	0.55	0.06
Protocatechuic acid	0.85	0.35
Pyrogallol	0.77	0.38
Quinol	0.88	0.69
Resorcinol	0.91	0.63
β-Resorcylic acid	0.93	0.54
Salicylic acid	0.95	0.84
Vanillic acid	0.92	0.81

TABLE 38. R_F values of phenols (Riley, *2684*)

Substance	n-Amyl alcohol sat. water	n-Butanol 1 Benzene 9 Water 10	n-Butanol 1 Benzene 19 Water 20
Hydroquinone	0.80	0.43	0.078
Resorcinol	0.83	0.54	0.13
Catechol	0.86	0.70	0.38
Toluhydroquinone	0.85		0.23
Phloroglucinol	0.61		0.00
Pyrogallol	0.65		0.025
α-Naphthol	0.90		0.83
β-Naphthol	0.92		0.85

gram with a 2% aqueous solution of phosphomolybdic acid and then exposing the strips to ammonia (see Table 38).

Roux (*2746*) has described the detection of di- and tri-hydroxyphenols by heating with a solution of sucrose in HCl (2 g sucrose, 10 ml HCl, 90 ml absolute ethanol). Durant (*834*) has separated phenols on buffered paper (pH 8) and detected the spots with Millon's reagent. Naphthols and poly-phenols have been separated by Barton *et al.* (*151*) with a single phase system of carbonic acid in water (pH 4,2); the phenols were detected by spraying with a mixture of equal volumes of 1% ferric chloride and 1% potassium ferricyanide, giving blue spots. Schleede (*2848*) has used benzene-cyclohexane (1:12) for the lower and benzene-cyclohexane (1:10) for higher phenols. Hudecek (*1471*) has used cyclohexane-CHCl$_3$-ethanol (27:3:0.6); both authors give extensive R_F tables.

Di- and trihydroxybenzenes have been separated by Mráz (*2300*), as well as Parke and Williams (*2444*), using benzene-acetic acid-water (40:50:10); development with diazotized *p*-nitroaniline and alkali gives different colours with the *o*-, *m*- and *p*-derivatives. See also Bray *et al.* (*368, 571*) and Wagner (*3442*). For the detection of phenols on paper chromatograms by ultra-violet absorption measurement, see Bradfield and Flood (*356*).

The two-dimensional chromatography of polyphenolic substances has been described by Cartwright and Roberts (*523*) with butanol-acetic acid-water (4:1:2.2) in one direction, 2% acetic acid in the second direction. Various new reagents for detecting the spots are described in this paper. Boscott (*325a*) has developed twodimensional methods for the analysis of phenolic constituents of urine. Barnabas (*141*) has used circular paper chromatography for the identification of phenolic substances.

o-, *m*- and *p*-Nitrophenols, α- and β-naphthols, *o*-, *m*- and *p*-hydroxybenzoic acids can be separated on paper, using butanol saturated with 5N NH$_4$OH (Lederer, *1917*). Nitrophenols have also been separated on paper by Robinson *et al.* (*2702*), aminophenols and their formyl and acetyl derivatives by Bray *et al.* (*367*) and chlorophenols by Johnson *et al.* (*1591*).

The paper chromatography of phenolic constituents of wood (pinosylvin and other stilbene derivatives, flavanols etc.) has been studied in detail by Lindstedt (2027), with Misiorny (*2028*) and Zacharias (*2029*).

Applications of paper chromatography to the separation of more complex phenolic substances concern: polymethylol phenols (Freeman, *1030*; Reese, *2646*), lignin (Bland and Gatley, *268*), depsides (Wachtmeister, *3435*), tea catechins and tannins (Bradfield and Bate-Smith, *355*; Roberts and Wood, *2694*; Zaprometov and Soboleva, *3668*; Oshima *et al.*, *2422, 2423*; Djemoukhadze and Chalneva, *791*); gallotannins (Schmidt and Lademann, *2864*); cocoa catechins (Forsyth, *1006*); tannins of the grape vine (Durmishidze and Nutsubidze, *835*).

Roberts and Wood (*2694*) seem to have achieved a separation of the optical antipodes of catechin, epicatechin, gallocatechin and epigallocatechin, using water as the mobile solvent. They have shown also that *epi*catechin has a substantially lower R_F in butanol-acetic acid than its epimer; Roberts (*2692*) explains this difference by hydrogen bonding of the 3-hydroxyl with the heterocyclic oxygen atom in catechin and its derivatives. Conformational analysis shows that such hydrogen bonding is impossible in *epi*catechin and its derivatives.

(ii) *Derivatives of phenols*

Hossfeld (*1451*) has used the azo-dyes obtained by coupling simple phenols with diazotized sulphanilic acid for separating *o*-, *m*- and *p*-cresol and guaiacol, using methyl ethyl ketone-water as mobile solvent. This method has been applied to a large series of phenols by Chang *et al.* (*545*); catechols are too sensitive to oxidation to be handled by this method. Rayburn *et al.* (*2637*) have separated several phenols after coupling with diazotized *p*-nitroaniline.

Schlögl and Siegel (*2850*) have separated mono-phenols as arylhydroxy-acetic acids (ROCH₂COOH); the spots were detected by the quenching of the fluorescence of a solution of umbelliferone.

CHAPTER 21

Aldehydes and ketones

1. SEPARATIONS OF FREE ALDEHYDES AND KETONES

LeRosen and May (*1975*) have studied the chromatographic behaviour of some aldehydes on silicic acid, from benzene solution. The positions of the zones were detected by streaking the extruded column with Schiff's reagent. The side chain of the aldehydes acts as a hindrance to adsorption and leads to an increase in the value of R (see p. 104) with increasing number of carbon atoms.

Similar work of Smith and LeRosen (*3018*) on the effect of the side chain on chromatographic behaviour of a series of aliphatic straight chain methyl ketones from C_3 to C_{19}, has shown that the relationship between adsorption and molecular weight is somewhat more complex. It was found that in general the ketones appeared to be adsorbed by interaction between the carbonyl group and the adsorbent, and that the heavier the side chain the smaller was the adsorption. Certain exceptions were however noted, where the addition of a CH_2 group to the side chain decreased the rate of zone movement. These decreases in the R value seemed to be periodic within the homologous series. Branched-chain compounds are adsorbed less, and cyclic aliphatic ketones more than the corresponding straight-chain methyl ketones; both of these facts point to the importance of steric effects.

On studying the effect of side chain on adsorption of some ketones on charcoal (Norit A), Smith and LeRosen (*3019*) found that adsorption becomes progressively stronger as the length of the side chain increases; chain branching diminishes adsorption.

As the carbonyl group has a lower affinity for alumina than the hydroxyl group, aldehydes and ketones may easily be separated from alcohols (e.g. separation of the components of the unsaponifiable material from wool fat, *693*).

Kirchner et al. (*1727*) have separated aldehydes and terpene ketones (cinnamaldehyde, pulegone, carvone, camphor) on "chromatostrips" (see p. 12 and Fig. 60) and Sease (*2927*) has separated aromatic aldehydes and ketones on fluorescent silica gel.

Stoll (*3128*) has purified unsaturated aliphatic aldehydes on a column of acid-activated alumina (pH 5.4).

The quantitative separation of vanillin and syringa-aldehyde has long been a problem to lignin chemists. Two solutions have been proposed: paper chromatography, by Bland (*267*) and adsorption on a column of magnesol, by Pearl and Dickey (*2481*). See also Pearl and Beyer (*2480*).

A patent of Distillation Products Inc. (*787*) describes the purification of β-ionone on sodium aluminium silicate.

Use of too active an alumina can lead to destruction of unsaturated aldehydes and ketones.

Gabrielson and Samuelson (*1066, 1067*) have studied the separation of aldehydes and ketones on the bisulphite form of ion exchangers. Aldehydes are thus adsorbed while ketones are readily washed out of the column. A quantitative separation of aldehydes and ketones was thus achieved.

Sjöström (*3005, 3006*) has shown that such columns can even separate ketones from one another.

Bisulphite treated paper can also be used (with hexane-$CHCl_3$, 9:1 as organic phase); Newcombe and Reid (*2361*) have thus separated aceto-veratrone, acetovanillone, veratraldehyde and vanillin.

Verzele (*3400*) has described the *partition chromatography* of humulone and related substances on columns of silica gel. Extensive work on the separation of the low molecular ketones and aldehydes by *gas-liquid partition chromatography* was carried out by James and Martin (*1538, 1539*) (Fig. 62).

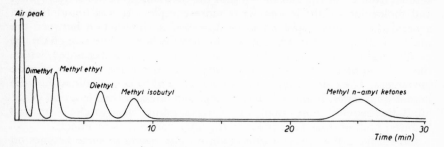

Fig. 62. Separation of ketones. Column length, 4 ft.; stationary phase, paraffin wax; temperature, 100°; rate of flow of nitrogen, 24.8 ml/min; pressure of nitrogen, 17.8 cm Hg (James and Martin, *1539*)

For the separation of vanillin, ethylvanillin, piperonal and *p*-hydroxybenzaldehyde on *paper*, see Stoll and Bouteville (*3129*), Mitchell (*2239*) and Ter Heide and Lemmens (*3221*).

2. SEPARATIONS OF DERIVATIVES OF ALDEHYDES AND KETONES

Di Modica and Spriano (*783*) have separated aliphatic *semicarbazones* by partition chromatography on silica gel.

The hydrosoluble hydrazones obtained by treating aldehydes or ketones with the Girard T and Girard P reagents can be separated by *paper chromatography*. The positions of the spots can be revealed by spraying with iodoplatinate. In this way quantities of the order of a few μg of ketosteroids have been separated by Zaffaroni *et al.* (*3658*).

Uno and Koyama (*3350*) treated volatile aldehydes with benzene sulfohydroxamic acid in alkaline solution and separated the *hydroxamic acid derivatives* of the aldehydes on paper.

The separation of simple methyl ketones after condensation with salicylaldehyde can not be recommended, each ketone giving several condensation products (Mentzer *et al.*, *2189*).

α-*Dicarbonyl-compounds* (glyoxal, diacetyl, etc.) also α-ketoacids can be separated on paper in the form of their *quinoxaline derivatives* (Harjanna, *1309*).

Conteben (*p*-acetylaminobenzaldehyde thiosemicarbazone) has been subjected to paper chromatography by Hinrichs *et al.* (*1395*).

2,4-Dinitrophenylhydrazones

Strain (*3136*) first recommended the separation of aldehydes and ketones as their 2,4-dinitrophenylhydrazones; since then many authors have described such separations. As adsorbents one can use alumina (*18, 1901, 2043, 2148, 2340, 2342*), silicic acid (*514, 2111, 2698*), activated $CaSO_4$ (*3052*), zinc carbonate (*3326*), talc or magnesium sulphate (*3444, 3082*), bentonite-Celite (*3535*); this latter adsorbent is especially recommended by Elvidge and Whalley (*891*).

Not only simple aldehydes or ketones, but also steroid ketones, such as oestrone and equilenin (*3386*) or androgens (*1592*) have thus been separated. Gordon *et al.* (*1174*) have separated the *syn* and *anti* forms of ethyl methyl ketone dinitrophenylhydrazones on silicic acid-Celite.

Onoe (*2415*) has applied the chromatostrip technique of Kirchner *et al.* (*1727*).

Kramer and Van Duin (*1777*) have used silica gel saturated with nitromethane as immobile and light petroleum as mobile phase for the separation of mixtures of 2,4-dinitrophenylhydrazones of *n*-aliphatic aldehydes and methylketones from C_1 to C_{18}. Up to C_{12}, the derivatives could be separated when differing only by one C atom; the higher homologs, up to C_{18}, only when differing by two C atoms. Identification is possible by measuring the retention volumes.

Howard and Tatchell (*1461*) have separated the 2,4-dinitrophenylhydrazones of volatile carbonyl compounds on a reversed phase partition column (kieselguhr pretreated with dimethyl-dichlorosilane, benzene as stationary, formamide as mobile phase).

The *filter paper chromatography* of 2,4-dinitrophenylhydrazones of aldehydes and ketones has been described by Rice *et al.* (*2676*). The spots are rendered visible by spraying with a 10 % solution of KOH in water, which produces blue, red or orange spots.

Sýkora and Procházka (*3179*) have used petroleum ether as stationary phase and 80 % ethanol or 65 % propanol as mobile phase. See also Meigh (*2180*), Forss *et al.* (*1004*).

Kirchner and Keller (*1724*) have obtained good separations of 2,4-dinitrophenylhydrazones of methyl ketones on paper impregnated with silicic acid (solvent: 5 % ethyl ether in light petroleum, boiling point 110°).

Kostiř and Slavik (*1771*) have described the reversed phase partition chromatography of 2,4-dinitrophenylhydrazones on acetylated paper. See also Neïman *et al.* (*2348*).

CHAPTER 22

Acids

The chromatographic separation of fatty acids by adsorption is generally difficult, but the methods of partition chromatography usually give excellent results.

Holman (*1434*), Boldingh (*301*) and Lugg (*2056*) have published reviews of the chromatography of fatty acids and related substances.

1. SEPARATIONS BY ADSORPTION

(a) *Free acids*

Kaufmann (*1651–1653, 1658–1660*) has published a series of papers on the chromatography of fatty acids on alumina and silica gel. Separations were almost always incomplete.

Kaufmann and Schmidt (*1659*) showed that chromatography on alumina or silica gel can be used to neutralise natural oils. This technique has the advantage of not destroying substances of dietary importance such as vitamins, provitamins, phosphatides, etc. (*1655*). See also Boekenoogen (*288*), Loury (*2051*), Sylvester *et al.* (*3180*).

Kaufmann and Wolf (*1660*) described the separation of stereoisomeric acids (fumaric and maleic; oleic and elaidic; erucic and brassidic acids).

Cassidy (*527*), after having determined the adsorption isotherms of the higher fatty acids (lauric, myristic, palmitic, stearic) on charcoal, alumina, magnesia, Fuller's earth and silica gel, investigated the possibilities of chromatographic separation. He found that comparison of the adsorption isotherms did not allow prediction of the possibilities of separation on a column. The acids in solution seem to exert a far from negligible effect on one another. Certain charcoals adsorb the higher acids more strongly whereas other charcoals and silica gel adsorb the lower acids more strongly.

Cassidy (*528*) also described the separation of lauric and stearic, and palmitic and stearic acids on a charcoal column using light petroleum as solvent.

In a later paper, Nestler and Cassidy (*2356*) report the separation of the lower fatty acids by frontal analysis and discuss the measurement of adsorption isotherms by the frontal analysis technique.

Cason and Gillies (524) have measured static adsorption isotherms on Darco-G 60 charcoal, with 95 % ethanol as solvent, for a series of saturated, unsaturated and branched-chain, acids. The isotherms were used for predicting separability by chromatography and a good separation of palmitic and stearic acid was reported.

Kiselev et al. (1729) have studied the effect of the desiccation temperature of silica gel on the adsorption of fatty acids. A gel dried at 485° adsorbs twice as much stearic acid as a gel dried at 800°.

Claesson (574) has studied the separation of the higher fatty acids by frontal analysis; with silica gel dried at 800° as adsorbent the unsaturated, saturated and branched chain acids could be separated. With charcoal as adsorbent, saturated acids could be separated, the adsorption increasing with chain length (for details see 579, also 3262). See also Dutton (842).

Holman and Hagdahl (1435) have carried out separations of the higher fatty acids using the displacement technique of Tiselius. Palmitic acid (0.5 % in alcohol) serves as displacer for myristic and lauric acids, but picric acid is the most useful displacing agent. The authors showed that the displacement of linoleic acid is difficult, as the acid leaves the column only very slowly, mixed with the displacing agent. The recoveries of this acid were only 70 %.

In further studies of Hagdahl and Holman (1257, 1436), separations of all the straight-chain fatty acids from C_1–C_{20} were described, using a mixture of one part Darco G-60 and two parts Hyflo Super Cel. The best displacer for a group of acids is the next higher acid, and as solvent, mixtures of water and ethyl alcohol are used (Fig. 63). Although under the correct conditions each acid gives a definite step, detectable optically as well as by titration, Holman (1431) in a later paper improves the separations and obtains quantitative recoveries by the use of carrier displacement (see Tiselius and Hagdahl, 3269). If small quantities of the acids are mixed with ten times the weight of their methyl esters in 95 % ethanol, the acids emerge immediately in front of their methyl esters. Titration of the effluent shows recoveries of 93.5–100.3 %. A displacer for the methyl esters must also be employed and Holman uses the next higher methyl ester. For example, in the separation of myristic, palmitic and stearic acids, methyl laurate, myristate and palmitate are used as carriers and 1 % methyl stearate as the displacer (Fig. 6, p. 8).

The separation of unsaturated acids on Darco G-60 mixed with two parts Hyflo was also studied by Holman and Williams (1438) and Holman (1433); non-conjugated double bonds diminish the adsorption, while acids with carboxylic conjugated double bonds are more strongly adsorbed. Oleic and linoleic acid can be separated.

Di Modica and Rossi (782) have separated oleic, linoleic and linolenic acids in acetone solution on a cooled alumina column.

Graff and Skau (1184) have separated oleic and stearic acid or myristic and stearic acid on magnesium oxide columns coloured by the addition of phenol red. The acids form visible zones which may be eluted by dissolving the magnesia in dilute HCl.

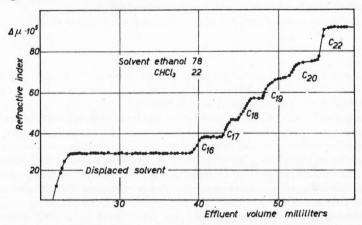

Fig. 63. Displacement diagram for acids of 16 to 22 carbon atoms. Sample, 12 mg of hexadecanoic, 15 mg of heptadecanoic, 20 mg of octadecanoic, 30 mg of nonadecanoic, and 35 mg of eicosanoic acids. Solvent, chloroform 22 and ethanol 78 volumes per cent. Filter, 2000, 800, 400, 200, 100, 50π c.mm = 11.2 ml. Displacer, 1.0 per cent docosanoic acid (Holman and Hagdahl, *1436*)

The *separation of saturated and unsaturated fatty acids* is possible in different ways: Simmons and Quackenbush (*2991*) have used the reaction of 2,4-dinitrobenzenesulphenyl chloride with double bonds:

$$R_2C\!\!=\!\!CR_2 + ArS\!\!-\!\!Cl \longrightarrow \begin{array}{c} R_2C\!\!-\!\!CR_2 \\ | \quad\; | \\ ArS \;\; Cl \end{array}$$

to prepare the 2,4-dinitrobenzenesulphenyl chloro derivatives of the unsaturated fatty acids; these can be separated from the saturated fatty acids by chromatography on magnesium sulphate.

Bergström and Pääbo (*210*) have hydroxylated the unsaturated fatty acids by performic acid and separated the esters of the saturated and hydroxylated acids on silicic acid; the former are eluted with CH_2Cl_2, the latter after adding 0.75 % methanol to the solvent.

Howton (*1463*) has shown that the addition of bromine to unsaturated fatty acid esters enhances the adsorption affinity contribution of each unsaturation site to such an extent that the resulting esters, differing in the number of pairs of vicinal bromine substituents are easily separated by gradient elution (increasing ether in pentane) on alumina columns. Methyl stearate, methyl *threo*-9,10-dibromostearate and methyl *threo,threo*-9,10,12,13-tetrabromostearate were cleanly separated.

English (900) has separated two series of dibasic acids HO_2C—$(CH_2)_n$—$CH{=}CH$—CO_2H and HO_2C—$(CH_2)_n$—$CH{=}CH$—CH_2CO_2H on charcoal. Acids of the first series having a conjugated double bond system are more strongly adsorbed than acids of the second series.

For the separation of cyclic acids on charcoal see Jeger *et al.* (1571).

(b) *Esters*

Swift *et al.* (3175) have prepared pure methyl linoleate by chromatography on alumina. The same technique has been used by White and Brown (3537) for the isolation of pure methyl arachidonate.

Dutton and Reinbold(843) have fractionated synthetic mixtures of the ethyl esters of stearic, oleic, linoleic and linolenic acids on alumina. Dugan *et al.* (826) and Bergström(207) have studied the products of autoxidation of methyl linoleate by chromatography on sodium aluminium silicates and alumina.

Sometimes even neutralized alumina saponifies esters rather easily; magnesium silicate is therefore to be preferred for the purification of such substances.

Pure samples of methyl linoleate and methyl linolenate were obtained from natural oils by chromatography on silicic acid by Riemenschneider *et al.* (2683). See also 1368.

Abu-Nasr and Holman (4) have isolated methyl eicosapentaenoate, docosapentaenoate and docosahexaenoate from cod liver oil esters by chromatography on charcoal and silicic acid.

The displacement analysis of oxidation products of methyl linoleate with a system of coupled filters packed with Darco G-60 and Hyflo Supercel (1:2), with ethanol as solvent and 1 % ethyl stearate as displacer has been described by Khan *et al.* (1706). Products with *cis-trans* and *trans-trans* configuration could be separated. See also p. 173 (separation of hydroxy- or bromoesters).

Kirchner *et al.* (1728) have reported the separation of the lower fatty acids on silicic acid in the form of their *p*-phenyl phenacylates (C_6H_5–C_6H_4CO–CH_2OCOR) which are fluorescent. The derivatives of higher fatty acids are less strongly adsorbed than those of the lower acids. Each straight-chain acid from C_2 to C_6 can be separated from its lower and higher homologues as can the *iso*-acids with 4, 5 and 6 carbon atoms, *iso*butyric esters could be separated from esters of *n*-butyric acid, but *iso*valeric esters could not be separated from esters of *n*-valeric acid. See also Klohs *et al.* (1741).

The separation of *p*-phenylazophenacyl esters, $C_6H_5N = NC_6H_4$ $COCH_2$ $OCOR$, of fatty acids (C_2 to C_{18}) on silicic acid has been described by Ikeda *et al.* (1499).

Sease (2926) has described the separation of esters of aromatic acids on fluorescent silica gel.

(c) *Hydroxy-acids*

Chromatography on alumina has proved a very valuable and efficient tool for the separation of the high molecular weight mycolic acids of Mycobacteria, which are branched-chain β-hydroxy acids. Thus Asselineau and Lederer (91) have separated two apparently isomeric α- and β-mycolic acids, $C_{88}H_{176}O_4$, by percolation of the free acids through alumina; the acids are eluted successively by increasing concentrations (0.1 to 0.5 %) of acetic acid in ether. Numerous derivatives of these acids have been purified on alumina and thus their formula established. The separation of complex mixtures of methoxylated and hydroxylated mycolic acids is better achieved by chromatography of the methyl esters on alumina (Demarteau, 733) or silicic acid (Tener and Asselineau, unpublished).

Chromatography on alumina has also been extensively used for the isolation of coryno-mycolic acid, $C_{32}H_{64}O_3$, from the lipids of the Diphtheria Bacillus and for the purification of its derivatives (Lederer and Pudles, 1910).

(d) Keto-acids

Chromatography of aliphatic β-keto-esters, $R_1COCH(R_2)COOCH_3$, on alumina can yield the corresponding ketones, $R_1COCH_2R_2$, because of saponification of the esters and decarboxylation of the β-keto-acids thus formed (2550).

Thompson et al. (3237) have used alumina for the purification of β-keto-esters of the type Aroyl-C(R_2)COOC$_2$H$_5$.

Datta et al. (705) have worked out a method for the determination of keto-acids in urine. A 1/10th of a 24 h specimen is treated with 2,4-dinitrophenylhydrazine in the presence of HCl and the 2,4-dinitrophenylhydrazones of the keto-acids extracted with ethyl acetate. The extract is chromatographed on 200 g alumina (Peter Spence Ltd., Type H) using ethyl acetate, alcohol, water and 0.5 M Na$_2$CO$_3$ as eluants. 1 mg amounts of the separated keto acids can be estimated with an accuracy of 5 %. See also Cremer and Beyer (652).

(e) Lactones

Great losses occur sometimes on chromatographing lactones on alumina; this is principally due to saponification of the lactone ring by the alkali of the adsorbent. Well washed, neutral alumina is mostly harmless. Reichstein and Shoppee (2658) mention 14β,21-oxido-23$\rightarrow$21-lactones ("iso"-compounds) (I) derived from heart poisons. They state: "the losses are decreased, but not completely eliminated by use of neutralised alumina; 20-keto-21$\rightarrow$14-lactones (II) suffer very large losses which cannot be avoided even by use of neutralised alumina".

(I)   (II)

Magnesium silicate is usually to be preferred for the purification of lactones (see for instance Polonsky and Lederer, *2550*).

Examples of successful purifications of hydroxylated and unsaturated lactones may be found in *1304, 1305, 1306, 2759, 2760*.

2. SEPARATIONS ON ION EXCHANGE COLUMNS

Kunin and Myers (*1837*) have studied the adsorption of organic acids (formic, acetic, oxalic, citric, benzoic and salicylic acids) on the anion exchange resin, Amberlite IR-4B, in the hydroxyl or chloride form. The anion exchange abilities of the various anions depend upon the structure, size and valence of the anion and upon the type of salt formed; ionisation does not seem to be an important factor.

Anion exchangers may be used for the adsorption of the bulk of organic anions present in solution and thus for the elimination of sugars and other impurities (Bryant and Overell, *436*; Hulme and Swain, *1485*; Hulme and Richardson, *1484*; Nordmann *et al.*, *2379, 2380*). The use of anion exchangers for the selective separation of complex mixtures of organic acids has however only been reported recently.

Thus Busch *et al.* (*479*) have used the formate form of Dowex-1 for the separation of the acids of the citric acid cycle; elution of the adsorbed acids was effected by continuously increasing the concentration of formic acid. Discrete, well separated peaks of lactic, succinic, malic, pyruvic, fumaric, α-ketoglutaric and *cis*-aconitic acids were obtained. Reproducibility of the positions of the peaks of known acids was fairly good.

The separation of citric, oxalic, malic, tartaric and glycolic acids on anion exchangers (Resin S of the Permutit Corp., or Dowex-1) has been described by Owens *et al.* (*2430*) and Schenker and Rieman (*2841*).

Davies (*711*) has shown that for the separation of two weak acids by elution the best separation is achieved when the pH of the solution applied to the resin is 1–2 units lower than the value of $\frac{1}{2}(pK + pK')$ where K and K' are the dissociation constants of the two acids. The weaker acid will then be taken up less readily by the resin and will be eluted preferentially. Hale *et al.* (*1274*) have used this principle for the quantitative separation of β- and γ-resorcylic acids on columns of Amberlite XE-76, Amberlite IRA-400, De-Acidite B and De-Acidite E. Separation of the two resorcylic acids was most readily achieved by adsorbing the mixture on an anion exchange resin in the chloride form and then selectively eluting the β- and the γ-acid with 0.01 N and 0.1 N HCl respectively in 75 % ethyl alcohol.

The authors used 75 % ethyl alcohol as mobile phase in order to obviate the use of large volumes of aqueous eluant which were required because of the tailing of the bands.

Berntsson and Samuelson (*228*) have studied the quantitative aspects of the elution of some organic acids from Dowex-2 and Dowex-3 columns.

Savary and Desnuelle (*2833*) have described the separation of glycerides from free fatty acids by passing an ether solution of the mixture on Amberlite IRA-400; the adsorbed acids can be eluted (in a partially esterified form) with a mixture of ethanol (50), 10 N HCl (10), ether (75).

3. SEPARATIONS BY PARTITION CHROMATOGRAPHY

(a) *Column separations*

(i) *Fatty acids*

Lester Smith (*3021*) first applied partition chromatography on silica gel to volatile fatty acids, and obtained qualitative separations of formic, acetic, propionic, butyric and valeric acids. The technique has been further developed by Elsden (*889*). As formic acid is strongly adsorbed, it must first be eliminated. The other four acids are dissolved in chloroform containing 5 % of butanol and passed through a column of silica gel, which acts as the support for the stationary phase of water, coloured with bromocresol green. The different acids are retained in order of their partition coefficients: acetic acid at the top of the column and the others distributed down the column in order of increasing chain length, each acid forming a yellow zone. The determination of the R_F values allows the identification of the acids. On washing the column with chloroform /20 % butanol, each zone passes out of the column in the filtrate. The zones are estimated quantitatively by titration of the eluates. The position of the zones, however, is not always constant but depends on the amount of acid present. Ramsey and Patterson (*2615*), Nijkamp (*2371*) and Kretovich *et al.* (*1801*) have described similar methods.

Several other workers have improved on the method of Elsden. Neish (*2350*) substituted benzene/butanol and Nijkamp (*2370*) CCl$_4$/butanol for Elsden's CHCl$_3$/butanol solvent, and obtained quantitative estimations of acetic acid.

Synge (*3183*) has discussed the reasons for variation of R_F with concentration. The existence in the organic phase of a partition chromatogram of associated molecules of the fatty acids and of ionic dissociation in the aqueous phase deforms the partition isotherm and causes the zones to "tail". By using a buffer in the aqueous phase, the pH is maintained constant and so the ratio $\dfrac{\text{ionised acid}}{\text{unionised acid}}$ is kept constant.

If a buffer system is used, such that the ions cannot pass into the organic phase, the following advantages are obtained:

1. "The bands of ionising substances may be rendered sharp by promoting linearity in the effective distribution isotherm."

2. "By controlling the pH of the buffer, the distribution coefficient of ionising substances may be varied over a wide range. Thus substances which in the un-ionised form would travel too fast for separation in the chromatogram may be slowed down to any desired rate."

3. "Ionising substances differing in pK but having similar partition coefficients in the un-ionised state may be separated on the basis of their different ionising strengths by partition chromatography at a suitable pH".

Scarisbrick et al. (2298, 2835) have obtained quantitative separations of the saturated volatile fatty acids (C_2 to C_8) by the use of silica gel partition columns having aqueous buffers as the stationary phase and chloroform-butanol mixture as the mobile phase. No indicators are used on the column but the eluate is titrated directly with 0.005 N KOH. Straight- and branched-chain isomers, however, cannot be separated.

Fairbairn and Harpur (932) used two columns of silica gel in series. The upper column retained C_2 and C_3 acids, and the lower separated the C_4–C_8 acids. By disconnecting the two columns after the C_4–C_8 acids have passed through the first, very fast quantitative estimations were carried out.

By using first an unbuffered column and re-chromatographing the higher acids on buffered columns, Gray et al. (1199) have achieved the separation of isomeric acids as well as obtaining quantitative results for all the acids present in rumen distillates. Ramsey (2614) has also obtained a partial separation of n- and isobutyric acid on a silica gel partition column. Isomeric C_5-acids were separated by Bueding and Yale (446) on Celite columns buffered to pH 6.5 with phosphate buffer. See also Brouwer and Nijkamp (405). Siegel (2980) has used a buffered celite column (pH 6.5) for the separation of acetic, propionic and butyric acids.

Ramsey and Patterson (2616) have used a column of silicic acid saturated with methanol (containing bromocresol green as indicator) and a mobile phase of isooctane to give good separations of the straight-chain fatty acids from C_5 to C_{10}. In a later publication (2618) the range was extended to the C_{11} to C_{19} acids by using a silicic acid column with a mixture of furfuryl alcohol and 2-aminopyridine as the stationary phase, and n-hexane as the mobile phase. The method of Ramsey and Patterson (2616, 2618) has been modified by Vandenheuvel and Hayes (3364) who use two different columns, the first for the acids C_2 to C_7, the second for the acids, C_7 to C_{12}.

Ramsey and Patterson's method was employed for the detection of n-pelargonic, n-caprylic and n-oenanthic acids in the degradation of a glycolipid by Jarvis and Johnson (1560), and for oenanthic acid in the degradation of trans-11-octadecenoic acid by Bumpus et al. (452). See also Brouwer and Nijkamp (404) for analysis of silage and Peterson et al. (2499) for fatty acids in cheese.

Quantitative separations of the straight-chain fatty acids from C_2 to C_{10} have been obtained by Peterson and Johnson (2498) using a Celite column with 27 to 35 N sulphuric acid as the stationary phase and benzene as the mobile phase.

Nijkamp (*2372*) has separated the fatty acids from C_{10} to C_{20} on a silicic acid column tainted with bromothymol blue and neutralized with methanolic ammonia solution (*isooctane* as mobile phase). Links and De Groot (*2030*) have purified C_{17} to C_{19} acids on a preparative scale by partition on silica gel.

New solvent systems for separating monocarboxylic acids from C_2 to C_{16} (and dicarboxylic acids from C_2 to C_{22}) have been described by Zbinovsky (*3671*) (Fig. 64).

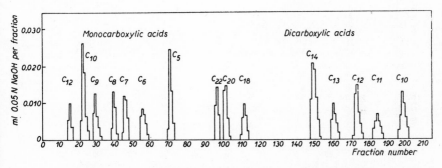

Fig. 64. Separation of monocarboxylic acids (1.8 mg of total acids) and dicarboxylic acids (2.0 mg of total acids) on 6-gram silicic acid column (Zbinovsky, *3671*)

Fractions 1 to 39, 0.5 ml; 40 to 69, 1.0 ml; 70 to 73, 2.0 ml; 74 to 171, 0.5 ml and 172 to 202, 1.0 ml. Solvents: Fraction 1 to 73, Skellysolve B; 74 to 127, Skellysolve B-*n*-butyl ether mixture (1:1); 128 to 202, *n*-butyl ether

a. Gas-liquid partition chromatography. Partition chromatography has been extended by James and Martin (*1537*) to columns using a stationary liquid phase and a mobile gas phase (nitrogen). The columns consist of 4 ft. or 11 ft. lengths of 4 mm internal diameter glass tube packed with kieselguhr (size graded Celite 545) which acts as the support for the liquid phase, which for separation of the volatile fatty acids is a mixture of DC 550 silicone and stearic acid (10 % w/w). The columns are held in a vapour jacket at a temperature such that the saturated vapour pressure of the acids to be separated lies between 10 and 1000 mm. The end of the column is drawn down to a capillary and by means of a thin rubber gasket fits into the bottom of a titration cell containing aqueous indicator (0.01 % phenol red). The mixture of acids is applied by means of a micro-pipette to a Fibreglass plug at the far end of the column packing. A stream of nitrogen gas from a manostat is then passed down the column. The acids move generally in order of molecular weight and are absorbed from the nitrogen gas stream by the water in the titration cell. The acids are continuously titrated by means of an automatic recording burette controlled by a green sensitive (unfiltered) photocell and an amplifier. A graph consisting of a series of steps (the integral of the more usual chromatogram peaks) is obtained, the straight line between each step denoting that no material is emerging from the column (Fig. 65). The height of each step gives directly the amount of each acid emerging from the column. By the use of this technique all the *n*- and *iso*-acids from formic to dodecanoic acid can be separated and quantitatively estimated in micro amounts. By the use of an 11 ft. column folded into three lengths so that it occupies no greater length than the 4 ft. column, all the isomers of valeric acid (trimethylacetic, *iso*valeric, methylethylacetic and

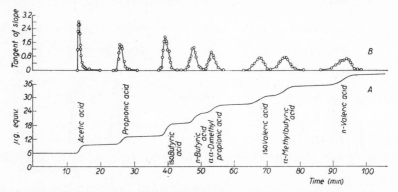

Fig. 65. Gas-liquid partition chromatography. The separation of acetic, propionic, *n*-butyric and *iso*butyric acids and the isomers of valeric acid, showing the complete resolution of all bands, and change in band shape in ascending series. *A*, experimental curve. *B*, differential of experimental curve. Column length, 11 ft.; liquid phase, stearic acid (10% w/w) in DC 550 silicone; nitrogen pressure, 74 cm Hg; flow rate, 18.2 ml/min; temp. 137° (James and Martin, *1537*)

n-valeric acid) can be separated. The reproducibility, speed (all the *n*- and *iso*-acids from formic to *n*-valeric acid can be qualitatively and roughly quantitatively estimated in 17 min) and general convenience of this technique make it especially valuable for work entailing a long series of routine estimations of the volatile acids. A technique has been worked out for starting with solutions of salts of the acids instead of dry samples of the free acids. El-Shazly (*890*) has applied this method to the analysis of volatile fatty acids of biological origin.

The separation of methyl or ethyl *esters* of fatty acids by gas-liquid partition has been described by James and Martin (*1539, 1539a*) (C_1–C_5) and by Cropper and Heywood (*662*) (C_{12} to C_{22}). Good separations of normal and *iso*- or *anteiso*-saturated acids were obtained as well as of mono- and di-unsaturated acids (*1539a*).

β. *Reversed phase partition chromatography.* Boldingh (*300*) showed that separations of the C_8–C_{18} acids can be carried out, if the lipid phase is stationary instead of the aqueous phase. Using a column of rubber saturated with benzene and eluting with aqueous methanol, he obtained complete separations of the higher fatty acids. However, the operation of the rubber columns is critical; for example the temperature must be kept between 21 and 23°. Mead *et al.* (*2178*) have modified this method for the separation of arachidic (C_{20}) and behenic (C_{22}) acids on Mealorub (vulcanized rubber powder).

A simpler reversed-phase column for the C_{12}–C_{18} acids was described by Howard and Martin (*1460*), who made Hyflo Supercel kieselguhr water-repellent by exposing it to vapours of dimethyldichlorosilane. The non-polar phase is readily supported by this material; the fractions of the eluant flowing out of the column were titrated and showed complete separations of the higher fatty acids.

Silk and Hahn (*2987*) have extended this method to cover the resolution of mixtures of higher fatty acids from 16 to 24 carbon atoms; they use paraffin oil fixed on non-wetting (dimethyldichlorosilane-treated) kieselguhr as stationary phase and acetone as developing solvent. They have also used this method for the analysis and preparative separation of unsaturated higher fatty acids of fish oils (*2988*). Hougen (*1453*) has used this method for the separation of all the normal fatty acids from six to eleven carbon atoms.

The use of polythene in acetone for the reverse phase chromatography of fatty acids has been reported by Green *et al.* (*1213*); aqueous acetone containing bromothymol-blue indicator (0.001 % w/v) was used as mobile phase.

Savary and Desnuelle (*2832*) have used the method of Howard and Martin (*1460*) to *separate saturated and unsaturated fatty acids* after hydroxylation of the latter with cold dilute alkaline permanganate; the di- or tetra-hydroxylated acids are rapidly eluted from the paraffin oil-cyclohexane column and can then be separated on a fresh column of celite with purified castor oil as stationary phase.

Crombie *et al.* (*658*) have used the system of Howard and Martin (*1460*) for estimating mg quantities of fatty acid mixtures, in which the saturated acids are determined separately by chromatography after removing the unsaturated acids, by oxidation by alkaline permanganate. They have studied the chromatographic behaviour of a large number of unsaturated acids and have found several rules: the occurrence of a double bond in the fatty acid increases the elution rate according to its position. Little change is noticed if the unsaturation occurs adjacent to the carbonyl group; if the unsaturation is further removed from the –COOH, the elution rate corresponds approximately to that of the saturated acid having a chain length of two CH_2 less (oleic acid is eluted with palmitic, erucic with arachidic acid). The *cis* and *trans* isomers appear to be inseparable. The presense of a second double bond, as in linoleic acid further increases the elution rate; triply unsaturated acids move still faster (linolenic and eleostearic acids move faster than myristic acid). The presence of a keto-group causes a marked increase in elution rate; a hydroxyl group causes very speedy elution.

(ii) *Hydroxy-acids, keto-acids, dicarboxylic acids*
A fluorometric determination of polycarboxylic acids following chromatography has been described by Frohman and Orten (*1052*); the method is

based on the reaction of the acids with resorcinol in sulfuric acid, to form fluorescent derivatives.

Isherwood (1512) has described the preparation of silica gel for partition chromatograms and an apparatus which allows separation of organic acids to be carried out on silica gel columns with water as the stationary phase. The column eluate is mixed with a dilute solution of a pH indicator which shows when an acid appears in the filtrate. Excellent quantitative separation of a number of acids (succinic, malonic, oxalic, malic, tartaric and citric) were obtained.

Gottlieb (1182) has separated tropic and atropic acids on silica gel columns using the indicator of Liddell and Rydon (2005).

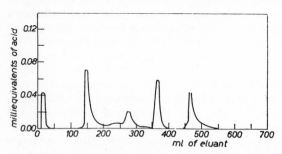

Fig. 66. Threshold volumes of *m*-hydroxybenzoic acid (15 ml), trichloroacetic acid (145 ml), cyanoacetic acid (275 ml), succinic acid (365 ml) and aconitic acid (465 ml) (Marvel and Rands, 2163)

The separation of non-volatile acids on silicic acid was also described by Marvel and Rands (2163), who eluted with successive mixtures of butanol and chloroform ranging from 5 % butanol-95 % $CHCl_3$ to 100% butanol. The peak effluent volume of a large number of aliphatic and aromatic acids was recorded; it varied somewhat depending on the solvent used for the mixture of acids (Fig. 66).

Marvel and Light (2162) utilised this technique for the identification of the acids produced by the oxidative cleavage of butadiene and polystyrenes, Begemann et al. (174) separated a number of aliphatic dicarboxylic acids obtained by the permanganate oxidation of monoethenoid fatty acids, and Frohman et al. (1053) employed silica gel partition columns for the determination of the acids of the citric acid cycle in tissues.

The method of Marvel and Rands (2163) can be improved by the use of a gradient of ethanol (or *tert*. amyl alcohol) (Donaldson et al., 802; Marshall et al., 2145; Shkol'nik, 2968) (separations of acetic, lactic, pyruvic, fumaric, succinic, malic, oxalacetic, ketoglutaric acids); see also Hulme (1479, 1480), Hulme and Richardson (1484).

The *dicarboxylic acids* from C_3 to C_{10} have been separated on columns of silicic acid; Vandenheuvel and Hayes (3364) use a water-alcohol mixture as stationary phase and benzene as mobile solvent, Higuchi et al. (1393) buffer the column with citrate at pH 5.4. Klenk and Bongard (1738, 1739) use a phosphate buffer for the separation of acids above adipic acid and water as stationary phase for the acids below adipic acid; good separations for dicarboxylic acids (C_2 to C_{22}) have also been described by Zbinovsky (3671) (Fig. 64).

A large number of di- and tri-carboxylic and keto- and hydroxy-acids has been separated on columns of Celite containing 0.5 N sulfuric acid, using two successive solvent systems: butanol in chloroform, followed by ethyl ether (Fig. 67) (Phares *et al.*, *2505*).

Bulen *et al.* (*448*) have used partition columns prepared by mixing 8 g silicic acid with 5.5 ml of 0.5 N sulphuric acid to separate a large number of organic acids from plant tissues; they use a "survey separation" on a first column, with a series of butanol-chloroform solvents and then separate

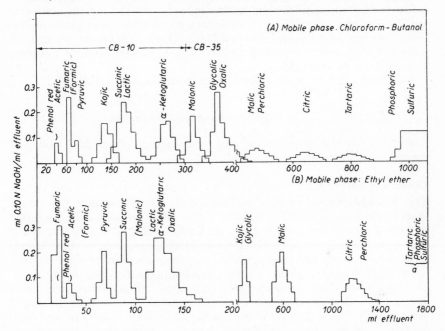

Fig. 67. Comparison of elution bands of acids with chloroform-butanol and with ethyl ether as the mobile phases on 45 × 1.0 cm Celite partition columns (20 ml per volume)

CB-10 = 10% butanol in CHCl$_3$, CB-35 = 35% butanol in CHCl$_3$, a) these acids not eluted at 2000 ml (Phares *et al.*, *2505*)

different groups on secondary columns. Kinnory *et al.* (*1722*) use 0.05 N sulphuric acid on silicic acid with a benzene-ethyl ether mixture as eluting solvent for similar acids (see also Hulme, *1479, 1480*). Scott (*2923*) has used a similar silicic acid-0.5 N sulphuric acid column, with elution by a gradient of 4-methyl-2-pentanone in methylene chloride for the separation of fumaric, succinic, oxalic, glycolic, malic and isocitric acids. Resnik *et al.* (*2670*) use acid washed silica gel containing 1 ml N sulphuric acid per gram and *tert.* butanol-chloroform as mobile phase.

The separation of pyruvic and ketobutyric acids by Isherwood's method was carried out by Wittenberg and Shemin (*3606*).

Drew *et al.* (*818*) have separated the 2,4-dinitrophenylhydrazones of pyruvic, oxaloacetic and α-ketoglutaric acids on cellulose columns (Solka Floc BW 100).

(iii) *Aromatic acids*

Bhargava and Heidelberger (*240*) have separated benzoic and naphthoic acids on columns of silicic acid with a mixture of 90% aqueous methanol and 0.5 N sulphuric acid (9:1) as stationary phase and ligroin as the mobile phase (see also Gordon and Beroza, *1176*). Ongley (*2414*) has studied the influence of hydrogen bonding on the degree of adsorption of hydroxy-benzoic acids on silica gel.

For other aromatic acids, see the chapter on phenols, p. 162.

(b) *Paper chromatography of acids*

Asselineau (*89*) has reviewed the paper chromatography of fatty acids and their derivatives.

Detection of acids on paper

The identification of organic acids depends usually on detection of the spots by spraying with pH indicators. Kalbe (*1619*) recommends a solution of 0.03% methyl red in 0.05 N borate buffer (pH 8.0) as a very sensitive reagent for the detection of acids (1–3 μg) which gives stable colours. Several authors have developed more specific spray reagents. Buch *et al.* (*442*) have used four reagents: silver nitrate-ammonium hydroxide, acetic acid-pyridine, ammonium vanadate and ceric ammonium nitrate (see Table 39). Godin (*1149*) has reported that a reagent prepared by mixing 7 volumes of pyridine and 3 volumes acetic anhydride detects *cis* and *trans* aconitic acids as brown-yellow spots, itaconic, citric and *iso*citric acids as yellow or pale yellow spots and fumaric acid as a brown spot, whilst other dicarboxylic acids do not show up.

Resnik *et al.* (*2670*) recommended the inclusion of 0.05% 8-quinolinol in the chromatographic solvent; the acids are thus disclosed in ultraviolet light as dark spots against a fluorescent background.

Martin (*2161*) has obtained distinctive colours with a ferrocyanide-ferric ammonium sulphate-ammonia spray. Barnabas and Joshi (*142*) recommend phenol-indo-2,6-dichlorophenol as a useful reagent for differentiating certain organic acids.

(i) *Lower fatty acids* (C_1–C_9)

The fatty acids C_1–C_9 were separated by various workers (Brown and Hall, *410*, *413*; Hiscox and Berridge, *1418*; Kennedy and Barker, *1689*)

Reagents and conditions of observation[a]

| Acid[b] | R values (Bromophenol blue) | | | Silver nitrate + ammonium hydroxide | | Acetic anhydride + pyridine | | Ammonium vanadate | | Ceric ammonium nitrate | |
	R_F	R_{lactic}	R_{malic}	Daylight	UV	Daylight	UV (20 hours)	Daylight	Daylight (20 hours)	Daylight or UV (10 min)	UV (20 hours)
Gluconic	0.02	0.04	0.07	Yellow	F	—	—	Yellow	Gray	Decolorized	+
Glucosaccharic	0.02	0.03	0.06	White	F	—	Yellow F	Yellow	Gray	Decolorized	—
Galacturonic	0.02	0.03	0.05	Yellow with red edge	Yellow F	—	F	Yellow	Gray	Decolorized	+
Sulphamic	0.08	0.14	0.26	White	Blue F	—	F	—	—	—	—
Glucuronic	0.10	0.15	0.29	Yellow with red edge	F	Yellow	F	Yellow	Gray	Decolorized	++
Ascorbic	0.12	0.19	0.37	Immed. black	Dark	Yellow	F	Blue gray	Gray	Decolorized	++
Tartaric	0.14	0.22	0.42	White	F	—	F	Red	Red	Decolorized	++
Glyceric	0.22	0.36	0.70	Yellow	F	—	—	—	Gray	Decolorized	—
Citric	0.23	0.36	0.70	Pink	F	—	F	Yellow	W. gray	Decolorized	++
α,γ-Dihydroxybutyric	0.29	0.46	0.91	White	F	—	—	Yellow	—	—	++
Malic	0.32	0.51	1.00	Gray	F	—	—	Yellow	Gray	Decolorized	++
α,β-Dihydroxybutyric	0.40	0.65	1.24	Yellow	Yellow F	—	—	—	Gray	Decolorized	—
Glycolic	0.43	0.67	1.30	Yellow	Yellow F	—	—	Yellow	Gray	—	—
Diglycolic	0.43	0.70	1.34	White	Yellow F	—	—	Yellow	—	—	—
Ketoglutaric	0.48	0.80	1.50	White	Pink F	—	W	Yellow	W. gray	Decolorized	—
Maleic	0.53	0.84	1.65	White	F	—	—	Yellow	—	—	—
Tricarballylic	0.53	0.85	1.65	White	F	—	—	Yellow	—	—	—
Malonic	0.53	0.86	1.67	White	F	—	—	Yellow	—	Decolorized	++
Succinic	0.61	0.98	1.93	White	White F	—	—	Yellow	—	—	—
Lactic	0.62	1.00	1.95	Yellow	Yellow F	—	—	W, yellow	Gray	Decolorized	+
Pyruvic	0.65	1.03	1.95	W. white	W. white F	Brown	F	Yellow	—	—	—
Aconitic	0.66	1.06	2.09	White	White F	—	—	Yellow	—	—	+
Citraconic	0.67	1.07	2.16	Gray	F	—	—	—	—	—	—
Itaconic	0.71	1.14	2.24	Gray	Yellow F	Orange	F	Yellow	—	Decolorized	++
α-Hydroxyisobutyric	0.75	1.20	2.32	Yellow	F	—	—	—	—	Decolorized	—
Fumaric	0.79	1.28	2.49	White	F	—	—	Yellow	—	—	—
Furoic	0.83	1.33	2.59	White	F	—	—	Yellow	—	Decolorized	—
Mesaconic	0.85	1.34	2.65	White	F	—	—	Yellow	—	—	—
Sorbic	0.88	1.35	2.80	White	F	—	—	Yellow	—	—	—

[a] F = fluorescent; W = weak; — = no reaction; + = weakly positive spot; ++ = strongly positive spot.

[b] Concentration: 15 mg/ml. Of this, 0.01 ml, or 150 γ, was used per test.

using alcohols containing ammonia. The detection of spots was carried out by spraying the paper with an indicator of the right pH range to show up the pH difference between the neutral background and the spots of the ammonium salts of the fatty acids. Bromocresol green and bromothymol blue (40 mg in 100 ml H_2O) (*413*) as well as bromophenol blue (50 mg and 200 mg citric acid in 100 ml H_2O) (*1689*) were recommended. R_F values are shown in Table 40.

Fig. 68. Separation of C_2–C_6 fatty acids from 5 μl samples of mixed solutions of their ammonium salts. Concentration of acids (μg/5μl) from left to right: acetic 93, 78, 62; propionic 77, 64, 51; *n*-butyric 76, 63, 50; *iso*valeric 68, 57, 45; *n*-hexanoic 64, 54, 43. The chromatogram was run for 18 h on Whatman no. 1 (special chromatographic) paper, with *n*-butanol-aqueous ammonia as the developing solvent (Reid and Lederer, *2662*)

The method of Hiscox and Berridge (*1418*) was further developed by Lindqvist and Storgårds (*2024*) and by Burton (*474*). Kalbe (*1619*) has used tetrahydrofuran–3N NH_4OH (4:1) for the separation of the fatty acids C_2 to C_8.

Quantitative estimation by measurement of the spot areas was achieved by Reid and Lederer (*2662*) after spraying the paper with alcoholic bromocresol purple and exposing for a short time to NH_3 vapour. Sharply defined yellow spots on a blue background appear, which can be traced onto graph paper, giving an accuracy of 2 to 4.5 % when three standards are run together with triplicate analyses on the same sheet.

This method was found satisfactory for the analysis of fatty acids in the rumen of sheep after distillation, also for the detection of fatty acids in nematode parasites (2662). However, the use of paper chromatograms seems limited by their relative inaccuracy when compared with columns, and their inability to separate isomeric acids.

TABLE 40. R_F values of volatile fatty acids

Acid		Butanol sat. with 1.5 N NH₄OH (Reid and Lederer, 2662)	100 ml 95% ethanol, 1 ml conc. NH₄OH (Kennedy and Barker, 1689)	Butanol/ethylamine H₂O (Hiscox and Berridge, 1418)
Formic	C_1	0.10	0.31	—
Acetic	C_2	0.11	0.33	0.20
Propionic	C_3	0.19	0.44	0.31
Butyric	C_4	0.29	0.54	0.44
Valeric	C_5	0.41	0.60	0.56
Caproic	C_6	0.53	0.68	0.77
Oenanthic	C_7	0.62	0.72	—
Caprylic	C_8	0.65	0.76	0.91
Pelargonic	C_9	0.67	—	—

(ii) *Derivatives of lower fatty acids*

Esters or acid chlorides can be converted quantitatively to the corresponding *hydroxamic acid* by heating with hydroxylamine (see Feigl, 940).

$$R \cdot COCl \text{ or } R \cdot COOEt + NH_2OH \xrightarrow{Alkali} R \cdot CONHOH$$

Hydroxamic acids produce an intensely red complex with ferric chloride.

$$3 R \cdot CONHOH + FeCl_3 \longrightarrow 3 HCl + Fe(R \cdot CONHO)_3$$

Fink and Fink (954) converted volatile and non-volatile fatty acids to the corresponding hydroxamates and separated them on paper. Thompson (3234) developed this technique especially for fatty acids, and by employing various solvents achieved good separation of all acids C_1–C_9 (see Table 41). A similar separation was also described by Inouye and Noda (1502). Fink and Fink's technique was employed for the identification of lactic acid by Schmid et al. (2856). (See also Lehninger and Greville, 1949.)

Another derivative of fatty acids, useful for paper chromatographic separation, is the *hydrazide*, as described by Satake and Seki (2824). The

hydrazides are formed by heating the acid with hydrazine hydrate at
100–130°.

$$\text{R} \cdot \text{COOH} + \text{NH}_2 \cdot \text{NH}_2 \cdot \text{H}_2\text{O} \xrightarrow[100-130°]{} \text{R} \cdot \text{CONHNH}_2 + 2\,\text{H}_2\text{O}$$

They can be readily detected since they reduce ammoniacal $AgNO_3$. The
R_F values are shown in Table 41; 10 μg is the detection limit for the acids.
There is usually a small difference in R_F values between isomeric acids,
but it is insufficient for separation.

TABLE 41. R_F values of derivatives of lower fatty acids

Derivative:	Hydroxamate	Hydroxamate	Hydrazide
Author:	Thompson (3234)	Thompson (3234)	Satake and Seki (2824)
Solvent:	Amyl alcohol 40 Acetic acid 10 Water 50	Benzene 75 Formic acid 75 Water 75	isoAmyl alcohol 10 Collidine 2 Water 1
C_1	0.26	0	0.11
C_2	0.34	0	0.18
C_3	0.51	0.01	0.37
C_4	0.67	0.04	0.54
C_5	0.78	0.11	0.70
C_6	0.86	0.26	0.77
C_7	0.89	0.51	—
C_8	0.89	0.77	—
C_9	0.89	0.88	—
C_{10}	0.90	0.92	—

Micheel and Schweppe (2204) have separated the hydroxamates of the
saturated fatty acids C_5 to C_{18} on cellulose acetate paper (22–26 % acetyl)
with ethyl acetate-tetrahydrofuran-water (0.6:3.5:4.7 v/v) as mobile
solvent; detection of the spots is carried out by spraying with ferric chloride
solution (see Table 42).

TABLE 42. R_F values of hydroxamates of fatty acids (Micheel and Schweppe, 2204)
        Solvent: ethyl acetate-tetrahydrofuran-water (0.6:3.5:4.7)
        Paper: acetylated Whatman No. 1 containing 25% acetyl

Acid	R_F value	Acid	R_F value
Valeric	0.84	Lauric	0.38
Caproic	0.72	Myristic	0.34
Oenanthic	0.64	Palmitic	0.30
Caprylic	0.57	Oleic	0.30
Pelargonic	0.51	Stearic	0.24
Capric	0.46	Erucic	0.22
Undecylic	0.40		

(iii) *Higher fatty acids* $(C_{10}-C_{22})$

The behaviour of high molecular weight fatty acids was studied by Kaufmann and Budwig (*1654, 1656*), who found filter paper a suitable medium for the radiometric determination of fatty acids by converting them to radioactive salts and carrying out a micro-determination of the iodine value.

Boldingh (*300*) achieved good separations of the ethyl esters on paper impregnated with rubber latex. On such rubber-coated paper strips, separations are possible with methanol and methanol/acetone as solvents. The R_F values are only approximate as they depend on the amount of rubber in the paper (see Table 43).

TABLE 43.   Approximate R_F values of ethyl esters of higher fatty acids on rubber-coated paper (Boldingh, *300*)

Ester	Methanol	Methanol - acetone 1 : 1
Ethyl stearate	0.20	0.35
Ethyl palmitate	0.28	0.60
Ethyl myristate.	0.39	—
Ethyl laurate	0.49	—
Ethyl oleate	—	0.52
Ethyl erucate	—	0.28

The *reversed phase paper chromatography* of higher fatty acids on paper impregnated with an apolar organic phase has been described by several authors. Inouye and Noda (*1504, 1505*) and Spiteri (*3072*) use paraffin oil, with methanol or methanol-acetone (3:1) or acetic acid as mobile solvent (see Table 44). Nunez and Spiteri (*2389*) and Kobrle and Zahradník (*1748*)

TABLE 44. R_F values of higher fatty acids on paper impregnated with petroleum hydrocarbons b.p. 140–170° (Inouye and Noda, *1504, 1505*)
Solvent: methanol 4, petroleum 1

Fatty acid	R_F value	Fatty acid	R_F value
Crotonic	0.88	Ricinelaidic	0.87
Hendecenoic	0.84	Stearolic	0.61
Oleic	0.52	Behenolic	0.40
Elaidic	0.54	Linoleic	0.55
Erucic	0.35	Linolenic	0.65
Brassidic	0.23	α-Eleostearic	0.60
Ricinoleic	0.85	β-Eleostearic	0.54

use triglycerides (purified olive oil) or chloronaphthalene; the spots can be developed by bromocresol green or by an ammoniacal silver spray.

Savary (*2831*) has separated the C_{12} to C_{18} acids on siliconated paper with mixtures of aqueous acetone and hexane or cyclohexane as mobile phase; the spots are detected by spraying with copper acetate and then Rhodamine B; this gives light green spots on a rose coloured background (violet and red in ultraviolet light).

Holasek and Winsauer (*1428*) have used paper impregnated by dipping in a 0.5 % aqueous potassium alum solution and then dried by 60–100°, for the separation of fatty acids from C_4 to C_{18} (mobile phase: CCl_4, methanol, conc. NH_4OH solution in volume ratio 81:81:1); the spots were developed with Rhodamine B and showed up in ultraviolet light. R_F values ranged from 0.11 to 0.80.

Savary (*2831*) has also separated the C_{12} to C_{18} acids on plain paper, with cyclohexane as solvent; the spots were detected by spraying with copper acetate solution, then with Rhodamine B.

Recently, Ashley and Westphal (*87*) have described a micromethod for the separation of 10–50 μg quantities of C_{12} to C_{24} acids on filter paper coated with paraffin oil or latex. Detection of the spots as lead sulfide or lead rhodizonate was found more sensitive than with bromothymol blue (Fig. 69).

Inouye and Noda (*1505*) have also studied the behaviour of unsaturated higher fatty acids on paper impregnated with paraffin oil; with acids of the same chain length the R_F values were in the following order: saturated < monoethenoid < diethenoid < triethenoid (conjugated) < monoethinoid < triethenoid (non conjugated) < hydroxymonoethenoid acids. Differences in R_F values between *cis* and *trans* forms were slight.

Fig. 69. Separation of 20 μg each of C_{18} and C_{20} acids. Whatman No. 1, coated with 21 % paraffin oil; mobile solvent: 90 % methanol; room temperature, development time $6\frac{1}{2}$ h (Ashley and Westphal, *87*)

Inouye *et al.* (*1506*) have described the separation of the mercuric acetate addition compounds of esters of unsaturated fatty acids formed by the reaction

$$-CH=CH- \xrightarrow{\text{Hg(OCOCH}_3)_2}_{\text{CH}_3\text{OH}} -CH(OCH_3)-CH(HgOCOCH_3)-$$

by a reversed phase paper chromatography: paper impregnated with tetralin or vaselin oil as stationary and 9 % (v/v) aqueous methanol-acetic acid-tetralin (30:1:3) as mobile phase; the spots were detected using the sensitive colour reactions with diphenylcarbazone. Good separations of a series of unsaturated C_{14} to C_{22} acids were obtained.

(iv) *Paper chromatography of non-volatile aliphatic hydroxy-, di- and poly-carboxylic acids*
The separation of the aliphatic polycarboxylic-hydroxy- and ketoacids on paper was first described by Lugg and Overell (*2057, 2058*). They suppressed tail formation by adding formic or acetic acid to the solvent, thus keeping the acids un-ionised. In view of the great biochemical interest of these acids, numerous solvents and techniques have been developed for improved or specific separations. For the detection of the acids the volatile acid of the solvent is completely evaporated and the paper sprayed with an indicator such as bromocresol green or bromophenol blue (0.04 g in 95 ml ethanol and 5 ml H_2O). When acetic acid is included in the solvent, the evaporation takes longer than with formic acid, however, in either case drying should be carried out slowly (for about 1–2 days) in a room at 40°. The only solvent used by Lugg and Overell (*2058*) which can be dried in a current of air in a few hours is the mixture of mesityl oxide-formic acid-water. Cheftel *et al.* (*556*) developed a solvent for fast evaporation of the "swamping" acid, containing a volatile constituent such as eucalyptol, which helps to evaporate the formic acid (see Table 45). Stark *et al.* (*3084*) investigated a series of solvents including chloroform, *iso*octane and phenol, all containing water and formic acid, for the separation of seventeen acids usually found in plant material.

Opieńska-Blauth *et al.* (*2418*) separated a whole series of acids with phenol-formic acid-water as solvent. Samples of pure lactic acid were found to give three spots on the chromatogram due to lactic, monolactyl lactic and polylactyl lactic acid (Loeb and Lichtenberger, *2039*; also Ohara and Suzuki, *2402*). Norris and Campbell (*2383*) separated gluconic, 2-ketogluconic and α-ketoglutaric acids with methanol and ethanol as solvents.

Bryant and Overell (*438*) have estimated the organic acids of carrot and apple extracts after adsorbing them on Amberlite IRA-400 for the elimination of non-acidic substances and elution with ammonium carbonate solution. The solvent used for the paper chromatography was 75 ml mesityl oxide, 75 ml water and 36 ml 85 % formic acid. For R_F values see (*436*); quantitative estimations were obtained by relating concentration to the weight of paper occupied by the spot.

For the separation of maleic and malonic acids see Airan and Barnabas (*25*), for the separation of succinic and *iso*succinic (methylmalonic) acid see Katz and Chaikoff (*1647*), for citric and *iso*citric acid, Cheftel *et al.* (*560*), for glyceric, malic and threonic acids, Isherwood *et al.* (*1514*), for citramalic acid, Hulme (*1479, 1480*), for quinic, shikimic acids, Hulme and Richardson (*1484, 2677*). Separations of lactic and polylactic acids have been described by Montgomery (*2256*).

Overell (*2429*) has reported the separation of hydroxy- and dicarboxylic acids by twodimensional chromatography.

Osteux and Laturaze (*2424*) have analyzed the non-volatile acids of urine; all acids were first adsorbed on Deacidite or Amberlite IR4-B and separated after elution on a twodimensional chromatogram (first solvent:

TABLE 45. R_F values of non-volatile acids

Author:	Lugg and Overell (2058)	Lugg and Overell (2058)	Cheftel, Munier and Macheboeuf (556)	Stark, Goodban and Owens (3084)	Opieńska-Blauth et al. (2418)	
Solvent:	Butanol Formic acid Water	Mesityl oxide Formic acid Water	Propanol 50 Eucalyptol 50 Formic acid 20	Phenol 3 g 90% Formic acid 1 ml Water 1 ml	Phenol-water in an atmosphere of formic acid	
Acid					Acid	
Malic	0.45	0.37	0.32	0.42	p-Amino-	
Tartaric	0.22	0.16	0.13	0.19	benzoic	0.87
Citric	0.37	0.30	0.24	0.26	Ascorbic	0.46
Succinic	0.74	—	0.64	0.66	Citric	0.26
Fumaric	0.89	0.90	0.83	0.63	Glucuronic	0.22
Pyruvic	0.66	—	—	—	Malic	0.43
Malonic	0.63	—	0.52	0.48	Mandelic	0.86
Lactic	0.67	—	0.58	0.72	Oxalic	0.33
Ketosuccinic	0.64	—	—	—	Picric	0.75
α-Ketoglutaric	0.58	—	—	—	Tannic	0.19
Monomethyl-		—	—	—	Benzoic	0.86
succinic	0.86	—	—	0.78	Aspartic	0.36
Glutaric	0.84	—	—	0.86	Fumaric	0.64
Adipic	0.89	—	—	0.18	Lactic	0.79
Oxalic	—	—	0.03	—	Malonic	0.54
Maleic	—	—	0.41	—	Nicotinic	0.89
Aconitic	—	—	—	0.36	Tartaric	0.21
Glycolic	—	—	—	0.59	Sulpho-	
Levulinic	—	—	—	0.91	salicylic	0.26
Syringic	—	—	—	0.95	Pyruvic	0.35
Tricarballylic	—	—	—	0.52		

absolute ethanol 80, conc. NH_4OH 5, water 10; second solvent: butanol 70, formic acid 30, water 120); by spraying with bromocresol green twenty seven acidic compounds were detected, eighteen of which could be identified.

For other solvent systems see Jones et al. (1597), Löffler and Reichl (2041), Scott (2923).

Cheftel et al. (557, 558) have also used a twodimensional separation: first with a hydro-alcoholic solvent made alcaline with ammonia, then with another solvent acidified with formic acid. They use a special spraying technique for dicarboxylic acids: the paper is first sprayed with bromo-cresol green then with a saturated solution of neutral lead acetate in 90 % ethanol; the acids show up as yellow-green spots on a violet background (559).

The method of Cheftel et al. (559) has been slightly modified by Nordmann et al. (2379, 2380) for the analysis of the non volatile acids of urine (which were first adsorbed on Dowex 2); R_F values for hundred and two acids are given, twenty one of which were found in normal urine.

"Rapid methods" have been described by Gore (*1178*) and Denison *et al.* (*738*).

For circular paper chromatography see Airan *et al.* (*27*), Giri *et al.* (*1119*), Scott (*2923*).

Smith and Spriestersbach (*3028*) have used filter paper coated with alginic acid for the separation of glyconic, glycaric and glycuronic acids; the incorporation of alginic acid suppresses the formation of streaked spots,

TABLE 46. R_F values of organic acids on paper chromatograms using mixtures of *n*-propanol and concentrated ammonia as developing solvents (Isherwood and Hanes, *1517*)

Temp. 20°, Whatman paper, No. 1, washed. The mixtures of *n*-propanol: conc. aqueous ammonia (sp. gr. 0.880) used as solvents were (in vol.): No. 1, 90:10; No. 2, 80:20; No. 3, 70:30; No. 4, 60:40.

Acids	x	R_F value in solvent *No.*			
		1	2	3	4
Monocarboxylic acids:					
$H \cdot (CH_2)_x \cdot COOH$	0	—	—	0.37	0.52
	1	0.13	—	0.37	0.52
	2	0.25	—	0.48	0.61
	3	0.33	—	0.57	0.69
	5	0.44	—	0.69	0.80
	7	0.52	—	0.73	0.84
	9	0.58	—	0.78	0.86
	13	0.62	—	0.82	0.88
Dicarboxylic acids:					
$HOOC \cdot (CH_2)_x \cdot COOH$	1	—	0.04	0.09	0.23
	2	—	—	0.13	0.30
	3	—	0.07	0.16	0.34
	4	—	0.083	0.19	0.39
	6	0.02	0.18	0.29	0.49
	7	0.04	0.24	0.33	0.58
	8	0.07	0.28	0.44	0.65
Other acids:					
Glycolic		—	—	0.26	0.39
Lactic (DL-)		—	—	0.395	0.48
Glyceric (DL-)		—	—	0.21	0.38
Maleic (*cis*)		—	—	0.08	0.21
Fumaric (*trans*)		—	—	0.11	0.23
Malic (DL-)		—	—	0.06	0.195
Tartaric (L-)		—	—	0.03	0.15
Tricarballylic		—	—	0.01	0.12
Aconitic (*trans*)		—	—	0.00	0.10
Citric		—	—	0.00	0.07

due to its acidity. Uronic acids, citric, tartaric and malic acids can be detected by the ammoniacal silver nitrate; the uronic acids also by reagents containing aromatic amines; spray reagents for other acids are also described.

Isherwood and Hanes (*1517*) have separated a large number of mono-, di- and tricarboxylic acids with different mixtures of propanol and aqueous ammonia (see Table 46).

For the quantitative estimation of the ammonium salts of the acids on paper, these authors use a thymol blue-glycine reagent* which is insensitive to free ammonia. The change in colour of the thymol blue is inversely proportional to the concentration of the salt from about 80 % to about 20 % neutralization of the reagent; the accuracy of the method is about ± 10 % when 50 μg of organic acid is present.

Higher aliphatic dicarboxylic acids up to C_{10} or C_{12} have been separated on paper by Isherwood and Hanes (*1517*) and by Kalbe (*1619*). The latter paper contains a very thorough study of experimental details and uses seven different solvent systems for the separation of normal and branched chain dicarboxylic acids.

(v) *Keto-acids*

α) *Separations of free acids.* The free keto-acids were separated by Wieland and Fischer (*3561*), also by Magasanik and Umbarger (*2103*), and were detected by spraying with either *o*-phenylenediamine and CCl_3COOH or 0.1 % semicarbazide and 0.15 % sodium acetate, and heating to 110°. Spots fluorescing in ultra-violet light are produced by both techniques (Table 47).

Umbarger and Magasanik (*3346*) have described a quantitative development of their method (see also Umbarger and Adelberg, *3345*). Hydroxy- and keto-acids with 5 and 6 carbon atoms were separated by Liberman *et al.* (*2004*) using toluene/acetic acid/water as solvent. As the solvent is allowed to run off the paper they record the distance travelled in 6 hours, which for the acids examined is: α-hydroxyvaleric acid, 5.3 cm; α-hydroxycaproic acid, 14.5 cm; β-hydroxycaproic, 10.5 cm; α-ketovaleric acid, 8.6 cm.

β) *Separations of the 2,4 dinitrophenylhydrazones and other derivatives.* The keto-acids present in blood and urine were separated by Cavallini *et al.* (*535*) as their 2,4-dinitrophenylhydrazones (see Table 47). In a later paper the same authors (*536*) describe the details for a clinical method by extraction of the spots and colorimetry.

The preparation and chromatographic determination of the 2,4-dinitrophenylhydrazone of oxalosuccinic acid was described by Cavallini *et al.*

---

* Add 1 ml of a 0.05 *M* solution of the Na salt of glycine (made by dissolving 3.75 g glycine in 1 l of 0.05 *N* NaOH, free from carbonate) to 450 ml of 35 % (v/v) aqueous ethanol containing 8 mg of thymol blue. Stir by a stream of CO_2-free air. To the solution add 1.4 ml of 0.1 *N* NaOH which should bring the intensity of the blue colour to about 80 % of the maximum value.

TABLE 47. R_F values of keto-acids

Author:	Magasanik and Umbarger (2103)	Cavallini et al. (535)		
Derivative:	Free acid	2,4-Dinitrophenylhydrazones		
Solvent:	Butanol 95 Formic acid 5 sat. with water	Butanol Water	Butanol 3% NH_3	Butanol 50 Ethanol 10 Water 40
α-Ketoglutaric	0.51	0.07	0.05	0.26
Oxaloacetic	0.08	0.13	0.12	0.28
Glyoxylic	—	0.17	0.24	0.32
Pyruvic	0.64	0.21	0.35	0.36
Acetoacetic	—	0.26	0.40	0.43
α-Keto-γ-methiobutyric	—	0.43	0.62	0.55
α-Ketobutyric	0.76	0.43	0.65	0.53
p-Hydroxyphenylpyruvic. . . .	—	0.60	0.64	0.55
Phenylpyruvic	—	0.59	0.80	0.66
α-Keto-isovaleric	0.83	—	—	—

(537). The keto-acids in the vitreous humour of animals and man were examined by Malatesta (2110), after previous reaction with 2,4-dinitrophenylhydrazine and two solvent extractions. The formation of oxaloacetate from pyruvate and CO_2 was investigated by Kaltenbach and Kalnitsky (1621).

Keup (1704) and Boscott and Bickel (326) have described the detection of phenylpyruvic acid in urine.

Cavallini and Frontali (534) have described their quantitative method more recently in detail and give the positions of eighteen keto-acids. Several authors have applied this or similar methods (Walker et al., 3458; Halliwell, 1282; Kulonen et al., 1829; Seligson and Shapiro, 2939; Hulme, 1480; El Hawary and Thompson, 880; Turnock, 3335). Fowden and Webb (1012) have isolated a new natural keto-acid, γ-methylene-α-oxoglutaric acid, from groundnut plants as the 2,4-dinitrophenylhydrazone, by chromatography on cellulose columns and paper sheets; see also Towers and Steward (3292).

Virtanen et al. (3416) have studied the keto-acids of green plants and described the preliminary purification of the 2,4-dinitrophenylhydrazones on a bentonite-celite column, prior to adsorption chromatography on paper in a glycine-NaOH buffer (pH 8.4).

Several of the above mentioned authors have found that pyruvic acid 2,4-dinitrophenylhydrazone gives *two spots*, when chromatographed on paper; this is due to *cis-trans* isomerism around the C=N bond, as shown by Stewart *(3112)* and Isherwood and Cruickshank *(1515)*. Stewart *(3112)* could separate the isomers on columns of bentonite or cellulose. Glyoxylic, ketobutyric, oxaloacetic and keto-glutaric acids also give two dinitrophenyl-hydrazones (see also Isherwood and Jones, *1519*; Kulonen, *1828*). The danger of confounding the second pyruvic 2,4-dinitrophenylhydrazone spot with the derivative of acetoacetic acid has been discussed by Markees *(2135)* and El Hawary and Thompson *(881)*.

Virtanen and Alfthan *(3407, 3408)* have isolated new α-ketoacids from green plants, using paper chromatography of 2,4-dinitrophenylhydrazones before and after *reduction to amino-acids* (α-keto-adipic, α-keto-pimelic, γ-hydroxy-α-keto-pimelic, hydroxy-pyruvic, β- and γ-hydroxy-α-keto-butyric acids). Towers and Mortimer *(3291)* have found 1-hydroxy-6-nitro-1,2,3-benzotriazole as an artefact on chromatograms of keto-acid 2,4-dinitro-phenylhydrazones. It is formed by the action of alkali on 2,4-dinitrophenyl-hydrazine.

Kulonen *(1827)* and Towers *et al.* *(3293)* have reduced the 2,4-dinitro-phenylhydrazones of keto-acids with aluminium amalgam or catalytically and have then separated the *amino acids* thus formed. Alfthan and Virtanen *(37)* have used tin in HCl for such reductions.

Hockenhull *et al.* *(1420, 1422)* have condensed *o*-phenylenediamine or 1,2-diamino-4-nitrobenzene with α-keto-acids to quinoxalinols which were separated on paper or alumina. The quinoxalinols derived from the latter reagent have the annexed structure in which the position of the nitro group (6 or 7) has not yet been determined.

(vi) *Aromatic acids*

Separations of aromatic hydroxy-acids were described by Evans *et al.* *(929)* in their study of phenols (see p. 163), by Bray *et al.* *(370, 371)* for the study of the fate of organic acids in the rabbit, also by Lederer *(1923)*, and Fewster and Hall *(945)*.

In general the non-ionised acids travel too fast and as a rule a solvent containing NH_3, collidine or pyridine is used to ionise the acids. Fewster and Hall *(945)* aim at buffering the paper with an ammonia-ammonium carbonate buffer, however, the R_F values are essentially the same as with butanol ammonia (see Table 48). Long *et al.* *(2045)* have used ethanol-ammonia-water and have determined the R_F values of a large number of acids (Table 50).

TABLE 48. R_F values of aromatic acids in various solvents

Acids	n-Butanol 1.5N NH₃ (NH₄)₂CO₃ buffer (Fewster and Hall, 945)	n-Butanol 1.5N NH₄OH (Lederer, 1923)
p-Aminobenzoic	0.09	0.12
m-Aminobenzoic	—	0.19
o-Aminobenzoic	0.26	0.38
p-Hydroxybenzoic	0.10	0.14
m-Hydroxybenzoic	—	0.23
o-Hydroxybenzoic	—	0.54
Benzoic	0.41	0.42
Phenylacetic	0.44	0.46
Hydrocinnamic	0.57	0.54
Cinnamic	0.62	0.54
Mandelic	0.28	—
Sulphanilic	0.11	0.13
Hippuric	0.31	—
o-Coumaric	0.27	
Gentisic	—	0.29
2,4-Dihydroxybenzoic	—	0.13
3,5-Dihydroxybenzoic	—	0.06
Phthalic	—	0.035
Anisic	—	0.35

Acids	Benzene 40 Acetic acid 40 Water 20 (Bray et al., 370)
p-Hydroxyphenylacetic	0.85
Salicyluric	0.75
2,4-Dihydroxybenzoic	0.7
m-Hydroxyhippuric	0.05
p-Hydroxyhippuric	0.05
2,3-Dihydroxytoluene	0.9–1.0
2,3-Dihydroxybenzoic	0.9–1.0
2,3-Dihydroxybenzamide	0.9–1.0
2,5-Dihydroxytoluene	0.7
2,5-Dihydroxybenzoic	0.5
3,4-Dihydroxybenzoic	0.2
Salicylic	1.0
p-Hydroxycoumaric	0.95
m-Hydroxybenzoic	0.90
p-Hydroxybenzoic	0.90

TABLE 49. R_F values of some aromatic acids and esters (Williams, *3575*)
Solvent: 2% acetic acid

	Main spot	Second spot		Main spot	Second spot
Chlorogenic acid	0.56	0.75	Methyl *p*-coumarate	0.49	0.73
Caffeic acid	0.30	0.63	Methyl *iso*ferulate	0.37	0.67
o-Coumaric acid	0.47	—	*o*-Hydroxybenzoic acid. . . .	0.72	—
m-Coumaric acid	0.44	0.80	*m*-Hydroxybenzoic acid . . .	0.70	—
p-Coumaric acid	0.40	0.73	*p*-Hydroxybenzoic acid. . . .	0.62	—
Ferulic acid	0.38	0.65	Protocatechuic acid	0.50	—
*iso*Ferulic acid	0.20	0.67	Phloretic acid	0.77	—
Methyl *m*-coumarate . . .	0.57	0.80	Phloroglucinol (R_F standard) .	0.56	—

TABLE 50. R_F values of aromatic and many other acids (Long *et al.*, *2045*)
Solvent: ethanol-ammonia-water (80:4:16)
Paper: Whatman No. 54

Acid	R_F value	Acid	R_F value	Substance	R_F value
Benzoic	0.58	Aconitic	0.16	Toluene-*p*-sulphonic	
o-Bromobenzoic . . .	0.76	Trimesic	0.0	acid	0.73
p-Bromobenzoic . . .	0.76			Catechol	0.76
m-Chlorobenzoic . . .	0.65	Formic	0.50	2,4-Dinitrophenol . .	0.76
Cinnamic	0.68	Acetic	0.52	*m*-Hydroxybenz-	
2,4-Dinitrobenzoic . .	0.60	Propionic.	0.56	aldehyde	0.85
3,5-Dinitrobenzoic . .	0.60	Butyric	0.64	*p*-Hydroxybenzaldehyde	0.69
α-Naphthoic	0.69	Valeric	0.65	*α*-Naphthol	0.84
β-Naphthoic	0.69	*iso*Valeric	0.67	*β*-Naphthol	0.82
m-Nitrocinnamic . . .	0.83	*n*-Hexanoic	0.70	*m*-Nitrophenol . . .	0.83
Protocatechuic	0.39	*n*-Octanoic	0.74	*p*-Nitrophenol	0.66
Salicylic	0.73	Trifluoroacetic . . .	0.75	Phloroglucinol . . .	0.61
o-Toluic	0.66	Chloroacetic	0.52	Picric acid	0.71
m-Toluic	0.63	Dichloroacetic . . .	0.60	Resorcinol	0.77
p-Toluic	0.63	Trichloroacetic . . .	0.70		
Phenylacetic.	0.60	Bromoacetic	0.54	Fluoride	0.27
Oxalic	0-0.15	*β*-Chloropropionic . .	0.60	Chloride	0.43
Malonic	0.26	Citric	0.11	Bromide	0.48
Succinic	0.29	Lactic	0.49	Iodide	0.53
Glutaric	0.32	Malic	0.25	Azide	0.54
Adipic	0.41	Tartaric	0.19	Chromate	0.09
Pimelic	0.44	Sulphanilic.	0.53	Nitrate	0.48
Maleic	0.31	1-Hydroxynaphthalene-		Nitrite	0.47
Phthalic	0.39	5-sulphonic	0.67	Phosphate	0.02
*iso*Phthalic	0.45	2-Hydroxynaphthalene-		Sulphate	0.09
Terephthalic	0.45	6-sulphonic	0.58	Sulphite	0.04

Reagents for aromatic acids are indicators, such as methyl red or bromo-phenol blue, diazotised sulphanilic acid, $FeCl_3$, 2,5-dichloroquinone-chloro-imide. For identification of the three hydroxybenzoic acids, see also Loebl et al. (2040). Sen and Leopold (2943) have separated a large number of aromatic acids with growth regulating properties using phenol-water, butanol-propionic acid-water or isopropanol-ammonia-water (Table 67). For urinary metabolites of phenylalanine see Boscott and Bickel (326).

Isomeric phthalic acids have been separated as hydroxamic acids on cellulose acetate paper by Micheel and Schweppe (2204); Williams (3575) has described the paper chromatography of cinnamic acid derivatives. The main spot was accompanied by a second weaker spot, with higher R_F, which was shown to be due to the *trans* isomer (see Table 49). For hippuric acid derivatives see Haberland et al. (1250), also Table 48.

See also the chapter on phenols.

(vii) *Aromatic lactones*

The paper chromatography of *coumarins* has been described by Swain (3168), Mitchell (2239), and Berlingozzi and Fabbrini (217). Jørgensen and Kofod (1605) have reported the paper chromatography of *podophyllin*.

(viii) *Heterocyclic acids and derivatives*

The detection and identification on paper of *isonicotinic hydrazide* (*isoniazide*) and other tertiary pyridine derivatives has been described by Cuthbertson and Ireland (681). Leuschner (1984) has also described a paper chromatographic method of determining isoniazide; sec.-butanol with water was used as solvent in the first dimension, then a mixture of 56 parts isoamyl alcohol, 24 parts acetone, 6 parts acetic acid and 14 parts water in the second dimension. Nicotinic acid and isonicotinic acid derivatives were differentiated by fluorescence after reaction with BrCN or BrCN-o-phenylenediamine colouring. Isoniazide was then eluted and determined with p-dimethyl-aminoazobenzene. See also Itai et al. (1523).

The separation of the isomeric pyridinecarboxylic acids has been achieved by Hashizume (1326); other natural pyridine acids detected on paper are piperidine 2-carboxylic acid (Hulme and Arthington, 1483), 4-hydroxy-piperidine-2-carboxylic acid (Virtanen and Kari, 3412) and 3-hydroxy-kynurenine (Makino et al., 2109).

CHAPTER 23

Nitrogen compounds

1. NITRO COMPOUNDS

Cruse and Mittag (*668*) have separated *m*-dinitrobenzene and *sym.* trinitrobenzene on MgO, with benzene as solvent.

Halfter (*1276*) has described the separation of di- and trinitrotoluenes on a column of MgO and Ca(OH)$_2$. Dinitrotoluene gave a green zone and trinitrotoluene a violet zone below the former.

Teague *et al.* (*3214*) have studied the order of adsorption of polynitrostilbenes; on silicic acid-Celite, adsorbability increases with the number of nitro groups, whereas on kaolin-Celite the order is almost completely reversed.

For the separation of the isomeric nitrophenols see Bunton *et al.* (*453*), and for the separation of vulcanisation accelerators see Bellamy *et al.* (*181*).

The separation of *o*- and *p*-nitroanilines and their monosubstituted derivatives on columns of alumina has been studied by Larson and Harvey (*1882*).

Schroeder (*2883*) has published a detailed study of the separation of various constituents of explosives, the mono-, di-, tri- and tetra-nitrodiphenylamines being studied in greatest detail. Many examples of the change of zone order with solvent were found with silicic acid as the adsorbent. Schroeder described a phenomenon which he called "double zoning", that is the formation of two zones with a pure substance (observed with 4-nitro-N,N-diethylaniline, N,N-diphenylurethan, carbanilide and various nitro derivatives of diphenylamine). The upper zone was usually the thinner of the two. One of the reasons envisaged for the explanation of this phenomenon was that the solvent used caused the precipitation of part of the substance (the nitro compounds having low solubilities). The author does not seem to have considered the possibility of the formation of complexes with impurities in the adsorbent (see also Schroeder *et al.*, *2890*).

Malmberg *et al.* (*2113*) have used chromatography on silicic acid to study the detection and estimation of various nitramines present as impurities in the high explosive hexogen (hexahydro-1,3,5-trinitro-*s*-triazine).

Öhmann (*2403*) has used the frontal analysis technique of Tiselius for the separation of the products obtained from the nitration of olefins (ethyl and butyl nitrates, diglycol dinitrates, and nitroglycol).

The *paper chromatography* of aromatic substances containing a NO_2 group has become of interest in relation to chloromycetin (see Chapter 37). NO_2 groups can be revealed on paper by reduction with titanous chloride, followed by diazotization and coupling with N-(1-naphthyl)-ethylene diamine dihydrochloride (Bratton-Marshall reagent) (Glazko *et al.*, *1136*).

See also: Dinitrophenylhydrazones, p. 169, Dinitrophenyl derivatives, p. 208 and Chapters 30 and 31.

2. AMIDES

The determination of *urea* on paper chromatograms has been described by Hübener *et al.* (*1469*) (detection with *p*-dimethyl amino-benzaldehyde) and by Rao and Giri (*2625*) (detection on circular chromatograms with phenol-hypochlorite).

Pyrazolone derivatives have been separated on paper by Dihlmann (*775*).

See also: Amino acids, Peptides, Proteins.

3. AMINES

(a) *Column chromatography*

Roberts and Selby (*2697*) have obtained good separations of aromatic amines (3,5-dinitroaniline, 3,5-dinitromethylaniline, 3,5-dinitrodimethylaniline) on Supercel kieselguhr impregnated with 6*N* HCl or 6 to 25*N* sulphuric acids

Sease (*2926*) has carried out the separation of various aromatic amines on fluorescent silica gel columns. The fluorescent acylating agents of Baker and Collis (*125*) should also prove valuable for separations of amines.

Aliphatic amines have been separated on ion exchange resins by Tsuda and Matsumoto (*3312*).

McIntire *et al.* (*2085, 2086*) have used cotton acid succinate for the quantitative separation of histamine.

Fuks and Rappoport (*1062*) have carried out a separation of ammonia, mono-, di- and trimethylamine, by *partition chromatography* on starch. The amines were run as the free bases and were detected and estimated by titration.

Clayton and Strong (*586*) have used Celite partition columns to separate the homologous series of volatile aliphatic amines (stationary phase: 18 ml CH_3OH, 3 ml H_2O, 2 ml of an ethanol solution of phenolphthalein; petroleum ether as mobile phase; pink bands are formed on the column). Van Duin (*3370*) has separated primary and secondary amines as *2,4-dinitrophenyl derivatives*, on columns of silica gel with nitromethane as stationary and petrol ether as mobile phase.

Gas-liquid chromatography. James *et al.* (*1540*) have described the quantitative separation and micro-estimation of ammonia and mono-, di- and trimethylamine by the technique of *gas-liquid partition chromatography* (see p. 111). 4 ft.

columns were used, packed with kieselguhr holding a mixture of undecanol and liquid paraffin (15 % w/w) as the stationary phase at a temperature of 78.6°. By varying the hydrogen bonding power of the solvent by changing the concentration of liquid paraffin, the position of the trimethylamine zone could be changed over a wide range.

In a later publication, James (1535) carried out separations of a large number of primary, secondary and tertiary aliphatic amines and also pyridine homologues (Fig. 70). By using two types of column, one having

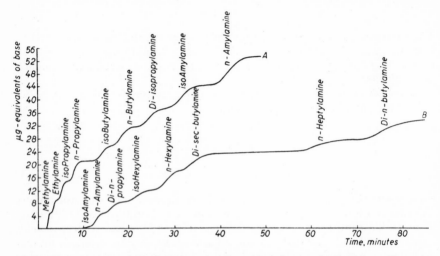

Fig. 70. The separation of 15 aliphatic amines on a 4 ft. column, with liquid paraffin as the stationary phase, at 100° (James, 1535)

Curve A, rate of flow of nitrogen, 5.7 ml/min; pressure of nitrogen, 7.5 cm Hg. Amines in order of appearance: methylamine, ethylamine, *iso*propylamine, *n*-propylamine, *iso*butylamine, *n*-butylamine, di*iso*propylamine, *iso*amylamine (overlapping with triethylamine). Curve B, rate of flow of nitrogen, 18.7 ml/min; pressure of nitrogen, 22.5 cm Hg. Amines in order of appearance: *iso*amylamine, *n*-amylamine, di-*n*-propylamine, *iso*hexylamine, *n*-hexylamine, di-*sec.*-butylamine (overlapping with di*iso*butylamine and cyclohexylamine), *n*-heptylamine (overlapping with tri-*n*-propylamine), di-*n*-butylamine

liquid paraffin as stationary phase (so that the only solution forces involved are Van der Waals forces), and the other a polyether (Lubrol MO, Imperial Chemical Industries Ltd.) so that both hydrogen bonding and Van der Waals forces operate as solution forces, distinction could be made between different types of aliphatic amines. Only primary and secondary amines are capable of hydrogen bonding with the polyether stationary phase, as an NH group is necessary for this bonding. The primary and secondary amines are held back relative to tertiary amines and show greater retention volumes than when run on a liquid paraffin column, whereas the tertiary amines show similar or lower retention volumes. By plotting the retention volumes (relative to ethylamine) of a series of amines when run on a paraffin column against those obtained on a polyether column, primary amines, secondary amines and tertiary amines are found to fall on straight lines of different increasing slope. Bifunctional amines such as hydroxyamines and diacidic amines lie on positions away from those of the monofunctional amine lines by virtue of their stronger hydrogen bonding. In this way it is possible to define closely the nature of an unknown amine. Further, with liquid paraffin as stationary phase, the position of an amine is defined only by the Van der Waals forces of interaction with the stationary phase. Since in compounds containing C, H, N and O atoms these forces depend mainly on the molecular weight it is possible to get a rough estimate of the molecular weight of an unknown amine simply from its position on the chromatogram. Of 25 different aliphatic amines studied, only two pairs (*iso*butylamine and *sec.*-butylamine, and also di*iso*butylamine and di-*sec.*-butylamine) could not be separated on either of the two 4 ft. columns.

(b) Paper chromatography

The *paper chromatography* of simple *aliphatic* and some *alicylic amines*, as hydrochlorides, in phenol, collidine or butanol has been described in detail by Bremner and Kenten (*376*). The spots can be detected by a ninhydrin spray (see Table 51). See also Bertetti (*232*) and Dihlmann (*776*).

Baker *et al.* (*126*) use *o*-acetoacetyl phenol to detect primary amines on paper. This reagent gives yellow, ultraviolet fluorescent spots.

Steiner and Stein von Kamienski (*3098*) have found the impregnation of the paper with sodium acetate, as recommended by Munier and Macheboeuf (*2323*) very useful for obtaining regular spots. Primary and secondary amines were detected with ninhydrin, tertiary amines with iodine vapours or with phosphomolybdate.

Bregoff *et al.* (*373*) have separated quaternary ammonium bases and related compounds (choline, carnitine, betaine, etc.) on paper (see Table 52); the spots are developed with $KBiI_4$. The *reineckates* of most of the bases are hydrolyzed during chromatography and give two distinct spots: one of the base, the other of free Reinecke reagent (pink spot before developement, $R_F = 0.6$ to 0.7 in 95 % ethanol + *conc*. NH_4OH (95 + 5)).

Roche *et al.* (*2709*) have described the separation of monosubstituted *guanidines* (octopine, arginine, glycocyamine, agmatine, arcaine, methylguanidine) on Whatman paper No. 1 treated with 0.06 to 0.2 N NaOH or KOH. The spots were revealed by the Sakaguchi reaction.

See also Tuppy (*3316*) and Lissitzky *et al.* (*2032*) who used paper electrophoresis in one direction and paper chromatography in a second direction.

Ekman (*876*, Table 53) has separated primary *aromatic amines* on paper, the spots being developed by diazotization (spray of 0.2 % $NaNO_2$ in 0.1 N HCl followed by a spray of N-ethyl-1-naphthylamine hydrochloride in alcohol). Burmistrov (*461*) has described similar separations for primary

TABLE 51. R_F values of amines in various solvents on Whatman No. 4 paper at room temperature (Bremner and Kenten, 376)

Solvent (% v/v):	n-Butanol	n-Butanol (40%) acetic acid (10%) water (50%)	"Collidine"	m-Cresol (50%) acetic acid (2%) water (48%)	Phenol	Phenol	Phenol	Colour with ninhydrin*
Addition:	—	—	Diethylamine	—	NH_3, HCN	Acetic acid	—	
Methylamine	0.10	0.37	—	0.52	0.96	0.91	0.72	P
Ethylamine	0.18	0.45	—	0.67	0.95	0.94	0.80	P
n-Propylamine	0.28	0.58	—	0.76	0.97	0.98	0.86	P
n-Butylamine	0.31	0.70	—	0.86	0.96	0.98	0.91	P
n-Amylamine	0.53	0.77	—	0.89	0.96	0.98	0.92	P
n-Heptylamine	0.58	0.85	—	0.93	0.95	0.98	0.94	P
isoPropylamine	0.27	0.57	—	—	—	—	—	P
isoAmylamine	0.52	0.77	—†	—	0.97	—	—	P
1,2-Diaminoethane	0.02	0.14	0.20	0.02	0.93	0.23	0.18	P
1,3-Diaminopropane	0.02	0.15	0.30	0.03	0.93	0.35	0.25	P
1,4-Diaminobutane (putrescine)	0.02	0.16	0.20	0.06(0.38)++	0.94	0.89	0.45	P
1,5-Diaminopentane (cadaverine)	0.02	0.17	0.22	0.09(0.50)+++	0.94	0.94	0.59	P
1,6-Diaminohexane	0.03	0.20	0.24	0.14(0.64)+	0.94	0.97	0.67	RP§
Benzylamine	0.38	0.68	0.79	0.86	0.95	0.98	0.91	B
β-Phenylethylamine	0.51	0.72	0.80	0.88	0.95	0.97	0.91	B
ββ-Diphenylethylamine	0.50	0.72	0.80	0.86	0.94	0.97	0.90	B
β-Phenyl-β-hydroxyethylamine	0.35	0.65	0.79	0.84	0.94	0.96	0.86	P
Adrenaline	0.14	0.45	Streak	0.58	Streak	0.91	0.74	P‖
Agmatine	0.00	0.05	0.16	0.06(0.64)++	0.94	0.60(0.93)++	0.42(0.75)+	P
Allylamine	0.03	0.50	—†	0.14	0.95	0.95	0.86	P
Ethanolamine	0.09	0.33	0.44	0.47	0.90	0.85	0.65	P
Dimethylamine	0.12	0.43	—†	—	—	0.97	0.95	P‖
Ephedrine	0.53	0.75	0.86	0.95	0.94	0.96	0.94	P
Glucosamine	0.04	0.24	0.50	0.03	0.67	0.42	0.30	P
Histamine	0.03	0.19	0.44	0.09	0.96	0.94	0.68	BG
Spermine	0.00	0.07	0.10	0.02	0.93	0.33	0.24	P
Tryptamine	0.35	0.67	0.84	0.86	0.95	0.96	0.91	GBr
Tyramine	0.30	0.62	0.84	0.70	0.95	0.96	0.85	GP

* Final colours on chromatograms run in n-butanol-acetic acid; P, purple; R, red; B, blue; G, grey; Br, brown.
† These amines were not revealed by ninhydrin in this solvent.
‡ Values in brackets refer to weaker spots detected,
+ Values in brackets refer to weaker spots detected,

TABLE 52. R_F values of methylated nitrogen bases (Bregoff et al., 373)
            Solvents: Bu-Acet = 100 ml of n-butanol + 30 ml of glacial acetic acid
            + 85 ml of water. Et-NH$_3$ = 95 ml of 95 per cent ethanol + 5 ml of con-
            centrated NH$_4$OH

| | R_F | | Reaction toward detection reagents | | | Precipitation with Reinecke solution; pH |
Compound	Bu-Acet	Et-NH$_3$	KBiI$_4$†	Levine-Chargaff	Ultraviolet, 250 mμ	
Tetramethylammonium bromide	0.49	—	I.	+		(Alkaline)‡
Cetyltrimethylammonium bromide	0.93	—	I.			(Alkaline)‡
Betaine hydrochloride	0.43	0.30	A.	—	—	Acid
Choline chloride	0.50	0.48	I.,§A.	+	—	Alkaline
Acetylcholine chloride	0.59	hydrolyzed	I.	+	—	Alkaline
Carnitine	0.48	—	A.	—	—	Acid
γ-Butyrobetaine bromide	0.52	0.13	A.	—	—	Acid
Carbomethoxypropyl-trimethylammonium bromide	0.64	—	I.	Faint	—	Alkaline
Carboethoxypropyl-trimethylammonium bromide	0.72	0.70	I.	Faint	—	Alkaline
Nicotine	0.65	0.93	I.‖		Absorbs moderately	(Alkaline)
Trigonelline	—	0.20	A.	Faint	Absorbs intensely	(Acid)
Hyoscyamine sulfate	—	0.94	I.	Faint	—	Alkaline
Hordenine sulfate	—	0.94	I.‖	Faint	Absorbs moderately	Alkaline

† I., immediate appearance of the spot; A., appearance of the spot only after drying.
‡ Reinecke precipitation was not carried out on these compounds, but the pH of the precipitation according to the scheme of Strack and Schwaneberg is given in parentheses.
§ Immediate reaction only at levels of > 50 γ. The immediate reaction is dark brown and often fades entirely or partially to give an orange spot on drying; lower levels of choline give only this latter reaction.
‖ Fades on drying, particularly if present at low levels.

arylamines, using paper exposed to vapours of HCl, formic or acetic acid, or pyridine.

See also Kariyone and Hashimoto (1630), Wankmüller (3474), Wickström and Salvesen (3543), Bertetti (233).

Vitte and Boussemart (3424) have separated local anaesthetics (novocaine, cocaine, etc.) on paper; they develop the spots with Dragendorff's reagent. See also Wagner (3441).

Erspamer and Boretti (920) have described the paper chromatography of

TABLE 53. R_F values of primary aromatic amines (Ekman, *876*)

Solvents	1	2	3	4	5
% MeOH	40	35	35	35	30.8
% Amyl alcohol	20	17.5	17.5	17.5	15.2
% Benzene	20	35	35	35	46
% H_2O	20	12.5	12.5	12.5	8
			(2N HCl)	(4% NH_3)	

Reagents: 0.2% $NaNO_2$ and 0.1N HCl, then 0.2% ethyl-α-naphthylamine·HCl in EtOH.

R_F values

	1	2	3	4	5
Sulphathiazole	0.86	0.78	0.53	0.64	0.64
Sulphanilamide	0.78	0.73	0.45	0.69	0.53
Sulphapyridine	0.87	0.82	0.53	0.73	0.72
Diaminodiphenyl sulphone	0.89	0.84	0.80	0.84	0.79
p-Aminobenzoic acid	0.88	0.56	0.79	0.41	0.71
o-Aminobenzoic acid	0.90	0.61	0.82	0.53	0.84
m-Aminobenzoic acid	0.81	0.42	0.78	0.45	0.76
p-Aminosalicylic acid	0.79	0.44	0.79	0.46	0.71
Sulphanilic acid	0.74	0.73	0.50	0.50	0.35
α-Naphthylamine	0.92	0.95	0.92	0.95	0.98
β-Naphthylamine	0.92	0.93	0.89	0.95	0.93
m-Phenylenediamine	0.62	0.75	0.35	0.71	0.50
p-Phenylenediamine	0.25	0.75	0.28	0.70	0.46
m-Toluidine	0.89	0.85	0.87	0.86	0.83
o-Toluidine	0.88	0.88	0.86	0.89	0.87
Aniline	0.88	—	0.71	0.91	0.84
p-Aminophenol	0.51	0.82	0.33	0.74	0.46

biogenic amines, such as enteramine (serotonine), octopamine, tyramine and histamine.

Paper chromatography of *histamine* has also been described by Urbach (*3352*), Bremner and Kenten (*376*) (see Table 51), Schayer (*2838*), Werle and Palm (*3507*), West and Riley (*3509*), Pénau *et al.* (*2487*), Stepanyan (*3099*).

Separations of biological arylamines on paper have been carried out by Tabone *et al.* (*3192*).

In presence of sulphate ions, diamines can give a "slow spot" due to the formation of a sulphate (Waldron-Edward, *3449*).

Pyridine bases have been separated on paper by Walker (*3459*) (see Table 63, p. 214); Jerchel and Jacobs (*1581*) have oxidized alkyl pyridines to the corresponding acids which were separated on paper; the latter authors also use the N-oxides of pyridines for paper chromatography. See Table 54. Baudet (*159*) has separated the copper complexes of picolines and lutidines on paper.

For quinoline derivatives see Smith and Williams (*3034*), Table 55.

TABLE 54. R_F values of N-oxides of pyridine bases (Jerchel and Jacobs, *1582*)
Solvent: butanol-formic acid-water (75:15:10)
Reagent: acridine in alcohol (0.005%); then viewed under short UVL

N-oxides of	R_F value	N-oxides of	R_F value
Pyridine	0.60	2,4-Lutidine	0.75
Quinoline	0.78	2,6-Lutidine	0.79
α-Picoline	0.71	2-Methyl-5-ethylpyridine	0.83
β-Picoline	0.69	Picolinic acid	0.55
γ-Picoline	0.66	Nicotinic acid	0.57
β-Ethylpyridine	0.78	*iso*Nicotinic acid	0.60
γ-Ethylpyridine	0.75	6-Methylpyridine-2-carboxylic acid	0.74

TABLE 55. R_F values of some quinoline derivatives (Smith and Williams, *3034*)
Whatman No. 1 or 4 paper was used and the chromatogram run until front had moved 12–15 in. from the origin. Solvent systems: *A*, *n*-Butanol-acetic acid-water, 4:1:5; *B*, saturated *n*-butanol-water; *C*, benzene-*n*-butanol-ammonia (sp.gr. 0.88), 2:5:2; *D*, ethyl methyl ketone saturated with water; *E*, ethyl methyl ketone-2*N* ammonia, 1:1; *F*, benzene-ethanol-water, 5:1:4; *G*, benzene-ethanol-water, 5:2:5.

	R_F in solvent system						
	A	B	C	D	E	F	G
Quinolyl-6-glucuronide	0.25	0.15	—	—	—	—	—
2-Quinolonyl-6-glucuronide	0.38	0.30	—	—	—	—	—
4-Quinolonyl-6-glucuronide	0.34	0.25	—	—	—	—	—
6-Hydroxyquinoline	0.86	0.87	0.88	—	—	—	—
2,6-Dihydroxyquinoline	0.78	0.80	0.59	—	—	—	—
4,6-Dihydroxyquinoline	0.78	0.80	0.41	—	—	—	—
6-Hydroxyquinolyl-5-sulphuric acid	0.53	0.33	—	0.68	0.20	—	—
3-Hydroxyquinoline	—	—	—	—	—	0.30	0.84
2,3-Dihydroxyquinoline	—	—	—	—	—	0.18	0.54
2,4-Dihydroxyquinoline	0.83	—	—	0.9	0.06	—	—

4. AMINO-ALCOHOLS

Pilgeram *et al.* (*2519*) have separated ethanolamine, diethanolamine and triethanolamine on a column of Dowex-50 (elution with 1.5 *N* HCl).

Amino-alcohols derived from the natural amino acids are important for the identification of C-terminal amino acids in natural peptides and proteins by the method of Fromageot *et al.* (*1055*); they are separated on paper with different solvent systems (see Table 56) and revealed with ninhydrin; quantitative determination can be achieved by oxidation with periodate.

TABLE 56. R_F values of amino-alcohols (Fromageot *et al.*, *1055*)

Solvents:	n-Butanol 77 Acetic acid 6 Water 17	n-Butanol sat. with 0.1% aq. NH_3	Phenol sat. with 0.1% aq. NH_3
Colamine.	0.18	0.15	0.74
Alaninol	0.25	0.23	0.83
Serinol	0.16	0.15	0.69
Threoninol	0.17	0.31	—
Valinol	0.40	0.49	ca. 1
Leucinol	0.53	0.63	ca. 1
Isoleucinol	0.50	0.63	ca. 1
Prolinol	0.28	0.26	ca. 1
Phenylalaninol . . .	0.54	0.70	ca. 1
Tyrosinol	0.40	0.54	0.81
Aspartidiol	0.19	0.23	0.82
Glutamidiol	0.21	0.23	0.85
Lysinol.	0.08	0.07	0.78
Argininol	0.08	0.07	—
Histidinol	0.08	0.12	—

Choline esters have been separated on Amberlite IRC-50 (Sheppard *et al.*, *2962*) and on Amberlite XE-97 (Gardiner and Whittaker, *1089*). Malyoth and Stein (*2117*) recommend ethyl acetate-pyridine-water (50:35:15, or 50:30:20) as solvent for the paper chromatography of acetylcholine and homologues; the spots can be detected with dipicrylamine in dilute acetone. The paper chromatography of acetylcholine and related substances has also been described by Augustinsson and Grahn (*103*), Whittaker and Wijesundera (*3540*) and Schümann (*2906*).

The paper chromatography of *serotonine* (5-hydroxy-tryptamine) and related compounds has been described by Erspamer and Boretti (*920*), Correale and Cortese (*636*) and Shepherd *et al.* (*2961*). Udenfriend *et al.* (*3339*) have identified 5-hydroxytryptamine in *Bufo* venom by their method of paper chromatography of the mixed ^{35}S and ^{131}I labeled pipsyl derivatives (see p. 320). Serotonine and related compounds from urine have been separated on alumina and on paper by Bumpus and Page (*451*).

2,4-Dinitrophenyl-amino-alcohols are used by several authors for the identification of aminoalcohols (obtained by reduction of amino acids at the carboxyl end of a peptide chain); they can be separated on columns of silicic acid-celite (Grassmann *et al.*, *1197*, *1198*) or of silicone-treated kieselguhr (Jutisz *et al.*, *1613*) or on paper (Jatzkewitz and Tam, *1562*), Grassmann *et al.* (*1198*).

See also Nitrogenous constituents of phosphatides (p. 272) and adrenaline (p. 405).

5. ALKALOIDS

(a) *Column chromatography*

The chromatography of alkaloids on alumina has been described in many papers (usually with benzene as solvent). (See *16, 293, 996, 1531, 2588, 2665, 3057*).

Beroza (*229*) has described the measurement of ultra-violet absorption ratios in the filtrate for control of the elution of the alkaloids. Trautner and Roberts (*3297*) used dimethylaminoazobenzene as a coloured indicator in the separation of hyoscine and hyoscyamine on silica gel columns. Jensen and Svendsen (*1578*) have separated strychnine and brucine on kieselguhr wetted with 0.2 *M* phosphate buffer using ether as the solvent. Verhaar (*3399*) has studied in detail the adsorption behaviour of the cinchona alkaloids.

For the separation of alkaloid *reineckates* see *1643* and *3546*. Prelog *et al.* (*2589*) described the separation of mixed crystals of alkaloids (dihydro-strychninolones).

Stolman and Stewart (*3131*) have separated morphine alkaloids on Florisil. Mukherjee and Sen Gupta (*2301*) and Saunders and Srivastava (*2827*) have used ion exchange columns (Zeo-Karb, Amberlite IR-100, Ionac C 284) for the purification of cinchona alkaloids.

Corynantheine can be separated from dihydrocorynantheine on a column of the sulfonated resin Duolite C 10 (Blumenthal *et al.*, *283*). Björling and Berggren (*259*) have studied the behaviour of free alkaloid bases and their salts on Decalso F.

Bottomley and Mortimer (*329*) have studied the column partition chromatography of tropane alkaloids (hyoscyamine, hyoscine, norhyoscyamine, tiglodine, valeroidine and apoatropine).

Culvenor *et al.* (*674*) have described partition chromatography of *Heliotropium* alkaloids on buffered kieselguhr.

(b) *Paper chromatography*

The *paper chromatography* of alkaloids has been reviewed by Munier (*2318*). It has been used with great success by Schmid and Karrer (*2857–2859*) and Kebrle *et al.* (*1670*) for the separation of the curare alkaloids (solvent: ethyl acetate-pyridine-water, or methyl ethyl ketone-cellosolve-water); the spots were detected with a ceric sulphate reagent or by observation of the fluorescence in ultra-violet light (Fig. 71). Munier and Macheboeuf (*2320–2323*) have published a series of studies on the paper chromatography of alkaloids, in which they described how the form of the spot depends on pK and solubility of the alkaloid, and on the pH of the solvent. Strong bases ($K > 10^{-3}$) and very weak bases ($K < 10^{-10}$) give round spots in neutral solvents; alkaloids with K between 10^{-3} and 10^{-10} give elongated trailing spots in neutral solvents; this can be overcome by choosing a suitable acid solvent. Munier *et al.* (*2324*) have obtained good separations on paper impregnated with KH_2PO_4 or $M/2\,KCl$ (*2325*). Some separations are possible with water-miscible solvents (*2322*).

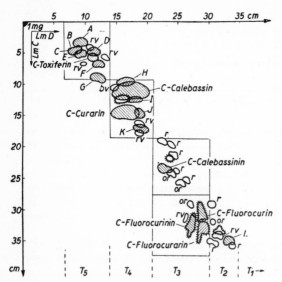

Fig. 71. Paper chromatogram of 1 mg of purified chlorides of calabash curare. The spots have been made visible by spraying with ceric sulphate or with iodine solution (Schmid *et al.*, *2859*)

◣ = fluorescence under UV light. r = red, bv = blue, rv = red-violet, or = orange ceric sulphate reaction. C-calebassinine can be only detected by spraying with iodine solution. The striped spots represent alkaloids which have been isolated. Solvents: LmC = water saturated methyl ethyl ketone + 1 to 3 % methanol. LmD = ethyl acetate-pyridine-water, 7.5:2.3:1.65

Carless and Woodhead (*509*) have used buffered filter paper ($M/15$ phosphate or citrate); the spots were detected with an iodine/KI spray. The authors point out that variations of pH and of solvent should make possible the separation of very complex mixtures. Werle and Koch (*3506*) have described the paper chromatography of nicotine alkaloids, in form of their hydrochlorides (solvent, butanol-10 % acetic acid; detection of the spots by the action of the vapour of cyanogen bromide followed by a 2 % aniline or 0.25 % benzidine solution in alcohol).

Thies and Reuther (*3229*) have described a modified $KBiI_4$ reagent for detecting alkaloids on paper.

Munier (*2319*) has separated alkaloids from their N-oxides.

Resplandy (*2671*) has studied the relation between R_F values and the concentration of electrolytes in the developing liquid; alkaloids of the same chemical group show very similar behaviour.

List (*2034*) has used the reduction of phosphomolybdate for the estimation of alkaloids on paper.

For *tobacco alkaloids* see also: Porter *et al.* (*2574*), Tso and Jeffrey (*3311*) (Table 57), Wegner (*3481*) and Tewari (*3224*).

Stoll and Ruegger (*3127*) have used paper impregnated with methyl phthalate for the separation of *ergot alkaloids*, which have also been separated on paper by Foster *et al.* (*1009*), Brindle *et al.* (*384*), Tyler and Schwarting (*3337*), Tanaka and Sugawa (*3201*), Stoll and Bouteville (*3130*), Pöhm (*2529*), Carless (*508*), and Pöhm and Fuchs (*2530*).

TABLE 57. R_F values of nicotine and related substances obtained with 50:50 *tert*-amyl alcohol and pH 5.6 buffer and colour produced with PABA-CNBr (Tso and Jeffrey, *3311*)

Substances	R_F	Colour
MAPB *	0.12	Pink
"N-Methylmyosmine" †	0.13	Lemon yellow
Nornicotine	0.17	Darker orange yellow
Anabasine	0.21	Orange yellow
Metanicotine	0.25	Rose lavender to peach red
Oxynicotine	0.30	Brick red
Nicotine	0.33	Lemon yellow to orange yellow
"N-Methylanabasine" †	0.36	Bright yellow
Dihydronicotyrine	0.36	Brownish orange
Isonicotinic acid	0.51	Lavender
3-Acetylpyridine	0.61	Light yellowish brown
Nicotinic acid	0.62	Yellowish orange
Nicotinamide	0.86	Orange yellow
β-Picoline	0.87	Rosy pink
Myosmine	0.92	Light yellow
2,3'-Dipyridyl	0.95	Light yellowish tan
Oxazine §	0.96	Pink
Nornicotyrine	0.97	Reddish lavender
Nicotyrine	0.98	Rose

* 4-Methylamino-1-(3-pyridyl)-1-butanol.

† The structure of these models has not been proved by chemical tests.

§ 2-Methyl-6-(3-pyridyl)-tetrahydro-1,2-oxazine hydrochloride.

For details of the separation of the alkaloids of the atropine, cocaine, nicotine, sparteine, strychnine and corynantheine groups, see Munier and Macheboeuf (*2323*), Drey and Foster (*819*), and Tables 58–64.

Curry and Powell (*678*) have separated twenty eight alkaloids and drugs liable to be of interest in *toxicology*, in the system *n*-butanol (50 ml), water (50 ml), citric acid (1 g) (spraying with the reagent of Munier and

TABLE 58. R_F values of atropine alkaloids (Munier and Macheboeuf, *2323*)
Paper: Durieux No. 122

	Amine acetates in n-Butanol 86, acetic acid 14 saturated with water	Amine hydrochlorides in n-Butanol 90, conc. HCl 10 saturated with water
Atropine	0.71	0.88
Hyoscyamine	0.72	0.88
Homatropine	0.64	0.82
Scopolamine	0.60	0.66
Tropine	0.43	0.51

Macheboeuf (*2323*). See also Curry (*677*), Jatzkewitz (*1561*), Dobro and Kusafuka (*795*), and Table 59.

Further papers to be mentioned concern: pomegranate alkaloids: Wibaut *et al.* (*3541*); opium alkaloids: Mannering *et al.* (*2122*) and Seibert *et al.* (*2932*); *Cytisus* alkaloids: Pöhm and Galinovsky (*2531*); *Strychnos* alkaloids: Denoël *et al.* (*739*), Jaminet (*1549*); coca alkaloids: Klementschitz and Mathes (*1736*); cinchona alkaloids: Lussman *et al.* (*2062*), de Moerloose (*734*) (Table 65). Aconitine and derivatives: Dybing *et al.* (*845*); alkaloids of *Schoenocaulon officinale* (esters of veratrine, cevacine, cevadine, veratridine, etc.): Macek *et al.* (*2078*), colchicine and derivatives: Mačák *et al.* (*2067*).

For the application of chromatography to pharmaceutical analysis (analysis of tinctures, etc.) see Van Espen (*3374*), and Lindhard Christensen and Jensen (*2017*).

TABLE 59. R_F values of narcotics (Dobro and Kusafuka, *795*)

	n-Butanol 15 Acetic acid 1 Water 4	n-Butanol 10 Acetic acid 1 Water 3
Morphine hydrochloride	0.39	0.45
Diacetylmorphine hydrochloride (heroin)	0.62	0.67
Codeine sulphate	0.49	0.53
Ethylmorphine hydrochloride (dionin)	0.60	0.65
Cocaine hydrochloride	0.64	0.77
Papaverine hydrochloride	0.77	0.82
Atropine sulphate	0.59	0.68
Quinine hydrochloride	0.82	0.88

TABLE 60. R_F values of strychnine alkaloids (Denoël *et al.*, *739*)

                Paper: Whatman No. 1

	n-Butanol 100, conc. HCl 15, saturated with water	iso-Butanol 100, conc. HCl 15, saturated with water
Strychnine	0.86	0.87
Genostrychnine	0.86	0.87
Brücine	0.70	0.66
α-Colubrine	0.87	0.87
β-Colubrine	0.84	0.84
Vomicine	0.86	0.87
Holstine	0.91	0.92
Retuline	0.94	0.96
Tubocurarine	0.46	0.32
Hypaphorine	0.85	0.77

TABLE 61. R_F values of various alkaloids at 22° on paper impregnated with $M/2$ KCl
                Solvent: butanol 98, conc. HCl 2, saturated with water
                (Munier *et al.*, *2325*)

Atropine	0.88	Hydrastinine	0.46
Scopolamine	0.66	Strychnine	0.78
Homatropine	0.76	Brucine	0.69
Tropine	0.32	Quinine	0.94
d-Cocaine	0.88	Quinidine	0.94
d-Pseudococaine	0.91	Cinchonine	0.94
Tropacocaine	0.91	Cinchonidine	0.94
Sparteine	0.86	Pilocarpine	0.44
Genisteine	0.91	Isopilocarpine	0.47
Morphine	0.49	Pilosine	0.55
Codeine	0.61	Nicotine	0.06
Thebaine	0.86	Pyridine	0.17
Heroine	0.84	Choline	0.07
Papaverine	0.94	Acetylcholine	0.08
Narcotine	0.94	Boldine	0.91
Cotarnine	0.70	Arecoline	0.41
Hydrastine	0.93		

TABLE 62. R_F values of the morphine, nicotine, sparteine and corynanthine alkaloids (Munier and Macheboeuf, *2321, 2323*)
Paper: Durieux No. 122

Solvent: butanol 80, conc. HCl 20, saturated with water

Nicotine 0.27
Pyrrolidine 0.53
Pyridine 0.36

Solvent: butanol 86, acetic acid 14, saturated with water

Sparteine 0.73
Genisteine 0.84

Solvent: butanol 70, acetic acid 30, saturated with water

Morphine 0.68
Codeine 0.75
Thebaine 0.84

Solvent: butanol 96, conc. HCl 4, saturated with water

Corynanthine 0.80
Corynantheine 0.93
Corynanthidine 0.72
Corynantheidine 0.89
Yohimbine 0.73

TABLE 63. R_F values of pyridine bases (Walker, *3459, 3460*)

Solvent: butanol/2N HCl

Pyridine 0.19
Quinoline 0.45
Isoquinoline 0.42
Collidine 0.53

TABLE 64. R_F values of various local anaesthetics, which yield coloured spots with iodobismuthate (Jaminet, *1548*)

Solvent: *iso*butanol 85, conc. HCl 15, saturated with water

Stovaine 0.96
Cocaine 0.83
Pantocaine 0.54
Novocaine 0.21

TABLE 65. R_F values of quinine alkaloids (De Moerloose, *734*)
Solvent: 5% aqueous ammonia

	Paper saturated with water	Paper air dried	Paper dried over H_2SO_4
Quinine	0.58	0.80	0.70
Quinidine	0.56	0.63	0.64
Cinchonidine	0.68	0.78	0.77
Cinchonine	0	0	0

Solvent: 5% aqueous pyridine

	Paper air dried	Paper saturated with water
Quinine	0.70	0.45
Quinidine	0.60	0.45
Cinchonidine	0.72	0.53
Cinchonine	comets in all cases	

6. OTHER NITROGEN COMPOUNDS

The chromatography of amino acids, peptides and proteins, purines, pyrimidines and their derivatives is described in special chapters. See also the sections on Synthetic dyestuffs, Phosphatides, Natural Pigments, Water-soluble Vitamins, Hormones and Antibiotics.

CHAPTER 24

Halogen compounds

(a) Column chromatography

Smith *et al.* (*3020*) have separated acetic, chloroacetic and dichloroacetic acids by adsorption on silicic acid-celite.

Haller *et al.* (*1281*) have described the chromatography of D.D.T. on alumina and have isolated a series of secondary products. Cristol *et al.* (*656*), after chromatography on alumina, succeeded in isolating the *o,o'*-isomer of D.D.T.

Alumina containing carbonate or bicarbonate gives rise to dehalogenations of steroids (Reichstein *et al.*, *2426*, *2650*) but this can be prevented by neutralisation of the alumina with acid (see also p. 61).

Ramsey and Patterson (*2616*) have described a *partition chromatographic* method for the separation of the isomeric *hexachlorocyclohexanes*. They used silicic acid to support the immobile phase of nitromethane, and *n*-hexane as the mobile phase. Good separations of the α-, β-, γ- and δ-isomers were obtained. This method has been used, with minor modifications, by Aepli *et al.* (*20*), Fouks and Chetverikova (*1061*), Contier *et al.* (*638*) and Kolka *et al.* (*1764*).

Granger and Zwilling (*1187*) have shown recently that the use of nitromethane is not essential, as moist silica gel also gives clear-cut separations. The authors state that this is a case of adsorption chromatography.

Chlorophenols can be separated on partition columns of Hyflo Super-cel with methanol-*N* sulfuric acid as immobile phase and light petroleum as moving phase (Freeman *et al.*, *1029*).

The separation of 2,4-dichlorophenoxyacetic and 2,4,5-trichlorophenoxyacetic acids (2,4-D and 2,4,5-T) by partition on silicic acid has been described by Gordon and Beroza (*1176*).

Beckman (*172*) has purified the chlorinated insecticides aldrin and dieldrin by partition on columns of silicic acid.

The *gas-liquid partition chromatography* of alkyl halides was described by James and Martin (*1538*, *1539*); isomeric aliphatic chlorides separate readily as shown in Fig. 72.

Mono-, di-, tri- and tetrachloromethanes were separated by Purnell and Spencer (*2603*).

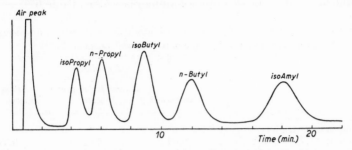

Fig. 72. The separation of alkyl chlorides. Column length, 4 ft.; stationary phase, benzyl-diphenyl; temperature, 100°; rate of flow of nitrogen, 11.8 ml/min; pressure of nitrogen, 12.5 cm Hg (James and Martin, *1539*)

(b) *Paper chromatography*

Schmeiser and Jerchel (*2855*) point out the possibilities of neutron irradiation of developed paper chromatograms for the detection of spots containing either S, Br or Cl.

Decay curves for these elements show that in all three cases distinctive radioactivities of reasonable half lives are produced (from S mainly. ^{32}P (14.3 days) with some ^{31}Si (2.8 h), from Br ^{82}Br (4.4 h) and from Cl ^{34}Cl (33 min)). A weak background of ^{14}C is also formed.

Alkyl halides can be identified by conversion into alkyl *iso*thiuronium salts which are identified by paper chromatography (*n*-amyl alcohol: acetic acid: water, 45:15:10) (Večeřa and Gasparič, *3385*). This method allows the identification of O-alkyl and N-alkyl groups which are easily transformed to alkyl halides by boiling HI (Table 66).

TABLE 66. R_F values of alkyl *iso*thiuronium salts (Večeřa and Gasparič, *3385*)
Paper: Schleicher and Schüll, 2034b; technique ascending; solvent *n*-amyl alcohol: acetic acid: water (45:15:10)

Alkyl	R_F
Methyl	0.33
Ethyl	0.46
Propyl	0.60
Butyl	0.72
Amyl	0.77

Chlorophenols can be separated on paper, either as free phenols or as chlorophenoxyacetic acid (Siegel and Schlögl, *2978*) or after coupling with sulphanilic acid (Johnson et al., *1591*).

Sen and Leopold (*2943*) have separated halogenated aromatic acids with growth-regulating properties on paper, using phenol-water or butanol-propionic acid-water, or *iso*propanol-ammonia-water (Table 67).

TABLE 67. R_F values of some halogenated aromatic acids with growth-regulating properties and allied compounds (Sen and Leopold, *2943*)

Compound	R_F value in		
	Phenol-water	Butanol-propionic acid-water	iso Propanol-ammonia-water
A. Benzoic acid derivatives:			
Benzoic acid	0.89	0.92	0.55
o-Bromobenzoic acid	0.86	0.93	0.55
m-Bromobenzoic acid	0.91	0.93	0.66
p-Bromobenzoic acid	0.86	—	0.59
o-Chlorobenzoic acid	0.76	0.93	0.58
m-Chlorobenzoic acid	0.85	0.94	0.69
p-Chlorobenzoic acid	0.90	0.91	0.64
2,4-Dichlorobenzoic acid . . .	0.82	0.95	0.73
3,4-Dichlorobenzoic acid . . .	0.75	0.94	0.78
2,3,5-Triiodobenzoic acid . . .	—	0.92	0.78
m-Hydroxybenzoic acid	0.78	0.90	0.21
p-Hydroxybenzoic acid	0.76	0.92	0.32
2,4-Dihydroxybenzoic acid . . .	0.60	0.91	0.37
o-Aminobenzoic acid	0.85	0.92	0.30
m-Aminobenzoic acid	0.86	0.86	0.14
p-Aminobenzoic acid	0.81	0.96	0.16
2,5-Dinitrobenzoic acid	0.80	0.80	0.52
3,5-Dinitrobenzoic acid	0.80	0.86	0.59
2-Amino-5-chlorobenzoic acid .	0.79	0.94	0.44
2-Chloro-5-nitrobenzoic acid .	—	—	0.59
B. Phenylacetic acid	0.83	0.92	0.51
C. Phenoxyacetic acid derivatives:			
Phenoxyacetic acid	—	0.91	0.67
o-Chlorophenoxyacetic acid . .	—	0.89	0.60
p-Chlorophenoxyacetic acid . .	—	0.89	0.56
2,4-Dichlorophenoxyacetic acid .	0.83	0.91	0.67
3,4,5-Trichlorophenoxyacetic acid	0.76	0.94	0.80
D. Naphthalene derivatives:			
α-Naphthaleneacetic acid . . .	0.93	0.94	0.58
β-Naphthoxyacetic acid	0.90	0.91	0.60

The paper chromatography of the isomeric benzenehexachlorides has been described by Mitchell (*2237*).

Moynihan and O'Colla (*2299*) have described the paper chromatography of chlorinated insecticides.

The paper chromatographic separation and identification of chlorinated organic pesticides (aldrin, isodrin, dieldrin, endrin, dilan, technical D.D.T., rhothane, methoxychlor and lindane) has been studied in detail by Mitchell (*2240, 2242, 2244*).

Winteringham *et al.* (*3602*) have described the reversed phase paper partition of D.D.T. and its derivatives, on paper impregnated with vaselin. (See also also Gruch (*1236*) and Fig. 56, p. 144.)

[82]Br-containing derivatives of D.D.T. have also been separated by Winteringham *et al.* (*3602*) on vaselin-impregnated paper.

Radioactive iodine (^{131}I) is used in a great number of chemical and biochemical studies on iodinated amino acids. The different iodinated spots are easily detected by radioautography (for further details see the chapters on halogenated amino acids (p. 324) and on thyroid hormones (p. 406).

CHAPTER 25

Sulphur compounds

(a) Column chromatography

The frontal analysis of sulphides has been described by Hurd *et al.* (*1488*).
Mercaptans can be selectively adsorbed on mercurated phenol-form-aldehyde resins (see p. 98).

Rovery and Desnuelle (*2748*) have described the chromatography of 3-phenyl-2-thiohydantoins (obtained in Edman's method of peptide degradation) on columns of silica gel (with 17 % *iso*propanol in cyclohexane as solvent).

Sease and Zechmeister (*2928*) separated di-, tri-, and tetra-thienyls (I) and their homologues, the position of the zones being detected by their fluorescence in ultra-violet light on alumina.

(I)

(b) Paper chromatography

The paper of Schmeiser and Jerchel (*2855*) concerning the detection of S, Br or Cl containing compounds on paper by neutron irradiation has already been mentioned (p. 217).

Many sulphur-containing substances (e.g. amino acids) can be detected on paper by their ability to catalyze the oxidation of sodium azide by iodine: $2NaN_3 + I_2 \rightarrow 2NaI + 3N_2$ (Feigl, *940*); after drying the paper for 30 min at 95°, it is sprayed with 0.05 N aqueous iodine containing 1.5 % NaN_3. White spots on a light brown background indicate positions of sulphur-containing substances (Chargaff *et al.*, *548*).

Kariyone and Hashimoto (*1629*) have described the paper chromatographic separation of potassium xanthates of simple alcohols (ROCSSK) as a means of separating and identifying the latter. The spots can be detected by their dark brown fluorescence in ultra-violet light, or by spraying with

Grote's reagent. This reagent is prepared by treating a solution of sodium nitro-ferricyanide in sodium bicarbonate, first with hydroxylamine and then with bromine.

The junior author of this book has found an acid solution of ammonium molybdate a very satisfactory spray for xanthates.

Naphthylamine-sulphonic acids, important intermediates for the synthesis of dyestuffs, have been separated on paper by Latinak (Table 68).

TABLE 68. R_F value of naphthylamine-sulphonic acids (Latinak, *1885*)

			R_F values	
Name	Position of NH_2	Position of SO_3H	Butanol 3 Pyridine 1 Water 1	Butanol 4 Acetic acid 1 Water 5
Koch acid	1	3,6,8	0	0
Freund acid	1	3,6	0.04	0.02
Amino-ε-acid	1	3,8	0.08	0.14
1-Naphthylamine-2- sulphonic acid	1	2	0.63	0.57
	1	3	0.47	0.41
Naphthionic acid	1	4	0.42	0.29
Laurent acid	1	5	0.44	0.26
Cleve acid	1	6	0.48	0.34
Cleve acid	1	7	0.54	0.46
Peri acid	1	8	0.63	0.69
1-Naphthylamine	1	0	0.92	0.95
Amino-R-acid	2	3,6	0.08	0.03
C-acid	2	4,8	0.06	0.04
	2	1,6	0.05	0.03
Amino-G-acid	2	6,8	0.04	0.03
Tobias acid	2	1	0.61	0.55
Dahl acid	2	5	0.47	0.34
Brönner acid	2	6	0.46	0.32
Amino-F-acid	2	7	0.47	0.33
Baden acid	2	8	0.51	0.48
2-Naphthylamine	2	0	0.92	0.94

Longenecker (*2047*) has described the paper chromatographic separation of sulphonamides, sulphones and their metabolic products on narrow strips of paper; the spots are developed with *p*-dimethylamino-benzaldehyde (Ehrlich's reagent) or by diazotizing. Paper chromatography of sulphonamides has also been described by Thomas *et al.* (*3231*), by Robinson (*2707*; Table 69), Steel (*3086*), San and Ultée (*2805*), Bray *et al.* (*368*; Table 70), Wagner (*3440, 3443*), Heinänen *et al.* (*1357*), De Reeder (*746*) and Řybár *et al.* (*2767*).

Schulman and Keating (*2903*) have studied the excretion of radioactive

TABLE 69. R_F values of sulphonamides (Robinson, 2707)

a) Acid solvent:

50 ml n-butanol is mixed with 15 ml glacial acetic acid and 60 ml distilled water. The mixture is well shaken in a separating funnel and allowed to separate. The lower aqueous layer is discarded and sufficient p-dimethylaminobenzaldehyde added to the remaining solvent to give a concentration of approximately 0.5%.

b) Basic solvent:

40 ml n-butanol is mixed with 10 ml ammonium hydroxide and 30 ml water. The mixture is shaken up as with the acid solvent, separated and p-dimethylaminobenzaldehyde added to the organic layer to give a concentration of approximately 0.5%.

Paper used: Whatman No. 1.

Sulphonamides	n-Butanol/ammonia	n-Butanol/acetic acid
Sulphaguanidine	0.43	0.36
Sulphathiazole	0.48	0.46
Sulphanilamide	0.56	0.42
Sulphadiazine	0.24	0.50
Sulphamezathine	0.48	0.48
Sulphamerazine	0.34	0.50
Sulphacetamide	0.71	0.44
Sulphapyridine	0.70	0.51

material after the injection of ^{35}S-labelled thiourea; radioautograms of paper chromatograms show that the major portion of thiourea is excreted unchanged.

Kjaer and Rubinstein (1730) have identified naturally occurring isothiocyanates by paper chromatography of the corresponding thioureas (see Table 71).

For R_F values of thiouracils, see Table 72 and Reinhardt (2668).

2-Thiohydantoins (II) derived from amino acids have been separated on paper by Edward and Nielsen (869) and Dautrevaux and Biserte (709). Sjöquist (3003, 3004) and Landmann et al. (1876) have separated on paper the 3-phenyl-2-thiohydantoins (III) obtained from various amino acids by the method of end group analysis of Edman (865).

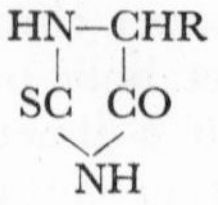

```
    HN—CHR          HN—CHR
    |    |           |    |
    SC   CO          SC   CO
      \  /             \  /
       NH               N
                        |
                        C₆H₅

      (II)             (III)
```

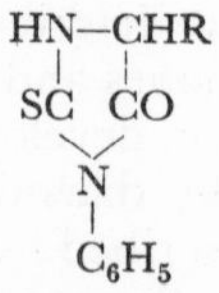

See also the sections on amino acids, synthetic dyestuffs, water soluble vitamins, and antibiotics.

TABLE 70. R_F values of sulphanilic acid, sulphamezathine and derivatives (Bray et al., 368)

Solvents: A n-Butanol 4 vol, pyridine 8 vol, sat. NaCl 5 vol, NH$_4$OH 3 vol.
B Collidine saturated with water.
C n-Butanol 20 vol, NH$_4$OH 3 vol.
D isoPropanol 4 vol, NH$_4$OH 2 vol, sat. NaCl 3 vol.
E Ethanol 96% 20 vol, NH$_4$OH 1 vol.
F n-Butanol 5 vol, acetic acid 1 vol, H$_2$O 4 vol.
G Ethyl acetate 10 vol, acetic acid 3 vol, H$_2$O 7 vol.
H Ethanol 96% 1 vol, acetone 3 vol.
J n-Butanol 1 vol, pyridine 1 vol, sat. NaCl 2 vol.

Paper: Whatman No. 4.

Compound	Colour with diazotized p-nitraniline	Colour after diazotizing and coupling with ethyl-naphthylamine	Reduction with ammoniacal AgNO$_3$	R_F values in solvents					
				A	B	C	D	E	F
Sulphanilic acid	pale red	red purple	—	0.63	0.51	0.10	0.44	0.50	0.21
Sulphanilamide	very pale red	red purple	—	0.96	0.95	0.77	0.88	0.90	0.71
2-Hydroxysulphanilic acid	blue purple	red purple	—	0.73	0.74	0.08	0.18	—	0.27
2-Hydroxysulphanilamide	blue	red purple	—	0.81	0.93	0.06	0.28	0.21	0.61
3-Hydroxysulphanilic acid	grey purple	orange	+	0.62	0.57	0.04	0.13	0.31	0.26
3-Hydroxysulphanilamide	grey purple	orange	+	0.86	1.00	0.22	0.41	0.40	0.63
Sulphamezathine	pale red	red purple	—	0.80	0.95	0.55	0.60	0.69	0.90

Compound	Colour with ammoniacal AgNO$_3$ after		heating at 110°	R_F values				
	5 min	30 min		A	C	G	H	J
2-Hydroxy-4,6-dimethylpyrimidine	—	pale grey	red brown	0.69	0.45	0.60	0.72	1.00
2,5-Dihydroxy-4,6-dimethylpyrimidine HCl	—	pale grey	bright silver	0.36	0.10	0.10	0.23	0.15
2-Amino-5-hydroxy-4,6-dimethyl-pyrimidine HCl	dark grey	dark grey	merge into background colour	0.74	0.35	0.91	0.83	0.90
3-Hydroxysulphanilic acid	dark grey	dark grey	merge into background colour	0.62	0.04	0.04	0.35	0.22
3-Hydroxysulphanilamide	dark grey	dark grey	merge into background colour	0.86	0.22	1.00	0.95	1.00

TABLE 71. R_F values of N-substituted thioureas (Kjaer and Rubinstein, *1731*)
Solvent: chloroform saturated with water.
Conditions: Whatman No. 1 at 23.5°, the paper being equilibrated for 16–18
hours before chromatography

N-substituents	R_{Ph} value*	N-substituents	R_{Ph} value*
Methyl	0.04	Allyl	0.26
Ethyl	0.15	Benzyl	0.90
n-Propyl	0.43	*sym.* Dimethyl	0.41
n-Butyl	0.84	Trimethyl	1.22

* $R_{Ph} = R$ with N-phenylthiourea as reference substance.

TABLE 72. R_F values of thiouracils (Lederer and Silberman, *1943*)

Solvent	Butanol/water	Amyl alcohol/water
Thiouracil	0.61–0.62	0.53–0.57
Methylthiouracil	0.69–0.72	0.64–0.70
Propylthiouracil	0.90–0.93	0.93–0.95

CHAPTER 26

Phosphoric esters

1. PHOSPHORIC ESTERS OF DIOLS, TRIOLS, POLYOLS AND ACIDS

Goodman (*1159*) has described the separation of various phosphates (of glycol, glyceric acid, glucose, etc.) by elution from a Dowex-1 column (see also Bartlett and Marlow, *149*).

Bublitz and Kennedy (*441*) have described the separation of α- and β-glycerophosphoric acid by adsorption on a column of Dowex-1 and elution with a gradient of acetate buffer pH 6.0 in distilled water. The α-isomer is eluted between 232 and 256 ml, the β-isomer from 256 to 288 ml.

The paper chromatography of glycero-phosphoric acid and 3-phosphoglyceric acid has been described by Hanahan (*1291*); Aronoff (*82*) has separated mono- and diphosphoglyceric acids labelled with ^{14}C (detection by autoradiography). Cowgill (*639*) has separated 2-phospho- and 3-phosphoglyceric acids.

Smith and Clark (*3016, 3017*) have separated *inositol phosphates* from soil on De-Acidite.

The paper chromatography of inositol phosphates has been reported by Anderson (*55*).

Stadtman and Barker (*3080*) have obtained a quantitative separation and estimation of the various *acyl phosphates* (acetyl, propionyl, butyryl, valeryl, caproyl phosphates) formed during fatty acid synthesis by enzyme preparations of *Clostridium*. They converted the acyl phosphates to their hydroxamic acid derivatives, which were chromatographed on paper (developing solvent: water-saturated butanol). After separation, the spots were located by spraying with a ferric chloride solution; the hydroxamic acids are thus converted to their highly coloured ferric complexes; these can be extracted from the paper and estimated quantitatively by colorimetry.

2. PHOSPHORIC ESTERS OF CARBOHYDRATES

McCready and Hassid (*2070*) have purified glucose-1-phosphate (Cori-ester) by passing the solution first through a cation exchanger (Amberlite

IR-100) to remove cations, then through an anion exchanger (Amberlite IR-4) on which the Cori-ester is adsorbed as it is a fairly strong acid; the impurities such as dextrins, proteins and weak organic acids pass out in the filtrate. The ester was then desorbed with alcohol.

Benson *et al.* (*188*) have purified fructose-6-phosphate, fructose-1,6-di-phosphate and 3-phosphoglyceric acid on Dowex A-1. Horecker *et al.* (*1446*) have separated pentose phosphates on Dowex-1.

Groth *et al.* (*1232*) have purified ribose-5-phosphate on a Dowex-1 column in the monochloracetate form; elution was carried out with a

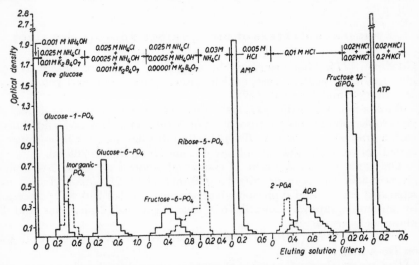

Fig. 73. Ion exchange separation of sugar phosphates, inorganic phosphate, adenosine phosphates and phosphoglyceric acid. Exchanger, Dowex-1, ca. 300 mesh, 0.86 sq.cm × 12 cm, chloride form; rate 3.5 ml/min (Khym and Cohn, *1709*)

gradient of 1 M monochloroacetic acid (see also Byrne and Lardy, *491*). Khym *et al.* (*1712*) have separated isomeric ribose phosphates on Dowex-1-sulphate in presence of borate. Khym and Cohn (*1709*) had shown previously that borate complexing can be used to effect separations of monophosphory-lated sugars on Dowex-1 in an alkaline chloride system. Diphosphates (e.g. fructose diphosphate) or acid monophosphates (e.g. phosphoglyceric acid) and the adenosine polyphosphates are separated from the sugar monophosphates and from each other by simple pH and ionic strength adjustment (Fig. 73).

Hanes and Isherwood (*1297*) were the first to study the separation of biologically important phosphoric esters on *filter paper* (Fig. 74). Various solvent systems were used, such as acidic water-immiscible mixtures (*tert.*-amyl alcohol, water, formic acid), acidic water-miscible solvents (*tert.*-butanol, water, picric acid), or basic water-miscible systems (ethyl acetate, pyridine, water). Tertiary alcohols were used to prevent esterification with the acid component. The authors recommended purifying the filter paper by washing with $2N$ acetic acid followed by rinsing with water; the elimination of metals was carried out either with H_2S or with oxine. The

phosphoric esters are detected by spraying the paper with an acid molybdate solution followed by heating under conditions causing hydrolysis of the esters without undue decomposition of the paper itself. The free orthophosphoric acid produced from the esters forms a phospho-molybdate complex with the molybdate present, and this is reduced to an intensely blue compound by treating the paper with H_2S. For quantitative use one can cut out the spot, ash the paper and determine the phosphorus by the method of Berenblum and Chain. Hanes and Isherwood (1297) have successfully separated glucose and fructose phosphates, as well as adenosine triphosphate. Using a similar method (detection with acid ammonium molybdate and ascorbic acid) Leuthardt and Testa (1985) have studied the phosphorylation of fructose in the liver. See also Cohen and Scott (595) and Table 73.

A two-dimensional method for the identification of phosphate esters of biological importance, including inorganic polyphosphates, has been developed by Bandurski and Axelrod (133); an acid solvent consisting of methanol, formic acid and water, and

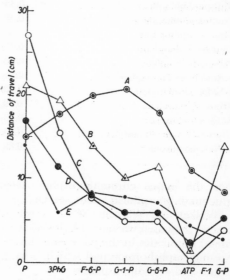

Fig. 74. Movement of free orthophosphate and phosphoric esters in different solvents. All chromatograms on No. 4 Whatman paper (purified) (Hanes and Isherwood, 1297).

A solvent: propanol/ammonia/water
     running time 15 h
B solvent: tert.butanol/water/picric acid
     running time 15 h
C solvent: isopropyl ether/90 % formic acid
     running time 6 h
D solvent: tert.amyl alc./water/formic acid
     running time 15 h
E solvent: ethyl acetate/pyridine/water
     running time 23 h
3PhG = 3-phosphoglyceric acid; F-6-P = fructose-6-phosphate; G-1-P = glucose-1-phosphate, etc.

a basic solvent containing methanol, ammonia and water were used. These authors have also described an improved method of colour development involving ultra-violet light, which permits hydrolysis of resistant esters and the minimization of background colour. A similar method has been described by Burrows et al. (464). See also Dulberg et al. (827).

TABLE 73. R_F values of sugar phosphates (Cohen and Scott, *595*)

	80% ethanol containing 0.8% acetate at pH 3.5	*80% ethanol containing 0.64% boric acid*
Glucose-6-phosphate	0.35	0
Fructose-6-phosphate . . . . :	0.38	0
Glucose-1-phosphate	—	0
Glucose-4-phosphate	0.45	0
Ribose-5-phosphate	0.50	0
D-Arabinose-5-phosphate . . .	0.54	0.25
D-Xylose-5-phosphate	0.55	0.0 and 0.25
Ribose-3-phosphate	0.50	0.0 and 0.19
Xylose-3-phosphate	0.53	0.0 and 0.23
Glyceraldehyde-3-phosphate . .	0.73	0.92
6-Phosphogluconate	0.89	—

For the paper chromatographic detection and separation of 2-keto-gluconate phosphoric esters, see De Ley (*730*).

Walker and Warren (*3454*) investigated phosphate esters in tissue extracts; they used versene (ethylenediamine tetra-acetic acid) for complexing metal impurities in the paper and have developed a spraying reagent containing naphthoresorcinol which is said to be specific for fructose-6-phosphate and -1,6-diphosphate.

Wade and Morgan (*3436*) detect phosphoric esters on paper by spraying with 0.1 % $FeCl_3 \cdot 6 H_2O$ in 80 % ethanol and then with 1 % sulphosalicylic acid in 80 % alcohol; this gives white spots on a pale mauve background and has the advantage over the usual method that even difficultly hydrolysable phosphates can be detected.

Another method which overcomes the difficulty of detecting the more acid-stable phosphates is described by Fletcher and Malpress (*982*) who use a solution of alkaline phosphomonoesterase to liberate orthophosphate from the esters; the spots of the phosphoric acids are then made visible by a spray with 5 % ammonium molybdate in 20 % HCl followed by benzidine and ammonia. Mortimer (*2286*) has used the following solvent systems for the separation of biologically important phosphate esters: ethyl acetate, acetic acid, water (3:5:1), methyl-cellosolve, methyl ethyl ketone, 3N NH₄OH (7:2:3) and ethyl acetate, formamide, pyridine (6:4:1). (The combinations of solvent 1 + 3 or 2 + 3 are used for twodimensional separation.) See also: Doman and Kagan (*801*), Opieńska-Blauth (*2416*), De Ley (*729*).

Wade and Morgan (*3436*) have described a two-dimensional fractionation of organic phosphates by a combination of paper ionophoresis (on N formic acid washed Whatman No. 3 paper impregnated with an aqueous solution of 9.2 % (v/v) n-butyric acid and 0.1 % NaOH at 400 V for 6 hours) and paper chromatography with an aqueous solution of 69 % (v/v) n-butyric acid and 0.85 % (w/v) NaOH in water saturated atmosphere at 20° for 3 days (Table 74).

TABLE 74. The movement of phosphates during paper electrophoresis and chromatography (Wade and Morgan, *3437*)
Electrolyte: 9.2% butyric acid – 0.1% NaOH with 400 V for 4.5 hours.
Solvent for chromatography: 69% (v/v) butyric acid – 0.85% (w/v) NaOH at 20° for 3 days.
M_o indicates the movement of phosphates relative to orthophosphates after ionophoresis.

	M_o	R_F
Inorganic phosphates		
1. Orthophosphate	1.00	0.36
2. Pyrophosphate	1.42	0.25
3. Triphosphate ($Na_5P_3O_{10}$)	1.63	0.16
4. Trimetaphosphate ($Na_4P_3O_9$)	2.05	0.10
5. Tetrametaphosphate ($Na_4P_4O_{12}$)	1.63	0.10
Nitrogen-free organic phosphates		
6. Acetyl phosphate	0.80*	0.52*
7. Glycolaldehyde phosphate	1.05	0.13
8. Glycolic acid phosphate	1.23	0.09
9. α-Glycerophosphate	0.78	0.35
10. β-Glycerophosphate	0.81	0.38
11. Glyceraldehyde phosphate	0.74	0.36
12. Dihydroxyacetone phosphate	0.81	0.36
13. Propane-2,3-diol α-phosphate	0.83	0.53
14. Glyceric acid 3-phosphate	1.07	0.26
15. Glyceric acid 2,3-diphosphate	1.32	0.15
16. Enolpyruvic phosphate	1.20	0.35
17. Ribose 5-phosphate	0.70	0.27
18. Glucose 1-phosphate	0.64	0.23
19. Glucose 6-phosphate	0.64	0.20
20. Gluconic acid 6-phosphate	0.80	0.19
21. Fructose 6-phosphate	0.62	0.24
22. Fructose 1,6-diphosphate	1.00	0.12
23. Mannose 6-phosphate	0.64	0.24
24. Galactose 6-phosphate	0.64	0.19
25. Sedulose 7-phosphate	0.64	0.20
Nitrogen-containing organic phosphates		
26. Ethanolamine phosphate	0.00	0.58
27. Choline phosphate	0.00	0.73
28. Creatine phosphate	0.61*	0.50*
29. Pantothenic acid 2-phosphate	0.64	0.62
30. Pantothenic acid 4-phosphate	0.64	0.57
31. Uridine 2'-phosphate	0.55	0.38
32. Uridine 3'-phosphate	0.55	0.39

* Dephosphorylates slowly during separation.

Continued p. 230

TABLE 74 *(Continued)*

	M_o	R_F
33. Uridine 5'-phosphate	0.55	0.28
34. Uridine 5'-triphosphate	1.12	0.16
35. Cytidine 2'-phosphate	0.13	0.65
36. Cytidine 3'-phosphate	0.13	0.59
37. Guanosine 2'-phosphate	0.42	0.38
38. Guanosine 3'-phosphate	0.42	0.36
39. Adenosine 2'-phosphate	0.23	0.72
40. Adenosine 3'-phosphate	0.23	0.70
41. Adenosine 5'-phosphate	0.21	0.63
42. Adenosine 5'-diphosphate	0.58	0.50
43. Adenosine 5'-triphosphate	0.86	0.41
44. Inosine 5'-phosphate	0.49	0.33
45. Inosine 5'-diphosphate	0.82	0.23
46. Inosine 5'-triphosphate	1.00	0.17
47. Pyridoxal phosphate	0.32	0.45
48. Pyridoxamine phosphate	0.20	0.78
49. Thiamine pyrophosphate	0.00	0.89
50. Flavin mononucleotide	0.36	0.43
51. Flavin-adenine dinucleotide	0.36	0.52
52. Diphosphopyridine nucleotide	0.13	0.56
53. Dihydrodiphosphopyridine nucleotide . .	0.48 †	0.40 †

† Oxidizes slowly during separation.

Ganguli *(1078)* has described the circular paper chromatography of phosphate esters with butanol, acetic acid, water (4:1:5).

Working with ^{32}P labelled phosphate esters, Benson *et al.* *(188)* have obtained autoradiograms on paper of different intermediates in photosynthesis.

3. NITROGEN-CONTAINING PHOSPHORIC ESTERS

Phosphoserine has been isolated from a Dowex-50 column by De Verdier *(761)* and from a Dowex-1 column by Kennedy and Smith *(1690)*. The paper chromatography of phosphoserine has been described by Dent *(741)* and by Ågren *et al.* *(22)*.

Aminoethyl phosphoric ester from rat organs and human tumours has been isolated and detected on paper chromatograms with ninhydrin (Awapara *et al.*, *105*; Ansell and Dawson, *70*; Dent, *741*). Glycerylphosphoryl-ethanolamine from liver and yeast extracts was isolated by Campbell *et al.* *(501)* using paper columns to separate it from taurine and ethanolamine.

Viscontini *et al.* *(3422)* have reported the paper chromatographic separation of thiamine, and its mono-, di- and triphosphates; the detection of the spots was made with ultra-violet light. Forrest and Todd (*1002*) have isolated pure synthetic riboflavin 5'-phosphate by using a paper chromatopile. Phosphates of pantothenic acid can be separated and detected by the method of Hanes and Isherwood (*1297*) (Baddiley and Thain, *120*). (See also Table 74 and the chapter on Vitamins.)

Numerous recent papers concern the separation and quantitative estimation of *nucleotides*; these will be referred to in Chapter 33.

CHAPTER 27

Synthetic dyestuffs

(a) Column chromatography

Chromatography can be of service in the determination of the purity of synthetic dyestuffs (see Williams, *3585*; Cropper, *661*; Mottier and Potterat, *2290*; Trubey and Christian, *3305*).

Karabinos and Hyde (*1626*) have described the chromatography of indicators using a mixture of silene EF and Celite as the adsorbent.

Galmarini and Deulofeu (*1077*) separated isomeric dimethoxy-azobenzenes (e.g. 2,2'-dimethoxy- and 3,3'-dimethoxy-) and also isomeric diethoxy-azobenzenes (2,2'-diethoxy- and 3,3'-diethoxy-) by adsorption on alumina. Mixtures of methoxyazobenzenes and ethoxyazobenzenes with the same number of carbon atoms could not be resolved.

Zechmeister and Pinckard (*3685*) have reported that a cyanine dye with four conjugated double bonds gives three dark blue zones on calcium carbonate. When any of the three zones was eluted and rechromatographed, exactly identical chromatograms, composed of three zones were obtained. This proves a rapid spontaneous interconversion of stereoisomers during elution and transfer operations.

For the separation of 1-amino-, 2-amino-, 1,2-diamino- and 1,4-diamino-anthraquinones etc., see Stewart (*3109*). Hoyer (*1464–1466*) has examined the influence of hydrogen bonding on the chromatographic behaviour of a great number of anthraquinone derivatives (see p. 47).

Saenz-Lascano-Ruiz *et al.* (*2779*) have studied the chromatographic separation on alumina of various groups of synthetic dyestuffs used in the food industry (azo derivatives of pyrazolone, naphthalene, xanthene and triphenylmethane, etc.). The solvents used were water, alcohol and mixtures of the two. A number of interesting observations were reported, concerning the influence of acidic groups (sulphonic and carboxylic) which increase adsorption, and of amino groups which cause a decrease in adsorption. The positions of these groups, as well as those of the hydroxyl group, exert a marked effect on chromatographic behaviour. Inversion of zone positions was frequently noticed when the solvents were changed. For example, seven derivatives of xanthene (rose bengal, erythrosin, cyanosin, phloxin, eosin, fluorescein and dinitrofluorescein) arranged themselves in the order given (of decreasing affinity) when water was the solvent; when alcohol was used the order was inverted (*2777-2779*).

King *et al.* (*1718*) purified the excretion products of mepacrine isolated from urine by chromatography on alumina. Mepacrine forms two distinct zones on alumina which the authors ascribe to varying degrees of solvation of the dye; it seems also possible that part of the mepacrine combines with an impurity in the alumina and that this compound forms a distinct zone.

Rao *et al.* (*2622*) have described the separation of direct dyes on cellulose acetate, nylon or vinyon and of vat dyes on cellulose paper.

A partition column of kieselguhr with $0.2 M$ sodium phosphate buffer has been used by De Repentigny and James (*747*) for the separation of aminofluorescein isomers. For the separation of fluorescein and halogenated fluoresceins see Graichen (*1185*).

(b) *Paper chromatography*

Paper chromatography can be a valuable tool for the analysis and identification of commercial dyestuffs and of universal indicators. Tables 75 to 77 give R_F values of indicators (Hough *et al.*, *1458*; M. Lederer, *1924*; Franglen, *1020*). Table 78 gives R_F values of different dyes, as published by Zahn (*3660*).

TABLE 75. R_G values of indicators (Hough *et al.*, *1458*)
Solvent: *n*-butanol 40%, ethanol 10%, water 50%

Auramine	1.00	Cresol red	0.41
Dimethyl yellow	0.95	Methyl red	0.38
Bromothymol blue	0.83	Bromocresol green	0.30
Brilliant cresyl green	0.73	Bromophenol blue	0.26
Metanil yellow	0.48	Methyl orange	0.23

Several other publications describe the paper chromatography of various dyes: Sudan III and IV (Christman and Trubey, *567*), azo dyes from arylamines and sulphanilamides (Zalokar, *3664*), mixtures of dyes (Emery and Stotz, *894*).

The analysis of dyestuffs used for foods, drugs and cosmetics is described in the following papers: Tilden (*3248*), Anderson and Martin (*59*), Thaler and Scheller (*3226*). Jakovliv and Colpé (*1534*), Mottier and Potterat (*2291, 2292*), Charro Arias (*554*), Deshusses and Desbaumes (*749*), Jax and Aust (*1563*), Mitchell (*2238*), Tilden (*3249*), Ishida *et al.* (*1520* and Table 79).

For the paper chromatography of inks see Somerford (*3048*), Brackett and Bradford (*354*) and Brown and Kirk (*406*). For the analysis of coloured papers see Navarro Sagristá (*2341*).

TABLE 76. R_F values of indicators (M. Lederer, *1924*)

	isoPropanol 90 conc. NH₄OH 10	Butanol sat. with 1.5N NH₄OH	Amyl alcohol saturated with 1.5N NH₄OH
Congo red	—	0.0	0.0
Indigo carmine	0.0	0.0	0.0
Chlorophenol red	—	0.17	0.01
Phenol red	—	0.18	0.01
Cresol red	—	0.41	0.12
Bromocresol purple	0.68	0.43	0.10
Bromocresol green	0.84	0.47	0.24
Bromophenol blue	—	0.55	0.19
Methyl orange	0.77	0.55	0.26
Methyl red	0.73	0.59	0.33
Neutral red	—	0.66	0.53
Bromothymol blue	0.93	0.79	0.63
Methyl violet	0.95	0.88	0.86
Thymol blue	1.0	0.90	0.75
Phenolphthalein	1.0	0.92	0.89
Thymolphthalein	1.0	0.92	0.92

TABLE 77. R_F values of Britisch Drug-House indicators (Franglen, *1020*)
The number of other components detected under UV are listed after each compound. Paper: Whatman No. 3MM

	tert.-Amyl alcohol 200 conc. NH₄OH 50		tert.-Amyl alcohol 200 glacial acetic acid 50 water 50	
	R_F	Number of other components	R_F	Number of other components
Bromophenol blue	0.27	1	0.74	0
Bromocresol green	0.32	1	0.79	0
Chlorophenol red	0.09	6	0.67	6
Bromophenol red	0.11	4	0.68	5
Bromocresol purple	0.19	2	0.76	5
Bromothymol blue	0.50	0	0.88	0
Phenol red	0.06	4	0.54	3
Cresol red	0.12	4	0.67	3
Metacresol purple	0.16	3	0.66	2
Thymol blue	0.70	0	0.85	0

TABLE 78. R_F values of synthetic dyes (Zahn, *3660*)
on Schleicher and Schüll paper 2043b

	Colour of spot	80% Ethanol	iso-Butanol	iso-Butyric acid
Kitonlichtgelb 3GRL	yellow	0.51	0.00	0.00
	yellow	0.75	0.12	0.13
	yellow	0.86	0.57	0.49
Anthralanorange GG		0.46	0.02	0.00
Orange II		0.85	0.45	0.41
Kitonlichtrot BGLE		0.73	0.19	0.18
Echtrot AV	violet	0.33	0.0	0.0
	red	0.75	0.48	0.41
Crystal ponceau 6R		0.65	0.14	0.10
Alizarin sapphire blue C	blue	0.0–0.34	0.00	0.00
	blue	0.58	0.10	—
Sulphocyanine 5R extra		0.75	0.41	0.07
Chromechtgelb O		0.72	0.41	0.34
Chromechtrot O		0.57	0.25	0.24
Chromechtblau 2RB		0.38–0.84	0.49	0.33
Acid alizarin black R		—	0.62	—
	violet	0–0.64	0.39	0.36
	brown	0.72	—	—
	blue	0.94	0.57	0.49
	blue	—	0.00	0.00
Chrome dyes				
Neolan yellow BE		0.54–0.86	0.08	0.02
Neolan pink BE		0.00	—	—
		0.62	0.37	0.24
Neolan red BRE		0.55	0.05	0.05
Neolan green BF		0.00	0.27	0.16
		0.53	—	—
Neolan blue 2RB		0.00	0.25	0.00
		0.53	—	0.08

M. Lederer (*1923*) has reported the separation of the acridine dyes atebrin,
acriflavin, proflavin and monacrin on paper, using water or dilute HCl as
developing solvents. This is a clear case of adsorption chromatography;
partition chromatography was achieved later with higher alcohols or
benzene/NH₃ (*1926*) (See Table 80).

TABLE 79. R_F values of food colours (Ishida *et al.*, *1520*)

Food colour	Solvent	R_F value
Orange I	10% HCl	0.05
Guinea green B	10% HCl	0.92
Erythrosin	25% ethanol 5% NH$_4$OH (1:1)	0.47
Eosin	25% ethanol 5% NH$_4$OH (1:1)	0.56
Rose bengal	25% ethanol 5% NH$_4$OH (1:1)	0.60
Phloxin	25% ethanol 5% NH$_4$OH (1:1)	0.70
Ponceau SX	80% phenol	0.25
Ponceau 3R	80% phenol	0.45
Ponceau R	80% phenol	0.45
Naphthol yellow S	80% phenol	0.52
Tetrazine	phenol:ethanol:water (40:25:40)	0.38
Amaranth	phenol:ethanol:water (40:25:40)	0.48
Indigo carmine	phenol:ethanol:water (40:25:40)	0.52
New coccine	phenol:ethanol:water (40:25:40)	0.59
Sunset yellow FCF	phenol:ethanol:water (40:25:40)	0.60
Lightgreen SF yellowish	phenol:ethanol:water (40:25:40)	0.78–1.00
Fast green FCF	phenol:ethanol:water (40:25:40)	0.78–1.00
Brilliant blue FCF	phenol:ethanol:water (40:25:40)	0.78–1.00

TABLE 80. R_F values of acridine dyes (M. Lederer, *1926*)

	I	II	III	IV	V	VI	VII
	Butanol, acetic acid, water 8:2:10	Caprylic alcohol, butanol, acetic acid, water 30:60:20:90	Caprylic alcohol, acetic acid, water 90:10:100	Caprylic alcohol, butanol, hydrochloric acid, water 30:60:30:80	Caprylic alcohol, ammonia 1.5 N 100:100	Caprylic alcohol, pyridine, water 10:20:90	Benzene, ammonia, water
Acriflavin	0.57	0.29	0	0.07	0.01	0.	0
Proflavin	0.56	0.33	0	0.09	0.17	0.025	0.04
Monacrin (5-amino-acridine)	0.82	0.73		0.47	0.81	0.14	0.51
Atebrin	0.78	0.54	0	0.47	0.93	0.59	1.0

CHAPTER 28

Carbohydrates

An excellent monograph of Binkley and Wolfrom (*250*), *Chromatography of Sugars and Related Substances*, gives a detailed account of the work done in this field up to 1948, and numerous examples and experimental details, mostly from the above authors' own experience. More recently, Binkley (*246a*) has reviewed the column chromatography of sugars and their derivatives; reviews on paper chromatography of carbohydrates are quoted p. 245.

1. SEPARATION OF CARBOHYDRATES ON ADSORPTION COLUMNS

(a) *Azoylated sugars*

Reich (*2652*) has described the preparation of p-phenylazobenzoyl chloride ($C_6H_5N=NC_6H_4COCl$) and the esterification of glucose and fructose with this reagent. The coloured esters (known as *"azoylates"* or *"azoates"*, Ia) were separated on columns of silica gel or alumina. The zones were separated mechanically and the esters eluted with a mixture of methanol and chloroform.

$$
\begin{array}{ll}
\text{O·COC}_6\text{H}_4\text{N=NC}_6\text{H}_5 & \text{O·C}_6\text{H}_4\text{N=NC}_6\text{H}_5 \\
| & | \\
\text{CH} & \text{CH} \\
| & | \\
(\text{CHOCOC}_6\text{H}_4\text{N=NC}_6\text{H}_5)_3 \quad \text{O} & (\text{CHOOCCH}_3)_3 \quad \text{O} \\
| & | \\
\text{CH} & \text{CH} \\
| & | \\
\text{CH}_2\text{OCOC}_6\text{H}_4\text{N=NC}_6\text{H}_5 & \text{CH}_2\text{OOCCH}_3 \\
\quad\quad\text{Ia} & \quad\quad\text{Ib}
\end{array}
$$

Coleman *et al.* (*606*) have prepared azoyl derivatives of various other sugars for chromatographic separation. The following pairs of esters were separated on Magnesol (hydrated magnesium silicate) with Dicalite as a filter aid, and also on silicic acid: α-lactose and α-D-galactose; trehalose and

β-D-glucose; α-lactose and saccharose, etc. The same methods were used for the separation of mono- and disaccharides, various disaccharides, and also mono- and trisaccharides. Coleman and McCloskey (607) developed the method further (especially the preparation and elution of the esters) and succeeded in separating two stereoisomeric derivatives, α-D-galactose penta-azoate and β-D-galactose penta-azoate. They also studied the separation of azoates of methylated and acetylated sugars; a mixture of four azoates (β-L-arabinose, β-D-glucose, α,α-trehalose and β-cellobiose) was completely resolved. The azoates are excellent compounds for the characterisation of sugars and their derivatives.

Mertzweiller et al. (2193) have applied the same technique to the separation of azoates of methylated sugars (on silica gel). They describe in detail the preparation of the azoates and the adsorbent, and also the separation of a mixture of the azoates of 2,3-dimethyl-glucose, 2,3,6-trimethyl-glucose and 2,3,4,6-tetramethyl-glucose.

Coleman et al. (608) have separated on silicic acid the azoates of the products of hydrolysis of methylated disaccharides (e.g. 2,3,4,6-tetramethyl-D-glucosyl azoate and 4-azoyl-2,3,6-trimethyl-D-glucosyl azoate).

Hurd and Zelinski (1489) have reacted p-hydroxyazobenzene ($HOC_6H_4N =$ $=NC_6H_5$) with the polyacetyl-α-glycosyl chloride derivatives of D-glucose, D-galactose, D-xylose, maltose and lactose and obtained crystalline yellow and orange glycosides (I b). Binary and ternary mixtures of these were resolved on columns of silica gel. This method has the advantage of introducing only one coloured group and uses a commercially available reagent.

(b) Methylated sugars

Dispensing with the necessity for coloured zones, Jones (1599) was able to separate fully methylated sugars, and sugars having one free hydroxyl group on alumina (light petroleum ether as solvent). Tetramethyl methylglucoside was obtained in 94 % yield from a mixture containing twenty times as much trimethyl methylglucoside. A certain degree of separation of the α- and β-forms of the methylglucosides was achieved, the β-form being least adsorbed.

Norberg et al. (2377) separated 2,3-dimethyl-glucose, 2,3,6-trimethyl-glucose, and 2,3,4,6-tetramethyl-glucose by adsorption on fibrous alumina (after Wislicenus, 3604). The position of the zones was detected by their fluorescence in ultra-violet light.

For the separation of methylated sugars obtained by the hydrolysis of dextrans and levans of microbiological origin, see Stacey and Swift (3078), Gilbert and Stacey (1106).

Methylated sugars can be separated by gradient elution (increasing alcohol concentration) from charcoal columns; glucose is first eluted, followed by the mono-, then the di-, tri- and tetramethyl-glucoses (Lindberg and Wickberg, 2014; Whelan and Morgan, 3526).

Boissonnas (294) has separated isomeric methylated sugars by reduction to the hexitols, which were then chromatographed as their azoates. The four isomeric trimethylglucopyranoses (2,3,4-, 2,3,6-, 2,4,6-, and 3,4,6-) have been quantitatively separated on alumina with a benzene/CHCl₃ solvent. The α-β-isomerism which complicates the chromatographic separation and identification is suppressed by reduction of the aldehyde group to a primary alcohol group. Azoylation of these alcohols provides coloured, readily crystallisable substances which can easily be identified and may be estimated by photometry. For the application of this technique in the study of the constitution of lichenin, see Boissonnas (295). Quantitative separations could not be obtained if the solvent was run too rapidly through the column.

(c) Acetylated sugars

Talley et al. (3198) purified the octa-acetate of 6-β-D-mannosido-β-D-glucose by adsorption on alumina (solvent: ether). The alumina must be freed from alkali by washing with acid, as the presence of alkali causes partial hydrolysis of the acetyl groups.

McNeely et al. (2093) separated a series of acetylated sugars (β-D-glucose penta-acetate from β-maltose octa-acetate, saccharose octa-acetate from keto-D-fructose penta-acetate, etc.). The position of the zones was detected by brushing the column with an alkaline permanganate solution. The acetates of di- and trisaccharides are much more strongly adsorbed than the acetates of monosaccharides.

Binkley and Wolfrom (250) give the following example of the influence of functional groups on the chromatographic behaviour of the sugar acetates: fully acetylated α- or β-D-glucopyranose (II) is loosely adsorbed on magnesol; 2,3,4,6-D-glucose tetra-acetate (III) is held more firmly: aldehydo-D-glucose (IV) penta-acetate is very strongly adsorbed.

(II) Ester   (III) Hemi-acetal   (IV) Aldehyde

By the chromatography of the acetyl derivatives of the constituents of sugar cane molasses, Binkley et al. (247) were able to isolate inositol, mannitol and glucose. See also Binkley and Wolfrom (248).

Chromatography of fully acetylated mono- and disaccharides on alumina can lead to deacetylation of the lactol hydroxyl; columns of silica or calcium carbonate are preferable (Bredereck et al., 372).

(d) Phenyl-osazones

Fischer Jørgensen (970) has separated mixtures of the phenyl-osazones of sorbose, glucose, galactose, altrose, xylose, arabinose, rhamnose, fucose and erythrulose on columns of $CaCO_3$.

(e) Free sugars

Lew et al. (1994) have described the chromatographic separation of sugars and various other polyhydroxycompounds by adsorption on different active earths, principally Florex XXX (Floridin Co., Warren, Pennsylvania). The solvents used were alcohols, ether, dioxan, ketones, acids, pyridine and water, alone and in mixtures. The zones were made visible by brushing the column with alkaline permanganate, pH indicators, or 2,6-dichlorophenol-indophenol. The authors give a list of substances: pentoses, hexoses, heptoses, disaccharides, oligosaccharides, glycosides, polyols, ethers, acids and lactones, arranged in a chromatographic sequence in order of adsorbability. This technique allows for instance the detection of 0.01 mg of D-mannitol on a column 0.35 cm × 4 cm, and the separation of a mixture containing 2 % of mannitol and 98 % of sorbitol. 1 mg of a mixture of D-galactose, D-glucose, D-xylose, L-rhamnose and a-D-galacturonic acid such as is obtained by the hydrolysis of a hemicellulose, can be resolved into its constituents. It is often difficult to extrude the column intact, but the authors have had less difficulty with aluminium tubes. Tapered glass columns can also be used (1099). When using isopropyl alcohol instead of ethyl alcohol, these authors often observed an inversion of zone order. See also the patent by Wolfrom and Lew (3617).

The number and position of the functional groups have a profound influence on the order of adsorption: thus, galacturonic acid is most firmly held because of its carboxyl group; galactose is more tightly held than glucose because of the position of the hydroxyl group on the fourth carbon atom. Glucose with 5 hydroxyl groups is higher on the column than xylose with 4 hydroxyl groups. Rhamnose is held most loosely because it has fewer polar groups per carbon atom (Binkley and Wolfrom, 250).

Georges et al. (1099) introduced a new adsorbent, "Silene EF", a synthetic hydrated calcium acid silicate (Columbia Chemical Division, Pittsburg Plate Glass Co., Barberton, Ohio). When mixed with 13 % of Celite this adsorbent forms readily extruded columns on which separations of sugars and their derivatives can be carried out. Detection of the zones with permanganate is particularly easy with this adsorbent. The authors give a list of thirty sugars and derivatives arranged in a chromatographic sequence for this adsorbent. In a later publication the separation of aldonamides was described (Wolfrom et al. 3615).

Binkley and Wolfrom (249) succeeded in recovering 74 % of saccharose from sugar cane molasses and 93 % of saccharose from sugar beet molasses by filtering the solution through Florex XXX (using dilute alcohol to develop the column).

Binkley and Wolfrom (251) have separated the constituents of cane blackstrap molasses and molasses fermentation residues on Florex XXX. After chromatography of the acetates, the following were obtained in crystalline form: sucrose, D-glucose, meso-inositol, D-fructose and D-mannitol.

Whistler and Durso (*3530*) have separated sugars on *charcoal* by successive displacement with water, 5 % ethanol and 15 % ethanol. Desorption was followed polarimetrically. Glucose, maltose and raffinose were separated in 90 % yield. Di- and trisaccharides were also separated. Salts do not interfere.

This method has found wide application, especially for the purification of oligosaccharides (see below); Whistler (*3527*) has given experimental details for the preparation of the columns.

Wolfrom *et al.* (*3619*) have isolated pure isomaltose (6α-D-glucopyranosyl-D-glucose) through chromatography on charcoal or silicate columns.

Tiselius (*3257*) has shown how his technique of *displacement development* (see p. 4) can be applied to the separation of sugars, e.g. the separation of saccharose and glucose on charcoal by displacement with 0.5 % phenol.

Tiselius and Hahn (*3270*) applied this technique to the separation of degradation products of starch, displacement being carried out with a solution of ephedrine. Losses due to irreversible adsorption were prevented by pre-treating the charcoal with a solution of the displacing agent at a tenfold dilution. This treatment increases the differences in adsorbability of the substances to be separated. Tiselius and Hahn were in this way able to detect and separate di-, tri-, tetra-, penta- and hexasaccharides.

Galactose was found to be specifically adsorbed on asbestos by Holzapfel and Engel (*1440*).

(f) Various sugar derivatives

Reichstein and his collaborators have separated numerous sugar derivatives by chromatography on alumina, using organic solvents. Thus for instance the separation of 10 g of a mixture of 4,6-benzal-α-methyl-D-galactoside (1,5), and its 2-mono-, 3-mono- and 2,3-ditosylate was effected on 300 g of alumina (52 elutions of one liter each) (*2639*). The purification of the following may also be mentioned: tosylated (*p*-toluenesulphonylated) sugars (*302, 2640*), tritylated (triphenylmethyl-) sugars and dinitrobenzoylated sugars (*2640*), derivatives of deoxy-sugars (*1566, 2101, 2594*), acetylated and methylated sugars (*3420*).

(g) Oligosaccharides

Separations of oligosaccharides can be effected on silene EF (Wolfrom and Dacons, *3616*), on Magnesol (Thompson and Wolfrom, *3233*; Thompson *et al.*, *3232*) and especially by the method of Whistler and Durso (*3530*) using charcoal (mostly Darco G-60)-Celite columns; applications of the latter method include the isolation of: oligosaccharides formed during the lactase hydrolysis of lactose (Roberts and McFarren, *2696*), a nitrogenous tetra-saccharide from human milk (Kuhn *et al.*, *1818*), isolation of a crystalline xyloheptaose (Whistler and Tu, *3532*), of maltotetraose from corn syrup (Whistler and Hickson, *3531*), of maltopentaose (Whistler and Duffy, *3529*), of the trisaccharide panose (Thompson and Wolfrom, *3233*), separation of lactotriose, lactobiose and galactobiose (Wallenfels *et al.*, *3468*), separation of disaccharides obtained by acid reversion of D-glucose (Thompson *et al.*, *3232*). Elution is best effected by a gradient of alcohol in water (Alm, *41*; Bacon and Bell, *119*; Bacon, *118*).

Barker *et al.* (*139*) suggest a new method for the fractionation of mixtures of oligosaccharides consisting in a mild treatment with methanolic HCl and subsequent charcoal chromatography; those components which form methyl furanosides can then be separated from others, as they are more strongly adsorbed on charcoal (example: separation of maltose and the disaccharide nigerose, of cellobiose and laminaribiose and of two tri-saccharides).

Cellulose columns have also been used for the separation of oligosaccharides (Whistler and Hickson, *3531*; Whistler and Duffy, *3529*).

Dickey and Wolfrom (*772*) have separated by chromatography on Silene EF and on magnesol a product obtained by the acetolysis of cellulose, into a series of homologous polymers of crystalline α-D-acetates, ranging in degree of polymerisation from 1 to 6 (glucopyranose penta-acetate to cellohexaose eicosa-acetate). The polymers are fixed on the column in the order of molecular weight, the hexaose on the top.

(h) *Polysaccharides*

Pascu and Mullen (*2433*) have separated α- and β-amylose by adsorption on cotton, and elution with hot water, the β-amylose being the most strongly adsorbed.

Ulmann (*3342*) has separated on alumina different dextrins obtained by digestion of starch with amylase. Fischer and Settele (*964*) showed that neutral alumina can separate amylose and amylopectin; amylose is eluted by an acid buffer as a single fraction, whereas amylopection eluted by a neutral buffer yields several fractions.

Kaval'skiï (*1662*) has studied the adsorption behaviour of glycogen on CaCO₃; differently coloured zones were obtained and glycogen was thought to be a mixture of closely related compounds.

That the zone order of polysaccharides depends on molecular weight, has also been shown for nitrocellulose (on corn starch; Brooks and Badger, *399*), and for the hydrolysis products of hyaluronic acid (on charcoal; Rapport *et al.*, *2627*).

Siegel *et al.* (*2979*) have separated polysaccharides from *Mycobacterium tuberculosis* on columns of silica gel.

2. SEPARATION OF CARBOHYDRATES ON ION EXCHANGE COLUMNS

The secondary reactions observed after the passage of carbohydrates on ion exchange columns have already been mentioned (p. 99). Roseman *et al.* (*2736*) have reported the retention of glucosamine and other reducing sugars on columns of strongly basic exchangers (Amberlite IRA-400 or Dowex-1); nonreducing substances, like sorbitol, mannitol, inositol and methyl-α-D-glucopyranoside are not retained. Retention was much less in columns saturated with chloride, acetate or carbonate ions. Losses approaching 50 % of reducing substances were encountered after elution. The adsorption of sugars on Dowex-2 and Dowex-50 from ethanol-water solutions has been studied by Rückert and Samuelson (*2752*). Reynolds (*2672*) has reported losses of about 1 % of glucose by passage through Amberlite IR-4B; repeated washing is necessary to recover the last traces of glucose. Laland and Overend (*1870*) report quantitative recovery of 2-deoxy-D-ribose from columns of mixed resins (Amberlite IR-120-H) + (Amberlite IR-4B-OH).

Acidic carbohydrates can be separated on anion exchangers. Thus, glucuronic and galacturonic acids have been separated on Dowex-1 in the acetate form (Khym and Doherty, *1711*). Epimeric aldonic acids have been separated on anion exchange resins by Karabinos (*1625*); the separation is dependent on the rate of hydrolysis of the lactones since the free acids and not the lactones are adsorbed. Derungs and Deuel (*748*) have obtained very nice separations of mono-, di-, tri-, and tetragalacturonic acids on Dowex-3 in the formate form. Andrews and Jones (*65*), Whistler *et al.* (*3528*) have used Amberlite IR-4B for the separation of glucuronic and galacturonic acid containing oligosaccharides.

Weissmann *et al.* (*3497*) have separated the products of enzymatic hydrolysis of hyaluronic acid on Dowex-1 formate, with gradient elution by dilute formic acid. Seven distinct oligosaccharide fractions were isolated, ranging from di- to tetradecasaccharide in increments of one disaccharide unit (the repeating unit being glucuronido-acetyl glucosamine).

Separation of neutral carbohydrates in presence of borate

Khym and Zill (*1713*) have developed an important new method of column separation of carbohydrates, by using the reaction of polyhydroxy compounds

with borate ions to form acidic borate complexes (I or II); these borate complexes are easily separated on columns of strong base anion exchange resins (Dowex-1) with boric acid-borate buffers (pH 8 to 9). The particular order of elution of the sugars suggests that several factors affecting the affinity of the sugar-borates for the exchanger are involved (mutarotation, furanose-pyranose interconversion, etc.). Disaccharides are readily separated from monosaccharides and the components of hexose and pentose mixtures are easily resolved. Zill *et al.* (*3697*) have applied this method to the separation of various mono- and oligosaccharides and polyols; after elution, boric

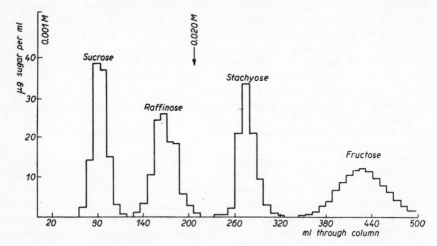

Fig. 75. Separation and recovery of 1 mg each of sucrose, raffinose, stachyose and fructose; Dowex-1 column (0.85 sq. cm × 2.5 cm). Flow ratio: 60 ml/h. Elution started with 0.001 *M* potassium tetraborate and changed to 0.02 *M* at 210 ml. Found: 1.01 mg sucrose, 1.02 mg raffinose, 0.89 mg stachyose, and 1.01 mg fructose (Noggle and Zill, *2374*)

acid can be eliminated as the volatile methyl borate. Noggle and Zill (*2374*) have applied the borate method to the analysis of sugars in plant extracts (Fig. 75). See also (*2375*). Lampen (*1874*) has used the method of Khym and Zill (*1713*) for the separation of pentoses.

3. SEPARATION OF CARBOHYDRATES ON PARTITION COLUMNS

The first application of partition chromatography to sugars was published by Bell (*177*), who separated 2,3,4,6-tetramethyl-glucose from 2,3,6-tri-methyl- and 2,3-dimethyl-glucose on columns of silica gel. The tetramethyl derivative was washed out of the column with $CHCl_3$, the trimethyl sugar with a mixture of $CHCl_3$ and butanol, and the dimethyl sugar was recovered by extrusion of the column. By a similar method, Bell and Palmer (*179*) later achieved quantitative separation of tetra-, tri- and dimethyl-fructose. This problem was also attacked by Schlubach and Heesch (*2853*). Bell and

Palmer (*179*) also separated 1,3,4,6-tetramethyl-, 1,3,4-trimethyl- and 3,4-dimethyl-fructoses quantitatively on silica gel columns.

Hough *et al.* (*1458*) used *columns of powdered cellulose* which were prepared by rubbing Whatman ashless tablets through an 80 mesh sieve. A column 20 in. long and 1½ in. diameter was used. A typical separation of simple sugars is as follows: a mixture of L-rhamnose hydrate (300 mg; R_G 0.30) and L-arabinose (200 mg; R_G 0.13) was resolved, using *n*-butanol saturated with water containing a little ammonia as the mobile phase. The eluate from the column was fractionated with an automatic fraction collector at 20 minute intervals (ca. 5 ml portions). The contents of the receivers were examined by sheet filter paper chromatography, which indicated complete separation. The two solutions thus obtained were filtered and evaporated under reduced pressure yielding (a) anhydrous L-rhamnose (260 mg; 96 % recovery) and (b) L-arabinose (193 mg; 96 % recovery). Each fraction crystallized on removal of the solvent and after recrystallization from methanol pure specimens of both sugars were obtained. Methylated sugars do not separate satisfactorily with butanol; better separations being obtained with a solvent composed of light petroleum (b.p. 100–120°) (60 % v/v) and *n*-butanol (40 % v/v). This solvent on paper sheets gives the following R_G values (*1458*):

6-Methyl-glucose	0.09	2,4,6-Trimethyl-glucose	0.71
Rhamnose	0.15	2,3,4-Trimethyl-xylose	0.95
2,4-Dimethyl-galactose	0.25	2,3,4,6-Tetramethyl-glucose	1.00
2,4-Dimethyl-xylose	0.56		

See also the review of Hough (*1455*).

Geerdes *et al.* (*1093*) have prepared hydrocellulose by dissolving cellulose powder in phosphoric acid followed by precipitation with water; they found that the resolving power and capacity of this hydrocellulose were superior to those of other cellulose powders, principally for the separation of methylated sugars; the anomeric forms of ethyl L-rhamnofuranoside were separated.

The separation of monosaccharides on columns of starch, with butanol-propanol-water (4:1:1) as moving phase was described by Gardell (*1087*).

Berenson *et al.* (*192*) have fractionated acid mucopolysaccharides (hyaluronic and chondroitinsulphuric acids) on columns of silicone-treated Celite. Several components with different viscosities were obtained.

4. PAPER CHROMATOGRAPHY OF SUGARS

The paper chromatography of carbohydrates and related compounds has been reviewed by Dedonder (*723*), Hough (*1455*), Isherwood (*1513*) and Kowkabany (*1773*).

Partridge (*2447–2449*) was the first to apply paper chromatography to the separation of sugars. The chromatograms were run with phenol and collidine and the spots were made visible by the reduction of a solution of ammoniacal silver nitrate sprayed on the paper. Partridge was able to separate the following sugars: rhamnose, xylose, mannose, galactose and galacturonic acid (see Fig. 76); he also published a fairly complete R_F value table (Table 81).

Hough *et al.* (*414, 1458*) extended paper chromatography to *methylated* sugars. For R_F values see Tables 84 and 86.

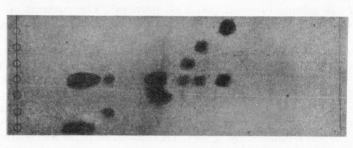

Ribose
Mannose
Fructose
A
Galactose
Lactose
Galacturonic Acid

Fig. 76. Paper chromatography of carbohydrates (Partridge, *2448*)
A = a mixture of ribose, mannose, fructose, galactose, lactose and galacturonic acid

Bayly *et al.* (*167*) have reported that glucose or other aldoses can give several spots on paper; this was later found to be due to the formation of glycosylamines during evaporation of the sugar solutions in presence of ammonium acetate (*168*).

(a) *The relation between structure and R_F values of sugars*

Isherwood and Jermyn (*1518*) discuss the relationship between the structure of sugars and their position on the chromatogram. The R_F values in various solvents vary with the water content of the solvent, with the exception of phenol, in which sugars appear to form complexes.

The sequence of R_F values for each group of sugars and methyl-glycosides (aldohexoses, aldomethylose, etc.) depends only on the configuration of the -OH of the ring. This is illustrated in Table 82 where the preferential formation of a pyranose ring in solution is assumed. A detailed analysis of the contribution of each -OH group to the observed R_F values was also made for the aldohexoses, aldopentoses and ketohexoses. If a substituent (-CH_2OH) is attached to carbon atom 5 of the pyranose ring, then the influence of a hydroxyl group on any particular carbon atom largely depends upon whether the hydroxyl is on the same side of the ring as the substituent or not. The effect on the R_F value is different for each carbon atom. In the absence

TABLE 81. R_F values of sugars in various solvents on No. 1. Whatman paper (Partridge, 2448)

Solvent	Phenol	s-Collidine	n-Butanol 40% Acetic acid 10% Water 50%	n-Butanol 45% Ethanol 5% Water 49%	n-Butanol	isoButyric acid	Methyl ethyl ketone
Additions	NH₃ 1% HCN	None	None	NH₃ 1%	NH₃ 1%	None	NH₃ 1%
D-Glucose	0.39	0.39	0.18	0.105	0.070	0.13	0.025
D-Galactose	0.44	0.34	0.16	0.090	0.060	0.14	0.015
D-Mannose	0.45	0.46	0.20	0.130	0.100	0.15	0.050
L-Sorbose	0.42	0.40	0.20	0.120	0.085	0.16	0.050
D-Fructose	0.51	0.42	0.23	0.135	0.100	0.18	0.045
D-Xylose	0.44	0.50	0.28	0.170	0.125	0.19	0.090
D-Arabinose	0.54	0.43	0.21	0.145	0.100	0.21	0.075
D-Ribose	0.59	0.56	0.31	0.210	0.180	0.22	0.165
L-Rhamnose	0.59	0.59	0.37	0.285	0.220	0.30	0.180
D-Deoxyribose	0.73	0.60	—	—	—	0.32	—
L-Fucose	0.63	0.44	0.27	—	—	—	0.095
Lactose	0.38	0.24	0.09	0.0	0.0	0.70	0.0
Maltose	0.36	0.32	0.11	0.15	0.01	0.85	0.0
Sucrose	0.39	0.40	0.14	—	—	—	—
Raffinose	0.27	0.20	0.05			—	—
D-Galacturonic acid	0.13	0.14	0.14			0.09	
D-Glucuronic acid	0.12	0.16 (0.72)+	0.12 (0.32)+			0.08 (0.22)+	
D-Glucurone	0.12	— (0.72)+	— (0.33)+			— (0.22)+	
D-Glucosamine HCl	0.62	0.32	0.13 (0.17)-			0.05 (0.20)-	
Chondrosamine HCl	0.65	0.28	0.12 (0.16)-			— (0.19)-	
N-Acetylglucosamine	0.69	0.50	0.26			0.25	
L-Ascorbic acid	0.24	0.42	0.38			0.19	
Dehydroascorbic acid	0.16	0.68	0.27			0.16	
i-Inositol	0.23	0.10	0.09			—	

+ Values in brackets due to lactone.

- Values in brackets due to free base (comet connects the two spots).

TABLE 82. Relationship between the configuration and the R_F value of sugars and glycosides in ethyl acetate–pyridine–water at 20° (Isherwood and Jermyn, *1518*)

(a) Sugars with the general formula ⟨ring⟩ H·OH, where R = —CH₃, —CH₂OH, —COOH, —CHOH·CH₂OH

Aldomethylose sugars R = —CH₃	M.p. (°)	R_F	Aldohexose sugars R = —CH₂OH	M.p. (°)	R_F	Aldohexuronic acids § R = —COOH	M.p. (°)	R_F	Aldoheptose sugars R = —CHOH·CH₂OH	M.p. (°)	R_F
Allomethylose*	146	0.36	Allose*	128	0.22				D-Gala-L-glucose†	190–194	0.16
Altromethylose	.	0.45	Altrose†	108	0.27	Glucuronic	165	0.15	D-Gluco-D-gulose†	210	0.19
Glucomethylose*	139–145	0.36	Glucose	146–150	0.195				D-Gala-L-mannose†	.	0.21
Gulomethylose*	.	0.38	Gulose	.	0.23	Mannuronic	165	0.18			
Mannomethylose*	123	0.51	Mannose	132	0.24				D-Gulo-L-galactose†	.	0.14
Idomethylose*	.	.	Idose	.	0.31				D-Manno-D-galactose†	134	0.15
Galactomethylose*	145	0.265	Galactose	165	0.175	Galacturonic	160	0.13	D-Gulo-L-talose†	.	0.23
Talomethylose*	.	0.48	Talose*	128	0.285						

Sugars marked * were a gift from Prof. T. Reichstein and those marked † were from Prof. C. S. Hudson.
§ These were run in ethyl acetate–acetic acid–water as solvent (Jermyn & Isherwood, *1586*).

(b) Sugars with the general formula ⟨ring⟩ R·OH, where R = —H, CH₂OH

Ketohexose sugars R = —CH₂OH	M.p. (°)	R_F	Aldo-pentose sugars R = —H	M.p. (°)	R_F
Allulose	Syrup	0.30	Ribose	95	0.33
Tagatose	124	0.27	Lyxose	106–117	0.30
Sorbose	164	0.24	Xylose	145	0.28
Fructose	102	0.24	Arabinose	159	0.23

(c) Sugars with the general formula ⟨ring⟩ R·OH, where R = —H, —CH₂OH and R₁ = —H, —CH₃

Ketopentose sugars R = R₁ = —CH₂OH	R_F	Aldotetrose sugars R = R₁ = —H	R_F	Keto-hexomethylose sugars R₁ = —CH₃, R = —CH₂OH	R_F
Araboketose	0.35	Erythrose	0.41	Allomethylo-ketose	0.44
				Galactomethylo-ketose	0.43
Xyloketose	0.37	Threose	0.43	Mannomethylo-ketose	0.43
				Idomethylo-ketose	0.47

(d) Glycosides with the general formula ⟨ring⟩ H·OCH₃, where R = —H, —CH₂OH

Aldohexose glycosides R = —CH₂OH	R_F	Aldopentose glycosides R = —H	R_F
Allose	0.23	Ribose	0.40
Altrose	0.36	Arabinose	.
Glucose	0.29	Xylose	0.46
Gulose	.	Lyxose	0.48
Mannose	0.37		
Idose	.		
Galactose	0.34		
Talose	0.29		

(e) Glycosides with the general formula ⟨ring⟩ H·OCH₃, where R = —CH₂OH, —CHOH·CH₂OH

Aldohexose glycosides R = —CHOH·CH₂OH	R_F	Aldopentose glycosides R = —CH₂OH	R_F
Allose	0.45	Ribose	0.54
Altrose	0.48	Arabinose	0.55
Glucose	0.43	Xylose	0.52
Gulose	.	Lyxose	0.55
Mannose	0.48		
Idose	.		
Galactose	0.47		

of a substituent (aldopentoses) or if the substituent is attached to carbon atom 1 of the pyranose ring (ketohexoses), it is the interaction between the hydroxyl groups on neighbouring carbon atoms which governs the R_F value, a *cis* disposition of hydroxyls giving a higher R_F value than a *trans*. A furanose has a higher R_F value than a pyranose. Methylation increases the R_F value, thus tetramethyl sugars generally travel faster than trimethyl sugars, etc.

An interesting correlation between melting point and R_F value was also noted by Isherwood and Jermyn (*1518*), the melting point being inversely proportional to the R_F value. This can be explained by the similarity between the hydration forces of the sugar and the intramolecular attraction forces in a crystal, both being due to the interactions of OH groups. Thus a high-melting sugar will also have a high affinity for the stationary aqueous phase and a correspondingly low R_F value.

Levy (*1993*) has tried to put into more quantitative terms the qualitative statements of Isherwood and Jermyn (*1518*). In comparing aldohexoses ($R = -CH_2OH$), aldomethyloses ($R = CH_3$) and aldohexuronic acids ($R = -COOH$) in the pyranose form the assumption was made that the various nuclei (allo-, altro-, gluco-, etc.) as well as the several R groups

$$R \atop \langle\!\!\!\!\!\!-O \atop \rangle H \cdot OH$$

mentioned, will make contributions to some mathematical function of the R_F value, characteristic of each nucleus and R group, and independent of each other. For the hexose series a rather good correlation was found, but for the aldoheptoses ($R = -CHOH \cdot CH_2OH$) and the methylfuranosides and pyranosides discrepancies were observed.

French and Wild (*1035*) have found that the regularity of the paper chromatographic mobilities of homologous *oligosaccharides* leads to a straight line characteristic for each series when the logarithm of a partition function, a' [where $a' = R_F/(1-R_F)$]* is plotted against molecular size. The partition function is obtained from single or multiple ascent paper chromatograms. It is concluded that the increase of the size of a saccharide by one hexose unit decreases the mobility by an amount which depends on the type of hexose and on its mode of attachment. Examples studied include oligosaccharides of the starch, dextran, levan, inulin and galactan type.

(b) *Various techniques*

Hirst and Jones (*1417*) reported the R_{TG} or R_G values of sugars and methylated sugars where R_{TG} or $R_G = \dfrac{\text{distance travelled by sugar}}{\text{distance travelled by 2,3,4,6-tetramethyl glucose}}$,

* See also R_M, p. 117.

thus any variation of R_F due to temperature or other variables are corrected for (see Table 83).

TABLE 83. R_G and R_F values of sugars, methylated sugars, esters, lactones, etc. (I) Hirst and Jones (1417); (II) Rafique and Smith (2612); (III) Boggs et al. (291)

	(I) R_G n-Butanol 50 Ethanol 10 Water 40 (top layer)	(II) R_G n-Butanol 40 Ethanol 10 Ammonia 1 Water 49	(III) R_F Methyl ethyl ketone saturated with water
Raffinose	0.001		
Lactose	0.016		
Maltose	0.021		
Sucrose	0.03		
D-Gluco-L-gala-octose	0.038		
D-Gulo-L-gala-heptose	0.050		
D-Gluco-D-gulo-heptose	0.053		
D-Manno-D-gala-heptose	0.058		
D-Gala-L-gluco-heptose	0.064		
Turanose	0.060		
Galactose	0.070		
D-Gluco-L-talo-octose	0.082		
Glucose	0.090		
Sorbose	0.10		
Mannose, manno-heptulose	0.11		
D-Gala-L-manno-heptose	0.11		
D-Gulo-L-talo-heptose	0.11		
Fructose, gluco-heptulose	0.12		
Gulose, arabinose, tagatose	0.12		
Xylose	0.15		
4-Methyl galactose	0.16		
Altrose	0.17		
Idose, 6-methyl galactose	0.18		
Talose, lyxose	0.19		
Ribose, fucose	0.21		
2-Methyl glucose	0.22		
2-Methyl galactose	0.23		
Riboketose, apiose	0.25		
2-Deoxygalactose	0.25		
3-Methyl glucose	0.26		0.40
Xyloketose	0.26		
6-Methyl glucose	0.27		
Quinovose	0.28		
Rhamnose	0.30		
α-Methylmannoside	0.30		
3,4-Dimethyl galactose	0.32		
4-Methyl mannose	0.32		0.40
β-Methylarabinoside	0.32		
2-Deoxyallose	0.33		
2-Methyl β-methyl altroside	0.34		
Rhamnoketose	0.37		
3,6-Anhydroglucose	0.37		

TABLE 83 *(continued)*

	(I) R_G n-Butanol 50 Ethanol 10 Water 40 (top layer)	(II) R_G n-Butanol 40 Ethanol 10 Ammonia 1 Water 49	(III) R_F Methyl ethyl ketone saturated with water
Talomethylose	0.37		
2-Methyl xylose	0.38		
2-Methyl arabinose	0.38		
2,4-Dimethyl galactose	0.41		
4,6-Dimethyl galactose	0.42		
2-Deoxyribose	0.44		
2,6-Dimethyl galactose	0.44		
4,6-Dimethyl glucose	0.46		
2-Methyl fucose	0.51		
3,6-Dimethyl glucose	0.51		
3,4-Dimethyl glucose	0.52		
4,6-Dimethyl altrose	0.52		
2,3-Dimethyl mannose	0.54	0.62	0.22
2,3-Dimethyl glucose	0.57	0.68	0.28
4-Methyl rhamnose	0.57		
4,6-Dimethyl mannose	0.57		
3,4-Dimethyl mannose	0.58		
3-Methyl quinovose	0.60		
3,4-Dimethyl fructose	0.61		
2-Deoxyrhamnose	0.61		
2,3-Dimethyl arabinose	0.64		
2,3,4-Trimethyl galactose	0.64		
2,4-Dimethyl xylose	0.66		
2,4,6-Trimethyl galactose	0.67		
3-Methyl altromethylose	0.68		
2,3,6-Trimethyl galactose	0.71		
2,3-Dimethyl xylose	0.74		
2,4,6-Trimethyl glucose	0.76		
3,4,6-Trimethyl mannose	0.79		
2,3,6-Trimethyl mannose	0.81	0.84	0.50
2,3,6-Trimethyl glucose	0.83	0.88	0.56
1,3,4-Trimethyl fructose	0.83		
3,4-Dimethyl rhamnose	0.84		
2,3,4-Trimethyl glucose	0.85		
3,4,6-Trimethyl fructose	0.86		
Cymarose	0.87		
Oleandrose	0.88		
2,3,4,6-Tetramethyl galactose	0.88	0.93	0.68
Tetramethyl fructopyranose	0.90		
2,3,4-Trimethyl xylose	0.94		
2,3,5-Trimethyl arabinose	0.95		
2,3,4,6-Tetramethyl mannose	0.96	1.00	
2,3,4,6-Tetramethyl glucose	*1.00*	1.00	0.82
2,3,5,6-Tetramethyl glucose	1.01		
2,3,4-Trimethyl rhamnose	1.01		
1,3,4,6-Tetramethyl fructose	1.01		
2,3,4-Trimethyl D-mannose	—	0.86	

Continued p. 252

TABLE 83 *(continued)*

	(I) R_G n-Butanol 50 Ethanol 10 Water 40 (top layer)	(II) R_G n-Butanol 40 Ethanol 10 Ammonia 1 Water 49	(III) R_F Methyl ethyl ketone saturated with water
	R_F values (Abdel-Akher and Smith, *1*)		
Esters and lactones			
D-Xylono-γ-lactone.	0.41		
L-Rhamnono-γ-lactone	0.50		
D-Glucono-γ-lactone	0.32		
D-Galactono-γ-lactone	0.35		
D-Mannono-γ-lactone	0.25		
D-Glucoheptono-γ-lactone	0.13		
D-Glucono-δ-lactone	0.22		
Methyl-β-D-glucofuronoside	0.30		
Mannitol hexa-acetate	0.85		
Arabitol penta-acetate	0.85		
Mannosaccharo-1,4-3,6-dilactone			0.60
3-Methyl-D-erythrono-γ-lactone			0.73
2,3,5-Trimethyl-L-rhamnono-γ-lactone			1.00
2,3,5-Trimethyl-D-galactono-γ-lactone			0.90
3,4,6-Trimethyl-D-mannono-δ-lactone			0.86
2,4-Dimethyl-D-galactosaccharo-3,6-lactone methyl ester			0.91
2,3,4-Trimethyl-D-glucosaccharo-1,5-lactone methyl ester			0.94
2,3,5-Trimethyl-D-glucosaccharo-1,4-lactone methyl ester			1.00

Jermyn and Isherwood (*1586*) attempted to increase the separation of sugars by the use of solvents containing ethyl acetate (Table 84).

TABLE 84. R_F values of sugars (Jermyn and Isherwood, *1586*)

Solvents	Ethyl acetate 2 Water 2 Pyridine 1	Ethyl acetate 3 Water 3 Acetic acid 1
Rhamnose	0.49	0.34
Xylose.	0.38	0.265
Arabinose	0.33	0.22
Mannose	0.32	0.195
Fructose	0.32	—
Glucose	0.28	0.17
Galactose	0.235	0.14
Galacturonic acid	0.025	0.13

However, the solvents are relatively unstable owing to the hydrolysis of the ethyl acetate.

Other solvents used are: butanol-formic acid-water (Wiggins and Williams, *3568*), benzene-butanol-pyridine-water (1:5:3:3) (Gaillard, *1072*), propanol-ethyl acetate-water (7:1:2) (Baar and Bull, *113*).

Ultramicrotechniques have been described for the enzymatic hydrolysis of carbohydrates prior to paper chromatography (Porter and Hoban, *2573*; Williams and Bevenue, *3579*).

Gaillard (*1072*) reports that it is not necessary to neutralize the N sulphuric acid used to hydrolyse polysaccharides prior to paper chromatography with benzene-butanol-pyridine-water (1:5:3:3) and using an acid spray (for instance aniline phosphate).

Baar and Bull (*113*) describe the interference of salts in paper chromatography of carbohydrates from biological liquids (lower R_F values being obtained); in neutral solvents (for instance propanol-ethyl acetate-water, 7:1:2) this interference is less.

The paper chromatography of radioactive (^{14}C containing) monoses and oligosaccharides has been described by Pazur and French (*2478*), Benson *et al.* (*189*), Putman and Hassid (*2604*).

The paper chromatography of carbohydrates in presence of boric acid has been described by Wachtmeister (*3434*), Barker and Smith (*137*); see also Micheel and Van de Kamp (*2206*).

Circular paper chromatography of carbohydrates has been recommended as a rapid method of analysis (Bersin and Müller, *231*; Giri and Nigam, *1122*; Parihar, *2442*; Venner, *3390*).

Jeanes *et al.* (*1564*) have recommended a multiple descent technique for obtaining higher R_F values for carbohydrates; this technique has been used by Koepsell *et al.* (*1756*) for the separation of oligosaccharides (the paper was irrigated five times with a solvent containing butanol-pyridine-water, 3:2:1).

Harris and MacWilliam (*1312*) have described dipping techniques for revealing sugars on paper chromatograms.

(c) *Reagents for sugars*

A great many, more or less specific, spray reagents have been used; they can be divided into two main classes:

1) The sugar reduces the reagent to give a coloured derivative.

2) Heating of the sugar with acid produces furfuraldehyde, methylfurfuraldehyde or ω-hydroxymethylfurfuraldehyde which can be condensed with an aromatic amine or phenol to give characteristically coloured compounds; different classes of sugars and sometimes even different sugars of the same class can give different colour reactions (Table 85, p. 254).

As a general method for the detection of sugars on a paper chromatogram, Partridge (*2447, 2448*) employed an ammonical solution of silver

TABLE 85. The properties of various spray reagents for carbohydrates. (+ positive reaction; — no reaction. Blank space signifies that reaction has not been described. The colours produced with different classes of sugars are indicated) (Isherwood, 1513).

Spray reagent	Reaction given by various classes of sugar															Comments	References
	Aldohexose	Aldopentose	Ketohexose	Ketopentose	Uronic acid	Methyl pentose	Methylated aldohexose	Methylated ketohexose	Methylated aldopentose	Methylated uronic acid	Amino sugar	Sugar alcohols	Deoxy sugars	Lactones, esters, amides	Glycosides		
CLASS I																	
0.1 M AgNO₃ in 5 M NH₄OH. Heated at 100° for 5–30 min. Addition of 2 N NaOH enables reagent to attack sugar alcohols, etc. Brown-black stains	+	+	+	+	+	+					+	±		±	±	This reagent is unspecific but is valuable because almost every compound on chromatogram either produces a stain or inhibits background colour. Sensitive to 0.001 M glucose. An elegant modification of this reagent has been described for reducing sugars	2447 3300
Alkaline solution of triphenyltetrazolium chloride reduced to give brilliant red triphenylformazan	+	+	+	+	+	+					+	—		—	—	In the author's hands this reagent was not sensitive to concentrations below 0.005 M glucose	3464
3,4-Dinitrobenzoic acid in ethanolic Na₂CO₃ reduced to 2-nitro-4-carboxylic phenylhydroxamine. Na salt, blue	+	+	+	+	+	+					+		+	—	—	Ketoses react more quickly than aldoses. Claimed to be as sensitive as the Ag reagent	3523
3,5-Dinitrosalicylic acid in N NaOH. Brown on yellow background	+	+	+	+	+	+					+	—	+	—	—	Sensitivity less than 0.01 M glucose	1565

TABLE 85 (continued)

Reaction given by various classes of sugar

CLASS II

Spray reagent	Aldohexose	Aldopentose	Ketohexose	Ketopentose	Uronic acid	Methyl pentose	Methylated aldohexose	Methylated ketohexose	Methylated aldopentose	Methylated uronic acid	Amino sugar	Sugar alcohols	Deoxy sugars	Lactones, esters, amides	Glycosides	Comments	References
$KMnO_4$ (1%) in 2% aqueous Na_2CO_3. No heating. Brown on purple background	+	+	+	+	+	+					+	±	+	±	±	Sensitivity less than 0.01 M glucose. Unspecific and not permanent	2432
Aniline phthalate (2%) in moist butanol (similarly its trichloroacetate or phosphate)	Brown	Red	—	+	Brown	Brown	Brown		Red	Red	—			—	—	Not quite as sensitive as the Ag reagent. Similar reagents based on di-	2448
p-Anisidine hydrochloride (3%) in moist butanol	Greenish brown	Yellow	—		Red	Green	Brown		Red				Light brown	—	—	phenylamine, dimethylamine, α-naphthylamine and benzidine have been described (1459). Reacts with aldoses	1459
Naphthoresorcinol (0.1%) in 0.4 N HCl in 80% ethanol	Bluish violet	Red violet	Red	—	Blue	Bluish violet					—					Similar reagents based on orcinol and resorcinol and anthraquinone. Sensitivity less than the Ag reagent. Reacts particularly with ketohexoses	1005
Urea (5%) in 0.4 N HCl in 80% ethanol	—	—	Intense brownish black	—	—	—		Brownish black								Reacts particularly with ketohexoses	724

nitrate, the reducing sugars giving rise to discrete brown-black spots after heating to 100°. This reagent was also employed for methylated sugars (Hirst *et al.*, *1412*), polyhydric alcohols and methyl glycosides of simple sugars (Hough, *1454*). Silver nitrate was included in the solvent by Mc-Farren *et al.* (*2080*) to ensure uniform distribution for subsequent quantitative estimation of the spots. A number of other materials, for example Cl⁻ and Na⁺, also give black spots with $AgNO_3$ (Westall, *3510*).

Chargaff *et al.* (*548*) employed *m-phenylenediamine* dihydrochloride in 76 % alcohol for reducing sugars, which produce spots which fluoresce in ultra-violet light. Patridge (*2450*) employed *aniline hydrogen phthalate* in butanol, which gives red spots with pentoses, brown spots with aldohexoses, methyl aldopentoses and hexuronic acids. The range of applicability of aniline hydrogen phthalate was further extended by Hough *et al.* (*1459*) who also obtained brown spots with methylated aldohexoses, ketohexoses, heptoses and octoses, as little as 1–5 μg being detectable. Partly methylated aldohexoses give rise to brown colours, with a tint of red; fully methylated aldohexoses a characteristic maroon colour; methylated aldopentoses a cherry red colour. Methylated uronic acids give crimson red colours of high brilliance.

Aniline phosphate was found by Bryson and Mitchell (*440*) to cover a wider range of sugars than the phthalate. The following phenols, in an alcoholic solution of H_3PO_4 are also used by these authors to detect sugars: resorcinol, naphthoresorcinol, naphthol and phloroglucinol.

Wallenfels (*3464*) employs alkaline *triphenyltetrazolium chloride* (TTC) yielding brilliant red spots (lower limit 5 μg) which can be extracted for subsequent colorimetry. Trevelyan *et al.* (*3300*) use a chloroform solution of TTC. Pacsu *et al.* (*2432*) spray the paper with 1 % $KMnO_4$ in 2 % Na_2CO_3, producing yellow spots with reducing and non-reducing carbohydrates on a purple background.

p-Anisidine hydrochloride in *n*-butanol yields a green-brown colour with aldohexoses, a brilliant lemon yellow with ketohexoses, an emerald green with methyl aldopentoses and a cherry red with uronic acids (*1459*). Methylated aldohexoses furnish brown colours and methylated aldopentoses intense red colours. 2-Deoxyaldoses give pale brown colours, which appear as characteristic white fluorescent spots in ultra-violet light thus providing a very sensitive means of detection. *p-Anisidine phosphate* in ethanol has been proposed by Mukherjee and Srivastava (*2302*). *Diphenylamine trichloracetate* yields brown spots with aldohexoses and purple with aldopentoses. *Dimethylaniline* yields purple and brown colours with highly methylated sugars. *a-Naphthylamine trichloracetate* is another general reagent (*1459*). Aldohexoses and their partly methylated derivatives yield brown spots, fully methylated aldohexoses red, ketohexoses yellow and aldopentoses green spots. A distinction between methylated arabinoses (faint green) and methylated xyloses (intense green) can be made.

A number of amines, notably *p-aminodimethylaniline* were also used by Boggs *et al.* (*291*); see also *926*.

Vámos (*3360*) recommends *p-aminophenol phosphate* in ethanol; Hirase *et al.* (*1396*) prefer *o-aminophenol phosphate* which yields a specific colour with each class of sugars.

Sattler and Zerban (*2825*) recommend a 0.3 % ethanolic solution of *p-aminohippuric* acid which after heating to 140° for 8 min gives stable orange spots with the common hexoses and pentoses (detection limit 1 μg).

Naphthylamine with HCl in alcoholic solution was employed by Novellie (*2385*). *Benzidine* in alcoholic acetic acid yields brown spots with reducing sugars (Horrocks, *1449*).

Several spray reagents allow to distinguish *aldoses* from *ketoses*: Orcinol and *resorcinol* react only with ketoses and their derivatives, giving red colours (*1459*); orcinol was also used by Klevstrand and Nordall (*1740*) who report blue-green spots with sedoheptulose and mannoheptulose and yellow spots with fructose and sucrose. Solutions of *urea hydrochloride,* and anthraquinone and HCl yield intense brown-black spots with ketoses and their methylated derivatives (*1459*). *3,4-Dinitrobenzoic acid* in alcoholic Na_2CO_3 produces, on heating, blue spots which on standing turn brown; ketoses react faster than aldoses (*3523*). *Alkaline hypoiodite* reveals only aldoses (Schneider and Erlemann, *2867*); *2,4-dinitrophenylhydrazine* in 95 % ethanol containing 1 % concentrated HCl gives, after heating on the paper for a few minutes, at 70°, orange spots on light yellow background, with ketoses only (Gray, *1202*). On spraying the paper first with *aniline oxalate,* then with a permanganate solution containing 3 % sulphuric acid and heating at 100°, ketoses give grey-black or black spots, aldoses brown spots (Malyoth, *2116*). The reagent of Godin (*1150*) (1 % *vanillin* in ethanol + 3 % perchloric acid in water, 1:1) gives, after heating at 85° for 3–4 min, pale blue to lilac spots with polyols, grey-green spots with ketoses, brick red with rhamnose; inositol and aldoses do not show up.

Lactones and esters were detected by Abdel-Akher and Smith (*1*) by spraying with alkaline *hydroxylamine* and detecting the hydroxamic acids thus formed with ferric chloride. Lanning and Cohen (*1881*) use a 2 % solution of *o-phenylenediamine* in 80 % ethanol as a specific spray for 2-ketohexonic acids.

(d) *Spray reagents for non-reducing carbohydrates, polyols, methylated sugars, etc.*

In absence of acid substances, carbohydrates and polyols can be detected on paper by a spray of 1 part 0.5 N sodium borate (pH 9.18), 2 parts phenol red (2 mg/ml ethanol) and 7 parts methanol (Hockenhull, *1419*).

The use of *periodate* for the detection of carbohydrates and polyols on paper has been recommended by various authors (Yoda, *3651*; Lemieux and Bauer, *1956*; Cifonelli and Smith, *572*); the chromatograms are sprayed with a dilute solution of periodate followed by a benzidine solution; the substances which react with periodate are located by the appearance of colourless spots on a blue blackground, which arises from the action of the periodate on benzidine (Cifonelli and Smith, *572*). Metzenberg and Mitchell (*2194*) detect periodate-oxidizable compounds (also several amino acids) by spraying first with aqueous KIO_4 (0.01 M), drying at room temperature for 8–10 min and then spraying with a solution of 35 % saturated sodium tetraborate containing 0.8 % KI, 0.9 % boric acid and 3 % soluble starch. Periodate reacts with the iodide ion to liberate iodine, which in turn gives a blue colour with starch. Where periodate has been reduced, white spots appear on a blue background. Lemieux and Bauer (*1956*) obtain greenish yellow spots with the following spray reagent: 2 parts of 2 % aqueous sodium metaperiodate and 1 part of 1 % permanganate in 2 % aqueous Na_2CO_3.

Sugar mercaptals, glycosides, hexonic acids, non-reducing oligo- and polysaccharides, can also be detected by exposing the developed and dried chromatogram to vapours of iodine (Greenway *et al.*, *1215*).

Schneider and Erlemann (*2867*) recommend a solution of tin chloride double salt of *p*-aminodimethylaniline hydrochloride which reveals all reducing and non-reducing carbohydrates. See also Table 85.

(e) *Quantitative estimation*

(i) *Spot area measurement*

Gustafsson *et al.* (*1244*) spray the paper with aniline phthalate and after heating to 105°, photograph the paper and evaluate a reduced film on a Zeiss-Schnellphotometer ($\pm$ 5 % accuracy). Direct measurement of spot intensity after reaction with $AgNO_3$ using a Standard Photovolt electron-transmission densitometer is described by McFarren *et al.* (*2080*) (accuracy 2–5 %). See also Wallenfels *et al.* (*3467*).

McCready and McComb (*2071*) have developed a reflectance method using the colour produced by 2 % aniline and 2 % trichloracetic acid in ethyl acetate.

The use of [110]Ag in quantitative paper chromatography of sugars was described by Jaarma (*1527*) using the method of Trevelyan *et al.* (*3300*) with [110]Ag incorporated in $AgNO_3$; metallic Ag is precipitated in the spot and the radioactivity measured; 3 to 40 μg can be estimated with precision.

(ii) *Elution and determination*

Laidlaw and Reid (*1866*) showed that complete extraction of carbohydrates from paper is effected with water at room temperature. Numerous workers describe techniques for elution and subsequent estimation. Somogyi's copper reagent was employed by Flood *et al.* (*983*) and in combination with arseno-molybdate (measurement of molybdenum blue) by Duff and Eastwood (*825*) (accuracy $\pm$ 1–10 %).

Roudier (*2743*) has described some modifications of the method of Flood et al. (*983*).

Hirst et al. (*1412*) later found the copper reagent not to be entirely satisfactory and employed alkaline iodine for the quantitative micro-determination of eluted sugars with very good results. Periodate oxidation followed by titration of formic acid with NaOH was used by the same authors (Hirst and Jones, *1416*), also by Weygand and Hoffmann (*3523*). Hawthorne (*1338*) used the titration of Willstätter-Schudel (5 % accuracy). See also Shu (*2971*).

The colorimetry of eluted formazan obtained by the reduction of triphenyltetrazolium chloride on the paper has been described in detail by Wallenfels et al. (*3467*), Fischer and Dörfel (*966*) and Giri and Nigam (*1122*), the latter in combination with circular paper chromatography. Blass et al. (*272*) and Baar (*112*) have described the quantitative estimation of sugars by elution and colorimetry of the colour obtained with aniline hydrogen phthalate. Morris (*2282*) used the anthrone reagent.

Wadman et al. (*3438*) have developed a quantitative method for estimation of reducing saccharides; the sugars are allowed to react with N-(1-naphthyl)-ethylenediamine and the resulting glycosylamines then separated by paper chromatography. After elution with 1 % $Na_3PO_4.12H_2O$ the fluorescence is determined in a fluorometer.

Estimation of reducing sugars in biological extracts by current methods usually gives too high values due to extraneous reducing substances; more exact values are obtained after paper chromatography and estimation of the eluted sugars (Wanner, *3476*).

Dimler et al. (*779*) have described the quantitative paper chromatography of hexoses and their oligosaccharides, using the anthrone reaction applied to the eluted spots.

(f) *Paper chromatography of various derivatives of carbohydrates*

The following papers concern separations of derivatives of carbohydrates on paper:

p-nitrophenylhydrazones: Stoll and Rügger (*3126*);
osazones: Barry and Mitchell (*148*);
uronic acids and derivatives: Rao et al. (*2624*): Macek and Tadra (*2077*),
    Roudier (*2744*), Roudier and Eberhard (*2745*);
saccharinic acids: Moilanen and Richtzenhain (*2247*);
anilides of saccharinic acids: Green (*1212*);
glycals: Edward and Waldron (*870*).

5. SOME APPLICATIONS OF THE PARTITION CHROMATOGRAPHY
OF SUGARS

As the technique is widely used in all branches of sugar chemistry, a complete survey of the field would be too extensive for this treatise. Some important fields of research made possible by the application of partition chromatography will be mentioned.

(a) *Identification of some special monosaccharides*

Ketopentoses: McGeown and Malpress *(2082)*; *ribose* and *ribulose*: Seegmiller and Horecker *(2931)*; *methylated riboses*: Anderson *et al.* *(54)*; *methylpentoses* and *deoxysugars*: Edward and Waldron *(870)*; *apiose*: Bell *et al.* *(178)*; *keto-heptoses*: Noggle *(2373)*; *cycloses*: Posternak *et al.* *(2576)*.

(b) *Separations of oligosaccharides*

The degradation of starch or partial hydrolysis products by amylase has been studied by several authors by paper chromatography, and very good separations of amylotriose, tetraose, up to hexaose, have been obtained; (Myrbäck and Willstaedt, *2331*; Pazur, *2476*; Bird and Hopkins, *253*; French and Wild, *1036*); Giri *et al.*, *(1123)*; galactosyl-oligosaccharides have been separated by Pazur *(2477)* and Montreuil *(2257)*; isomaltose and isomaltobiose by Jeanes *et al.* *(1564)*, inulobiose, triose and tetraose by Pazur and Gordon *(2479)*; radio-active oligosaccharides by Pazur and French *(2478)*; circular chromatography has been applied to similar problems by Giri and Nigam*(1121)*, Giri *et al.* *(1123)* (see also Fig. 77).

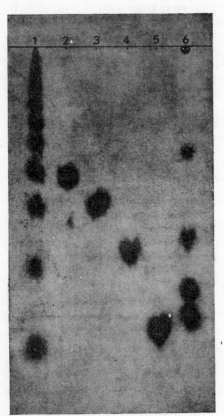

Fig. 77. Paper chromatography of oligo-saccharides; the spots were revealed with aniline phthalate + urea.

1 = natural mixture of oligosaccharides from an extract of Jerusalem artichoke.
2 = a tetrasaccharide isolated from the extract;
3 = a trisaccharide from the same extract;
4 = saccharose;
5 = fructose;
6 = an artificial mixture of inulin, raffinose, saccharose, glucose and fructose (Dedonder; cf. *724*)

Dedonder (*724*) has separated on cellulose columns a series of gluco-fructosans from *Helianthus tuberosus* extracts; the first fraction eluted was identified as saccharose, the second was a trisaccharide containing two molecules of fructose and one of glucose and the third was a tetraholoside containing three molecules of fructose for one of glucose (Fig. 77).

Difficulties are however encountered with the *paper chromatography of higher oligosaccharides*, as the distances moved by the spots grow smaller and smaller; development for several days is rarely desirable; the movement can be increased by working at 37° (*1459*) or by multiple development. The latter technique, described by Jeanes *et al.* (*1565*) gave discrete spots of amylosaccharides ranging from maltose to the decaose (solvent: equal parts fusel oil, pyridine, water; two to four developments, each of 22 hours). Bayly and Bourne (*166*) have described a new approach, by which considerable increases in R_F values (about eight-fold for disaccharides) can be obtained; the method consists in the conversion of the oligosaccharides, on the paper, into the corresponding N-benzyl-glycosylamines, thus reducing hydro-solubility; separation of the glycosylamines of dextrins containing up to ten sugar units are described. The spots can be detected with ninhydrin. A similar method has been adopted by Wadman *et al.* (*3438*) for the quantita-tive estimation of reducing oligosaccharides; the sugars are allowed to react with N-(1-naphthyl)-ethylenediamine and are then separated on paper; the amount of each sugar may be estimated from the brightness of the fluorescence under UV light.

(c) *Analysis of polysaccharides*

The determination of the structure of oligo- and polysaccharides is usually carried out by methylation and subsequent hydrolysis. The position of the linkages in the polysaccharide can be recognised from the positions of the free OH groups present in the methylated monosaccharides.

For the separation and identification of the constituents of the hydroly-sates, partition chromatography on columns of cellulose and on paper is far better than the previously employed techniques. Hirst, Hough, Jones and their collaborators using this technique determined the structures of numer-ous gums and mucilages, for example slippery elm mucilage (*1414, 1457*), cherry (*1600*), peach, lemon (*1417*), grapefruit (*611*) and sterculia gums (*1411, 1413, 1456*), inulin from Dahlia tubers (*1415*), esparto grass xylan (*544*). Nordal and Klevstrand (*2378*) examined the constituents of crassu-laceous plants. Guar gum was studied by Rafique and Smith (*2612*).

Frog spawn mucin was studied by Folkes *et al.* (*998*), the capsular poly-saccharide of *Cryptococcus neoformans* by Evans and Mehl (*926*).

(d) *Paper chromatography of polysaccharides*

Leitner and Kerby (*1953*) have separated acid mucopolysaccharides on Whatman No. 1 with propanol or ethanol in 0.066 M phosphate buffers of pH 6.5. The spots are detected with 0.06 % toluidine blue 0 in 0.5 % aqueous acetic acid. Kerby (*1698*) has described the separation of heparin and chondroitin sulphate, using the above mentioned solvent. See also Molho and Molho-Lacroix (*2250*).

(e) *Study of enzymatic reactions*

Enzymatic reactions with carbohydrates are readily examined by placing samples obtained after various reaction times side by side on one sheet of paper. The action of pectinase on pectin was observed in this manner by Jermyn and Tomkins (*1588*). The rate of change of sugar transformation during fermentation was examined by Barton-Wright and Harris (*152*). Wallenfels (*3466*) studied the enzymatic synthesis of oligosaccharides from disaccharides. The enzymatic synthesis and degradation of starch was examined also by paper chromatography by Barker *et al.* (*140*). Edelman and Bacon (*858*) studied the trans-fructosidation reaction in extracts of *Helianthus tuberosus*. Leuthardt and Testa (*1985*) studied the phosphorylation of fructose in the liver.

The separation of mono- and oligosaccharides from malting and brewing liquors has been described by Montreuil and Scriban (*2261*); Opieńska-Blauth *et al.* (*2417*) have studied the metabolism of hexoses in liquid cultures of *E. coli*. A paper by Pazur (*2475*) contains photos of "progress chromatograms" showing transfructosidation reactions.

(f) *Clinical applications*

Wallenfels (*3465*), using paper chromatography, was able to demonstrate that fructose is a normal constituent of human blood (0.5–5 mg %). Sugars in urine were detected by Odin and Werner (*2395*), Eastham (*849*), Horrocks and Manning (*1450*), Montreuil and Boulanger (*2259*) and Robertson (*2699*); the biochemistry of lactose was studied by Caputto and Trucco (*504*) and Malpress and Morrison (*2115*).

(g) *Various applications*

The isolation of ^{14}C uniformly labelled fructose prepared by photosynthesis was carried out in 75 paper strip chromatograms by Udenfriend and Gibbs (*3340*). The stimulant involved in the germination of *Orobanche minor* was found by Brown *et al.* (*423*) to travel with the solvent front and thus to differ from the usual carbohydrates. The analysis of sulphite waste liquors on starch columns is described by Mulvany *et al.* (*2314*). Patton and Chism (*2467*) observed the formation of 10 distinct fluorogens by the condensation of glycine and glucose (Maillard's browning reaction).

The reaction of carbohydrates with ammonia was studied by Raacke-Fels (*2608*); for the analysis of sugars in honey, see Goldschmidt and Burkert

(*1156*). The stability of sugars in boiling aqueous solution has been studied by Soutar and Hampton (*3053*).

6. AMINOSUGARS

The separation of glucosamine and galactosamine on columns of Dowex-50 has been achieved, by elution with 0.3 N HCl and detection by recording the electrolytic conductivity (Drake and Gardell, *814*) or with the Elson-Morgan colour reaction (Gardell, *1088*). A similar separation on Dowex-50 using the method of Moore and Stein (*2268*) and detection with ninhydrin, has been described by Eastoe (*850*).

The preparation of pure galactosamine with the aid of Dowex-1 in the carbonate form has been described by Roseman and Ludowieg (*2737*).

Charcoal-Celite columns have been used for the purification of a glucosamine containing tetrasaccharide from human milk (Kuhn *et al.*, *1818*, *1819*) and for the separation of N-acetyl-lactosamine (I) from N-acetyl-isolactosamine (II) (Kuhn and Kirschenlohr, *1820*); the mixture of these two latter substances had been previously separated from non aminated sugars by adsorption on Amberlite IR-120.

$$
\begin{array}{ll}
\text{HCO} & \text{CH}_2\text{—NHCOCH}_3 \\
| & | \\
\text{HC—NHCOCH}_3 & \text{CO} \\
| & | \\
\quad\text{(I)} & \quad\text{(II)}
\end{array}
$$

The separation and quantitative estimation of glucosamine and galactosamine has been effected by Leskowitz and Kabat (*1982*) by reduction of the aminosugars with $NaBH_4$ to the corresponding hexosaminitols which were then dinitrophenylated by the method of Sanger; the DNP-aminohexitols are first purified on a silicic acid column and then separated on a borate buffered column of silicic acid-Celite, with 20 % ethanol-chloroform.

Aminoff and Morgan (*49*) were the first to separate hexosamines on *paper*; they revealed these substances by spraying with ninhydrin, or with ammoniacal silver nitrate or with Ehrlich's reagent.

The R_F values of glucosamine and chondrosamine are given in Partridge's table (Table 81); the aminosugars can be detected by the Elson-Morgan reaction. Although the R_F differences are small, Aminoff *et al.* (*50*) were able to identify them both in hydrolysates of the blood group A substance on two-dimentional chromatograms with collidine and phenol as solvents and using the red colouration with acetylacetone and Ehrlich's reagent as confirmation. Kaye and Stacey (*1668*) separated glucosamine from glucuronic acid on paper with butanol/ethanol/water or butanol/acetic acid/water mixtures as the solvents. See also Klenk (*1737*), Masamune and Maki (*2164*).

The identification of glucosamine in an antigenic polysaccharide from tuberculin was carried out by Kent (*1694*, *1695*) by chromatographing the hydrolysate and the 2,4-dinitrophenyl (DNP) derivative (see p. 330). In another paper Kent *et al.* (*1696*) list the R_F values of a large number of DNP derivatives of aminosugars (and amino acids) (Table 86). The separation of 10–20 mg of mixtures of glucosamine and chondrosamine was carried out by Annison *et al.* (*69*). The DNP derivatives were separated on a Hyflo Supercel column using a 0.2 M borate/KOH buffer (pH 9.9) as stationary

TABLE 86. R_F values of methylated sugars and of N-(2,4-dinitrophenyl) derivatives of aminosugars and amino acids (Kent *et al.*, *1696*)

Solvent: 40% butanol/10% ethanol/50% water

| | | | Reaction with | |
	R_F	$AgNO_3$	aniline phthalate	ninhydrin
3,4-Dimethyl-D-mannose	0.53	+	+	—
3,4,6-Trimethyl-D-mannose	0.70	+	+	—
2,3,4,6-Tetramethyl-D-mannose	0.68	+	+	—
2,3-Dimethyl-D-glucose	0.51	+	+	—
2,3,4-Trimethyl-D-glucose	0.70	+	+	—
2,3,6-Trimethyl-D-glucose	0.66	+	+	—
2,3,4,6-Tetramethyl-D-glucose	0.85	+	+	—
1,3,4-Trimethyl-D-fructose	0.67	+	—	—
1,3,4,6-Tetramethyl-D-fructose	0.75	+	—	—
D-Glucosamine HCl	0.10	+	—	+
N-Acetyl-D-glucosamine	0.24	+	—	+
N-Carbobenzoxy-α-methylglucosamidine	0.82	—	—	—
3-Methyl-N-acetylglucosamine	0.34	+	—	—
4,6-Dimethyl-N-acetylglucosamine	0.50	+	+	—
3,4,6-Trimethyl-N-acetylglucosamine	0.72	+	+	—
Glucosaminic acid	0.02	—	—	+
α-Methyl-2-acetamido-glucopyranoside	0.31	—	—	+
β-Methyl-2-acetamido-glucofuranoside	0.42	±	±	—
N-DNP-D-Glucosamine	0.75	+	+	—
1,3,4,6-Tetraacetyl-DNP-D-glucosamine	0.93	+	+	—
DNP-glucosamine diethyl mercaptal	0.89	—	—	—
Tetra acetyl-DNP-glucosamine diethyl mercaptal	0.96	—	—	—
DNP-glucosaminic acid	0.60	—	—	—
DNP-chondrosamine	0.61	+	+	—
1,3,4,6-Tetraacetyl-DNP-chondrosamine	0.92	+	+	—
DNP-glycine	0.42	—	—	—
DNP-tyrosine	0.77	—	—	—
DNP-histidine	0.64 & 0.85	—	—	—
DNP-arginine	0.65	—	—	—
DNP-valine	0.72	—	—	—
DNP-aspartic acid	0.28	—	—	—
DNP-alanine	0.58	—	—	—
DNP-tryptophan	0.69	—	—	—

phase and a 30 % amyl alcohol/CHCl$_3$ mixture as the mobile phase. 75 % yields of highly pure products were obtained.

In presence of sulphate ions, aminosugars can give a "slow spot" due to the formation of a sulphate (Waldron-Edward, *3449*).

Payne *et al.* (*2474*) have studied the chromatographic determination of glucosamine with ninhydrin, using the butanol-pyridine solvent of Jeanes *et al.* (*1565*); they found that aminosugars do not respond to aniline hydrogen phthalate under these conditions; N-acetyl-glucosamine does respond, except in presence of galacturonic acid.

A method allowing the determination of microgram amounts of aminosugars described by Stoffyn and Jeanloz (*3120*) uses a pretreatment with ninhydrin followed by the identification of the pentoses formed. See also the relevant chapter in the book of Kent and Whitehouse, *Biochemistry of the aminosugars* (*1697*).

Kuhn *et al.* (*1826*) have found that the α and β forms of N-acetyl-methylglucosaminides can be separated on paper columns or sheets with ethyl acetate, pyridine, water (2:1:2). The substances were detected by the –CO–NH– reagent of Rydon and Smith (*2769*).

For the separation and detection on paper of neuraminic, lactaminic and gynaminic acids see Kuhn and Brossmer (*1817*) and Zilliken *et al.* (*3698*).

7. GLYCOSIDES

(a) *Column chromatography*

Mowery Jr. and Ferrante (*2297*) have separated α- and β-methyl-D-galactosides on Florex XXX.

Augestad and Berner (*102*) have described the separation of anomeric methyl glycosides (methyl furanosides of galactose, arabinose and xylose) on columns of powdered cellulose.

The behaviour of several glycosides on Dowex-1 in the chloride and borate forms has been studied by Chambers *et al.* (*543*). Some of the separations achieved were ascribed to the adsorption of the aromatic aglycon moiety, rather than to exchange of borate complexes of the glycosides.

Cardiotonic glycosides can be purified on neutralized alumina (solvent CHCl$_3$–CH$_3$OH) (Schindler and Reichstein, *2844*). Acetylated glycosides are better chromatographed on magnesium trisilicate (Siegfried, Zofingen) because alumina easily splits off the acetyl groups (Aebi and Reichstein, *19*; Schindler and Reichstein, *2842*).

Stoll *et al.* (*3124, 3125*) employed columns of cotton linters, diatomaceous earth and silica gel for preparative scale separations of cardiac glycosides by partition chromatography. For 10 g quantities, an aluminium tube 4 m long and 5.6 cm wide containing 5 kg of diatomaceous earth, 1250 ml water and a mobile phase of ethyl acetate/butanol was used, and separations of mixtures of ten glycosides were carried out. Silica gel columns (36 g) with ethyl acetate/0.5 % methanol/water as solvent gave excellent separations of 5–100 mg quantities of mixed glycosides. The eluates are collected in small fractions which are evaporated, weighed and identified. Excellent separations were obtained.

Very good separations of cardiotonic glycosides were also obtained by Hegedüs et al. (1350) on kieselguhr-water (1:1) columns, with a series of organic solvents (petroleum ether-benzene, benzene, benzene-chloroform, chloroform, chloroform-butanol) as mobile solvent. The columns must be relatively large (500 g for 1 g of substance).

(b) *Paper chromatography*

Cifonelli and Smith (572) have described a method for the detection on paper of non-reducing glycosides: the chromatogram is first sprayed with a dilute periodate solution whereby the α-glycol groupings of the sugar moiety are split and the periodate ion is reduced to iodate; upon spraying with benzidine, the glycosides are located by the appearance of white spots on blue background, the latter arising from the oxidizing action of periodate on benzidine.

Janot et al. (1559) have given the R_F values for several solvents of the following glycosides and their aglycones: calycanthoside, fraxoside, esculoside, methylesculoside, salicoside, populoside, coniferoside, syringoside, colchicoside, monotropitoside, aucuboside, asperuloside, loroglossoside.

The paper chromatography of *cardiac glycosides* has been described by several authors. Schindler and Reichstein (1377, 2843, 2846) first used filter paper impregnated with formamide (see Zaffaroni on sterols, p. 278) and chloroform or benzene-chloroform as mobile solvent. As spray reagent they used *m*-dinitrobenzene and alkali, which only shows up the steroid glycosides having an unsaturated γ-lactone ring. Jaminet (1547, 1548) as well as Svendsen and Jensen (3161) heat the paper with trichloroacetic acid for detecting the spots by their fluorescence (limit 0.25–1 μg). Lawday (1887) has recommended $SbCl_3$ in chloroform for detection of digitalis glycosides (limit 0.5 μg per cm²). Heftmann and Levant (1349) have described the use

TABLE 87. R_F values of cardiac glycosides and genins in chloroform-tetrahydrofuran-formamide 50:50:6.5 (Kaiser, 1615).
The numbers refer to the chromatogram, Fig. 78, p. 267.

Substance	R_F	Substance	R_F
1 Digitalinum verum	0.16	6 Desacetyldigilanide B	
2 Digilanide C	0.33	(Purpureaglycoside B) . .	0.19
3 Digilanide B	0.47	7 Desacetyldigilanide A	
4 Digilanide A	0.85	(Purpureaglycoside A) . .	0.50
5 Desacetyldigilanide C	0.13	8 Strospeside	0.075

of $Ag(NH_3)_2{}^+$ in alkaline medium (Tollens' reagent) and give the mobilities of a series of cardiac glycosides and aglycones. See also Heftmann *et al.* (*1347*). Quite recently, Bernasconi *et al.* (*225a*) have described the detection of glycosides and aglycones of the scilla-bufo-type (which have a strong absorption at 300 mμ) by direct photocopy with ultraviolet light.

Jensen (*1576*) has also used $CHCl_3$-methanol-water (5:5:1) and Silberman and Thorp (*2984*) have used ethyl acetate-benzene-water (86:14:50). See also Okada *et al.* (*2410*) and Bally *et al.* (*130*).

Tschesche *et al.* (*3308*) have described similar separations using water saturated organic alcohols (octanol, pentanol) as stationary and aqueous formamide as mobile phase. Reaction with an appropriate reagent and colorimetry of the separated substances yields quantitative values.

The paper chromatography of cardiac glycosides and aglycones of *Digitalis purpurea* has been described in detail by Jensen (*1577*); for glycosides and aglycones of *Digitalis lanata*, see Jensen and Tennöe (*1579*); the detection of scilliroside has been described by Dybing *et al.* (*846*).

For the paper chromatography of strongly polar cardiac glycosides, Schenker *et al.* (*2840*) recommend water impregnated paper with butanol or butanol-toluene (1:1) as mobile phase.

Kaiser (*1615*) uses formamide impregnated paper with xylene-methyl ethyl ketone (1:1) or chloroform-tetrahydrofuran-formamide (50:50:6.5) for separating a large number of cardiac glycosides (see Fig. 78 and Tables 87 and 88).

The α and β forms of acetyldigitoxin, acetylgitoxin and acetyldigoxin could be separated.

Kaiser also gives the R_F values in 8 solvents of the most important sugars found in the cardiac glycosides (Table 89).

For the paper chromatography of saponins see Dutta (*841*) and Gootjes and Nauta (*1162*).

For N-glycosides of arylamines see Inouye *et al.* (*1507*).

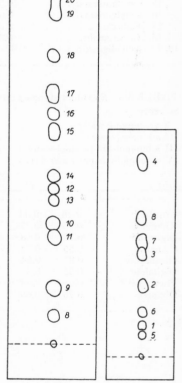

Fig. 78. Paper chromatograms of cardiac glycosides and genins. For explanation of the numbers, see Tables 87 and 88

For separations of purine and pyrimidine glycosides, see Chapter 33. See also the sections on Anthoxanthins and Anthocyanins.

TABLE 88.　R_F values of cardiac glycosides and genins in xylene-methyl ethyl ketone 1 : 1 saturated with formamide (Kaiser, *1615*). The numbers refer to the chromatogram, Fig. 78, p. 267.

Substance	R_F	Substance	R_F
8 Strospeside	0.075	Oleandrin	0.93
9 Digoxin	0.16	Oleandrigenin	0.72
10 Digoxigenin	0.32	Neriifolin	0.55
11 Gitoxin	0.29	Cymarin	0.55
12 Gitoxigenin	0.42	Allocymarin	0.48
13 Acetyldigoxin α . . .	0.39	Strophanthidin	0.40
14 Acetyldigoxin β . . .	0.45	Allostrophanthidin	0.32
15 Acetylgitoxin α . . .	0.59	Isostrophanthidin	0.76
16 Acetylgitoxin β . . .	0.64	Strophanthidol	0.26
17 Digitoxin	0.69	Cymarol	0.36
18 Digitoxigenin	0.81	Emicymarin	0.11
19, 20 Acetyldigitoxin $\alpha + \beta$.	0.90–0.95	Alloemicymarin	0.07

TABLE 89.　R_F values of sugars occurring in cardiac glycosides (Kaiser, *1615*).

Solvents:

I *sec.*-butanol-water
II *n*-butanol-water
III *n*-butanol-*tert.*-butanol-water 1 : 1 : 1
IV *n*-butanol-ethanol-water 4 : 1 : 5

V *n*-butanol-acetic acid-water 4 : 1 : 5
VI ethyl acetate-acetic acid-water 5 : 2 : 2
VII *n*-butanol-pyridine-water-benzene
　　50 : 30 : 30 : 4.5
VIII ethyl acetate-pyridine-water 25 : 10 : 35

Solvent	I	II	III	IV	V	VI	VII	VIII
Glucose	0.28	0.08	0.25	0.12	0.17	0.25	0.36	0.10
Rhamnose	0.55	0.27	0.36	0.34	0.42	0.53	0.65	0.44
Digitoxose	0.70	0.58	0.59	0.65	0.66	0.81	0.87	0.88
Cymarose	0.85	0.74	0.89	0.82	0.85	1.0	0.95	1.0
Oleandrose	0.81	0.84	0.76	—	0.87	0.95	1.0	1.0
Digitalose	0.55	0.37	0.43	0.51	0.52	0.60	0.71	0.52
Thevetose	0.61	0.48	0.55	—	0.62	0.70	—	—
Diginose	0.78	0.68	—	0.74	0.76	0.91	—	1.0

CHAPTER 29

Lipids

Asselineau (*89*) has published a review of the paper chromatography of lipids.

Fillerup and Mead (*950*) have separated plasma lipids into several fractions (sterol esters, triglycerides, sterols, fatty acids and phospholipids) by chromatography on silicic acid and elution with a series of solvents of increasing eluting power (see Table 90).

TABLE 90. Chromatographic separation of an artificial lipid mixture (Fillerup and Mead, *950*)

50 mg of lipid, 13 g silicic acid in a tube of 2 cm diameter

Fraction (in order of elut.)	Compounds used	Solvent mixture eluting fraction	Column vol. of eluate	% recovery
1. Sterol ester	Cholesterol palmitate, cholesterol oleate	1% ether in petroleum ether	20	98.4–100
2. Triglyceride	Tripalmitin, triolein	4% ether in petroleum ether	20	96 –100
3. Sterol	Cholesterol	10% ether in petroleum ether	15	99 –100
4. Acid	Stearic, oleic	50% ether in petroleum ether	15	97
5. Phospholipid	Lecithin	25% methanol in ether	15	100

Hack (*1251*) has described a technique for the analysis of lipids in quantities of $10^{-2} \mu$moles by means of spot tests applied to radial filter paper chromatograms (see also Kaufmann and Budwig, *1657*).

Casselman (*526*) has recommended acetylated Sudan black B for the localization of lipids on paper (as saturated solution in 70 % ethanol, or ethylene- or propylene glycol).

Dieckert and Reiser (*774*) have separated various lipids on glass-fibre paper impregnated with silicic acid; developing with 2 % ethyl ether in *iso*octane, the following R_F values were found: 1-monopalmitin, 0.05; dipalmitin, 0.27; cholesterol, 0.41; tripalmitin, 0.79; cholesterol acetate, 1.0.

1. GLYCERIDES

Trappe (*3294*) has carried out a systematic study of the behaviour of various lipids on alumina columns; the order of increasing adsorbability was found to be: hydrocarbons, cholesterol esters, triglycerides, free cholesterol, fatty acids, phosphatides. See also Table 90. Williams (*3577*) has published a review of the applications of chromatography to the analysis of oils. A review by Weil (*3483*) on "industrial oil and fat chromatography" contains 220 references.

Kaufmann (*1651*) has studied the separation of glycerides on columns of alumina, acid earths and silica gel, and has described the inversion of chromatographic behaviour on changing from alumina to silica gel columns. The first adsorbs high molecular weight glycerides least, and the latter adsorbs the low molecular weight glycerides the least. Separation of mono-, di- and triglycerides were described, the monoglycerides being adsorbed most strongly and the triglycerides the least. See also Desnuelle *et al.* (*751*).

Hamilton and Holman (*1285*) have studied the displacement analysis of mixtures of glycerides using as adsorbent a mixture of 1 part Darco G-60 charcoal and 2 parts Hyflo Supercel; ethanol was used for the separation of monoglycerides and triglycerides up to including trilaurin, benzene for the higher triglycerides; triglycerides were used as displacer. The adsorbabilities of homologous glycerides on Darco G-60 increase with molecular weight. The order of adsorbability of the mono-, di- and triglycerides depends upon the polarity of the solvent. In ethanol (polar solvent) the least polar (tri) glycerides are adsorbed most strongly. In the non-polar benzene, the most polar (mono) glycerides are adsorbed most strongly (see also Hamilton, *1283*).

Savary and Desnuelle (*2833*) have described a method for the fractionation of mixtures of free acids, mono-, di- and tri-glycerides; in ether solution, the free acids are adsorbed on Amberlite IRA-400, then the glycerides are fractionated on a column of silicone treated kieselguhr (cyclohexane as stationary, aqueous ethanol as mobile phase); mono-, di- and triglycerides are thus easily separated; some difficulties are encountered when the glycerides contain short (C_{12}) and long chains (C_{16}, C_{18}) simultaneously.

Borgström (*316*) has also separated free acids from glycerides and cholesterol esters by adsorbing the free fatty acids on Amberlite IRA-400; the remaining neutral esters are fractionated on silicic acid. More recently, Borgström (*317*) has described the separation of tri-, di- and monoglycerides on silicic acid; the first are eluted with benzene, the second with benzene-chloroform (85:15), the last with chloroform. 2-Monoglycerides are partially isomerized to 1-monoglycerides on the column.

For the neutralisation of fats and oils see *288, 1653, 1655, 2051.*

The chromatography of oils oxidized by air blowing allows separation into fractions of different viscosity (*1658*). For a study of oxidized oils, or oils combined with sulphur, see Williams (*3577*).

Walker and Mills (*3457*) have described the separation of the glycerides of flax oil and linseed oil (*3456*) into fractions containing 4,5,6 and 7 double bonds per glyceride molecule (alumina as adsorbent and hexane as solvent).

Natural oils, such as *Stillingia* oil or *Perilla* oil can easily be separated on alumina into fractions having widely different properties; e.g. mono-, di- and tri-glycerides, sterol esters and waxes (Huang *et al.*, *1468*; Tischer and Tögel, *3251*). Some fractions can be distinguished by their fluorescence.

Koch (*1749*) has separated on clay or alumina the sodium sulphonates of the lipoid-soluble substances formed during the sulphonation of petroleum. In this way the emulsions which interfere in the more usual analytical methods are avoided.

Fisk (*974*) and Mills (*2227*) have reviewed the applications of chromatography to the chemistry of drying oils.

2. PHOSPHATIDES

(a) *Column chromatography*

Thannhauser *et al.* (*3227*) have described the desalting of phosphatides on Amberlite XE-58 and XE-64 columns.

Macpherson (*2096*) has removed cadmium chloride from the $CdCl_2$ complex of lecithin by filtration on a column of Amberlite MB3, in ethanol-water-$CHCl_3$ (7:3:1).

The separation of phospholipids from neutral fats and fatty acids by chromatography on silicic acid has been described by Borgström (*315*). Nitrogenous impurities of phospholipids can be removed by filtration of a chloroform solution through a column of cellulose powder (Lea and Rhodes, *1890*).

Choline-containing phospholipids can be separated from other phospholipids by adsorption on MgO, only the former substances are eluted with methanol (*3209*).

Rice and Osler (*2675*) have purified commercial beef heart lecithin by adsorption on magnesol, and have thus obtained pure lecithin.

Hanahan *et al.* (*1294*) have used alumina for the isolation of phosphatidyl choline from mixed egg phospholipids. A 3 % solution of the phospholipids in 95 % ethanol is passed over the column; the lecithins are eluted nearly quantitatively by several portions of ethanol. This method has been used with success for the preparation of pure lecithins from yeast (Hanahan and Jayko, *1293*) and other sources (Hanahan, *1292*).

Silicic acid-Celite mixtures can also be used for the separation of phospholipids, as described by McKibbin and Taylor (*2088*), Lea and Rhodes (*1891*); the latter authors have also used paper impregnated with silicic acid and achieved good separations of phosphatidyl-ethanolamine, lecithin, lysolecithin and sphingomyelin (*1892*). The method of McKibbin and Taylor (*2088*) combined with a gradient elution (methanol in chloroform) has been used by Vilkas and Lederer (*3405*) and Michel and Lederer (*2207*) for the separation of phosphatides of Mycobacteria in several groups differing markedly in their chemical properties.

The purification of sphingomyelins and cerebrosides on alumina columns has been described by Marinetti and Stotz (*2131*).

Svennerholm (*3162*) has described the purification of gangliosides by partition on columns of cellulose powder.

The separation of choline containing phospholipids from the other phosphatides by chromatography on columns of cellulose powder as described by Bevan *et al.* (*238*) could not be confirmed by Lea and Rhodes (*1890*).

(b) *Paper chromatography*

Amelung and Böhm (*47*) have described a good method for the separation of lecithin, phosphatidyl-ethanolamine and phosphatidyl-serine on paper, using phenol as mobile solvent. The detection of the different constituents can be made with ninhydrin, phosphomolybdic acid, Reinecke salt, the plasmal reagent, a modified phosphoric acid reagent and with Sudan black; the latter only detects phosphatides containing unsaturated acyl radicals. Phosphatidyl-ethanolamine and phosphatidyl-serine can be distinguished by the use of the copper carbonate method of Crumpler and Dent (*666*).

Huennekens *et al.* (*1473*) have used paper chromatography for the separation and identification of lecithins and their hydrolysis products, such as lysolecithins, mono- and diglycerides, phosphatidic acids, glycerylphosphorylcholine etc. (Table 91). Specific spray reagents for each of these groups are described. See also Douste-Blazy *et al.* (*806*).

Kornberg and Pricer (*1768*) have studied the enzymatic synthesis of phosphatides using ^{32}P and ^{14}C-stearic acid. The intermediate phosphatidic acids were separated on paper with di*iso*propyl ether as mobile phase.

The *nitrogenous constituents* of the phosphatides can easily be separated by paper chromatography; colamine and serine are detected by a ninhydrin spray, choline with a phosphomolybdic reagent (Chargaff *et al.*, *548, 549, 1989*; Munier, *2317*; Schulte and Krause, *2905*; Hecht and Mink, *1343*). Levine and Chargaff (*1990*) have also studied the paper chromatographic behaviour of analogues of nitrogenous lipid constituents. It is worth noting that 1-hydroxy-2-aminopropanol and 1-amino-2-hydroxypropanol have exactly the same R_F value in five different solvents. (See also Cooley *et al.*, *633, 885*.) This fact shows again the necessity of confirming by some independent method identifications of "unknown substances" by paper chromatography.

TABLE 91. R_F values of some phosphatides and their components (Huennekens *et al.*, *1473*)

Solvent 1, *n*-butanol saturated with water; Solvent 2, ethanol-water (8:1); Solvent 3, *n*-propanol-water (8:1); Solvent 4, *n*-propanol-acetic acid-water (8:1:1).

Substance	Solvent 1	Solvent 2	Solvent 3	Solvent 4
Unsaturated lecithin	0.81	0.89	0.78	0.86
Saturated lecithin	0.87	0.89	0.78	0.86
Unsaturated lysolecithin	0.70	0.87	0.80	0.80
Saturated lysolecithin	0.67	0.89	0.61	0.85
α-Monopalmitin	0.93	0.92	0.91	0.90
α,β-Dipalmitolein	0.93	0.88	0.91	0.92
Phosphatidic acid	Streak (0.00–0.50)	Streak (0.00–0.50)	Streak (0.00–0.30)	Streak (0.03–0.10)
Oleic acid	0.95	0.92	0.94	0.94
Glycerylphosporylcholine . . .	0.00	0.70	0.25	0.76
β-Glycerophosphate	0.00	0.46	0.03	0.07 / 0.09
Phosphorylcholine	0.00	0.45	0.00	0.08
Choline	0.11	0.72	0.30	0.40·

Barbier and Lederer (*135*) have detected *hydroxylysine* by paper chromatography of a hydrolysate of a phosphatide of *Mycobacterium phlei*.

The principal nitrogenous constituent of many phosphatides of Mycobacteria is L-*ornithine*, which has been isolated by Gendre and Lederer (*1097*) after chromatography on Dowex-50 and elution with a gradient of aqueous ammonia.

The paper chromatography of dihydrosphingosin and related substances has been described by Gregory and Malkin (*1220*) (development of the spots with ninhydrin).

3. WAXES

Tischer and Ilner (*3250*) have reported a chromatographic procedure for the neutralisation and decoloration of crude beeswax. Daniel *et al.* (*693*) have chromatographed the unsaponifiable fraction of wool wax on alumina.

The waxes of tubercle bacilli have been extensively purified in recent years by chromatography on alumina, magnesium silicate and silicic acid (Asselineau and Lederer, *92*).

The "cord factor", a toxic lipid of tubercle bacilli, which is a 6,6'-dimycolate of α,α-trehalose ($C_{186}H_{366}O_{17} \pm 10\ CH_2$) has been obtained pure after chromatography on magnesium silicate, silicic acid and silica gel (Noll *et al.*, *2376*).

4. STEROIDS

Reviews on chromatography of steroids and sterols have been recently published by Bush (*480*) and Heftmann (*1346a*); the chromatography of the steroids of the adrenal glands has been described in detail by Haines and Karnemaat (*1266*), Romanoff and Wolff (*2733*); the uses and limitations of adsorption chromatography for the separation of urinary ketosteroids have been reviewed by Robinson (*2701*).

Pure samples of steroids for chromatographic identification can be obtained from the United States Pharmacopeial Convention, see *Nature*, 1954, *173*, 1170.

(a) *Separation on adsorption columns*

Many papers have been published on the subject of the chromatography of the unsaponifiable material from plant or animal extracts (*693, 1907, 2580, 2585–2587, 2762, 2764*). Chromatography on alumina usually gives a series of fractions, amongst which hydrocarbons, ketones and sterols can be identified.

Aluminium silicate strongly adsorbs free sterols or sterol glycosides directly from the natural oils without the use of a solvent (*1793, 3246*).

In the sterol field, chromatography has become an indispensable technique for separation and purification. For this reason only those papers presenting some new addition to the technique are dealt with here. The most usual adsorbent is *alumina* (whose free alkali can give rise to secondary reactions, see page 61), and a system of solvents of progressively greater eluting power (the "liquid chromatogram" of Reichstein) is used. The reader is referred to the series of papers by Reichstein and his school on the constituents of the adrenals, on sterols, bile acids, and cardiotonic glycosides (e.g. *925, 1645, 2200, 2426, 2650, 2656, 2657, 3089*) as well as those of Ruzicka, Prelog, Plattner *et al.* on sterols and sex hormones and on the unsaponifiable fractions of extracts of organs (e.g. *1306, 2527, 2582, 2586, 2760, 2763, 2764, 3193*). See also *150, 693, 784, 785, 946, 2786, 3296*.

Bretschneider (*380*) has described the separation of cholestanone and cholestenone on *silica gel*, as well as the isolation of androstenedione and progesterone. The separation of free and esterified cholesterol from blood plasma or cerebrospinal fluid has been described by Hess (*1378*) and by Selbach and Trappe (*2937*).

Mention has already been made of the separation of sterols as their coloured esters (*594*). Sterol azoylates can be separated on silica (Idler and Baumann, *1497*), 3,5-dinitrobenzoates on alumina and silica (Kellie and Wade, *1676, 1677*), 2,4-dinitrophenylhydrazones of progesterone, pregnenolone and other ketosteroids on alumina (Reich *et al.*, *2651*; Zwingelstein *et al.*, *3704*).

Dobriner, Lieberman *et al.* (*794, 2006*) have published a systematic study of the ketosteroids of urine. Forty-two different substances were isolated after repeated chromatography on alumina and *magnesium silicate* (No. 34 from the Philadelphia Quartz Co., Berkeley, Calif.) mixed with equal parts of Celite. Magnesium silicate is a weaker adsorbent than alumina and allows fractionation of those steroids too strongly retained·by alumina. One chromatographic fractionation involves 100 to 200 elutions. These authors insist that despite the considerable variations in the composition of the urinary steroids, the order of elution always remains the same; this greatly assists identification. Diastereoisomers are separable by the use of two or three columns (e.g. androsterone and etiocholan-3a-ol-17-one). The separation of saturated compounds from their analogues having one double bond is more difficult (e.g. dehydro*iso*androsterone and *iso*androsterone). In this case the epoxides are easier to separate (Seebeck and Reichstein, *2929*).

Table 92 (Lieberman *et al.*, *2006*) shows the elution order of sterols after chromatography on alumina.

Lieberman *et al.* (*2008*) have continued their study of the separation of urinary steroids and have indicated the following correlations between chemical constitution and relative order of elution during chromatography on alumina:

(1) "C_{21} pregnane derivatives are eluted before the corresponding C_{19} androstane derivatives."

(2) "Compounds with ring A–B having a *trans* juncture (androstane and allopregnane) are eluted before their corresponding isomers with rings A–B *cis* (etiocholane and pregnane). One exception to this rule has been noted: etiocholanol-3β-one-17 is eluted from alumina before androstanol-3a-one-17. This inversion in the expected sequence of elution may be a special case because these substances also differ in their configuration at C_3. The configuration of the 3-hydroxyl group seems to be more important in determining the chromatographic sequence than does the steric juncture of rings A and B."

(3) "Saturated compounds are eluted before their unsaturated analogues except in the case of etiocholanolone, which is eluted after $\Delta^{9,11}$-etiocholenolone. Δ^1-Androstenedione-3,17, as expected, is eluted after its saturated analogue androstanedione-3,17. The unsaturated Δ^1-compound (XI) precedes its Δ^4-isomer (XII) in the sequence of elution."

(4) "The diketomonohydroxy compounds are eluted before their corresponding dihydroxymonoketo compounds with the important exception that the dihydroxymonoketones having one hydroxyl group on C_{11} are eluted before those diketomonohydroxy compounds with one ketonic group on C_{11}."

"Thus, 11-hydroxyandrosterone and 11-hydroxy-etiocholanolone are eluted before their corresponding C_{11} carbonyl derivatives 11-keto-androsterone and 11-keto-etiocholanolone."

TABLE 92. Elution order of ketosteroids after chromatography on alumina (**Lieberman** *et al., 2006*)

	Example
1. Monoketones	I
2. Monoketones containing one acetoxy group .	II
3. Diketones a) of the allopregnane series . . .	III
b) of the pregnane series	IV
c) of the androstane series	V
d) of the etiocholane series . . .	VI
4. Mono-hydroxy-monoketones:	
a) of the allopregnane series . . .	VII
b) of the pregnane series	VIII
c) of the androstane series	IX
d) of the etioch)lane series . . .	X

O

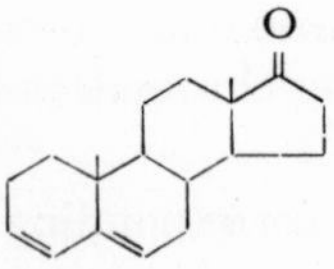

(I)

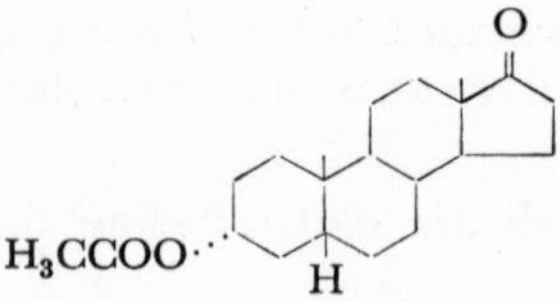

O

H₃CCOO···   H

(II)

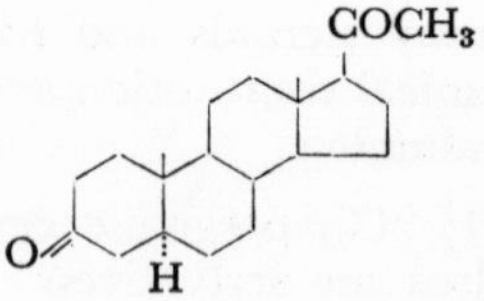

COCH₃

O   H

(III)

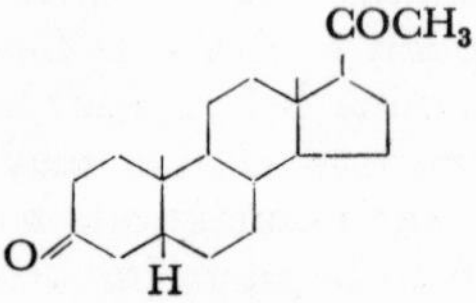

COCH₃

O   H

(IV)

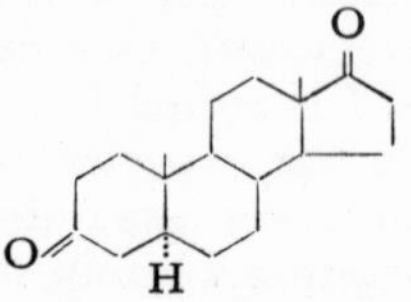

O

O   H

(V)

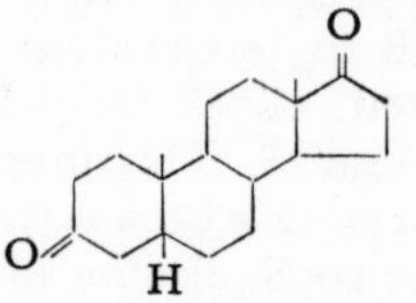

O

O   H

(VI)

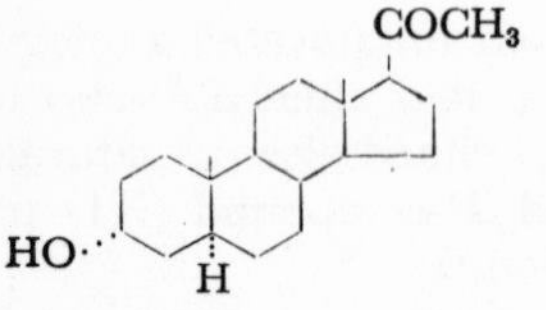

COCH₃

HO···   H

(VII)

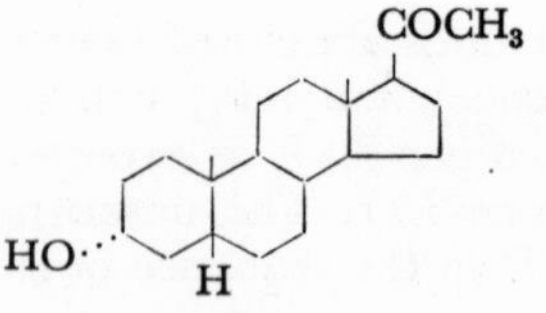

COCH₃

HO···   H

(VIII)

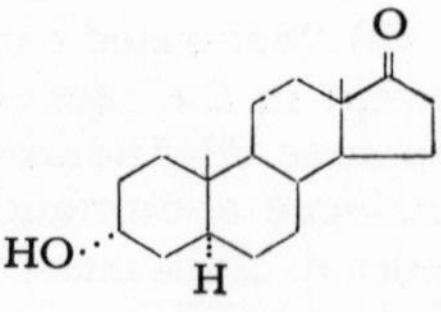

O

HO···   H

(IX)

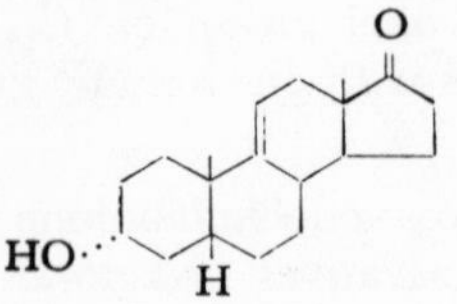

O

HO···   H

(X)

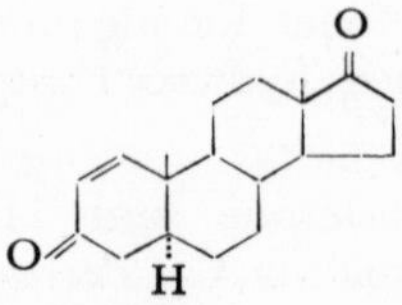

O

O   H

(XI)

O

O

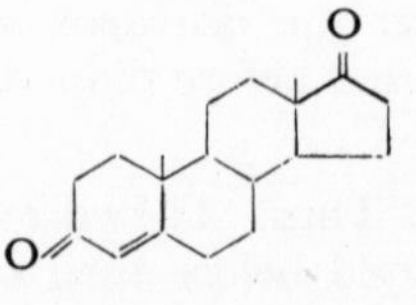

(XII)

On changing from alumina to magnesium silicate-Celite, several inversions of the order of elution were observed.

A notable improvement in alumina chromatography of steroids has been recently introduced by Lakshmanan and Lieberman (*1867*) who have standardized their alumina by exposure to the atmosphere of a salt solution of known vapour pressure and have used a carefully studied gradient elution (see p. 42).

(see p. 42)

The *displacement analysis* of steroids on Darco G-60-Hyflo (2 : 1), with 0.5 % cholesterol as displacer has been described by Hamilton and Holman (*1284*).

Charcoal has been used by Lombardo *et al.* (*2044*) to remove the major portion of urinary pigments which interfere with the usual chromatographic techniques for the separation of steroids. Activated charcoal, *magnesol* and *florisil* have been used by Peterson *et al.* (*2495*), Nelson and Samuels (*2353*) and Meister *et al.* (*2183*) for the separation of ketosteroids.

A *silica gel microcolumn* for separation of cortical steroids has been described by Sweat (*3171, 3172*); silica gel Davison was used and the steroids eluted with increasing concentrations of alcohol in chloroform; mixtures of several steroids in amounts as small as 0.25 μg can be analyzed.

Brooks *et al.* (*400, 401*) have recently shown that good separations of steroids can be obtained by chromatography of their benzoates on alumina.

Stokes *et al.* (*3121*) have separated p-^{131}I-benzoates of sterols on columns of silicic acid-Celite, following the separation by scanning of the column at hourly intervals with a collimated scintillation counter.

Epimeric pairs of steroids have been separated by Galinovsky and Vogl (*1075*) and Ruzicka *et al.* (*2763*) on alumina, and by Brooks *et al.* (*400, 401*) as the corresponding benzoates.

Fieser (*947*) reports the easy splitting of a *molecular complex* of epicholesterol and $\varDelta^4$-cholestene-6β-ol-3-one by chromatography on alumina.

For *aromatic* steroids, see the chapter on *sex hormones*.

For di- and triterpenes see the papers by Ruzicka *et al.* (e.g. *445, 1568, 1569, 1570, 2759, 2764*) of Spring *et al.* (e.g. *1207, 2293, 2360*) as well as *1907, 2277, 2549*.

(b) *Partition columns*

Jones and Stitch (*1601*) have separated urinary ketosteroids by partition on silicic acid using nitromethane as stationary phase and petroleum ether-chloroform as mobile phase (see also Cook *et al.*, *632*). Katzenellenbogen *et al.* (*1648*) have used silica gel impregnated with ethanol or formamide and methylene chloride containing 1 to 5 % ethanol as mobile phase for the separation of highly polar steroids. They describe in detail the influence of lenght of column and of flowrate on the degree of separation obtained.

Columns of kieselguhr impregnated with aqueous methanol or ethanol with hexane, toluene or benzene as mobile phases have been used by Morris and Williams (*2281*) for corticosteroids, by Cox and Marrian (*640*) for pregnanetriol.

Mosbach *et al.* (*2287*) have separated the air-oxidation products of cholesterol on columns of Celite partially saturated with aqueous methanol or propyleneglycol, using cyclohexane or petroleum ether as stationary phase.

Baker *et al.* (*124*) have separated 17-hydroxycorticosteroids on a column of Solca-Floc (cellulose powder) impregnated with propylene glycol (toluene as mobile solvent).

A special mention should be made of the work of Simpson *et al.* (*2996*) who isolated crystalline aldosterone from adrenal extracts after chromatography of purified extracts on a column of Celite saturated with water, using petroleum-ether-benzene and then benzene alone as mobile phase (140 fractions were obtained during the experiment which took 70 days); further purification was achieved on a column of cellulose.

See also Chapter 36 on *oestrogens*.

(c) *Paper chromatography*

The paper chromatography of steroids has been reviewed by Bush (*480, 481*) and Heftmann (*1346a*).

The paper chromatography of keto-steroids was first described by Zaffaroni *et al.* (*3658*). The substances were transformed into their water-soluble hydrazones with Girard's reagent T, and then separated on paper; the spots were detected by treatment with an iodoplatinate or iodobismuthate reagent. The R_F values of a large number of keto-steroids are given in this paper. On applying this technique to the adrenocortical hormones, the same authors found, however, that the R_F values of the individual steroid-reagent-T-hydrazones were not sufficiently different for identification. Moreover, some compounds gave rise to two spots corresponding to the mono- and dihydrazones respectively. They finally developed a more satisfactory method, based on the paper chromatography of the *free* keto-steroids, using paper impregnated with formamide or propyleneglycol, and with benzene or toluene as the mobile solvents. The spots were detected by spraying with alkaline silver nitrate, sodium thiosulphate, triphenyltetrazolium chloride, iodine in aqueous KI, or potassium permanganate (*476*).

Zaffaroni and Burton (*3657*) have developed a detailed procedure for the fractionation of complex mixtures of adrenocortical steroids by paper chromatography. The crude chloroform extract is first chromatographed on paper in toluene–propylene glycol and thus divided into three main fractions: fraction A, the overflow of the first 4 hours; fraction B, the overflow of 4–48 hours; fraction C, steroids remaining on the paper. Each fraction is then again chromatographed on paper in different solvent systems (benzene-formamide with or without methanol, etc.) and thus further divided into subfractions. Analysis of a commercial extract of beef adrenal glands revealed the presence of twelve α-ketols of which eight, including 11-deoxycorticosterone, were identified.

Burton *et al.* (477) have described an analogous method for the analysis of adrenocortical steroids in urine extracts; see also Hofmann and Staudinger (1426, Table 93).

TABLE 93. R_F values of cortical steroids (Hofmann and Staudinger, 1426)

Solvent: water saturated with *n*-butanol

Substance	R_F
Deoxycorticosterone	0.09
11-Deoxy-17-hydroxycorticosterone	0.15
Corticosterone	0.27
11-Dehydro-17-hydroxycorticosterone	0.35
17-Hydroxycorticosterone	0.36

ABLE 94.   Methods of detection of steroids (Bush, 480)

thod	Substances detected	Approximate limit of sensitivity (μg)*	Reference
llens reagent	α-Ketols	5	Zaffaroni *et al.* (3658)
ueous iodine	Cortisone and others	5	Zaffaroni & Burton (3657)
line in petroleum	$\alpha\beta$-Unsaturated ketones, etc.	1–2	Brante (362), Bush (486)
icotungstic acid	Mainly 3-hydroxysteroids	5	Kritchevsky & Kirk (1804)
osphomolybdic acid	Mainly 3-hydroxysteroids	5	Kritchevsky & Kirk (1804)
timony trichloride	Mainly 3-hydroxysteroids	2–5	Kritchevsky & Kirk (1804)
tra-violet light	$\alpha\beta$-Unsaturated ketones	5	Haines & Drake (1265), Bush (486)
iphenyltetrazolium chloride	α-Ketols	3–5	Zaffaroni & Burton (3657)
lue tetrazolium"	α-Ketols and others	1–5	Chen & Tewell (561), Simpson & Tait (2995)
mic acid vapour	$\alpha\beta$-Unsaturated ketones	2–5	Mancuro & Zygmuntowicz (2119)
OH fluorescence	Δ^4-3-Ketones	0.25–1.0	Bush (481, 486), Simpson & Tait (2995)
osphoric acid	A wide range of steroids	1–10	Neher & Wettstein (2345)
Dinitrobenzene and alkali	Cardiac aglycones, etc.	5	Schindler & Reichstein (2843)
Dinitrobenzene and alkali	17-Ketosteroids	2–5	Savard (2828, 2829)
Dinitrobenzoic acid and alkali	Cardiac aglycones	5	Bush & Taylor (484)
4-Dinitrophenylhydrazine	Ketosteroids	3–5	Kochakian & Stidworthy (1751)
azotized sulphanilic acid	Oestrogens	2–5	Mitchell (2233)
lin and Ciocalteu's reagent	Oestrogens	2	Mitchell (2233)
richloroacetic acid	Cardiac glycosides, etc.	5–10	Svendsen & Jensen (3161), Hassall & Martin (1333)
fra-red spectra	$\Delta^{9:11}$-Androstenolone, etc.	50	Dobriner & Lieberman (794)

This is stated in terms of the usual size of zone on a paper chromatogram, since figures in terms of antity per unit area are mostly unavailable. Most zones are 1–3 cm² in area in practice.

TABLE 95. R_F values of various steroids on activated alumina paper (Bush, *486*) These values are taken from medium activity papers with a surface reaction to "Universal" indicator (B.D.H.) between pH 7.5 and 7.8. Solvents: (a) benzene-$CHCl_3$ 3:1 (by vol.); (b) benzene-$CHCl_3$ 2:1; (c) benzene-$CHCl_3$ 1:1; (d) benzene-acetone 19:1

Compound	Solvents			
	(a)	(b)	(c)	(d)
Progesterone	0.75	0.95	1.0	0.8
Testosterone	0.40	0.60	—	—
Oestrone	0.30	0.40	0.9	0.6
Oestradiol	0.05	0.15	0.5	0.25
Pregnenolone	—	0.80	—	—
Dehydroandrosterone	—	0.41	—	—
Androsterone	0.25	0.39	—	—
11-Deoxycorticosterone 21-acetate	0.70	0.85	—	0.70
11-Dehydrocorticosterone 21-acetate	0.35	0.60	0.95	0.40
11-Dehydro-17-hydroxycorticosterone 21-acetate	0.05	0.10	0.40	0.10
17-Hydroxycorticosterone 21-acetate	0.02	0.06	0.30	0.05
Corticosterone 21-acetate	—	0.15	0.67	0.20

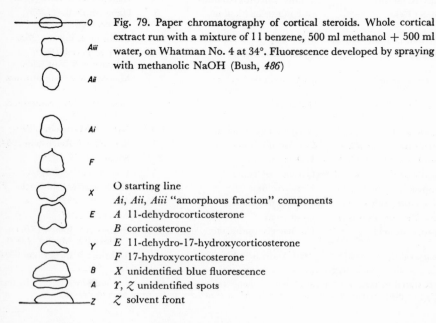

Fig. 79. Paper chromatography of cortical steroids. Whole cortical extract run with a mixture of 1 l benzene, 500 ml methanol + 500 ml water, on Whatman No. 4 at 34°. Fluorescence developed by spraying with methanolic NaOH (Bush, *486*)

O starting line
Ai, Aii, Aiii "amorphous fraction" components
A 11-dehydrocorticosterone
B corticosterone
E 11-dehydro-17-hydroxycorticosterone
F 17-hydroxycorticosterone
X unidentified blue fluorescence
Y, Z unidentified spots
Z solvent front

Reich *et al.* (*2649*) have used a similar method to identify adrenosterone; the spot was detected with 2,4-dinitrophenylhydrazine in dilute acid or an alkaline solution of *m*-dinitrobenzene in alcohol.

Steroids can also be detected on paper by a $ZnCl_2$ spot test (*2390, 2391*). Neher and Wettstein (*2344*) have found that corticosteroids can be detected on paper by treatment with dilute phosphoric acid, followed by heating; characteristic ultraviolet fluorescent spots are obtained. $SbCl_3$ and anisaldehyde-sulphuric acid can also give useful colorations. (See also Table 94.)

Axelrod (*107*) has described a specific test for differentiating between the a-ketol and the dihydroxyacetone groups of C_{21} steroids on paper chromatograms. The detection of steroid 17,20-diols on paper can be made by oxidation on paper with lead tetra-acetate, which converts them to 17-ketosteroids (Bush, *482*).

TABLE 96. R_F values of various steroids obtained with partition systems using aqueous methanol as stationary phase (Bush, *486*)

The values given are taken from 44 cm flows on No. 4 Whatman paper at 34°. With higher temperatures the values increase. Errors of more than $\pm$ 0.05 are rare and are usually due to faulty equilibration. Mixtures of compounds III and IV, or IV and V can only be separated completely by overrunning with system A or B_3

Compound	(A)	(B_1)	(B_2)	(B_3)	(B_4)	(B_5)	(C)
17-Hydroxycorticosterone (I)	0.0	0.03	0.10	0.01	0.15	0.32	0.4
11-Dehydro-17-hydroxycorticosterone (II)	0.01	0.05	0.20	0.02	0.30	0.50	0.7
Corticosterone (III)	—	0.23	0.65	0.15	0.70	0.85	
11-Deoxy-17-hydroxycorticosterone (IV)	—	0.27	0.70	0.20	0.72	0.87	
11-Dehydrocorticosterone (V)	—	0.36	0.75	0.25		0.96	
11-Deoxycorticosterone (VI)	0.39	0.86	0.96	0.75		1.0	
21-Acetate of I	—	0.25					
21-Acetate of II	0.04	0.39					
21-Acetate of IV	0.25						
21-Acetate of VI	0.70						
Testosterone	0.40						
Androst-4-ene-3,11,17-trione	0.26						
Androst-4-ene-3,17-dione	0.70						
Progesterone	0.85						

Solvent systems:

A. 1000 ml light petroleum, 800 ml methanol, 200 ml water.

B. (1) 500 ml toluene, 500 ml light petroleum, 700 ml methanol, 300 ml water.
    (2) 667 ml toluene, 333 ml light petroleum, 600 ml methanol, 400 ml water.
    (3) 667 ml light petroleum, 333 ml benzene, 800 ml methanol, 200 ml water.
    (4) 1000 ml toluene, 500 ml methanol, 500 ml water.
    (5) 1000 ml benzene, 500 ml methanol, 500 ml water.

C. 800 ml toluene, 100 ml ethyl acetate, 500 ml methanol, 500 ml water.

Axelrod (*109*) has used the solvent systems decalin-formamide, methyl-cyclohexane-propylene glycol, cyclohexene-formamide and methylcyclo-hexane-1,3-butanediol for the separation of $C_{21}O_3$, $C_{21}O_2$ and $C_{19}O_2$ steroids and hydriodic acid, fuming sulfuric acid, antimony pentachloride, *m*-dinitrobenzene, 2,4-dinitrophenylhydrazine and triphenyltetrazolium chloride for detection.

A one phase solvent system (xylene 225, methanol 75) has been used by Sakal and Merrill (*2784*) for the separation of corticosteroids. For aldo-sterone see Schmidt and Staudinger (*2862*).

Pechet (*2484*) has described solvent systems for the separation of very polar steroids (cortisone, hydrocortisone and their dihydro and tetrahydro-derivatives). See also Meyerheim and Hübener (*2201*).

For the analysis of adrenocortical steroids from biological fluids, see also Pechet (*2483*), Bush and Sandberg (*483*), and Bloch *et al.* (*274*).

Neher and Wettstein (*2345*) have described a new solvent system which is particularly applicable to the separation of weakly polar steroids or their esters and ethers. The stationary phase is formed by ethylene glycol mono-phenyl ether (phenyl cellosolve) on Whatman paper No. 7, the mobile phase by heptane.

Bush (*486*; Fig. 79) has described an adsorption method using alumina impregnated paper (Table 95) and partition systems with aqueous methanol as stationary phase (Table 96). Relatively non-polar steroids, such as progesterone, testosterone, deoxycorticosterone, could be separated with benzene-chloroform mixtures as mobile solvents.

Alumina impregnated paper has also been used by Shull *et al.* (*2973*), with non-aqueous solvents as developers; adrenal steroids were separated on plain paper with benzene-water.

Boscott (*325*) has used paper impregnated with hydrotropic agents, such as sodium *p*-toluenesulphonate for the separation of relatively non-polar steroids.

Androgens have been separated in a propylene glycol-methanol system by Kochakian and Stidworthy (*1751*, see Table 97); the separation of $C_{19}O_3$ steroids has been described by Arroyave and Axelrod (*84*).

Neutral 17-ketosteroids from urine were separated by McDonough (*2075*).

Savard (*2828*) has published a detailed paper concerning the migration of ketosteroids in a system of propylene-glycol impregnated paper with ligroin as mobile solvent. The relative mobility, or R_T, was calculated from the following formula:

$$R_T = \frac{\text{movement of unknown steroid}}{\text{movement of standard}} \times \text{rate of movement of standard} = \frac{D_u}{D_{st}} \times K_{st}$$

where D_u = the distance traversed by the steroid of unknown mobility,

D_{st} = the distance traversed concurrently by the standard, and

K_{st} = the rate of movement of the standard (androsterone or 11-β-hydroxy-androsterone) in cm per hour in the same solvent at room temperature.

TABLE 97.   Mobility of androgens in paper chromatograms* (Kochakian and Stidworthy, *1751*)

	R_t†	Ultraviolet,‡ 254 mμ	DNB§	DNPH§
Δ^4-Androstene-3,17-dione-6α-ol	0.11	+	V.	Y.-O.
Androstane-3β,17β-diol	0.15	—		
Δ^5-Androstane-3β,17β-diol	0.20	—		
Androstane-3α,17β-diol-16-one	0.31	—	T.	Y.
Δ^5-Androstene-3β,17α-diol	0.32	—		
Δ^4-Androstene-3,17-dione-6β-ol	0.33	+	V.	Y.-O.
$\Delta^{1,4}$-Androstadiene-17β-ol-3-one . . .	0.42	+	V.	O.
Androstane-3α,17β-diol	0.45	—		
Testosterone	1.00	+	Bl.	R.-O.
Epitestosterone	1.35	+	Bl.	O.
Etiocholan-3α-ol-17-one	1.57	—	V.	Y.
Epiandrosterone	1.66	—	V.	Y.
Dehydroepiandrosterone	1.73	—	V.	Y.
$\Delta^{1,4}$-Androstadiene-3,17-dione	2.07	+	V.	O.-Y.
Adrenosterone	2.08	+	V.	Y.-O.
Androstane-17β-ol-3-one	2.08	—	Bl.	Y.-O.
Androstane-3,7,17-trione	2.57	—	V.	Y.
Androsterone	2.72	—	V.	L.-Y.
$\Delta^{4,6}$-Androstadiene-3,17-dione	2.85	+	Bl.-V.	Br.-O.
Δ^5-Androsten-3β-ol	4.60	—		
Δ^4-Androsten-3-one	4.81	+	Bl.	O.
Δ^4-Androstene-3,17-dione	5.80	+	V.	O.-Y.
Androstan-3-one	6.25	—	Bl.	Y.-O.
Androstan-17-one	6.30	—	V.	Y.
Androstan-3α-ol	6.85	—		
Androstan-3β-ol	8.34	—		
Androstane-3,17-dione	8.40	—	V.	Y.
Androstan-17β-ol	∞	—		

* Whatman No. 1 paper impregnated with propylene glycol-methanol (1:1). Developed with benzene-cyclohexane (1:1) saturated with propylene glycol for 17 hours for compounds moving slower than testosterone and 6 hours for those moving faster.

† R_t = (movement of steroid)/(movement of testosterone).

‡ Ultraviolet absorption in the Haines-Drake fluorescence scanner.

§ Colours developed with alkaline *m*-dinitrobenzene test (DNB) and 2,4-dinitrophenylhydrazine (DNPH). V. = violet; T. = tan; Bl. = blue; Y. = yellow; O. = orange; R. = red; L. = lemon; Br. = brown.

Savard has thus established the chromatographic sequence of some 50 ketosteroids (see Table 98). When ketosteroids of very similar relative mobilities were chromatographed together, displacement effects were observed, whereby the faster moving components moved even faster than the R_T would have indicated; the "displaced R_T" values are also listed in Table 98. This displacement effect which Savard only observed in the ligroin-

TABLE 98. Relative mobilities (R_T) of α-ketosteroids in ligroin-propylene glycol (Savard, *2828*

Class No.	Compounds	C atoms	C=O	—OH	R_T	Displaced R_T	Other solvents, R_T	Ag	DNPH	Zin
V	6α-Hydroxy-11-deoxycortico-sterone	21	2	2	0.008		0.27*	+	Red	Bl.
V	6β-Hydroxy-11-deoxycortico-sterone	21	2	2	0.016		0.53*	+	Red	Bl.
IV	Δ⁴-Androsten-6α-ol-3,17-dione	19	2	1	0.024		0.80*	—	Red	V.
V	Corticosterone	21	2	2	0.025	0.027	1.00*	+	Red	Bl.
IV	6α-Hydroxyprogesterone	21	2	1	0.026	0.033		⌐	Red	Bl.
V	11-Deoxy-17α-hydroxy-corticosterone	21	2	2	0.029	0.035	0.83*	+	Red	Bl.
IV	Δ⁴-Androsten-14α-ol-3,17-dione	19	2	1	0.039			—	Red	V.
V	Etiocholane-3α,11β-diol-17-one	19	1	2	0.048			—	—	V.
V	21-Deoxycortisone	21	3	1	0.061		2.00*	—	Red	Bl.
V	11-Dehydrocorticosterone	21	3	1	0.074		3.46*	+	Red	Bl.
IV	11α-Hydroxyprogesterone	21	2	1	0.079		2.54*	—	Red	Bl.-*
IV	Δ⁴-Androsten-6β-ol-3,17-dione	19	2	1	0.080	0.080		—	Red	V.
V	Pregnane-3α,17α-diol-20-one	21	1	2	0.090			—	Y.	Br.
V	Androstane-3α,11β-diol-17-one	19	1	2	0.10	0.10	3.60*	—	—	V.
	Δ⁴-Androsten-11β-ol-3,17-dione	19	2	1			4.60*	—	Red	Bl.-*
III	Δ¹,⁴-Androstadien-17β-ol-3-one	19	1	1	0.12				Red	Bl.-
IV	6β-Hydroxyprogesterone	21	2	1	0.14	0.15			Red	Bl.
IV	Androstan-11β-ol-3,17-dione	19	2	1	0.17				Red	V.
IV	11β-Hydroxyprogesterone	21	2	1	0.20	0.22 ·			Red	Bl.
III	Δ⁴-Pregnen-20α-ol-3-one	21	1	1	0.24				Red	Br.
IV	Etiocholan-3α-ol-11,17-dione	19	2	1	0.26				—	V.
IV	17α-Hydroxyprogesterone	21	2	1	0.28	0.28			Red	Bl.
IV	Androstan-3α-ol-11,17-dione	19	2	1	0.30				—	V.
III	Testosterone	19	1	1	0.33	0.36			Red	Bl.-
IV	Adrenosterone	19	3		0.36	0.41			Red	Bl.-
III	Epitestosterone	19	1	1	0.40	0.56			Red	Bl.-
IV	Deoxycorticosterone	21	2	1	0.45	0.60	6.60*	+	Red	Bl.
II	Δ¹,⁴-Androstadiene-3,17-dione	19	2		0.62				Red	Bl.-*
III	Δ⁹-Etiocholen-3α-ol-17-one	19	1	1	0.70	0.70		—	—	V.
	Etiocholan-3α-ol-17-one	19	1	1				—	—	V.
	Etiocholan-17β-ol-3-one	19	1	1				—	—	Bl.
III	Δ⁹-Androsten-3α-ol-17-one	19	1	1	1.00	1.00	0.12†	—	—	V.
	Androsterone	19	1	1				—	—	V.
IV	Allopregnan-21-ol-3,20-dione	21	2	1	1.06	1.06		+	Y.	Br.

LE 98 *(Continued)*

Compounds	No. of C atoms	C=O	—OH	R_T	Displaced solvents, R_T	Other solvents, R_T	Ag	DNPH	Zimm.
Pregnan-3α-ol-20-one	21	1	1	1.30			—	Y.	Br.
Allopregnan-3α-ol-20-one	21	1	1	1.40			—	Y.	Br.
i-Androstanolone	19	1	1	1.5			—	—	V.
Δ4-Androstene-3,17-dione	19	2		1.7			—	Red	Bl.-V.
Δ1-Androstene-3,17-dione	19	2		2.0			—	Red	Bl.-V.
Δ16-Dehydroprogesterone	21	2		2.3	2.3		—	Red	Bl.
Progesterone	21	2		2.3	2.4		—	Red	Bl.-V.
{ Etiocholane-3,17-dione	19	2)		2.5	2.5	0.13†	—	Y.	V.
{ Androstane-3,17-dione	19	2)				0.13†	—	Y.	V.
{ Pregnane-3,20-dione	21	2)		3.0	3.0	0.16†	—	Y.	Bl.
{ Allopregnane-3,20-dione	21	2)				0.16†	—	Y.	Bl.
Δ4,16-Androstadien-3-one	19	1		6.0		0.48†	—	Red	Bl.
{ Δ5-3β-Chloroandrosten-17-one	19	1		9.0)	9.0	0.53†	—	—	V.
{ Δ2 or 3-Androsten-17-one	19	1		9.2)		1.15†	—	—	V.
Δ4-Cholesten-3-one	27	1		9.5			—	Red	

The following abbreviations are used: Ag = silver diamine reaction; DNPH = dinitrophenylhydrazine; Zimm. = modified Zimmermann reagent; Bl. = blue; Br. = brown; Y. = Yellow; V = violet.

* In toluene-propylene glycol.

† In heptane-ethylene glycol phenyl ether.

The "α-ketosteroids" of this table are ketosteroids which are not precipitated with digitonin; for a similar table of "β-ketosteroids", see Savard (*2828*). The substances are arranged in order of increasing R_T.

    Class   I = monoketo- and monohydroxysteroids

            II = diketosteroids

        III and IV = monohydroxymonoketosteroids, monohydroxydiketosteroids and triketosteroids

            V = Dihydroxymonoketones, monohydroxytriketones, dihydroxydiketones.

propylene glycol system was even more pronounced when steroid mixtures from natural extracts were chromatographed and are attributed in some instances to greater concentrations of the slower moving components, as well as to non-steroid material present. This apparent distortion, however, in no way impaired the separation of the steroids and the relative order remained the same (see Fig. 80).

The following generalizations could be drawn by Savard (*2828*): C_{21}-ketosteroids (oxygen function at C-20) move faster than the corresponding C_{19}

steroids (with similar oxygen function at C-17); derivatives of the androstane (*allo*) series move faster than the corresponding derivatives of the etiocholane (*normal*) series; this appears to be equally so in the few C_{21}-ketosteroids studied. The mobility sequence of steroids with oxygen function at a fixed position (particularly in C-3 position) is: saturated ketone > α,β-un-

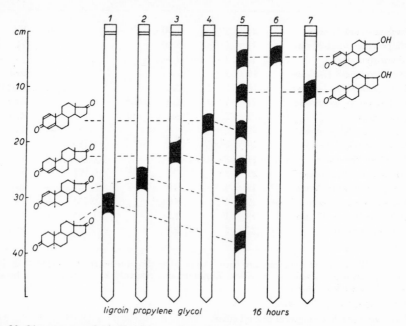

Fig. 80. Chromatography in ligroin-propylene glycol; 50 γ of steroid on each strip visualized by the Zimmermann reagent or dinitrophenylhydrazine or both (Savard, *2828*)

saturated ketone > hydroxyl in the polar conformation > hydroxyl in the equatorial conformation.

Savard was the first to note that steroids possessing the equatorial (less hindered) hydroxyl have lower mobility than the corresponding compounds with the axial hydroxyl.

Savard *et al.* (*2830*) described the precautions to be taken during drying or irradiation of the chromatograms to reduce to a minimum secondary reactions and destructions of the steroids, which are particularly labile when spread in a very thin layer over a great area of paper.

Kritchevsky and Calvin (*1803*) have described the *reversed phase paper chromatography* of steroids on paper treated with Quilon (stearato chromic chloride); mobile solvent: methanol, ethanol, or ethanol-water 8:2; for ease of location tritiated cholesterol was used and detected by scanning the paper with a windowless counting tube designed for locating weakly radiating substances on paper. The position of cholesterol was confirmed by the red colour developed after the paper was treated with a solution of silicotungstic acid, or with iodine in KI. See also Kritchevsky and Kirk (*1805*).

Kritchevsky and Tiselius (*1807*) have described the reversed phase partition chromatography of steroids on silicone-treated paper. Strips of paper (Swedish Munktell 20,150G) were drawn through a 5 % solution of Dow Corning Silicone No. 1107 in cyclohexane, then blotted and dried at 110° for 1 hour. The mobile solvent was a mixture of water, ethanol and chloroform. The spots were revealed with alkaline *m*-dinitrobenzene. See also Edgar (*861, 862*).

The paper chromatography of *cardiac aglycones* and of *steroidal sapogenins* has been described by Jensen (*1575, 1576*), Bush and Taylor (*484*), Heftmann and Hayden (*1348*), Sannié and Lapin (*2814*). Mills and Werner (*2226*) have described the reversed phase paper chromatography of constituents of natural resins (triterpene alcohols and acids) using kerosene as stationary and aqueous *iso*propanol as mobile phase. See also the section on glycosides, p. 265.

Steroid amines and steroid alkaloids can be separated on paper in moist butyl acetate or in a reversed phase system of paraffin oil-aqueous ethanol (Procházka *et al.*, *2599*).

Quantitative methods

After elution from the paper, sterols can be estimated by the Zimmermann reaction (Rubin *et al.*, *2751*; Oertel, *2397, 2398*) or by ultraviolet photometry (Edgar, *861, 862*). Zaffaroni and Burton (*3657*), Hübener *et al.* (*1470*), Hoffmann and Staudinger (*1425*) treat the paper strips with triphenyl tetrazolium chloride and then elute the formazans for colorimetry.

Schwarz (*2917*) has described a quantitative method for the estimation of reducing steroids in which the chromatograms are incubated with arsenomolybdate reagent; sterols containing the ketol side chain or having an α,β-unsaturated 3-keto group quantitatively reduce the reagent to molybdenum blue.

The estimation of plasma cholesterol after elution from paper has been described by Comfort (*609*).

5. BILE ACIDS

Trickey (*3301*) has transformed an amorphous mixture of bile acids into the acid chlorides of the bile acid formates and reacted the mixture with aminoazobenzene to obtain the azoylamide formates; these were chromatographed on alumina; the original bile acids can be recovered after saponification of the eluates.

Bergström and Sjövall (*211, 212*) have separated bile acids by reversed phase partition chromatography on Hyflo-Supercel made hydrophobic with dimethyl dichlorosilane; $CHCl_3$ was used as stationary phase and aqueous methanol as mobile phase.

Mosbach *et al.* (*2288*) have used Celite columns with 7 % aqueous acetic acid as stationary and petroleum ether-*iso*propyl ether as mobile phase for the separation of various bile acids.

The separation of the taurine and glycine conjugates of bile acids on kieselguhr columns has been reported by Norman (*2382*).

The *paper chromatography* of bile acids has been described by Kritchevsky and Kirk (*1804, 1806*), Sjövall (*3007, 3008*), Beyreder and Rettenbacher-Däubner (*239*), Haslewood (*1329*), Siperstein *et al.* (*3000*). The spots can be detected by spraying with phosphomolybdic acid in ethanol.

Bile salts have been separated on paper by Haslewood (*1329*), Haslewood and Sjövall (*1330*).

Eriksson and Sjövall (*913*) have made quantitative determinations of bile acids after elution from paper chromatograms (ultraviolet spectrophotometry of the colour formed after heating with 65 % sulfuric acid) (see also Arima, *78*).

Procházka (*2598*) has described the paper chromatography of 32 steroid acids, their methyl esters and acetates.

CHAPTER 30

Amino acids

The great progress made in the field of amino acids, peptides and proteins since 1940 has largely been due to chromatography.

Many original techniques have been developed and remarkable quantitative results have been obtained. This work has been the subject of reviews by Jutisz (*1609, 1610*), Martin and Synge (*2157*), Sanger (*2811*), Tiselius (*3259*), Turba (*3320*) and Wieland (*3552, 3555*). Many experimental details of chromatographic procedures for the separation of amino acids are given in the book of Block and Bolling (*281*), which also contains illustrations of apparatus for paper chromatography and charts showing the R_F values of amino acids in various solvents.

The excellent book of Turba, *Chromatographische Methoden in der Protein-chemie* (*3320a*) covers in detail all interesting aspects of chromatography in the field of amino acids, peptides and proteins. A recent review of Thompson and Thompson entitled *Paper chromatography in the study of the structure of peptides and proteins* (*3238*) is also to be recommended.

1. ADSORPTION CHROMATOGRAPHY

(a) *Charcoal*

Tiselius has studied the adsorption of the aliphatic monoamino acids on *charcoal* using the "frontal analysis" technique (see p. 4). Table 99 gives a

TABLE 99. Retention volumes in cm³/g of charcoal (Tiselius, *3259*)

Alanine	0.3	Histidine	15.0	Leucyl-glycine	18.2
Hydroxyproline	2.0	Arginine	40.4	Leucyl-glycyl-glycine	29.8
Proline	2.5	Tryptophan	76.5	Valyl-alanine	22.0
Valine	3.2	Phenylalanine	62.5	Alanyl-leucyl-glycine	34.4
Leucine	7.7	Glycyl-glycine	3.5	Glycyl-leucyl-alanine	42.5
Isoleucine	9.2	Glycyl-alanine	4.0	Glycyl-leucyl-glycine	38.0
Methionine	12.4				

list of retention volumes (see p. 4) of various amino acids and peptides and Table 100 gives a list of retention volumes of various substances used as displacing agents (Tiselius, *3259*). Good separations of a number of mixtures of neutral amino acids have been obtained.

TABLE 100. Retention volumes of displacing agents (in cm³/g charcoal) (Tiselius, *3259*)

Saccharose.	23	Raffinose	46
Methyl propyl ketone	31	Phenol	53
Ethyl acetate	32	Ephedrine	83
Pyridine	33	Picric acid	83

Wachtel and Cassidy (*3432, 3433*) have used adsorption on charcoal to separate a mixture of glycine, leucine, phenylalanine and tyrosine. In order to speed up the rate of solvent flow though the column, the charcoal was mixed with one or two parts of paper pulp. The aliphatic amino acids passed rapidly through the column while phenylalanine and tyrosine were retained. A partial separation of these two acids was obtained by washing the column with 5 % aqueous acetone. The mixture of glycine and leucine was resolved on a second charcoal column, glycine being eluted with water and leucine with aqueous ethyl acetate.

Schramm and Primosigh (*2880*) used charcoal for the separation of the neutral amino acids into an aliphatic and an aromatic group, the latter (phenylalanine, tyrosine and tryptophan) being retained on the column. The charcoal was pre-treated with a trace of KCN in order to inactivate the traces of metals that catalyse the oxidation of amino acids. The aromatic amino acids were eluted with a 5 % solution of phenol in 20 % acetic acid. As the eluate fractions were analysed only for nitrogen it is by no means certain that the amino acids were eluted as such. It was found (*1054*) that a large part of the tryptophan and some of the tyrosine could not be estimated colorimetrically when eluted according to the method of Schramm and Primosigh.

Turba *et al.* (*3325*) also used charcoal for the adsorption of tyrosine and phenylalanine, elution being carried out with a mixture of pyridine and acetic acid. A series of solvents for elution of aromatic amino acids from charcoal has been studied by Moreira de Almeida (*2272*).

Hall and Tiselius (*1278*) have described an elegant micro-analytical separation of the three aromatic amino acids by carrier displacement on charcoal. 104 μg tyrosine, 120 μg of phenylalanine, 140 μg of tryptophan and 97 μg of methionine dissolved in 0.1 M sodium carbonate-bicarbonate buffer (pH 9.7) were placed on a column composed of a filter having a total volume of 14.1 ml and containing 3 parts of Super Cel to 1 part of Carbo-raffin Supra. On passing through a mixture of *n*-butanol, 2-methyl-2-butanol, 3-methyl-1-butanol and benzyl alcohol in 1 % concentration in the same buffer, the amino acids were eluted as follows: methionine at the foot of the first step, i.e. between the buffer and the n-butanol; tyrosine at the interface butanol/2-methyl-2-butanol; phenylalanine was displaced by the 3-methyl-1-butanol and tryptophan by the benzyl alcohol.

Separations of valine and leucine or of valine and methionine can also be carried out by the use of charcoal (*3325*).

Robson and Selim (*2708*) have described a method of isolation of the basic amino acids in protein hydrolysates: the aromatic amino acids are first removed by adsorption on charcoal, then the basic amino acids in the filtrate of the charcoal are transformed into their picrates, which can be quantitatively adsorbed on charcoal, while the free monoamino acids and the acidic diamino acids pass out in the eluate. Treatment of the charcoal column with 0.5 N HCl or H_2SO_4 decomposes the adsorbed picrates, the bases appearing in the percolate and the picric acid being retained by the charcoal.

(b) *Other adsorbents*

Ando and Yonemoto (*63*) have described brush methods for detection of some amino acids on alumina columns.

Ergothioneine has been purified on alumina columns by Melville *et al.* (*2185*).

Hamoir (*1290*) used a new adsorbent for amino acids consisting of a precipitate of silver sulphide carrying an excess of silver ions. Silver was chosen as many of the amino acids form silver salts of low solubility. The dicarboxylic acids, histidine, and the aromatic amino acids were adsorbed on such columns. Methionine had a high affinity for the column but could be eluted with dilute acetic acid; cystine and cysteine were irreversibly adsorbed (formation of a silver complex?).

Thompson *et al.* (*3239*) have described an interesting method for separating the bulk of the α-amino acids from other similar compounds, consisting in the use of mixed columns of copper carbonate and alumina, which adsorbs the α-amino acids as their copper complexes (see also Grobbelaar *et al.*, *1228*).

O'Connor and Bryant (*2393*) have studied the adsorption of amino acids on sparingly soluble inorganic salts, such as ZnS, PbS, $BaSO_4$, etc.

(c) *Separations of coloured derivatives of amino acids*

The paper of Karrer *et al.* (*1642*) describing the separation of the p-phenylazobenzoyl derivatives of amino acids on basic zinc carbonate has already been mentioned (p. 54).

Various authors have investigated the possibility of separating amino acids as the azo-urea derivatives, obtained by reaction of the amino acids with p-phenylazophenyl *iso*cyanate $C_6H_5N=NC_6H_4N=C=O$ (Turba and

Schrader-Beielstein, *3327*; Zeile and Oetzel, *3695*; Kruckenberg, *1810*). Masuyama (*2168*) has coupled the ethyl esters of amino acids with the same reagent.

Flowers and Reith (*993*) have prepared the sulphazoyl-derivatives

$$(C_6H_5N = NC_6H_4SO_2NHCH—COOH)$$
$$|$$
$$R$$

of amino acids and have separated them on alumina. In some cases a pure sulphazoylate gave two zones. Evans and Reith (*928*) have prepared the red 4-dimethylamino-3,5-dinitrophenyl *iso*cyanate which gives deep orange coloured carbamyl derivatives with amino acids; these were separated on alumina. See also Dinitrophenylamino acids, p. 330.

Another way to use coloured derivatives is to heat the amino acids with ninhydrin, then to prepare the 2,4-dinitrophenylhydrazones of the aldehydes thus formed and to chromatograph them on zinc carbonate (Turba and Schrader-Beielstein, *3326*).

2. ION EXCHANGE CHROMATOGRAPHY

Block (*277*) has reviewed the use of ion exchangers for the separation of amino acids.

(a) *Basic amino acids*

Turba (*3319*) was the first to show that the basic amino acids (lysine, arginine and histidine) were adsorbed on acidic earths (Filtrol-Neutrol or Floridin XXF, H. Bensmann, Bremen). In this way the other amino acids may be separated from a mixture. Histidine is adsorbed only by Filtrol-Neutrol and not by Floridin XXF. By the use of Filtrol-Neutrol, lysine and arginine can be separated.

Englis and Fiess (*899*) have studied the behaviour of amino acids on various synthetic resins. Cleaver *et al.* (*589*) examined in detail the conditions under which various amino acids could be adsorbed on the two Amberlite resins, IR-100 and IR-4. Particular attention was paid to the effect of the type of resin, grain size, column length, rate of solvent flow, amino acid concentration, pH and also to the behaviour of many binary and tertiary mixtures. The neutral amino acids are only slightly retained by cationic exchangers, the adsorption of the dicarboxylic acids on anion exchangers presents no difficulty (see below), whereas the quantitative separation of the hexone bases is more difficult. Block (*275*) however, has carried out such separations on Amberlite-IR-4 and IR-100, and Wieland (*3553*) on Wofatit C.

Rauen and Felix (*2631*) obtained a quantitative separation of the three

basic amino acids on Wofatit C. Schramm and Primosigh (*2882*) have
studied the conditions for separation of these amino acids on alumina,
Wofatit C and acid earths. Histidine could be separated quantitatively from
arginine and lysine on Floridin XS.

Sheehan and Bolhofer (*2952*) describe the isolation of natural hydroxylysine by adsorbing it on Amberlite IRC-50 and then on alumina; excess
of flavianic acid and HCl (originating from the previous precipitation of
arginine flavianate) was first eliminated by filtration through the anion-
exchanger Amberlite IR-4B.

Arginine and lysine are retained by a column of "basic alumina" i.e.
one containing sodium ions (Wieland, *3549*). Arginine can be estimated
after adsorption on Permutit and elution with sodium chloride (Dubnoff,
822) and it can be separated from glutamine by filtration through Decalso
(Permutit) (*74*).

Archibald (*73*) has developed a method of estimation of citrulline in blood
after adsorption on a column of Amberlite IR-100.

Bergdoll and Doty (*198*) carried out quantitative adsorption of the basic
amino acids on Fuller's earth mixed with a filter aid (Hyflo-Super Cel) and
eluted lysine with N HCl, histidine with 0.125 M bicarbonate, and arginine
with a 10 % solution of pyridine in 0.7 N HCl.

More recently, Dowex-50 has been used by several workers for separations
of basic amino acids.

The separation of homoarginine and hydroxy-homoarginine from hydrolysates of guanidinated gelatin, on Dowex-50 has been reported by Eastoe
and Kenchington (*851*), the isolation of 3-methylhistidine from human
urine by Tallan *et al.* (*3197*). The method of Moore and Stein (*2268*) has
been modified by Hamilton and Anderson (*1287*) to obtain a better separation of the basic amino acids, especially of histidine from hydroxylysine and
ornithine from lysine. The same authors (*1288*) have described the isolation
of hydroxylysine from gelatin and the resolution of its diastereoisomers on
columns of sodium Dowex-50. (See also Piez, *2517*.)

The British patent No. 568,369 describes the preparation of histidine
from protein hydrolysates by adsorption on a column of activated Fuller's
Earth. A series of other patents describes the adsorption and elution of the
diamino acids on cation exchangers (*80, 224, 225, 276, 278, 1462, 1508,
1836* and *2464*).

(b) *Dicarboxylic amino acids*

Wieland (*1824, 3549, 3550*) has shown that alumina treated with N HCl
takes up Cl⁻ ions and behaves as an anion exchanger. This "acid alumina"
quantitatively adsorbs glutamic and aspartic acids and also cystine, the
latter being due either to the insolubility of the cystine at the pH of the
column (4–5) or to its acid isoelectric point (pH 5). Cystine is eluted by
water saturated with H_2S (reduction to cysteine), and the two dicarboxylic

acids with alkali (or acid, *1054*). Schramm and Primosigh (*2881*) avoided the adsorption of cystine by working in a medium saturated with H_2S.

Turba and Richter (*3324*) activated alumina by a buffer of N acetate of pH 3.3 (the isoelectric point of the dicarboxylic acids) and separated aspartic and glutamic acids, the latter being eluted with the acetate buffer and the aspartic acid with 0.05 N alkali (see also Wieland and Wirth, *3566*; Wieland and Paul, *3563*; Rauen and Wolf, *2634*). Prescott and Waelsch (*2591*) described a microtube for the adsorption of quantities of glutamic acid of the order of 50 μg, the adsorbent being packed by centrifuging. The glutamic acid was eluted with 0.5 N HCl (as described by Wieland), the aspartic acid remained adsorbed and was eluted with alkali.

Kretovich and Bundel (*1800*) separated aspartic and glutamic acids on alumina which had been heated previously with aluminium chloride at 700°.

Boulanger and Osteux (*344*) describe a method for the separate determination of D- and L-glutamic acid. They first isolate both by adsorption on, and specific elution from, acid alumina; the L-isomer was then determined by decarboxylation with the specific L-glutamodecarboxylase of *Clostridium Welchii* SR-12. Paper chromatography was used to control the separation on alumina.

Several papers have described the adsorption of the dicarboxylic acids on synthetic resins (Amberlite IR-4, or De-Acidite); Cleaver *et al.* (*589*) have studied in detail the effect of pH and salts on the adsorption of glutamic acid on Amberlite IR-4. See also Englis and Fiess (*899*) and Drake (*811*). Consden *et al.* (*621*) have eluted glutamic and aspartic acids separately after adsorption on Amberlite IR-4. The latter authors also isolated optically active cysteic acid from a hydrolysate of chlorinated wool, after adsorption on Amberlite IR-4. Rauen (*2630*) has quantitatively separated aspartic and glutamic acids and tryptophan on Wofatit M.

(c) *Neutral amino acids*

As the neutral amino acids carry a negative charge in 90 % alcohol, they can be retained by a column of "acid alumina" and can thus be separated from other neutral substances (e.g. glucose). Elution is carried out with water (Wieland, *3551*).

Schramm and Primosigh (*2880*) showed that glycine, serine and threonine were sufficiently acidic in the presence of formaldehyde to be adsorbed on "acid alumina". In this way these authors were able to separate these three amino acids from the other neutral amino acids by elution from the column with alkali. Amino acids with the amino group in the ω-position, e.g. β-alanine, are also adsorbed on "acid alumina" from a 10 % formaldehyde solution (*1911*). The elution of substances adsorbed under these conditions is simply carried out with warm water, as this causes dissociation of the formaldehyde-amino acid complexes (*1609, 1611*).

Turba *et al.* (*3325*) carried out some selective separations of neutral amino acids on alumina activated by an acetate buffer of pH 3.3, or on Filtrol-Neutrol (cystine, alanine, valine, leucine and proline).

Turba *et al.* (*3325*) found that the synthetic resin Wofatit M was a specific adsorbent for tryptophan (eluted by aqueous pyridine or by alcohol). For adsorption of other amino acids on the Wofatit resins, see Freudenberg *et al.* (*1038*).

The neutral amino acids are weakly adsorbed by cation exchangers. Cleaver and Cassidy (*588*) found that adsorption of neutral amino acids by ion exchange resins increases with increasing chain length, phenylalanine being especially strongly retained. Different ion exchange resins differ in the degree of adsorption, Dowex-50 and Ionac A-300 exhibiting the lowest adsorption capacities for the neutral amino acids.

Carsten and Cannan (*516*) have studied exchange equilibria of neutral amino acids on Amberlite IR-100, IR-120 and IR A-400 and Dowex-50 of several neutral amino acids and have determined retardation volumes.

Pipecolic acid has been isolated from natural sources by ion exchange by Zacharius *et al.* (*3656*) and by Morrison (*2285*).

Taurine, hypotaurine and cysteinsulphinic acid have been separated on Amberlite IR-400 and Permutit 50 (Bergeret and Chatagner, *200*).

Schram *et al.* (*2878*) have developed a method of determination of cystine by performic acid oxidation to cysteic acid which can then be estimated after chromatography on Dowex-50.

Campbell *et al.* (*503*) isolated the toxic substance of agenized wheat flour, methionine-sulphone-imine

$$CH_3-\overset{\overset{O}{\uparrow}}{\underset{NH}{S}}-CH_2CH_2\underset{NH_2}{CH}-COOH$$

by adsorption on Zeo-Karb 215 and then on paper.

Haworth *et al.* (*1337*) isolated sarcosine from ground nut protein by adsorption on Zeo-Karb, then on cellulose columns and finally on paper sheets.

Cation exchangers have recently been used with success for the separation of diastereoisomeric mixtures of amino acids.

The separation of threonine and *allo*threonine (labeled with ^{14}C) on a column of Dowex-50, by elution with 1.5 N HCl has been described by Shulgin *et al.* (*2972*). Work *et al.* (*3630*) have recently reported the separation of the three isomeric components of synthetic α,ε-diaminopimelic acid; the action of hog kidney amidase on the diamide of the synthetic acid (a mixture of the LL, DD and DL acid) gave a mixture of LL-diaminopimelic acid, DD-diamide and L-diaminopimelic acid D-monoamide; these could be separated on an Amberlite XE-64 column; after hydrolysis of the separated amides the three isomers were obtained in pure form.

Piez (*2517*) has reported the separation of the diastereoisomers of *iso*-leucine and hydroxylysine on Dowex-50 and Blackburn and Lee (*264*) have separated the diastereoisomers of lanthionine on Dowex-50.

A patent of Morris (*2283*) describes the isolation of tryptophan with the use of columns of Amberlite IR-100 H or Dowex-50.

(d) *The separation of amino acids into groups*

Schramm and Primosigh (*2881*) have suggested a method of separation of the amino acids into five groups: (1) phenylalanine, tyrosine and tryptophan adsorbed on charcoal; (2) arginine, lysine and histidine on silica gel; (3) aspartic and glutamic acids on acid alumina; (4) serine, threonine, glycine and cystine on acid alumina (in the presence of formaldehyde), and (5) alanine, valine, leucine, isoleucine, proline and hydroxyproline which are not adsorbed on any of these adsorbents.

This method has been used by Schramm and Braunitzer (*2879*) for the analysis of β-lactoglobulin and the protein of tobacco mosaic virus (Wofatit C being used for the adsorption of the basic amino acids).

Tiselius *et al.* (*3268*) have described a separation of amino acids into four groups by adsorption of the aromatic amino acids on charcoal, the basic acids on Wofatit C, and the acidic amino acids on Amberlite IR-4. The filtrate from the Wofatit C column contains the neutral and acidic amino acids, and on passing the mixture onto a column of Wofatit KS all the acids are adsorbed so that foreign substances are eliminated in the filtrate. Elution is carried out with N HCl. The same authors have described an assembly of three superimposed columns (for charcoal, Wolfatit C and Wofatit KS) which allows the operation to be carried out with the minimum of manipulation and loss. The three columns can be detached and eluted separately (Fig. 10, p. 15). See also the patent of Roberts (*2695*).

Fromageot *et al.* (*1054*) have described an analogous method using silica gel to adsorb the bases, and acid alumina for the dicarboxylic amino acids. Silica gel prepared as described by Gordon *et al.* (*1172*), must be washed with 20 % acetic acid and 0.1 N HCl in order to remove nitrogenous impurities (see p. 107). The technique suggested by Fromageot *et al.* (*1054*) can also be used to separate peptides into four groups (basic, acidic, neutral aromatic, and neutral aliphatic) (*1612*). Separations on silica gel are not quantitative if salts are present. If the acid used for hydrolysis is neutralised with lithium hydroxide and the amount of silica gel used is increased, satisfactory separations may be achieved. The salts may be removed by electrodialysis as described by Joseph (*1606*) and also by Consden *et al.* (*620*). (See the section on desalting, p. 127.) The synthetic resins used by Tiselius *et al.* (*3268*) are also affected by the presence of salts.

Brenner and Frey (*378*) have also obtained group separations by passing a solution of leucine, glutamic acid and arginine continuously for twenty-three hours through two columns, the first being Amberlite IR-4B, the second Amberlite IRC-50.

(e) Displacement chromatography on ion exchange resins

A detailed description of these methods and their application to the isolation of-gram quantities of crystalline amino acids from hydrolysates of proteins is given by Partridge and Brimley (2460), and Partridge (2454, 2456).

Partridge and Westall (182, 2463) followed separations of amino acids on ion exchange columns by observing changes in the effluent using four methods:

1. continuous measurement of electrical conductivity,
2. continuous measurement of pH,
3. titration of successive small fractions,
4. qualitative analysis of successive fractions by means of paper chromatography.

Data for the adsorption of various bases and amino acids on Zeo-Karb 215 are given in the form of "retention isotherms".

Effective separations of two components of a mixture of ampholytes were obtained on the column provided that the isoelectric points of the components differed by more than 0.5–1.0 pH unit.

Fine grinding of the resin was found essential for production of sharp bands, 40–60 mesh being the best grade.

Partridge and Brimley (2457) report the preparative isolation of glutamic and aspartic acids by adsorption on the anion exchanger De-Acidite B, followed by development with acetic acid (see also Drake, 811). Neutral amino acids are only slightly adsorbed. Acetic acid is adsorbed to about the same extent as glutamic acid, but less strongly than aspartic acid; separation is thus achieved by formation of two distinct amino acid bands with an acetic acid band between them. A complete separation was obtained with a mixture of 5 g of the amino acids on a column containing 125 g of resin.

Fractionation on Zeo-Karb 215 of a protein hydrolysate using displacement by dilute ammonia gave seven distinct but slightly overlapping fractions (2452) (Fig. 81).

After evaporating the effluent fractions, the amino acids or amino acid mixtures contained in each band were recovered as salt-free crystalline solids. Individual amino acids could be obtained by further fractionation on a second column.

Glucosamine and histidine are eluted together from a Zeo-Karb 215 column by 0.1 N NaOH, but can be separated by displacing the glucosamine with 0.5 N NaCl, histidine remaining on the column (2453).

Partridge *et al.* (*2461*) have studied the separation of the basic amino acids on a sulphonated cross-linked polystyrene prepared by them. Mixtures of leucine, histidine, lysine and arginine were successfully separated.

Separation of certain mixtures (for instance, leucine + methionine) which cannot be effected at room temperature, are possible at higher temperatures

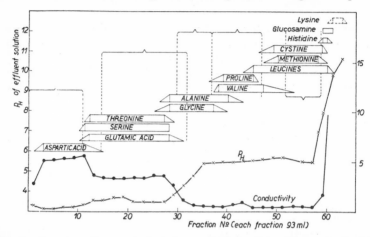

Fig. 81. Fractionation of a protein hydrolysate. Analysis of the effluent obtained from a column of Zeo-Karb 215 by displacing the mixture of amino acids with dilute ammonia solution. The upper portion of the figure shows an analysis carried out by means of filter-paper chromatography, while the lower portion shows the corresponding variations in pH and conductivity. The brackets include fractions collected together for treatment on secondary columns (Partridge, *2451*)

•—•—•—•—• conductivity   —×—×—×— pH

(Partridge and Brimley, *2458*). In general, the position (in the order of displacement) taken up by an amino acid at higher temperature approaches more closely the position expected from its dissociation constant pK_1. The increased reaction rate at a higher temperature is a further advantage.

The mechanisms operative in separations of amino acids on ion exchange resins have been discussed by Davies (*711*). Three principal factors are responsible for the separations achieved:

(1) differences in the extent of ionisation of the individual amino acids in their passage through the column,
(2) differences in charge,
(3) adsorptive (Van der Waals') forces.

A first inspection of the order of amino acids eluted by displacement (*2452*) shows that except for proline and cystine, they are in the order of pK_1 (COOH) (or pK_2 for the diamino acids). The two exceptions can be explained by the cyclic structure of proline and by the existence of bivalent cations of cystine (Table 101).

TABLE 101. Order in which amino acids appear in the effluent from columns packed with Zeo-
Karb, sulphonated polystyrene or Dowex-2 (Partridge and Brimley, *2459*)
The brackets include groups of amino acids which form mixed bands. Values of pK_1 and
pK_2 are included for comparison.
* marks amino-acids occupying anomalous positions in the order of displacement

Column no. (1)	(2)	(3)		(4)	
Zeo-Karb 215 at 20°	Zeo-Karb 215 at 60°	Sulphonated polystyrene at 20°		Dowex-2 at 20°	
Order of displacement	Order of displacement	Order of displacement	pK_1 (COOH)	Order of displacement	pK_2 (NH_3^+)
Aspartic acid	Aspartic acid	Aspartic acid	1.88	Arginine	12.48 (Guanidine)
Glutamic acid	⎰ Glutamic acid	Threonine	—	Lysine	10.53 (ε NH_3^+)
Serine	⎱ Serine	⎰ Serine	2.21	Proline*	10.60
Threonine	⎱ Threonine	⎱ Glutamic acid	2.19	Alanine	9.69
Glycine	⎰ Glycine	Proline*	1.99	⎰ Valine	9.62
Alanine	⎱ Alanine	⎰ Glycine	2.34	⎰ Leucine	9.60
Valine	Proline*	⎰ Alanine	2.34	⎱ Glycine	9.60
Proline*	⎰ Valine	⎱ Valine	2.32	⎰ Threonine	—
Methionine*	⎱ Methionine*	⎰ Methionine*	2.28	⎱ Serine	9.15
Leucine	Leucine	⎱ Leucine	2.36	Histidine*	9.17
Phenylalanine*	Phenylalanine*	Phenylalanine*	1.83	Methionine*	9.21
Bivalent cations	Bivalent cations	Bivalent cations	—	Bivalent anions	—
				Phenylalanine*	9.13

As for the separation into fraction (iii) (iv) and (v) (see Fig. 81), Davies
(*711*) points out that the pK values are too close to expect a separation
depending on this factor, but that they are eluted in the order of their
molecular weight.

(iii) glycine, mol. wt. 75; alanine, mol. wt. 89
(iv) valine, mol. wt. 117; proline, mol. wt. 115
(v) leucines, mol. wt. 131; methionine, mol. wt. 148.

Here adsorption forces seem to play an important part.

Davies *et al.* (*713*) also showed that strongly *basic* resins (Dowex-2) can
effectively separate mixtures of amino acids which are not separated by
elution from *acid* exchangers. Thus leucine and methionine, with closely
similar dissociation constants ($pK = 2.36$ and 2.28) for the reaction $A^+ \rightleftarrows$
$H^+ + A^\pm$ are not separable on Dowex-50. For the reaction $A^\pm \rightleftarrows H^+ + A^-$
however, the pK values are: leucine 9.60, methionine 9.21; as expected, these
two substances could be separated on Dowex-2 (see also Partridge, *2451*).

Partridge and Brimley (*2459*) have studied in detail the separation of
neutral amino acids on Dowex-2, with 0.1 N HCl as displacement developer.
Complete separations of alanine-glycine, and proline-valine mixtures were

obtained. The order of displacement was substantially the same as the order of pK_2 (NH_3^+) for the monocarboxylic amino acids; exceptions are again due to adsorption forces.

Table 101 lists the order in which amino acids appear in the effluent from different ion exchange columns.

Complete separations of the fourteen amino acids listed in Table 101 are possible by the use of two columns. Partridge has suggested that the primary fractionation could be carried out on a strongly acidic resin and the mixed bands so obtained could be separated on a strongly basic resin. The only exception to complete separation would be the leucine-isoleucine mixture which forms a mixed band on both types of resin (2459).

Displacement chromatography on ion exchange resins has been used by Westall (3511, 3512) for the isolation of glutamine, betaine and γ-amino-butyric acid from beetroot and of an S-hydroxyalkyl cysteine ("felinine") from the urine of cats (3514) and for a general survey of amino acids and other ampholytes of urine (Westall, 3513) and by Shewan et al. (2963) for the analysis of the basic extractives in haddock muscle.

(f) Complete amino acid analysis on Dowex-50

Moore and Stein (2268) described in full detail a procedure for the analysis of amino acid mixtures which marked an important step forward in protein analysis. The drawbacks of amino acid separations on starch columns had been stressed by the authors themselves (see p. 306). The Dowex-50 resin (a sulphonated polystyrene resin) they used later, possesses higher resolving power and is not adversely affected by the presence of inorganic salts; accordingly, blood plasma dialysates and urine may be fractionated directly. In preliminary experiments, Stein and Moore (3095) used Dowex-50 columns operated in the hydrogen cycle and eluted the amino acids with dilute HCl. Later, the use of the sodium or ammonium form of the resin and elution with buffers was shown to be more advantageous. Moore and Stein (2268) use a 0.9 × 100 cm column of Dowex-50 and elute the amino acids with buffers of progressively increasing pH (from pH 3.4 to pH 11). In experiments with synthetic mixtures of amino acids simulating the composition of protein hydrolysates, a sequence of buffers has been developed which yields in a single chromatogram an effluent curve in which every component emerges as a discrete peak. For analytical work, 3 to 6 mg of the amino acid mixture are required. Integration of the curves has given quantitative recoveries (100 ± 3 %) for all except the basic amino acids. These may be determined with the aid of a second column (0.9 × 15 cm) using buffers in the pH range 5 to 6.8.

This method of amino acid analysis has been extensively used by workers in many laboratories.

Stein (3091) has analyzed the amino acids in normal urine, Schram et al. (2878) have analyzed amino acids in foods, Koechlin and Parish (1755) the amino acid composition of a protein isolated from lobster nerve. Waksman and Bigwood (*3446*) have described the identification and estimation of γ-aminobutyric acid from yeast, Rogers *et al.* (*2730*) have separated hydroxyproline and proline of oxypolygelatin, Carsten (*515*) has described a modified method for separation of amino acids of urine. See also *2877*.

More recently, Moore and Stein (*2270*) have described the use of a longer column, a resin with 4 % cross-linking (Dowex-50-X4 instead of Dowex-50-X8) and an eluent of continuously changing pH and ionic strength. The results obtained with this procedure, when a synthetic mixture of 50 components is chromatographed, are shown in Fig. 82. The use of a 150 cm column and the introduction of gradient elution improves the resolving power of the system. The lower cross-linked resin gives satisfactorily sharp peaks with small peptides as well and thus permits mixtures of amino acids and oligopeptides to be fractionated on the same column. The method permits the recovery of amino acids within 3 % of theory in a single chromatogram of one week's duration at pH 3.1 to 5.1, by elution with sodium acetate-citrate buffers of gradually increasing pH and ionic strength.

The eluted amino acids and related compounds can be estimated photometrically by a modified ninhydrin reagent described by Moore and Stein (*2271*).

While the new Dowex-50-X4 method represents a considerable improvement in respect to the previous one and has been used by the above authors for the quantitative analysis of the free amino acids of human blood plasma (Stein and Moore, *3097*), of free amino acids and related compounds in the tissues of the cat (Tallan *et al.*, *3195*) and of the amino acid composition of ribonuclease (Hirs *et al.*, *1405*), certain difficulties of reproducibility have been noted by the authors themselves, due to variations in the properties of different batches of Dowex-50-X4 resin. Moore and Stein (*2270*) describe in detail the way to avoid these difficulties, mainly by adjusting the properties of the resin through admixture with Dowex-50-X5 resin.

Van der Schaaf and Huisman (*3366*) have used this method for the estimation of the amino acid composition of human adult and foetal carbonmonoxyhaemoglobin.

Campbell *et al.* (*500*) have described a similar method for amino acid analysis using Zeo-Karb 225; these authors consider it more satisfactory to define a resin by its water regain (W.R.) as defined by Pepper (*2490*) rather than by the amount of cross-linking agent added at the polymerization stage, since the former can be measured accurately and the latter does not necessarily give a true indication of the degree of cross-linking in the final product. (W.R. 1.5, as used by Campbell *et al.* corresponds to about 5 % cross-linking.)

Hirs *et al.* (*1402*) have described a procedure which permits the isolation of about 100 mg quantities of amino acids from protein hydrolysates. They employ four columns of ion exchange resins: a 7.5 × 15 cm column of Dowex-50 is used to separate the basic amino acids (the other amino acids

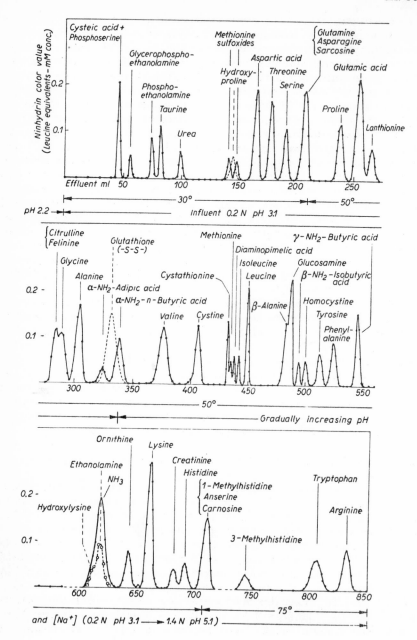

passing into the filtrate); a 7.5 × 15 cm column of Amberlite IR-4B (XE-59) to separate the acidic amino acids (the neutral amino acids passing into the filtrate) and two Dowex-50 columns, 7.5 × 120 cm and 7.5 × 60 cm, to separate the neutral amino acids. Elution is effected with ammonium formate or acetate buffers of pH range 3 to 7 which can be removed by sublimation at 40° *in vacuo*, the residual amino acids being readily recovered in crystalline form. Hirs *et al.* applied their fractionation scheme to the isolation of the amino acids of an acid hydrolysate of 2.5 g of bovine serum albumin. All the amino acids (methionine excepted) were obtained in analytically pure form in an average yield of 66 %. The use of triethyl-ammonium buffers, which are very volatile, is recommended by Porath (*2560a*).

More recently Hirs *et al.* (*1403*) have developed an improved method using volatile acids for elution. The protein hydrolysate is first passed over a Dowex-1-X8 column, which retains aspartic acid, glutamic acid and tyrosine, elutable by 0.5 N acetic acid; the remaining components of the mixture are separated on a 150 cm column of Dowex 50-X4 by elution with a gradient of HCl (1 to 4 N). Recoveries of glutamic and aspartic acid were only of 85 and 70 % respectively; the recoveries from the Dowex-50-X4 column were quantitative except for methionine which is partially oxidized to the sulfoxide during hydrolysis and chromatography. These authors give detailed prescriptions for the isolation of the amino acids from the eluates in a crystalline form.

3. PARTITION CHROMATOGRAPHY ON COLUMNS

(a) *Silica gel*

Partition chromatography was invented and developed by Martin and Synge (*2156*) in 1941, for the separation of acetylated amino acids.

The principle of the method has already been described (p. 103). Martin and Synge (*2156*) originally used silica gel as the support for the aqueous phase of the solvent system, the organic phase chloroform-butanol-water, being passed through the column.

Fig. 82. Separation of amino acids and related compounds from a synthetic mixture containing 50 components. The column of Dowex-50-X4 (150 × 0.9 cm) was operated with sodium acetate and citrate buffers at the temperatures indicated. The amino acids were each present in 0.05 to 0.20 mg quantities. The effluent was collected in 1 ml fractions, except in the range from cystathionine through leucine, in which 0.5 ml fractions were collected. The positions of hydroxyproline and glutathione, which were not included in the mixture, are indicated by dotted curves. The concentrations of hydroxylysine and ethanolamine were determined after removal of NH_3 by evaporation. The dotted base line at 600 and 725 effluent ml indicates the positions at which there is a change in blank resulting from the elution of traces of NH_3 contained in the influent buffers (Moore and Stein, *2270*)

Preliminary work having established that the N-acetylamino acids possessed sufficiently different partition coefficients, Martin and Synge (*2156*) developed the technique into a precise analytical method for the separation and estimation of a dozen neutral amino acids. The zones on the column were made visible by the addition of an indicator of suitable pH range to the column (methyl orange, anthocyanins, or the indicator of Liddell and Rydon, *2005*).

Gordon, Martin and Synge used this method for the analysis of the amino acids of wool (*1168*), gelatine (*1170*), gramicidin (*1169*), tyrocidin (*1171*) and gramicidin S (*3182*).

The preparation of suitable silica gels is rather difficult and not always successful (*1172*). See Tristram (*3303*), Isherwood (*1512*), Harris and Wick (*1315*).

Tristram (*3303*) has described in detail the preparation of silica gels suitable for partition chromatography; the gels should be free from Fe and Al, as far as possible, and should be able to absorb up to 75 % of water. In applying the method of Gordon, Martin and Synge to other proteins, difficulty in the estimation of methionine and proline is encountered owing to the fact that the rate of movement of the zones is affected by the presence of other acetylamino acids, and by variations in the percentage of the stationary phase in the silica gel. Blackburn *et al.* (*263*) used this technique to purify the S-methylcysteine produced by the methylation of wool treated with bisulphite. See also Berlingozzi *et al.* (*223*).

Wieland and Fremery (*3562*) used columns of silica gel imbibed with water for the separation of the copper salts of the neutral amino acids; blue zones were obtained, and valine, alanine, etc. were quantitatively separated.

In their study of the capsular substance of *Bacillus anthracis* composed of D-glutamic acid residues, Hanby and Rydon (*1295*) described a specific method for the determination of glutamic acid which consists in the formation of pyrrolidone carboxylic acid which was purified by partition chromatography on silica gel. The acid was eluted with $CHCl_3$-17% butanol. Chromatography showed the absence of other amino acids in the capsular polypeptide.

Kieselguhr columns (Hyflo Supercel) were used by Smith (*3029*) for the separation of acetylated amino acids.

(b) *Starch and cellulose*

Synge (*3181*) applied partition chromatography to the separation of free amino acids on *columns of starch*, containing water. Moore and Stein (*2265, 3093*) have developed this method with considerable success. Fig. 83 shows the separation of phenylalanine, leucine and isoleucine (solvent, *n*-butanol-benzyl alcohol, 1:1, saturated with water). The filtrate was collected in 1 ml fractions by the use of an automatic fraction collector. Quantitative determinations were carried out by a specially developed photometric ninhydrin method (*2266*). Traces of metals upset the separations by distorting the first band (e.g. phenylalanine); they can be removed by pretreating the column with a solution of 8-hydroxyquinoline.

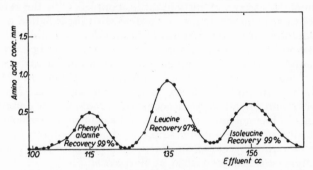

Fig. 83. Separation of phenylalanine, leucine and isoleucine on starch columns (Moore and Stein, *2265*)

A comparison of the partition coefficients of the amino acids and their behaviour on the column led Moore and Stein (*2265*) to conclude that the starch was not only acting as a support for the stationary phase but also as an adsorbent for the aromatic amino acids. When starch was used as an adsorption column with water as solvent, they obtained good separations of alanine, glycine, leucine, phenylalanine and tryptophan.

Nevertheless, Moore and Stein have found that, in contrast to adsorption chromatography, where the ease of elution of a substance depends on the quantity of impurity present in solution, a given substance always behaves in the same way on a starch chromatogram independently of the amount of other substances present.

The rate of movement of certain amino acids was found to be increased by the addition of propanol or butanol (serine, glycine) or by acidification of the organic solvent (propanol-0.5 N aqueous HCl, 3:1, for the hexone bases).

In a subsequent paper, Moore and Stein (*2267*) examined in detail all the experimental conditions influencing resolution of unknown amino acid mixtures. By the use of different solvent mixtures, such as *n*-butanol, *n*-propanol, 0.1 N HCl (1:2:1), propanol, 0.5 N HCl (2:1) and butanol, benzyl alcohol, water (1:1:0.288) most separations could be accomplished; with three separate columns, all the usual 18 constituents of proteins can be determined quantitatively. The average recoveries in duplicate and triplicate determinations were 100 ± 3 %. The same authors (*3094*) applied their method to the analysis of the composition of lactoglobulin and bovine serum albumin. Total nitrogen recoveries were 99.6 and 101.2 % respectively. For further details and applications of this method see also Stein and Moore (*3095*) who state that the separations obtained are not due to partition but to adsorption. They finally point out that "Starch columns possess the disadvantage that fluids of high salt content, such as blood plasma or urine, require desalting prior to chromatography. Moreover, contamination of the effluent by traces of carbohydrate, coupled with the relatively low capacity of these columns, tends to interfere with the isolation of pure compounds. Methods in which columns of Dowex-50 are employed have been found to be free from these difficulties." These are described on p. 300.

The method of starch chromatography of amino acids developed by Moore and Stein has been used by Pierce and Du Vigneaud (*2516*) for the analysis of the amino acids of oxytocin, by Turner *et al.* (*3334*) for the analysis of vasopressin and by Schroeder *et al.* (*2889*) for the determination of amino acids in haemoglobin of normal and sickle cells. Borsook *et al.* (*320*) separated glutamic and α-aminoadipic acids on a column of starch. The analysis of the amino acids from bacitracin (an antibiotic peptide) using starch columns has been described by Barry *et al.* (*147*).

Lens and Evertzen (*1959*) have extended Stein and Moore's method of amino acid analysis on starch columns to protein digests; polypeptides present generally remain on the column, only the free amino acids being eluted.

Sluyterman and Veenendaal (*3011*) have described the preparative separation of glycine, sarcosine and dimethylglycine on starch columns.

Dobyns and Barry (*796*) have separated iodinated amino acids from thyroid tissue on starch columns.

Cellulose columns have been used by Blackburn (*262*) for the fractionation of α-amino acids.

4. PAPER CHROMATOGRAPHY

For reviews of the paper chromatography of amino acids see Boulanger *et al.* (*339*), and Thompson and Thompson (*3238*). Dent and Walshe (*744*) have reviewed the applications of paper chromatography to the study of amino acid metabolism.

(a) *Technical details of procedure*

(i) *Effects of pH on chromatography of amino acids*

Amino acids are generally applied to the paper as hydrochlorides, obtained from protein hydrolysates, after removing most of the acid by repeated evaporation with water. Landua *et al.* (*1878*) found that the pH of the sample affected the spread of the spot and also its position on the chromatogram. Variability in R_F and area of the spot at pH values near the isoelectric point was observed for most of the substances studied. This implies that in certain cases the suitable pH for the sample has to be chosen, to obtain good separations.

McFarren (*2074*) has studied in detail the use of buffered filter paper for chromatography of amino acids, and claims to have obtained very reproducible R_F values. By employing several solvents buffered at a chosen pH between 1.0 and 12.0 it is possible to separate each amino acid from all the others by one-dimensional chromatography.

Best results are obtained by buffering both the paper and the solvents. Difficulties are encountered in presence of large concentrations of tyrosine and cystine, as these are only soluble at a very low pH, which gives R_F values quite different from those usually encountered.

Berry and Cain (*230*) incorporate buffer salts into the phenol (20 ml of a solution containing 6.3 % sodium citrate and 3.7 % KH_2PO_4 for 100 g phenol) so as to eliminate interference by extraneous salts.

Buffered solvent systems have also been proposed by Hais and Horešovský (*1268*) (*tert.* butanol-borate, pH 8.5).

(ii) *Solvents*

Water-saturated phenol and collidine are probably still the most widely used solvents for two-dimensional chromatograms. Phenol can be purified by distillation in presence of 0.1 % aluminium turnings and 0.05 % $NaHCO_3$; the water used to saturate this solvent should be metal free (triple distilled) (Draper and Pollard, *816*).

Collidine has certain disadvantages, such as the formation of double spots and haloes (Bentley and Whitehead, *190*), varying chemical composition, unpleasant smell and toxic effects, as well as, in some countries, difficulties of supply. Alternative solvents have been proposed, such as: butanol-acetic acid (Phillips, *2510*), *iso*butanol-acetic acid-water (4:1:5) (Dakshinamurti, *682*), and pyridine-amyl alcohol (Verdier and Ågren, *762*; Bryant, *435*). Mesityl oxide, which gives an extremely wide distribution of R_F values, has also been proposed, but has the inconvenience of not being very stable (Bryant and Overell, *437*).

The following solvent systems have been described for *two-dimensional separations*: methanol-water-pyridine (40:10:2) and butanol-methyl ethyl ketone-water-diethylamine (20:20:10:2) (Sisakyan *et al.*, *3002*), phenol saturated with a solution of 6.3 % sodium citrate and 3.7 % NaH_2PO_4 and butanol-acetic acid-water (4:1:5) (Wahi and Nigam, *3445*), methanol-water-pyridine (80:20:4) and *tert.*butanol-methyl ethyl ketone-water-diethylamine (40:40:20:4) (Redfield, *2641*), butanol-acetic acid-water (4:1:5) and *m*-cresol-phenol, pH 9.3 borate buffer (1:1) (Levy and Chung, *1992*), butanol-acetic acid-water (4:1:5) and either α-picoline-acetic acid-water (75:2:23) or α-picoline-pyridine-water (72:5:23) (Pfenning, *2503*).

Thompson and Thompson (*3238*) stress that butanol-acetic acid in the first direction followed by phenol in the second direction gives better resolution and spread of amino acids than does the reverse order.

Rockland and Underwood (*2729*) have described a "rapid two-dimensional procedure" for the separation of amino acids on 5 inch square paper sheets, with *tert*.butanol-formic acid-water as the first and phenol-ammonia-water as the second solvent; this was developed after a thorough study of several factors affecting separation and sequence of amino acids (Underwood and Rockland, *3349*).

Decker *et al.* (*722*) have proposed three solvents for "multidimensional chromatography":

A: 30 % liquid phenol + 70 % liquid phenol saturated with water + NH$_3$ + HCN in the atmosphere

B: 70 % α-picoline + 2 % conc. NH$_3$ + 28 % water

C: 70 % *iso*propanol + 20 % acetic acid + 10 % H$_2$O

Methylcellosolve (ethylene glycol monomethyl ether) gives R_F values slightly higher than collidine (Bender, *183*).

The *salting out chromatography* of amino acids on paper has been studied by Hagdahl and Tiselius (*1258*).

Water-miscible solvents have been studied by Bentley and Whitehead (*190*); they suggest that aqueous acetone can replace collidine for use, in conjunction with phenol, in two-dimensional chromatograms; mixtures of furfuryl alcohol and tetrahydrofurfuryl alcohol give small sharp spots. Fisher *et al.* (*972*) have also used furan derivatives (furfuryl alcohol-water, 25 % v/v and tetrahydrofurfuryl alcohol-water, 20 % v/v) for separations and quantitative estimations of amino acids. Boissonnas (*296*) has developed a combination of two-dimensional ascending chromatograms which is claimed to separate all common natural amino acids in 26 hours. The leucines, valine, methionine, phenylalanine, tyrosine and tryptophan are separated by the system: *tert*.-butanol/methyl ethyl ketone/water, 4:4:2; *tert*.butanol/methanol/water, 4:5:1. The other amino acids do not migrate at all and can then be separated with the system phenol/water, 7:3; *n*-butanol/water, 7:3.

The theoretical implications of the use of water-miscible solvents have been discussed already (see p. 115).

Clayton and Strong (*587*) have proposed a mixture of completely miscible solvents (methyl ethyl ketone-propionic acid-water, 75:25:30), the composition of which is not altered by temperature fluctuations and which gives very good resolution.

Hardy *et al.* (*1308*) have studied a great number of different one-phase solvent mixtures and could separate all the constituents of a mixture of 18 amino acids by a single chromatography with the following solvent: butanol-butanone-water (10:10:5), or butanol-butanone-water-cyclohexylamine (10:10:5:2) or ethanol-butanol-water-dicyclohexylamine (10:10:5:2) or butanol-butanone-water-dicyclohexylamine (10:10:5:2).

Drying of the paper is best effected at as low a temperature as possible; overheating causes losses of amino acids, especially after use of phenol as solvent. See the section on quantitative estimation (p. 139).

(iii) *Detection of spots*

The most widely used reagent for amino acids is ninhydrin. It can be purified by a single crystallisation from water in presence of Norit (Moore and Stein, *2266*) or, if necessary, by dissolving in 5 parts of 2 N HCl and boiling with 0.1 part of Norit (*1289*). The ninhydrin is generally used as a 0.1 to 0.25 % solution in butanol or butanol-water (Berry and Cain, *230*). Nicholson (*2369*) has proposed the incorporation of 0.1 % ninhydrin in the collidine used as one of the solvents. Williams and Kirby (*3581*) prefer to use a 0.25 % solution of ninhydrin for developing the spots instead of the more usual 0.1 % solution, as this is claimed to double the sensitivity of the method (See also Table 102).

A chloroform solution of ninhydrin (0.1 % w/v) containing 0.1 % collidine is recommended by Woiwod (*3610*).

Instead of spraying the ninhydrin solution, Boissonnas (*297*) applies the reagent with a special flat "pin cushion" which places the reagent on the paper in uniformly spaced needle point spots. Instead of spraying, one can dip the paper in an acetone solution of ninhydrin (*3277*).

Amino acid chromatograms developed with ninhydrin can be preserved by dipping the paper in a suitable preserving varnish (Adams and Stuart, *15*) or a solution of Perspex in chloroform (Kawerau and Wieland, *1664*).

Porath and Flodin (*2561*) have described a method of detecting amino acids, peptides, proteins and other buffering substances on paper. They partly hydrolyze the cellulose by heating the paper with dilute acid; the spots of buffering substances protect the paper against hydrolysis. By spraying the whole paper with an orcinol reagent, the buffered spots appear white on a red-violet background.

$KMnO_4$ (1 % solution containing 2 % Na_2CO_3) can be used as spraying reagent for the detection of certain amino acids, especially methionine, tryptophan, tyrosine and (to a lesser extent) proline and histidine (Dalgliesh, *684*).

Various other spray reagents have been described for amino acids, such as: an alkaline bromothymol blue solution containing formaldehyde (Kemble and Macpherson, *1683*), aromatic aldehydes (Curzon and Giltrow, *680*), isatin (Saifer and Oreskes, *2782*), Folin's reagent (sodium 1,2-naphtho-quinone-4-sulphonate) (Müting, *2327*; Giri and Nagabhushanam, *1120*).

Jepson and Smith (*1580*) recommend a multiple dipping procedure (for instance ninhydrin or isatin followed by Ehrlich's reagent and then Sakaguchi's reagent); individual amino acids can thus be identified with greater precision.

A general spray reaction which detects also peptides and proteins has been described by Rydon and Smith (*2769*) and further developed by Reindel and Hoppe (*2667*). The former authors treat the chromatograms with gaseous chlorine which transforms the amino compounds to chloroimides. These show up as blue-black spots after treatment with starch-potassium iodide. Reindel and Hoppe (*2667*) oxidize with ClO_2 and detect the spots with o-tolidine or benzidine; less then 1 μg of substance can thus be detected.

TABLE 102.   Sensitivity of the ninhydrin reaction (Pratt and Auclair, *2579*)
             Method: 28 h in phenol, 60 h 1:1 mixt. collidine and 2,4-lutidine at 23–26°.
             Sprayed with 0.1 % ninhydrin in *n*-butanol and heated in oven for 5 min. at
             80–100°. Best illumination through X-ray illuminator.

Compound	Minimum quantity (µg)	Colour of spot	R_F Phenol 28 h	R_F Collidine- lutidine 60 h
Alanine	0.2	purple	0.63	0.41
β-Alanine	0.2	blue	0.71	0.33
Alanylglycine	3	pink purple	0.56	0.36
α-Amino-*n*-butyric acid . .	0.2	purple	0.77	0.46
ε-Amino-*n*-caproic acid . . .	0.5	purple	0.91	0.34
Arginine.HCl	4	blue purple	0.66	0.14
Asparagine	1	brown yellow	0.42	0.29
Aspartic acid	0.4	blue	0.19	0.24
Citrulline	0.5	purple	0.67	0.31
Cysteic acid	8	blue	0.10	0.43
Glucosamine.HCl	4	purple brown	0.52	0.65
Glutamic acid	0.1	purple	0.32	0.26
Glutamine	2	purple	0.62	0.32
Glutathione	10	blue purple	0.10	0.16
Glycine	0.1	pink purple	0.42	0.33
Histamine.2 HCl	12	yellow brown	0.92	0.46
Histidine.HCl	25	brown	0.77	0.34
Homocystine	4	purple	0.38	0.28
Hydroxyproline	1	brown yellow	0.72	0.42
Isoleucine	0.5	purple	0.88	0.62
Leucine	0.5	purple	0.88	0.65
Lysine.HCl	3	purple	0.56	0.14
Methionine	1	purple	0.85	0.61
Methionine sulphone . . .	5	brown purple	0.66	0.51
Methionine sulphoxide . . .	1	purple	0.84	0.34
Norleucine	0.4	purple	0.89	0.69
Norvaline	0.5	purple	0.84	0.56
Ornithine.HCl	3	purple	0.42	0.13
Phenylalanine	5	grey brown	0.90	0.67
Proline	1	yellow	0.90	0.41
Serine	0.3	brown red	0.37	0.37
Taurine	1	purple	0.39	0.50
Threonine	2	pink purple	0.53	0.43
Tryptophan	2	yellow brown	0.79	0.66
Tyrosine	3	brown	0.63	0.74
Valine	0.2	purple	0.82	0.53

(iv) *Fluorescence of amino acids on paper*

Phillips (*2507*) has described how amino acids can be detected on paper by
their fluorescence in ultra-violet light transmitted through a Woods filter.

Approximately 20 μg of each acid is the lower limit of detection. The fluorescence, however, develops only after heating the amino acid spots with the paper and is due to a "browning reaction", i.e. reaction of the free amino groups of the amino acids with free aldehyde groups of the paper (Patton et al., 2469; Gal, 1073; Woiwod, 3611). Only a few per cent of the amino acids seem to react, so that extraction of the fluorescent spots yields about 90 % of ninhydrin-reacting amino acids.

(v) *Circular paper chromatography*

The separation of amino acids by circular paper chromatography has been described by several authors (Giri, 1111; Giri and Rao, 1131; Giri et al., 1115; Rao and Wadhwani, 2621; Berlingozzi et al., 222; Oreskes and Saifer, 2419; Ray et al., 2635). Giri et al. (1128) have studied the factors influencing the quantitative determination of amino acids by circular paper chromatography. Amino acid analysis of proteins has been reported by Giri et al.

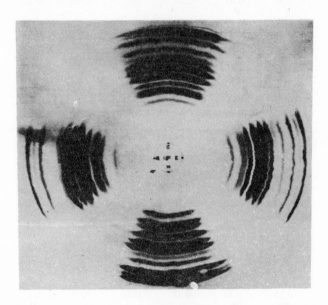

Fig. 84. Circular paper chromatogram showing the amino acid composition of casein (top), gelatin (left) and edestin (right); below, a synthetic mixture of amino acids (Giri, *1111*)

(1116) (see Fig. 84), and Giri and Rao (1132) have described a technique for the separation and estimation of "overlapping amino acids" by cutting out the bands of amino acids not well separated and using the cut strip as wick for developing another chromatogram with another solvent.

TABLE 103. R_F values of amino acids in various solvents on Whatman No. 1 paper, at ro

Solvent	Phenol	Phenol	Phenol	Phenol	Phenol	Collidine	n-Butanol	B
Addition	HCN	3% NH₃*	0.1% NH₃*	Coal gas	Cupron	—	Cupron	J
Glycine	0.40 RP	0.40 RP	0.41	0.40	0.42	0.25 P	0.05	
Alanine	0.54 P	0.54 P	0.55	0.57	0.59	0.32 P	0.08	
Norvaline	0.81 P	0.79 P	0.78	0.78	0.80	0.48 P	0.26	
Valine	0.77 P	0.76 P	0.76	0.78	0.77	0.45 P	0.20	
Norleucine	0.88 P	0.85 P	0.87	0.85	0.89†	0.60 P	0.47	
Isoleucine	0.86 P	0.81 P	0.87	0.82	0.86†	0.54 P	0.37	
Leucine	0.85 P	0.83 P	0.86	0.84	0.88	0.58 P	0.38	
Phenylalanine	0.89 BP	0.87 P	0.90	0.86	0.93†	0.59 G	0.43	
Tyrosine‡	0.64 GP	0.63 B	0.66	0.59	0.62	0.64 G	0.28	
Serine	0.36 P	0.33 P	0.35	0.33	0.36	0.28 G	0.05	
Threonine	0.50 P	0.41 P	0.47	0.50	0.50	0.32 P	0.07	
Hydroxyproline	0.67 O	0.50 O	0.66	0.66	0.67	0.34 OY	0.07	
Proline	0.91 Y	0.85 Y	0.89	0.87	0.86†	0.35 Y	0.12	
Tryptophan	0.83 B	—	—	0.76	0.86	0.62 P	0.35	
Histidine‡	0.69 RP	0.68 B	0.70	0.72	0.69	0.28 G	0.06	
Arginine‡	0.59 P	0.89 P	0.85	0.67	0.62	0.16 P	0.03	
Ornithine‡	0.33 P	0.73 P	0.61	0.40	0.37	0.13 BG	0.01	
Lysine‡	0.46 P	0.82 P	0.73	0.50	0.48	0.14 BG	0.01	
Aspartic acid‖	0.15 BP	0.12 B	0.12	0.14	0.17	0.22 B	0.01	
Glutamic acid	0.25 P	0.13 P	0.19	0.24	0.28	0.25 P	0.01	
Lanthionine‡	0.20 RP	0.19 G	0.21	0.18	0.20	0.12 G	0.01	
Cystine‡	0.30 RP	0.24 YG	0.13§	0.13§	0.29	0.14 G	0.01	
Methionine	0.80 P	0.76 P	0.83	0.82	0.81†	0.57 GP	0.26	

Coulour given by ninhydrin = B, blue; G, grey; O, orange; P, purple; R, red; Y, yellow.
* The % figure refers to the strength of NH₃ solution present in quantity in the tray.
† These acids were applied to the paper after the solvent had already run some distance.

:rature (Consden *et al.*, *618*)

.- vl hol -	Benzyl alcohol —	1:1 n-Butanol benzyl alcohol HCN	o- Cresol Cupron	o- Cresol Cupron + 0.1% NH₃*	m- Cresol Cupron	m- Cresol Cupron + 0.1% NH₃*	p- Cresol Cupron	p- Cresol Cupron + 0.1% NH₃*	iso- Butyric acid —
7	0.02	0.03 P	0.07	0.07	0.12 P	0.14 G	0.15 RP	0.19	0.36
0	0.03	0.05 P	0.13	0.15	0.23 P	0.24 B	0.28	0.32	0.44
3	0.12	0.19 P	0.45	0.45	0.60 B	0.62 B	0.66 BP	0.66	0.71
8 •	0.11	0.15 P	0.41	0.38	0.52 P	0.54 P	0.59 BP	0.60	0.65
2	0.27	0.36 P	0.72	0.67	0.80 B	0.78 P	0.82 BP	0.80	0.79
1	0.18	0.27 P	0.58	0.55	0.70 B	0.70 B	0.73 P	0.74	0.76
6	0.21	0.31 P	0.61	0.62	0.73 B	0.73 B	0.73 BP	0.77	0.78
6	0.36	0.38 BG	0.81	0.74	0.82 GB	0.82 P	0.82 P	0.83	0.80
4	0.14	0.19 G	0.24	0.23	0.35 P	0.35 P	0.39 B	0.38	0.58
8	0.01	0.02 P	0.05	0.05	0.08 BP	0.08 B	0.11 P	0.13	0.34
9	0.02	0.04 P	0.11	0.08	0.14 B	0.14 P	0.19 P	0.21	0.43
0	0.04	0.05 O	0.27	0.23	0.32 GY	0.32 Y	0.40 BG	0.42	0.42
1	0.12	0.12 Y	0.69	0.66	0.73 Y	0.75 Y	0.76 Y	0.78	0.57
-	—	0.30 B	0.65	0.58	0.76	0.69	0.70 B	0.68	—
7	0.02	0.03 RG	0.25	0.33	0.34 P	0.44 GP	0.36 B	0.52	0.45
2	0.01	0.01 P	0.02	0.44	0.07 P	0.76 P	0.12 BP	0.81	0.40
2	0.00	0.00 P	0.01	0.21	0.03 P	0.47 P	0.04 P	0.59	0.24
2	0.00	0.01 P	0.01	0.20	0.04 P	0.54 P	0.07 BP	0.66	0.27
2 P	0.00	0.00 P	0.00	0.00	0.01 RP	0.01 B	0.02 BP	0.02	0.31
2	0.00	0.01 P	0.01	0.01	0.01 P	0.02 B	0.03 BP	0.03	0.38
3 G	0.00	0.00 P	0.01	0.01	0.01 R	0.03 RP	0.03 RP	0.05	0.21
2 G	0.00	0.00 P	0.02	0.01	0.04 P	0.04	0.02 RP	0.02	0.25
7	0.17	0.21 P	0.58	0.49	0.64 B	0.63 P	0.71 BP	0.68	0.69

‡ These acids were put on as the hydrochlorides and neutralized with NH₃ before development.
§ Cystine is decomposed by these additions. R_F value refers to the decomposition product.
‖ Colour initially green. Only final colour shown in table.

(b) Special reactions of groups of amino acids

(i) α-Amino acids

Crumpler and Dent (666) describe a distinctive test for α-amino acids consisting in the formation on the paper of the copper complexes of these substances, which no longer react with ninhydrin. β-Amino acids survive only when present in larger amounts; γ-, δ- and ε-amino acids are resistant.

(ii) N-Substituted amino acids

Plattner and Nager (2525) describe the separation and detection of N-methylamino acids using a specific colour test involving reaction with *p*-nitrobenzoylchloride and pyridine. Novellie and Schwartz (2388) find however that sarcosine does not give the reaction of Plattner and Nager. Gal and Greenberg (1074) have separated N-ethyl, N-propyl, N-*iso*propyl and N-phenyl-amino acids with phenol-water and collidine-water; free amino acids give a blue fluorescence, the N-substituted amino acids dark spots; the latter give only weak spots with ninhydrin. Kuhn and Ruelius (1822) have described the retention analysis of glycine, sarcosine and di-methylglycine; see also p. 138 for detection of amphoteric substances.

Kiessling and Porath (1714) have described the detection of N-dimethyl-amino acids on paper by conversion to betaines by means of methyl iodide. The iodide remaining in the spot after evaporation of excess CH_3I may be detected either by oxidation with chlorine in vapor phase yielding iodine which can be detected with a starch solution (see Rydon and Smith, 2769) or by precipitation of silver iodide followed by reduction with a photographic developer.

(iii) ω-Amino acids

Synge (3185) has described the paper chromatography of β-alanine, γ-aminobutyric acid, δ-aminovaleric acid and ε-aminocaproic acid, after having separated them from α-amino acids by ionophoresis in silica gel (Table 104).

TABLE 104. R_F values of ω-amino acids (Synge, 3185)

Paper, Munktell OB	Butanol: water: acetic acid (by volume)			Butanol-water
	8 : 10 : 1	4 : 5 : 1	2.67 : 3.33 : 1	
β-Alanine	0.17	0.32	0.65	0.05
γ-Aminobutyric acid	0.24	0.39	0.70	0.05
δ-Aminovaleric acid	0.32	0.50	0.75	0.05
ε-Aminocaproic acid	0.38	0.61	—	—
Alanine	0.19	0.33	0.65	—
Valine	0.41	0.53	0.77	0.24
Leucine	0.58	0.68	0.82	0.42

(iv) *Hydroxy-amino acids*

These can be detected by their reaction with periodate and can be identified after oxidation with periodate by the R_F values of the oxidation products (Block and Bolling, *281*; Barbier and Lederer, *135*).

(v) *Sulphur-containing amino acids*

These can be detected by their ability to catalyse the oxidation of sodium azide with iodine (Chargaff *et al.*, *548*), by their reaction with potassium iodoplatinate (Winegard *et al.*, *3593*, Table 105), and also by incorporation of ^{35}S (Tomarelli and Florey, *3279*).

TABLE 105. R_F values of sulphur-containing amino acids (Winegard *et al.*, *3593*). K_2PtI_6 is used as reagent

Amino acid	R_F in Phenol-NH_3-coal gas	Bleaching time	Amount detected (μg)
Cystathionine	0.30	immediate	24
Cysteine	*	2 minutes	12
Cysteine sulphinic acid . . .	0.21	immediate	12
Cysteic acid	0.10	does not bleach	
Cystine	0.25	2 minutes	12
Cystine disulphoxide	0.21	immediate	12
Djenkolic acid	0.30	immediate	12
Lanthionine	0.27	immediate	24
Methionine	0.76	immediate	12
Methionine sulphone	0.65	immediate	18
Methionine sulphoxide . . .	0.81	immediate	18

* Could not be detected when phenol was used as solvent.

Cysteine can best be detected after oxidation to cysteic acid with bromine water or hydrogen peroxide (*619, 3050*). Van Halteren (*3376*) has described artefacts formed after the bromine oxidation of mixtures of amino acids containing cysteine.

Toennies and Kolb (*3277*) describe three reagents for the detection of sulphur compounds: a nitroprusside, a platinum and a palladium reagent.

(vi) *Detection of D-amino acids on paper*

Following a suggestion of Synge, Jones (*1602*) showed that by treating a paper chromatogram with D-amino acid oxidase, the D-amino acids could be distinguished from the L-amino acids. By spraying the paper with a solution of the D-amino acid oxidase in a pyrophosphate buffer and incubating at 37° for 2½ hours, the D-amino acids were destroyed. Jones

(*1602*) applied this technique to a study of the configuration of the amino acids obtained from the antibiotic aerosporin (polymyxin).

Auclair and Patton (*101*) detected D-alanine in the haemolymph of insects by incubating the paper chromatogram with D-amino acid oxidase and by detecting the α-keto-acids formed with a 2,4-dinitrophenylhydrazine spray. Using a similar method, Stevens *et al*. (*3103*) found D-aspartic acid in some natural materials.

Bonetti and Dent (*308*) have described experimental details for the determination of the optical configuration of amino acids in biological fluids; this method uses D-amino oxidase of sheep's kidney and L-amino oxidase of snake venom and is applicable to the following amino acids: alanine, histidine, isoleucine, leucine, methionine, phenylalanine, proline, tryptophan, tyrosine and valine. After incubation with the enzymes, the amino acids are subjected to paper chromatography (see also Berlingozzi *et al*., *222*).

(c) *Special spraying reagents for detection of individual amino acids*

Glycine can be detected by a reagent containing *o*-phthalaldehyde (Patton and Foreman, *2468*).

Creatine, creatinine and derivatives give colours with a 3,5-dinitrobenzoate reagent (Epshteïn *et al*., *904*; Barker and Ennor, *138*).

Histidine and *tyrosine* give only weak colour reactions with ninhydrin; a more sensitive spot test was described by Dent (*741*), who coupled these amino acids with diazobenzene-*p*-sulphonic acid to reveal their presence on a one-dimensional chromatogram. DeVay *et al*. (*760*) have shown that the azo-dyes of histidine and tyrosine obtained by coupling with diazotized *p*-nitroaniline in presence of 4 % Na_2CO_3 can be separated and identified on paper.

Sanger and Tuppy (*2812, 2813*) use a modification of the Pauly test (with diazotised *p*-anisidine) for the detection of *histidine* on paper. Bolling *et al*. (*303*) recommend diazotised *p*-bromoaniline (see also Block and Bolling, (*281*). For monoiodohistidine, see Roche *et al*. (*2712*).

Tyrosine can also be detected with an α-nitroso-β-naphthol spray (Acher and Crocker, *6*).

Arginine and other substances containing the *guanidine* group can be revealed on paper with a Sakaguchi spray (*2709*). See also Thoai *et al*. (*3230*), Roche *et al*. (*2722*), Tuppy (*3316*), Jepson and Smith (*1580*).

Proline gives a specific blue spot with an isatin-containing reagent (Acher *et al*., *7*); hydroxyproline shows up at much higher concentrations only. Turner *et al*. (*3334*) have used this test for the detection of proline in vasopressin.

Friedberg (*1041*) has separated hydroxyproline and *allo*-hydroxyproline on paper. A "new" amino acid in apple peel was tentatively identified as methyl-hydroxyproline (Hulme, *1481*). Giri and Nagabhushanam (*1120*) have discussed the use of Folin's reagent for the estimation of proline and hydroxyproline. Jepson and Smith (*1580*) recommend a first spray with isatin, followed by Ehrlich's reagent as a specific test, for hydroxyproline (change of colour from blue to purple red).

Tryptophan derivatives can be detected by their characteristic fluorescence as well as with Ehrlich's reagent (Tabone *et al.*, *3190, 3191*; Mason and Berg, *2166*; Block and Bolling, *281*).

(d) *Quantitative estimation of amino acids*

Reviews of the methods of quantitative estimation after paper chromatography were made by Schwerdtfeger (*2919*) and Chibnall (*566*).

Polson *et al.* (*2553*) have described a *semi-quantitative* method consisting in comparing the intensity of the spots obtained with ninhydrin with those given by a series of known dilutions of amino acids. The authors applied their technique to the analysis of a hydrolysate of silk and of bacteriophage (Polson and Wyckoff, *2554*).

Woiwod (*3607, 3608*) described a colorimetric method for the estimation of amino-nitrogen in the amino acids obtained by eluting the spots from the paper, based on the method of Pope and Stevens (*2556*) (colorimetry of the amyl alcohol-soluble copper complexes formed by treating the amino acids with copper phosphate). The position of the spots was determined by comparison with a duplicate strip on which the spots were developed with ninhydrin, or by treating the paper with a dilute solution of ninhydrin so that only an inappreciable amount of amino nitrogen was destroyed.

In a later paper, Woiwod (*3609*) reported results obtained with 25 amino acids; recoveries of 81–95 % for uni-dimensional, and 63.9–86.5 % for two-dimensional chromatograms were obtained. It was also noted that the losses increased with the distance of travel of the spot and with the time of heating of the chromatogram. The Pope-Stevens determination was criticized and found unsuitable for paper chromatograms by Rauen *et al.* (*2632*) and by Kofranyi (*1761*). In this as in other methods of estimation after paper chromatography the value of the results depends on the problem and the accuracy required. Satisfactory results with techniques based on Woiwod's method were recorded by Wellington (*3501*), Giri *et al.* (*1128*), Landucci and Pimont (*1879*), Fischer and Dörfel (*965*), Pernis and Wunderly (*2493*), Gerok (*1101*) and Bode (*287*). Heyns and Anders (*1385*) describe an ultra-micro Kjeldahl determination of nitrogen in eluates from paper chromatograms. See also Klatzkin (*1732*).

Naftalin (*2333*) describes a colorimetric method using ninhydrin, carried out in test tubes containing the paper strips cut from the chromatogram. Quantities of the order of 0.2 μg of amino-nitrogen can be estimated with some accuracy. This method was used by Polonovski *et al.* (*2548*) (order of accuracy 10 %).

Fisher *et al.* (*973*) showed experimentally that the area of the spots on the chromatogram was proportional to the logarithm of the concentration of solution used. Fisher *et al.* (*973*) claimed an accuracy of ± 2 % for the measurement of the spot areas obtained from known dilutions of amino acids. The measurement of spot area was carried out more easily on reproductions of the chromatogram on a special photographic paper.

Åkerfeldt (*29*) also employs the spot area method and uses an equation for the spot area-concentration relationship involving the molecular weight of the amino acid.

Martin and Mittelmann (*2153*) recommend the estimation of amino acids on paper by polarography of the copper complexes. The direct photometry of the spots has been developed by Bull *et al.* (*449*). Detailed instructions are given of the development procedure necessary for quantitative results. Other direct photometric methods were described by Rockland and Dunn (*2727*) and also by Block (*279*); the latter moves the paper strip at a constant rate through a photovolt densitometer and records the galvanometer deflection (accuracy ± 12 %). For details, see Block and Bolling (*281*). For further work see also McFarren and Mills (*2081*), Redfield and Barron (*2642*), Christomanos (*568*), Roland and Gross (*2732*) and Grassmann *et al.* (*1196*).

The usual order of accuracy in these estimations ranges from ± 3 % to ± 10 %. Brüggemann and Drepper (*426*) and Hiller *et al.* (*1394*) dip the paper into an oil (example bromonaphthalene-paraffin oil mixture) to increase the paper transparency in the photometer. Darmon and Faucquembergue (*699*), also Keil (*1675*) prefer to scan a photographic copy of the chromatogram, as a better range for accurate results is thus obtained. Other methods proposed are the colorimetry of the DNP-derivatives (Levy, *1991*, and Isherwood and Cruickshank, *1516*), the reaction with chloramine T and measurement of the evolved CO_2 in a Warburg apparatus (Kemble and Macpherson, *1684*), reaction of the eluted spots with a suspension of copper phosphate labelled with ^{64}Cu and measurement of the dissolved radioactivity (Blackburn and Robson, *266*). See also Boser (*327*) and Barrollier (*146*).

Wieland (*3555*) has described the quantitative determination of amino acids by "retention analysis"; the amino acid spots retard the flow of a dilute copper acetate solution by complex formation (see p. 142).

(e) *Sources of error in the determination of amino acids*

The colorimetric determination of the ninhydrin colour developed in solutions of eluted amino acids was used by Moore and Stein (*2266*), who recommend a 2 % ninhydrin solution in methyl cellosolve, mixed with an equal volume of pH 5 citrate buffer containing $SnCl_2$. The same reagent was employed by Landua and Awapara (*1877*) who carried out the determination in the following steps:

(i) The paper is sprayed with 0.05 % ninhydrin, which reveals the position of the spots without reacting with the bulk of the amino acid present. Another method for detection is to view the dry paper under ultraviolet light (Pereira and Serra, *2491*).

(ii) The spots are cut out and fully reacted with 1 ml of 2 % aqueous ninhydrin in the presence of pyridine in a boiling water bath for 20 min. The presence of pyridine is very important, as also shown by Atkinson *et al.* (*97*).

(iii) The colour developed is measured at 570 mμ with a Beckmann Spectrophotometer. For purple spots, wavelengths of 550 mμ or 570 mμ were used. for orange brown spots 330 mμ. The greatest error found in this procedure was the blank due to filter paper ($\pm$ 7 %), on the same sheet 10 μg samples gave an accuracy of $\pm$ 1–2 %. The paper blank was reduced by Fowden and Penney (*1010, 1011*), by placing the excised spots in contact with 0.1 N NaOH (0.5 ml) in a desiccator containing H_2SO_4 overnight. Novellie (*2386*) preferred to pre-treat the paper with 1 % alkali and to wash the paper free of alkali before use. Heating with methanolic KOH was used by Boissonnas (*298*) to remove NH_3. Pereira and Serra (*2491*), on the other hand, found satisfactory blanks (not requiring any corrections) with a large variety of papers, when using ninhydrin buffered to pH 7 with veronal (accuracy $\pm$ 3–5 %).

Heating of the paper to dry off the solvent, however, appears to be another important source of error. Fowden and Penney (*1011*) showed that the yields depended on the temperature (36 % at 75° and 74 % at 110°) also that on drying below 50°, i.e. at room temperature, the losses are negligible, which was also confirmed by Novellie (*2386*). Contrary to Woiwod's (*3609*) previous observation, Fowden (*1010*) was unable to observe a decrease of yield with the distance moved by a spot. Brush *et al.* (*434*) who studied the losses by heating, after using phenol as solvent, recommend 27° as suitable for drying, losses occurring at the other temperatures examined (60°, 85° and 110° C). In another study of the sources of error in quantitative estimations Thompson *et al.* (*3241*) recommend heating in an atmosphere of CO_2 at 60° to drive off the solvent without loss of amino acid. In all other respects the above authors used techniques similar to those of Landua and Awapara (*1877*) with only small variations. Thus Fowden (*1010*) employed ninhydrin saturated with hydrindantin and Thompson *et al.* (*3241*) an ethanolic solution of ninhydrin containing 2 % of a collidine-lutidine mixture.

(f) *Paper chromatography of amino acids containing radioactive isotopes*

Radioactive tracers are extensively used in marking amino acids. No complete survey of the field will be presented, as the general techniques have already been mentioned (page 143). Below are several examples which illustrate the possibilities of the combination of tracers with chromatographic methods. (See also the sections on iodinated amino acids and on thyroid hormones.)

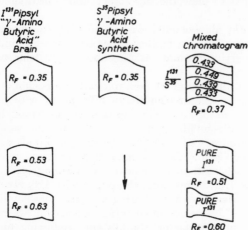

The pipsyl method. Keston *et al.* (*1700*) chromatographed on paper the N-*p*-¹³¹I-phenylsulphonyl amino acids (I). The position of the spots was detected by the radioactivity of the ¹³¹I.

¹³¹I—⟨　⟩—SO₂—NH—CHR
　　　　(I)　　　　　　　CO₂H

This technique was also applied to the *quantitative estimation* of amino acids (*1699, 1700*).

Fig. 85 a. Chromatograms of the ¹³¹I-labelled pipsyl derivative obtained from extracts of mouse brain and the ³⁵S-labelled pipsyl derivative of an authentic sample of γ-aminobutyric acid (Udenfriend, *3338*)

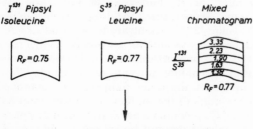

Fig. 85 b. Chromatograms of ¹³¹I-labelled pipsyl isoleucine and ³⁵S-labelled pipsyl isoleucine

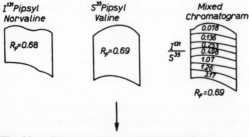

Fig. 85 c. Chromatograms of ¹³¹I-labelled pipsyl norvaline and ³⁵S-labelled pipsyl valine

Udenfriend *et al.* (*1701, 3338, 3340, 3341*) also employed the N-*p*-iodophenylsulphonyl (abbr. *pipsyl*) derivatives for the following problems:

(I) Detection of amino acids having only small R_F differences (0.01). For example, norvaline (R_F 0.68) cannot usually be detected in the presence of valine (R_F 0.69). If synthetic ¹³¹I-pipsylnorvaline and synthetic ³⁵S-pipsylvaline are chromatographed together a mixed band is obtained, which, however, does not give a uniform ¹³¹I/³⁵S ratio throughout (Fig. 85b, c). This is determined by a count with a Geiger counter with and without an Al absorber. One application of this method was the identification of γ-aminobutyric acid (from brain) (*3338*, Fig. 85a). This amino acid was converted to the ¹³¹I-pipsyl-derivative and mixed with synthetic ^{35}S-pipsyl-γ-aminobutyric acid. The band obtained was cut into four sections and gave the same ^{131}I/^{35}S ratio in all four portions, thus proving the identity with the added ^{35}S marked derivative.

(II) A quantitative method for the estimation of amino acids from protein hydrolysates. The mixed amino acids are converted to their ^{131}I-pipsyl derivatives and known amounts of ^{35}S-pipsyl-derivatives of the suspected amino acids added. The mixture is then separated by counter-current distribution and paper chromatography, and the initial amounts of amino acids calculated from the ^{131}I/^{35}S ratio, not from the yields of amino acids; thus all losses during the separations cancel out (Velick and Udenfriend, *3387*, Table 106). A complete scheme for the isotope derivative analysis of eleven amino acids in a single mg sample of protein was developed.

TABLE 106. R_F values of N-p-iodo-phenylsulphonyl derivatives of amino acids (Velick and Udenfriend, *3387*)

Solvent	n-Butanol – 0.1 N NH_3	n-Butanol – 0.1 N HCl
p-Iodophenylsulphonic acid	0.65	0.70
ε-N-Pipsyl lysine	0.45	0.55
Pipsyl aspartic acid	0.03	—
Pipsyl glutamic acid	0.03	—
Pipsyl hydroxyproline	0.30	—
Pipsyl serine	0.40	—
Pipsyl threonine	0.50	—
Pipsyl glycine	0.40	—
Pipsyl alanine	0.50	0.70
Pipsyl proline	0.40	0.70
Pipsyl methionine	0.65	0.60
Pipsyl phenylalanine	0.65	0.35
Pipsyl valine	0.70	0.60
Pipsyl leucines	0.75	0.60
Pipsyl methionine sulphone	0.40	—

(III) Amino end group analysis of proteins (*3341*). Free amino groups readily react with pipsyl chloride and the sulphonamide bond so formed is much more stable than a peptide bond and is not hydrolysed by boiling for four hours at 115° with 6 N HCl. 200 μg quantities of proteins (for example, bovine insulin) sufficed for identification of the amino end groups as their pipsyl derivatives. Quantitative determinations using the ^{131}I/^{35}S ratio were also carried out. The R_F values of a number of pipsyl derivatives are given in Table 106. Velick and Udenfriend (*3388*) have also applied this method to the analysis of salmine and of phosphorylase (*3389*).

(g) *Some details of the paper chromatography of individual amino acids*

(i) *Basic amino acids*

Lysine can give rise to more than one spot on paper, in a saturated water-phenol system, depending on the pH. Lysine ions seem to associate with phenol to give a new complex behaving as an independent substance (Aronoff, *81*). This effect is absent in basic solvents, such as collidine-water, or butanol-water-pyridine. Landua *et al.* (*1878*) have observed similar multiple spots at certain pH values for arginine in phenol, and glutathione in phenol and lutidine. The multiple spots were connected in all cases by a diffuse area of colour.

Waldron-Edward (*3449*) reports that in neutral or mildly acidic or basic solvents, the diamino carboxylic acids give a "slow" spot in presence of small quantities of sulphate ion, due to the formation of the sulphate.

Hydroxylysine has been identified in a phosphatide of *M. phlei* (Barbier and Lederer, *135*). Hydroxylysine phosphate was detected by Gordon (*1166*) in the intracellular fluid of calf embryo muscle.

The estimation of *histidine* by paper chromatography has been described by Frank and Petersen (*1021*).

The paper chromatography of *ergothioneine* has been described by Lawson *et al.* (*1889*) and by Work (*3627*).

Lugg and Weller (*2059*) have observed that on chromatographing the *picrates* of basic amino acids on paper in an NH_3 atmosphere, picric acid is liberated and migrates separately as ammonium picrate.

(ii) *Dicarboxylic amino acids*

Koch and Hanson (*1750*) report that when glutamic and aspartic acids are kept in hydro-alcoholic solution and are then chromatographed on paper, secondary spots corresponding to the ethyl esters of these amino acids are observed.

Ellfolk and Synge (*883*) have described the detection of pyrrolidone-carboxylic acid (which is easily formed from glutamic acid and glutamine and does not react with ninhydrin) by the chlorine-starch-iodide reaction of Rydon and Smith (*2769*).

The isolation of ^{14}C-glutamine by paper chromatography has been described by Bidwell *et al.* (*242*).

γ-Hydroxy-glutamic acid (Virtanen and Hietala, *3410*) and other higher amino-dicarboxylic acids have been isolated from green plants (Virtanen and Berg, *3409*).

(iii) *Neutral amino acids*

Consden *et al.* (*624*) have proved that norleucine did not occur in the hydrolysate of spinal cord; the three leucines involved were well separated on the chromatogram. The isomeric leucines can also be separated with pyridine-amyl alcohol (Heyns and Walter, *1387*).

Cornforth *et al.* (*635*) have described the paper chromatography of tryptophan, hydroxytryptophan, kynurenine and kynurenic acid; the Folin-Denis reagent for phenols (phosphomolybdic-tungstic acid) was found useful for detecting the spots; Dalgliesh (*686*) has used this reagent for the detection of dihydroxyphenyl compounds related to tyrosine.

Citrullin (N-carbamoyl-ornithine) and other N-carbamoyl-amino acids (H_2N—CO—NH—CH(R)—COOH) can be detected on paper with 4 % *p*-dimethylaminobenzaldehyde-*N* HCl (Phillips, *2511*).

(iv) *Neutral hydroxy-amino acids*

Hardy and Holland (*1307*) have described an improved separation of DL-threonine from DL-*allo*threonine on paper. Homoserine has been found in the free state in peas (Miettinen *et al.*, *2211*). A cyclic homoserine derivative has been found in some *Liliaceae* (Virtanen and Linko, *3414*).

Lambooy (*1872*) has described the preparation and paper chromatographic separation of 2,3-, 2,4-, 2,5-, 2,6- and 3,5-dihydroxyphenylalanines. See also Dalgliesh (*686*). 4-Hydroxy- and 5-hydroxy-piperidine-2-carboxylic acids have been identified in green plants by Virtanen and Kari (*3411*).

(v) *Creatine and creatinine*

Maw (*2174*) has described the separation of creatine and creatinine on paper (collidine as solvent); the spots were detected by treatment with alkaline sodium picrate after conversion of creatine to creatinine on the paper with acid. See also Ames and Risley (*48*).

Creatine, creatinine and creatine phosphate have been separated on paper by Epshteïn and Fomina (*904*). Creatinine can be determined by a special 3,5-dinitrobenzoate-containing reagent. Barker and Ennor (*138*) have also identified creatine phosphate by paper chromatography.

(vi) *Sulphur-containing amino acids*

Consden *et al.* (*619*) have studied the use of paper chromatography as a means of identifying the substances derived from cystine by various chemical treatments of wool (cysteic acid, lanthionine, djenkolic acid, thiazolidine-carboxylic acid, etc.). Small amounts of cysteic acid and lanthionine were discovered in natural wool. Cysteic acid was found to give a zinc salt which formed a distinct spot on paper similar to the copper salts of the amino acids. The addition of ammonia diminished the intensity of such secondary spots; with collidine as solvent no metallic salts could be detected. See also Smith and Tuller (*3027*).

Dent (*741*) has given details of the paper chromatography of cystathionine, cysteic acid, cysteine and cystine, penicillamine, djenkolic acid, homocysteic acid, homocystine, lanthionine, methionine, its sulphone and sulphoxide. For the detection of ^{35}S-containing amino acids, see Schlüssel *et al.* (*2854*).

(vii) *Halogenated amino acids*

It has already been mentioned that these substances have been extensively studied with radio-iodine (^{131}I) as tracer. Roche *et al.* (*2714*) have described in detail the techniques of separation and detection of substances containing radioactive iodine.

Monoiodotyrosine has been detected in thyroid tissue and determined quantitatively (Taurog *et al.*, *3208*, *3210*); mono-, di-iodotyrosine and thyroxine have been separated (Fink *et al.*, *957*; Lemmon *et al.*, *1957*) and the metabolism of the thyroid gland studied (Tishkoff *et al.*, *3272*) (Fig. 102, p. 407). Roche *et al.* (*2710*) have described a quantitative paper chromatographic method for the separation and determination of radioactive iodinated substances and have applied it to the study of marked thyroglobulin. The iodination of tyrosine and of histidine has been examined (Roche *et al.*, *2711*, *2712*). The metabolism of ^{131}I and labeled diiodotyrosine has been studied by Tong *et al.*, (*3286*). Detailed methods for the separation and estimation of radioactive compounds isolated from the living organism after administration of ^{131}I have been published (Brown and Jackson, *415*; Roche *et al.*, *2716*). Roche *et al.* (*2719*) have studied the formation of iodinated amino acids in casein. See also the chapter on thyroid hormones.

Roche *et al.* (*2723*) have identified monobromotyrosine as a natural amino acid in a hydrolysate of Gorgonines. Roche and Eysseric-Lafon (*2709*) have made a comparative study of the iodinated and brominated amino acids of the scleroproteins of Corals.

The bromination of tyrosine and thyronine in presence of ^{82}Br and the paper chromatographic separation of the various bromination products has been described by Yagi *et al.* (*3642*).

(viii) *New amino acids*

Paper chromatography is an excellent method for revealing new and unknown constituents of complex mixtures; a whole series of "new" natural amino acids has been discovered as "unknown spots" on papergrams.

α-Aminobutyric acid. This substance was found by Dent *et al.* (*743*) during their study of the free amino acids in potatoes, and by Polson (*2551*) and Woiwod and Proom (*3614*) in hydrolysates of bacteria (see also Dent, *741*; Heyns and Walter, *1388*).

γ-Aminobutyric acid. An unknown substance, occurring in the phenol-collidine-lutidine chromatogram below valine and having the R_F of γ-aminobutyric acid was first found in potato extracts by Dent *et al.* (*743*, *3106*), then in yeast (Reed, *2643*), in bacteria (Woiwod and Proom, *3614*), in plant tissue (Hulme and Arthington, *1482*), and in brain (Awapara *et al.*, *106*; Boulanger and Biserte, *335*). Reed (*2643*) and Awapara *et al.* (*106*) isolated it on starch columns and identified it with synthetic γ-aminobutyric acid by infra-red spectra, elementary analysis and mixed paper chromatography; the isomeric aminobutyric and amino*iso*butyric acids have different R_F

values. Westall (*3511*) has isolated γ-aminobutyric acid from beetroot by adsorption on Zeo-Karb 215 and displacement elution with 0.15 N aqueous NH₄OH. Boulanger and Biserte (*335*) have found that γ-aminobutyric acid is present as a peptide in brain tissue. See also Auclair *et al.* (*100*).

β-*Amino*isobutyric acid was discovered by Crumpler *et al.* (*667*) in human urine forming an "unknown spot" in the position occupied by methionine sulphoxide. That it was not the sulphoxide was shown by the fact that it could not be further oxidised to the sulphone. It did not form a copper complex and therefore was not an α-amino acid. The authors were able to isolate the substance by adsorption on Zeo-Karb 215 and Dowex-2. See also Fink *et al.* (*955*).

α-*Aminoadipic acid*. Boulanger and Biserte (*334*) have described the identification of this substance from urine. It occupies a place near that of glutamic acid, its lower homologue (see also Bergström and Pääbo, *209*). Windsor (*3592*) used starch columns for the detection and isolation of α-aminoadipic acid; radioactive α-aminoadipic acid was added as tracer.

α,ε-*Diaminopimelic acid* was discovered by Work (*3628*) in hydrolysates of *Corynebacterium diphtheriae*, isolated by chromatography on acid alumina and on cellulose columns, and finally identified with a synthetic product (*3629*). Diaminopimelic acid is a neutral amino acid, but is adsorbed on acid alumina if three times the usual quantity of adsorbent is used. It can thus be separated from other amino·acids. Asselineau *et al.* (*90*) had found the same compound in a lipopolysaccharide of *Mycobacterium tuberculosis*. A semi-quantitative method of determination of diaminopimelic acid has been developed, based upon its behaviour on acid alumina and on estimation after paper chromatography (Gendre and Lederer, *1096*). Diaminopimelic acid has until now been found only in material of bacterial origin (see also Bremner, *374*; Blass *et al.*, *271*).

Methylhistidine. Normal human urine contains a substance giving a green spot with ninhydrin (near proline, R_F 0.86 in phenol, 0.28 in collidine-lutidine) (Dent, *741*). Searle and Westall (*2924*) isolated the substance by displacement chromatography on Zeo-Karb 215 and Dowex-2 and identified it with authentic methylhistidine. Datta and Harris (*703*) have shown that it arises from the anserine of the diet.

β-*Methyl-lanthionine*. This substance was isolated from the antibiotic Subtilin, where it is accompanied by lanthionine. The two can be separated on paper with a mixture of acetone, water, acetic acid and urea (Lewis and Snell, *1997*). It is also found in the antibiotic nisin (Newton *et al.*, *2363*).

Monoiodotyrosine and *monobromotyrosine* have been mentioned on p. 324.

Numerous papers contain references to "unknown spots"; some may be artefacts, formed by destruction of cysteine (*3376*), others may be peptides hydrolysed only with difficulty.

(ix) *Enzymological studies*
Enzymatic reactions involving amino acids can easily be followed on paper.

Hird and Rowsell (*1398*) have described "progress chromatograms" (Fig.

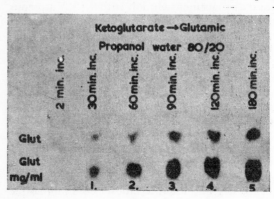

Fig. 86. Progress chromatogram showing glutamate synthesis from phenylalanine and α-ketoglutarate (both *M*/20); separate chromatograms of known amounts of glutamate are set underneath for comparison (Hird and Rowsell, *1398*)

86) showing the formation of phenylalanine from glutamate and phenylpyruvate in the presence of transaminase, and the formation of glutamate from phenylalanine and α-ketoglutarate by insoluble particle preparations of rat liver (see also Rowsell, *2750*).

Feldman and Gunsalus (*941*) have studied specific transaminases producing glutamic acid from α-ketoglutarate and Müller and Leuthardt (*2304*) the formation of citrulline in tissue slices. Gladner and Neurath (*1134*) have published very nice photographs showing the course of the action of carboxypeptidase on DFP-α-chymotrypsin.

(x) *Microbiological studies*

Woiwod and Linggood (*3613*) have described the analysis of the amino acids of diphtheria toxin and toxoid by means of two-dimensional chromatograms (butanol-acetic acid and phenol as solvents), Fromageot *et al.* (*1056*) the evolution of amino acids in corn-steep liquor during culture of *Penicillium chrysogenum*. Proom and Woiwod (*2601, 3614*), have studied in detail the ninhydrin-reacting substances of bacterial culture filtrates. Amino acids of bacteria have been studied by Asselineau *et al.* (*90*), Polson (*2551*) and Work (*3628, 3629*).

(xi) *Clinical studies*

Dent (*740*) has studied the amino acids excreted in human urine in a case of Fanconi syndrome. A substance forming a spot below alanine was identified as a tripeptide consisting of serine and glycine ("under alanine", see Fig. 87). Another substance occurring above glycine on the chromatogram ("overglycine", see Fig. 87) was identified as taurine.

Urinary amino acids have been studied by paper chromatography by Boissonnas and Lo Bianco (*299*), Boulanger *et al.* (*338*), Datta and Harris (*704*), Giddey (*1104*), Westall (*3515*); numerous unidentified spots are still observed.

Dent and Schilling (*742*) have studied the amino acid pattern in portal blood, by two-dimensional paper chromatography. Amino acids in blood plasma were studied by Walker (*3455*) and Gordon and Nardi (*1177*). Boulanger and Biserte (*336*) have studied amino acids and polypeptide fractions of blood plasma after adsorption on Permutit-50.

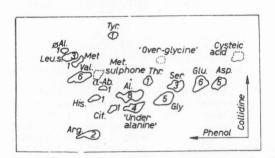

Fig. 87. Tracing showing the positions of the amino acids in a typical urine. The figures inset represent relative colour strengths on an arbitrary scale. "Overglycine" is usually very weak. Methionine sulphone and cysteic acid are regularly seen if the urine is previously treated with H_2O_2, but the former is very weak except after methionine feeding. Methionine sulphoxide occurs just to the left of histidine (Dent, *740*). ∅ Al = phenylalanine; α-Ab = α-aminobutyric acid

(xii) *Botanical studies*

A great many papers describe the analysis of the free, ninhydrin reacting substances extracted from plant material; a few of these can only be quoted here. Thus, Hulme and Arthington (*1482, 1483*) have found γ-amino-butyric acid, β-alanine and a methylhydroxyproline in apples, Virtanen and Linko (*3413*) have found piperidine-2-carboxylic acid, free ornithine and its N-acetyl derivative in the leaves of *Asplenium nidus*, Virtanen and Miettinen (*3415*) have found citrulline to be the predominating acid in alder, whereas it is absent from peas. The free amino acids in the pea plant have been studied by Miettinen (*2210*).

Steward *et al.* (*3107*) have described a quantitative chromatographic study of nitrogenous components of shoot apices and Steward *et al.* (*3108*) have published a detailed survey of the paper chromatography of nitrogenous compounds found in plants (see Table 107).

Thompson and Steward (*3240*) have described a detailed quantitative analysis of the amino acids of zein, edestin, cucumber seed globulin and squash seed globulin by paper chromatography.

Amino acids, organic acids and sugars can be identified on the same chromatogram by successive sprays with ninhydrin, methyl red and 3,5-dinitrosalicylic acid (Woodward and Rabideau, *3624*).

TABLE 107. R_F values of amino acids and amides (Steward *et al.*, *3108*)

Compound	Ninhydrin Colour	R_F Values phenol-water (pH 5.0–5.5)	collidine-lutidine (1 : 3)-water	n-butanol-acetic acid (9 : 1)-water
		(A)	(A)	
Cysteic acid . . . . . . violet	violet	0.069 (0.024)	0.187 (0.026)	0.006
Aspartic acid . . . . . blue blue-violet	blue blue-violet	0.179 (0.022)	0.106 (0.023)	0.030
Glutamic acid . . . . . violet	violet	0.311 (0.023)	0.116 (0.028)	0.050
Serine . . . . violet	violet	0.363 (0.027)	0.234 (0.014)	0.053
Glycine . . . . red-violet	red-violet	0.385 (0.032)	0.219 (0.007)	0.057
Asparagine . . . . . . orange-brown	orange-brown	0.442 (0.020)	0.172 (0.015)	0.024
Threonine . . . . violet	violet	0.492 (0.018)	0.299 (0.009)	0.086
Alanine . . . . violet	violet	0.585 (0.013)	0.297 (0.013)	0.110
Glutamine . . . . . . violet	violet	0.598 (0.021)	0.209 (0.010)	0.035
α-Amino-n-butyric acid . violet	violet	0.689 (0.020)	0.354 (0.014)	0.160
Histidine . . . . grey-violet	grey-violet	0.643 (0.025)	0.253 (0.018)	0.031
Lysine·HCl . . . . . . violet	violet	0.476 (0.030)	0.083 (0.010)	0.017
Arginine·HCl . . . . . violet	violet	0.561 (0.030)	0.135 (0.023)	0.031
Methionine sulphoxide . violet	violet	0.791 (0.007)	0.246 (0.013)	0.050
Proline . . . . yellow	yellow	0.860 (0.032)	0.310 (0.014)	0.135
Valine . . . . violet	violet	0.777 (0.017)	0.435 (0.026)	0.241
Methionine sulphone . . violet	violet	0.600 (0.052)	0.396 (0.015)	0.265
Leucine . . . . violet	violet	0.842 (0.025)	0.554 (0.038)	0.337
Isoleucine . . . . violet	violet	0.842	0.554	0.355
Phenylalanine . . . . . grey-violet	grey-violet	0.836 (0.019)	0.580 (0.068)	0.285
Tryptophan . . . . . . grey-violet	grey-violet	0.802 (0.014)	0.656 (0.060)	0.220
Tyrosine . . . . grey-violet	grey-violet	0.665 (0.036)	0.640 (0.040)	0.155
β-Alanine . . . . blue	blue	0.654 (0.012)	0.226 (0.006)	0.90
γ-Aminobutyric acid . . violet	violet	0.751 (0.018)	0.220 (0.014)	0.150
Hydroxyproline . . . . orange-red	orange-red	0.671 (0.007)	0.310 (0.007)	0.068
Glutathione . . . . . . violet	violet	0.109 (0.005)	0.098 (0.060)	—
α-Aminoadipic acid . . violet	violet	0.444	0.154	0.113
Ornithine·HCl . . . . violet	violet	0.327	0.151	0.024
Ornithine . . . . violet	violet	0.334	0.507	—
α,ε-Diaminopimelic acid . violet	violet	0.273	0.114	0.011
Hydroxylysine . . . . . violet	violet	0.291	0.058	0.020
Isoserine . . . . brown	brown	0.391	0.209	—
α,γ-Diaminobutyric acid violet	violet	0.334	0.174	0.025
Leucanol . . . . brown-violet	brown-violet	0.487	0.028	—
Taurine . . . . grey-violet	grey-violet	0.423	0.368	0.060
3,4-Dihydroxyphenyl-alanine . . . . grey-violet	grey-violet	0.217	0.527	0.083
Cystathionine . . . . . violet	violet	0.367	0.202	—
Homoserine . . . . . . violet	violet	0.582	0.297	0.073
Ethanolamine . . . . . violet	violet	0.542	0.436	0.240
δ-Hydroxy-α-amino-valeric acid . . . . . violet	violet	0.648	0.296	0.093
Canavanine . . . . . . violet	violet	0.509	0.199	0.029

(A) Standard deviation of the *mean R_F values* shown in brackets.

TABLE 107 *(Continued)*

Compound	Ninhydrin Colour	R_F Values phenol-water (pH 5.0–5.5)	collidine-lutidine (1:3)-water	n-butanol-acetic acid (9:1)-water
γ-Methylproline	yellow	0.873	0.393	0.259
Glucosamine	violet	0.639	0.455	0.058
Putrescine· di-HCl . . .	violet	0.398	0.118	—
Putrescine	violet	0.398	0.537	—
Citrulline	violet	0.643	0.229	0.037
Sarcosine	red-violet	0.775	0.258	0.087
δ-Aminovaleric acid . .	violet	0.789	0.246	0.258
β-Aminoisobutyric acid .	grey-violet	0.757	0.325	0.204
Glutamic acid diamide .	violet	0.763	0.392	—
Allo-δ-hydroxyleucine . .	brown to violet	0.770	0.378	0.200
δ-Hydroxyleucine . . .	violet	0.770	0.378	0.167
Norvaline	violet	0.761	0.481	0.254
Kynurenin· SO₄	violet	0.829	0.479	—
γ-Hydroxyleucine . . .	violet	0.826	0.400	0.186
Pyrrolidine· HCl . . .	yellow	0.868	0.472	0.259
Nipecotic acid	yellow-brown	0.882	0.270	0.171
γ-Pipecolic acid	brown-yellow	0.878	0.254	0.178
Pipecolic acid	red-violet to brown-yellow	0.895	0.384	0.214
Baikiain· HCl	yellow-brown	0.878	0.374	0.192
Histamine	grey-violet	0.934	0.465	0.045
Copellidine	grey-green	0.922	0.667	—
Aspartophenone	orange-yellow	0.843	0.602	—
Diiodotyrosine	brown-violet	0.813	0.624	0.350
Tryptamine· HCl	red-violet	0.868	0.752	0.376
Tyramine· HCl	grey-violet	0.881	0.717	0.354
Leucinamide	violet	0.819	0.785	0.309
Phenethylamine	violet	0.813	0.788	0.445
Piperidine	red-violet	0.931	0.396	0.267
β-Aminobutyric acid . .	none (reacts at 100°)	0.780	0.308	0.168
Allo-β-hydroxyglutamic acid	violet	0.140	0.163	0.051
β-Hydroxyglutamic acid .	violet	0.146	0.175	0.056
Theanine (γ-glutamylethylamide) . . .	violet	0.894	0.431	0.175
(γ-Glutamyl)-2-aminopropionitrile	violet	0.828	0.417	0.091
Methioninemethylsulphonium iodide . .	violet	0.839	0.161	0.031
γ-Methylglutamic acid. .	violet	0.346	0.226	0.151
Homoglutamine (α-aminoadipamic acid)	violet	0.675	0.264	0.054
Ethylamine· HCl	violet	0.802	Volatile	0.230
Glycocyamine	none [1]	0.714	0.300	0.118
γ-Guanidinobutyric acid	none [1]	0.920	0.337	0.247

(Continued p. 330)

TABLE 107 *(Continued)*

Compound	Ninhydrin Colour	phenol-water (pH 5.0–5.5)	collidine-lutidine (1 : 3)-water	n-butanol-acetic acid (9 : 1)-water
			R_F Values	
Piperidine-2,6-dicarbo-xylic acid	none [2]	0.577	0.305	0.082
Norleucine	violet	0.881	0.537	0.344
Glycylglycine	yellow-brown turning violet slowly	0.467	0.268	0.024
Alanylglycine	violet	0.643	0.319	0.060
Isoasparagine	violet	0.466	0.237	0.048
Isoglutamine	violet	0.601	0.252	0.087
Pyrrolidonecarboxylic acid (pyroglutamic acid)	none [2]	0.660	0.413	0.315

[1] Reacts after alkaline treatment.
[2] Detected by Rydon-Smith Test.

5. DNP-AMINO ACIDS

(a) *Column chromatography*

In his classical study of the terminal amino acids of insulin, Sanger (*2807*) treated insulin with 2,4-dinitrofluorobenzene and isolated, after hydrolysis, the terminal amino acids as the N-2,4-dinitrophenyl (*DNP*) derivatives which were purified by partition chromatography on silica gel. The addition of an indicator was unnecessary as the dinitrophenylamino acids are themselves yellow. In a second paper Sanger (*2808*) described the isolation by the same technique of 2,4-dinitrophenylornithine from gramicidin-S after treatment with dinitrofluorobenzene.

The α- and δ-DNP-ornithines were easily separated by partition chromatography (solvent: methyl ethyl ketone-ether 2:1).

Details for the preparation of pure DNP-amino acids and for the separation of mixtures of these substances on silica gel have been given by Porter and Sanger (*2570*). These, as well as subsequent authors: Blackburn (*260*), Desnuelle *et al.* (*752*), Middlebrook (*2209*), Phillips and Stephen (*2512*), have noted that adsorption effects play an important part in the chromatography of DNP-amino acids on silica gel.

Heikens *et al.* (*1353*) have observed that polymerization reactions may occur during the preparation of DNP-amino acids; the polymers (up to the tetramer) were identified by their R_F values. Excess dinitrophenol and dinitroaniline which are formed during the preparation of the DNP-derivatives can be eliminated by chromatography of a chloroform solution on moist silicic acid (Li and Ash, *1999*) or on acid alumina, from aqueous solution; 2% acetic acid elutes the dinitrophenol, whereas the DNP-amino acids are eluted with dilute NaHCO$_3$ (Turba and Gundlach, *3322*).

Blackburn (*260*) has described the separation of DNP-amino acids on buffered columns of silica-gel, and Perrone (*2494*) has used buffered Celite-545 columns.

A satisfactory and reproducible method for the separation of sixteen ether-soluble DNP-amino acids on columns of silicic acid-Celite has been elaborated by Green and Kay (*1203*).

Partridge and Swain (*2462*) obtained good separations of DNP-amino acids on chlorinated rubber (commercial "Alloprene") with butanol as the stationary phase and butanol-saturated buffers of pH 3, 4 or 5 as mobile phase. Knessl *et al.* (*1744*) have described the partition of DNP-amino acids on buffered kieselguhr and on siliconized materials.

Fletcher *et al.* (*981*) have separated the methyl ester of various dinitrophenyl amino acids on alumina columns.

(b) *Paper chromatography*

Paper chromatography of DNP-amino acids is only possible by using special solvent systems to prevent tailing, due to the strong adsorption effects of the DNP-radicals. Phillips and Stephen (*2512*) describe the difficulties thus encountered; Blackburn and Lowther (*265*) obtained good separations with paper buffered with a phthalate buffer of pH 6.0 and with 30 % propanol-cyclohexane as mobile solvent. Monier and Penasse (*2255*) used solvents containing potassium benzoate and Biserte and Osteux (*256*) either phenol-*iso*amyl alcohol (1:1) or toluene-glycol monochlorhydrin-pyridine-ammonia (5:3:1:3). The latter solvent has been widely used by other authors. It is probable that the aromatic constituents of these systems diminish the adsorption of the dinitrophenyl radicals on paper.

Rovery and Fabre (*2749*) have separated DNP-amino acids on paper in aqueous solution with a citrate-HCl buffer, pH 6.2. See also Kent *et al.* (*1696*) and Table 88, p. 268, Williamson and Passmann (*3588*), Felix and Krekels (*942*).

Iwainsky (*1525*) has studied the influence of various buffers on the separation of DNP-amino acids on paper and recommends the following solvent: butanol-*iso*amyl alcohol-ethanol-buffer (20:20:6.5:30); the buffers used were those of McFarren (*2074*).

DNP-tryptophan is not stable to acid hydrolysis; after 18 hours boiling with 5.5 N HCl, only 30 % were recovered (Desnuelle *et al.*, *752*).

Thompson (*3235*) has reported that DNP-amino acids are partially destroyed during hydrolysis in the presence of excess tryptophan.

Mills (*2225*) has shown that DNP-amino acids can be hydrolyzed to the

TABLE 108.   Paper chromatography of DNP-amino acids (Biserte and Osteux, *256*)

D.NP-amino acids	Solvent I	Solvent II	Solvent III
DNP-cysteic acid	0.02	—	—
DNP-aspartic acid	0.08	0.05	0.05
DNP-glutamic acid	0.09	0.05	0.06
DNP-asparagine	0.24	0.20	—
DNP-glutamine	0.27	0.23	—
DNP-hydroxyproline	0.28	0.23	0.40
DNP-serine	0.30	0.23	0.45
Di-DNP-cystine	0.32	0.20	0.41
DNP-glycine	0.35	0.33	0.45
DNP-threonine	0.36	0.31	0.53
DNP-proline	0.43	0.50	—
DNP-alanine	0.45	0.44	0.55–0.60
2,4-Dinitrophenol	0.42–0.45	0.25	0.62
DNP-valine	0.56	0.60	0.70
DNP-methionine	0.59	0.57	0.71
DNP-methionine oxide	{ 0.35 { 0.48	{ 0.23 { 0.36	{ 0.40 { 0.66
DNP-tryptophan	0.62	0.63	0.75
DNP-leucine	0.64	0.68	0.80
DNP-phenylalanine	0.66	0.68	0.80
Di-DNP-lysine	0.72	0.75	0.78
Di-DNP-tyrosine	0.74	0.70	0.83
2,4-Dinitroaniline	0.95–1	0.95–1	0.95–1
ε-Mono-DNP-lysine	0.40	0.65	0.40
α-DNP-arginine	0.48	0.74	0.57
Di-DNP-histidine	{ 0.43 { 0.70	{ 0.32 { 0.70	{ 0.47 { 0.70

I = toluene-pyridine-glycol monochlorohydrin-0.8 N NH$_4$OH (5:1:3:3).
II = phenol-*iso*amyl alcohol-water (1:1:1).
III = pyridine-*iso*amyl alcohol-1.6 N NH$_4$OH (6:14:20).

free amino acids, preferably with baryta, and the amino acids identified in the usual way by chromatography on paper. All commonly occurring amino acids could be recovered except histidine and cystine. DNP-arginine gives, after hydrolysis, several ninhydrin reacting substances.

The paper chromatography of DNP-amino acids has also been used for the quantitative estimation of N-terminal residues in proteins (Levy, *1991*; Anfinsen *et al.*, *67*; Jollès and Fromageot, *1594*).

CHAPTER 31

Peptides

1. INTRODUCTION

In this chapter we will be concerned with the chromatographic behaviour of the peptides themselves; the application of chromatography to the qualitative and quantitative analysis of their component amino acids has already been described in detail in the previous chapter.

Reviews of Sanger (*2811*), Campbell and Work (*502*), Schroeder (*2885*), and Tiselius (*3266*) describe the latest developments in this field, a great many of which have been achieved by the combined use of various chromatographic methods.

Waldschmidt-Leitz *et al.* (*3450, 3451*) carried out the first separation of peptides by a chromatographic technique. By passing an enzymatic hydrolysate of clupein through a column of an acid earth, they obtained peptide fractions containing 2, 3 or 4 molecules of arginine, the peptides forming clear zones on the adsorbent.

The most successful methods of separating peptides, which have been widely used in recent years, are: ion exchange on columns of synthetic resins, partition chromatography on columns of starch or paper powder and on sheets of paper. Tiselius's methods of frontal analysis, displacement elution and carrier displacement have also proved very efficient. Detailed references to these methods will be given below. The paper chromatography of peptides and proteins has been reviewed by Boulanger and Biserte (*337*).

2. GROUP SEPARATIONS

Natural peptide mixtures and the products of the partial hydrolysis of proteins are often very complex. Thus, Consden *et al.* (*623*) have stated that a two-dimensional chromatogram of a partial hydrolysate of wool showed at least forty different peptides to be present. It is therefore often profitable to separate such mixtures first into three or more groups: basic, acid and neutral peptides. Such separations are achieved fairly easily, as lower peptides usually behave like the corresponding amino acids: thus basic peptides (e.g. carnosine) are adsorbed on silica gel or on cation exchangers like basic amino acids, glutathione and other acidic peptides are adsorbed on acid alumina or on anion exchange resins, like the dicarboxylic amino acids (*617, 622, 1054*), and aromatic peptides containing at least one aromatic group are specifically adsorbed on charcoal (*1054, 3188*).

The separation of peptides into four groups, basic, acidic, neutral aromatic and aliphatic, by filtration through columns of silica gel, acid alumina, and charcoal has been described by Fromageot *et al.* (*1054*). The presence of salts (from the neutralisation of the acid used for the partial hydrolysis) causes some difficulties but this can be circumvented by using lithium hydroxide for the neutralisation and a larger amount of silica gel. This method has been used by Acher *et al.* (*9, 11*) in their study of peptides obtained from lysozyme.

The separation of acidic peptides by adsorption on Amberlite IR-4 has been described by Consden *et al.* (*622*).

Group separations of polypeptides on ion exchangers have also been described by Biserte and Boulanger (*255*); basic peptides are adsorbed on Zeo-Karb or on Permutit-50, acid peptides on De-Acidite.

A partial hydrolysate of steer hide collagen was fractionated by Kroner *et al.* (*1809*) by adsorption of the aromatic part on charcoal, the basic fraction on Amberlite XE-64 and the acidic one on Amberlite IR-4B. The neutral fraction was found in the filtrate.

The curious behaviour of the "neutral" peptide lysyl-glycyl-glycyl-glutamic acid has been observed by Jutisz (*1608*); it is adsorbed on a column of acid alumina, like acid peptides, and also on a column of silica gel, like basic peptides.

The specific separation of neutral *cystine peptides* is possible by a method of Consden and Gordon (*617*). Acidic amino acids and peptides of a partial hydrolysate (of wool) were first eliminated by adsorption on Amberlite IR-4; the filtrate containing neutral and basic amino acids and peptides was then oxidized with bromine water, the SH groups of free and peptide-bound cysteine being thus oxidised to SO_3H. A second passage through an Amberlite IR-4 column retained all the cysteic acid and cysteic acid peptides formed.

Acidic peptides have been purified on Dowex-2: cysteic acid peptides from oxidized ovalbumin (Flavin and Anfinsen, *980*), acidic peptides from performic acid oxidized vasopressin (Popenoe and du Vigneaud, *2557*), peptides from partial hydrolysis of tyrocidin (Paladini and Craig, *2435*).

Dowex-50 has been used with success for purification of peptides: peptides of partial hydrolysis of gelatine (Schroeder *et al.*, *2887*), phosphopeptides from partial hydrolysates of several proteins (Flavin, *978*).

For larger peptides the polyacrylic acid resin Amberlite IRC-50 can be used, as illustrated by several examples, p. 353.

1 % Formaldehyde can be used for the separation of glycyl-, seryl- and threonyl-peptides from other neutral peptides as only the former are

sufficiently acidic to be adsorbed on acid alumina. The other neutral peptides as well as glycine, serine and threonine pass through the column. In this way glycyl-leucine and leucyl-glycine can easily be separated (*1611*).

3. ADSORPTION CHROMATOGRAPHY

Tiselius (*3258*) used his frontal analysis technique to separate alanyl-glycine from alanyl-glycyl-glycine and leucyl-glycine from leucyl-glycyl-glycine. Table 99 (p. 289) shows a comparison of the retention volumes of some neutral peptides with those of the component amino acids.

Synge and Tiselius (*3187*) studied the behaviour on charcoal of antibiotic peptides of the gramicidin group; tyrocidin was shown to be heterogenous; they also separated aromatic from non-aromatic peptides on charcoal (*3188*).

Tiselius and Hagdahl (*3269*) separated methionine and leucyl-glycyl-glycine by carrier displacement with butanol.

Li *et al.* (*2000*) have purified adrenocorticotropic peptides by carrier displacement on Darco G-60, using normal alcohols with 8 to 10 carbon atoms as displacer. Similar experiments with pretreated charcoal have been carried out by Porath and Li (*2562*) (see also p. 36). Porath (*2558, 2559*) has described the carrier displacement of bacitracin on charcoal.

For work with such peptides, Porath and Li (*2562*) use plastic-lined columns to avoid contact of the solvent with the stainless steel usually employed for sectional columns; they thus obviate the danger of inactivation of the peptides by traces of metals.

Ågren and Glomset (*23*) have fractionated phosphopeptones from casein on charcoal.

Selective removal of glutathione (and cysteine) can be achieved by adsorption on lead carbonate (Bessman *et al.*, *235*).

Moring-Claesson (*2279*) has carried out a frontal analysis of ovalbumin and its partial hydrolysates on alumina and charcoal.

4. ION-EXCHANGE SEPARATIONS

The use of displacement chromatography for the separation of lower peptides on ion exchange resins has been reviewed by Partridge (*2456*).

Cook and Levy (*630*) separated alanyl-glycine from alanine and glycine by adsorption on Zeo-Karb 215; the adsorption zones were clearly visible as the resin changed from orange to yellow brown; development with $0.1 \, N$ ammonia eluted first the amino acids, then the dipeptide.

The behaviour of a series of glycyl-, leucyl-, glutamyl-, and tyrosyl-peptides on acid and basic ion exchangers has been studied by Brenner and Burckhardt (*377*). Tripeptides are generally more strongly adsorbed than the analogous dipeptides; long side chains (e.g. those of leucine) also increase adsorption. This effect seems to be even stronger than that of an additional peptide bond; thus, leucyl-glycine is more strongly adsorbed than triglycine, and leucyl-leucine more strongly than leucyl-glycyl-glycine. This is ascribed to molecular adsorption, a factor which could also explain the retardation of leucyl-leucine on the basic resin Amberlite IR-4B. Tyrosine peptides are particularly strongly adsorbed.

Moore and Stein (2268) have discussed the importance of using synthetic resins with the proper degree of cross-linking. The standard Dowex-50 resin used in their work was prepared from styrene co-polymerized with about 8 % divinylbenzene and is considered to have a molecular structure, the pores of which correspond to 8 % cross-linking. Amino acids chromatographed on a 16 % cross-linked resin gave only very poorly resolved zones. This indicates that the pore size of this resin was too small to permit rapid penetration into the resin particles. For amino acids and dipeptides, the 8 % cross-linked resin was shown to be the best; the tetrapeptide leucyl-leucyl-glycyl-glycine already gave a broader peak. Moore and Stein (2268) conclude that for higher peptides, resins of 4 or 2 % cross-linking may be required. "In general, it appears preferable to use a resin with the highest degree of cross-linking compatible with the separations to be achieved."

The method using Dowex-50-X4 described more recently allows the separation of oligopeptides up to octa- or deca-peptides; Dowex-50-X2 is preferable for larger peptides (Moore and Stein, 2270).

Dowmont and Fruton (810) have separated several synthetic di- and tri-peptides on Dowex-50-X4 with a buffer of pH 4.72, and Thompson (3236a) has separated oligopeptides from a partial acid hydrolysate of lysozyme on Dowex-50-X4, using gradient elution with ammonium formate and acetate buffer from pH 3.4 (0.2 M) to pH 6.8 (0.6 M).

5. PARTITION CHROMATOGRAPHY ON COLUMNS

(a) Silica gel

Gordon et al. (1170) who studied the partition chromatography of neutral peptides, stated that the rate of movement of the peptides on a silica gel column was approximately the mean of the rates of the component amino acids of each peptide. The following peptides were isolated from a partial hydrolysate of gelatine: leucyl-glycine and glycyl-leucine, as well as di-peptides of proline and alanine, and of proline and glycine. In addition, a micro-chromatographic method for the determination of the arrangement of amino acids in the peptides was elaborated.

The fractionation of DNP-peptides on silica gel is mentioned below (p. 348).

(b) Starch

Synge (*3181*) has used *starch columns* for the study of the products of partial hydrolysis of gramicidin; L-valyl-glycine was isolated and later also D-leucyl-glycine, L-alanyl-D-valine and L-alanyl-D-leucine (*3184*). Synge (*3184*) observed that dipeptides such as leucyl-leucine and· leucyl-tryptophan were difficult to purify because they either run very fast in the more polar solvents, or do not run at all in less polar solvents (CCl_4). They also give poor colours with ninhydrin (especially tryptophyl-tryptophan; Synge and Tiselius, *3188*).

Moore and Stein (*2265*) separated the isomers leucyl-glycine and glycyl-leucine on a starch column; displacement development was used with a 0.25 % solution of phenol as displacer. These authors (*3095*), as well as Li and Pedersen (*2002*) and Fels and Tiselius (*943*) have found starch unsuitable for efficient fractionation of the higher peptides.

A peptide fraction isolated by Borsook et al. (*318, 319, 321*), and Fuerst et al. (*1057*) from different biological sources, which emerges from the starch column before the least adsorbed amino acid (leucine), was shown to be a mixture by Fels and Tiselius (*943*). These authors also observed that the behaviour of starch as an adsorbent for higher peptides is different from that of filter paper (reversal of zone of a mixture of polypeptides and leucine).

Ottesen and Villee (*2427*) have separated on starch the oligopeptides formed by the action of a proteinase of *Bacillus subtilis* on ·ovalbumin.

Lindquist and Storgards (*2025*) found a *molecular sieve action* in starch columns developed with aqueous buffers. A separation between peptides of molecular weight 200 up to 500 from amino acids and peptides of molecular weight 75–200 was obtained; the high molecular peptides were eluted ahead of the more retained smaller molecules.

(c) Cellulose

Wieland and Schmidt (*3565*) have purified the toxic peptides amanitin and phalloidin on columns of cellulose powder; such columns have also been used by Ågren and Glomset (*23*) for the fractionation of phospho-peptones from casein. Kihara et al. (*1715*) have purified oligopeptides on a chromatopile, with pyridine-water (4:1).

(d) Kieselguhr

The peptides representing the oxidized A and B chains of insulin were separated by Andersen (*51*) on silane treated Hyflo-Supercel with 2-butanol-0.01 N aqueous trichloracetic acid, the organic layer of which was the stationary phase. Bumpus et al. (*450*) have purified angiotonin on a Celite column with a butanol-propanol-NaCl-HCl solvent system.

The work of Porter is mentioned p. 356.

6. PAPER CHROMATOGRAPHY OF PEPTIDES

Consden *et al.* (*620*) have described in detail a method for the identification of lower peptides present in partial hydrolysates of proteins, employing ionophoresis followed by partition chromatography on paper. Peptide spots obtained after chromatography on paper can be eluted from duplicate sheets (see Fig. 88), then hydrolyzed, and the resulting amino acids identified by chromatography on paper. The structure of a dipeptide can be determined by deamination with nitrous acid and hydrolysis; the deaminated amino acid will no more show up on a paper chromatogram.

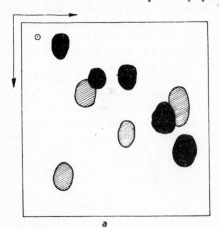 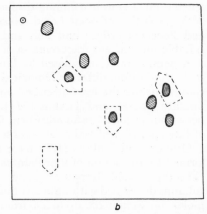

Fig. 88. Diagram showing method of cutting a two-dimensional chromatogram; (a) and (b) represent duplicate sheets, the solution to be analyzed being applied at the open circles. The arrows indicate the direction of development of the chromatograms in the two solvents. In the guide chromatogram (a), amino acids are shown as black spots and peptides as shaded spots. After treatment with dilute ninhydrin, the duplicate sheet (b) shows faint spots (some may be barely visible) and is cut as shown by the dotted lines (Consden *et al.*, *620*)

Consden and Gordon (*616*) recommend that the hydrolysis of peptides eluted from paper chromatograms should be carried out in Pyrex capillaries in order to avoid the introduction of large amounts of mineral salts. They used a solution of ninhydrin in butanol containing 1 % acetic acid to neutralise alkaline spots present on the paper chromatograms.

(a) *Detection of peptides on paper*

The position of the peptide spots on paper is generally revealed by spraying with a ninhydrin solution and heating; this causes only slight destruction of the peptides, which can then be eluted for further study.

On heating the peptide spots on paper, fluorescence is produced, which is also very useful for revealing the spots. For the nature of this reaction see p. 310. Peptides can also be detected by their amphoteric properties (see p. 138.)

Rydon and Smith (2769) have shown that peptides (including proteins, diketopiperazines, acetylated amino acids and peptides) can be detected by chlorination in an atmosphere of gaseous chlorine and subsequent spraying with starch-potassium iodide; the colour formed is due to the liberation of iodine by the N-chloropeptide (-CO-NCl-) formed in the chlorination stage. Reindel and Hoppe (2666) have modified this method, by treating the paper with chlorine dioxide, followed by benzidine.

(b) Solvents for separating peptides on paper

The most generally used solvents are phenol, collidine and butanol-acetic acid. The latter mixture is especially useful for larger peptides which tail in other solvents (Jones, 1603; Phillips, 2507).

Grassmann and Deffner (1191) stress that phenol has a special affinity for the peptide bond; this explains why, contrary to the behaviour in other solvents, the R_F values of peptides increase with increasing chain length, when they are chromatographed in phenol.

(c) R_F values of oligopeptides

Martin (2150) has shown by a theoretical treatment that the partition coefficient of a dipeptide divided by the product of the partition coefficients of the constituent amino acids is a constant for any given phase, i.e. that for a peptide A.B, $a_{A.B}/a_A\,a_B$ = constant, where a_A, a_B and $a_{A.B}$ are the respective partition coefficients.

As the R_F values are related to the partition coefficients by the equation

$$a = \frac{A_L}{A_S}\left(\frac{1}{R_F} - 1\right) \quad \text{(Consden et al., 618)}$$

and as A_L/A_S is a constant for any solvent system (see p. 104),

$$\frac{\dfrac{1}{R_F^{A.B}} - 1}{\left(\dfrac{1}{R_F^A} - 1\right)\left(\dfrac{1}{R_F^B} - 1\right)} = \text{constant} = K$$

Then

$$R_F^{A.B} = \frac{1}{K\left(\dfrac{1}{R_F^A} - 1\right)\left(\dfrac{1}{R_F^B} - 1\right) + 1}$$

Thus the R_F value of a peptide may be calculated in terms of the R_F values of the constituent amino acids and of a constant K, which is best determined experimentally for each solvent system.

As Sanger (*2811*) pointed out "this relationship is not absolutely accurate, since for instance peptides containing the same amino acids in different order may be frequently separated. It has, however, been found to apply satisfactorily in most cases and to be a useful check on the identity of a peptide".

Pardee (*2441*) has developed a similar approach for calculation of R_F values of oligo-peptides; from the values of the component amino acids and two general, experimentally determined constants, the R_F value of any desired peptide can be calculated with an average accuracy of about ± 0.05, as compared with the figures of Knight (*1746*) (Table 110). Some deviations are however as high as 0.17.

(d) *Methods for determining end-groups*

Chromatography is extensively used in the different methods employed for the determination of the N- and C-terminal amino acids of peptides. The use of DNP-derivatives introduced by Sanger (*2807*) has already been reviewed in detail (p. 330). Ingram (*1501*) methylated the N-terminal amino acid with formaldehyde; the N-dimethylamino acid thus formed does not give a colour with ninhydrin; the corresponding amino acid will thus be absent from the hydrolysate. This method seems to be limited to oligo-peptides.

The "pipsyl" method of Udenfriend *et al.* has been mentioned on p. 320.

The method of Edman (*863–865*) which is frequently used for stepwise degradation of peptides, consists in treating the peptide or protein with phenyl*iso*thiocyanate ($C_6H_5N{=}C{=}S$) which reacts with the N-terminal amino acid forming a phenylthiocarbamyl $C_6H_5N{-}C(S){-}NHCH(R)CO$- derivative which is then cyclised to a phenylthiohydantoin (I); this can be identified after hydrolysis of the peptide chain;

$$
\begin{array}{c}
C_6H_5N{-}CS \\
 \mid \mid \\
OC NH \\
\diagdown C \diagup \\
\diagup \diagdown \\
\text{(I)} H R
\end{array}
$$

the relevant methods have already been quoted on p. 222; for applications, see Acher *et al.* (*10*), Harris and Li (*1313*), Rovery and Desnuelle (*2748*).

The treatment of a carbobenzoxypeptide with hydrazine transforms the N-terminal amino acid into a dihydrazide which can be cyclised to a

3-aminohydantoin; these can be identified by paper chromatography (Schlögl *et al.*, *2851*).

$$
\begin{array}{c}
\underset{\substack{|\\ \text{COOCH}_2\text{C}_6\text{H}_5}}{\text{NH}-\overset{\overset{\displaystyle R_1}{|}}{\text{CH}}-\text{CO}-\text{NH}-\overset{\overset{\displaystyle R_2}{|}}{\text{CH}}-\text{CO}-} \longrightarrow \underset{\substack{|\\ \text{CONHNH}_2}}{\text{NH}-\overset{\overset{\displaystyle R_1}{|}}{\text{CH}}-\text{CO}-\text{NHNH}_2} + \text{H}_2\text{N}-\overset{\overset{\displaystyle R_2}{|}}{\text{CH}}-\text{CO}-
\end{array}
$$

$$
\longrightarrow \begin{array}{c} R_1\text{--CH--CO} \\ | \qquad\qquad \rangle\text{N--NH}_2 \\ \text{NH--CO} \end{array}
$$

Fromageot *et al.* (*1055*) worked out a method for the determination of the C-terminal residue of peptides, based on the reduction of the free carboxyl with LiAlH$_4$ to the corresponding amino alcohol. After hydrolysis the latter can be isolated and identified by paper chromatography (see p. 207).

(e) *Synthetic oligopeptides*

Table 109 shows R_F values of oligopeptides determined by Consden *et al.* (*620*). Brockmann and Musso (*392*) studied the paper chromatography of glycine, glycyl-glycine, triglycine, up to hexaglycine. Complete separation was obtained in cresol by running the chromatogram for 336 hours (with paper pads at the bottom of the sheets, see Miettinen and Virtanen, *2213*). In butanol, butanol-acetic acid and benzyl alcohol the R_F values decrease regularly with increasing number of glycine radicals; in phenol they increase. In cresol, the peptides are obtained in the following order (increasing R_F): tri-, di-, tetraglycine, glycine, pentaglycine. The intensity of colour obtained by spraying and heating with ninhydrin solution decreases with chain length; the peptides give a violet colour only after heating for 20 minutes.

Alanine- and serine-peptides and the methyl esters of several of these peptides were also studied.

Cook and Levy (*630*) have measured the R_F values of 24 glycylpeptides and a number of alanylpeptides (*631*), in butanol-acetic acid; considerable variations of R_F values with different papers, temperatures, etc. were observed, although their relative values remained rather constant; glycine was selected as arbitrary standard with R_F 0.10 and run with every chromatogram.

Heyns and Anders (*1384*) also studied the paper chromatography of polyglycylpeptides. Those containing more than six glycyl residues cannot be revealed with ninhydrin. The same authors studied alanylpeptides up to tetra-alanine; R_F values increase in phenol and decrease in butanol and pyridine-*iso*amyl alcohol. Tetra- and higher peptides autolyze easily in water solution even in the ice box after 1 or 2 weeks as shown by the appearance of amino acids and di- and tripeptide spots on the paper.

Knight (*1746*) has determined the R_F values in phenol-water and in pyridine-*iso*amyl alcohol for lower peptides (see Table 110).

TABLE 109. R_F values of some peptides* in various solvents on Whatman No. 4 paper (Consden et al., 620)

Addition \ Solvent	Phenol 0.1% cupron	Phenol coal-gas 0.3% NH₃†	s-Collidine	n-Butanol 0.1% cupron	Benzyl alcohol 0.1% cupron	m-Cresol 0.1% cupron	m-Cresol 0.1% cupron 3% NH₃†
Glycylglycine	0.53	0.57	0.28	0.01	0.00	0.10	0.18
Glycyl-DL-alanine	0.64	0.63	0.32	0.03	0.01	0.20	0.29
Glycyl-DL-valine	0.78	0.74	0.47	0.09	0.02	0.37	0.51
Glycyl-DL-leucine	0.87	0.79	0.53	0.24	0.05	0.59	0.66
Glycyl-L-proline	0.77	0.69	0.34	0.05	0.03	0.68	0.69
Glycyl-L-hydroxyproline	0.57	0.59	0.28	0.02	0.01	0.49	0.25
Glycyl-L-phenylalanine	0.78	0.70	0.79	0.18	0.08	0.65	0.72
Glycyl-L-tryptophan	—	0.66	0.85	0.29	—	0.44	0.63
Glycylglycylglycine	0.58	0.59	0.32	0.01	0.02	0.08	0.19
Glycylglycyl-L-leucylglycine	0.86	0.75	0.71	0.11	0.04	0.67	0.67
DL-Alanylglycine	0.68	0.65	0.32	0.03	0.01	0.21	0.29
L-Alanylglycylglycine	0.67	0.63	0.46	0.01	0.00	0.19	0.29
DL-Valylglycine	0.83	0.73	0.53	0.11	0.04	0.48	0.48
DL-Leucylglycine	0.86	0.78	0.61	0.23	0.08	0.63	0.62
DL-Leucyl-DL-leucine	0.95	0.84	0.89	0.65	0.30	0.87	0.85
L-Leucyl-L-tryptophan	0.92	0.83	0.95	0.60	0.34	0.87	0.83
L-Leucylglycylglycine	0.88	0.75	0.55	0.11	0.04	0.61	0.60
L-Prolylglycine	0.86	0.68	0.49	0.03	0.03	0.67	0.55
L-Tyrosylglycine	0.59	0.59	0.95	0.09	0.03	0.27	0.33

Peptides in which glycine carries a free amino group, on heating with ninhydrin give first a yellow colour, then grey and finally purple. All the other peptides in the table give a purple colour, except prolylglycine, which is yellow at first, then orange and finally grey. Diketopiperazines do not give colours with ninhydrin.

* For the R_F values of peptides related to Gramicidin S, see Consden, Gordon, Martin and Synge (625).

† The % figure refers to the concentration of the NH₃ solution present in the tray at the bottom of the chamber.

TABLE 110. R_F values* of some peptides on Whatman No. 1 paper at 25° (Knight, *1746*)

Peptide	Phenol-water	Pyridine-isoamyl alcohol	With ninhydrin ‡
L-Leucylhexaglycylglycine		0.17	Purple
Glycyl-L-aspartic acid	0.09	0.04	Yellow
DL-Asparagylglycine§	0.14	0.04	Pink
Glycyl-DL-asparagyldiglycine‖.	0.15	0.04	Yellow
L-Seryl-L-serine	0.25	0.11	Yellow
L-Leucyl-L-aspartic acid	0.37	0.11	Purple
L-Leucylglycyl-L-aspartic acid.	0.46	0.10	Purple
L-Leucyl-L-glutamic acid	0.48	0.12	Purple
Triglycine	0.48	0.08	Yellow
Glycyl-L-tyrosine	0.55	0.27	Yellow
Tetraglycine	0.57	0.07	Yellow
Glycyl-L-tyrosylglycine.	0.57	0.26	Yellow
DL-Alanylglycine	0.58	0.12	Purple
D-Alanylglycylglycine	0.63	0.11	Purple
L-Alanyl-L-alanine. . . . . , . . .	0.66	0.15	Purple
Leucylcystine†	0.69	0.34	Purple
Glycyl-L-tryptophan	0.70	0.34	Yellow
Glycyl-L-leucine. . . . . , .	0.72	0.29	Yellow
Glycyl-L-asparaginyl-L-leucine	0.73	0.19	Yellow
DL-Alanyl-DL-leucine.	0.76	0.34	Purple
DL-Leucylglycine	0.78	0.31	Purple
Phenylglycyl-DL-alanine . . . . , .	0.80	0.32	Purple
Glycyl-L-alanylglycyl-L-tyrosine . . . . . , . .	0.81	0.26	Yellow
Triglycylglycine amide.	0.82	0.15	Yellow
DL-Leucyl-DL-alanine	0.82	0.37	Purple
Leucylalanylglycine †	0.86	0.28	Purple
L-Leucyl-D-leucine. . . . . ,	0.87	0.54	Purple
DL-Leucylglycyl-DL-alanine	0.87	0.29	Purple
DL-Leucyldiglycylglycine	0.90	0.24	Purple
Leucylalanylalanine †	0.94	0.34	Purple
L-Alanyldiglycyl-L-alanylglycylglycine	0.94	0.09	Purple
L-Leucyl-L-histidine	0.95	0.31	Purple
L-Leucylglycyl-L-leucine	0.96	0.51	Purple
L-Prolyl-L-phenylalanine	0.97	0.39	Yellow

* These values respresent the averages of duplicate or triplicate determinations. Individual values differed from the averages by about $\pm$ 0.015 R_F units.

‡ Initial colours after spraying the chromatograms with ninhydrin and heating.

§ It is not known whether this is α- or β-aspartylglycine.

‖ Glycyl-α, β-DL-aspartyldiglycine.

† Configuration of the optical centres not known.

Brenner *et al.* (*379*) separated on paper enzymatically formed di- and tri-peptides of methionine.

Trypsin- and chymotrypsin-catalysed transpeptidations were studied by paper chromatography by Waley and Watson (*3452*), Blau and Waley (*273*). Ekstrand (*877*) has used the same method for the estimation of peptidase activity.

7. PURIFICATION OF NATURAL PEPTIDES AND THEIR PARTIAL HYDROLYSIS PRODUCTS

(a) *Oligopeptides*

Yudaev (*3655*) described the paper chromatographic detection of *carnosine*; Lindan and Work (*2013*) studied the behaviour of *glutathione* on paper. See also Miller and Rockland (*2224*). Glutathione and γ-glutamylpeptides were separated on paper by Hird and Springell (*1399*) with propanol-water (8:2). Enzymic transpeptidation reactions involving γ-glutamylpeptides have been studied by Hanes *et al.* (*1296*) using the same solvent.

Chattaway *et al.* (*555*) isolated on paper a peptide (probably composed of serine, glycine and glutamic acid) which is a growth factor for *Coryne-bacterium diphtheriae*.

The acidic peptides formed on partial hydrolysis of wool were separated by Consden *et al.* (*622*) by ionophoresis in a slab of silica gel, and were then extensively studied by paper chromatography.

One of the triumphs of paper chromatography was the establishment of the chemical structure of the antibiotic gramicidin S by Consden *et al.* (*625*).

The peptides obtained by partial hydrolysis were separated on paper, each peptide was separately eluted, hydrolysed and examined on a new paper chromatogram. By comparing the spots obtained from the hydrolysis of the peptide with those obtained from the hydrolysis after deamination of the peptide with nitrous acid, the structure of the dipeptide was established unequivocally.

Šorm and Keil (*3050*) studied the composition of phalloidin, the toxic crystalline peptide of *Amanita phalloides*. Chromatography of the peptide on a starch column shows its homogeneity, total hydrolysis gave four spots, iden-tified as cysteic acid (C), hydroxytryptophan (T), alanine (A), and *allo*-hydroxyproline (P). Partial hydrolysates were studied by two-dimensional paper chromatography and different di-, tri- and tetrapeptides identified. For more recent work on phalloidin, see Wieland and Schön (*3565a*).

(b) *Polypeptides*

(i) *Insulin*

a) *Chromatography of the intact hormone.* The elution and displacement analysis of insulin on pretreated charcoal has been described by Porath and Li (*2562*).

Porter (*2565*) has made a detailed study of the partition chromatography of insulin on silane treated Hyflo-Supercel, in a system containing water, ethyl cellosolve, butyl cellosolve and a sodium phosphate buffer; separation of inactive proteins was easily achieved and crystalline material of uniform activity obtained; the recovery from the columns was quantitative. Robinson and Fehr (*2704*) have described the separation of insulin and protamine on paper sheets, by development with the upper phase of a mixture of *n*-butanol-glacial acetic acid-water (3:1:4); bromocresol green was used for locating the zones. Asaoka and Higashi (*86*) have used this method for the determination of insulin.

β) *Chromatography of degradation products of insulin.* Sanger's extensive work on insulin has culminated in the establishment of the whole amino acid sequence and the complete structure of this hormone (*2807–2813*).

Insulin contains two open polypeptide chains joined together by –S–S– bridges. These can be broken by oxidation with performic acid and two main products are obtained: an acidic fraction *A* with a glycyl terminal residue, and a basic fraction *B* with a phenylalanyl terminal residue (*2809*).

Peptide *A* gives no colouration with ninhydrin; the most satisfactory method of detecting it on paper is by the Pauly reaction; both peptides give fast moving tailing spots with phenol or collidine and do not move with butanol or butanol-acetic acid. Tiselius and Sanger (*3271*) have studied these two fractions by frontal analysis. Andersen (*52*) purified them on silane-treated Hyflo-Super Cel.

Oligopeptides from partial acid hydrolysates were separated on paper with phenol-0.3 % NH_3 in the first direction and butanol-acetic acid-water in the second direction (Brown et al., *419*). See also Ryle et al. (*2770*).

Progress in the study of the structure of peptides *A* and *B* was made by extensive fractionation of the *DNP-peptides* obtained after partial hydrolysis (see p. 348).

Phillips (*2510*) also separated the peptides from an enzymic digest of insulin by paper chromatography.

(ii) *Lysozyme*

Acher et al. (*9*) have separated the whole fraction of basic peptides of a partial hydrolysate of lysozyme on silica gel and then separated the individual peptides by two-dimensional paper chromatography (phenol and butanol). Most of these are tri-, tetra- and pentapeptides having arginine or lysine at the amino-end of the peptide chain. Acher et al. (*11*) used the method of Fromageot et al. (*1054*) to separate lysozyme peptides into four groups. Schroeder (*2884*) has recently isolated a DNP-tetrapeptide from lysozyme and shown it to be a,ε-di-DNP-lysyl-valyl-phenylalanyl-glycine. Thompson (*3236*) digested lysozyme with carboxypeptidase and identified leucine as C-terminal amino acid. The chromatography of lyzosyme itself is mentioned in the chapter on proteins (p. 353).

(iii) *ACTH*

The elution and displacement analysis of ACTH on pre-treated charcoal has been described by Porath and Li (*2562*).

Adrenocorticotropic hormone can be purified by adsorption on powdered cellulose followed by elution with 0.1 N HCl (Payne *et al.*, *2473*); oxidized cellulose is a stronger adsorbent, because of the persence of carboxyl groups which adsorb the basic hormone (Astwood *et al.*, *94*).

Brink *et al.* (*386*), Geschwind *et al.* (*1102*), Bazemore *et al.* (*169*) have also used oxycellulose for the purification of ACTH; after pepsin digestion, the corticotropin peptides can also be purified on oxycellulose (Bazemore *et al.*, *169*; Richter *et al.*, *2680*).

A nearly hundredfold concentration of the hormone has been achieved by Dixon *et al.* (*789*) by passing it through a column of Amberlite IRC-50. Using a sodium phosphate buffer (pH 6.65), 50–100 % of the activity of the hormone was recovered in a fraction corresponding to only 2–5 % of the starting solution, on a ninhydrin colour base. Mendenhall and Li (*2186*) have studied the behaviour of ACTH-peptides on starch and diatomaceous earth.

Hess and Carpenter (*1376*) have studied the partition of ACTH on kieselguhr columns.

Li and Pedersen (*2002*) studied the paper chromatography of the pepsin-digested hormone; six spots were obtained, one of which was active. Li *et al.* (*2003*) applied the methods of Tiselius to such peptides. Displacement development was used with a mixture of 1 part Carboraffin Supra and 9 parts Hyflo-supercel as adsorbent. The column was made of 8 sections and was equipped with the mixers described by Hagdahl (*1254*), thus assuring straight fronts of the zones. Displacement of the adsorbed peptides was effected with 0.4 % zephiran chloride in 0.1 N HCl. Carrier displacement was also tried, using octyl, nonyl and decyl alcohols as carriers; this gave the purest preparations.

More recently, spectacular success has been achieved by several groups of workers in the purification of ACTH peptides from hog or sheep glands (corticotropins of molecular weight 5000 to 7000), by chromatography on Amberlite IRC-50 (or XE-97) (Richter *et al.*, *2680*; White and Fierce, *3538*; Li *et al.*, *2001*); in the latter case elution was obtained with a gradient of NaHCO₃. See also Otsuka and Kimura (*2425*).

(iv) *Miscellaneous oligo- and polypeptides*

Bell *et al.* (*180*) studied the homogeneity of the antibiotic *polymyxins* on buffered Hyflo-Supercel columns. The polymyxins A and D were purified and separated on paper by Jones (*1603*; Fig. 89) (solvent: butanol-acetic acid). Nash and Smashey (*2339*) used buffered paper strips; on development with butanol, the R_F values for polymyxin hydrochloride were controlled within wide limits by simply varying the concentration of the buffer.

Heyns *et al.* (*1386*) studied partial hydrolysates of *gelatine* by paper chromatography. Some di- and tripeptides were identified.

Ottesen and Villee (*2427*) separated on starch the peptides released in the enzymatic transformation of ovalbumin to plakalbumin. Klungsøyr *et al.* (*1743*) isolated on paper some microbiologically active peptides from partial hydrolysates of meat. The separation on paper of some peptides produced by papain hydrolysis of wool has been described by Blackburn (*261*). Acidic

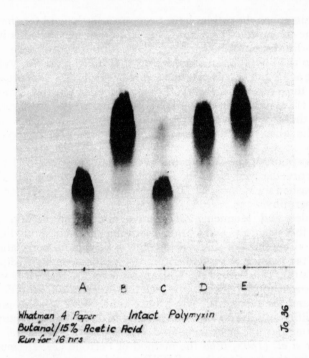

Fig. 89. Paper chromatography of five different polymyxins (Jones; cf. *1603*)

peptides from partial hydrolysates of wool have been extensively studied by Consden *et al.* (*622*).

Urinary polypeptides have been studied by Boulanger *et al.* (*338*) (group separations on ion exchangers, paper chromatography of the hydrolysates).

Peptides resulting from the partial hydrolysis of *salmine* have been studied by Monier and Jutisz (*2254*).

A ferriporphyrin-C peptide obtained from cytochrome-c by digestion with pepsin has been purified on a column of Hyflo-Supercel in a butanol-acetic acid-water system (Tuppy and Bodo, *3317*; Tuppy and Paléus, *3318*).

Taylor (*3211*) has described the complete separation of oxytocin, arginine-vasopressin and lysine-vasopressin on Amberlite XE-64 with a 0.2 *M* phosphate buffer of pH 6.95.

Two biologically active peptides "hypoglycin A and B" from *Blighia sapida* have been purified on the Amberlites IR-120, IRC-50 and IR-4B (Hassall and Reyle, *1334*).

(c) *DNP-peptides*

N-2,4-dinitrophenyl peptides have been extensively used for the determination of the structure of peptides. The peptide to be studied is first coupled with 2,4-dinitrofluorobenzene and then submitted to partial hydrolysis leading to the liberation of a series of N-2,4-dinitrophenyl (DNP) peptides. These are acids and can be extracted from acid solution with organic solvents and are thus separated from other products of partial hydrolysis. The methods of fractionation of the DNP peptides on silica are essentially the same as those used for the separation of DNP-amino-acids. Details concerning the solvent systems used can be found in Sanger's papers (*2810, 2812, 2813*).

By adsorption of the DNP-peptides on talc (Sanger, *2810*) or talc-celite (1:1) (Schroeder and Honnen, *2886, 2888*) the amino acids and peptides can be washed away and the DNP-derivatives eluted separately for further chromatography.

Schroeder and Honnen (*2887*) have established certain correlations between the structure and chromatographic behaviour of dinitrophenyl-peptides on columns of silicic acid-celite. There is no correlation between length of the peptide chain and chromatographic behaviour (dipeptides can be more strongly adsorbed than tetrapeptides etc.). There is however a very definite correlation between the relative adsorption affinities of DNP-peptides and the type of the constituent amino acids. In this paper a large series of different solvents has been investigated (generally mixtures of acetic acid-acetone-ligroin, or benzene in different proportions).

Kroner *et al.* (*1809*) used the method of Perrone (*2494*) (buffered celite column) for the fractionation of DNP-peptides. For the isolation of DNP-peptides see also Askonas *et al.* (*88*), Desnuelle and Fabre (*750*).

Šorm and Šormová (*3051*) have analysed partial hydrolysates of clupein with the DNP-method, and identified DNP-prolyl-alanine, DNP-prolyl-alanyl-serine and DNP-seryl-alanyl-serine.

The separation of DNP-peptides on paper has been studied by Phillips and Stevens (*2512*), Rovery and Fabre (*2749*), and Desnuelle and Fabre (*750*). Generally, the same solvent systems are used as for DNP-amino acids (see p. 330).

Identification of amino acids present in the isolated DNP-peptides can be made by paper chromatography after hydrolysis with 5.7 N HCl (see p. 330).

CHAPTER 32

Proteins

The relative instability and ease of denaturation of proteins, as well as the difficulty of detecting minor changes in their structure, has precluded until recently the extensive use of chromatography in their purification. The chromatographic behaviour of hormones and enzymes, the activity of which can be specifically controlled, has been more fully studied. Ion exchange methods are actually the most promising in this field.

Turba (*3320*), Schroeder (*2885*) and Porter (*2568*) have reviewed the chromatography of proteins; Zechmeister and Rohdewald (*3688*) have discussed in detail the chromatography of enzymes, and Boulanger and Bizerte (*337*) the paper chromatography of proteins.

Drake (*813*) described the use of polarography for the automatic recording of the presence of proteins in chromatographic filtrates; all proteins which contain cysteine or cystine can thus be detected.

1. ADSORPTION CHROMATOGRAPHY

(a) *Adsorbents*

Columns of alumina as well as acid earths and silica gel were used by Zechmeister and Rohdewald (*3687*), in combination with the brush technique for the detection of the zones of proteins.

Tricalcium phosphate was used by Sumner et al. (*3155*) for the purification of catalase, on which it forms a coloured zone; Swingle and Tiselius (*3176*) recommend this adsorbent for chromatography of proteins, as being readily prepared in a reproducible form with desirable physical characteristics and showing reversible adsorption for many proteins. Tiselius (*3265–3267*) has described the preparation of the adsorbent and several applications of this method; amongst these the separation of the chromoproteins phycocyanin and phycoerythrin is the most conspicuous; elution is effected by phosphate buffers of increasing concentration. Phycoerythrin is eluted with 0.005–0.008 M Na_2HPO_4, phycocyanin with 0.03–0.04 M Na_2HPO_4. The separation of colourless proteins (serum albumin, egg albumin) can be observed by using a quartz column and photographing the column in reflected ultraviolet light. CO-haemoglobin, haemocyanin and tobacco mosaic virus could also be purified by the calcium phosphate column. Haxo et al. (*1339*) have applied the same method for the separation of several chromoproteins of red and bluegreen algae.

Displacement chromatography on tricalcium phosphate and then on silicic acid-Celite was used by Polis and Shmukler (*2534*) for the isolation of crystalline lactoperoxidase.

Price and Green field (*2593*) precipitated a calcium phosphate gel on cellulose and used a column of this adsorbent for the purification of catalase.

Bergdoll et al. (*199*) have purified a staphylococcal enterotoxin by adsorption on Hyflo Supercel in presence of a citrate-phosphate buffer of pH 6.35 and 0.02 ionic strength. Elution was performed with a nitrate buffer of pH 7.8 and 0.12 ionic strength.

Clauser and Li (*585*) have described the successful purification of hypophyseal growth hormone, ACTH, ovalbumin and serumalbumin on Hyflo-Supercel columns; the adsorbed proteins could be eluted into separate peaks by increasing, either continuously or discontinuously, the pH of the developing solvent.

Chiba (*564*) has reported the separation of a crystalline chlorophyll lipoprotein into two components by chromatography on a column of filter paper pulp; with 55 % picoline, two zones were obtained, one blue green above, the second yellow-green below.

Starch columns can be used for the specific adsorption of amylase (Hockenhull and Herbert, *1421*; French and Knapp, *1034*). Cellulose columns have been used by Hurst and Butler (*1491*) for the separation of phosphatases in snake venoms; fractional elution with NaCl solution resulted in a preparation of phosphodiesterase free of 5-nucleotidase. See also the section on specific adsorbents, p. 32.

(b) *Salting-out adsorption*

The difficulties encountered in the chromatography of proteins are in the main due to the slowness of equilibration, to the frequent existence of irreversible adsorption, or of denaturation following elution. Tiselius (*3260*) found one solution to the problem by showing that many proteins were more strongly adsorbed in the presence of salts at concentrations lower than that necessary for precipitation; this phenomenon is called "salting-out adsorption". In the absence of salts, 1 g of silica gel adsorbs only 0.5 mg of ovalbumin, but in the presence of 1.2 M ammonium sulphate, 9 mg of the protein are adsorbed. Salting out of ovalbumin does not occur below an ammonium sulphate concentration of 2.5 M. Details of this technique have been given by Shepard and Tiselius (*2958*).

Mitchell et al. (*2235, 2236*) showed that considerable enzyme resolution could be attained by use of the filter paper "chromatopile"; variations in salt concentration and pH have a great influence on the behaviour of the proteins on the paper column. See also Kritsman and Lebedeva (*1808*).

Brandenberger (*361*) has studied the adsorption of blood plasma proteins on cellulose powder columns, in presence of salt.

Schwimmer (*2921*) has described column procedures for salt fractionation of enzymes: the mixture of proteins is precipitated with ammonium sulphate, then mixed with Celite and filled into a column which is then eluted by ammonium sulphate; alternatively a solution of the proteins can be filtered in presence of ammonium sulphate on a column of $CaSO_4 \cdot \frac{1}{2}H_2O$, which hydrates itself and thus increases the salt concentration in the solution.

Riley (*2688*) has shown the effect of salt concentration on the chromatography, on Celite, of the virus of the Rous sarcoma. In this and subsequent papers (*2685, 2686, 2687*), Riley has shown that chromatography can be extended to the separation of particulates ranging from viruses to mitochondria and bacteria. Riley *et al.* (*2689*) have reversibly adsorbed melanized granules (size 0.2 to 0.6 μ or more) of mouse melanomas on Celite columns. Other constituents of the tumour homogenates were not adsorbed. The particulate elements separated by chromatography were found to be very homogenous under the electron microscope, and to possess high dopaoxidase and succin-oxidase activities. Adsorption was proportional to the salt concentration; elution was effected with distilled water.

Riley (*2686*) has described in detail salting-out adsorption for the purification of the virus from chicken tumours on Celite columns. A 20-fold increase in purity was obtained in a single adsorption.

Leyon (*1998*) has shown that Theiler's virus could be adsorbed on filter paper at a concentration of ammonium sulphate too low to precipitate the virus and that it could be eluted with distilled water.

Shepard (*2957, 2959*) has studied the chromatographic behaviour of *E. coli*-phages in HCl-treated paper strips (solvent: 0.1 % bovine plasma albumin + 0.1 M NaCl).

More recently, Albertsson (*32a*) has described experiments of chromatography of *Chlorella* cells on calcium phosphate and has separated cell walls and starch grains of *Chlorella* in a specially constructed apparatus.

(c) *Enzymes*

Some applications have already been mentioned above. For details concerning the adsorption chromatography of enzymes and the application of the brush streak technique, see Zechmeister and Rohdewald (*2731, 3687*). Cleavage of an enzyme into co- and apo-enzymes has been observed by Weygand and Birkofer (*3521*), by passing a solution of a yellow enzyme onto a column consisting of an upper section of Franconite SB, and a lower section of Granosil; the unaltered apo-enzyme was found in the filtrate. Von Euler and Fonó (*921*) obtained reversible inactivation of glycerophosphatase by adsorption on alumina.

Partial purifications of enzymes have been reported by Zechmeister *et al.* (*3692–3694*; glycosidases, chitinase on bauxite), by Schöberl and Rambacher (*2869*; rennin), by Enselme *et al.* (*901*; *ortho*-diphenolase), by Tóth and Bársony (*3290*; tannase), by Schormüller (*2875*; trypsin) and by McQuarrie *et al.* (*2097*; penicillinase). For details see the review of Zechmeister and Rohdewald (*3688*).

(d) *Chromoproteins*

Altschul *et al.* (*45*) observed that solutions of haemoglobin formed two distinct red zones when passed through alumina.

This was confirmed by Kruh (*1811*) using rabbit haemoglobin marked with ^{59}Fe; one fraction is eluted with water, the second with a phosphate buffer.

(e) *Miscellaneous*

Competition for the dye azorubin between serum proteins and alumina has been used by Westphal *et al.* (*3518*) to distinguish some of these proteins. Spies *et al.* (*3071*) purified an allergenic fraction from cotton seed embryo by adsorption of its picrate on alumina and eluting with 0.5 N NaOH.

Moring-Claesson (*2279*) investigated the frontal analysis of crystalline egg albumin and its partial hydrolysis products, on charcoal and aluminium hydroxide (Mataki-gel). The intact egg albumin was only slightly adsorbed on charcoal (to the same extent as alanine), probably because of its large molecular size, precluding entrance into the pores of the adsorbents; it was strongly adsorbed on alumina.

2. ION-EXCHANGE CHROMATOGRAPHY

Permutit has been used for the purification of urinary gonadotrophins (Katzman *et al.*, *1649*) and of the hormones of the posterior lobe of the hypophysis (Potts and Gallagher, *2578*).

Schlubach and Hauschildt (*2852*) have described the removal of proteins from plant juices by means of ion exchange resins.

Gilbert and Swallow (*1105*) studied the use of ion exchange resins for deionisation of proteins; denaturation was prevented by taking special precautions. Aminized and phosphorylated cotton fabrics have been used for de-ionisation of oil seed proteins (Hoffpauir and Guthrie, *1424*).

Amberlite IRC-50 has recently been used by different authors for the successful purification of basic proteins of low molecular weight.

Paléus and Neilands (*2346, 2438*) obtained a good separation of *cytochrome c* from several iron-containing impurities, by eluting with 0.1 M NH$_4$OH-ammonium acetate at pH 10.8. Margoliash (*2129, 2130*) has described in detail a method for obtaining pure cytochrome-c by elution with an ammonium acetate buffer by ion exchange on Amberlite IRC-50 and has separated the reduced and oxidized forms. The variables involving the purification of cytochrome-c on this resin have also been studied by Boardman and Partridge (284, 285).

Talboys (3194) used the same resin for partial purification of bacterial *pectinase*. Hirs *et al.* (1404) obtained a separation of crystalline *ribonuclease* into two peaks by chromatography on Amberlite IRC-50, thus confirming the presence of the two active components found by Martin and Porter (2154) under quite different conditions (see below). Elution of the enzyme was carried out with a 0.2 M sodium phosphate buffer of pH 6.45, and proceeded without loss of enzymatic activity. Crystalline *lysozyme* carbonate was chromatographed under similar conditions and also yielded two active peaks, though one of these seemed to be an artefact (Tallan and Stein, 3196; Stein, 3090). A lysozyme-like protein from spleen of rabbit has been separated into two active fractions on columns of Amberlite XE-64 (Jollès and Fromageot, 1593).

Sober *et al.* (3042) separated an egg white albumin fraction into three main components by percolation through a column of Dowex-50. They obtained a good qualitative agreement between electrophoretic and chromatographic results.

Chymotrypsinogen a has been purified on Amberlite IRC-50 by Hirs (1401), by elution with a sodium phosphate buffer of pH 6, *hyaluronidase* by Rasmussen (2628) and by Högberg (1427), *thrombin* by Rasmussen (2629), *calf thymus histone* by Crampton *et al.* (645). In the latter case difficulties were encountered in the elution; even with calcium or barium acetate buffers (pH 6.2 or 6.7) of continously increasing ionic strength (0.1 to 0.2 M) only the less basic part of the total histone mixture could be eluted. This part was found to consist of five or more proteins. For ACTH see p. 346.

Neumann (2358) has separated crystalline *crotoxin* into two active components, on Amberlite IRC-50 at a pH of 6.87.

Boardman and Partridge (285) have shown recently that Amberlite IRC-50 can also be used for the purification of neutral proteins; they have separated sheep foetal carbon monoxide *haemoglobin* from sheep maternal CO haemoglobin and from bovine CO haemoglobin and bovine CO haemoglobin from bovine methaemoglobin; these separations are sharply dependent on the pH and sodium ion concentration of the eluting buffer. The proteins can be recovered unaltered from the column and are readily crystallized, if the necessary precautions are taken (working at 0°, using freshly prepared proteins). Jollès and de Repentigny (1595) have purified metmyoglobin on Amberlite XE-64. Prins and Huisman (2595) have described a very good separation of different kinds of *human haemoglobin*, eluting with sodium citrate buffers (Fig. 90).

Boman (306) has obtained good separations of *serum proteins* on columns of the anion exchange resin Dowex-2 in the chloride form, eluting by stepwise increase of buffer concentration and keeping the pH constant at 7.2; several other proteins (phosphatases, haemoglobins, phycoerythrin and phycocyanine) could also be purified. Boman (*305*) has also purified *prostatic phosphatase* on Dowex-50 by elution with a pH gradient.

Two patents by Reid (*2659, 2660*) describe ion exchange processes for the separation of proteins; see also Reid and Jones (*2661*) who used ion exchangers for fractionating blood plasma proteins.

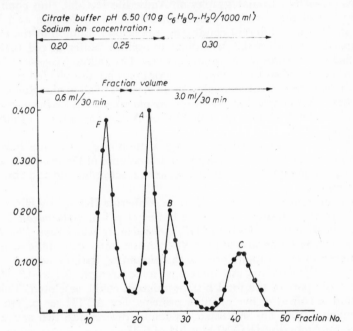

Fig. 90. Elution curves of four different kinds of human haemoglobin, chromatographed on Amberlite IRC-50 (XE-64). 48.5 mg carboxyhaemoglobin (in 0.5 ml) was chromatographed; from each fraction, after dilution to 4 ml, the extinction was measured at 5700 A. Temp. 10°; recovery 85 per cent. The carboxyhaemoglobins are indicated by *F* (fœtal); *A* (adult), *B* (sickle cell), and *C* (Prins and Huisman, *2595*)

Sober and Peterson (*3043*) have described a new approach to chromatography of proteins, by preparing special cellulose ion exchangers. A cation exchanger (CM-cellulose) was obtained by treating strongly alkaline cellulose with chloroacetic acid. An aspartic-glutamic transaminase could be successfully purified on a column of this powder, the enzymatic activity being associated with a tan band (second from bottom) (Fig. 91). An anion exchanger (DEAE-Polycel) obtained by treating strongly alkaline cellulose with 2-chloro-N,N-diethylamine was used to separate amidases (Fig. 92).

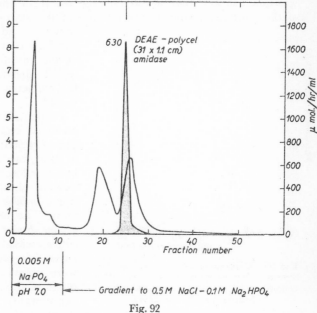

Fig. 91

0.005 M
NaPO₄
pH 7.0 ◄──────── Gradient to 0.5 M NaCl - 0.1 M Na₂HPO₄

Fig. 92

Fig. 91. 3.5 g CM-cellulose column, 21 × 1.1 cm, buffered initially at pH 5.1 with 0.02 M sodium phosphate; flow rate 2 ml/h (Sober and Peterson, *3043*)

Fig. 92. 5.0 g DEAE-cellulose buffered at pH 7.0 with 0.005 M sodium phosphate; load was 270 mg of dialyzed, lyophilized kidney fraction in 2.7 ml of same buffer. Fraction volume was 6 ml. Gradient was produced by continuous introduction of 0.1 M Na_2HPO_4 – 0.5 M NaCl into a constant volume reservoir initially containing 100 ml of 0.005 M sodium phosphate, pH 7.0, and was begun at fraction 11. Left ordinate is optical density at 280 mμ and represents protein (solid line). Right ordinate is amidase activity and is represented by the shaded area. Specific activity of the amidase preparation was 80 μmole leucine amide split/h/D_{280}. Fraction 25 had a specific activity of 630 (Sober and Peterson, *3043*)

3. PARTITION CHROMATOGRAPHY ON COLUMNS

The partition chromatography of proteins has been reviewed by Porter (*2567*). Herbert and Pinsent (*1370*) were the first to show that proteins could be purified by partition methods. They used the system ammonium sulphate-water-ethanol for purification of bacterial catalase. Recently, Martin and Porter (*2154*) used the system ammonium sulphate-water-cellosolve (ethylene glycol monoethyl ether) for partition chromatography of *ribonuclease* on kieselguhr. The crystalline enzyme was found to contain two enzymatically active components (Fig. 93).

Martin and Porter (*2154*) found that the rate of movement of the bands was much slower than would be expected from the partition coefficients of the substances. Since in the absence of the stationary phase kieselguhr shows no adsorption, the authors conclude that adsorption must occur at the interfaces between the two liquid phases. This might explain the efficiency of the separation obtained.

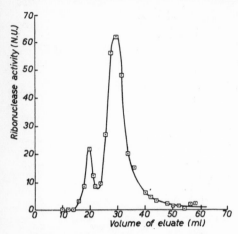

Fig. 93. Chromatogram of ribonuclease. System: 20 g ammonium sulphate, 20 g cellosolve, 56 g water, 6 g silica column (Martin and Porter, *2154*)

Porter (*2565*) has studied in detail partition systems for chromatography of proteins, and has used three of these, containing water, ethyl and butyl cellosolve and sodium or potassium phosphate for the purification of *insulin* on silane treated kieselguhr. The behaviour of *ribonuclease, avidine, bovine serum albumin* and *penicillinase* was also studied; in the latter cases, however, some irregularities were observed, the rate of movement of the principal peaks varying considerably from one chromatogram to the next. *Glucagon, chymotrypsin* and *chymotrypsinogen* could also be purified (Porter, *2566*).

More recently, Porter (*2569*) has studied the partition chromatography of *γ-globulin*, using the same system as described before, but working at —3° to avoid denaturation. It was concluded that a large number of components was present whose chromatographic behaviour ranges about a mean value.

4. PAPER CHROMATOGRAPHY

The separation of proteins by paper chromatography is rendered difficult by the general properties of proteins. Thus partition between the usual chromatographic solvents is ruled out since most solvents other than water denature proteins. The solvents that have been used with some success are aqueous acetone and alcohol, aqueous solutions of salts, sugars, buffers or combinations of these; all protein chromatograms on paper are hence due rather to adsorption than to partition.

Preliminary experiments of Tiselius (*3260*) and of Wolfson *et al.* (*3620*) indicated that salting-out chromatography can be successful under conditions of controlled evaporation of water from the paper. Separations by means of concentration gradients of salts were also obtained by Legge (*1948*).

Due to adsorption on the paper and/or the fast flow of these aqueous solutions, most proteins do not move as distinct round spots, but as long comets. The temperature during development should be kept between 0–5° to avoid denaturation.

Macek (*2076*) discusses the effects of the various gradients established on the paper during development and their effects on protein adsorption-desorption equilibria. Analyses revealed considerable differences between the composition of the bulk solvent and the water-alcohol ratio of the ascending solvent front. Similar observations, more particularly with salt gradients were recorded by Boman (*304*) who showed that round spots may be obtained with proteins in paper chromatography if these are placed on the paper behind the advancing salt gradient.

(a) *Separation of enzymes*

The identification of enzymes after separation is usually carried out by reaction with appropriate substrates. Franklin and Quastel (*1023*) were the first to show movement of proteins, including enzymes, on filter paper, without denaturation, using sucrose or buffer solutions as developing solvents. Reid (*2663*) separated the pectin esterase and the polygalacturonase of microfungi on Whatman No. 1 or No. 3 paper with buffered salt solutions as developers. The paper strips were laid on agar plates containing substrates, or cut into sections and reacted in test tubes with substrates. An almost complete separation of the two enzymes was obtained. Fungal amylases were separated into four fractions by development with acetone-water mixtures by the same author (*2664*). R_F values were found to be considerably affected by slight variations of conditions.

Giri and Prasad (*1125–1127*) reported similar separations with aqueous alcohol and acetone, and used agar plates containing substrates for the detection of the spots. They found that the β-amylases of sweet potato and of *Aspergillus niger* could be distinguished by their positions on the paper chromatogram. Trypsin (2 % aqueous solution) was studied with 50 % aqueous acetone containing 1.5 % NaCl, which separates it completely from amylase. The trypsin was detected by leaving the paper in contact with a gelatin film strip which was liquified in the region of the trypsin spot. Simonart and Chow (*2992, 2993*) studied the enzymes of *Aspergillus tamarii*. A complete separation of the amylase and proteinase was effected with 35 % saturated $(NH_4)_2SO_4$ buffered to pH 6.5. Detection of the enzymes was again carried out with substrates. Cabib (*492*) separated invertase from glucomutase (of *Saccharomyces fragilis*) with 20–25 % alcohol and also obtained two fractions from the invertase of brewer's yeast.

The method of Cabib was employed by Reese and Gilligan (*2645*) for the examination of cellulytic components of a filtrate from cellulytic organisms. Jermyn (*1584, 1585*) studied the separation of the enzymes of *Aspergillus oryzae* both by paper chromatography and electrophoresis. Horse radish peroxidase was also examined by Jermyn et al. (*1108*) who found more active fractions by chromatography than by electrophoresis. A separation of pepsin from rennet using NaH_2PO_4 solution as solvent on Whatman No. 4 paper was reported by Simonart and Chow (2994). Tauber and Petit (3206) used the two-dimensional technique of Quastel (see p. 359) for the study of cytochrome-c, cow-liver catalase etc.

(b) Separation of non-enzymatic proteins

(i) Reagents for proteins on paper

Jones and Michael (1598) recommend the dye Solway purple, which is strongly adsorbed by proteins and not by cellulose. Dyeing at 50–90° with 0.05 % dye in 0.5 % H_2SO_4 and washing with warm water produces purple spots with serum proteins. Papastamatis and Wilkinson (2440) spray the paper with bromothymol blue (0.1 %) and, after drying, immerse it in 0.2 N acetic acid, which results in blue spots on a yellow background. Cabib (492) employed bromophenol blue for plasma proteins. Franklin and Quastel (1023) used haemin as a "marker" for proteins which combine with haemin. The haemin-protein complexes were chromatographed and detected with a benzidine-H_2O_2 reagent. Other reagents for dyeing proteins that have been used in paper electrophoresis, are azocarmine B (Turba and Enenkel, 3321) and bromophenol blue in saturated ethanolic $HgCl_2$ (Durrum, 837). See also Wunderly (3633) and McDonald (2072).

(ii) Separation of blood proteins by uni-dimensional development

Jones and Michael (1598), as well as Papastamatis and Wilkinson (2440) separated albumin from globulins either with half-saturated $(NH_4)_2SO_4$ or with veronal buffer at pH 6.5. In both cases two comets were produced; one from the starting point and one from the liquid front. Fractionation of the globulins into components α, β, and γ is not readily achieved. Papastamatis and Wilkinson (2440) also noted a difference in the speed of travel of egg and bovine albumins.

Cabib (493) separated human serum and rabbit serum into constituents using 30 % ethanol as solvent. Albumin travels with R_F 0.68 while the globulins leave an elongated trail. With radial development, Giri (1113) reports constant and considerable differences between normal and cirrhosis sera using 40 % ethanol as solvent. Zimmermann and Kludas (3701) use a mixture of butanol saturated water, sodium dodecanesulphonate solution and propanol (4:2:1) with radial development and claim a good resolution of human sera in only 15 minutes. See also Thomson (3245) and Berlingozzi et al. (218).

Sansone and Cusmano (2815) were able to distinguish between foetal and adult haemoglobin by developing with pyridine containing 10 % water or with 0.2 N NaOH. Adult haemoglobin travels from the point of application, while foetal haemoglobin remains at the starting point with a small trail forward.

Further work by Sansone and Cusmano (*2816, 2817*) on foetal and adult haemoglobins was carried out with pyridine as solvent in horizontal development. Kruh *et al.* (*1812*) separated adult rabbit haemoglobin into two bands. See also Berlingozzi *et al.* (*219*).

The possibility of proteins interfering with amino acid separations was examined by Wynn and Rogers (*3639*), and it was found that proteins do not split off amino acids when developed in phenol (protein $R_F = 1$) or collidine (protein $R_F = 0$).

The usefulness of phenolic solvents for protein separations was studied by Grassmann and Deffner (*1191*).

(iii) *Two-dimensional separation of plasma proteins*

Franklin and Quastel (*1024, 1025*) discussed the patterns obtained by two-dimensional treatment of "haemin-marked" proteins using, for example, aqueous sucrose and tartrate as the two solvents and adding surface active agents (Tween 85 and Tween 81) to the proteins. Although no distinct spots were produced (merely a pattern of comets and elongated trails), it is possible to utilise these patterns for the examination of certain pathological (*3170*) and physiological conditions resulting from heparin administration, fat diets (*3169*), multiple sclerosis; also to study the combination of radio-active thyroxine with plasma proteins (*1230*) and the non specific agglutinating substances of bovine sera.

Swank *et al.* (*3169, 3170*) have for example shown that the patterns consistently obtained in normal individuals are significantly altered during or just previous to the exacerbation in multiple sclerosis, indicating a change in the blood constituents. The altered patterns revert to normal when the patient is relieved. The administration of heparin as well as of high fat meals to dogs bring about large changes in the patterns obtained in these paper chromatograms. These results are taken to indicate that the patterns are functions not only of the protein constituents of the plasma, but of other substances (e.g. lipids) which are also present.

Franklin *et al.* (*1025*) have pointed out that the paper chromatography patterns of plasma proteins exhibit marked changes in multiple myeloma, hepatitis, liver cirrhosis and other pathological conditions. The patterns revert to normal on recovery of the patient.

It is also possible to obtain a semi-quantitative evaluation of artificial protein mixtures, e.g. cytochrome-c and albumin, and to distinguish the patterns of various plasma fractions (e.g. the fractions of the Cohn method), and of snake venom (*1025*).

The techniques of Quastel and co-workers were employed by McKerns (*2087*) for the examination of commercial rennet and its action on casein. Hall and Wewalka (*1279*) criticized the interpretation of Franklin and Quastel's protein patterns and suggested that most patterns obtained were artefacts, a view which is difficult to reconcile with the distinct and varied patterns obtained.

More recently, Quastel and Van Straten (*2606*) have demonstrated that a separation of plasma proteins takes place during unidimensional ascent on paper, the proteins being distinguished by electrophoresis in the second dimension. They were also able to show by the same technique that interactions occur between proteins, accounting for the difficulty in obtaining clear-cut separations in plasma protein mixtures.

For further work see also Quastel *et al.* (*2606*) and Kalz *et al.* (*1622*). Kazmeier and Gassen (*1669*) examined critically the method of Quastel and found it unreliable for clinical work.

(c) *Miscellaneous*

Plant viruses were separated by Gray (*1201*) using 40 % ethanol as solvent. See also Ragetli and Van der Want (*2613*). *Viper venom* (*Ammodytes*) was separated into several constituents by Piantanida and Muic (*2514*). *Antibody reactions* can be revealed by the chromatographic pattern as was shown by Spalding and Metcalf (*3056*).

Insulin was separated from *protamine* using butanol-CH_3COOH-water (3:1:4) as solvent (Robinson and Fehr (*2704*). Kohn (*1762*) separated *intermedin*, the melanophorotropic hormone of the pituitary intermediate lobe, into a band of R_F 0.28 and another of R_F 0.04 using water-butanol-CH_3COOH (4:4:1) as solvent.

CHAPTER 33

Purines, pyrimidines and related compounds

Authoritative reviews of this field can be found in articles by Markham (*2136*) ("Chromatography of nucleotides and related substances"); Cohn (*600*) ("Separation of nucleic acid derivatives by chromatography on ion exchange columns") and Wyatt (*3637*) ("Separation of nucleic acid components by chromatography on filter paper").

1. PURINES AND PYRIMIDINES

(a) *Column methods*

Smith and Wender (*3039*) separated xanthine and guanine on columns of Amberlite IR-100H but only xanthine could be eluted quantitatively.

Cohn (*597*) obtained good separations of uracil, cytosine, guanine and adenine by anion exchange on Dowex-A-1, as well as by cation exchange on Dowex-50 (*596*). See also Volkin *et al.* (*3429*).

Soodak *et al.* (*3049*) separated cytosine from uracil by adsorption of the former on Amberlite IR-100-H or on Decalso.

Wall (*3462*) has studied the simultaneous separation of purines, pyrimidines, amino acids and other nitrogenous compounds extracted from bacteria using a modification of the method of Moore and Stein (*2268*) with Dowex-50. Cohn (*600*) ascribes the stronger retention of purines (relative to pyrimidines) to a non polar attraction of the resin. This "family difference" is also noted in the nucleotides.

Starch columns were used by Edman *et al.* (*866*) for the quantitative separation of adenine and guanine (solvent: butanol-ethylene glycol mono-methyl ether).

Daly and Mirsky (*690*) were the first to describe the quantitative separation of all six bases: thymine, uracil, cytosine, adenine, guanine and hypoxanthine. They prepared starch columns according to Moore and Stein (*2265*) and used *n*-propanol $+ 0.5 \, N$ HCl $(2:1)$ as solvent. Eluate fractions of 0.5 ml were collected and analysed spectrophotometrically (Fig. 94). This method has been applied to the analysis of some deoxypentose nucleic acids (Daly *et al.*, *689*).

Hayaishi and Kornberg (*1340*) have separated the products of enzymatic oxidation of uracil on Dowex-1; barbituric acid was thus identified. See also Wang and Lampen (*3473*); Johnson (*1590*) has separated methylated derivatives of uric acid on Zeo-Karb 215.

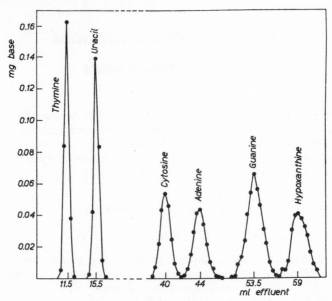

Fig. 94. Chromatography of purines and pyrimidines on starch columns
(Daly and Mirsky, *690*)

(b) *Paper chromatography*

Vischer and Chargaff (*3417*, *3418*) were the first to elaborate techniques for the separation and quantitative analysis of purine and pyrimidine bases by paper chromatography (with quinoline-collidine 3:1 as the mobile phase). The position of the purines was determined by treating the paper with a mercury salt. The mercuric salts of the purines and pyrimidines formed in this way were rendered visible by treatment with ammonium sulphide. By using duplicate chromatograms and making the spots visible on one, the other chromatogram could be cut up in such a way that the purines could be eluted separately and estimated by their ultra-violet absorption. The purines studied included hypoxanthine and xanthine. Vischer and Chargaff applied this technique to the analysis of the nucleic acids from yeast and pancreas (*3419*), thymus and spleen (*553*), as well as the nucleoproteins from an avian tubercle bacillus (*552*, *3421*).

Hotchkiss (*1452*) used Whatman No. 1 paper with *n*-butanol as solvent. The chromatogram was cut up into strips from 5 to 20 mm wide and the ultraviolet absorption spectrum of the material eluted was measured. Precise details of the procedure for the spectral identification of the five bases (cytosine, thymine, uracil, adenine and guanine) were given. The nucleotides cytidine, guanidine, adenosine and thymidine were also separated and estimated quantitatively.

Chargaff *et al.* (*551*) and Holiday and Johnson (*1429*) showed that the purine and pyrimidine spots on paper were visible as dark spots on a fluorescent background in ultraviolet light. Concentrations of the order of 0.5 to 1.0 μg/ml could be detected by this technique.

Carter (*517*) has described chromatographic techniques for resolving mixtures of purines, pyrimidines, nucleosides and nucleotides on paper, and locating the compounds by ultra-violet fluorescence. He used buffered aqueous systems with an overlying thin layer of *iso*amyl alcohol. The optimal amount of compounds for analysis is between 20 and 50 μg.

Markham and Smith (*2137, 3032*) have developed a method for the quantitative analysis of ribonucleic acids. They hydrolyse the latter quantitatively to a mixture of purine bases and pyrimidine nucleotides, and separate the resulting mixture on paper. Detection of the spots is made by an ultra-violet printing technique on photographic paper (*2137*). The spots are then cut out, extracted with 0.1 N HCl and the ultra-violet absorption measured quantitatively. The amount necessary for one analysis is about 0.1 mg. This method has been applied to the analysis of the nucleic acids of several strains of virus (*2138, 2139*). Edström (*867*) has described a similar method. Goeller and Sherry (*1154*) give details of the technique of ultra-violet photography in the study of nucleic acids.

MacDonald (*2074*) has described a method for the estimation of adenine and guanine from nucleic acids, based on the precipitation of the silver purines followed by separation of the purines by paper chromatography and estimation by spectrophotometry.

Löfgren (*2042*) has reported four solvent systems permitting the separation of mixtures of natural purines, pyrimidines and their nucleosides. See also Fig. 95 and Table 111.

Paper chromatography of hydrolysates of deoxypentose nucleic acids has led Wyatt (*3635*) to the discovery of an amino-pyrimidine which on deamination gives rise to thymine, and which he identified as *5-methylcytosine*. This pyrimidine base occurs in all animal deoxy-ribonucleic acids and the one nucleic acid of plant origin examined, and is apparently lacking in those from bacteria and viruses.

Chargaff *et al.* (*550*) and Wyatt (*3636*) described micromethods for the estimation of the purine and pyrimidine bases in *deoxypentose* nucleic acids. Both purines and pyrimidines are obtained by hydrolysis with formic or perchloric acid and separated on a single paper chromatogram.

Smith and Wyatt (*3033*) studied the composition of some microbial

TABLE 111. R_F Values of Purine and Pyrimidine Bases and Nucleosides (Wyatt, *3637*)

	n-Butanol[a]	n-Butanol-NH₃[b]	n-Butanol-NH₃[c]	n-Butanol-formic acid[d]	Isopropanol-NH₃[e]	Isopropanol-HCl[f]	Collidine-quinolines	Isobutyric acid-NH₃[h]	Na₂HPO₄-iso-amyl alc.[i]	Water (pH 10)[j]
Adenine	0.38	0.28	0.40	0.33	0.37	0.32	0.34	0.83	0.44	0.37
Guanine	0.15	0.11	0.15	0.13	0.16	0.22	0.22	0.70	0.02	0.40
Hypoxanthine	0.26	0.12	0.19	0.30	0.16	0.29	0.44	0.69	0.57	0.63
Xanthine	0.18	0.05	0.01	0.24	0.11	0.21	0.62	0.60	0.49	0.62
Uracil	0.31	0.19	0.33	0.39	0.38	0.66	0.74	0.67	0.73	0.76
Thymine	0.52	0.35	0.50	0.56	0.52	0.76	0.84	0.78	0.73	0.74
Cytosine	0.22	0.24	0.28	0.26	0.32	0.44	0.21	0.80	0.73	0.70
5-Methylcytosine	0.29	0.27	0.36	—	0.37	0.52	—	—	—	0.73
5-Hydroxymethylcytosine	0.13	0.12	—	—	0.25	0.44	—	—	—	0.75
Adenosine	0.20	0.22	0.33	0.12	0.31	0.34	—	0.91	0.54	0.49
Guanosine	0.15	0.03	0.10	0.17	0.13	0.30	—	0.59	0.62	0.68
Inosine	—	0.03	0.08	—	0.14	0.30	—	—	—	0.81
Uridine	0.17	0.08	—	0.25	0.31	0.64	—	0.60	0.79	0.84
Cytidine	0.12	0.11	0.15	0.18	0.28	0.45	—	0.73	0.76	0.76
Adenine DR[k]	0.35	—	0.41	—	—	—	—	0.91	0.55	0.47
Guanine DR	0.21	—	0.18	—	—	—	—	0.67	0.62	—
Hypoxanthine DR	0.23	—	0.17	—	—	—	—	0.70	0.70	0.80
Uracil DR	0.38	—	0.34	—	—	—	—	0.67	0.79	0.83
Thymine DR	0.51	0.40	0.48	—	0.57	0.81	—	0.75	0.78	0.77
Cytosine DR	0.23	—	0.26	—	—	0.60	—	0.83	0.77	0.75
5-Methylcytosine DR	0.25	—	—	—	—	—	—	—	0.76	—

[a] 86% (vol./vol.) aq. n-butanol; Whatman No. 1 paper, descending; Markham and Smith, *Biochem. J.* (1949) **45**, 294. Values for deoxyribosides are from Buchanan (*443*), using n-butanol satd. with water; values for methylcytosine and hydroxymethylcytosine detd. by Wyatt.

[b] 86% (vol./vol.) aq. n-butanol, with 5% by vol. of conc. NH₃ soln. (sp. gr. 0.880) added to solvent in bottom of tank; Whatman No. 1, descending; Markham and Smith, *op. cit.* Values for methylcytosine, hydroxymethylcytosine, inosine, and thymine deoxyriboside detd. by Wyatt.

[c] n-Butanol satd. with water at about 23° 100 ml., 15 N NH₄OH 1 ml.; Whatman No. 4, ascending; Mac-Nutt, *Biochem. J.* (1952) **50**, 384. Values for xanthine, adenosine, and guanosine are from Hotchkiss, *J. Biol. Chem.* (1948) **175**, 315, with a similar solvent system.

[d] n-Butanol 77%, water 13%, formic acid 10% by vol.; Whatman No. 1, descending; Markham and Smith, *op. cit.*

[e] Isopropanol 85 ml., water 15 ml., conc. (28%) NH₃ soln. 1.3 ml.; Whatman No. 1, descending; Hershey *et al.*, *J. Gen. Physiol.* (1953) **36**, 777. R_F values at 20–23° detd. by Wyatt.

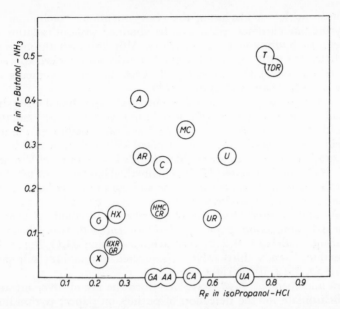

Fig. 95. Diagram of the positions of nucleic acid derivatives on a two-dimensional chromatogram run on Whatman No. 1 paper by the descending technique first in solvent f (Table 111), then in solvent c (Table 111), however in the large chromatogram tank used here, the effective NH_3 concentration is reduced. A, adenine; AA, yeast adenylic acid; AR, adenosine; C, cytosine; CA, cytidylic acid; CR, cytidine; G, guanine; GA, guanylic acid; GR, guanosine; HMC, 5-hydroxymethylcytosine; HX, hypoxanthine; HXR, inosine; MC, 5-methylcytosine; T, thymine; TDR, thymidine; U, uracil; UA, uridylic acid; UR, uridine; X, xanthine (Wyatt, 3637)

TABLE 111 *(continued)*

f *Iso*propanol 170 ml., conc. HCl (sp. gr. 1.19) 41 ml., water to make 250 ml.; Whatman No. 1, descending; Wyatt, *Biochem. J.* (1951) **48**, 584. Values redetd. at 20–23° by Wyatt.

g Collidine 1 vol., quinoline 2 vol., mixt. satd. with 1.5 vol. water; Schleicher and Schüll No. 597 paper, descending,at about 22°; Vischer and Chargaff, *J. Biol. Chem.* (1948) **176**, 703.

h *Iso*butyric acid 400 ml., water 208 ml., 25% NH_3 soln. 0.4 ml.; Whatman No. 4, descending, at 22°; Löfgren, *Acta Chem. Scand.* (1952) **6**, 1030; excepting deoxyribosides, for which solvent and conditions are as in footnote e, Table 113, and R_F values are calcd. from the relative mobilities given by Tamm *et al.*, *J. Biol. Chem.* (1953) **203**, 673, taking the R_F of thymidine as 0.75.

i 5% aq. Na_2HPO_4 satd. with *iso*amyl alc., both aq. and nonaq. phases being present in the trough; Whatman No. 1, descending; Carter, *J. Am. Chem. Soc.* (1950) **72**, 1466. Values for deoxyribosides are from Buchanan, *Nature* (1951) **168**, 1091.

j Water adjusted to pH 10 with N NH_4OH; Whatman No. 1, ascending, 22–23°; Levenbook, personal communication to Dr. Wyatt, 1953. Values for methylcytosine and hydroxymethylcytosine detd. by Wyatt.

k DR = deoxyriboside.

deoxypentose nucleic acids, using the above mentioned technique. Good purine-pyrimidine chromatograms can be obtained without isolation of the nucleic acids by hydrolysis of whole viruses. Although such chromatograms are heavily loaded with amino acids, the latter do not interfere appreciably with the estimation of the nucleic acid bases. For improvements in the technique, see Markham and Smith (*2139*).

Marshak and Vogel (*2144*) have described a method, based on hydrolysis with perchloric acid, which permits a quantitative micro-determination of nucleic acid purines and pyrimidines without prior isolation of the nucleic acid in lipid-free biological materials extracted with trichloroacetic acid. The chromatographic procedures are slight modifications of those of Vischer and Chargaff (*3418*) and Hotchkiss (*1452*). This method has been used by Marshak (*2143*) for the determination of the purine and pyrimidine content of nucleic acids of cell nuclei and cytoplasm. See also Kariyone and Inouye (*1632*).

Dorough and Seaton (*803*) have described a method of separating nucleic acid fragments on paper by development with two solvents (first butanol-acetic acid-H_2O, 8:2:2, then acetone-butanol-H_2O, 8:1:1, in the same direction, then a third solvent (saturated ammonium sulphate-*iso*-propanol-H_2O, 79:2:19) at right angles.

Reguera and Asimov (*2648*) have described the use of silver nitrate and sodium dichromate for the detection of purines on paper; pyrimidines do not show up.

Michl (*2208*) has described colour reactions for the detection of purine derivatives on paper, which can be useful where extraneous absorbing substances interfere with the usual spectrophotometric detection: (1) the mercuric salts of adenine, guanine, xanthine and hypoxanthine give coloured compounds with eosin, (2) adenine and guanine can be detected with an alcoholic solution of bromophenol blue saturated with mercuric chloride, (3) caffeine, theobromine, theophylline give fluorescent compounds after treatment with chlorine gas.

Wood (*3622*) has described a silver reagent for the detection of certain purines and pyrimidines (adenine, guanine, xanthine, hypoxanthine, cytosine, but not uracil, thymine, uric acid).

Boser (*328*) has used Folin's reagent (sodium 1,2-naphthoquinone-4-sulphonate) for the detection of adenine on paper (dark violet spot). Gerlach and Döring (*1100*) have described a specific colour test for adenine and adenosine derivatives on paper: the paper is first sprayed with $KMnO_4$ (70 mg in 100 ml water), then brought for 15 sec in an atmosphere of Cl_2 and dried 5 min at 100°. Adenine or adenosine and derivatives show up as yellow or orange spots which become red on spraying with 3 N KOH.

Edström (*868*) has described the analysis of nucleic acids in microgram amounts, using Munktell OB paper and spectrophotometric detection; the purines and pyrimidines of less than 2 μg of nucleic acid could be estimated.

The detection of acid degradation products of purines, by circular paper chromatography had been described by Giri *et al.* (*1118*).

Chargaff and Kream (*547*) have studied bacterial cytosine deaminase by incubating bacterial suspensions, or extracts, with cytosine on paper, and by subsequently developing a chromatogram to identify the reaction products. Similarly the conversion of adenine to hypoxanthine by *E.coli* and of guanine to xanthine by the guanase of rabbit liver were studied. Roth *et al.* (*2740*) have separated the radioactive metabolites of ^{14}C-labelled pentobarbital on paper.

Johnson (*1590*) has separated mono- and di-methylated uric acids from urine on paper using butanol, ethylene glycol mono-methyl ether, acetic acid, water (7:7:2:4) as solvent. In a similar study, Weissmann *et al.* (*3496, 3497*) identified the following purines in the urine of normal persons: hypoxanthine, xanthine, 7-methylguanine, adenine, 1-methylguanine and guanine, accompanied by five unidentified spots giving a purine absorption spectrum. After administration of caffeine, 1-methylxanthine, 7-methylxanthine and paraxanthine were found. See also Table 112.

TABLE 112. R_F values of xanthine bases (Munier and Macheboeuf, *2320, 2321*)
Paper: Durieux No. 122

	Butanol 90, conc. HCl 10, saturated with water	Butanol 100, conc. NH$_4$OH 2, water 16
Caffeine.	0.65	0.73
Theophylline	0.49	0.36
Theobromine	0.33	0.25
Xanthine	0.13	—
Trigonelline	0.21	0.10
Nicotinamide	—	0.78

Albert and Brown (*31*) have determined the R_F values of a great number of purines (containing methyl, phenyl, chloro, hydroxyl, methoxyl, amino, methylamine, dimethylamino, mercapto, methylthio and other groups).

The separation of methylated xanthines and uric acids on paper is described by Dikstein *et al.* (*776a*) (Fig. 101, p. 378).

The paper chromatography of barbituric acids has been described by Grieg (*1222*), Wickström and Salvesen (*3542*) and Riebeling and Burmeister (*2681*). Barbiturate metabolites in man have been studied by paper chromatography (Allgén, *38*).

See also the review of Boulanger and Montreuil (*342*).

2. NUCLEOSIDES

(a) *Column methods*

The nucleosides cytidine and uridine can be separated on ion exchangers; cytidine with its free amino group is retained on Zeo-Karb 215, whereas uridine passes through the column (Elmore, *887, 888*; Harris and Thomas, *1316, 1317*).

Dekker and Elmore (*727*) have isolated 5-methylcytosine deoxyriboside from wheat germ deoxyribonucleic acid by adsorption on Zeo-Karb 215 and elution with aqueous ammonia. Further purification was achieved by adsorption on a column of powdered cellulose. The *picrate* of the nucleoside gave the free nucleoside by percolation of an aqueous solution through a column of Amberlite IRA-400.

Andersen *et al.* (*53*) have described a method using Dowex-2 columns in the chloride form for the large-scale preparation of deoxyribonucleosides; they report that these compounds are hydrolysed when chromatographed on cation exchange resins (Dowex-50), even in the formate form. Friedkin and Roberts (*1045*) have separated thymine and thymidine on Dowex-2.

Nucleosides can also be separated on anion exchange columns in the presence of borate; weakly sorbed nucleosides as cytidine and adenosine become more strongly sorbed in the presence of borate and thus more easily separable (Jaenicke and Dahl, *1533*; Cohn, *600*). The difference between deoxyribonucleosides and ribonucleosides with respect to borate complex formation facilitates the separation of these two groups.

The separation of ribonucleosides by partition on starch columns has been described by Reichard (*2653*).

Beale *et al.* (*170*) used a similar method for the chromatography of the pentose nucleosides from a fowl sarcoma. Dekker and Todd (*728*) separated thymidine and uracil deoxyriboside on a cellulose column, using butanol-water as solvent. See also Drell (*817*).

(b) *Paper chromatography*

Hotchkiss (*1452*) was the first to separate the nucleosides cytidine, guanosine, adenosine and thymidine from each other and from the purine and pyrimidine bases (in butanol); detection and identification was achieved by ultra-violet spectrophotometry. Pentose nucleosides can also be detected on paper by periodate oxidation; deoxypentose nucleosides cannot be oxidised (Buchanan *et al.*, *444*). A Dische reaction with diphenylamine or the Feulgen reaction has been used by Buchanan *et al.* (*444*) for the detection of the latter. They can also be detected by a Dische reaction with cysteine-sulphuric acid; deoxyribose and deoxyribosides, but not ribose or ribose nucleosides, can be detected by this method (Buchanan, *443*).

The microbiological detection of thymidine by its activity as a growth factor in a vitamin B_{12} test has been used by Smith and Cuthbertson (*3026*), as well as by Winsten and Eigen (*3600*).

Carter (*517*) has used 5% dibasic sodium phosphate in *iso*amyl alcohol for separating the nucleosides of yeast nucleic acid; complete separation of nucleosides, nucleotides, purines and pyrimidines can be obtained by two-dimensional chromatography.

Carter (*518*) has also studied the products of enzymatic degradation of uridine by paper chromatography. Greenberg (*1214*) has separated ^{14}C inosine, inosine-5-phosphate and hypoxanthine. The radioactive fractions were located with a Geiger-Müller counter. (See also *2124*.)

For the separation of uridine, cytidine and thymidine, the solvent systems of Löfgren (*2042*) can also be used. Rose and Schweigert (*2734*) have separated nucleosides on paper in presence of borate.

3. NUCLEOTIDES

(a) *Column methods*

(i) *Ribonucleotides*

Ion exchange has been extensively used for the separation and preparation of nucleotides. Cohn (*596*) separated the four mononucleotides of yeast nucleic acid, first by cation exchange on Dowex-50, then by an improved method by anion exchange on Dowex-1 or Dowex-2, by successive elution with weak acids and buffers of controlled pH and anion concentration

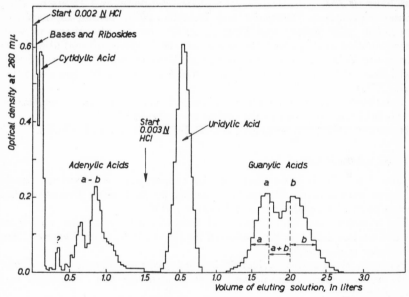

Fig. 96. Anion-exchange chromatography on Dowex-1 of rat liver ribonucleic acid mono-nucleotides (Volkin and Carter, *3427*)

(*596, 597*). Three isomeric adenylic and inosinic acids were separated. This method was applied by Volkin and Carter (*3427*) to the analysis of the mononucleotides of rat liver ribonucleic acid. Two adenylic acid and two guanylic acid peaks were obtained (Fig. 96). See also Hurst *et al.* (*1492, 1493*). Heterogeneity of pyrimidine nucleotides was also reported, as indicated by the formation of two peaks in ion-exchange elution diagrams (598).

Cohn and Carter (601) have described the preparation of uridylic and cytidylic acids from yeast ribonucleic acid by anion exchange on Amberlite IRA-400. Loring et al. (2050) have separated isomeric cytidylic acids on a preparative scale by chromatography on Dowex-1 or Dowex-2.

The enzymic degradation of ribonucleic acid by crystalline ribonuclease has been studied by Carter and Cohn (519), by applying ion exchange and paper chromatographic methods to the products of hydrolysis.

More recently, gradient elution methods using Dowex-1 in the formate form with elution by increasing concentrations of formic acid or formic acid + formate have been described by Bergkvist and Deutsch (202), Hurlbert et al. (1490); Schmitz et al. (2865, 2866); these methods permit the successive elution of most of the mono-, di- or triphosphates of adenosine, inosine, guanosine, cytidine and uridine. These and similar methods have been largely used in recent years for the isolation of different nucleotides: inosine-phosphates (Schulman and Buchanan, 2904; Deutsch and Nilsson, 758), guanosine-triphosphate (Bergkvist and Deutsch, 204), uridine-triphosphate (Bergkvist and Deutsch, 203), uridine-diphosphate-glucuronic acid (Storey and Dutton, 3132), cyclic phosphates of adenosine, uridine and cytidine (Brown et al., 407), pyrimidine-diphosphates (Cohn and Volkin, 604), guanosine-diphosphate-mannose from yeast (Cabib and Leloir, 494), cytidine-phosphates (Whitfeld et al., 3539). See also Sacks et al. (2776).

A detailed discussion of physico-chemical factors governing separation in this field can be found in the review by Cohn (600).

Reichard (2653) has separated ribonucleotides on a *starch* column.

(ii) *Deoxyribonucleotides*

Volkin et al. (3429) have described the preparation of *deoxyribonucleotides* by anion exchange on Dowex-1. A similar method using Dowex-A1 has been reported by Sinsheimer and Koerner (2999).

The isolation and identification of deoxy-5-methyl-cytidylic acid from thymus nucleic acid has been effected by Cohn (599), using Dowex-1. See also Potter et al. (2577).

Permutit-50 separates nucleotides into two groups; adenylic and cytidylic acids are adsorbed, guanylic and uridylic acids are not (Montreuil and Boulanger, 2258).

(iii) *Oligonucleotides*

Merrifield and Woolley (2192) have used the methods of Cohn (596–598) for the separation of dinucleotides obtained by partial acid hydrolysis of yeast ribonucleic acid.

Ribonucleotides can be separated from deoxyribonucleotides in presence of borate (Khym and Cohn, 1710).

In a study of the products of ribonuclease action on ribonucleic acid, Volkin and Cohn (*3428*) have used Dowex-1 with only 2 % cross-linking (400 mesh); the low degree of cross-linking seems essential for the equilibration of the larger oligonucleotides with the active groups of the resin (Fig.97); in this experiment the concentration of HCl was held at 0.01 N with stepwise increments in NaCl to remove the more strongly adsorbed oligonucleotides. Cohn (*600*) has discussed the relative positions of the latter in regard to the positions of the corresponding mononucleotides.

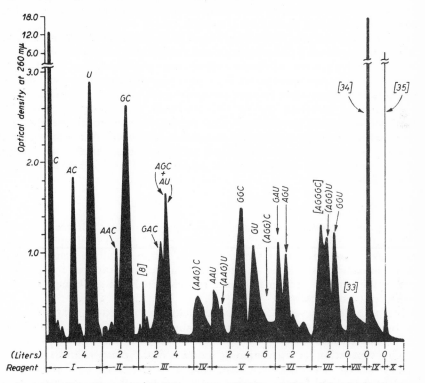

Fig. 97. Separation of the products of ribonuclease digestion of ribonucleic acid. Exchanger: Dowex-1 (2 % DVB)-chloride, 400 mesh, 15 cm × 3.7 cm². Solution: HCl + NaCl as follows: I, 0.005 N HCl; II, 0.01 N HCl; III–IX, 0.01 N HCl + 0.0125, 0.025, 0.05, 0.1, 0.2, 0.3, and 1 N NaCl, respectively; X, 2 N HCl. Sorbed material: 700 mg calf liver RNA + 10 mg ribonuclease in 105 ml H₂O, 22 hr at 37°, + NaOH as required to keep at pH 7.0; pH lowered to 2.0; chilled; centrifuged; supernatant made alkaline with NH₄OH; sorbed (Volkin and Cohn, *3428*)

A = adenosine, C = cytidine, G = guanosine, U = uridine

Sinsheimer (*2998*) has separated a mixture of nineteen mono- and dinucleotides obtained by the action of deoxyribonuclease on highly polymerized deoxyribonucleic acid, by successive elutions from Dowex-1-8X and Dowex-1-2X.

(iv) *Adenosine phosphates*

Cohn and Carter (*602*) have reported a very clearcut separation of adenine, adenosine mono-, di-, and triphosphates by adsorption on Dowex-1 and successive elution with HCl or HCl + NaCl.

Turba *et al*. (*3323*) have also used Amberlite IRA-400 for such separations. Marrian (*2141*) has isolated adenosine tetraphosphate by chromatography of commercial samples of ATP on Dowex-2.

Crane and Lipman (*646*) have found that Norit A adsorbs specifically adenosine containing phosphates; those compounds tested which were not adsorbed at all were inorganic ortho- and pyrophosphate, fructose diphosphate, glucose-1-phosphate, fructose-6-phosphate and acetylphosphate; it is presumed that other nucleotidic compounds are also adsorbed.

Coenzyme A has been purified by Stadtman and Kornberg (*3081*) on charcoal and then on Dowex-1 X2 in the formate form. Basford and Huennekens (*154*) have separated commercial coenzyme A into four distinct forms (–SH form, two disulfides and an "anomalous form") on cellulose powder columns or on paper sheets (R_F 0.42, 0.35, 0.15, 0.00 in ethanol-water, 7:3).

(v) *Pyridine nucleotides*

LePage and Mueller (*1963*) used chromatography on a charcoal column (Nuchar C) for the purification of triphosphopyridine nucleotide; elution was carried out with successive portions of dilute pyridine. Feigelson *et al*. (*939*) adapted this method to the quantitative estimation of the pyridine nucleotide content of animal tissue, the charcoal eluates being analysed spectrophotometrically.

Neilands and Åkeson (*2347*) have described a method for the preparation of diphosphopyridine nucleotide using adsorption on Norit and chromatography on Dowex-2.

The separation of nicotinamide and its diphosphopyridine nucleotide (DPN) on Dowex-1 (formate form) has been described by Zatman *et al*. (*3670*). A method for the purification of nicotinamide mononucleotide by chromatography on charcoal and Dowex-1 has been described by Plaut and Plaut (*2528*).

(vi) *Flavin nucleotide*

The isolation of *flavin* nucleotides on florisil, celite or dicalcium phosphate has been described by Dimant *et al*. (*778*). Siliprandi and Bianchi (*2985*) have purified the flavin-adenine dinucleotide by chromatography on Amberlite IRC-50 followed by electrophoresis on a column packed with cellulose powder.

(b) *Paper chromatography*

(i) *Purine and pyrimidine nucleotides*

Markham and Smith(*2139*) have studied a series of solvents for the separation of nucleotides on paper and recommend the following: saturated ammonium sulphate 79 %, water 19 % and *iso*propanol 2 % (v/v). With this solvent they separated two adenylic and two guanylic acids, thus confirming the findings of Carter (*517*). Cytidylic and uridylic acids were not separated from each other in this system.

Chargaff *et al.* (*551*) and Magasanik *et al.* (*2104*) have separated the ribonucleotides on paper in an atmosphere of ammonia, with *iso*butyric acid as solvent. The components were revealed by inspection under ultra-violet light. The separated nucleotides were eluted with *M* phosphate buffer of pH 7.1 and the ultra-violet spectra of the extracts determined. This method has been applied to the study of the structure of ribonucleic acids (Magasanik and Chargaff, *2102*).

Boulanger and Montreuil (*340*) have used solvents containing phenol or *iso*propanol for the quantitative separation of ribonucleotides (Fig. 98). The spots were detected by

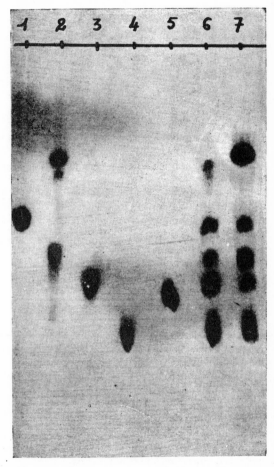

Fig. 98. Unidimensional chromatogram of ribonucleotides (Solvent: phenol/*iso*propanol/formic acid (85:10:100) (Boulanger and Montreuil, *340*)
1 uridylic acid
2 guanylic acid with a spot of phosphoric acid (above)
3 cytidylic acid
4 3′-adenylic acid
5 5′-adenylic acid
6 a mixture of the four nucleotides
7 yeast ribonucleic acid hydrolysed with ammonia. The fifth spot is due to free phosphoric acid.

TABLE 113. R_F values of nucleotides (Wyatt, 3637)

	Solvent							
	tert-Butanol-HCl[a]	*Isopropanol-HCl*[b]	*Acetone-tri-chloroacetic acid*[c]	*Isoamyl alc.-tetrahydrofurfuryl alc.-buffer*[d]	*Isobutyric acid-NH₃*[e]	*Phenol-tert-butanol-formic acid*[f]	*Na₂HPO₄-isoamyl alc.*[g]	*(NH₄)₂SO₄-buffer-isopropanol*[h]
Adenosine-2'-phosphate[i]⎫ Adenosine-3'-phosphate ⎬	0.50	0.48	—	0.35	0.49	0.70	⎧ 0.74 ⎨ 0.67	0.26 0.16
Guanosine-2'-phosphate ⎫ Guanosine-3'-phosphate ⎬	0.46	0.43	0.20	0.67	0.24	0.46	0.79	⎧ 0.50 ⎨ 0.40
Uridylic acid	0.80	0.77	0.51	0.43	0.24	0.35	0.85	0.73
Cytidylic acid	0.56	0.58	0.34	0.26	0.37	0.57	0.85	0.73
Adenosine-5'-phosphate .	—	0.43	0.37	0.28	0.43	—	0.69	—
Adenosine diphosphate .	—	—	0.10	0.07	—	—	0.77	—
Adenosine triphosphate .	—	—	0.04	0.08	—	—	0.83	—
Deoxycytidylic acid . . .	—	0.64	—	—	—	—	—	—
Thymidylic acid	—	0.81	—	—	—	—	—	—
Orthophosphate	0.90	0.84	0.61	—	—	0.22	—	—

[a] *tert.*-Butanol at 26° 700 ml, const.-boiling HCl 132 ml., water to make 1 liter; Whatman No. 1, descending; Smith and Markham, *Biochem. J.* (1950) **46**, 509. R_F values (20–22°) are from Boulanger and Montreuil, *Bull. soc. chim. biol.* (1951) **33**, 784, 791.

[b] See footnote *f*, Table 111. The value for orthophosphate is from Markham and Smith, *Biochem. J.* (1951) **49**, 401.

[c] Acetone 75 vol., 25% (wt./vol.) trichloroacetic acid 25 vol.; Whatman No. 1 paper washed in 2 N HCl and water, ascending, at 4°; Burrows *et al.*, *Nature* (1952) **170**, 800.

[d] *Iso*amyl alc. 1 vol., tetrahydrofurfuryl alc. 1 vol., 0.08 M potassium citrate buffer (pH 3.02) 1 vol.; Whatman No. 1, descending, 20–25°; Carpenter, *Anal. Chem.* (1952) **24**, 1203.

[e] *Iso*butyric acid 10 vol., 0.5 N NH₄OH 6 vol., pH 3.6–3.7; Schleicher and Schüll No. 597, descending, 21–25°; Magasanik *et al.*, *J. Biol. Chem.* (1950) **186**, 37. R_F values calcd. from the published relative mobilities and figures.

[f] 90% aq. phenol 84 vol., *tert.*-butanol 6 vol., formic acid 10 vol., water 100 vol.: after sepn. at a temp. 2-3° below that of the chromatography room, the nonaq. phase is used and the aq. phase is placed in the bottom of the tank; Whatman No. 1 paper, descending, 20–22°; Boulanger and Montreuil, *op. cit.*

[g] See footnote *i*, Table 111. Values for adenosine-5'-phosphate, diphosphate, and triphosphate are from W. E. Cohn and C. E. Carter, *J. Am. Chem. Soc.* (1950) **72**, 4273.

[h] Satd. (NH₄)₂SO₄ in water 79 vol., 0.1 M buffer soln. (pH 6) 19 vol., *iso*propanol 2 vol.; Whatman No. 1 paper, descending; Markham and Smith, *op. cit.* R_F values are measured from the published figure.

[i] The "a" and "b" nucleotides of Carter, *J. Am. Chem. Soc.* (1950) **72**, 1466, and subsequent authors are now identified as the 2'-phosphates and 3'-phosphates, respectively, of the ribosides.

spraying with the molybdate reagent of Hanes and Isherwood (*1297*). The above authors have applied this method to the analysis of the ribonucleotides of yeast and pancreas (*341*).

Montreuil and Boulanger (*2260*) have developed a rapid quantitative procedure for the analysis of the ribonucleotides of tissues. After hydrolysis with $N/2$ NaOH the nucleotides are adsorbed on Deacidite 200 (formate form), eluted with formic acid-formate and then separated on paper with phenol-*iso*propanol-formic acid-water (85:5:10:100); they are then estimated by the method of Boulanger and Montreuil (*340*).

Boulanger *et al.* (*343*) and Szafarz and Paternotte (*3189*) have described the paper chromatography of ^{32}P containing ribonucleotides; the content of the spots can be estimated by their radioactivity.

The separation of several uridine-phosphates has been described by Paladini and Leloir (*2437*), Kenner *et al.* (*1693*), Munch-Petersen *et al.* (*2315*), Herbert *et al.* (*1371*) and Storey and Dutton (*3132*).

Oligonucleotides have been separated by Smith and Allen (*3036*).

For the separation of cyclic (2′,3′)-cytidine phosphate and di- and trinucleotides of cytidine see Heppel and Whitfeld (*1366*), Heppel *et al.* (*1367*); for cyclic phosphates of adenosine, guanosine, uridine and cytidine, see Brown *et al.* (*407, 408*).

See also Table 113.

(ii) *Adenosine phosphates*

Carter (*517*) resolved, on paper, adenylic acid derived from yeast nucleic acid into two components which were identified as adenosine-2-phosphoric and adenosine-3-phosphoric acids. The synthetic isomers were separated on paper by Brown and Todd (*409*) using 5 % aqueous Na_2HPO_4-*iso*amyl alcohol or butanol-acetic acid-water (4:1:5).

Cohn and Carter (*602*) have described the paper chromatographic separation of adenine, adenosine, adenosine mono-, di- and triphosphate. Turba and Turba (*3328*) have separated ATP from its hydrolysis products (adenosine-diphosphoric and adenosine-5-monophosphoric acids) using the water-soluble mixture of *iso*butyric acid-water-acetic acid (100:50:1). Spots were printed on photographic paper by ultra-violet light (See Table 113). Snellman and Gelotte (*3040*) have obtained good resolutions of adenosine 2- and 3-phosphoric acid by using 0.5 % laurylamine in n-amyl alcohol.

Hersey and Ajl (*1375*) proved by paper chromatography that ^{32}P is incorporated into adenosine triphosphate by cell-free extracts of *E.coli*.

Eggleston and Hems (*872*) have described a method for the separation of the adenosine phosphates using two different solvents: 1) 90 ml *iso*propyl ether + 60 ml 90 % (v/v) formic acid; 2) 240 ml propanol, 120 ml ammonia (sp. gr. 0.880), 40 ml 0.002 M ethylenediamine tetraacetic acid, used successively to develop the material in the same direction, the first ascending, the second descending on the reversed strip. The spots are said to be more compact than with the usual twodimensional methods.

This method has been used by several authors (e.g. Paecht and Katchalsky, *2434*); Krebs and Hems (*1796*) have modified the method for the separation of the mono-, di- and tri-phosphates of adenosine and inosine. See also Turba *et al.* (*3323*), Lindberg and Ernster (*2015*). Adenine nucleotides can be detected specifically by the oxidation test of Gerlach and Döring (*1100*).

Giri *et al.* (*1118*) have described the circular paper chromatography of adenosine derivatives.

Deutsch and Nilsson (*757*) have separated the three phosphates of adenosine and inosine by a twodimensional method, using first propanol-ammonia-water (60:30:10), then saturated ammonium sulphate solution-water-*iso*propanol (79:19:2). "Pure" preparations of ATP obtained from

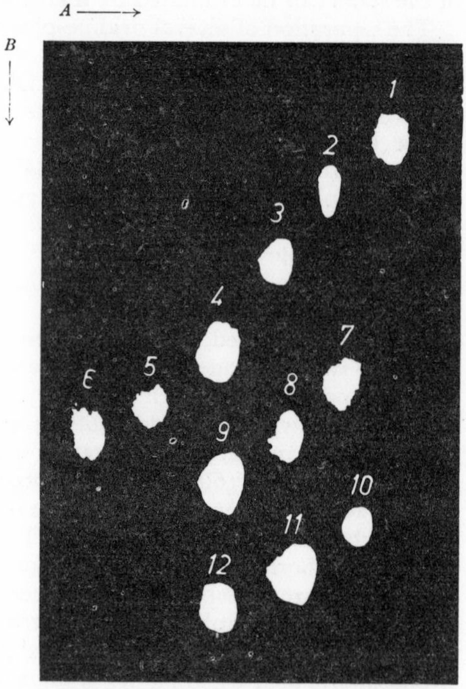

Fig. 99. Two-dimensional chromatogram (Whatman No. 1 paper) of a mixture of ATP, ADP, AMP, GTP, GDP, GMP, ITP, IDP, IMP, UTP, UDP and UMP, approximately 50 μg of each compound (Bergkvist and Deutsch, *205*)

(A) first solvent: *n*-propanol-ammonia-water (60:30: 10), 60 hours. (B) second solvent: saturated ammonium sulphate solution - *iso*propanol - water (79:2:19), 10 hours:

(1) AMP   (4) GMP   (7) IMP   (10) UMP
(2) ADP   (5) GDP   (8) IDP   (11) UDP
(3) ATP   (6) GTP   (9) ITP   (12) UTP

muscle contain guanosine triphosphate and uridine phosphate as contaminants; these can be detected by paper chromatography (Bergkvist and Deutsch, *201*). Quite recently, Bergkvist and Deutsch (*205*) have described a twodimensional system for the separation of all the mono-, di- and tri-phosphates of adenosine, guanosine, inosine and uridine (see Fig. 99).

Guérin (*1241*) has used propanol - butanol - conc. ammonia - water (36:30:10:24) for the separation of adenosine and inosine phosphates in one dimension.

(iii) *Pyridine and flavin nucleotides*

Hummel and Lindberg (*1486*) have described the simultaneous separation of adenylic acid, adenosine triphosphate, diphosphopyridine nucleotide, flavin mononucleotide and flavin-adenine dinucleotide, using a series of solvents. The location of the flavin nucleotides was observed under a "black light" mercury lamp and the areas showing a yellow green fluorescence were outlined with pencil. Compounds containing adenine were located by measurement of the ultra-violet absorption at 260 mμ. The diphosphopyridine nucleotide was localized in two ways: in the first method the strips were cut into 1 cm sections, each section clipped into pieces of 4 sq. mm and the activity determined enzymatically. The second procedure was the fluorometric measurement against a standard quinine solution, after destroying the fluorescence of the flavins by strong irradiation during alkaline incubation.

TABLE 114. R_F values of oxidised and reduced pyridine nucleotides (Burton and San Pietro, *478*).

The chromatograms were run with pyridine – water (2:1) for 16–17 hours at room temperature (ca. 23°) with a front movement of 33–35 cm

Compound	R_F
Desaminodiphosphopyridine nucleotide	
oxidized form	0.65
reduced form	0.39
Diphosphopyridine nucleotide	
oxidized form	0.63
reduced form	0.76
Triphosphopyridine nucleotide	
oxidized form	0.51
reduced form	0.24
Nicotinamide mononucleotide	0.33
Nicotinamide riboside	0.77
Adenosine diphosphoribose	0.80
2-Phosphoadenosine diphosphoribose . .	0.13
Adenosine.	0.89
2-Adenylic acid (adenylic acid a). . . .	0.28
3-Adenylic acid (adenylic acid b). . . .	0.28
5-Adenylic acid	0.41

The separation of diphospho and triphosphopyridine nucleotides on paper has been described by Pfleiderer and Schulz (*2504*) and by Boone et al. (*313*). See also Gerlach and Döring (*1100*). Burton and San Pietro (*478*) have separated the oxidized and reduced forms of pyridine nucleotides (see Table 114). See also the section on flavins, p. 395.

4. NUCLEIC ACIDS

Zamenhof and Chargaff (*3665*) have separated pentose nucleic acids from highly polymerized deoxypentose nucleic acids by adsorption of the former on charcoal. The latter is not adsorbed, due to its high molecular weight.

The heterogeneity of deoxyribonucleic acids was proven by Brown and Watson (*417*) by chromatography on "specific columns" of kieselguhr covered with histone (see also p. 34) and by Bendich *et al.* (*184*) with a synthetic cellulose exchanger ("*ECTEOLA* cellulose" prepared by heating alkaline cellulose with epichlorohydrin and triethanolamine); in both cases discontinuous and gradient elution gave very different results (Figs. 14a and 14b, p. 34). Brown and Martin (*416a*) have separated two chemically distinct fractions of a bacteriophage nucleic acid on a cellulose-histone column (Fig. 100). Lerman (*1969*) has purified the transforming principle of the pneumococcus on a column consisting of methylated serum albumin mixed with 20 parts of Celite; elution was effected with a gradient of NaCl or phosphate buffer.

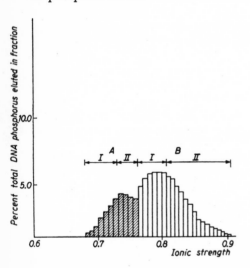
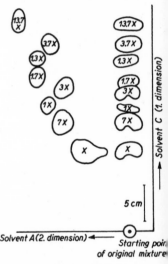

Fig. 100   Fig. 101

Fig. 100. Elution of *T2r* bacteriophage deoxyribonucleic acid from a column of cellulose-histone by a concentration gradient of sodium chloride at pH 6.8. The percentage of total deoxyribonucleic acid phosphorus eluted in a fraction is plotted against the ionic strength of the fraction (Brown and Martin, *416a*)

Fig. 101. Two-dimensional paper chromatogram of methylated xanthines. Location of the homologous xanthines after application of Solvent C is shown on the right side. Although 3-methyl- and 1,7-dimethylxanthines are closely connected, the two spots are distinguishable by the different colour of their complexes with mercury and diphenylcarbazone (Dikstein *et al.*, *776a*). Solvent A: 95% ethanol 85, acetic acid 5, water 10; Solvent C: pyridine 94, 25% NH$_3$ 6.

CHAPTER 34

Natural pigments

1. CAROTENOIDS

Chromatography has been an indispensable technique in the field of the carotenoids since 1931 (*1821, 1825*); it is used for the isolation of the pigments and for the separation and estimation of stereoisomers (*1899, 3674, 3679*). For details see the book of Karrer and Jucker (*1640*).

A special mention should be made of the work of Karrer and his collaborators concerning the synthesis and transformations of epoxides (*1638, 1639*), the discovery of naturally occurring epoxides (*1640, 1642*), and the synthesis of β-carotene (*1637*) which was achieved after chromatography of an intermediary product on alumina and of the final product on Ca(OH)$_2$ and ZnCO$_3$. Zechmeister and his co-workers have published an important series of papers on the numerous *cis-trans* isomers of the carotenoids, leading also to the isolation of the natural *cis*-forms of lycopene and γ-carotene (*1981, 3674, 3676, 3681, 3690, 3691*).

Porter and Zscheile (*2564*) were able to isolate, on MgO-Supercel, twelve isomers C$_{40}$H$_{56}$ from an extract of tomatoes (see also Porter and Lincoln, *2563*). Moore (*2264*) used an alkali-activated dicalcium phosphate for the separation of chlorophylls, xanthophylls and carotenes, Mackinney *et al.* (*2089*) a mixture of MgO and silicic acid whilst Mann (*2121*) used bone meal. Kemmerer and Fraps (*1685*) used Ca(OH)$_2$ to separate the crude carotene from plant extracts into "impurity A", "carotenoid X", β-carotene, neo-β-carotene and α-carotene; losses of pigments occurring during adsorption varied from 0 to 17 %. Wilkes (*3570*) used kieselguhr (Celite, Hyflo, etc.) as the adsorbent for the separation and estimation of carotene, which passed through into the filtrate, the xanthophylls remaining on the column.

Bickoff (*241*) has published a detailed study of solvents which can be used for the quantitative separation of various stereoisomers of β-carotene on chalk. The best separations were obtained with light petroleum containing 3 % anethole (Fig. 20, p. 44). Minor polyene pigments of plants have recently been studied by Goodwin (*1160, 1161*). Chromatography, controlled by following the bands by their fluorescence, enabled Zechmeister and Polgar (*3686*) to isolate *phytofluene*, a colourless carotenoid occurring in plants.

TABLE 115. R_F values of flavonoid compounds (Gage *et al.*, *1069*)

Compound \ Solvent	Ethyl acetate sat. with water	Phenol sat. with water	m, p-Cresols sat. with water	Chloroform sat. with water	Butanol Acetic acid Water
Flavonol aglycones					
Gossypetin	0.59	0.09	0.07	—	0.2
Kaempferol	0.90	0.74	0.53	0.17	0.8
Morin	0.71	0.65	0.17	—	0.8
Nortangeretin	0.79	0.92	0.72	—	0.7
Patuletin	0.81	0.13	0.43	—	0.7
Quercetagetin	0.17	0.06	0.09	—	0.2
Quercetin	0.81	0.42	0.22	0.05	0.7
Rhamnetin	0.92	0.71	0.63	0.96	0.8
Robinetin	0.41	0.25	0.10	—	0.5
Flavonol glycosides					
Gossypin	0.02	0.81	0.25	—	0.8
Gossypitrin	0.11	0.39	0.08	—	0.4
Isoquercitrin	0.40	0.51	0.24	0.05	0.7
Quercemeritrin . . .	0.50	0.52	0.27	—	0.7
Quercitrin	0.50	0.56	0.33	0.05	0.8
Robinin	0.21	0.55	0.35	0.14	0.5
Rutin	0.15	0.45	0.21	0.07	0.5
Xanthorhamnin . . .	0.02	0.70	0.27	0.08	0.5
Flavone aglycones					
Acacetin	0.94	0.96	0.95	0.94	0.9
Apigenin	0.87	0.89	0.87	—	0.9
Auranetin	0.92	0.93	0.99	0.94	0.9
Chrysin	0.86	0.93	0.95	—	0.9
Genkwanin	0.93	0.96	0.98	0.94	0.9
Isowogonin	0.92	0.74	0.99	—	0.9
Norwogonin	0.85	0.94	0.99	—	0.8
Oroxylin A	0.97	0.98	0.98	—	0.9
Wogonin	0.94	0.97	0.99	0.94	0.9
Flavanone aglycones					
Homoeriodictyol . . .	0.97	0.95	0.93	0.08	0.9
3,3′, 4′, 5, 7-Penta-hydroxyflavanone . .	0.79	0.62	0.42	—	0.8
Flavanone glycosides					
Hesperidin	0.77	0.85	0.40	0.17	0.4
Naringin	0.51	0.75	0.58	0.16	0.7
Neohesperidin	0.38	0.75	0.67	0.16	0.6
Flavane aglycones					
d-Catechin	0.91	0.35	0.16	0.11	0.7
l-Epicatechin	0.80	0.59	0.83	0.00	0.6
Miscellaneous compounds					
Hesperidin methyl-chalcone	0.79	0.96	0.96	—	0.8
Phloretin	0.93	0.81	0.84	0.81	0.9
2′,3,4-Trihydroxy-chalcone	0.93	0.85	0.88	0.77	0.9
Esculetin	0.78	0.85	0.76	—	0.8
Pomiferin	0.96	0.98	0.96	0.06	0.9

rm 20% alc. 40% · 40%	isoPropyl alcohol 22% Water 78%	isoPropyl alcohol 60% Water 40%	Acetic acid 15% Water 85%	Heptane 29% Butanol 14% Water 57%	Acetic acid 60% Water 40%
—	0.06	0.51	0.12	0.07	0.43
05	0.09	0.77	0.10	0.04	0.50
,12	0.26	0.58	0.27	0.13	0.68
—	0.08	0.60	0.10	0.04	0.54
—	0.10	0.76	0.10	0.06	0.50
—	0.24	0.61	0.19	0.13	0.63
,05	0.06	0.67	0.07	0.04	0.40
07	0.08	0.73	0.08	0.03	0.60
,05	0.07	0.58	0.08	0.03	0.32
—	—	—	—	—	—
,16	0.13	0.54	0.14	0.17	0.44
27	0.40	0.79	0.46	0.24	0.74
28	0.42	0.80	0.45	0.27	0.73
54	0.55	0.79	0.46	0.45	0.74
,77	0.72	0.76	0.77	0.71	0.84
,54	0.60	0.83	0.62	0.45	0.75
69	0.66	0.83	0.68	0.58	0.82
—	0.00	0.80	0.00	0.00	0.71
—	0.12	0.89	0.15	0.00	0.66
—	—	0.99	0.63	0.34	0.90
—	0.00	—	0.00	0.00	0.75
—	0.00	0.78	0.00	0.00	0.72
,02	0.00	0.87	0.00	0.00	0.81
,11	0.21	0.83	0.26	0.00	0.73
—	0.00	0.88	0.00	0.00	0.84
02	0.00	0.88	0.00	0.00	0.79
,32	0.49	0.92	0.55	0.29	0.80
—	0.57	0.87	0.13	0.00	0.73
80	0.79	0.63	0.82	0.17	0.89
81	0.75	0.86	0.80	0.77	0.88
80	0.79	0.88	0.81	0.72	0.90
—	0.65	0.79	0.41	0.51	0.68
—	0.54	0.72	0.55	0.38	0.68
—	0.82	0.95	0.92	0.14	0.89
—	0.45	0.91	0.42	0.00	0.73
08	0.15	0.81	0.19	0.06	0.68
—	0.66	0.73	0.60	0.38	0.74
—	0.00	0.89	0.00	0.00	0.78

Meunier (*2195*) has dealt with the relationship between chromatography and mesomerism; he considers the green or blue colour given by carotenoids and Vitamin A on certain bleaching earths to be analogous to the blue colour given with antimony trichloride. (See also Zechmeister and Sandoval, *3689*.) Meunier and Vinet (*2199*) dealt in detail with the ionisation adsorption of Vitamin A and carotene from the point of view of the theory of mesomerism.

Bauer (*160*) has recently described the *paper chromatography* of carotenoids and chlorophylls. The carotenes migrate with the solvent front (benzene).

Several papers describing the separation of chloroplast pigments on paper are mentioned in the chapter on chlorophyll. See also Grangaud and Garcia (*1186*).

The industrial preparations of carotene from alfalfa meal by chromatography on charcoal (using columns 60 cm in diameter and 2.5 metres high) has been described by Shearon and Gee (*2951*) and by Williams and Hightower (*3580*).

2. ANTHOXANTHINS AND ANTHOCYANINS

Ice and Wender (*1495*) reported the successful separation of flavonoid compounds by adsorption on magnesol columns (from acetone solution, elution with ethyl acetate saturated with water).

Gage et al. (*1070*) have purified flavonoid pigments on columns of Amberlite IRC-50(H); adsorption in aqueous, neutral or alkaline solution was followed by elution with alcohol.

Williams and Wender (*3576*) have isolated kaempferol and quercetin from strawberries on Amberlite IRC-50 and on paper. Freudenberg and Hartmann (*1037*) have used the same exchange resin, as well as chromatopacks for isolation of flavones and dihydroflavonols.

Spaeth and Rosenblatt (*3055*) separated anthocyanins by partition on columns of silicic acid.

Gage, Wender et al. (*1069, 1070, 3505*) have described the chromatography of flavonoid pigments on *paper*; the location of the spots can be facilitated by observation in ultra-violet light, or by spraying with Na_2CO_3, lead acetate, etc. (Table 115). The glycosides of quercetin and isoquercetin can also be separated (Douglass et al. *804*; see also Naghski et al., *2334*). In an analysis of flavonol-3-glycosides, the pigments were located under UV-light, leached from the paper with $AlCl_3$, and the absorption intensities of the flavonoid aluminium complexes measured on a Beckmann spectrophotometer (Gage and Wender, *1071*). For paper chromatography of *flavonoids* and their glycosides see also: Lindstedt (*2027*), Oshima and Nakabayashi (*2422*), Nordström and Swain (*2381*), Krebs and Wankmüller (*1797*), Geissman et al. (*1095*), Gripenberg (*1227*), Paris (*2443*), Puri and Seshadri (*2602*), Siegelman (*2981*).

Bate-Smith (*157*) has given numerous R_F values of anthocyanins (see Table 116).

TABLE 116. R_F values of anthocyanins and anthocyanidin chlorides (Bate-Smith, *157*)

		R_F values in		
		Butanol-acetic acid	*2N HCl*	*m-Cresol-acetic acid*
Apigeninidin		0.82	—	1.0
Cyanidin:		—	0.69	—
-3-glucoside	Chrysanthemin	0.33	0.27	Trail
-3-rhamnoglucoside	Antirrhinin	0.37	0.28	0.25
-3-gentiobioside	Mekocyanin	0.29	0.22	0.18
-3,5-diglucoside	Cyanin	0.16	0.08	0.19
Delphinidin:		—	0.35	—
-3-monoside	ex Verbena	0.16	0.14	0.11
-3,5 diglucoside	Delphin	0.11	0.06	0.03
Hirsutidin:		—	0.72	—
-3-glucoside		0.61	—	—
-3,5-diglucoside	Hirsutin	0.38	0.07	0.69
Malvidin:		—	0.53	—
-3-glucoside	Oenin	0.40	0.23	0.75
-3-galactoside	Primulin	0.40	0.24	0.76
-3,5-diglucoside	Malvin	0.22	0.07	0.54
Pelargonidin:		—	0.80	—
-3-glucoside	Callistephin	0.59	0.52	0.67
-3,5-diglucoside	Pelargonin	0.34	0.20	0.42
Peonidin:		—	0.72	—
-3-glucoside	Oxycoccicyanin	0.47	0.31	0.72
-3,5-diglucoside	Peonin	0.26	0.10	0.48

Bate-Smith and Westall (*158*) have studied the behaviour on paper of a large number of simple polyphenols, of natural and synthetic C_{15} compounds (flavanols, flavones, anthocyanidins, and chalcones), as well as their glycosides. The R_F values are related to the nature and number of substituent groups, in such a way that in many instances a straight line is obtained when $\log (1/R_F - 1)$ is plotted against the number of substituent groups of any one kind. This relationship, already predicted by Martin (*2150*), follows from the relationship between constitution and partition coefficient. The symbol R_M is suggested for the function $\log (1/R_F - 1)$ (see p. 117).

Several instances of departure from regularity in the above relationship can be attributed to constitutional factors, especially *ortho* or *vicinal* arrangement of substituent groups.

For paper chromatography of *anthocyanins*, see also Forsyth and Simmonds (*1007*), Ribéreau-Gayon and Ribéreau-Gayon (*2674*), Dupuis and Puisais (*833*).

3. QUINONES

(a) *Naphthoquinones*

The echinochromes and spinochromes, naphthoquinone pigments of sea urchins, can be adsorbed and separated on calcium carbonate; they are eluted by dissolving the adsorbent in dilute HCl (*1823, 1902, 1905, 3463*). The degree of adsorption depends on the alkalinity of the calcium carbonate(*1902*).

Green and Dam (*1211*) have reported the adsorptiochromism of certain hydroxynaphthoquinones on alumina; it is possible that the free alkali of the adsorbent causes the colour shift.

The *paper chromatography* of substituted naphthoquinones has been described by Nakanishi and Fieser (*2337*) and by Sproston and Bassett (*3074*) (Table 117); Green (*1209*) has stated that nineteen alkyl, hydroxyl, or chloro substituted 1,4 naphthoquinones (including vitamin K) could be detected on paper by their red fluorescence; in solution no fluorescence is visible. (See also Vitamin K, p. 395.)

(b) *Anthraquinones*

Brockmann *et al.* (*395*) used calcium oxalate or silicic acid for the purification of naturally occurring anthraquinones, Seebeck and Schindler (*2930*) isolated a bioside of emodine from the bark of *Rhamnus frangula*, after chromatography of its acetate on silicic acid or Floridin XXF. See also Bandemer (*132*). The behaviour of hydroxyanthraquinones on silica gel columns has been studied by Hoyer (*1465–1467*) in relation to hydrogen bonding (see p. 47).

Shibata *et al.* (*2966*) described the *paper chromatography* of anthraquinone pigments; the yellow spots can be sprayed with 0.5 % alcoholic magnesium acetate and acquire pink, red or violet shades depending on the number and position of the hydroxyl groups in the pigments. See also Tsukida and Yoneshiga (*3313*). Danilovič (*695*) has described the paper chromatography of anthraquinone galenicals.

4. PORPHYRINS

Falk (*933*) has recently reviewed chromatographic methods used for separation of porphyrins, and outlined the possibilities of separating natural and "unnatural" isomers.

The esters of these pigments lend themselves well to purification by chromatography on talc or alumina (Fischer and Conrad, *967*; Grinstein and Watson, *1225, 1226*; Comfort and Weatherall, *610*). For the purification of uroporphyrins, see Watson *et al.* (*3480*), for the separation of protoporphyrin and mesoporphyrin, see Lederer and Tixier (*1913*); Rimington

TABLE 117. R_F values of some substituted naphtho- and benzoquinones (Sproston and Bassett, *3074*)

Solvent: amyl alcohol-pyridine-water (3:2:1.5)

The colour reactions are obtained by spraying with 5% aqueous NaOH

Compound	R_F	Color[a]
1,4-Naphthoquinones		
Unsubstituted.	0.85	OB
2-Methyl-	0.87	OB
2-Méthoxy-	0.81	O
2,3-Dimethoxy-	0.84	PuP
2-Methoxy-3-hydroxy-	0.37	PuP
2-Hydroxy-	0.47	O
2-Methyl-3-acetyl-	0.83	PaB
3-Methyl-6-succino-	0.53	P
2,3-Dichloro-	0.41	Y
2-Chloro-3-methoxy-[b]	0.33	O
	0.86	PaO
2-Chloro-3-ethoxy-[b]	0.27	PaRO
	0.89	RO
Fe+++ salt of 2-chloro-3-hydroxy-	0.42	Op
2-Amino-	0.82	YB
2-Amino-3-chloro-	0.83	O
2-Acetylamino-	0.83	POB
2-Dimethylamino-	0.83	O
2-Chloro-3-dimethylamino-	0.87	OP
2-Chloro-3-ethylamino-	0.88	PO
2-Chloro-3-*n*-decylamino-	0.91	PaY
2-(*N*-Acetanilido)-3-chloro-	0.89	PaOB
2-Methylmercapto-	0.84	Y
2-Mercapto-3-chloro-[b]	0.43	Y
	0.95	PaY
2-Methyliminonaphtho(2,3)-1,3-dithiole-4,9-dione-metho-chloride monohydrate	0.92	PaYB
1,2-Naphthoquinone[b]	0.46	P
	0.84	PaY
Benzoquinones		
p-Benzoquinone	0.87	PaB
p-Toluquinone	0.85	YB
p-Xyloquinone	0.97	PaB
2,5-Dichloro-*p*-benzoquinone	0.28	PaY
	0.93	PaY
2,5-Dichloro-3,6-dicarbethoxy-*p*-benzoquinone	0.94	PaY
2,3,5,6-Tetrachloro-*p*-benzoquinone	0.22	PaY
3,4,5,6-Tetrachloro-*o*-benzoquinone	0.90	PaYB
Others		
2,2,3,4,4-Pentachloro-1-ketotetrahydro-naphthalene. . . .	0.90	PaPB

[a] Pa, pale; P, pink; B, brown; Y, yellow; O, orange; R, red; Pu, purple.

[b] Impure compounds producing two spots on chromatographing.

(*2690*) has separated free uro- and copro-porphyrins on kieselguhr (solvent 1 % HCl, development with a pH 5 buffer). Tixier (*3275, 3276*) has purified on alumina different porphyrin pigments of invertebrates.

Nicholas (*2364*) has studied the chromatographic behaviour of porphyrin esters, under standardized conditions, on columns of MgO, MgCO$_3$ and CaCO$_3$.

Lucas and Orten (*2053*) have described a method for the separation and crystallisation of the natural porphyrins by partition chromatography on hydrated silica gel. The porphyrins were eluted in the order proto-, meso-, copro-, uro-porphyrin, by increasing the chloroform concentration in the solvent from 20 to 50 %.

Morrison *et al.* (*2284*) have separated two different *haemins* (obtained from pure cytochrome-c) on silicic acid columns.

Nicholas and Rimington (*2366, 2367*) have applied *paper chromatography* to the separation of free porphyrin using a lutidine-water system; their method separates the pigments into groups corresponding to the number of carboxyl functions in the molecule. Using a graph established with porphyrins having 2, 4, and 8 carboxyl groups, they have revealed the presence, in different animal excretions, of hitherto unrecognised porphyrins with three, five, six, or seven carboxylic functions (Table 118). Nicholas and

TABLE 118. R_F values of porphyrins (Nicholas and Rimington, *2366*)
Solvent: lutidine at 19° in an atmosphere of ammonia vapour

Porphyrin	R_F value	Number of carboxylic groups
Uroporphyrin	0.3	8
Coproporphyrin	0.6	4
Protoporphyrin.	0.8	2
Deuteroporphyrin	0.8	2
Mesoporphyrin.	0.8	2
Haematoporphyrin	0.8	2
Phylloerythrin	0.9	1
Monazoprotoporphyrin	0.8	2
All porphyrin esters.	1.0	0

Comfort (*2365*) used this method for analysing the porphyrins of molluscan shells; they could find no support for the existence of the pentacarboxylic "conchoporphyrin". McSwiney *et al.* (*2098*) have studied the porphyrins of acute porphyria.

Eriksen (*912*) has modified the method of Nicholas and Rimington (*2366*) using a mixture of 2,6-lutidine and water, which gives a single phase system with water thus diminishing the temperature sensitivity of the system. See also Kehl and Stich (*1674*), Comfort and Weatherall (*610*), Kennedy (*1691*). Falk and Benson (*935*) have described the paper chromatography of highly carboxylated porphyrins.

Nicholas and Rimington (*2368*) have examined the porphyrin mixture resulting from the removal of copper from turacin, by chromatography on MgO and paper and have isolated the methyl ester of uroporphyrin III.

The chromatographic behaviour of a "Waldenström porphyrin" (a mixture of uroporphyrin I and a type III porphyrin) on MgO, CaCO₃ and paper has been studied by Watson *et al.* (*3478, 3479*).

Chu *et al.* (*571*) have described the paper chromatography of methyl esters of porphyrins. By using ascending movement and two successive developments with chloroform-kerosene and then with *n*-propanol-kerosene, they could separate the methyl esters of coproporphyrin I and III, as well as others; the spots were detected by their fluorescence in ultra-violet light (Table 119).

TABLE 119. R_F values of methyl esters of porphyrins (Chu *et al.*, *571*)
Solvent: (1) 4 ml pure alkane and 2.6 ml chloroform (25 mins) followed in the same direction by (2) 5 ml alkane and 1 ml of *n*-propyl alcohol

Methyl ester of	n-Decane	n-Dodecane	n-Tetradecane	n-Hexadecane
Uroporphyrin I	0.14	0.20	0.13	0.15
Coproporphyrin I . . .	0.42	0.52	0.45	0.47
Coproporphyrin III. . .	0.70	0.76	0.74	0.66
Protoporphyrin IX . . .	0.86	0.92	0.92	0.89
Mesoporphyrin IX . . .	0.92	0.96	0.95	0.93

Chu and Chu (*569*) have described several solvent systems for the separation of the dimethyl esters of dicarboxylic porphyrins. Falk and Benson (*934*) have separated uroporphyrin esters I and III, Kehl and Günther (*1672, 1673*) coproporphyrins I and III. For radial chromatography of porphyrins see Serchi and Rapi (*2946*) as well as Rappoport *et al.* (*2626*) who also separate the methyl esters of coproporphyrins I and III. (Solvents: petroleum ether-CHCl₃, 3:1, or heptane-ethylene dichloride-*tert*.butanol, 20:1:1,5.)

The paper chromatography of the iron complexes of porphyrins has been described by Chu and Chu (*570*).

5. BILE PIGMENTS

Siedel *et al.* (*2975–2977*), as well as Schwartz and Watson (*2916*) have described the purification of both natural and synthetic bile pigments. Tixier (*3274*) has purified the methyl ester of biliverdin on alumina, and With

(*3605*) attempted the chromatographic purification of bilirubin. For the fractionation of the pigments of Aplysia, see (*1906*) and for the fractionation of the green pigments of fish skeletons, see Willstaedt (*3589*).

Kench *et al.* (*1687*) separated protoporphyrin from biliverdin on columns of silica gel, with butanol/CHCl₃ as solvent. Cole *et al.* (*605*) have described the separation of bile pigments on siliconized kieselguhr. Tixier (*3276*) has described the paper chromatography of bile pigments; here, as with the porphyrins, the pigments are separated into groups, depending on the number of carboxyl groups in the molecule. See also Stich *et al.* (*3116*). Gries *et al.* (*1223*) have described the separation of direct and indirect bilirubin on paper (see also Mendioroz *et al.*, *2187*).

For the paper chromatography of porphobilinogen, see Ågren and Verdier (*21*).

6. CHLOROPHYLLS

Strain *et al.* succeeded in isolating new chlorophylls by chromatography on powdered sugar: chlorophyll C from diatoms and dinoflagellates (*3143, 3145*); chlorophyll D from red algae (*2123*); chlorophylls a' and b' (*3144*). Wendel (*3504*) has used starch columns. Fischer and Gibian (*968*) purified some chlorophyll derivates on alumina. Bonelline, the green pigment of *Bonellia viridis* was obtained crystalline after adsorption on CaCO₃ (*1900*).

Paper chromatography of chloroplast pigments has been developed with success by several authors (Asami, *85*; Bauer, *160*; Sapozhnikov *et al.*, *2819*; Sironval, *3001*; Douin, *805*; Chiba and Naguchi, *565*). Strain (*3142*) states that the interface of two immiscible liquids, such as water and petroleum ether exhibits selective affinity for chloroplast pigments. Chlorophylls and carotenoids were separated by sorption on moist or dry glass or cellulose paper, by partition between methanol (80 %) on the paper and petroleum ether, and by partition between petroleum ether plus vaselin on paper and 80 % methanol.

Lind *et al.* (*2012*) develop chloroplast extracts in the first direction with a succession of solvents, then in a second direction with another solvent and obtain six well separated areas: carotenes, xanthophylls, chlorophylls a and b, an unknown pigment and a colourless fluorescing compound.

Sporer *et al.* (*3073*) and Freed *et al.* (*1028*) have separated chlorophylls a and b on paper impregnated with sucrose.

Chiba (*564*) has reported the separation of two chlorophyll lipoproteins on paper, with picoline-water; the chlorophylls isolated from these two fractions were identified after separation on paper.

7. URINARY PIGMENTS

Lepkovsky and Nielsen (*1964*) isolated a green pigment from the urine of rats deficient in pyridoxin, after chromatography on paper pulp. Lepkovsky *et al.* (*1965*) isolated xanthurenic acid from the urine of rats deficient in tryptophan after chromatography on paper pulp or cotton. Rimington (*2691*) separated indigotin and indirubin on alumina; the pigments were identified by means of mixed chromatograms (See also *884*).

Urochrome B has been analysed by Stich and Stärk (*3117*); Heikel *et al.* (*1352*) have described the paper chromatography of uroerythrin. Urinary indicans and indicanoids have been separated by Decker (*721*). For urine melanogens see Leonhardi (*1962*).

8. PTERINS

Uropterin (xanthopterin) has been purified by Koschara (*1770*). Polonovski *et al.* (*2547*) isolated fluorescyanin (ichthyopterin) from fish scales by adsorption on Franconite and separated it from riboflavin, which was less adsorbed.

Good and Johnson (*1157*), as well as Tschesche and Korte (*3309*, Table 120) have described the paper chromatography of natural pterins and

TABLE 120. R_F values of pterins (Tschesche and Korte, *3309*)
   Paper: SS 2043a; Solvent: 3% aqueous ammonium chloride solution

Xanthopterin	0.68
9-Methylxanthopterin	0.39
Pteroylglutamic acid	0.19
9-Hydroxypteroylglutamic acid	0.38
6,9-Dihydroxy-2-aminopteridine-8-aldehyde	0.40
6-Hydroxy-2-aminopteridine-8-aldehyde	0.61
6-Hydroxy-2-aminopteridine-8-carboxylic acid	0.60
6-Hydroxy-2-aminopteridine-9-carboxylic acid	0.57
Xanthopterin carboxylic acid	0.40
Isoxanthopterin carboxylic acid	0.30
6,9-Dihydroxy-2-amino-8-acetonylpteridine	0.14
Erythropterin	0.33
Leucopterin	0.27
Pterorhodin	0.25

some of their derivatives, as well as of synthetic analogues (solvent: butanol/acetic acid). Albert *et al.* (*32*) have used paper chromatography for identification and as a criterium of purity of synthetic pteridines. They found, for instance, that 2,4-dihydroxypteridine, previously described as yellow needles, gave 5 spots on paper and when sufficiently purified consisted of white needles (see Table 121, also Günder, *1242*). (See also Folic acid, p. 400.)

TABLE 121. R_F values of pteridines (Albert *et al.*, *32*)
Solvent: *n*-butanol, 2 vol., 5*N* acetic acid, 1 vol.

Compound	R_F value
Pteridine	0.25
2-Aminopteridine	0.70
4-Aminopteridine	0.70
2-Dimethylaminopteridine.	0.90
2-Hydroxypteridine.	0.80
4-Hydroxypteridine.	0.50
4-Hydroxy-6,7-dimethylpteridine	0.60
2,4-Hydroxypteridine	0.50
4-Mercaptopteridine	0.70

Wald and Allen (*3448*) have chromatographed the red eye pigments of Drosophila on powdered talc. Heymann *et al.* (*1383*) have described the partition chromatography of these pigments on silica; ten closely related pigments were separated. Forrest and Mitchell (*1001*) have used "Filtrol Grade 58" for the isolation of pteridines from Drosophila.

9. MISCELLANEOUS

Willstaedt (*3590*) purified lactaroviolin, $C_{15}H_{14}O$, the violet azulene pigment of a lactary as well as its oxime and the corresponding nitrile by adsorption on alumina. For *azulenes*, see papers by Plattner *et al.*, for instance (*2524*). The paper chromatography of azulenes has been described by Knessl and Vlastiborová (*1745*).

For the isolation of Evans Blue from blood plasma, see Morris (*2280*).

Schmidli (*2861*) has obtained two well defined dark brown zones on filtering a 0.1 *N* NaOH solution of natural *melanin* on Floridine XXF.

CHAPTER 35

Vitamins

Schoen (*2870*) has published a review on the paper chromatography of the vitamins and their derivatives, with detailed R_F tables.

1. LIPOSOLUBLE VITAMINS

(a) Vitamin A

The calorimetric method used by Müller (*2305, 2306*) to prepare and control the activity of alumina has already been mentioned (see p. 27). The main application of this technique has been in the separation of the natural esters of vitamin A in fish liver oils. As vitamin A is destroyed by alumina of too high an activity, Müller used a column consisting of alumina of three different activities: the top section had $Q = 50$ cal and held back strongly adsorbed impurities; then came a section $Q = 56.5$ which adsorbed the vitamin A ester; then a section of still higher activity ($Q = 83.5$ cal) which retained all the substances remaining in solution. The light petroleum solvent used left the column completely pure and was used for about twenty subsequent separations. This method could be modified for the separation of vitamin A and carotene, etc. More recently, Müller (*2307*) has used columns consisting of alumina of five different activities for the purification of vitamin A and D and their separation from esters of vitamin A and from carotene. Wilkie and Jones (*3572*) have described the standardization of adsorbents for chromatography of vitamin A. For the separation of vitamin A and carotene on Na_2CO_3 see *634*.

Many fish liver oils of low vitamin A content do not show the characteristic 328 mμ absorption of vitamin A but chromatography of these oils on partially deactivated alumina allows the separation of spectroscopically pure vitamin A ester (*1912*). See also Servigne et al. (*2948*) and *809*.

Eden (*860*) has described a micro-method for the quantitative separation of free and esterified vitamin A. The ester is not adsorbed on alumina and the alcohol can be eluted with 20 % acetone in petroleum ether. Defatted bone meal has been used with success for the separation of free and esterified vitamin A (Glover et al., *1139*).

A patent of Robeson (*2700*) describes the isolation of neo-vitamin A by chromatography on Doucil.

Green and Singleton (*1208*) have described a rapid chromatographic method for determination of vitamin A in whale liver oils.

Meunier and Vinet (*2198*) have studied the red pigment formed during the chromatography of vitamin A on alumina, which seems to be bis-axerophthyl ether, formed by the elimination of water from two molecules of vitamin A. Wald (*3447*) has shown that vitamin A is oxidized to retinene by chromatography on MnO_2. For the chromatography of retinene, see *1336*.

Green (*1204*) has studied the adsorption and destruction of vitamin A on Floridin. A single adsorption on this earth can quantitatively separate the vitamins A and D, the latter being recovered without loss.

A patent of Embree and Hawks (*892*) describes the concentration of the vitamins A and D by selective adsorption on sodium aluminium silicate; vitamin A was eluted with light petroleum, vitamin D with acetone.

Vitamins A_1 and A_2, having respectively five and six conjugated double bonds, are inseparable by chromatography; the observation of Embree and Shantz (*893, 1574*) that the two "anhydrovitamins" are separable, is now explained by the finding of Henbest *et al.* (*1362a*) that "anhydrovitamin A_2" is in reality ethoxyanhydrovitamin A_1.

For the chromatography of hepaxanthine, see Karrer and Jucker (*1640*). Barua and Morton (*153*) have separated vitamin A esters from kitol esters by chromatography on alumina.

Datta and Overell (*706, 707*) have studied the separation and detection of vitamin A and derivatives (esters, anhydro-vitamin A, retinene) by chromatography on alumina-treated *filter paper*. Amounts of vitamin A down to 1 μg may be readily identified.

J. A. Brown (*420*) has described the determination of vitamins A and E on silicone impregnated paper, after separation with acetonitrile-water as mobile phase; the vitamins are located by means of an automatic spectrophotometer; quantitative results are obtained by measuring the zonal areas on the graph and comparing with standards.

(b) *Vitamin D*

DeWitt and Sullivan (*767*) purified vitamin D from fish liver oils by adsorption on a column of MgO/Celite (1:1) and controlled the progress of adsorption by examining the column in ultra-violet light (see also *892*).

Ewing *et al.* (*930*) have described a two-step chromatographic process for the separation of vitamin D from other nonsaponifiable oil components. The solution is first filtered on a column of Superfiltrol (activated bentonite, Filtrol Corp., Los Angeles) which removes vitamin A, carotenoids and some sterols and then on activated alumina which retains certain polyenes, as well as other substances; vitamin D can then be estimated in the filtrate by spectrophotometry at 265 mμ (see also Burnett, *462* and Schlabach, *2847*).

Brüggemann *et al.* (*427*) have used a talc column impregnated with $SbCl_3$ to separate vitamins A and D (the former being retained on the column in a dark blue zone).

Thibaudet (*3228*) has reported the transformation of calciferol to tachysterol by acid earths.

Green (*1205*) has studied in detail the estimation of vitamin D in irradiated products and in fish liver oils by adsorption on floridin earth. This adsorbent must be specially prepared for the determination of vitamin D (*1206*).

Rossi (*2738*) has reported the *paper chromatography* of vitamin D_2 and McMahon *et al.* (*2092*) have separated on paper the vitamin D_2 and D_3 and 7-dehydrocholesterol (solvents: phenol and methanol-water; detection by spraying with $SbCl_5$ in chloroform).

Davis *et al.* (*716*) have separated vitamin D from other steroids on paper impregnated with quilon (solvent: methanol-water-ether 65:10:25; detection of the spots with a 40 % solution of $SbCl_3$ in $CHCl_3$). Vitamin D can also be detected on paper by ultraviolet photography (Fischer *et al.*, *969*). Kodicek and Ashby (*1752*) have used paper impregnated with liquid paraffin (Table 122). The same authors (*1753*) have described the estimation of vitamin D after preliminary adsorption on MgO-kieselguhr (1:1) and subsequent chromatography on paper.

TABLE 122. R_F values of vitamin D, related sterols and tocopherols (Kodicek and Ashby, *1752*)

Development: Ascending method for 16–17 hours.

Paper impregnated with liquid paraffin B.P.; Whatman No. 2 sheets dipped into a 5 % solution of liquid paraffin in light petroleum.

	Ethylene glycol monoethyl ether 35 n-Propanol 10 Methanol 30 Water 25	n-Propanol 15 Methanol 82 Water 3	Methanol 95%
Ergocalciferol	0.44	0.66	0.69
Cholecalciferol	0.46	0.68	0.68
7-Dehydrocholesterol.	0.39	0.57	0.56
Ergosterol	0.0	0.56	0.0
Lumisterol	0.27	0.59	0.52
Suprasterol II.	0.36	0.52	0.57
Cholesterol	0.29	0.49	0.50
Sitosterol	0.0	0.51	0.42
Zymosterol	0.41	0.61	0.48
α-Tocopherol	0.15	0.58	—
γ-Tocopherol	0.26	0.72	—
δ-Tocopherol	0.45	0.77	—
α,β,γ-Carotenes	0	0	0
Tachysterol.	0.37	0.61	0.58

(c) *Vitamin E*

Meunier and Vinet (*2197*) removed the carotenoids which interfere with the estimation of tocopherols by filtering the unsaponifiable material to be estimated through a column of Jagolite (montmorillonite from Bezenet, Allier, France). The carotenoids were adsorbed by the column in a green zone (see also Emmerie and Engel, *897*; Tošić and Moore, *3289*).

Kofler (*1757*) removed the carotenoids by filtration on alumina; the carotenes were not adsorbed, the tocopherols were eluted with benzene and the xanthophylls with methanol. In order to estimate the tocopherols, they were condensed with *o*-phenylenediamine to give a phenazine derivative which was estimated by fluorescence. This phenazine derivative can also be purified by adsorption on alumina, the elution being controlled by observation in Wood's light (see also Lieck and Willstaedt, *2010*). More recently Kofler (*1759*) described the separation of α-, β-, γ-, and δ-tocopherols by chromatography on alumina treated with $SnCl_2$ (this treatment prevents the autoxidation of the tocopherols on the column). The adsorbability increases in the series α-tocopherol (5,7,8-trimethyltocol), 5,7-dimethyltocol, β-tocopherol (5,8-dimethyltocol), γ-tocopherol (7,8-dimethyltocol) and δ-tocopherol (8-methyltocol). Kofler (*1759*) has suggested an explanation for this order of adsorbability: the adsorption depends mainly on the phenolic hydroxyl in position 6 (I). Methyl groups in neighbouring positions decrease the adsorption by steric hindrance. As would be expected, trimethyltocol is the least adsorbed and monomethyltocol the most strongly adsorbed. For the isolation of natural δ-tocopherol, see (*3100*).

(I)

Quaife (*2605*) has described the determination of the individual tocopherols by separation of the nitroso derivatives of the β, γ and δ compounds on zinc carbonate-celite; the concentration of α-tocopherol is determined by difference. Emmerie (*895*) has shown that α- and γ-tocopherols may be separated from one another by chromatography on alumina or floridin.

A patent of Baxter and Robeson (*165*) describes the separation of the different tocopherols on activated clays, kaolin, silica gel, Na-Al silicate, or oxides of Al, Ca and Mg.

F. Brown (*411*) has described the separation of the α, β, and δ-tocopherols by *reversed phase paper chromatography* on vaselin coated Whatman No. 1 paper (mobile phase 70 % aqueous ethanol, detection of the spots by spraying with 2,2′-dipyridyl in ethanol, followed by ferric chloride; bright red spots are formed on a white background). This method has been developed for the quantitative estimation of synthetic and natural mixtures of tocopherols (Brown, *412*). See also Eggitt and Ward (*871*). As already mentioned in the section on vitamin A, J. A. Brown (*420*) has reported a method of separating vitamins A and E on silicone impregnated paper, with acetonitrile-water; the substances are located and estimated by ultraviolet spectrophotometry. See also Kodicek and Ashby (*1752*) and Table 122.

Guerillot *et al.* (*1240*) have used paper impregnated with ferric undecylate to separate the tocopherols.

(d) *Vitamin K*

Karrer *et al.* (*1636*) have isolated natural vitamin K from alfalfa after chromatography on $MgSO_4$ and on $ZnCO_3$. Kofler (*1758*) has described a fluorimetric method for the estimation of 2-methyl-1,4-naphthoquinone after condensation with *o*-phenylenediamine and purification of the product of condensation on alumina. For the adsorption of Vitamin K on Decalso, see (*246*). Dam and Glavind (*691*) describe in a patent the purification of vitamin K on $CaSO_4$, $BaSO_4$ or alumina. Dam and Lewis (*692*) have drawn attention to the destruction of vitamin K on alumina and MgO. (See also *2682*.)

Green and Dam (*1211*) have studied in detail the *paper chromatography* of vitamin K and related compounds on siliconized paper (see Table 123).

Hais (*1267*) has described the paper chromatography of haemorrhagic derivatives of 4-hydroxycoumarin.

2. WATER-SOLUBLE VITAMINS

Brown and Marsh (*421*) have separated thiamine salts, riboflavin, nicotinamide and pyridoxine hydrochloride by paper chromatography (solvent: upper phase of *n*-butanol-acetic acid-water,40:5:55) and have used quantitative spectrophotometry for estimation of the four vitamins.

Radhakrishnamurty and Sarma (*2611*) use butanol, methanol, benzene, water (2:1:1:1) for the separation of most of the B vitamins on paper.

Harrison (*1318*) has given a general description of the microbiological assay of growth factors after their separation by paper chromatography.

(a) *Thiamine and riboflavin*

Conner and Straub (*612*) have separated thiamine and riboflavin on a composite column, the former being held on the top section of Decalso and the latter on the lower section of Supersorb. Herr (*1374*) adsorbed thiamine on Amberlite IR-100-H, riboflavin not being adsorbed.

Siliprandi and Siliprandi (*2986*) have separated thiamine, its mono-, di- and triphosphoric esters by chromatography on starch and ion exchange columns and on paper. See also Malyoth and Stein (*2118*). The paper

TABLE 123. R_F values and fluorescence of 2-methyl-1,4-naphthoquinone and its derivatives (Green and Dam, 1210)

Formula of group in 3-position or designation of compound	Solvent system			Colour of fluorescence		
	Ethyl alcohol-acetic acid-water (750:25:225)	Isopropyl alcohol-acetic acid-water (600:25:375)	n-Propyl alcohol-acetic acid-water (600:25:375)	Before activation	After activation	After activation and spraying with KOH
—H (menadione)	0.71	0.82	0.89	red	blue	green
(dimer of menadione)	0.73	0.88	0.97	red	blue	green
—OH (phthiocol)	0.63	0.72	0.80	red	red	cherry red in visible light
—CH₂—CH=C—CH₂—CH₂—CH=C—CH₃ / CH₃	0.55	0.66	0.73	red	green	orange
—CH₂—CH=C—[CH₂—CH₂—CH₄—CH=C]₂—CH₃ / CH₃	0.49	0.58	0.68	red	green	orange
—CH₂—CH=C—[CH₂—CH₂—CH₂—CH]₂—CH₃ / CH₃	0.42	0.46	0.54	red	green	orange

Structure					
—CH₂—CH=C—[CH₂—CH₂—CH=C]₃—CH₃ (with CH₃)	0.31	0.40	0.47	red	green	orange
—CH₂—CH—C—(CH₂)₃—CH—(CH₂)₃—CH—CH₃ (with CH₃ groups) (vitamin K₁)	0.19	0.28	0.36	red	green	orange
the 2,3-oxide of vitamin K₁	0.19	0.27	0.38	red	green	orange
—CH₂—CH—C—(CH₂—CH₂—CH₂—CH)₄—CH₃ (with CH₃)	0.12	0.16	0.26	red	green	orange
—CH₂—CH—C—CH₂—(CH₂—CH=C—CH₂)₄—CH₂—CH=C—CH₃ (with CH₃ groups) (vitamin K₂)	0.14	0.18	0.30	red	green	orange

chromatography of thiamine and its mono-, di-, and triphosphates has been described by Viscontini *et al.* (*3422*).

Fujiwara and Shimizu (*1059*) have described a microanalytical estimation of riboflavin using cation exchange resins. Yagi *et al.* (*3641*) report the separation of flavins by ion exchange resins. Forrest and Todd (*1002*) have purified riboflavin-5′-phosphate on a paper chromatopile.

Crammer (*644*) has described the paper chromatography of free riboflavin, its phosphate and its nucleotides. Hais and Pecáková (*1271*) have studied riboflavin decomposition products by chromatography on paper, and Whitby (*3533*) the enzymatic formation of new riboflavin derivatives. (See also Woiwod and Linggood, *3612*, and p. 377).

(b) *Pantothenic acid*

Kuhn and Wieland (*1824*) have purified pantothenic acid by adsorption on acid alumina. Ackermann and Kirby (*13*) have separated pantothenic acid from α-aminobutyric acid and α-amino*iso*butyric acid on paper. Baddiley and Thain (*120, 121*) have separated phosphates of pantothenic acid on ion exchangers, on cellulose columns and on paper, using the method of Hanes and Isherwood (*1297*) for detection of the spots.

For paper chromatography of *Lactobacillus bulgaricus* factor, pantethine, etc., see Vitucci *et al.* (*3425*), Brown and Snell (*418*). For the separation of coenzyme A and hydrolysis products, see Stadtman (*3079*), Gregory and Lipmann (*1221*). See also p. 372.

(c) *Niacine and derivatives*

The paper chromatography of nicotinic acid in *n*-butanol saturated with 0.2 N NH$_4$OH followed by a microbiological assay has been reported by Leifer *et al.* (*1950*). Huebner (*1472*) showed subsequently that all nonvolatile pyridine derivatives giving the colour reaction of König can be de-

TABLE 124. R_F values of derivatives of nicotinamide (Huebner, *1472*)
Reagent: Expose to vapours of cyanogen bromide, then spray with 25% benzidine in 50% ethanol. Detection limit, 5 γ.
Solvent: butanol/1.5 N NH$_4$OH

	R_F	Colour
Nicotinic acid	0.24	Red violet
Isonicotinic acid	0.25	Grey lavender
Nicotinamide	0.65	Red violet
Diethylnicotinamide (Coramine)	0.84	Violet

tected on paper, after exposure to a few crystals of CNBr and spraying with a 0.25% solution of benzidine in 50% alcohol (Table 124). See also Munier

(2316), Allouf and Munier *(40)* and Wollish *et al.* *(3621)*. Kodicek and Reddi *(1754)* have described a similar method, using *p*-aminobenzoic acid instead of benzidine, in the König colour reaction. They have also reported the behaviour of a number of quaternary pyridinium compounds, which can be detected by their fluorescence; 60 or 80 % propanol was the best solvent for these compounds. For di- and triphosphopyridine nucleotides, 60 % acetone, 60 % *n*-propanol or 60 % ammonium sulphate in 0.1 *M* phosphate buffer (pH 6.8) with 2 % *n*-propanol were used (see also p. 372).

Human urine contains, especially after the administration of nicotin-amide, a highly fluorescent substance that Huff and Perlzweig *(1474)* identi-fied as N-methylnicotinamide (trigonellamide) after isolation by adsorption on Decalso, followed by elution with KCl. The presence of this substance causes trouble in the estimation of thiamine by the thiochrome method but it can be removed by adsorption on Permutit (Najjar and Ketron, *2336*).

Six urinary metabolites of nicotinic acid were found by Leifer *et al.* *(1951)* after the injection of radioactive nicotinic acid (with a ^{14}C carboxyl group) or its amide. Five of these were identified by paper chromatography: N^1-methylnicotinamide, nicotinuric acid, nicotinic acid, N^1-methyl-6-pyri-done-3-carboxylamide and nicotinamide.

(d) *Biotin*

The resolution of biotin components in acid- and pepsin-digested liver by paper chromatography was described by Bowden and Peterson *(348)*. Bound forms of biotin can be separated in several components by partition on celite (with butanol-Skellysolve C-water as solvent) (Chang and Peterson, *546*).

Wright *et al.* *(3632)* have isolated biocytin (ε-N-biotinyl-L-lysine) from yeast extract, using chromatography on Superfiltrol-Celite and then on alumina.

Wright *et al.* *(3631)* have described the paper "bioautography" of biotin and related compounds (see Table 125).

(e) *Vitamin B_6*

Winsten and Eigen *(3599)* have separated the components of the vitamin B_6 group with butanol; the spots containing fractions of a μg only were detected by "bioautography" on agar.

Snyder and Wender *(3041)* have separated these compounds on paper with *iso*amyl alcohol - pyridine - water (2:1:2); (R_F: pyridoxamine 0.21, pyridoxal 0.62, pyridoxine 0.70).

(f) *p-Aminobenzoic acid*

Lemberg *et al.* *(1954, 1955)* identified *p*-aminobenzoic acid after diazotisation and coupling with dimethyl-α-naphthylamine, the coloured product being adsorbed on alumina where it formed a yellow-orange zone (solvent: ether-2 % acetic acid). The coloured isomer derived from o-aminobenzoic acid remained at the top of the column giving a purple zone (eluted with ether-20 % acetic acid). The coloured compounds derived from the p- and m-aminobenzoic acids could not be separated.

For the separation of the three aminobenzoic acids from one another on paper, see Ekman (*876*), M. Lederer (*1917*), Tabone *et al.* (*3192*).

(g) *Folic acid*

Stokstad *et al.* (*3122*) described the purification and isolation of folic acid. The method consists in adsorption of the methyl ester on a column of Superfiltrol, followed by elution with aqueous acetone.

Chromatography has played an important part in the purification of various derivatives of folic acid; details will not be given here. Amongst recent work the purification of synthetic compounds with folic acid activity on starch and Florisil columns (Flynn *et al.*, *994*) and the purification of the Citrovorum factor on Florisil and Dowex-1 (Silverman and Keresztesy, *2989*) may be mentioned. See also Sauberlich (*2826*).

For paper chromatography, see *2633*, *3548* and *3663*.

(h) *Thioctic acid* (*lipoic acid, protogen*)

Protogen A was isolated from beef liver by counter current distribution and chromatography on silicic acid; paper chromatography was used to study the interconversion of protogen A (dithiooctanoic acid) to protogen B (thiosulfinyloctanoic acid) (Patterson *et al.*, *2466*).

Reed and DeBusk (*2644*) have described the bioautography of α-lipoic acid conjugates on paper.

(i) *Cobalamin* (*vitamin* B_{12})

The isolation of *vitamin* B_{12} would never have been possible without chromatography. It was first recognised to be a pigment because it gave a red zone on columns of silica or alumina. The paper of Fantes *et al.* (*936*) contains the description of adsorption chromatography of vitamin B_{12} on alumina, silica gel and charcoal, as well as partition chromatography on silica gel and on paper. (See also Lester Smith, *3023*). Purification of vitamin B_{12} on alumina is described by Schindler and Reichstein (*2845*), by Lens *et al.* (*1960*) and in a patent of Folkers and Shavel (*997*).

The combination of paper chromatography and microbiological assay (bioautography) can be used for the detection of different B_{12} vitamins, as well as substitute growth factors (Winsten and Eigen, *3600*; Woodruff and Foster, *3623*; Kon, *1765a*).

Paper chromatography has been widely used for the isolation and the study of the interrelation of the various cobalamins, as well as for the identification of their degradation products (see Table 125); it is impossible to give adequate details of this work here; the recent papers of Armitage et al. (79), Brown et al. (416), Ford et al. (999), will serve as a guide for further references. The latter authors have described autoradiography of ^{60}Co labelled cobalamins.

TABLE 125. R_F values of microbiologically active vitamin B_{12}-like factors (Kon, 1765a)

Factor	State*	Base of nucleotide	Absorption maxima $(m\mu)$	R_F sec.-Butanol-acetic acid-water	R_F sec.-Butanol-ammonia-water
Factor B	A	None	276, 315, 355, 503, 530	0.50	0.45
Pseudo-vitamin B_{12}	C	Adenine	278, 308, 320, 361, 518, 548 50	0.11	0.085
Factor A	C	2-Methyl-adenine	280, 320, 361, 520, 548	0.13	0.12
Factor C_1	?	?	?	0.02	0.04
Factor C_2	?	?	?	0.04	0.06
Factor E	?	?	?	0.35	0.40
Factor F	C	?	?	0.21	0.17
Factor G	C	Hypoxanthine	359, 516, 540		
Factor H	C	2-Methylhypo-xanthine	358.5 517, 540		
Vitamin B_{12}III	C	?	295, 361, 518, 550	0.13	0.14
Cyanocobalamin	C	5,6-Dimethyl-benziminazole	278, 361, 520, 550	0.25	0.30

* C = crystalline; A = amorphous

(j) Carnitine

Carnitine (γ-hydroxy-β-butyro-betaine), or vitamin B_T can be estimated after paper chromatography (Bregoff et al., 373; Strack and Lorenz, 3134).

(k) Vitamin C

Jackel et al. (1528) described the recovery of small amounts of ascorbic acid from urine by adsorption on Amberlite IR-4B and elution with HCl.

Probst and Schultze (*2596*) have reported the isolation of the pure 2,4-dinitrophenylhydrazone of dehydroascorbic acid from the urine of rats after chromatography on magnesium phosphate tetrahydrate.

Mapson and Partridge (*2126*) have separated ascorbic acid and related substances on paper. HCN has to be added to the solvent to prevent the traces of Cu present in the paper from oxidizing ascorbic acid. The R_F values are given in Table 126. Another separation of reductone and ascorbic acid was described by Weygand (*3520*), who employed dichlorophenol-indophenol as a spraying reagent (white spots on blue background).

Patschky (*2465*) has described the determination of ascorbic acid by paper chromatography of its 2,4-dinitrophenylhydrazone and colorimetry of the separated spots, after elution.

TABLE 126. R_F values of substances related to ascorbic acid (Mapson and Partridge, *2126*)

	Colour reactions			Solvents		
	Indo-phenol	$AgNO_3$ NH_3 Room t.	$AgNO_3$ NH_3 100°	Butanol 40 AcOH 10 Water 50 (KCN)	Phenol 1% AcOH sat. water (KCN)	Collidine sat. with water (KCN)
Ascorbic acid . .	White	Black	Black	0.37	0.35	0.40
isoAscorbic acid .	White	Black	Black	0.38	0.40	0.41
Hydroxytetronic acid	White	Black	Black	0.63	0.62	0.49
Reductone . . .	White	Black	Black	0.63	0.66	0.46
Reductic acid . .	White	Black	Black	0.64	0.78	0.40
Dehydroascorbic acid	White	Light brown	Dark brown	0.41	0.38	0.44

Weygand and Csendes (*3522*) have described a colour reaction with $TiCl_3$ which is specific for enediols and enols of 1,3-diketones; 1 μg of ascorbic acid can thus be detected on paper. Ogawa (*2400*) detects vitamin C through the violet-indigo fluorescence obtained after reaction of dehydroascorbic acid with o-phenylenediamine. Chen et al. (*562*) use butanol or phenol saturated with water and oxalic acid to separate ascorbic acid and related compounds; the enediols are estimated by colorimetry after reaction with 2,6-dichlorophenolindophenol. Tegethoff (*3215*) uses a cacotheline solution to detect ascorbic acid, whereas Heimann et al. (*1355*) prefer a molybdate solution. Mitchell and Patterson (*2245*) use paper impregnated with a mixture of acetonitrile, water, acetone and acetic acid; the spots are detected by spraying with ammoniacal $AgNO_3$.

Schmidt and Staudinger (*2863*) determine separately ascorbic and dehydroascorbic acid after separation on paper. See also Ulmann (*3343*).

For paper chromatography of bound forms of ascorbic acid (ascorbigen) see Kořístek and Procházka (*1767*).

CHAPTER 36

Hormones

1. STEROIDAL HORMONES

(a) *Oestrogens*

The chromatographic fractionation and identification of oestrogens and related compounds has been reviewed by Axelrod (*110*).

Stimmel (*3113–3115*) has carried out a systematic study of the separation of oestrone, oestriol and oestradiol by adsorption on alumina. Breuer (*381*) has also used alumina columns and eluted oestrone with 0.3 to 0.5 % ethanol in benzene, β-oestradiol with 0.6 to 0.7 % and oestriol with 15 to 20 % ethanol in benzene.

Veitch and Milone (*3386*) have separated the 2,4-dinitrophenylhydrazones of oestrone and equilenin (see also Johnston, *1592*). For the isolation of 3-deoxyequilenin from pregnant mares urine, see Prelog and Führer (*2582*).

Umberger and Curtis (*3347*) have described a method for the quantitative esterification of the natural oestrogens with *p*-phenylazobenzoyl chloride. The oestrogen azoates may be separated quantitatively into monoazoates and diazoates on Florisil.

Stern and Swyer (*3101, 3178*), Braunsberg (*365, 366*), Haenni *et al.* (*511, 1253*) and Bitman and Sykes (*258*) have used partition columns of Celite containing NaOH or NaHCO₃ for the separation of various oestrogens, with benzene as mobile solvent. Marrian and Bauld (*2142*) have described the isolation of 16-*epi*-oestriol from urine on a column of Celite mixed with equal parts of 70 % methanol, with ethylene dichloride as mobile phase. A discussion of the relative merits of the different methods, as well as a description of two partition chromatograms for the quantitative separation of oestriol, oestrone and oestradiol in extracts of hydrolysed urine can be found in a paper of Bauld (*161*).

Nyc *et al.* (*2391*) have separated oestrone, oestradiol and oestriol on specially prepared powdered vulcanized rubber; the hormones were eluted successively with aqueous methanol of different concentrations.

Bosch (*323*) has recommended the use of rubber columns for the isolation of oestriol from pregnancy urine.

The *paper chromatography* of oestrogens after coupling with diazotised *p*-nitrobenzene-azo-dimethoxyaniline (Fast black Salt K) has been carried out by Heftmann (*1345, 1346*); in this way oestrone, oestriol, oestradiol-17α and -17β, equilin and equilenin were separated (see Table 127).

TABLE 127. R_F values of oestrogens and some other steroids coupled with diazotised *p*-nitrobenzene-azo-dimethoxyaniline (Heftmann, *1345*)

Solvent: 200 toluene, 100 petroleum ether, 30 ethanol and 70 water

	Colour of spot	R_F value
Oestrone	purple	0.95
Oestradiol.	purple	0.81
Oestriol	purple	0.07
Equilin	purple	0.96
1-Methyloestradiol diacetate . .	purple	0.81 and 0.96
Dehydrocorticosterone acetate .	yellow	0.95
Progesterone.	yellow	0.74
cis-Testosterone.	yellow	0.21
Δ^4-Androstene-3,17-dione . . .	yellow	0.60 and 0.97

Solvent: 200 petroleum ether, 100 toluene, 10 ethanol and 90 water
(Heftmann, *1346*)

Oestradiol-17α	purple	0.32
Oestradiol-17β	purple	0.09
Equilin	purple	0.49
Equilenin	blue	0.05
Oestrone	purple	0.29
Oestriol	purple	0.00

Boscott (*324*) separated phenolic acids and oestrogens from urine, using different solvent systems (glycols as stationary phase and benzene, ethylene-dichloride or aliphatic ketones as mobile phase). Paper chromatograms employing aqueous alkali as the mobile phase, and no organic solvent have also proved useful.

The paper chromatography of oestrogens has also been described by Heusghem (*1381*), Jellinek (*1572*), Markwardt (*2140*), Zbudovská and Hais (*2974, 3672*).

The isolation and estimation of steroid oestrogens in placental tissue has been studied in detail by Mitchell and Davies (*2234*) who have compared the various methods available (chromatography on paper impregnated with alumina, glycerol and propylene glycol or silicone, or partition on columns of Celite or of rubber). They finally recommend paper chromatography with methanol as stationary phase and petroleum ether as mobile phase for the separation of oestrone and oestradiol, or methanol-water (1:1) as stationary phase and benzene as mobile phase for the separation of oestriol. The spots are detected by ultraviolet absorption or by the reagent of Folin and Ciocalteu (995).

Axelrod (108) has described a modification of the method of Burton et al. (476) (paper impregnated with formamide, using o-dichlorobenzene or methylene chloride or cyclohexane as mobile solvents). Color reagents for detection of the spots were studied in detail.

(b) Progesterone

Haskins et al. (1328) have separated progesterone from the oils, in which it is commercially supplied, by utilizing paper chromatography. The hormone is located on the paper strip by applying m-dinitrobenzene and KOH. After elution, it can be determined quantitatively by spectrographic analysis. See also the chapter on steroids.

Male sex hormones and cortical hormones: see the section on steroids, p. 275.

2. WATER-SOLUBLE HORMONES

(a) Adrenaline

Bergström and Hansson (208) have used Amberlite IRC-50 for the purification and isolation of adrenaline and histamine from dilute aqueous solutions.

Bergström and Sjövall (213) have described the separation of adrenaline and noradrenaline on partition columns of Hyflo-Supercel (with dilute HCl as stationary and phenol as mobile phase).

James (1545) was the first to show that adrenaline and noradrenaline could be separated by paper chromatography; the spots were detected by spraying with potassium ferricyanide; oxidation to adrenochrome and similar pigments produces brown spots. Using this method, Holton (1439) found that noradrenaline was the main pressor component of extracts of adrenal medullary tumours, adrenaline being present only in much smaller amounts. Goldenberg et al. (1155) detected noradrenaline in commercial adrenaline preparations; the spots can also be developed with ninhydrin.

Von Euler et al. (923) have also separated on paper adrenaline and noradrenaline from suprarenal extracts; in urine they found a third .base, hydroxytyramine (924).

Shea (2950) found that the adrenaline spot could be detected by its fluorescence in ultra-violet light, after holding the paper over NH_3 vapour. Crawford (649) has given a list of R_F values of a series of derivatives of adrenaline (see Table 128, also James and Kilbey, 1546).

Gregerman and Wald (1216), Van Espen (3375) and Shepherd and West

(*2960*) have observed that adrenaline and similar compounds gave two spots when chromatographed on paper in presence of trichloracetic acid. This is ascribed to the formation of unstable compounds between the phenolic amines and trichloracetic acid.

Schayer (*2837*) has separated on paper the urinary excretion products of DL-adrenaline labelled with ^{14}C in the β-position. The radiograms (Fig. 58, p. 145) show the presence of at least five metabolites.

TABLE 128. R_F values of adrenaline and related compounds (Crawford, *649*)
Spray reagent: 0.44% (w/v) $K_3Fe(CN)_6$ in 0.2 M phosphate buffer pH 7.8

Substance	Formula						Phenol sat. with H_2O and SO_2	Butan sat. w 0.5 N
	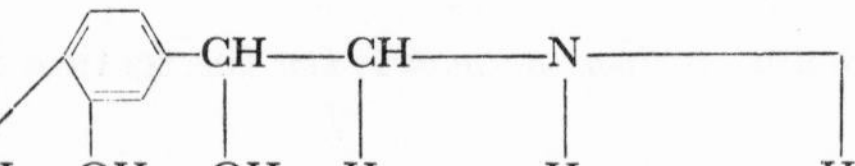		—CH—	—CH—	—N—			
Noradrenaline	OH	OH	OH	H	H	H	0.28	0.10
3,4-Dihydroxy-phenylalanine	OH	OH	H	COOH	H	H	0.29	0.21
α-Ethylnoradrenaline	OH	OH	OH	C_2H_5	H	H	0.37	0.25
Hydroxytyramine	OH	OH	H	H	H	H	0.43	0.24
Corbasil	OH	OH	OH	CH_3	H	H	0.50	0.33
Adrenaline	OH	OH	OH	H	CH_3	H	0.51	0.17
Isoprenaline	OH	OH	OH	H	$CH(CH_3)_2$	H	0.66	0.37
Epinine	OH	OH	H	H	CH_3	H	0.67	0.30
N-Ethylnoradrenaline	OH	OH	OH	H	C_2H_5	H	0.68	0.24
Adrenalone	OH	OH	=O	H	CH_3	H	0.69	0.24
N-Methyladrenaline	OH	OH	OH	H	CH_3	CH_3	0.72	0.16
Sympatol	OH	H	OH	H	CH_3	H	Not visible	Not vis
Metasympatol	H	OH	OH	H	CH_3	H	Not visible	Not vis

(b) *Thyroid hormones*

The biochemistry of thyroxine has been studied by several groups of workers with the aid of radioactive iodine.

This work has been recently reviewed by Gross (*1229*); Roche *et al.* (*2714*) have given a detailed description of the technical aspects of the preparation and chromatographic analysis of radioactive iodine compounds from the thyroid gland and body fluids.

Gross *et al.* *(1230)* examined by filter paper chromatography the iodinated organic substances in both hydrolysed and unhydrolysed thyroid gland after administration of radioactive iodine. They showed the presence of free thyroxine, diiodotyrosine and an unknown compound which has since been identified as triiodothyronine.

Roche *et al.* *(2715)* studied the formation of thyroxine from iodinated proteins and detected thyroxine in proteins of *Gorgonides* *(2723)*. Chaikoff *et al.* *(3287)* studied the non-thyroglobulin iodine of the thyroid and the iodinated excretion products of thyroxine *(3207)*. In all these experiments clearcut separations of thyroglobulin, diiodotyrosine, thyroxine and organic iodide were obtained as shown by autoradiographs (Fig. 102).

3,5,3'-Triiodothyronine has been isolated from thyroid glands by chromatography on a column of kieselguhr (with 0.5 N NaOH as stationary and 20 % $CHCl_3$ in butanol (v/v) as mobile phase) (Gross and Pitt-Rivers, *1231*); the paper chromatography of this and related compounds has been described by the latter authors, as well as by Roche *et al.* *(2713)* who have also studied the metabolites of ^{131}I containing triiodothyronine by paper chromatography *(2718)* (see also Albright *et al.*, *33*, and Table 129, from Gross, *1229*). The starch chromatography of ^{131}I containing thyroxine, tri-, di- and monoiodotyrosine has been described by Dobyns and Barry *(796)*. A source of error in column and paper chromatography lies in the possibility of exchange between radioactive inorganic iodide and organic iodine compounds, which occurs easily in acid solvents, due to the oxidation of iodide to elemental iodine. This can be prevented by adding a reducing reagent, such as thiosulphate (Dobyns and Barry, *796*).

Roche *et al.* *(2720, 2721)* have detected 3,3',5'-triiodothyronine and 3,3'-diiodothyronine in thyroglobuline, accompanying

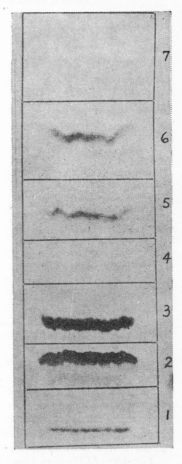

Fig. 102. Radioautogram of ascending chromatogram from hydrolysate of thyroid of chicken injected for 24 hours with [131]I: exposure time 5 hours. Solvent: collidine 125 ml, water 44 ml; atmosphere of ammonia (Taurog *et al.*, *3210*)

1 unhydrolysed thyroglobulin (?); 2 diiodotyrosine; 3 monoiodotyrosine; 5 thyroxine; 6 inorganic iodine

TABLE 129. R_F values of iodinated derivatives of tyrosine (Gross, *1229*)

(The upper layer is taken for solvent, and the substances in parentheses indicate what the solvent was equilibrated with in each case)

Compound	n-Butanol (2 N acetic acid)	n-Butanol (2 N formic acid)	n-Butanol (4 vol.), dioxan (1 vol.) (2 N NH₄OH)	n-Butanol (2 N NH₄OH)	Collidine (conc. NH₄OH)	isoPentanol (6 N NH₄OH)	n-Butanol (1 vol.), pentanol (1 vol.) (2 N NH₄OH)	Collidine (1 vol.), lutidine (1 vol.) (water)
Thyroxine	0.88	0.91	0.45	0.58	0.45	0.15	0.41	0.79
Triiodothyronine	0.91	0.91	0.65	—	0.65	0.27	—	—
Diiodothyronine	0.85	0.85	0.60	0.73	—	0.38	0.63	—
Thyronine	—	—	—	—	—	—	0.28	—
Diiodotyrosine	0.70	0.69	0.11	0.08	0.11	—	0.20	0.56
Monoiodotyrosine	0.57	0.51	0.17	0.13	0.28	—	0.20	0.68
Tyrosine	0.50	—	0.12	—	0.25	—	0.20	—
Iodide	0.27	0.29	0.40	0.37	0.90	0.08	—	0.89
Iodine	1.00	1.00	0.40	—	—	—	—	—
Thyroxine "glucuronide"	0.45	—	0.33	—	0.20	—	—	—
Triiodothyronine "glucuronide"	—	—	—	—	0.28	—	—	—
Monoiodohistidine	0.10	—	—	0.21	—	—	—	—
Diiodohistidine	—	—	—	0.13	—	—	—	—

the 3,5,3'-isomer isolated previously; these three hormones are readily separated on paper.

The separation of the phenylthiohydantoins of mono- and diiodotyrosine, and diiodo- and triiodothyronine and of thyroxine on paper has been described by Roche *et al.* (*2717*); the method of Edman has been used by these authors for the detection of tyrosine, 3,5-diiodotyrosine and thyroxine as N-terminal amino acids in thyroglobulin.

Maclagan *et al.* (*2090*) have detected thyroxine and triiodothyronine on paper with a ceric sulphate-arsenious acid reagent.

Gleason (*1137*) uses a new solvent (*tert.*-amyl alcohol saturated with 2 N ammonium hydroxide) for separating thyroxine derivatives.

See also Fink *et al.* (*957*), Hird and Trikojus (*1400*), Tishkoff *et al.* (*3272*), Gross *et al.* (*1230*), and the section on halogenated amino acids, p. 324.

(c) *Miscellaneous*

The chemistry of the *adrenocorticotropic hormone* (ACTH) and of *insulin* has been intensively studied by chromatographic methods; these papers were referred to in the chapter on *peptides* (pp. 345, 346).

The purification of bovine *thyrotropic hormone* on Amberlite IRC-50 has been described by Heidemann (*1351*).

Potts and Gallagher (*2578*) separated the *pressor hormone* and the *oxytocic hormone* of the posterior lobe of the hypophysis by chromatography on Decalso; only the former was adsorbed. Elution of the pressor substance was accomplished with sodium chloride solution. The *chorionic gonadotrophin* from urine has been purified by adsorption on Permutit, elution being carried out with alcohol containing ammonium acetate (Katzman *et al.*, *1649*). Vignes *et al.* (*3404*) have used kaolin and tricalcium phosphate.

Helmer (*1361*) has reported the purification of *angiotonin* (hypertensin) on a wet paper pulp column with aqueous phenol as developer. Separation of free amino acids, salts and inactive polypeptides was thus achieved. Kuether and Haney (*1816*) used a partition column of silica for the purification of angiotonin.

The composition of *hypertensin* was studied by Edman (*863*) using two-dimensional paper chromatograms. For the adsorption of hypertensin on ion exchangers, see Cruz-Coke *et al.* (*670*).

(d) *Plant hormones*

Jerchel and Müller (*1583*) and Pachéco (*2431*) have studied the paper chromatography of indoleacetic acid. Luckwill (*2054*) has used paper chromatography for the identification of auxins and growth inhibitors. After development, the paper was cut transversely into segments 1 cm long, eluted with water and the growth effects of the eluate on the wheat coleoptile measured. Bennet-Clark *et al.* (*187*) have described similar experiments and have detected unidentified growth factors (Bennet-Clark and Kefford, *186*).

Detailed studies of the paper chromatography of indole derivatives in connection with growth hormones have been published by Denffer and Fischer (*737*), Müller (*2308*), Fischer (*961*), Stowe and Thimann (*3133*), Weller *et al.* (*3500*) and Sen and Leopold (*2943*). See also Table 130.

TABLE 130. R_F values of indole-derivatives in different solvents (Sen and Leopold, 2943)

Substance	Structure	R_F in Solvents							
		Phenol-water	Butanol-propionic acid-water	Isopropanol-ammonia-water 10:1:1	Butanol-ethanol-water 4:1:1	Butanol-ethanol-ammonia 1:1:2	70% ethanol	Pyridine-ammonia 4:1	Water
1. Indole	NH	0.99	0.99	0.99	0.93	0.99	0.86	0.97	—
2. 2-Methylindole	CH₃ NH	0.97	0.98	0.98	0.92	—	0.88	0.99	—
3. 3-Methylindole (skatole)	CH₃ NH	0.98	0.99	0.98	0.93	—	0.88	0.98	—
4. Indole-3-aldehyde	CHO NH	—	—	0.86	0.92	0.98	0.86	0.97	0.47
5. Indole-3-acetaldehyde	CH₂·CHO NH	—	—	—	—	—	0.45	—	—
6. Indole-3-carboxylic acid	COOH NH	—	—	0.22	0.82	—	0.81	0.56	0.92
7. Indole-3-acetic acid	CH₂·COOH NH	0.80	0.93	0.37	0.66	0.81	0.77	0.56	0.89
8. Ethyl indoleacetate	CH₂·COOC₂H₅ NH	0.96	0.99	0.97	0.91	0.99	0.80	0.97	0.59

Compound	Structure								
9. Indole-3-propionic acid	indole, $CH_2 \cdot CH_2 \cdot COOH$, NH	0.82	0.96	0.44	0.76	0.85	0.91	0.61	0.85
10. Indole-3-butyric acid	indole, $CH_2 \cdot CH_2 \cdot CH_2 \cdot COOH$, NH	0.87	0.98	0.56	0.86	0.85	0.84	0.63	0.89
11. Tryptophan	indole, $CH_2 \cdot CH \cdot NH_2 \cdot COOH$, NH	—	—	0.19	0.26	—	0.40	0.45	0.63
12. Tryptamine	indole, $CH_2 \cdot CH_2 \cdot NH_2$, NH	—	—	0.75	0.42	0.96	0.71	0.86	0.28
13. Gramine	indole, $CH_2 \cdot N(CH_3)_2$	—	—	0.88	0.60	0.98	0.76	0.93	0.35
14. Isatin	indole (=O =O), NH	0.92	0.95	0.79	0.91	0.86	0.80	0.96	0.59
15. N-Acetylisatin	indole (=O =O), $NCOCH_3$	0.32	0.99	0.42	—	—	0.92	0.98	—
16. Dihydroxyindole	indole, OH OH, NH	0.80	0.96	0.79	—	0.97	0.79	0.93	0.63
17. Indican (urinary)	indole, OSO_3K, NH	0.44	0.32	0.68	0.34	0.94	0.73	0.95	0.91
18. Indican glucoside	indole, $OC_6H_{12}O_5$, NH	0.00	0.98	0.00	0.00	0.00	0.00	0.90	0.00

(Table continued)

TABLE 130 (continued)

Substance	Structure	R_F in Solvents							
		Phenol-water	Butanol-propionic acid-water	Isopropanol-ammonia-water 10:1:1	Butanol-ethanol-water 4:1:1	Butanol-ethanol-ammonia 1:1:2	70% ethanol	Pyridine-ammonia 4:1	Water
19. N-Acetylindoxyl	OH, NCOCH₃	0.93	098	0.87	0.87	0.87	0.80	0.97	0.63
20. Indoxyl acetate	OCOCH₃	—	0.98	0.86 / 0.66	0.94	0.88 / 0.96	0.89	0.95	0.57
21. Indigotin		0.00	0.00	0.00	0.00	0.00	0.00	0.00	0.00
22. Indigo disulfonate		0.00	0.00	0.16	0.00	0.00	0.34	0.00	0.81
23. Indigo tetrasulfonate		0.00	0.00	0.06	0.00	0.00	0.48	0.00	0.88
24. Indirubin		0.97	0.99	0.89 / 0.96	0.89	0.99	0.69 / 0.77	0.97	—
25. Indoleacetonitrile	CH₂·CN	—	0.97	0.99	0.94	0.99	0.86	0.95	0.41
26. Indolebutyronitrile	CH₂·CH₂·CH₂·CN	—	0.97	0.95	0.95	0.99	0.88	0.96	0.54

CHAPTER 37

Antibiotics

Jones (*1604*) has given a short account of the paper chromatography of antibiotics.

1. PENICILLIN

Penicillin was purified by repeated chromatography on alumina by Abraham and Chain (*2*); in this way the yellow pigments and pyrogenic substances were removed. The purification of penicillin salts on alumina and of free penicillin on magnesium silicate has been patented (*1423*).

Catch *et al.* (*533, 627*) achieved considerable purification by partitioning crude penicillin on columns of silica gel impregnated with an alkali or alkaline earth carbonate.

Levi (*1986, 1988*) has introduced the use of silica gel buffered with concentrated potassium phosphate buffer and has described the theory of separations carried out on such columns. Chromatography of crude penicillin on these columns, using ether as the mobile solvent, showed the presence of five penicillins (Boon *et al.*, *312*). Fischbach *et al.* (*958, 960*) have independently developed a similar method, using silica gel columns having a phosphate buffer (pH 6.4) as the stationary phase. The chromatogram was developed with ether saturated with water and the filtrate collected in 25 ml fractions; the penicillin in each fraction was estimated by iodometry. Penicillin K, being the least hydrophilic, appeared first in the filtrate. For the preparation of silica gel suitable for the purification of penicillin, see Harris and Wick (*1315*).

Behrens *et al.* (*176*) used partition chromatography on silica gel to separate new penicillins obtained by biosynthesis.

James *et al.* (*1541*) separated *n*-heptyl, *n*-amyl, *n*-pent-2-enyl and benzyl penicillins on a column of Hyflo Supercel mixed with sodium citrate buffer of pH 5.5. The mobile solvent was a 1:1 mixture of ethyl ether and di-*iso*propyl ether and the separations were followed automatically by recording the changes in electrical conductivity of the stationary phase of the column.

Leigh (*1952*) has given details of the partition chromatography of tertiary amine salts of penicillins on phosphate-buffered silica gel.

Goodall and Levi (*1158*) separated the penicillins by partition chromatography on *paper* impregnated with a phosphate buffer of pH 6 to 7 (mobile

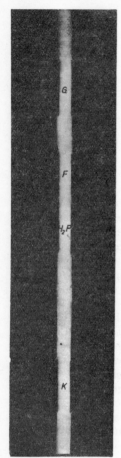

phase: ether saturated with water). The paper strips were placed in contact with agar plates inoculated with a micro-organism sensitive to penicillin. After a few hours contact the paper strips were removed and the agar plates showed zones of inhibition of growth in the positions corresponding to the penicillin spots on the paper (Fig. 103). Winsten and Spark (*3601*) using the same method were able to show the existence of at least eight different penicillins. Kluener (*1742*), as well as Karnovsky and Johnson (*1633*) obtained good separations using slight modifications of Goodall and Levi's technique. Lester Smith and Allison (*3024*) have checked the method of Goodall and Levi with ^{35}S-penicillin and found it to be very satisfactory.

Baker *et al.* (*123*) have separated penicillin mixtures by chromatography of the hydroxamic acid derivatives on heavily buffered paper; quantitative results were obtained by extraction and colorimetry of the iron complexes. It is very difficult to control the humidity of buffer-loaded paper and this difficulty is accentuated when a volatile solvent is used. Goodall and Levi (*1158*) tried to overcome this by covering the walls of the chamber with cloth soaked in water. Baker *et al.* (*123*) obtained better results by pumping both phases continuously over the walls of the chamber; see also Albans and Baker (*30*).

Burton and Abraham (*475*) have separated the related cephalosporins on paper.

Fig. 103. Paper chromatography of penicillins by the method of Winsten and Spark (*3601*). Solvent aqueous ether, development at 4°, 18 h. The four spots correspond to penicillin G, F, dihydro-P and K (Ziéglé and Rolovick, S.I.F.A., Paris)

Drake (*815*) has described how clear, high contrast photographs of antibiotic paper chromatograms can be obtained.

2. STREPTOMYCIN

Streptomycin can be purified by adsorption on acid-washed alumina; the antibiotic is retained by the column from a solution in acid methanol and can be eluted with aqueous methanol (*520*); see also Vander Brook *et al.* (*3365*). The streptomycin culture media contain another antibiotic *streptothricin*, which Peck *et al.* (*2486*) purified on alumina or on charcoal (Darco-G-60, mixed with paper pulp). Streptomycin can also be purified on Darco-G-60 (Kuehl *et al.*, *1815*). Fried and Wintersteiner (*1044*) purified the picrates of streptomycin and streptothricin by chromatography. Peck (*2485*) has published a detailed article devoted to the chromatography of streptomycin.

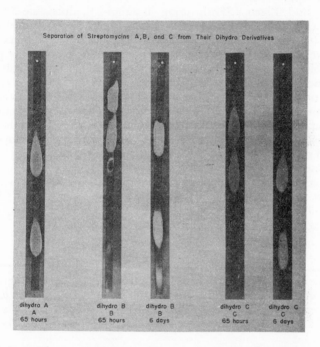

Fig. 104. Paper chromatography of streptomycins and their dihydro derivatives
(Stodola *et al.*, *3119*)

Steptothrycine has been purified by chromatography of neutral aqueous solutions on columns of active charcoal and elution by very dilute acetone (1 % v/v in water, Carter *et al.*, *521*).

Brockmann and Musso (*394*) have described several solvent systems which can be used to distinguish antibiotics of the streptomycin-neomycin-geomycin group on radial chromatograms.

Fried and Titus (*1042*) separated mannosido-streptomycin (streptomycin B) on alumina columns on which it is more strongly adsorbed than streptomycin. The same authors obtained evidence for the existence of tautomeric modifications of streptomycin which were partly separable by chromatography on alumina (see also *3273*).

An industrial plant for the purification of streptomycin with alumina columns 80 cm in diameter and 4 m high has been described by Williams and Hightower (3580).

Horne and Pollard (1447) described the identification of streptomycin by *paper chromatography*, the position of the antibiotic being shown by means of the Sakaguchi reagent. 3 % Ammonium chloride was found to be necessary for good separations.

Winsten and Eigen (3598), using butanol-2 % piperidine-2 % toluene sulphonic acid monohydrate as the developing liquid, have obtained evidence for the existence of 5 antibiotics in the streptomycin complex. Some of these may be due to the previously mentioned tautomeric forms. The positions of the antibiotics were revealed by "bioautography" on agar. Their method has been used by Solomons and Regna (3046) and by Stodola et al. (3119) for other separations in the streptomycin field (Fig. 104). Peterson and Reineke (2496) prefer the use of 2 % toluenesulphonic acid monohydrate in wet butanol. Separations of the antibiotics of crude streptomycin broths are best when the salt concentration is not in excess of 50 μg per streptomycin unit. A modified solvent containing NaCl is used for low potency culture filtrates. In the presence of salts, the antibiotics studied (streptomycin, dihydrostreptomycin, mannosidostreptomycin, streptothricin, neomycin A) gave two distinct zones; it is probable that this is also due to tautomeric forms.

Brink et al. (385) have made extensive use of chromatography in their study of degradation products of streptomycin.

3. MISCELLANEOUS

The isolation of *neomycin* has been accomplished by chromatography on charcoal columns and on Amberlite IRC-50 (Leach et al., 1893, 1894). The paper chromatography of neomycin is carried out with 2 % toluene sulphonic acid monohydrate in water with saturated butanol as the mobile solvent.

Haskell et al. (1327) have separated the hydrolysis products of the basic polypeptide *viomycin* on Zeo-Rex, a phenolic methylenesulphonic acid type exchanger (Permutit Co., New York) and on paper. Serine, α,β-diaminopropionic acid and isolysine were isolated in a pure state.

The paper chromatography of *chloromycetin* has been reported by Smith and Worrel (3030) and Hais et al. (1270). Glazko et al. (1136) have studied urinary metabolites of chloromycetin by paper chromatography; the spots were detected after reduction with titanous chloride.

Vining and Waksman (3406) have described several solvent systems for the separation of the different *actinomycins*. Brockmann and Gröne (389) have separated several actinomycins on columns of cellulose and by radial paper chromatography. For the paper chromatography of the amino acids of actinomycins, see Brockmann et al. (390) and Dalgliesh et al. (687).

Gliotoxin, $C_{13}H_{14}O_4N_2S_2$, from *Penicillium obscurum* is accompanied by another sulphur-containing antibiotic, $C_{14}H_{16}O_4N_2S_2$, which was purified by Mull *et al.* (*2303*) on alumina.

Clavacin (patulin, $C_7H_6O_4$) from *Aspergillus clavatus* is extracted from the culture media by passage through a charcoal column and then purified in chloroform solution on Permutit (Katzman *et al.*, *1650*).

Hays *et al.* (*1342*) studied the complex mixture of antibiotics produced by *Pseudomonas aeruginosa*; one of these substances, pyo II, is an acid and is retained by a column of oxalic acid (solvent: benzene, elution carried out by dissolving the column in potassium bicarbonate). The neutral antibiotics, pyo Ib, Ic, II and IV, can be separated by repeated chromatography on Permutit. Wells (*3502*) has shown quite recently that these substances are 2-alkyl-4-quinolinols.

The papers of Synge *et al.* dealing with the structure of *gramicidin* (*1169*), *tyrocidin* (*1171*) and *gramicidin S* (*3182*) have already been mentioned. Synge and Tiselius (*3187*) studied the homogeneity of these polypeptides by means of frontal analysis.

Grisein (a red amorphous powder, $C_{40}H_{61}N_{10}O_{20}SFe$) has been purified by Kuehl *et al.* (*1814*), by partition on silica columns, with 17 % phenol in chloroform as the mobile solvent. On hydrolysis, two ninhydrin-reacting spots were obtained after paper chromatography. Iron-free grisein can be purified by chromatography on Magnesol.

Hickey and Phillips (*1390*) have described the separation of *aureomycin* from *terramycin* by countercurrent distribution and paper chromatography. See also Dohnal and Bialá (*799*). Sokolski *et al.* (*3044*) have described spray reagents for differentiating these two antibiotics on paper.

Bird and Pugh (*252*) have reported the separation of tetracycline, chlorotetracycline (aureomycin) and oxytetracycline (terramycin) by paper chromatography.

Pettinga *et al.* (*2501*) have described the separation of *erythromycin A* and *B* on a column of powdered cellulose, developed with 0.01 N NH_4OH saturated with methyl *iso*butyl ketone. *Azaserine*, the tumor-inhibitory antibiotic from *Streptomyces* is destroyed by chromatography on a series of ion exchange resins (see also p. 99), but can be purified on activated charcoal, silica gel or alumina (Fusari *et al.* (*1065*).

CHAPTER 38

The separation of high polymers

1. COLUMN CHROMATOGRAPHY

Claesson (*573, 584*) has carried out the frontal analysis of high polymers such as polymethyl methacrylate, polyvinyl acetate, nitrocellulose etc. and has found that with molecular weights above 5 to 10 thousand the adsorption decreases with increasing molecular weight. The rate of adsorption is low, principally due to the low diffusion constant; thus the rate of flow must be extremely low to reach adsorption equilibrium.

Landler (*1875*) obtained fractionation by passing solutions of synthetic elastomers (a butadiene-styrene co-polymer, G.R.S.; a butadiene-acrylonitrile copolymer, Perbunan; and a polyisobutylene, Vistanex) through columns packed with carbon black deposited on activated charcoal. At low concentrations the polymers of low molecular weight are adsorbed first; at higher concentrations the order is reversed. This phenomenon can be explained by assuming that at high concentrations the molecules are held at several positions; this multiple adsorption will increase with chain length. At low concentrations, on the contrary, the fixing of a molecule at only one position will not be sufficient to hold large molecules. For the chromatography of *rubber* see Cajelli (*496*).

Brooks and Badger (*399*) have found that the extent of adsorption of nitrocellulose on starch increases with molecular weight; using a starch column they were able to demonstrate fractionation of nitrocellulose according to molecular weight.

Williams (*3582*) has studied the fractionation of two *nylon polymers* (average molecular weight 350,000 and 150,000) on columns of Celite by washing with a gradient of increasing formic acid in water. Contrary to the observations of Claesson, lower molecular weight fractions were eluted first. Solubility may have been the principal factor in this sort of experiment.

Silicones (polymethylsiloxanes) have been studied by Bannister *et al.* (*134*) using charcoal as adsorbent and a gradient of ether in methanol for elution; it was found that the molecular weight of the fractions increased with effluent volume.

2. PAPER CHROMATOGRAPHY

The paper chromatography of *polyethylene oxide* (molecular weight range 1000–5000) with fluorescein as indicator has been described by Kume *et al.* (*1831*). Surface active agents (Postonal, an ethylene oxide polymer, Tween 61, Triton X 100) have been studied by Gallo (*1076*). With butanol-acetic acid, medium R_F values were obtained. Polyvinyl pyrrolidone has been chromatographed on paper by Doebbler (*798*); either base-line or solvent front spots were obtained which were revealed by BiI_3 or K_2PtI_6.

CHAPTER 39

The separation of stereoisomers

Zechmeister (*3676*) has published a review on the subject of stereochemistry and chromatography.

1. OPTICAL ISOMERS

Two methods of using a column of adsorbent for the resolution of a racemic mixture into the optical antipodes can be visualised: firstly, passing a racemic mixture through a column of an optically *active* adsorbent; secondly, transforming the racemic compound into a mixture of two diastereoisomers by combination with a *d*- or *l*-compound, followed by passing the mixture through a column of optically *inactive* adsorbent. In the second case the column replaces the usual operation of fractional crystallisation.

The *first method* (resolution with the aid of an optically *active* adsorbent) has been applied by Karagunis and Coumoulos (*1628*): a column of quartz was used for the partial resolution of triethylenediammine chromic chloride, (Cr en₃) Cl₃·3.5H₂O. Henderson and Rule (*1364*) used lactose as the optically active adsorbent for the resolution of *dl-p*-phenylene-bis-iminocamphor (30 mg of substance on 6 kg of adsorbent). Lecoq (*1895*) succeeded in resolving *dl*-ephedrine on a column of lactose.

Prelog and Wieland (*2590*) effected the first resolution of an organic compound containing asymmetric, trivalent nitrogen, "Tröger's Base" (I), by the use of a D-lactose column. By an appropriate treatment they prepared a highly active lactose, 450 parts of which were sufficient to resolve partially one part of base (fractions of $[a]_D = +75°$ to $-52°$ were obtained, the pure base having $[a]_D = \pm 287°$).

$$CH_3 \text{---} \begin{array}{c} N \text{---} CH_2 \\ CH_2 \\ CH_2 \text{---} N \end{array} \text{---} CH_3$$

(I)

Karagunis (*1627*) has tried, without success, to separate the optically active components of the free radical phenyl-biphenyl-*a*-naphthyl-methyl by adsorption on D-lactose, sucrose, *l*-quartz, *d*-sodium chlorate and alumina coated with D-alanine or D-lactose.

Leonard and Middleton (*1961*), used chromatography on D-lactose hydrate to decide whether a given base (hexahydrojulolidine) was a *meso* or a racemic compound.

Krebs and Rasche (*1795*) have obtained a partial resolution of triethylene-diamine cobalt(III) chloride on starch.

Di Modica and Angeletti (*781*) have shown that in cases where substances without an asymmetric carbon atom are resolvable because of restricted rotation, chromatography on an optically active column can also be applied. Thus they resolved completely the enantiomorphs of *dl*-2-amino-2'-nitro-6,6'-ditolyl on a column of D-lactose prepared according to Prelog and Wieland (*2590*).

Proteins could probably also be used as optically active adsorbents; Bradley and Easty (*359*) have shown that both wool and casein selectively adsorb (+)-mandelic and (+)-α-naphthyl-glycolic acids from an aqueous solution of the racemates. The resolution of mandelic acid on wool occurs at the L-arginine and L-lysine residues (*360*). α-Methoxyphenylacetic acid could be resolved on wool, the + form combining preferentially with the protein in dilute aqueous ethanol (Bradley and Brindley, *358*).

Grubhofer and Schleith (*1234*) have combined Amberlite XE-64 with quinine and have resolved *dl*-mandelic acid on the asymetric column thus obtained.

The new approach tried recently by Curti and Colombo (*679*) (resolution of racemic mixtures on columns of "specific silica gel") has already been mentioned on p. 33.

The *second method* of resolution using an *inactive* adsorbent was adopted by Jamison and Turner (*1550*) who partially separated *l*-menthyl *d*-mandelate from *l*-menthyl *l*-mandelate on a column of alumina with light petroleum as solvent.

Hass *et al.* (*1331*) obtained a partial resolution of *dl*-α-phenylethyl ammonium bi-D-tartrate on alumina, calcium sulphate, magnesia, charcoal or Fuller's earth, with 60 % alcohol or water as solvent. In the same way a partial resolution of brucine *dl*-mandelate was achieved on calcium sulphate, charcoal, cotton, dextrose or magnesia with benzene as solvent. Cook *et al.* (*628*) succeeded in resolving the amide of *dl*-N-methylvaline and D-α-hydroxyvaleric acid, using alumina as adsorbent.

Macbeth *et al.* (*2068*) have applied column techniques to the resolution of acid esters. "If a racemic acid is combined with an active base B, and the mixture of salts A +, B and A—, B is washed through a column of excess B, with a solvent which dissolves the salts and free base only sparingly, a series of exchange reactions is possible. Some dissociation of the salts in dilute solution is to be expected, and the free acids A + and A— may then react with the surfaces of solid B in the column, re-forming A +, B and A—, B which in part may be held on the surface of solid B at the moment of formation but will later re-dissolve. This process may be repeated through the column and, if there are differences in solubility and stability of the salts and in the degree of their adsorption on solid B, one salt should be preferentially removed from the column by continued washing".

On testing this hypothesis by filtering a solution of the brucine salt of ($\pm$)-2-octyl hydrogen phthalate in light petroleum-acetone through a column of brucine mixed with filter aid, the first fractions to be eluted were highly active ($-$)-2-octyl hydrogen phthalate. Up to 80 % of the theoretically possible amount of the ($-$) isomer could be recovered, of 97 % optical purity. Comparable results were obtained with the strychnine salt of ($\pm$)-*trans*-3-methylcyclohexyl hydrogen phthalate on a column of strychnine, the salt first eluted being that of the ($+$) enantiomorph, of 98 % purity. The separations described afforded in each case the enantiomorph giving the more soluble salt, i.e. the one which is more difficult to purify by fractional crystallisation.

The importance of the solubility factor was shown by the observation that other, similar salts could not be resolved with such success and, finally that in the case of 2-octyl hydrogen phthalate the brucine could be omitted from the column altogether: a 97 % resolution was still obtained even when the mixture of the brucine salts was passed through a column of pure filter aid.

Resolution of optical antipodes by *paper chromatography* seems possible only under particular circumstances. Thus, Flood *et al.* (*983*) have observed no resolution of DL-arabinose chromatographed on paper in the presence of *laevo*-menthol or *laevo*-amyl alcohol. Bonino and Carassiti (*309, 310*) however, have succeeded in separating the optical antipodes of racemic 1-(2-hydroxy)-naphthylbenzylamine (β-naphtholbenzylamine) (II) on

$$\begin{array}{c} H \\ C_6H_5-C-NH_2 \\ | \\ C_{10}H_6OH \end{array}$$

(II)

paper impregnated with phenol, using a solution of D-tartaric acid as the mobile solvent. The *l*- and *d*-bases, chromatographed separately, give different R_F values, the racemic base gives two spots, corresponding to the position of the *d*- and *l*-spots respectively. The R_F values are strongly influenced by the concentration of D-tartaric acid (Table 131); this confirms that the separation is due to the formation of the D-tartaric acid salts of the *d*- and *l*-bases, which of course can have different R_F values.

No resolution was obtained in the presence of *meso*-tartaric acid.

Using *l*-methyl-(β-phenyl-*iso*propyl)-amine as the mobile phase, Kotake *et al.* (*1772*) found slightly different R_F values for D- and L-tyrosine or D- and L-glutamic acid. In the case of the D- and L-isomers of tyrosine-3-sulphonic acid the difference in R_F values was 0.21. A preparative separation of 250

TABLE 131. R_F values of optical antipodes in dependence of concentration of D-tartaric acid (Bonino and Carassiti, *309, 310*)

	50% D-*tartaric acid*	13.04% D-*tartaric acid*
(l) β-Naphtholbenzylamine (II) . . .	0.17–0.19	0.14–0.16
(d) β-Naphtholbenzylamine (II) . . .	0.45–0.50	0.35–0.40

mg on a chromatopile gave two well defined fractions (at discs 180–255 and 275–330). Elution gave a fraction with $[\alpha]_D = -4.14°$ and another with $[\alpha]_D = +4.47°$. The authors conclude: "It is most reasonable to consider that these resolutions are due at least in part to the asymmetric character of the cellulose".

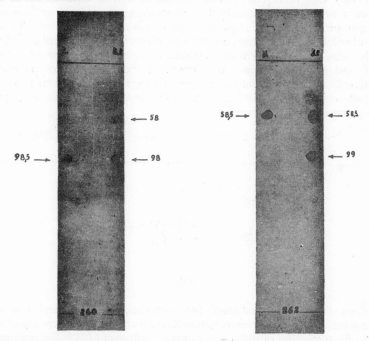

Fig. 105. Separation of the optical antipodes of α-aminophenylacetic acid on paper impregnated with *d*-camphor-10-sulphonic acid. Left: *l* and *dl*, right: *d* and *dl* compounds (Berlingozzi *et al.*, *221*)

Berlingozzi *et al.* (*221*) have separated the optical antipodes of α-aminophenylacetic acid on paper impregnated with *d*-camphor-10-sulphonic acid. The *l*-acid had an R_F of 0.378, the *d*-acid of 0.223 (Fig. 105).

More recently, Berlingozzi *et al.* *(216)* have also separated the optical antipodes of DL-phenylalanine in presence of D-tartaric acid in the mobile solvent and have described modified conditions for such resolutions.

Dalgliesh *(685)* has shown that some aromatic amino acids, e.g. D,L-kynurenine (III), 3-hydroxy-DL-kynurenine (IV), 2,3- and 2,5-dihydroxy-phenyl-DL-alanine (V), and 3,4-dihydroxy-2-methylphenyl-DL-alanine (VI), but not 3,4-dihydroxyphenyl-DL-alanine or its 5-methyl derivative, can be resolved on paper using *optically inactive solvents* (the organic phase of butanol-acetic acid-water). Resolution was shown by the appearance of two spots on the chromatogram of a chemically pure racemate, these being of equal size, shade and intensity. It is evident that in these cases, resolution must be due to *adsorption on the optically active cellulose.* The ratio of the R_F values of the two optical isomers was about 0.9, suggesting a common mechanism operative in all cases. Dalgliesh *(685)* points out that substances containing flat areas, such as an aromatic ring, and hydrogen bonding groups, such as the α-amino acid grouping, should be capable of being adsorbed on the long flat molecules of cellulose, which contain large numbers of groups forming hydrogen bonds. The structural features necessary for resolution are discussed in detail and a "three point" attachment of the molecule to the cellulose postulated: at the α-amino group, the carboxyl and one substituent in the ring, allowing a closer "fit" with the cellulose surface, and hence a greater adsorption of one of the antipodes.

(III R=H)
(IV R=OH)

(V)

(VI)

Lambooy *(1872)* has confirmed the observations of Dalgliesh concerning the resolution of the dihydroxyphenylalanines. See also Ogawa and Ohno *(2401)* for resolution of sodium glutamate, Fujisawa *(1058)* for resolution of kynurenine and tyrosine-3-sulphonic acid, Closs and Haug *(591)* for resolution of tryptophan, Weichert *(3482)* for resolution of histidine, Alessandro and Caldarera *(35)* for resolution of sympathicomimetic drugs (pervitin, ephedrine, noradrenaline, adrenaline, *iso*propyladrenaline, sympanine). No rule seems to have been found until now to predict which of the antipodes moves further than the other (Weichert, *3482*).

Roberts and Wood *(2693)* have adsorbed the catechins of tea leaf on paper with water as the mobile solvent and concluded that separation of optical isomers "appears likely".

2. CIS-TRANS ISOMERS

Zechmeister and McNeely (*3683*) studied the separation of *cis*- and *trans*-stilbenes on Superfiltrol (Filtrol Corp., Los Angeles, U.S.A.). Quantitative separations of the *cis* and *trans* isomers of *p*-methyl- and *p*-methoxy-stilbenes were achieved. In each case the *trans* isomers were more strongly adsorbed than the *cis* isomers. The position of the zones was detected by streaking the column with permanganate. This chromatographic separation also allowed the removal of fluorescent impurities which were retained at the top of the column.

Sandoval and Zechmeister (*2806*) separated the stereoisomeric diphenyl-butadienes on alumina; a detailed study of the chromatographic behaviour of the three stereoisomeric 1,4-diphenylbutadienes has been published by Pinckard *et al.* (*2521*). Zechmeister and LeRosen (*3682*) obtained three distinct zones with diphenyloctatetraene (all *trans*; *trans*, *cis*, *trans*, *trans*; and *trans*, *cis*, *cis*, *trans*) on a column of chalk. Splitting of the zones was observed if the column was irradiated for a short time due to the formation of new isomers.

Kaufmann and Wolf (*1660*) separated stereoisomeric acids on silica or charcoal (maleic-fumaric; oleic-elaidic and erucic-brassidic acids). For the use of chromatography in the separation of the various *cis-trans* isomers of the carotenoids, the review of Zechmeister (*3674*) should be consulted as well as the section on Carotenoids (p. 379).

Bickoff (*241*) studied the relationship between the separation of the β-carotene stereoisomers as a function of solvent composition. The best separations were obtained with low concentrations (0.8 to 1.7 %) of anethole or *p*-methoxytoluene in light petroleum ether (Fig. 20, p. 44).

Zechmeister *et al.* (*3684*) separated the *cis*- and *trans*-oximes of benzoin and anisoin by adsorption on Neutrol Filtrol (Filtrol Corp., Los Angeles, U.S.A.). The position of the zones was determined by streaking with an ammoniacal solution of copper sulphate which gave a green colour with the *trans*-oxime (the upper zone) and a brown colour with the *cis* isomer. The Neutrol Filtrol column had the disadvantage of causing slight isomerisation of the *cis* to the *trans* form.

For separations in the azobenzene series, see Cook and Jones (*629*).

Frankel and Wolovsky (*1022*) have separated *cis*- and *trans*-azobenzene on paper. Elenkov (*879*) has proposed to use paper chromatography to differentiate ethylenic isomerism from physical polymorphism.

3. STERIC FACTORS IN ALICYCLIC SUBSTANCES

Vavon and Gastambide (*3381*) separated cyclic *cis* and *trans* isomers on alumina (menthol and *neo*menthol, borneol and *iso*borneol), the *trans* isomers being the more strongly adsorbed.

The importance of the steric factor has been demonstrated by Vavon *et al.* in subsequent notes (*3382–3384*); they have shown that with ketones and their oximes, with amines and their derivatives, the less adsorbed substance is the one which has the sterically more hindered group, e.g. fenchone (VII) is less adsorbed than camphor (VIII); and *neo*bornylamine (IX) less adsorbed than bornylamine (X).

(VII)   (VIII)   (IX)   (X)

This is true not only for alumina, but also for other adsorbents, such as activated charcoal, silica gel and activated clays (*3383*). For further examples see Gastambide (*1090*).

Ruzicka *et al.* (*2763*) separated the epimers of 7-hydroxycholesterol in the form of their acetates. For the separation of coprosterol and epicoprosterol see *1907*. For the separation of *cis-* and *trans-*testosterone and other diastereoisometric sterols see the chapter on steroids, p. 274.

4. SEPARATIONS OF DIASTEREOISOMERS

Diastereoisomers are usually easily separated by various chromatographic methods. It may suffice to quote some examples of successful separations:

Adsorption chromatography: the diastereoisomeric corynomycolic acids (XI)

$$CH_3(CH_2)_{14}CH(OH)\underset{|}{C}HCOOH$$
$$(XI) \qquad C_{14}H_{29}$$

on alumina (Polonsky and Lederer, *2550*) the diastereoisomers of the heterocyclic bases (XII and XIII) on alumina (Prelog *et al.*, *2581, 2583*).

Ion exchange chromatography: isoleucine and *allo*-isoleucine on Dowex-50 (Piez, *2517*), the diastereoisomeric hydroxylysines on Dowex-50 (Hamilton and Anderson, *1288*), threonine and *allo*threonine on Dowex-50 (Shulgin *et al.*, *2972*).

Partition chromatography: For numerous examples in the carbohydrate field, see the relevant chapter.

The D,D-, L,L- and *meso*-forms of α,ε-diaminopimelic acid can be separated on paper with certain solvent systems (Rhuland *et al.*, *2673*).

(XII)   (XIII)   COC₆H₅

DIVISION V

CHROMATOGRAPHY OF INORGANIC SUBSTANCES

CHAPTER 40

Inorganic adsorbents

1. INTRODUCTION

The use of inorganic adsorbents such as Al_2O_3, MgO and $CaCO_3$, yields the best results for the separation of closely related organic compounds hence the same adsorbents were tried for similar separations of inorganic cations and anions by Schwab *et al.* (*2908–2915*). These workers effected separations on columns of alumina by the same techniques as are employed for organic separations although it was realized that a separation of metallic cations on alumina columns must depend on a mechanism differing from that of the fractional adsorption process of the organic chromatograms.

In this chapter the developments of chromatography of inorganic ions on adsorption columns will be described as distinct from the separation on ion-exchange resins and by fractional solvent-solvent extraction (partition).

A review of this field is also found in books by L. Zechmeister (*3673*), Smith (*3038*) and Pollard and McOmie (*2539*).

As will be seen in the following pages, the "adsorption" columns in use for inorganic separations, function mainly as fractional precipitation, hydrolysis or complexing columns rather than as true adsorption columns, and in some cases the mechanism of the separation is not clear at all.

2. ALUMINA

(a) *Separation of cations*

The mechanism of the separation of cations from aqueous solutions was reviewed by Sacconi (*2775*), who lists the following hypotheses:

(i) Na^+ ion-exchange of adsorbed Na^+ ions present as sodium aluminate.
(ii) Precipitation of basic salts caused by the alkaline reaction of technical alumina (*2914*).
(iii) Precipitation of basic carbonates due to alkali carbonates and bicarbonates present in the alumina (*2982*).

(iv) Al^{+++} ion-exchange.
(v)   Amphoteric adsorption by alumina.
(vi) Preferential adsorption of ions, which then binds the ion of opposite
      charge by preferential secondary adsorption.
(vii) Molecular adsorption (*1530*).

Sacconi rejects hypotheses (i), (iii) and (iv) on experimental evidence
and suggests a theory of hydrolytic adsorption associated with Al^{+++}–H^+
and Na^+–H^+ exchange processes. The adsorption affinity is proportional to
the polarizing power exerted by the ions as well as to the polarisability of
groups coordinated around the ions; this conclusion was also reached by
Venturello (*3391*).

Na^+ ions are not necessary for adsorption, according to Fricke and Neugebauer (*1039*), who prepared Na-free Al_2O_3 from Al foil. The same observation was made by Grasshof (*1189*) and by Tanaka and Shibata (*3203*).

The formation of zones of hydrolysed ions was demonstrated by Schwab
and Issidoridis (*2913*) by measurement of the light absorption of the adsorbed ions by reflected light. The adsorbed layer showed a light absorption
shift intermediate between that of the free hydrated ions and the hydroxide.

Venturello and Burdese (*3393*) also showed that there exists a reversible
and irreversible adsorption of Cu^{++} and SO_4^{--} ions on alumina. The irreversible adsorption increases with time and temperature. It is suggested that
the irreversible adsorption involves the loss of H_2O molecules of hydration
from the $Cu(H_2O)_4^{++}$ ion.

A series of cobaltammines (*2775*) gave a sequence of adsorption bands
on the chromatogram depending on their ionic potentials. The sequence
of metallic ions correlate well with their ease of hydrolysis and their ionic
potentials, a conclusion also reached by Shibata (*2964*) and Tanaka and
Shibata (*3203*).

Jacobs and Tompkins (*1529, 1530*) carried out studies on the kinetics of
the adsorption of ions on alumina. They studied a few variables for the
stationary adsorption of ions and also examined the rate of attainment of
equilibrium when solutions are run through columns of alumina. They found
that the zones, which by visual examination appear to be separated, overlap
to a considerable extent and that chromatography on alumina is unsuitable
both for separation and quantitative analysis.

Heinrich (*1358*) examined the influence of temperature on the adsorption
of Cu^{++} and SO_4^{--} ions. The amount adsorbed rises steeply from 20° to 80°.

Many of the more recent papers on alumina chromatography deal
primarily with explanations of the mechanism rather than with practical
applications. Precipitation owing to buffering by the alumina is convin-
cingly suggested by Hayek and Lorenz (*1341*), also by Schäfer and Neugebauer (*2836*). The importance of the OH layer on alumina was demonstrated
by Specker and Hartkamp (*3058*) who find adsorption of Cu^{++} to be in
relation to the solubility of the alumina in NaF. Further studies of the
equilibria of adsorption were made by Umland and Fischer (*3348*) who

claim that a mass action equation holds for very pure γ-Al_2O_3. See also D'Ans et al. (698), Yasunaga and Shimomura (3650) and Shemyakin et al. (2956).

The separability of ions is usually expressed as the order of zones obtained with the slowest moving band placed first, since the conditions are not sufficiently standardized to obtain comparable rates of travel for the ions. Schwab and colleagues (2914) give the following order for cations from aqueous solution on technical alumina:

$$As^{+++},\quad Sb^{+++},\quad Bi^{+++},\quad \begin{matrix}Cr^{+++}\\ Fe^{+++}\\ Hg^{++}\end{matrix},\quad UO_2^{++},\quad Pb^{++},\quad Ag^+,\quad Zn^{++},\quad \begin{matrix}Co^{++}\\ Ni^{++}\\ Cd^{++}\end{matrix},\quad Tl^+,\quad Mn^{++}.$$

This order is supplemented by Sacconi (2771–2775) with other metals:

$$\begin{matrix}Th^{++++}\ Zr^{++++}\ Fe^{+++}\\ Al^{+++},\quad Cr^{+++},\quad Ti^{++++},\ Hg^{++},\ UO_2^{++},\ Pb^{++},\ Cu^{++},\ Ag^+,\ Zn^{++},\quad \begin{matrix}Fe^{++}\ Ni^{++}\\ Co^{++},\ Cd^{++}\end{matrix},\ Tl^+,\ Mn^{++}.\\ U^{++++}\ Ce^{++++}\ Ce^{+++}\end{matrix}$$

Venturello (3392), who examined still more metals, gives a slighty different order:
Ca, Pd, Pt; Au; Mo; As; Bi, Ga, Os, Th, ZrO; Co, Cr, Cu, Fe, Hg, Ni, Pb, UO_2, V; Ag; Be, Cd, Mg, Mn, Tl, Zn; see also Tanaka and Shibata (3203, 3204).

In dilute solutions, Cu–Ag–Zn is reversed to Cu–Zn–Ag as shown by Srikantan and Krishnan (3075). From ammoniacal solutions (2914) the complex ammines give the order

$$\begin{matrix}Co^{++}\ Zn^{++},\ Cd^{++},\ Ni^{++},\ Ag^+\\ Cu^{++}\end{matrix}$$

From a solution containing tartaric acid (2914) the tartrate complexes adsorb in the order:

$$\begin{matrix}Zn^{++}\\ Pb^{++}\\ Mn^{++},\ Cd^{++},\ Cu^{++},\ Co^{++},\ Ni^{++}\\ Bi^{+++}\\ Fe^{+++}\\ Cr^{+++}\end{matrix}$$

From a solution of dioxan (2035): $\begin{matrix}Fe^{+++}\ Ag^+ \qquad\qquad Cd^{++}\\ Ni^{++},\ Co^{++},\ Zn^{++},\ Cu^{++},\ Ce^{+++}.\\ Cr^{+++}\end{matrix}$

Karschulin and Svarc (1644) separated As, Sb and Sn as the tartrate complexes. Pinterović (2522) separated Bi, As, Sb and Sn tartrate complexes on HCl-washed alumina. On development with H_2S water the sequence is Bi^{+++}, Sb^{+++}, Sn^{++++}, Sn^{++} and As^{+++} if the column is first moistened with tartrate solution and developed with tartaric acid solution.

The platinum metals were separated by Schwab and Ghosh (*2912*) giving the order Ir, Ru, Pt, Pd, Rh from aged solutions. The platinum metals can also be separated from Pb, Cu, Zn and Ni but not from Fe^{+++}.

Venturello and Saini (*3398*) also separated the platinum metals on alumina. Benzidine may be used as a reagent for all the metals, Na$_2$S and thiobarbituric acid for Rh and thiourea for Os.

Venturello (*3392*) separated mixtures of Al–Mg–Be (in this order) by developing with quinalizarin (1,2,5,8-tetrahydroxyanthraquinone). *Rare earth mixtures* were examined by Erämetsä (*906, 908*) in aqueous as well as tartrate and citrate solutions. Complete separations were not obtained but purification is feasible by repeated chromatography combined with other methods. Croatto (*657*) who worked independently on the separation of rare earth metals also failed to achieve complete separations, but obtained practical separations of Ce–La, Ce–Nd, Ce–Sm and Ce–Pr mixtures. Different behaviour was noted on alumina prepared by different methods.

Fischer and Niemann (*970a*) use chromatography on alumina for the purification of yttrium. The final products contained less than $2 \cdot 10^{-3}\%$ of other rare earth elements. The sequence of elution of the nitrates from a column of γ-alumina is given as:

$$\text{Y–La–Pr–Nd–till Tb–Dy–Er–Yb–Lu (Y being eluted first).}$$

(i) *Radiometric analysis*

The analytical problems associated with fission products were approached in the U.S.A. by the use of ion-exchange resins. In Germany, similar work was carried out at the same time with columns of Brockmann's alumina (*2018–'23*).

The submicro amounts of radioactive material were found to behave identically to macro amounts. Thus the thallium isotope ThC″ could be separated from lead (ThB) and bismuth (ThC) without a carrier (*2018*). Separations of radium and barium were also achieved on alumina with better results than previously known (*905, 907*). Some results were also recorded with radioactive rare earths. The technique used in this work is to run a solution through a column and to determine the change of activity of the column, the filtrate and the original solution over ca. 20 days.

By comparison with equilibrium experiments, the separation factor between La and Pr was found to be 2.2; the number of theoretical plates of the column was found to be 20. The separation factors of all the rare earths were obtained by this method (*2019, 2023*). However only La is readily separated from other rare earths. Several articles by Lindner (*2020–2022*) review this work and indicate proposed developments, which have become obsolete due to the better separations obtained with ion-exchange resins.

The behaviour of "masurium" on an alumina column was studied by Flagg and Bleidner (*977*).

Kamieński detects elution of ions from a column with sealed-in Sb-electrodes (*1624*).

(ii) *Separation of metal complexes*

The cobaltammines (*2775*) are adsorbed from aqueous solution in the order:

$(Co(NH_3)_3(H_2O)_3)^{+++}$ $(Co(NH_3)_5Cl)^{++}$
$(Co(NH_3)_4(H_2O)_2)^{+++}$ $Co(H_2O)_6^{++}$ $(Co(NH_3)_6)^{+++}$ $(Co(NH_3)_4CO_3)^{+}$
$(Co(NH_3)_3(H_2O)_2Cl)^{++}$ $(Co(NH_3)_4Cl_2)^{+}$ $(Co(NH_3)_5NO_2)^{++}$

The dithizone complexes can be adsorbed from CCl_4 and $CHCl_3$ solutions. Erämetsä (*907*) gives the following order for a CCl_4 solution:
Sb^{+++}, Sn^{++}, Ni^{++}, Mn^{++} are completely adsorbed.
Cu^{++} forms a grey-green ring at the top of the column.
Cd^{++} forms an orange ring below Cu^{++}.
Fe^{++} also forms an orange ring in this region, as does excess dithizone.
Co^{++} gives a typical, rapidly wandering ring of blue-violet, while between Co^{++} and Cd^{++} nothing is adsorbed.
Zn^{++} forms a bright carmine-red ring below Co^{++}.
Hg^{++} and Hg^{+} form an orange-yellow ring below Zn^{++}.
For elution of dithizonates with acetone see Bach (*117*). Tanaka *et al.* (*3202, 3204*) found that as little as 0.001 γ Co and 0.05 γ Cu could be detected by chromatographing dithizonates. Paulais (*2471*) separated the α-nitroso-β-naphtholates of Co, Fe, Ni and Cu by chromatographing a chloroform solution on alumina columns; Bach (*117*) used a benzene solution. Dean (*719*) developed a quantitative estimation of Co, by separating the nitroso-R salt of Co on acid-washed alumina. Cu, Cr, Ni and Fe are eliminated for subsequent colorimetric determination. The method is employed for cast iron, steels and Ni base alloys. Berkhout and Jongen (*214*), also King *et al.* (*1721*) have employed this method successfully, the latter workers for Co in animal feeds. Blair and Pantony (*266b*) describe a method for the determination of *chromium*. Cr(III) is precipitated with 8-hydroxy-quinaldine from a suitable solution. The dried precipitate is dissolved in $CHCl_3$, diluted once with benzene and passed over a column of activated alumina. The eluate containing the chromium complex only is used for a spectrophotometric determination. For further separations of organo-metallic complexes see Al-Mahdi and Wilson (*43, 44*). Bach (*117*) tried ethereal solutions of the thiocyanate complexes of Fe and Co on alumina. The Fe complex hydrolyses and turns yellow.

(b) *Separation of anions*

Schwab and Dattler (*2909*) found that mixtures of anions could be separated on alumina pre-treated with nitric acid, and suggested ion exchange to be mainly responsible for the effects observed. They give the following order for the bands:

$$Fe(CN)_6^{----} \quad Fe(CN)_6^{---}$$
$$OH^-, \quad PO_4^{---}, \quad F^-, \quad CrO_4^{--}, \quad SO_4^{--}, \quad Cr_2O_7^{--}, \quad Cl^-, \quad NO_3^-, \quad MnO_4^-, \quad ClO_4^-, \quad S^{--}.$$

Depending on the pH, chromate forms a chromate and dichromate band. This series was extended by Kubli (*1813*) with additional anions:

$$SO_3^{--} \qquad\qquad Fe(CN)_6^{---} \ NO_3^-$$
$$OH^-, \ PO_4^{---}, \ C_2O_4^{--}, \ F^-, \ Fe(CN)_6^{----}, \ S_2O_3^{--}, \ SO_4^{--}, \ Cr_2O_7^{--}, \ CNS^-, \ I^-,$$
$$CrO_4^{--}$$
$$Br^-, \ Cl^-, \ NO_3^-, \ MnO_4^-, \ ClO_4^-, \ CH_3COO^-, \ S^{--}.$$

(c) $Al(OH)_3$ as adsorbent

Fricke and Schmäh (*1040*) report the use of $Al(OH)_3$ as adsorbent instead of Al_2O_3. They separate Fe, Cu and Co in that order.

(d) Applications to qualitative and quantitative analysis

Complete analysis of unknown mixtures is difficult but separations of isolated groups are readily carried out. Schemes for such separations have been suggested by Schwab and Ghosh (*2911*) and by Fillinger (*951*). See also Olshanova and Chmutov (*2413*) and Yasunaga and Shimomura (*3649*).

The sensitivity of such analyses can equal that of spot tests and by combining the use of organic reagents with micro-columns of 1–2 mm diameter, better sensitivities are obtained than with the usual methods. The sensitivity of colours obtained varies between 0.005–0.5 γ for various cations (*2908*).

Combination of spectrography with chromatography was employed by Dunabin et al. (*829*). Cadmium was separated on an alumina column and the portion containing the Cd^{++} spectrographed; as little as $^1/_{50}\ \gamma$ could be determined.

Quantitative results obtained by Schwab (*2915*) were only approximate owing to uneven packing of columns. Lecoq (*1896*) determined Pb in drinking water by passing it over an Al_2O_3 column and developing with sodium sulphide. Quantitative results are obtained by comparison with known amounts. A determination of copper in brass and bronze was worked out by Srikantan and Krishnan (*3076*). The copper is adsorbed on an alumina column from 0.02–0.00125 M solutions forming sharp bands, and developed with ammonium sulphide or potassium ferrocyanide. The solution to be analysed is prepared by dissolving 1 g of the alloy in HNO_3 and diluting to 100 ml. The alumina is prepared from $KAl(SO_4)_2 \cdot 12H_2O$ and precipitated with NH_4OH and dried at 120°. The column is prepared by keeping an alumina suspension at pH 9.4 overnight before packing the column. 99 % accuracy for Cu determinations has been obtained. In a later paper Srikantan and Venkatachallum (*3077*) consider buffering of the column (at pH 9.4) necessary for a linear concentration-zone length relationship in the determination of Cu^{++}. Pinterović (*2523*) determined iron in common salt by adsorption on a column 3–6 mm in diameter and development with $K_4Fe(CN)_6$. A zone – length – concentration curve is constructed and enables the estimation of 0.02 % Fe in 1 g of salt.

A quantitative method for *Ba and Sr* was published by Ballczo and Schenk (*128*). *Nb and Ta* were separated by Tikhomiroff (*3247*) by dissolving the oxides by a $KHSO_4$ fusion and then in 4 % ammonium oxalate at pH 6.4.Ta is eluted with 4 % ammonium oxalate at pH 7 from an alumina column. Nydahl (*2392*) has worked out a chromatographic separation of sulphuric acid for the determination of S in iron and steel. The sulphate is adsorbed on acid washed alumina and eluted quantitatively with ammonia for subsequent determination.

For the detection of heavy metals in pharmaceuticals see Khoklova (*1707*).

(e) *Separation on paper strips impregnated with Al_2O_3*

Blotting paper can be impregnated with Al_2O_3 by dipping it into a sodium aluminate solution and, after drying, immersing it in a sodium bicarbonate solution; Na^+ ions are removed from the paper by washing in water for several days. Flood (*984–991*) employed such paper in preference to columns of alumina. With the exception of the sequences of Cd–Co and Ag–Cu, the same order of adsorption is given by alumina-impregnated paper as by alumina columns. After pre-treating the paper with HNO_3, the same order as is obtained on columns is also given by anions.

Since the "packing" is very uniform, a better mathematical correlation of concentration and zone length is obtained, which Flood expresses as

$$h = h' + k c$$

where

h = zone length
c = concentration of ion adsorbed
h' = 0.3–0.5 mm connected with the smaller adsorption capacity of the zone at the lower end, where the paper is secured by rubber bands.

The chromatography of glycine complexes on alumina-impregnated paper was also studied by Flood (*987*). Depending on the glycine concentration, separations of Cu and Ni, Cu and Co etc. can be carried out. Flood developed a "chromatographic titration"; by varying the amount of glycine added to unknown copper solutions, different chromatograms are obtained and an approximate quantitative analysis of copper can be made by comparison with standard chromatograms.

Iijima *et al.* (*1498*) and Goto and Kakita (*1180*) also developed methods of separation on alumina-impregnated filter paper. Limiting concentrations were found to be 0.1–0.05 mg per ml for metals and 0.5–0.05 N for acids. Paper impregnated with $Cr(OH)_3$ was found better for some cations than alumina paper.

Zolotavia (*3703*) established the following order for some cations on $Al(OH)_3$ impregnated paper: Hg^+, Hg^{++}, Bi^{+++}, Fe^{+++}, (Pb^{++} and

UO_2^{++}), Cu^{++}, Ag^+, (Cd^{++} and Zn^{++}), Tl^+, Fe^{++}, (Ni^{++} and Co^{++}), Mn^{++}, Cr^{+++}.

Vanyarkho and Garanina (*3379*) employed alumina paper for numerous separations of pairs of the metals Fe, Cu, Co, Mo, Ni, Cr and Mn as well as for some plant extracts. Oka and Murata (*2404–2407*) employed alumina papers for qualitative and quantitative analyses. See also Murata (*2326a*). For the combination with polarographic methods see Kemula (*1686*).

Okáč and Černý (*2408*) separate Bi, Sb and Sn by impregnating the paper with a solution of these ions and treating with drops of $(NH_4)_2S$ followed by $(NH_4)_2S_2$. Concentric rings of Bi_2S_3, SnS_2 and Sb_2S_3 are thus formed. Cd–Cu are separated by adding alkali and KCN to the solution and placing a few drops on filter paper impregnated with $(NH_4)_2S$.

For further work see also Okáč and Černý (*2409*). Essentially the same principle is used by Teige (*3218*) who separates halides on paper impregnated with silver chromate.

(f) *Surface chromatography*

T. I. Williams (*3585*) suggested disc chromatograms for the selection of suitable solvents. The method employed is to place some adsorbent between two glass plates and to feed the solvent and sample centrally through a hole in the upper plate. This technique was evolved from that of Brown (*425*) and Crowe (*663*). Meinhard and Hall (*2181, 2182*) developed it further. Uniformly spread layers of adsorbent on micro-slides are used and pipettes of known volume for additions of eluant. A mathematical treatment of the rate of travel of zones was also carried out. However, there appears to be no marked advantage of this technique over impregnated paper or micro-columns.

3. OTHER INORGANIC ADSORBENTS

Besides alumina, various adsorbents and adsorbent mixtures have been described, some of which work essentially by chemisorption forming definite compounds with the adsorbed substance.

Gapon and Chernikova (*1082*) examined the ability of soil clays such as kaolinite, bentonite and montmorillonite to separate Cu and Co. When a column is developed with water, copper is strongly adsorbed as a pale blue band in all cases and cobalt travels down as a pink band.

Hansen and Gunnar (*1298*) separated Zr and Hf as the chlorides on a silica gel column using a solution of the salts in methanol. This method was further elaborated (*1299*) and the details for a large scale process worked out. From natural mixtures a 30–60 % concentrate of Hf can be produced. Complex Ag and Cd cyanides were separated from most other metals by passage through a silica gel column (*1766*). Milone and Cetini (*2229a*) carried out extensive studies with basic silica gel as adsorbent. Numerous separations of the transition elements were recorded such as $Hg(II)$–Cu, Cu–Zn, Cd–Co, Zn–Cd, Cd–Ni, Zn–Ni, Fe–Ni, Pb–Cd, Ag–Ni, Cu–Co, Al–Cu and $Hg(II)$–Fe.

As little as 6 γ of Fe^{+++} in concentrated H_2SO_4 were detected by Shemyakin and Mitselovskiï (*2955*) by running the acid through a micro column of silica gel and developing with ferrocyanide.

Using slices of asbestos millboard Sen (*2940a*) separated Cu–Cd and As–Sb–Sn mixtures with the technique of Flood.

In later papers Sen (*2941*) describes separations of Hg^+–Pb^{++}, Hg^{++}–Pb^{++}, and Hg, Cu, Pb, Fe, Zn, Co, Ni, Mn and Zn. Sen (*2942*) also describes the use of $CaSO_4$ pencils as a medium for inorganic chromatography.

Gapon and Gapon (*1085*) employ inert supports impregnated with various reagents. Silica gel with sodium silicate as the precipitant gives separations of the following pairs of metals: Cu^{++}–Co^{++}, Cu^{++}–Ni^{++}, Cu^{++}–Ag^+, Fe^{++}–Cu^{++}, UO_2^{++}–Cu^{++}, UO_2^{++}–Ni^{++}. Dry Al_2O_3 mixed with 10 % powdered potassium iodide separates the mixtures Bi^{+++}–BiI_4^-, Bi^{+++}–Pb^{++}, Hg^{++}–Bi^{+++}–Pb^{++}, Ag^+–Pb^{++}.

Dry Al_2O_3 mixed with 10 % ammonium thiocyanate gives two zones with cobalt Co^{++} and $Co(CNS)_4^{--}$, and a wet column of Al_2O_3 with $FeCl_3$ as precipitant separates $Fe(CN)_6^{---}$ from CNS. See also Cetini and Ricca (*538a*).

Khomlev and Tsimbalista (*1708*) employ a silver sulphate column with an inert diluant for the separation of the halides.

For further work on precipitation chromatography see also Gapon and Belenkaya (*1081a*).

Bach (*114–117*) employs columns of zinc sulphide for the separation of copper and cadmium from many other metals. Columns containing zinc sulphide mixed with Supercel pulp at pH 4–6.8 are employed for Cu^{++}. Apart from Cu^{++}, only Bi^{+++}, Ag^+ and Hg^{++} form zones and only Bi^{+++} interferes with a reasonably accurate quantitative estimation by measurement of zone length. Cadmium forms a band on a ZnS column from a solution containing Zn^{++} and cyanide ions in excess of the cadmium. Again good zone-length concentration relationships are reported. Bach also employs a column consisting of diatomaceous earth impregnated with mercuric bromide in place of the paper strips impregnated with $HgBr_2$ in the Gutzeit test for arsenic.

In a later paper, Bach (*117*) records experiments with numerous other adsorbents such as $Zn_2Fe(CN)_6$ and $Sn(OH)_4$.

Separation of the inert gases

Janak (*1554, 1558*) describes the separation of H_2, CO, He, Ne, A, saturated and unsaturated hydrocarbons using columns of activated charcoal or silica and CO_2 as eluant. Separations with an accuracy of 0.3–0.5 % were achieved in 5–30 minutes. See also p. 58.

4. SEPARATION OF OPTICAL ISOMERS

Karagunis and Coumoulos (*1628*) separated triethylenediammine chromic chloride isomers on columns of *d-* or *l*-quartz. The quartz column was activated by heating it in a vacuum and the racemic mixture was run over it and eluted with 85 % alcohol. *l*-Quartz adsorbed the dextro form of the complex more strongly than the laevo form and the first portions of the filtrate were laevorotatory. With *d*-quartz the opposite occurs.

Krebs and Rasche (*1795*) obtained only partial resolution of triethylene-diammine-cobaltic chloride when passing over a column of starch. The first fraction showed 32 % resolution; further fractions only 10 %. See also page 420.

CHAPTER 41

Organic adsorbents

1. INTRODUCTION

No detailed studies of the mechanism of the separation of metal ions on organic adsorbents have been reported up to date. However, as the adsorbents used are well known chelating compounds, it appears certain that the separation depends on compound formation between the adsorbent and the metal ion. Thus the chromatograms obtained have the same appearance as those obtained by Gapon and Gapon (*1085*) with inorganic adsorbents, which separate by compound formation.

2. 8-HYDROXYQUINOLINE

Erlenmeyer and Dahn (*914*) first employed columns of powdered 8-hydroxy-quinoline, or 8-hydroxyquinoline mixed with two parts of kieselguhr as an inert support. The columns were 5–8 cm long and 0.3 cm in diameter. From aqueous solutions the following order and colour of bands was obtained:

VO_3^-	WO_4^{--}	Cu^{++}	Bi^{+++}	Ni^{++}	Co^{++}	Zn^{++}	Fe^{+++}	UO_2^{++}
black	yellow	green	yellow	green	pink	yellow	black	orange

In solutions acidified with acetic acid the Fe^{+++} band precedes the Zn^{++}. The sensitivity of the band colours was found to be good, as little as 2γ of Fe^{+++} giving a black band.

G. Robinson (*2705, 2706*) employed a column of 8-hydroxyquinoline for the analysis of cupro-nickel-zinc alloy and copper-zinc-silver alloy. By using as little as 2–6 mg dissolved in a solution buffered to pH 5–6 with sodium acetate, and 8-hydroxyquinoline of grain size 60–80 mixed with 50 or 60 % starch, it was possible to obtain quantitative results by comparing the zone lengths of the metals with those of known amounts under the same conditions.

Considerable difficulties were encountered in an attempt to employ this method for alloy steels since the large amounts of Fe^{+++} present prevented the formation of bands by other metals. It was found that separations with solutions containing acetic acid produced channelling in the adsorbent; thus iron could not be removed in an acid solution. A large number of metals were examined giving the order of adsorption:

VO_3^-	Mn^{++}	WO_4^{--}	Ag^+	Cu^{++}	Bi^{+++}	Ni^{++}	MoO_4^{--}	Al^{+++}
grey	green	yellow	yellow	green	yellow	yellow	orange	green

Co^{++}	Zn^{++}	Mg^{++}	Fe^{+++}	UO_2^{++}
reddish	yellow	green	black	orange

Since there are already a number of yellow zones adjacent to each other, further extension of this series would not be of analytical value.

3. OTHER ADSORBENTS

Columns of violuric acid, mixed with barium sulphate, diatomaceous earth or starch were investigated by Erlenmeyer and Schoenauer (*917*) for the separation of alkali metals, alkaline earths and some pairs of other metals. Distinct coloured zones were obtained with the following pairs: K–Na, NH₄–Na, K–Mg, Mg–Na, NH₄–Mg, Ba–Ca, Sr–Ca, Cu⁺⁺–Hg⁺⁺, Pb⁺⁺–Hg⁺⁺, Cu⁺⁺–Pb⁺⁺, and poor separations of K-NH₄, and Ba-Sr. Robinson (*2706*) also separates Na–K on a violuric acid column. Erlenmeyer and Schmidlin (*916*) separate Na and K semiquantitatively on columns of 5-oxo-4-oximino-3-phenylisoxazoline.

Shemyakin and Mitselovskii (*2954*) employed columns of 8-hydroxy-quinoline, naphthoquinoline and cupferron mixed with potato starch for separations of Fe, Cu, Ni, Co mixtures; only Co and Ni form distinct bands.

Dimethyl glyoxime, either alone or mixed with calcium carbonate or magnesium carbonate, was studied by Burriel-Marti and Pino-Perez (*463*) for the detection of nickel. 5γ of Ni were detectable as a red band on the top of the column even in presence of 10,000 times as much Co. Pd⁺⁺ gives an upper yellow band. In presence of large amounts of both Co⁺⁺ and Fe⁺⁺, the band of Ni⁺⁺ does not form.

Dithio-oxamide, nitrosonaphthols and anthranilic acid were tried by Bach without great succes (*117*).

Activated charcoal can also be employed for some separations. Columns of activated carbon remove noble metals from solution by reduction on the surface of the carbon (Dubrisay, *824*). Alexander (*36*) separated perrhenic acid from molybdic acid by eluting with 1.95 N H_2SO_4 from a column of Norit for several days. The adsorption of complex cyanides on charcoal was examined by Labruto and d'Alcontres (*1842*). From a solution of $K_3Cu(CN)_4$ all the Cu⁺⁺ and all but 0.5 % of the cyanide is adsorbed. A $K_2Cd(CN)_4$ solution is adsorbed except for 2.6 % of the cyanide. The following complexes are completely adsorbed: K_2HgI_4, $K_4Fe(CN)_6$ and $K_3Co(NO_2)_6$.

4. FILTER PAPER IMPREGNATED WITH ORGANIC ADSORBENTS

Under this heading the work of Feigl *et al.* must receive adequate attention. Feigl developed the field of spot tests with special emphasis on filter papers impregnated with reagents. By placing a drop of solution on such a paper, more sensitive reactions could be obtained than in solution. A great deal of information was obtained from so-called capillary pictures, which formed under these conditions. For example, on filter paper impregnated with rubeanic acid, a drop containing Cu, Co and Ni will precipitate Cu^{++} in the centre of a spot, while Co^{++} and Ni^{++}, having more soluble rubeanates, will form a ring around the centre spot of copper rubeanate.

Extensive studies on such capillary phenomena are described in Feigl's books *"Specific and Special Reactions"* (New York, 1940) and *"Spot Tests"* (Amsterdam, 1956) (*940*).

Pfau and Bergt (*2502*) detected 0.05 mg of $CuSO_4 \cdot 5H_2O$ in 20 ml of water by capillary analysis on filter paper strips with 1 % $K_4Fe(CN)_6$. Miličević (*2215*) found that cations may be separated on filter paper with water as solvent. The order of increasing R_F values is Fe^{+++}, UO_2^{++}, Pb^{++}, Hg^{++}, Cu^{++}, Ag^+, Cd^{++}, Co^{++} and Ni^{++}; for anions MnO_4^-, $Cr_2O_7^{--}$, I^- and Br^-. (Compare Lederer, *1914*, who found all metals except Sb^{+++} to travel with the liquid front on Whatman No. 1 paper.)

A number of chromatographic spot tests were described by Hopf (*1443*). While separations on columns seem to have little future owing to the difficulties of getting reproducible, quantitative results, filter paper impregnated with organic adsorbents shows promise as a general analytical tool. Hopf confirms the equations of Flood (*987*), i.e. the relation between zone length and concentration. As Hopf uses "radial development", by placing three drops of the solution on the same point on the impregnated paper, the equation becomes:

$$a = (r_1^2 - r_2^2) K_a \pi$$
$$b = (r_3^2 - r_4^2) K_b \pi$$

where
a = concentration of constituent A
b = concentration of constituent B
r_1, r_2, r_3 and r_4 = radii bounding the zones
K_a and K_b = constants for the substances

If, however, one test is denoted by suffix a and the other by β, the variation may be obtained by direct measurement:

$$\frac{a_a}{a_\beta} = \frac{r_{1_a}^2 - r_{2_a}^2}{r_{1_\beta}^2 - r_{2_\beta}^2}$$

Further, from the above:

$$\frac{a}{b} = \frac{r_1^2 - r_2^2}{r_3^2 - r_4^2} K_{a/b}, \text{ where } K_{a/b} = \frac{K_a}{K_b}$$

The paper employed must be free from impurities; the following methods are recommended for some separations:

Ferromanganese

A paper prepared with peptised alumina and formaldoxime hydrochloride is employed and the solution to be tested containing a metal concentration of 3 % is neutralised to incipient precipitation. A standard stain with a ratio of Fe/Mn = 0.01081 gives $r_1 = 39.17$ mm, $r_2 = 34.00$ mm, $r_3 = 28.8$ mm, $r_4 = 0$. Thus $(r_1{}^2 - r_2{}^2)/r_3{}^2 = 0.456$, i.e. $K_{Fe/Mn} = 0.0237$.

A metal stain of unknown composition gave the radial ratio 0.363, i.e. Fe/Mn = 0.363 × 0.0237 = 0.0086, agreeing with the theoretical figure of 0.00889.

Ferromolybdenum

Paper impregnated with peptised alumina and 8-hydroxyquinoline is used and a solution prepared by dissolving the alloy in HCl, fuming HNO_3 and H_2SO_4, and neutralizing with NH_4OH. On spotting, this solution gives an outer black zone of Fe and an inner yellow spot of Mo, quantitative results being within 1 % of the theoretical value.

Molybdenum and vanadium

On paper prepared with alumina and 8-hydroxyquinoline and at pH 8–9 vanadium is in the centre.

Alkali metals

A paper treated with starch and violuric acid is spotted with a mixture of K^+, $NH_4{}^+$ and Na^+ containing a little acetic acid. A red sodium zone is close to the centre, adjacent to a pink potassium zone and separated from a blue ammonium zone; zone measurements are again satisfactory.

TABLE 132. R_F values of cations with various solvents on paper impregnated with 8-hydroxyquinoline (Laskowski and McCrone, *1883*)

Solvent:	Di-oxan	Pyri-dine	Chloro-form	Ace-tone	Metha-nol	Etha-nol	Propa-nol-1	Propa-nol-2	Buta-nol-1
Al^{+++}	0.79	0.96	0.65	0.88	0.65	0.97	1	1	1
Sb^{+++}	0.0	0.90	—	0	0.69	0.57	0.34	0.25	0.24
Ba^{++}	0.68	0	—	0	0.46	0.06	0	0	0
Cd^{++}	0.0	0.91	0.01	0	0.14	0	0	0	0
Ca^{++}	—	0.11	0	0.08	0.42	0.08	0	0	0
Co^{++}	0.94	0.87	0.70	0.99	0.86	1	1	1	1
Cu^{++}	0.81	0.91	0.62	0.68	0.65	0	0	0	0
Fe^{+++}	0.99	0.92	0.74	—	—	—	—	—	—
Pb^{++}	0.69	0.83	0	0.55	—	—	—	—	—
Mg^{++}	—	0.53	0	0	0.6	0.1	0	0	0
Ni^{++}	0.79	0.93	0.71	0.84	0.79	0.75	1	1	1

Boric acid

Boric acid was separated on turmeric paper by Flood and Risberg (*988*).

Development with organic solvents

Laskowski and McCrone (*1883*) employ filter paper impregnated with 8-hydroxyquinoline. By developing with pure organic solvents, numerous separations are obtained for the common metals. See Table 132.

5. QUANTITATIVE USE OF CHROMATOGRAPHIC COLUMNS WITHOUT SEPARATION

Displacement chromatograms in which a dye such as methylene blue is displaced by alkaline earth metals on starch columns were used as a quantitative method for these ions by Dykyj and Černý (*847*).

6. INDUSTRIAL SEPARATIONS

Hopf, Lynam and Weil (*1445*) patented an apparatus for large scale separation by central feeding and radial development through a column of silica containing one third of oxine. Zinc was separated from iron and nickel by this method. See also page 17.

7. CHROMATOGRAPHY USING GELS AS ADSORBENTS

Milone and Cetini (*2228, 2229*) worked out interesting separations of inorganic ions using gels containing such salts as sodium arsenite, borax and sodium silicate. The very striking and sharp separations obtained with such mixtures as Fe–Cu, Hg–Co, Fe–Pb, Fe–Bi, Pb–Co and Hg–Cu are based on the precipitation of the cations with the salt contained in the gel. The gels employed were either gelatine or agar with about 1 % of the precipitating salt added. In another paper (*2230*) the same authors also show excellent separations of ternary mixtures of cations and tabulate the separations obtained between ten common cations on columns containing some of the precipitating salts mentioned above. While there is no obvious advantage over other known chromatographic methods gel-chromatography presents another variant which may in some cases be of use, for example where optical scanning of the transparent column is desired.

CHAPTER 42

Inorganic separations on ion exchange columns

1. APPARATUS

As pointed out in chapter 10, a particle of an ion exchange resin is completely permeable and reacts in its entirety, not only on the surface as with adsorbents. Thus it is not so important to have the resin in tightly packed columns, consisting of fine grains, but larger particles may be used, permitting faster flow of liquid. The columns used for ion exchange are therefore equipped with arrangements to slow down the flow of liquid, while suction or pressure is necessary for the Tswett adsorption columns. Fig. 106 shows the type of column used by Tompkins et al. (3284) for experimental separations. Column lengths are termed "resin bed depths" in ion exchange chromatography. Fig. 106 has a "resin bed" 1 sq.cm × 10 cm, with a porous glass disc to support the resin. The stopcock allows regulation of the flow rate, which is adjusted so as to keep the liquid level above the resin. Fig. 107 is a column set-up adaptable for deeper resin beds. Instead of employing a glass swan neck, rubber tubes are used. By raising or lowering the bottle of solvent, the flow rate can be changed. For an improved column design, see Brunisholz (430).

With ion exchange resins, liquid chromatograms are invariably carried out. The out-flowing solvent is collected in convenient portions and analysed. In most of the work radioactive tracers were used and the solvent evaporated in cylindrical capsules. The activity of the residue was determined with a conventional Geiger-Müller counter.

Alternatively, the radio-activity of the effluent was continuously recorded by Ketelle and Boyd (1702), employing a Geiger-Müller tube. Fig. 108 is a schematic representation of such an apparatus, which also includes a jacket for temperature control of the resin bed. Ketelle and Boyd carried out separations at 100° and for this purpose the solvent was pre-heated by passage through an Allihn condenser before the admission to the resin bed.

After flowing trough the resin bed A, the solvent passes through the counting cell B. The counting cell B consists of a cylindrical block of Lucite, with a spiral groove cut in one face. This channel is covered by a window of polystyrene 3.2 mg/sq.cm thick, which is sealed to the Lucite block with Amphenol cement. Access to each end of the spiral channel is attained through holes and threaded nipples.

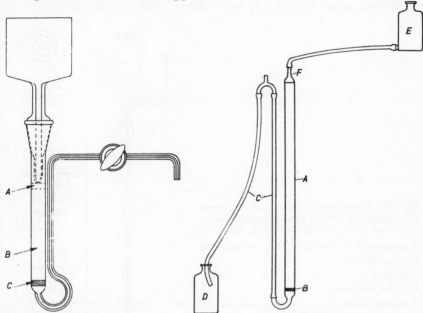

Fig. 106　　　　　　　　　　　　　　Fig. 107

Fig. 106. Experimental column. The funnel is removable to facilitate resin addition and removal. The resin bed, *B*, in this column is 1 sq. cm × 10 cm and rests on the porous glass disc, *C*. A stopcock in the outlet tube allows regulation of the flow rate. The opening in the outlet tube is above the top of the resin bed, thus maintaining a liquid layer, *A*, above the resin at all times (Tompkins *et al.*, *3284*)

Fig. 107. Apparatus after Harris and Tompkins (*1311*). The resin bed, *A*, rests on a porous discs *B*, in the glass column *F*. The flow rate is adjusted by varying the height of the bottle, *E*, which contains the influent solution. The effluent is collected in a bottle, *D*. The vent in tube *C* ensures a continuous liquid layer over the resin bed

In pilot scale separations of rare earths, Spedding (*3059*) employed 8 ft. columns with 4 in. diameter and collected the eluate in 45 litre bottles, which were changed every twelve hours; see also Gilwood (*1110*) and Brunisholz (*431*).

2. PREPARATION OF THE RESIN BED

Spedding *et al.* (*3069*) recommend the following treatment for Amberlite IR-1:

The resin was first screened to remove large particles ($>$ 20 mesh) and then soaked in water. The resin slurry was poured in the column and then backwashed with water to remove the fine resin particles, which tend to plug the column and thus cut down the flow rate. The resin was conditioned as follows: A solution of 5 % hydrochloric acid was passed through the

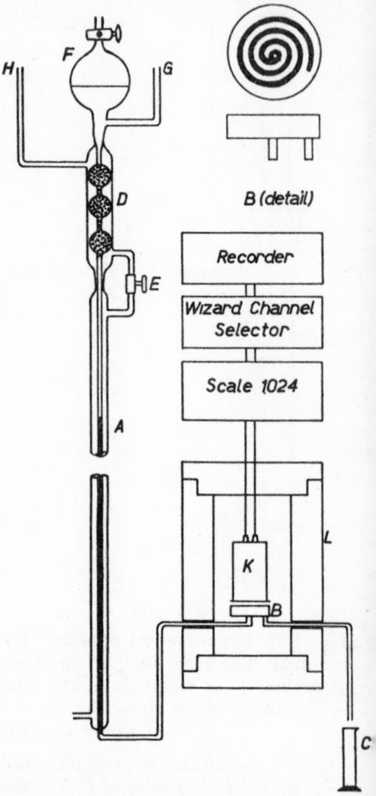

Fig. 108. Experimental arrangement employed in ion exchange column separations with radioactive rare earths (Ketelle and Boyd, *1702*)

A adsorbent bed, Amberlite IR-I or Dowex-50

B counting cell

C receiver

D Allihn condenser

E throttle valve

F gas entrainment bulb

G eluant inlet

H thermostat fluid inlet

K mica end-window Geiger-Müller counting tube

L two-inch lead radiation shield

column converting the resin to the acid form. The time of contact was at least 30 minutes to insure complete conversion. This was followed by a 2 % sodium chloride solution to convert the resin to the sodium form. Treatment with acid and salt was repeated at least two more times, the resin being finally left in the acid form.

In large scale separations the resin beds had to be re-used repeatedly and the regeneration was carried out by Spedding *et al.* as follows:

4 litres of 5 % citrate solution at pH 5.0, 4 litres of 5 % NaCl and 4 litres of 5 % hydrochloric acid were run through the bed. After removing the excess acid with distilled water, the new sample, dissolved in HCl, was poured onto the column. For a review of ion exchange technique see also Abrahamczik (*3*).

3. SEPARATION OF RARE EARTH METALS

(a) *Factors influencing separation*

Spedding (*3059*) reviewed the separation of rare earths; in a later article Spedding and Powell (*3066a*) present their procedures for obtaining rare earths of high purity on a pilot scale. The important variables which influence purity, zone shape and degree of separation are: (1) pH; (2) citrate concentration; (3) column load or rare earth concentration; (4) flow rate; (5) dimensions of the column; (6) particle size; (7) temperature; (8) mould growth in the citrate solution.

(i) *pH*

Considerable amount of work has been carried out to determine the optimum pH. There is only one optimum pH for a given resin and a given

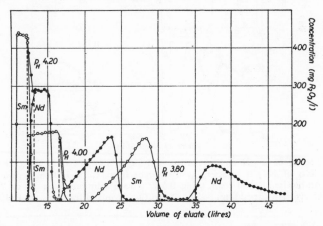

Fig. 109. The effect of pH on the elution of 1.713 g samples of equimolar Sm_2O_3-Nd_2O_3 mixtures from 2.2 × 120 cm beds of —30 +40 mesh size Amberlite IR-100 using 0.5% citrate solutions at a linear flow-rate of 0.5 cm/min (Spedding, *3059*). ○ Sm_2O_3, ◑ mixed fractions, ● Nd_2O_3

citrate concentration. Figs. 109 and 110 show the effect of pH on the separation of Nd and Sm, Nd and Pr at a citrate concentration of 0.5% and Fig. 111 the separation of Sm and Nd at a citrate concentration of 0.1%. In large scale separations a lower degree of purity is sometimes more economical, if it reduces the elution time. In earlier work (*3284*), higher concentrations of citrate (5%) gave much lower optimum pH values. For various resins the optimum pH also changes; thus for 5% citrate the optimum pH on Amberlite IR-100 is 2.6–2.9, while for Dowex 50 it is 2.9–3.3 (*1702*).

Nervik (*2355a*) employed gradient elution (see page 41) in rare earths separations using a hot lactate solution with a pH gradient.

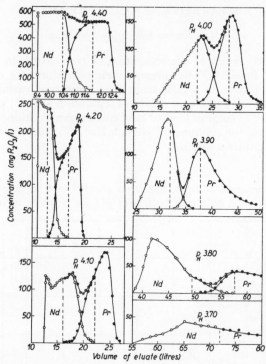

Fig. 110. The effect of pH on the elution of 1.695 g samples of a mixture containing equal amounts of Nd_2O_3 and Pr_6O_{11} by weight from 2.2×120 cm beds of $-30 +40$ mesh size Amberlite IR-100 resin, using 0.5% citrate solutions at a linear flow-rate of 0.5 cm/min (Spedding, 3059). ○ Nd_2O_3, ◑ mixed fractions, ● Pr_6O_{11}

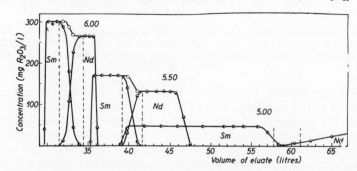

Fig. 111. The elution of 1.713 g equimolar mixtures of Sm_2O_3 and Nd_2O_3 from 2.2×120 cm beds of $-30 +40$ mesh size Amberlite IR-100 resin, using 0.1% citrate solutions at pH values of 5.0, 5.5 and 6.0 at a linear flow-rate of 0.5 cm/min (Spedding, 3059).
○ total R_2O_3, ◐ Sm_2O_3, ◓ Nd_2O_3; broken vertical lines indicate amount of overlap between Sm and Nd bands

(ii) *Citrate concentration*

The citrate concentrations reported to give good separations are 5.0 %, 0.75 %, 0.5 % and 0.1 % (expressed as grams of hydrated citric acid per 100 ml of water). In the elution with 5.0 % citrate solutions gently sloping elution curves are formed which have a considerable zone of cross contamination.

At lower concentrations (0.5 % and 0.1 %) elution bands with steep front

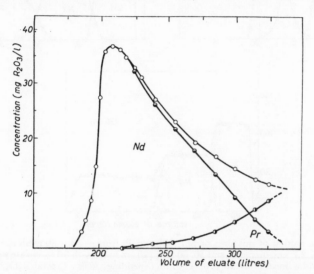

Fig. 112. The elution of a sample, 0.25 g/sq.cm, analyzing 51.7% Pr_6O_{11} and 48.3% Nd_2O_3, from a bed of Amberlite IR-1 resin, 6.4 × 175 cm, using 5% citrate solution at a pH value of 2.55 and a flow-rate of 1.5 cm/min. (Spedding, *3059*)

and rear edges and flat tops are produced. One earth immediately follows the next, indicating displacement of one by the next. Low citrate concentrations are preferable because good yields of very pure elements are obtained and also for economical reasons. Fig. 112 shows a Nd band with 5%, Fig. 110 with 0.5% and Figs. 111 and 113 with 0.1% citrate.

(iii) *Column load*

Spedding *et al.* (*3070*) recommend an optimum of 0.25–0.5 g of rare earth oxide per square cm of resin bed for the separation of Nd and Pr on a macro scale. Overloading gives poorer separations; however, column load and column length are interdependent variables; see Fig. 114.

(iv) *Flow-rate*

With slow flow-rates, equilibrium is more easily established; thus band overlap is reduced and better separations are obtained. Channelling is also greatly

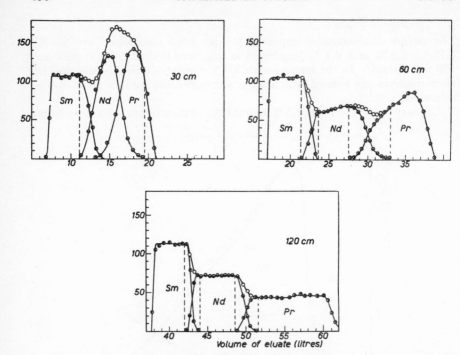

Fig. 113. The elution of mixtures of Sm, Nd and Pr from $-30 + 40$ mesh size Amberlite IR-100 beds, 2.2 cm diameter and 30, 60 and 120 cm long, with 0.1 % citrate solution at a pH of 5.30 and a linear flow-rate of 0.5 cm/min (Spedding, *3059*). ○ total R_2O_3, ◑Sm_2O_3, ◒ Nd_2O_3, ◐ Pr_6O_{11}; broken vertical lines indicate overlap between bands

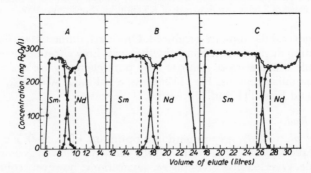

Fig. 114. The effect of increasing column load and column length proportionately (Spedding, *3059*)

A 1.65 g of equimolar Sm_2O_3-Nd_2O_3 mixture on a 2.2 × 30 cm bed of Amberlite IR-100 resin; *B* 3.30 g on 2.2 × 60 cm bed; *C* 4.95 g on 2.2 × 90 cm bed. ○ total R_2O_3, ◑ Sm_2O_3, ◒ Nd_2O_3; broken vertical lines indicate amount of overlap between bands

reduced at lower flow rates (*3064*). Fig. 115 shows the effect of varying the flow rate for two different **particle sizes.** The flow rate is **here varied from** 0.5 cm/min **to 2.0 cm/min.** Another method of **expressing** the flow rate is in ml per cm² of resin bed. Usual rates are 0.2 ml–2.2 ml per cm².

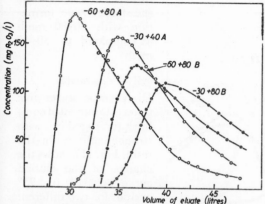

Fig. 115. The effect of particle size of resin on the elution of 1.683 g samples of pure Nd₂O₃ from 2.2 × 120 cm beds of Amberlite IR-100 resin, using 0.5% citrate solution at a pH of 3.80 and linear flow-rates of 0.5 cm/min and 2.0 cm/min (Spedding, *3059*). ○ 0.5 cm/min; ● 2.0 cm/min

(v) *Dimensions of the column*

The effect of the length of the column on the shape of the elution curve is shown in Fig. 113 and 116. A 30 cm column is too short, the widths of the elution bands for the 60, 120 and 240 cm columns are the same. Once maximum separation has been reached, there is no advantage in increasing the column length any further.

The width of the column has no influence on the separations obtained.

(vi) *Particle size*

The commercial ion exchangers are screened; the usual sizes used are 30–40 mesh, 40–60 mesh, or 60–80 mesh.

Fig. 116. The elution of 1.713 g equimolar Sm₂O₃-Nd₂O₃ mixtures from — 30 + 40 mesh size Amberlite IR-100 resin beds, 2.2 cm diameter and 30, 60, 120 and 240 cm long, respectively, using 0.1% citrate solution at a pH of 5.50 and a linear flow-rate of 0.5 cm/min (Spedding, *3059*).

○ total R₂O₃, ◐ Sm₂O₃, ◑ Nd₂O₃; broken vertical lines indicate amount of overlap between Sm and Nd bands: solid line across figures gives the pH of the eluate, reading on the right-hand scale

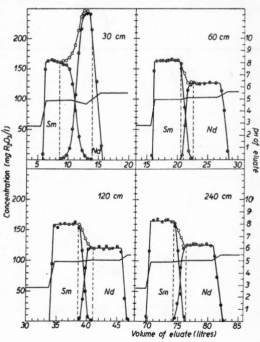

Smaller particles produce a more uniform column and hence better separations. Fig. 115 shows the steeper zone formation with 60–80 mesh as compared to 30–40 mesh.

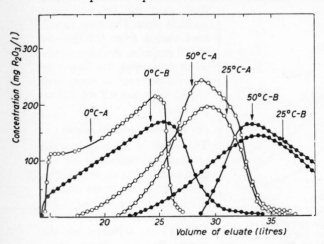

Fig. 117

(vii) *Temperature*

High temperatures give faster attainment of equilibrium together with stronger affinity for the resin. Fig. 117 shows the change of the elution curve between 0°, 25° and 50° for two different flow rates. Ketelle and Boyd (*1702*, *1703*) separated rare earths on Dowex-50 at 100° and obtained much improved separations as compared to room temperature (20°) (see Fig. 118). Thus for analytical separations involving many different ions, high temperatures are of advantage, while

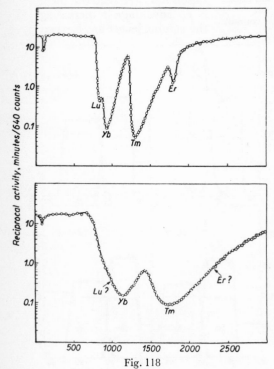

Fig. 118

Fig. 117. The effect of temperature on the elution of 1.714 g samples of pure Sm_2O_3 from 2.2 × 120 cm beds of −30 +40 mesh size Amberlite IR-100 resin, using 0.5% citrate solution at a pH value of 3.80 (measured at room temperature) and linear flow-rates of 0.5 cm/min and 2.0 cm/min (Spedding, *3059*). ○ 0.5 cm/min, ● 2.0 cm/min.

Fig. 118. Effect of temperature on the separation of the yttrium group earths with a 270/325 mesh Dowex-50 column: bed dimensions, 91 cm × 0.26 sq.cm.
Upper curve for 100°, flow rate 0.35–0.45 ml/sq. cm/min, and 5% citrate buffer at pH 3.2; lower curve for 20°, flow-rate 0.55 ml/sq.cm/min and pH 2.98
(Ketelle and Boyd, *1702*)

in large scale separations this is uneconomical, since re-fractionation of impure fractions is cheaper.

(viii) *Mould growth in the citrate solution*

Growth of a mould in the citrate solution was found to prevent separations either by blocking the exchange points or by poisoning the cyclic processes due to the formation of oxalic acid or other by-products of the mould growth. 0.1 % Phenol added to the citrate solution prevented the mould from being troublesome for about two years.

(ix) *Further variables*

In later papers Spedding and Powell (*3065, 3066*) discuss further factors encountered in a successful attempt to provide a mathematical basis for rare earth elutions. The quantitative theory developed is applicable for citrate elutions in the pH range of 5.0 to 8.0; the reader is referred to the original paper for a thorough mathematical treatment.

One impurity encountered is colloidal iron which was found in condensed steam used as pure water and which had only been passed through de-ionising exchange beds. For its removal a filtration through Pyrex glass wool was necessary. Another factor was the infection with bacteria which produce acetic acid. These could not be inhibited by phenol like the mould growth mentioned in (viii) but required sterilisation of all equipment used and increasing the phenol to 2g/litre.

(b) *Separation of rare earths on a micro-scale*

Ketelle and Boyd (*1702*) developed methods for the separation of both the yttrium and cerium group rare earths with as little as 0.1 mg of each element. 97 cm long columns of Dowex-50 (270–325 mesh) were used and the temperature for elution kept at 100° by means of a steam jacket. The rare earth mixture was first bombarded with neutrons, then placed on the column and eluted with 4.75 % citrate, first at pH 3.25 and then at pH 3.33, with a flow-rate of 2 ml/cm²/min. The radioactivity was continuously recorded in the set-up shown in Fig. 108. Typical results are shown in Fig. 119.

The method was used for the analysis of spectrographically pure erbium oxide which was shown to contain 10 p.p.m. of Tm. Good separations of intermediate and light rare earth mixtures containing as many as eleven elements were carried out.

Higgins and Street (*1392*) found that the "Gd peak" obtained by Ketelle and Boyd was actually due to terbium; this was confirmed by Ketelle and Boyd (*1703*) in a more recent note; also the band previously labelled as Y was actually Dy. Better separations of the rare earths were also realized (see Fig. 120).

Numerous papers dealing with the nuclear chemistry of rare earth isotopes contain methods similar to the above designed to separate adjacent

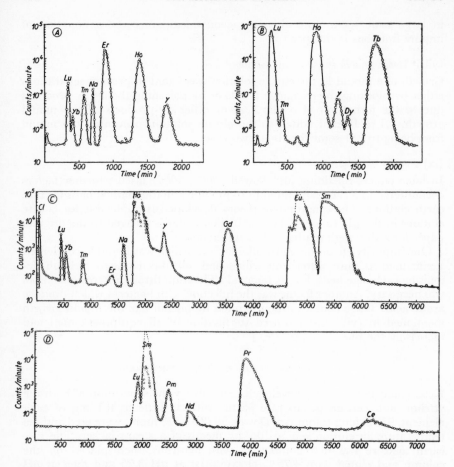

Fig. 119. Elution curves of rare earth separations effected with a 270/325 mesh Dowex-50 column at 100°. Bed dimensions 97 cm by 0.26 sq.cm; flow-rate 1.0 ml/sq. cm/min, except in A where 2.0 ml/sq.cm/min was used (Ketelle and Boyd, *1702*).

A fractionation of activities produced by neutron irradiation, 0.8 mg spectrographic grade Er_2O_3 (Hilger) (pH 3.20)

B fractionation of heavy rare earth mixture consisting of 0.1 mg each of Lu_2O_3, Yb_2O_3, Ho_2O_3 and Tb_2O_3 (Tm, Er, Y and Dy present as impurities, pH 3.20)

C fractionation of intermediate rare earth mixture consisting of 0.1 mg Ho_2O_3 and 1.0 mg each of Dy_2O_3, Gd_2O_3, Eu_2O_3 and Sm_2O_3 (Cl, Lu, Yb, Tm, Er and Na present as impurities, pH 3.25 for 4550 minutes, then pH 3.33)

D fractionation of light rare earth mixture consisting of 0.1 mg each of Sm_2O_3 and Nd_2O_3 plus 0.01 mg each of Pr_2O_3, Ce_2O_3 and La_2O_3 (Eu present as impurity, Pm produced by 1.7 h $^{149}Nd \rightarrow$ 47 h ^{149}Pm; pH 3.33 for 1610 minutes, then pH 3.40)

rare earths as quickly as possible. Freiling and Bunney (*1033*) described a method for rare earth fission products employing lactic acid at pH 3 and 87° as eluant. The separations obtained in 49 hours are claimed to be "virtually complete". For example a mixture of Nd, Pm, Sm, Eu, Gd, Tb and Y (with

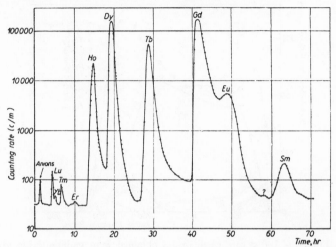

Fig. 120. Yttrium earth separations effected with a 97 cm column of 270/325 mesh Dowex-50 at 100° using 5% citrate buffer (pH 3.28) at a flow-rate of 0.64 cm/min; bed area 0.26 cm². Rare earth radioactivities produced by neutron irradiation of 2 mg of Gd_2O_3, 0.2 mg of Dy_2O_3 and 1 μg of Eu_2O_3. Samarium activity peak due to impurity in Eu; Tb peak from decay of ¹⁶¹Gd; Lu, Yb, Tm, Er and Ho peaks from impurities in Dy (Ketelle and Boyd, *1703*)

Ce, Pr, Nd, Sm carriers) yielded pure rare earths with the exception of Tb (which contained 0.15% Y) and Sm (which contained 0.3% Eu). Brooksbank and Leddicotte (*403*) used ion exchange for the separation of trace impurities which were radioactivated before chromatography. Citrate at pH 3.26 on a 100 cm Dowex-50 column gave satisfactory qualitative separations.

Other fast methods for tracer amounts were described by Stewart (*3111*) and Cuninghame *et al.* (*675*).

(c) *Large scale separation of rare earth salts*

Spedding and co-workers successfully prepared pound quantities of pure rare earth oxides. The approach to the large scale techniques involves economic considerations; thus one operator can control a large number of columns as well as only one column. Hence repeated fractionations at room temperature and low citrate concentrations are preferred to higher temperatures and high citrate concentration. Fig. 121 shows the elution curve of the first run of rare earth oxides obtained from gadolinite ore on Amberlite IR-100.

Figs. 122, 123 and 124 show similar fractionations of various crude rare earth mixtures. Adjacent pairs of metals can then be separated to 99.9 % purity in separate runs.

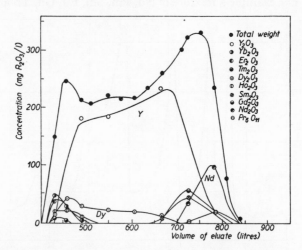

Fig. 121. Elution of 100 g rare earth oxides obtained from gadolinite ore using 0.1% citrate solution at a pH of 6.00 and a linear flow-rate of 0.5 cm/min. The column consisted of a —30 + 40 mesh size Amberlite IR-100 bed, 4 in. diameter and 4 ft. long (Spedding, *3059*)

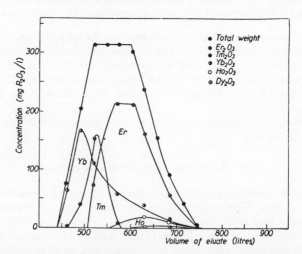

Fig. 122. Elution of 67 g of rare earth oxides originally obtained from gadolinite ore using 0.1% citrate solution at pH 6.00 at a flow-rate of 0.5 cm/min. Original composition was 20% Yb_2O_3, 13% Tm_2O_3, 60% Er_2O_3, 5% Ho_2O_3 and 2% Dy_2O_3 (Spedding, *3059*)

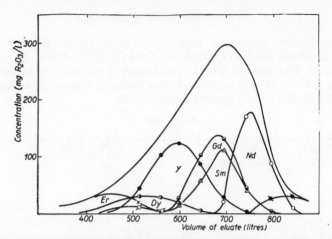

Fig. 123. Elution curve for 100 g R₂O₃ from acid-soluble fraction of Lindsay Light and Chemical Company "yttrium oxalate" from an Amberlite IR-100 bed, 4 in. diameter and 8 ft. long, using 0.5% citrate solution at a pH of 3.9 and a linear flow-rate of 0.5 cm/min (Spedding, *3059*)

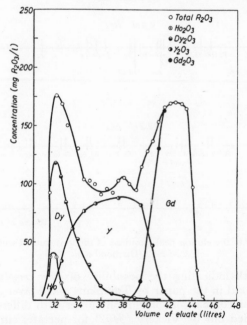

Fig. 124. Elution of 1.5 g mixture of Ho, Dy, Y and Gd oxides from an Amberlite IR-100 resin bed, 2,5 cm diameter and 90 cm long, using 0.1% citrate solution at a pH of 5.4 (Spedding, *3059*)

(d) *Other separations of rare earths*

For several other important papers describing citrate elutions and not mentioned above see (*1311, 3060, 3061, 3062, 3063, 3284*). *Anion exchange* with 0.0125 *M* citrate at pH 2.1 on Dowex-1 successfully separated Pm from Eu (*1477*). *Frontal analysis* without complex formation attempted by Lister and Smith (*2036*) proved unsuccessful.

3.2 M HCl

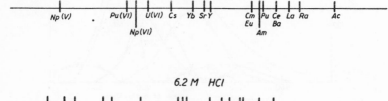

6.2 M HCl

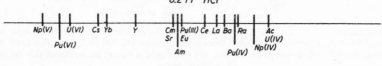

9.3 M HCl

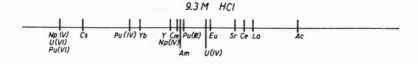

12.2 M HCl

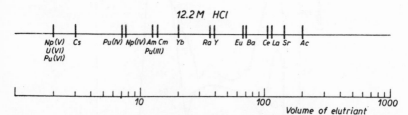

Volume of elutriant

Fig. 125. Summary of the elution peak positions of the ions studied with 3.2, 6.2, 9.3 and 12.2 *M* HCl (Diamond *et al.*, *770*)

Preliminary work indicating the possibility of *other complexing agents* than citric acids is found in the paper by Tompkins and Mayer (*3285*) (see also page 84). Actual comparison of elution curves with different complexing agents was carried out by Vickery (*3402*) for acetate, citrate, "Trilon"*

* Enta = ethylenediaminetetraacetic acid   Trilon = nitrilotriacetic acid
  TTA = thenoyltrifluoroacetone

and "Enta"* and by Mayer and Freiling (*2175*) for "Enta" > lactate = glycolate > malate ≥ citrate which were placed in the above order of preference. TTA* complexes in dioxan were separated on Dowex-50 by James and Bryan (*1544*). See also Fitch and Russell for aminoacetic acids (*975, 976*), Loriers and Quesney (*2049*) for a preparation of Y using an "Enta"* elution and Loriers and Carminati (*2048*) for a Ce earth elution using "Trilon"*.

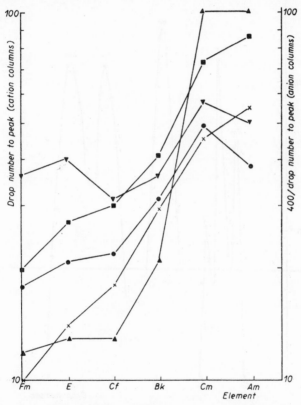

Fig. 126. Summary of elution positions of actinide elements as a function of Z (Thompson *et al.*, *3243*)

  × Dowex-50, lactate   ▲ Dowex-1, 1 *M* SCN⁻

  ● Dowex-50, 20% alcoholic HCl   ▼ Dowex-50, citrate

         ■ Dowex-1, saturated HCl

Choppin and Silva (*566a*) employ α-hydroxy*iso*butyric acid as complexant.

Transition elements may be used either to place them between the rare earths (Vickery, *3403*) or to pass the rare earth mixture over the resin

* See footnote on preceding page

saturated with the cation. For example Spedding *et al.* (*3065, 3068*) pass the "Enta" complexes of rare earths over a Cu form of Nalcite NCR. They record good results with what they term the "troublesome" rare earth mixtures (Lu–Yb, Dy–Y–Tb and Gd–Eu–Sm). A *series of short columns* was

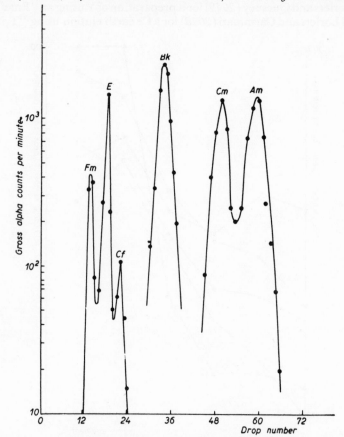

Fig. 127. Elution of tripositive actinides from Dowex-50 with ammonium lactate eluant (∼ 2 minutes per drop with 35λ drop size) (Thompson *et al.*, *3243*)

Element	Drop number
100, Fm	13.6
99, E	18
98, Cf	22
97, Bk	33
96, Cm	49
95, Am	58.5

employed by several workers to decrease the time necessary to elute the slow-moving bands through a long column (*3304, 3066*).

(e) *Separations of Ac and Sc*

The separation La–Ac was carried out by Yang and Haissinsky (*3646, 3647*) who were able to show that Ac is eluted after all the rare earths. See also McLane and Peterson (*2091*). Separations of Sc, La and Y were described by Rhadakrishna (*2610*) and by Iya and Loriers (*1526*).

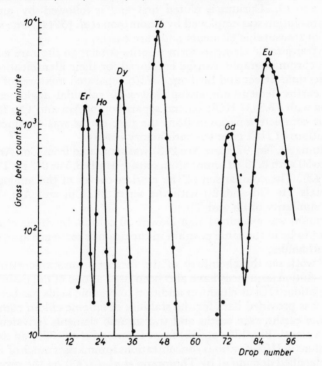

Fig. 128. Elution of homologous lanthanides with ammonium lactate (conditions same as for actinides in Fig. 127) (Thompson *et al.*, *3243*)

Element	Drop number × 0.721
68, Er	13
67, Ho	17.2
66, Dy	23
65, Tb	33
64, Gd	52
63, Eu	62.6

Diamond *et al.* (*769*) separate Y–Sr–Ce,La–Ac (in that order) on Dowex-50 with 6.2 M HCl as eluant. See also the section on transuranic elements, p. 462. Hagemann and Andrews (*1260, 64*) describe the separation of Ac from AcX on Dowex-50 with aqueous HNO_3 as eluant (2 N HNO_3 for elution of AcX, then 4 N HNO_3 for Ac).

4. SEPARATION OF TRANSURANIC ELEMENTS

Werner and Perlman (*3508*) separated mixtures of americium and curium on Dowex-50 with 5 % citrate at pH 3.05. Spongy palladium was mixed with the resin since the strong radiation of Am and Cm causes the formation of H_2 and O_2; the palladium acts as a catalyst for the formation of H_2O from H_2 and O_2. Curium is eluted first and is followed by americium. Fluosilicate elution was employed by Thompson *et al.* (*3244*), who separated mixtures of transuranic elements and rare earths.

As the transuranic elements form a series similar to the rare earths, ion exchange chromatography can be employed for their identification as well as for detection. Street and Seaborg (*3150*) separated mixtures of Am, Cm and rare earths elements employing Dowex-50 colloidal agglomerates and by eluting with 13.3 M HCl; partial separation of Am and Cm from each other, but complete separation from the rare earths was obtained. Lower concentrations of HCl gave less satisfactory results.

Berkelium (At. No. 97, mass No. 243) was separated from Cm, Am and Pu on Dowex-50 with 0.25 M ammonium citrate at pH 3.5 at 87° by Thompson *et al.* (*3242*). The comparison of the elution curves of the analogous rare earth metals Tb, Gd and Eu with the series Bk, Cm and Am shows the expected similarity of Bk and Tb.

Californium was examined on a Dowex-50 column by Street *et al.* (*3151*) and shown to be in the same position in the uranide sequence as dysprosium in the lanthanides.

In the work on the chemistry of the further transuranic elements ion exchange elution sequences have also been used as one of the main methods of investigation. HCl as eluant in addition to organic acids has been widely used and has provided data for the stability of anionic chloro complexes of the alkaline earths, rare earths and transuranic elements (trivalent as well as higher valent). Fig. 125 (p. 458) from Diamond *et al.* (*770*) on the elution peak positions in HCl of various concentrations summarises much of this work.

An unclassified document by Thompson *et al.* (*3243*) on the properties of elements 99 and 100 gives Fig. 126 (p. 459) comparing elution sequences of elements 95–100 with various eluants. Without going into the chemistry of the transuranic elements in detail, the important point to note is that the sequence is not always that of the atomic numbers as is the case with citric acid.

In general, lactic acid was preferred to citric as far as clean separations and sharpness of bands is concerned, at rapid flow rates. Typical separations of transuranic elements and lanthanides are shown in Fig. 127 and 128 (pp. 460, 461). Here columns only about 5 cm long were employed with Dowex-50 (200–400 mesh) and kept at 87° with a jacket of trichlorethylene. For citrate elution data of Am and Cm at temperatures ranging from 10–100°, see Hyde (*1494*). For a separation of Np from U and fission products, see Johansson (*1589*). For neptunium see Magnusson *et al.* (*2105*).

5. SEPARATION OF ZIRCONIUM FROM OTHER METALS

Separation of Zr and Nb from fission mixtures of U

Both Zr and Nb can be removed from an Amberlite column by washing with 0.5 % oxalic acid. Zr and Nb both form very stable oxalate complexes and thus can be separated from other metals (Russell *et al.*, see *3284*).

Purification of zirconium

In reasonably concentrated solutions tetravalent zirconium exists mainly in colloidal form and hence passes through a column of an ion exchanger. Ayres (*111*) separated Zr from Fe, Be and the rare earths by collecting the effluent from an Amberlite IR-100 column. However, hafnium, which is also colloidal, can not be separated from Zr and [186]Hf can be used as a tracer to determine the break-through of Zr.

Separation of zirconium and hafnium

Street and Seaborg (*3149*) prepared a solution of Zr and Hf and adsorbed it on 250–300 mesh Dowex-50 in presence of $HClO_4$ under conditions preventing colloid formation. When the Dowex-50 with adsorbed Zr and Hf was then placed on top of a column of Dowex-50 (1 $cm^2 \times$ 30 cm) and developed with 6 M HCl, about 66 % of Hf containing less than 0.1 % Zr was obtained. Zirconium and hafnium were also separated by Kraus and Moore (*1780*) as the oxalate or fluoride complexes on a strong base type anion exchanger.

Huffman and Lilly (*1476*) separated the fluoride complexes of Zr and Hf by eluting with 0.2 M HCl and 0.1 M HF from an Amberlite IRA-400 column.

See also U.S. Patent 2,546,953 for the separation of Zr and Hf on a cation exchange resin with 3–6 M HCl (*3148*), also Newnham (*2362*) and Forsling (*1003*).

Benedict *et al.* (*185*) describe an interesting separation without the use of HF by employing an HNO_3-citric acid mixture (0.091 M citric acid-0.45 M HNO_3) as eluant with a Dowex-50 column. Zr is eluted first (0–1250 ml under the conditions described) and then Hf (2000–5000 ml).

See also Rajan and Gupta (*2613a*).

6. SEPARATION OF Nb, Ta, Pa AND Zr

Kraus and Moore studied the equilibria of Nb, Ta and Pa in HCl solution, containing small amounts of HF, with the anion exchanger Dowex-1.

Although the metals Nb, Ta and Pa are very similar in their properties, a separation is very readily achieved as under certain conditions complexes of different charges are formed.

For a complete separation of Nb, Ta, Pa and Zr, Kraus and Moore (*1785*) elute Zr and Pa with 9 M HCl-0.004 M HF, then Nb with 9 M HCl-0.18 M HF and Ta with 1 M HF-4 M NH$_4$Cl.

Nb-Ta were separated by Kraus and Moore (*1779*) using 9 M HCl-0.05 M HF on Dowex-1; also Zr-Nb as fluoride complexes on Dowex-1. For the equilibrium studies of Ta, Nb, Zr, and Pa on exchange resins, see Kraus and Moore (*1781, 1783, 1784*).

Huffman and Iddings (*1475*) note the formation of three peaks when Nb in 7 M HCl is eluted from a Dowex-2 column (8 cm × 3 mm). Either slow equilibration or formation of several basic compounds is suggested as explanation. With 6 M HCl only one peak was observed.

For the separation Zr-Pa see also Kahn and Hawkinson (*1614a*); for a separation of Ta and Nb, see Cabell and Milner (*491a*).

7. SEPARATION OF THORIUM FROM OTHER ELEMENTS

The separation of Zr, Ti and Th on Dowex-50 with citrate (pH 1.75) was described by Brown and Rieman (*424*). Ti produces a very irregular elution curve; according to Haissinsky (private communication) the "Ti tracer" in this work is not a Ti isotope.

Radhakrishna (*2609*) separated La and Th on Amberlite IR-100; the La is eluted with 10 % citrate at pH 3 and then the Th with 6 N H$_2$SO$_4$.

A method for the separation of UX$_1$ ([234]Th) from UO$_2^{++}$ and Fe^{+++} is described by Dyrssen (*848*) using Wofatit KS and 2 M HCl.

8. SEPARATION OF URANIUM FROM OTHER ELEMENTS

Dizdar (*790*) separates traces of Cd^{++} from UO$_2^{++}$ on Amberlite IR-120 eluting U with 0.5 N oxalic acid and Cd with N HCl. Dolar and Draganic (*800*) separate traces of rare earths from UO$_2^{++}$ eluting the U with N oxalic acid and then the rare earths with 5 N HCl. The total anion analysis of aqueous solutions of uranyl ions was described by Day et al. (*717*); see also section 18. See also section 12 for the separation of fission products.

9. SEPARATION OF ALKALI METALS

While it is necessary to convert rare earths into their complexes in order to produce a difference in their equilibrium constants with ion exchange resins, this is not necessary with alkali metals. Table 24, p. 82, shows that the difference in free energies is sufficient to permit separation of the simple hydrated ions. Cohn and Kohn (*603*) obtained almost complete separations of a mixture of Na$^+$, K$^+$, Rb$^+$ and Cs$^+$ by the use of a Dowex-50 column and 0.15 N HCl as eluant. Radioactive tracers were used for the recording of the composition of the effluent. The efficiency of their separations is shown in Table 133.

TABLE 133.   Elution of alkali metals from Dowex-50 (Cohn and Kohn, *603*)

	% of total			
Fraction	Na	K	Rb	Cs
I.	73	0	0	0
II	27	0	0	0
III.	0	44	1	0
IV.	0	48	11	0
V	0	8	16	0
VI.	0	0	31	2
VII	0	0	40	34
VIII	0	0	1	64

Kayas (*1666*) reported the first complete separation of KCl (65 mg) and NaCl (45 mg) on a column of Amberlite IR-100 20 cm long and 11 mm in diameter, and using 0.1 N HClO$_4$ at the rate of 0.12 ml per min as eluant. The first 190 ml contain all the Na while the K is completely retained in the column. Bouchez and Kayas (*330*) obtained a complete separation of mixtures of carrier-free ^{22}Na and Mg on Amberlite IR-100, using either N/10 HCl or N/10 HClO$_4$ as eluant.

Complete details of an analytical procedure for the separation of mixtures of Na, K and Mg are given by Beukenkamp and Rieman (*236*). A column of 3.8 sq.cm cross section and containing 59.5 g of oven-dried Dowex-50 is used and eluted with 0.70 M hydrochloric acid with a flow-rate of 0.6 ml per min per sq.cm. The first 370 ml of eluate are discarded, the next 160 ml contain the Na and the following 190 ml contain the K. Mg does not appear until 1100 ml of eluate have passed; the fractions are evaporated and heated in an oven at 140° and the residues of alkali halides analysed by means of Mohr titrations to determine the total chloride. The errors were within 0.6 %.

More recently, Kayas (*1667*) reported the complete separation of mixtures of Na, K, Rb and Cs on Amberlite IR-100 eluting Na and K with 0.1 N HCl, and Rb and Cs with N HCl.

Kakihana (*1616, 1617*) records the equilibrium constants of the alkali ions and prepares KCl containing less than 10^{-7} % Cs and 10^{-6} % Rb by the use of a cation exchange column. For the separation of alkali ions as complexes see Buser (*485*).

Li, Be and Al were separated by Honda (*1441*) who eluted Li with 0.1 N HCl, Be with 0.05 N CaCl$_2$, and Al with 4 N HCl.

For a method of determination of Na, K, Mg and Ca in milk ash see Sutton and Almy (*3160*); for Na and K in human blood Arons and

Solomon (*83*). For another separation of K, Na and Li, see Okuno *et al.* (*2411*). For alkali metals in insoluble silicates, see Sweet *et al.* (*3174*). For Na in urine, see Vanatta and Cox (*3361, 3362*). For Rb and Cs in sea water, see Cabell and Thomas (*491b*); for an *anion exchange* separation on Dowex-1 with EDTA, see Nelson (*2353a*); for an automatic analysis, see Simanek and Janak (*2989a*).

10. SEPARATION OF ALKALINE EARTHS

The first separation of the alkaline earths is described in the paper by Tompkins *et al.* (*3284*) on the separation of fission products. On a column of Amberlite IR-1 or IR-100 (40–60 mesh) with 5 % citric acid adjusted to pH 5.0 with NH_4OH and at a flow rate of 1–2 ml per cm^2 per min, good separations of radioactive isotopes of Sr and Ba containing their daughters ^{90}Y and ^{140}La were obtained.

Tompkins (*3280*) describes the separation of radioactive Sr, Ba and Ra, and points out the advantages of this method over the separation by fractional crystallisation. A column of colloidal agglomerates of Dowex-50 1 cm^2 × 15 cm was employed. The starting solution contained 20 μg of Ra, 20 mg of Ba with ^{140}Ba tracer (and ^{140}La daughter) and 20 mg of Sr with $^{89,90}Sr$ tracer. The eluant used was 0.5 M ammonium citrate at pH 7.8 at a rate of 0.3 ml per min, and the activity of the effluent was continuously recorded on a Geiger-Müller counter.

Table 134 shows the composition of the effluent; after fraction VI the flow of eluant was stopped overnight; the concentration of La had thus increased during that time. See also U.S. Patent 2,554,649 (*3283*).

Separation of Be, Ca and Al on a phenol sulphonic resin is described by Kakihana (*1616, 1617*); Be was eluted with 0.01 N $CaCl_2$.

Honda (*1441*) separated Be-Al, using 0.01–0.1 N Ca^{++} or Mg^{++} to elute the Be^{++}.

A detailed analytical method was worked out by Lerner and Rieman (*1970*) using a column of colloidal Dowex-50 equilibrated with 1.2 M ammonium lactate. (Column dimensions: 19 cm × 2.5 cm^2; flow rate 0.56 cm/min.) A solution (6 ml or less) containing about 1 millimole of each metal is placed on the column. The elution is carried out with 1.2 M ammonium lactate; the first 74 ml contain all the Ca, the next 58 all the Sr and the next 325 all the Ba. Ca is then determined as oxalate with $KMnO_4$ and both Sr and Ba estimated gravimetrically as the sulphates. With 1000 micromolecules the accuracy is 0.2–0.8 %.

Ca and Mg in limestones and dolomites were separated quantitatively by Campbell and Kenner (*498*). A Dowex-50 column 6.3 cm long is used with a flow rate of 8 ml/min. Mg is eluted with 1.21 M HCl (in 375 ml) or with 1.07 M HCl (in 475 ml). Ca is left on the column and estimated by deducing Mg from the total Mg–Ca. See also Honda (*1442*). Brauns *et al.* (*364*) determine S and Ba in organic Ba sulphonates using an ion exchange separation. Kindt *et al.* (*1717*) separate Ca from phosphates.

TABLE 134.   Elution of alkaline earths from Dowex-50 (Tompkins, *3280*)

Fraction	La	Sr	Ba	Ra
I.	16.7	2.9		
II	75.0	4.4		
III.	1.1	6.3		
IV	0.0	78.5		
V	1.0	7.8		
VI	0.4	0.1		
VII	5.8		82.5	
VIII			15.9	
IX.			1.6	
X				
XI				
XII				
XIII				9.4
XIV				24.8
XV				40.6
XVI				22.5
XVII.				2.6
XVIII				0.1

11. SEPARATION OF TRANSITION ELEMENTS

Kozak and Walton (*1776*) attempted separations of Cd-Zn and Ag-Cu on Zeo-Karb and Glueckauf (*1142*) attempted a separation of Cu-Mn, also on Zeo-Karb resin. Ni, Cu, Zn as their ammines were examined by Nelson and Walton (*2355*).

Schuler *et al.* (*2902*) describe the separation of Cu^{++} and Ag^+ on Amberlite IR-100H with M $NaNO_3$ as eluant, as a class experiment. A 65 % yield of Cu^{++} free of Ag^+ is obtained. Cu–Cd separation was used by Peterson (*2500*), also as a class experiment.

Separation of Fe^{+++} from other metals was achieved by Kraus and Moore (*1782*) with an anion exchanger (Dowex-1) and with 9 M HCl as eluant. Fe then exists as the anion $FeCl_4^-$ and is strongly retained by the resin, while Al^{+++}, Cr^{+++} and the rare earths are readily eluted.

Teicher and Gordon (*3216*) separate Al^{+++} and Fe^{+++} by using a thiocyanate solution as the eluant on an anion exchanger (Amberlite IRA-400). An anionic iron-thiocyanate complex is formed under the conditions employed. See also Teicher and Gordon (*3217*); for a separation of Fe^{+++} and Cr^{+++}, see Venturello and Gualandi (*3397b*).

A complete separation of Mn^{++}, Fe^{++}, Co^{++}, Ni^{++}, Cu^{++} and Zn^{++}, also of Zn-Cd and As-Sb using a strong base anion exchanger and HCl of varying concentration as an eluant was discussed by Atteberry *et al.* (*99*). Blasius and Negwer (*269*) found that the concentration of HCl necessary for elution is much lower at 80° than at room temperature. Using Permutit ES or Dowex-2, Fe^{+++}–Mn^{++} and Fe^{+++}–Al^{+++} were separated with 10–15 % HCl at 80°; Co^{++}–Ni^{++} could be separated by eluting Ni^{++} at 80° with 120 ml of 20 % HCl and then Co^{++} with 300 ml H_2O at room temperature. For the separation of Fe and Ti, see Yoshino and Kojima (*3654*). Salmon and Tietze (*2785*)

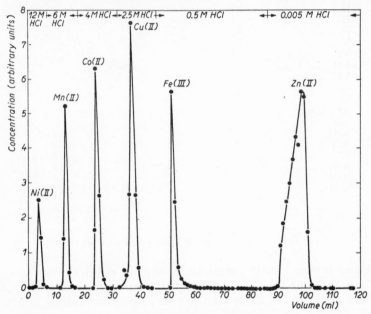

Fig. 129. Separation of transition elements Mn to Zn. Dowex-1 column, 26 cm × 0.29 cm²; flow rate 0.5 cm/min (Kraus and Moore, *1786*)

separate Fe^{+++} from V^{++++} in presence of phosphate ions. Honda (*1441*) separates Fe from silicates on a cation exchanger. For a separation of Cr^{++} and Cr^{+++}, see Haissinsky (*1272*).

The most important recent contribution to the separation of the transition elements is a series of papers by Kraus *et al.* (*1786, 1788, 1790, 1791, 2263*) dealing mainly with their equilibria in HCl solutions and their elution with HCl of varying concentration.

Kraus and Moore (*1786*) determined the elution constants of Ni, Mn, Cu, Fe, Co and Zn in 0–14 M HCl with Dowex-1. Fig. 129 shows a separation of these elements with decreasing concentrations of HCl as eluant. In a further paper Kraus *et al.* (*1791*) discuss the behaviour of a large number of elements in HCl solutions. Alkalis, alkaline earths and group III elements with the exception of Sc are not adsorbed appreciably in 12 M HCl.

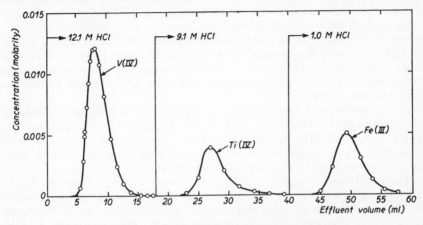

Fig. 130. Separation of V(IV), Ti(IV) and Fe(III) by anion exchange. Dowex-1 column, 18.6 cm $\times$ 0.49 cm^2; 1 ml = 0.05 M V(IV), 0.018 M Ti (IV), 0.025 M Fe(III) in 12 M HCl (Kraus *et al.*, *1781*)

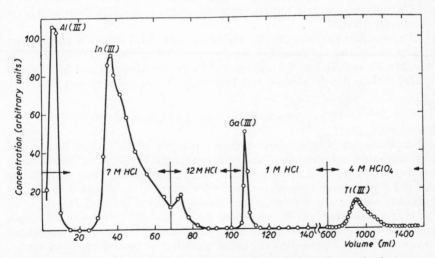

Fig. 131. Separation of Al(III), Ga(IIII), In(III) and Tl(III). Dowex-1 column, 20 cm $\times$ 0.4 cm^2, room temperature, flow rate 0.3–0.8 cm/min (Kraus *et al.*, *1791*)

V(IV), Ti(IV) and Fe(III) can be separated as shown in Fig. 130. Platinum metals, also gold (III) (see Kraus and Nelson, *1788*), are usually in the anionic form in all dilutions of HCl. Ir(III) being an exception can be readily separated from Pt(IV) and Pd(II).

The separation of Al, In, Ga and Tl may also be carried out with a HCl elution from Dowex-1 columns as shown in Fig. 131. Tl(III) being strongly anionic at all HCl concentrations must be eluted with $HClO_4$. For equilibrium studies with indium halides see also Schufle and Eiland (*2901*). Kraus *et al.* (*1791*) discuss the comparison of anion exchange and solvent extraction and overlook the much better correlation between ion-exchange sequences and paper chromatographic sequences which for all the above metals are identical (see p. 486).

Anion exchange studies with *bromide complexes* were recorded by Herber and Irvine (*1369*). The elutions are essentially identical with HCl elutions. More than 7 M HBr cannot be employed without attacking the resin (Dowex-1). Separations of Co, Cu, Zn, Ga and Ni were reported with carrier-free and macro amounts behaving identically.

The possibilities of *sulphate elutions* from Dowex-1 were explored by Kraus *et al.* (*1787*). Fe(III) is adsorbed on neutral Dowex-1 sulphate while Al(III) passes through. Also sulphuric acid and $CuSO_4$ can be separated on a neutral Dowex-1 sulphate column (*1789*). For phosphoric acid as complexing agent see Genge *et al.* (*1098*).

For a specific separation of Zn, see Miller and Hunter (*2216*). For further Fe(III) separations, see Kojima and Kakihana (*1618, 1763*). Kimura *et al.* (*1716*) use tartaric acid elution for a separation of Sb(III) from Sn(II) on Amberlite IR-120.

Yoshino and Kojima (*3654a*) separate traces of Cd from Co and Zn and traces of Zn and Cu from Cd. Ryabchikov and Osipova (*2765a*) separate Cr, Mn, Fe and Ni. For the separation Mo-Fe, see Alimarin and Medvedeva (*37a*) and for Mo-V Matsuo and Iwase (*2171b*). See also section 19.

12. SEPARATION OF METALS INTO GROUPS

Tompkins *et al.* (*3284*) showed that a mixture of fission products could be separated into divalent and trivalent ions by washing the column with oxalic acid to remove colloidal ions (Zr and Nb), then with citrate at pH 3 to remove rare earths, and then with citrate at pH 5 to remove alkaline earths. For separations of fission products see also Honda *et al.* (*1442a*). Samuelson (*2789*) separated Fe^{+++} and Al^{+++} from other metals as well as from phosphates by adsorption on an exchanger. Subsequent elution with HCl permitted the usual quantitative procedures to be used.

13. SEPARATION OF PLATINUM METALS

In addition to the studies of Kraus *et al.* (*1791*) (see above) several other workers also effected ion exchange separations of platinum metals at about the same time. Blasius and Wachtel (270), like Kraus, elute from anion exchangers. Cation exchangers were used by Stevenson *et al.* (3104) as well as by MacNevin and Crummett (2094, 2095). Stevenson *et al.* (3104) prepare a solution free of chlorides by evaporation with HNO_3–$HClO_4$; then adsorb on Dowex-50 and elute Pt(IV), then Pd(II) with 0.05–0.5 M HCl, Rh(III) with 2 M HCl and Ir with 4–6 M HCl. Sulphate ions must be absent from the starting solution. This method was used to separate Rh from fission products. See also Berg and Senn (195).

14. PRECIPITATION OF METALS INSIDE AN ANION EXCHANGER SATURATED WITH H_2S

Amberlite IR-4 saturated with H_2S was used by Gaddis (1068) for separating the hydrogen sulphide group metals from the other metals.

15. DETERMINATION OF PHOSPHATES

Various workers besides Samuelson (see section 19 on separation of anions and cations) employed exchange resins to separate elements which interfere in the determination of phosphates. A method for the determination of phosphate in rock was developed by Helrich and Rieman (1362). Usatenko and Datsenko (3354) first pass a solution through an exchange resin, when determining phosphorus in Phosphorus copper and Ferrophosphorus. Klement (1733) and Klement and Dmytruk (1735) employ Wofatit M for the separation of cations previous to the determination of phosphates. See also Hahn *et al.* (1263) and Yoshino (3652). For a purification of sodium tetrametaphosphate see Barney and Gryder (144). See also section 21e, p. 475.

16. TOTAL SALT CONCENTRATION

By passing solutions of salts through ion exchangers, total salt concentrations may be determined by acidimetry. Fedorova (938) developed a method for estimating total salt in natural waters, Polis and Reinhold (2533) for total base in serum.

Kakihana (1616, 1617) determines the following in natural waters: acidity after passage through a cation exchange column, alkali chlorides by elution with 0.1 N HCl; the other ions are eluted with N HCl for further analysis.

Burton and Lee (472) determined the total salt in vegetable tanning liquors. Investigations by Erler (918, 919), Tolliday, Thompson and Foreman (1000, 3278) and Lee (1946) have shown that erroneous results may be obtained due to the adsorption of the liberated acids on certain exchangers. A nonphenolic polystyrene resin gives less trouble than other resins.

17. ANALYSIS OF CHROME TANNING LIQUORS

Exchange resins were used to determine the various complex and ionised states of chromium (*17, 1244–1249*). These results were subject to discussions as decomposition of complexes occurs during chromatography. For comparisons with paper electrophoretic results see Gustavson (*1249a*), Kawamura *et al.* (*1662a, 1662b*) and Inouye *et al.* (*1501a*).

18. TRACE ELEMENTS

Traces of transition metals were found to adsorb strongly on cation exchange resins. Traces of Cu in milk (*647*) and traces of Cd, Cu, Mn, Ni and Zn can be removed from solutions (*2679*). Radioactive yttrium in large volumes of urine can also be recovered (*2897*). Traces of sulphates can be adsorbed on a column of anion exchange resin (*1441*) and interfering elements (Ca and Fe) in the determination of traces of silica can also be removed (*1864*).

Ions in rainwater were collected and concentrated by placing an ion exchange column before the collection reservoir (*873*).

19. SEPARATION OF ANIONS AND CATIONS

Samuelson and Runneberg (*2753–2755, 2787–2798, 2802*) developed numerous applications of organic and inorganic exchangers as aids to quantitative analyses. Elimination of interfering elements in $BaSO_4$ precipitations, of Fe and Al from phosphates in rock analyses, removal of Al and K from solutions was described by the above authors. Goehring and Darge (*1152, 1153*) also developed methods for the separation of Ca, Fe and Cr on Wofatit previous to the precipitation of $BaSO_4$. Samuelson further examined numerous salts such as the bromides, iodides, chlorates, bromates, iodates, hydrogen phosphates, nitrates, perchlorates of the alkali metals, alkaline earths and Zn, Mn, Ni, Co, Al and Cr and achieved satisfactory separations of the anions from the cations except with BrO_3^- and IO_3^- which oxidized the resin; see also *792, 793, 1863*.

Lindqvist (*2026*) adsorbed Na on a resin and thus removed many anions which interfere in the determination of sodium. Usatenko and Datsenko (*3355*) separated molybdenum in ferromolybdenum by adsorbing the Fe on a cation exchanger.

Separation of molybdenum from iron-chromium-molybdenum alloys for subsequent determination of Mo was reported by Shemyakin *et al.* (*2953*).

Lur'e and Filippova (*2061*) remove Ni and Cu previous to determinations of S, P and As in alloys. Odencrantz and Rieman (*2394*) employ a resin column in the estimation of As in insecticides. Erler (*918, 919*) describes the alkalimetry of salts after passing solutions through Wofatit Ks, and Frizzell (*1051*) after passing over Zeo-Karb.

Elements interfering with the titration of borates in presence of sugar were removed with a cation exchange column by Martin and Hayes (*2160*). Honda (*1441*) removes Al, Fe and Be before determining fluorides with Al-haematoxylin. Removal of F^- and SO_4^{--} from solution for the estimation of K is described by Dean (*718*).

The preparation of otherwise unstable acids is made possible by the use of cation exchange resins. Klement (*1734*) prepared HCNS and some phosphorus acids and Baker et al. (*122*) some heteropolyacids.

For Wofatit K and a sulphite furfural resin, particle size of 0.2–0.4 mm diameter was found most satisfactory by Samuelson (*2799*). The acidity of the solution decreases the adsorption capacity of the resin for the metal. The following generalisations regarding the *regeneration* of adsorbed ions were made:

Small particles give fast regeneration. An optimum acidity is reached with 3–4 M HCl. Experiments show that the flow-rate has little influence upon the speed of regeneration. It is the time of the regeneration process that determines its efficiency. Equilibrium studies of ions in various concentrations of HCl showed that the adsorption is at a minimum at the highest HCl concentration with the exception of Fe^{+++} which reaches a minimum at 5.47 M and is more strongly adsorbed at higher concentrations.

For further work see also Samuelson et al. (*1319, 2799–2801, 2804*).

20. SEPARATION OF COMPLEX COMPOUNDS

King and Dismukes (*1719*) were able to separate $Cr(H_2O)_6^{+++}$, $Cr(H_2O)_5CNS^{++}$ and $Cr(H_2O)_4(CNS)_2^+$ on a column of Dowex-50 using $HClO_4$ of various concentrations (0.15 M, 1.0 M and 5 M) as eluant.

Cis-trans isomers were successfully separated by King and Walters (*1720*) namely the two isomeric dinitrotetramminecobalt (III) ions using 1 M NaCl and 3 M NaCl as eluants.

21. SEPARATION OF INORGANIC ACIDS

(a) Halides

A mixture of ammonium fluoride, chloride, bromide and iodide was separated by Atteberry and Boyd (*98*) on Dowex-2, initially in the nitrate form, by eluting at a flow rate of 1.3 ml/min with a 1 M $NaNO_3$ solution adjusted to pH 10.4 with NaOH. Radioactive indicators were employed; the three close-lying curves of F^-, Cl^- and Br^- appeared at some distance from I^-, fluoride being eluted first. An analytical method for chloride-bromide mixtures based on the above work was elaborated by Rieman and

Lindenbaum (*2683*). Later DeGeiso *et al.* (*725*) suggest a method for the determination of a chloride-bromide-iodide mixture. A column of Dowex-1-X10 7 cm long equilibrated with 0.5 *M* NaNO₃ was used; the sample containing not more than 2.6 meq of any halide was poured on the column

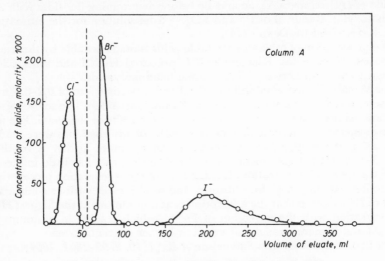

Fig. 132. Elution graph of halides (De Geiso *et al.*, *725*)

and eluted with 0.5 *M* NaNO₃ at a flow rate of 1.0 cm per min. Fig. 132 shows the elution curves with a column 6.7 cm long and a cross-section of 3.4 cm². After the elution of the chloride band (i.e. after 55 ml) 2.0 *M* NaNO₃ is used as eluant. The halides were then determined argentometrically with an average error of 0.08 % of the total halide.

For periodate, see Smith and Willeford (*3037*).

(b) *Technetium and rhenium*

Mixtures of radioactive Re and Tc as NH_4ReO_4 and NH_4TcO_4 were separated on Dowex-2 in the sulphate form by eluting with a 1 *M* solution of $(NH_4)_2SO_4$ and NH_4SCN adjusted to pH 8.3–8.5 with NaOH. The elution curve gave a peak of an unidentified impurity and well separated peaks of Re and Tc (*46*). Hall and Johns (*1280*) use an Amberlite IRA-400 column to separate pertechnetate from Mo, Co and Ag. The separation of ReO_4^- from Mo was achieved on cation as well as anion exchangers using 0.1–1 *M* HCl as eluant. Re is then anionic while Mo is cationic (*2765*).

(c) *Ferricyanide and ferrocyanide*

Ferricyanide and ferrocyanide were separated on Amberlite IR-4 in the carbonate form simply by running the mixture through the column (Cobble and Adamson, *593*).

(d) *Selenite and tellurite*

Atteberry *et al.* (*99*) worked out a separation of selenium and tellurium using 3 M HCl and a mixture of 1 M HCl with 1 M NH$_4$CNS as eluant. See also Aoki (*70a*).

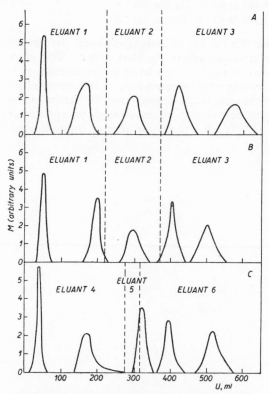

Fig. 133. Separations of the five phosphates (Lindenbaum *et al.*, *2016*)
  A. Dowex 1-X10, lot 2330–15, pH = 5.0
  B. Dowex 1-X10, lot 1195, pH = 5.0
  C. Dowex 1–X10, lot 2330–15, pH = 7.0
    For details see p. 476

(e) *Condensed phosphates*

In an excellent theoretical study Beukenkamp *et al.* (*237*) and Lindenbaum *et al.* (*2016*) measured the equilibrium constants of condensed phosphates

on Dowex-1-X10 with different Cl⁻ concentrations and different pH. For
the theoretical part see p. 89. The best separations were obtained at a pH
between 5.0 and 7.0. As eluant the following six solutions were used:

1. 0.250 M KCl buffered to pH 5.0 with 0.005 M acetate buffer.
2. 0.400 M KCl buffered to pH 5.0 with 0.005 M acetate buffer.
3. 0.500 M KCl buffered to pH 5.0 with 0.005 M acetate buffer.
4. 0.225 M KCl buffered at pH 7.0.
5. 0.400 M KCl buffered at pH 7.0.
6. 0.500 M KCl buffered at pH 7.0.

Fig. 133 shows the elution of orthophosphate, pyrophosphate, tri-, tetra-
meta- and trimeta-phosphate (in that order). The elution volumes are
sufficiently constant to elute each band as one fraction so that a subsequent
quantitative analysis may be carried out with a mean error of 0.2%. These
authors claim that duplicate elutions may be carried out in as little as five
hours. For gradient elution of condensed phosphates see Grande and
Beukenkamp (*1185a*); for the separation of orthophosphate and arsenate
Yoshino (*3653*).

CHAPTER 43

Inorganic paper chromatography

1. INTRODUCTION

Pollard *et al.* (*2543*) suggested a complete scheme of several uni-dimensional separations in conjunction with various spot tests. This scheme, although permitting detection of the common metals, does not yield separations for subsequent determinations.

Other schemes of complete qualitative analysis were proposed by Harasawa (*1301*), Tamura (*3199*) and Surak *et al.* (*3156*). For trace elements, see (*1869*).

Reeves and Crumpler (*2647*) suggest precipitation of the 8-hydroxy-quinoline complexes of a number of metals for their subsequent detection with paper chromatography.

Two-dimensional separations have not been employed widely, though very good results were obtained both by Lacourt *et al.* (*1851*) and Pollard *et al.* (*2542*). A number of specific tests for metals have been developed which in selectivity and reliability equal or exceed other spot tests.

The study of a large number of ions in one solvent has yielded numerous useful separations as well as giving comparative data for a theoretical approach. For experiments for school laboratories, see Surak and Schlueter (*3157*) and Frierson and Ammons (*1046*).

For reviews, see Pollard (*2535, 2538*), Wells (*3503*), Irving and Williams (*1511*) and O. Kaufmann (*1661*).

2. SOLVENTS

(a) *Division of solvent systems into groups*

The chemical basis of the separation of inorganic substances will not be discussed here in detail; for general discussions see pages 115–125 and Lederer (*1939*). Tables 142–148 list the R_F values in those solvents where numerous ions have been chromatographed. We have grouped some of these differently to the first edition owing to supplementary material being available.

For separations of inorganic ions a difference in complexation or a diffierence in polarity must exist (see Lederer, *1921*, Pollard *et al.*, *878*, and Harasawa, *1300*; also separations of alkalis and alkaline earths, pages 499–501). The solvent systems may be divided into three groups:

(i) Solvents containing strong acids. These yield constant R_F values even at tracer levels and it has been shown that no ion exchange is possible with the paper.

(ii) Solvents containing weak acids and complexing agents or weak bases. Here tracers often behave diffierent to macro amounts (see section 13 on radioelements); also ion exchange with COOH groups on the paper is possible.

(iii) Separations under non-equilibrium conditions such as in an atmosphere saturated with other solvents etc., see Linstead *et al.* (*467*) and Lacourt *et al.* (*1859, 1860*). Here excellent separations may be obtained with strict observance of the specified conditions. The spots do not necessarily have a Gauss curve distribution and the separations are known to vary with ratio and quantity of the substances to be separated.

To mention now some further solvent systems employed: Martin (*2158*) used butanol-HCNS as solvent obtaining the R_F values shown in Table 135 using Rutter's radial technique. Ether containing various acids was studied by Anderson and Whitley (*60*), alcoholic KCN by Pickering (*2515*). Reeves

TABLE 135. R_F values in butanol-HCNS mixture (Martin, *2158*)

Cation	R_F value	Cation	R_F value
Fe^{+++}	1.00	As^{+++}	0.54
Ni^{++}	0.42	Sb^{+++}	0.96
Co^{++}	1.00	Bi^{+++}	0.96
Cr^{+++}	0.23	Hg^{++}	1.00
Cu^{++}	1.00	Pb^{++}	0.0
Cd^{++}	1.00	Ag^{+}	1.00

and Crumpler (*2647*) examined butanol containing various concentrations of 12 N HCl and obtained the R_F values shown in Table 136. For solvents employing complexing agents see Table 142. Hartkamp and Specker (*1318a*) recommend tetrahydrofuran-HCl (50:15) in which the following R_F values were recorded: UO_2^{++} 0.976, Cu^{++} 0.911, Co^{++} 0.777, VO_2^{++} 0.555, Mn^{++} 0.506 and Ni^{++} 0.274. Szarvas *et al.* (*3189a*) used nicotine as solvent for cations. For butanol-N HCl containing 10% H_2O_2 especially for the separation of Ti, V and Mo, see (*1929*). A study of butanol-HNO$_3$ mixtures not mentioned in Table 146 was made by Bergamini and Rovai (*197a*).

TABLE 136. R_F values in butanol-12 N HCl mixtures (Reeves and Crumpler, *2647*)

% 12 N HCl	15	20	25	30
Ag$^+$	0.0	0.0	0.0	0.0
Al^{+++}	0.03	0.03	0.04	0.07
Ni^{++}	0.03	0.04	0.04	0.09
Co^{++}	0.09	0.19	0.39	0.51
Cu^{++}	0.30	0.40	0.51	0.55
Bi^{+++}	0.44	0.51	0.60	—
Zn^{++}	0.86	0.78	0.79	0.79
Cd^{++}	0.88	0.83	0.83	0.83
Hg^{++}	0.88	0.89	0.83	0.83
Fe^{+++}	0.92	0.93	0.92	0.92

TABLE 137. R_F values with butanol - 1 N HCl on No. 302 D'Arches paper (*1927*)

	Not impregnated	Impregnated with N NaCl
Ag$^+$	0.0	0.0
Pb^{++}	0.0 tail	0.04
Hg^{++}	0.79	0.36
Cu^{++}	0.09	0.13
Cd^{++}	0.61	0.23
Bi^{+++}	0.59	0.09
Sb^{+++}	0.70 tail	0.70
Fe^{+++}	0.19	0.74
Co^{++}	0.08	0.09
Ni^{++}	0.08	0.08
Mn^{++}	0.11	0.07
MoO$_4$$^{--}$	0.32	0.62
Ti^{++++}	0.08	0.04
V^{+++++}	0.22 tail	0.39 tail
UO$_2$$^{++}$	0.15	0.39
Ba^{++}	0.01	0.01
Rh^{+++}	0.15	0.06
Pd^{++}	0.65	0.07
Pt^{++++}	0.70	0.22
Ru^{+6}	0.14	0.04
Au^{+++}	0.82	1.0

TABLE 138. Organic reagents for detecting ions by ultra-violet fluorescence (Elbeih et al., 878)

Fluorescent Substance	Bright Spots			Dark Spots			Invisible Spots	
	+++	++	+	– –	–	–	–	
o-Amino-benzoic acid			Sb	Ag, Cu, Cd, Mn, Co	Hg^{++}, Pb, Hg^+, Bi, Al, Fe, Zn, Ni	–	Cr, Ca, Sr, Ba, Mg, K, Na	As, Sn^{++}, Sn^{++++}
8-Hydroxy-quinoline	Cd, Al, Zn, Mg	Ag, Sn^{++}, Sn^{++++}, Ca	Sr, Ba	Fe	Hg^+, Hg^{++}, Bi, Cu, Sb, Mn, Co, Ni		Pb, Cr, K, Na	As
Kojic acid	Ag, Hg^+, Hg^{++}	Ca	Sr, Ba, Mg	Cu, Fe, Ni	Bi, Al, Mn, Co		Pb, Cd, Sb, Sn^{++}, Sn^{++++}, Cr, Zn, K, Na	As
Naphthionic acid				Hg^{++}, Fe, Mn	Ag, Hg^+, Pb, Bi, Cu, Cd, Al, Cr, Zn, Co, Ni, Ca, Sr, Ba, Mg, Na, K			As, Sb, Sn^{++}, Sn^{++++}
Morin	Al, Zn	Pb, Cd, Sn^{++}, Sn^{++++}, Mg		Cu, Fe	Ag, Hg^+, Hg^{++}, Bi, Cr, Mn, Co, Ni			As, Sb, Ca, Sr, Ba, K, Na

(b) *Paper impregnated with salt solutions*

It was shown (*1927*) that the "salting-out" effect extensively employed in solvent–solvent extraction may also be used in inorganic paper chromatography by dipping paper sheets in salt solutions and drying before chromatography.

The salting-out effect of various cations on uranyl nitrate was thus studied and found to be identical to that observed by solvent extraction. Also a large number of cations was chromatographed with butanol-1 N HCl on paper dipped into N NaCl solution. The R_F values shown in Table 137 do not resemble those found with any butanol-HCl mixture studied and seem to be strongly influenced by the salting-out effect of Na$^+$ ions. This effect should be considered, when applying concentrated solutions to the paper as a zone with a salting-out effect is thus formed and may considerably influence the movement of the ions by arresting them on the point of application.

3. METHODS OF DETECTION FOR METALS AND ACIDS

(a) *General reagents*

In a chromatographic analysis of mixtures of metals a universal reagent is just as necessary as a suitable solvent pair. Up to now the attempts to develop organic reagents for metals were linked with the desire to obtain specificity as well as sensitivity for a given metal. Exactly the opposite is required for paper chromatography, and the less specific an organic reagent is towards metals the more useful it will be as a reagent for chromatographic analyses.

Among the inorganic reagents, H_2S, passed over the paper directly or after exposure to NH_3, is a good reagent for most of the heavy metals. Spots are obtained with Ag$^+$, Pb^{++}, Hg$_2^{++}$, Tl$^+$, Hg^{++}, Tl^{+++}, Cu^{++}, Cd^{++}, Bi^{+++}, As^{+++}, Sb^{+++}, Sn^{++}, Sn^{++++}, Fe^{+++}, Co^{++}, Ni^{++}, Mn^{++} and MoO$_4^{--}$.

A number of organic chelating compounds were employed by Elbeih *et al.* (*878*), who, after spraying with a reagent, view the chromatogram under ultraviolet light. The reagents and their reactions are shown in Table 138. The most satisfactory spray is a solution of 5 g 8-hydroxyquinoline and 1 g kojic acid in 1 litre of 60 % alcohol. After spraying, the paper is exposed to ammonia vapour.

Reeves and Crumpler (*2647*) observed the reactions listed in Table 139 by viewing the paper chromatograms in daylight and UV light. Weiss and Fallab (*3489*) recommend quercetin as a useful general reagent. Table 140 shows the limits of detection and the colours in ordinary and ultraviolet light for a number of cations. See also Michal (*2201a*).

Naito and Takahashi (*2335*) suggest phenylthiosemicarbazide for Cu, Cd, Bi, Ni, Co, Zn, Hg(II) and Fe(III), thiogallein for Sn(II), Sb(III), Sb(V),

TABLE 139. Some reagents for cations (Reeves and Crumpler, *2647*)

Reagents	8-Hydroxyquinoline		Diphenylcarbazide		Resorcinol	
	Daylight	UVL	Daylight	UVL	Daylight	UVL
Al⁺⁺⁺	yellow	yellow	red	—	—	—
Ni⁺⁺	yellow	red	red	—	blue	red
Co⁺⁺	yellow	red	purple	—	brown	red
Cu⁺⁺	yellow	red	green	—	black	purple
Bi⁺⁺⁺	yellow	red	orange	red	—	—
Zn⁺⁺	yellow	yellow	red	—	—	—
Cd⁺⁺	yellow	yellow	tan	—	—	—
Hg⁺⁺	yellow	red	red	—	yellow	red
Fe⁺⁺⁺	black	purple	green	red	—	—

TABLE 140. Reaction with quercetin in acid and alkaline media (Weiss and Fallab, *3489*)

Cation	Quercetin in acid		Quercetin in NH_3		Limit of detection in gamma
	daylight	UVL	daylight	UVL	
K	—	light yellow	—	—	6
Na	—	pale yellow	—	—	12.5
Mg	—	yellow green	yellow	—	6
Ca	—	yellow green	pale brown	dark grey	25
Sr	—	pale yellow	yellow	—	100
Ba	—	yellow green	—	—	100
Al	yellow green	green	yellow	orange	2.5
Ag	grey	grey	grey brown	dark violet	6
Hg(II)	—	—	grey	violet	6
Pb(II)	—	—	orange	dark violet	6
Sn(II)	yellow green	green	yellow orange	orange	2.5
Sn(IV)	pale yellow	pale green	yellow	violet	6
Cd	—	—	yellow	yellow	12.5
Zn	—	grey green	yellow	green	12.5
Fe(III)	grey green	dark violet	brown	dark violet	32
Mn	—	pale yellow	brown violet	dark violet	6
Cr(III)	grey green	grey	grey	grey	6
Ni	—	—	yellow	dark violet	6
Co	—	grey brown	grey brown	dark violet	6
Cu	—	—	grey	violet	9
UO₂⁺⁺	—	—	light brown	dark violet	6
As(III)	—	—	—	—	
Sb(III)	pale yellow	green	yellow brown	dark violet	6
Bi	—	—	yellow brown	grey violet	6

As(III) and As(V) and gallein for $Sn(IV)$, $Sb(III)$ and $Sb(V)$. Benzidine as general reagent was employed by Miller and Kraemer (*2219*).

Other reagents which have been employed successfully to detect cations on the paper are listed in Table 141. In this table only the name of the reagent is stated, the conditions being those of the ordinary spot test; for example for K, dipicrylamine is used in a solution of Na_2CO_3 and after spraying the paper is dipped into dilute nitric acid to discolour the unreacted sodium salt (*1921*).

TABLE 141. Some spot test reagents (*1921*)

Metal	Reagent	Metal	Reagent
Ni^{++}	rubeanic acid or dimethyl glyoxime	Didymium Yttrium and	alizarin (ammoniacal)
Co^{++}	dimethyl glyoxime	Erbium	alizarin (ammoniacal)
Mn^{++}	ammoniacal $AgNO_3$	Ti^{++++}	H_2O_2
K^+	dipicrylamine	V^{+++++}	H_2O_2 and H_2SO_4
Cs^+	dipicrylamine	Pt^{++++}	H_2S or KI
Rb^+	dipicrylamine	Pd^{++}	H_2S or KI
UO_2^{++}	potassium ferrocyanide	Rh^{+++}	H_2S
Tl^+ and Tl^{+++}	potassium iodide	Al^{+++}	aluminon or alizarin (ammoniacal)
MoO_4^{--}	H_2S or $SnCl_2$		
Th^{++++}	alizarin	Cr^{+++}	aluminon or alizarin (ammoniacal)
Be^{++}	quinalizarin		
In^{+++}	dithizone, quinalizarin or alizarin	Zn^{++}	ferrocyanide (white spot on blue background)
Ga^{+++}	aluminon	Ca^{++}	alizarin
Ce^{+++}	alizarin (ammoniacal)	Ba^{++} and Sr^{++}	rhodizonic acid
Praseodymium	alizarin (ammoniaacl)	Mg^{++}	azoresorcinol

(b) *Radiochemical detection*

Van Erkelens (*3373*) proposes exposure of the developed chromatogram to radioactive $H_2{}^{35}S$. The sensitivity of the H_2S reaction is thus considerably enhanced and special precautions such as thorough washing of the paper are necessary to permit detection of as little as 1 μm. A limited number of suitable solvents are also suggested. For other radiochemical methods such as neutron activation, see p. 143.

(c) *Instrumental methods*

Hashimoto *et al.* (*1323, 1325*) have suggested detection of substances on the paper chromatogram by passing the paper strip between the plates of a condenser in a tuned-grid-type high frequency circuit; large changes in the grid current being noted for each "spot". An inorganic example is shown in Fig. 134. For further data on inorganic salts see also Mori (*2273*). Conductivity measurements with electrodes along the paper were used by De Vries (*764*).

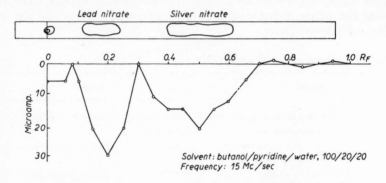

Fig. 134. Detection of $Pb(NO_3)_2$ and $AgNO_3$ by passing the chromatogram between condensor plates (Hashimoto and Mori, *1323*)

4. TWO-DIMENSIONAL SEPARATIONS

Lacourt *et al.* (*1851*) described the two-dimensional separation of Fe, Ti and Al, by developing the mixture dissolved in 50 % HCl, first for 1 hour with 6.5 N formic acid and then at right angles to the direction of the first development with dioxan. In formic acid, Al moves away from Ti and Fe and in dioxan, Fe moves away from Ti. McOmie *et al.* (*878*) separate the mixture Al, Fe, As, Cu and Cd on a square sheet of paper by developing first with collidine shaken with 0.4 N HNO_3, washing the paper with carbon tetrachloride and developing at right angles with butanol, containing benzoylacetone. All five metals give well separated spots. In a later paper, Pollard *et al.* (*2542*) describe additional two-dimensional separations of various mixtures of metals. To describe R_F values the x-axis of the paper sheet is lettered from A to J in ten regular intervals and the y-axis numbered from 1 to 10.

Two-way separations with (a) collidine, x-axis and (b) butanol-benzoyl-acetone, y-axis, produce spots with the following ordinates:
Al, Bi, Pb (A,1); Ba (C,1); K, Sr (D,1); or (E,1); Na (E,1); Ca (F,1); Mg (G,1); Mn, Cd (H,1); Zn (a) (I,1); As (a) (A,4); As (b) (F,4); Hg_2 (A,6); Cu (H,8); Fe (A,9).

Good separations were also produced by running first one solvent, then another at right angles and then a third in the same direction as the second.

TABLE 142. R_F values of cations with complex-forming mixtures (Pollard *et al.*, *2541*)

	Butanol/2 N HNO₃ containing 0.1% dibenzoylmethane	Butanol/0.1 N HNO₃ containing 0.5% benzoylacetone	Butanol/2 N HNO₃ containing 1% acetylacetone	Butanol/N HNO₃ containing 5% acetoacetic ester	Butanol/N HNO₃ containing 1% antipyrine	Dioxan 100 ml, antipyrine 1 g, conc. HNO₃ 1 ml, water 2.5 ml	Pyridine 60% Water 40%	Butanol 50% Acetic acid 10% Acetoacetic ester 5% H₂O 35% (2540)
Ag⁺	0.18	0.10	0.15	0.18	0.19	0.08	0.88	0.1
Hg⁺	0.23	0.24	0.43	0.50	0	0.43	0.86	0.09
Pb⁺⁺	0.11	0.03	0.09	0.09	0.10	0.15	—	0.18
Hg⁺⁺	0.23	0.31	0.43	0.50	0	0.42	0.86	0.84
Bi⁺⁺⁺	0.15	0.02	0.23	0.20	0.20	0.63	0	0.34
Cu⁺⁺	0.13	0.22	0.12	0.15	0.17	0.24	0.85	0.65
Cd⁺⁺	0.13	0.05	0.12	0.15	0.13	0.18	0.87	0.29
As⁺⁺⁺	0.42	0.43	0.43	0.45	0.50	0.18	0.75	0.17
Sb⁺⁺⁺	0	0	0.02	0	0	0.65	0.72	0.16
Sn⁺⁺	0.73	0.58	0.82	0.70	0.73	0.77	0	0.16
Sn⁺⁺⁺⁺	0.65	0.55	0.81	0.65	0.68	0.58	0	0.18
Al⁺⁺⁺	0.13	0.03	0.09	0.10	0.09	0.03	0	0.17
Cr⁺⁺⁺	0.13	0.03	0 09	0.10	0.09	0.01	0	0.64
Fe⁺⁺⁺	0.20	0.95	0.43	0.13	0.14	0.10	0	0.73
Zn⁺⁺	0.14	0.05	0.10	0.12	0.10	0.08	0.86	0.30
Mn⁺⁺	0.16	0.07	0.11	0.13	0.12	0.09	0.87	0.23
Co⁺⁺	0.13	0.06	0.10	0.12	0.11	0.05	0.88	0.22
Ni⁺⁺	0.13	0.03	0.09	0.12	0.09	0.05	0.87	0.22
Ca⁺⁺	0.11	0.05	0.08	0.12	0.09	0.10	—	0.20
Sr⁺⁺	0.08	0.04	0.07	0.08	0.05	0.04	—	0.18
Ba⁺⁺	0.06	0.02	0.09	0.06	0.04	0.02	—	0.16
Mg⁺⁺	0.11	0.06	0.10	0.13	0.10	0.04	0.85	0.20
K⁺	0.10	0.05	0.10	0.11	0.06	0.03	—	0.25
Na⁺	0.10	0.06	0.10	0.11	0.06	0.04	—	0.23

With collidine (x-axis), benzoylacetone in butanol, followed by antipyrine in dioxan (y-axis) the following values were obtained:
Al (A,2); Mg (E,2); Mn (F,2); Cd (G,3); Bi (A,5); Cu (G,8).

Two solvents used consecutively in the same direction also produced good separations; thus butanol containing benzoylacetone followed by dioxan containing antipyrine, separated a mixture of Fe, Bi and Al.

TABLE 143. R_F values in alcohol-HCl mixtures (Lederer, *1925*, and De Carvalho, *719a*)

	Ethanol + 10% 5 N HCl	isoPropanol + 10% 5 N HCl	Butanol N HCl	Butanol 2 N HCl	Butanol 4 N HCl	Butanol 6 N HCl	Butanol 8 N HCl	Butanol 10 N HCl	Butanol 12 N HCl	Amyl alcohol 2 N HCl
Ag+	0.02	0.06	0.0	0	0	0	0	0	0	0.0
Hg+	0.08T	0.05		—	—	—	—	—	—	—
Pb++	0.16	0.03	0.0T	0.17	0.39	0.55	0.52	0.45	0.40	0.0
Hg++	1.0	1.0	1.05	0.84	1.0	0.99	0.97	0.90	0.83	0.55
Bi+++	0.94	0.84	0.65	0.63	0.76	0.79	0.68	0.58	0.50	0.22
Cu++	0.47	0.28	0.10	0.11	0.31	0.52	0.58	0.58	0.57	0.02
Cd++	1.0	0.84	0.60	0.67	0.97	0.99	0.97	0.90	0.81	0.23
As+++	0.50	0.66	0.70	0.58	0.83	0.94	0.92	0.84	0.78	0.41
Sb+++	0.85	0.77	0.8T	0.76T	0.99	0.95	0.83	0.77	0.73	
Sn++	0.97	0.88	0.95	0.76	1.0	0.98	0.89	0.83	0.78	0.52
Sn++++				0.79	1.0	0.92	0.89	0.83	0.78	
Al+++	0.37	0.35	0.07	0.03	0.21	0.41	0.38	0.34	0.28	
Cr+++	0.47	0.28	0.07	0.09	0.22	0.44	0.37	0.29	0.22	
Fe+++	0.56	0.35	0.12	0.18	0.42	0.72	0.98	0.98	0.97	0.02
Zn++	0.93	0.87	0.76	0.71	0.98	1.0	0.89	0.84	0.74	
Mn++	0.36	0.37	0.09	0.08	0.24	0.48	0.43	0.44	0.36	
Co++	0.32	0.27	0.07	0.05	0.22	0.42	0.40	0.46	0.63	0.03
Ni++	0.34	0.23	0.07	0.05	0.22	0.40	0.36	0.28	0.24	0.0
Ca++			0.03	0.02	0.18	0.35	0.33	0.23	0.18	
Sr++	0.11	0.11	0.0	0.01	0.09	0.30	0.25	0.16	0.10	
Ba++	0.04	0.05	0.0	0.01	0.09	0.17	0.09T	0.05T	0.03T	
Mg++	0.33	0.23	0.11	0.06	0.24	0.43	0.41	0.36	0.29	
Li+				0.16	0.29	0.52	0.49	0.45	0.41	
Na+			0.07	0.06	0.24	0.39	0.35	0.25	0.20	
K+	0.08	0.15	0.08	0.05	0.19	0.38	0.34	0.25T	0.18T	
Rb+		0.13	0.08	0.06	0.22	0.41	0.39	0.30	0.29	
Cs+	0.05	0.13	0.08	0.06	0.25	0.45	0.44	0.37	0.37	
UO$_2$++	0.57	0.36	0.20	0.19	0.36	0.49	0.50	0.55	0.57	
U++++			0.0	0.02	0.16	0.31	0.24	0.16	0.17	
Tl+	0.09	0	0.0	—	—	—	—	—	—	0.0
Tl+++	1.0	1.0	1.11	0.94	1.0	1.0	0.98	0.95	0.95	0.75
MoO$_4$--	0.37	0.28	0.05	0.47	0.79T	0.93	0.88	0.83	0.76	
Be++	0.70	0.62	0.30	0.18	0.45	0.59	0.60	0.52	0.42	
In+++	0.65	0.44	0.33	0.28	0.46	0.55	0.54	0.50	0.46	
Ce++++			0.26							
Ga+++			0.21	0.34	0.76	1.0	1.0	1.0	0.91	
La+++			0.02	0.02	0.13	0.28	0.25	0.20	0.17	
Ce+++	0.07	0.06	0.03	0.01	0.16	0.31	0.27	0.22	0.19	
Y+++			0.03	0.03	0.13	0.32	0.27	0.23	0.21	
Rare earths	0.11	0.07	0.03	0.02	0.16	0.30	0.28	0.21	0.17	
	(Pr Nd)	(Pr Nd)	(Pr Nd)							
Ti++++	0.50	0.33	0.07	0.11	0.29	0.46	0.46	0.42T	0.43T	0.02
Zr++++	0.0	0.02	0.0	0.0	0.12T	0.28T	0.23T	0.13T	0.11T	
Th++++	0.12	0.14	0.03	0.02	0.17	0.32	0.24	0.16	0.13	
V^{5+}	0.38	0.30	0.17T	0.14T	0.26T	0.48T	0.43T	0.36T	0.31T	0.05
Au+++	0.95	1.0	1.1	0.80	1.0	0.99	0.99	0.97	0.96	0.36T
Pt++++	0.90	0.95	0.76	0.73	0.92	0.92	0.88	0.75	0.71	0.30
Pd++	0.90	0.85	0.60	0.62	0.80	0.80	0.73	0.62	0.54	0.26
Rh+++			0.07	0.15T	0.35T	0.52	0.55	0.48	0.42	

(*Continued in Table 144*)

TABLE 144. R_F values of some further ions in butanol-HCl mixtures (De Carvalho, *719a*)

	Butanol shaken with HCl :						
	1 N	*2 N*	*4 N*	*6 N*	*8 N*	*10 N*	*12 N*
W^{6+}	0	0	0	0	0	0	0
BO_3^{3-}	0.56	0.62	0.76	0.74	0.70	0.66	0.60
Ra^{++} (AcX)	0	0	0.10	0.14	0.07	0.05	0.03
PO_4^{---}	0.54	0.66	0.91	0.92	0.89	0.77	0.77
SeO_3^{--}	0.59	0.66	0.82	0.82	0.82	0.97	0.96
TeO_3^{--}	0.15T	0.32	0.84	0.96	0.94	0.90	0.90
Po^{4+}	0.63	0.79	0.99	0.96	0.95	0.87	0.87
SO_4^{--}	0.31	0.45	0.62	0.80	0.87	0.78	0.71
Br^-	0.63	0.78	0.94	0.91	0.86	0.81	0.75
I^-	0.68	0.83	0.99	0.96	0.89	0.84	0.77
Ac^{+++}	0	0.03	0.16	0.30	0.27	0.21	0.17
Sc^{+++}	0.02	0.04	0.18	0.41	0.35	0.31	0.29
ReO_4^-	0.64	0.74	0.99	0.99	0.93	0.88	0.82
TcO_4^-	0.62	0.77	0.98	0.98	0.90	0.73	0.66
Hf^{4+}	0.01	0.02	0.15	0.30	0.27	0.21	0.12

Note: each solvent including the one-phase mixtures is prepared by shaking 100 ml of butanol with 100 ml of the HCl solution.

TABLE 145. R_F values in binary alcohol mixtures containing HCl (Walker and Lederer, *3461*)

	45 Ethanol 45 isoPropanol 10 5 N HCl	45 isoPropanol 45 Butanol 10 5 N HCl	50 Butanol 50 Amylalcohol sat. N HCl
Ag^+	0.02	0.02	0.01
Pb^{++}	0.03	0.03	0.0
Hg^{++}	0.82	0.85	
Bi^{+++}	0.67	0.65	0.30
Cu^{++}	0.26	0.12	0.03
Cd^{++}	0.75	0.73	0.22
As^{+++}	0.43	0.44	0.41
Sb^{+++}	0.77	0.76	
Sn^{++}	0.83	0.81	0.42
Al^{+++}	0.25	0.08	0.0
Cr^{+++}	0.18	0.07	0.0
Fe^{+++}	0.42	0.11	0.02
Zn^{++}	0.90	0.80	0.31
Mn^{++}	0.21	0.08	0.01
Co^{++}	0.15	0.08	0.01
Ni^{++}	0.15	0.08	0.01
Sr^{++}	0.02	0.04	0.0
Ba^{++}	0.05	0.04	0.0
UO_2^{++}	0.30	0.14	0.05
Tl^{+++}	0.98	0.94	0.75
MoO_4^{--}	0.20	0.23	0.2
Ti^{++++}	0.32	0.04	0.01
Au^{+++}	0.96	0.89	0.61
Pt^{++++}	0.70	0.77	0.36
Pd^{++}	0.73	0.67	

TABLE 146. R_F values of cations in alcohol-HNO_3 mixtures

	Ethanol 90 8 N HNO_3 10 (1862a)	Isopropanol 180 Water 10 HNO_3 20 (M.Lederer, unpublished)	Isopropanol 90 8 N HNO_3 10 (1862a)	Butanol saturated with 10% aqu. HNO_3 (1921)	Butanol saturated with 1 N HNO_3 (878)	Butanol saturated with 2 N HNO_3 (878)	Butanol saturated with 3 N HNO_3 (878)	Butanol saturated with 10% aqu. HNO_3 (1862a)	Amyl alcohol saturated with 20% aqu. HNO_3 (1862a)
Ag^+	0.18	0.24	0.13	0.23	0.12	0.13	0.19	0.12	0.06
Hg^+		0.30							
Pb^{++}	0.11	0.07	0.02	0.15	0.01	0.08	0.15	0.06	0.02
Bi^{+++}	0.67	0.39	0.03T	0.27	0.18	0.19	0.25	0.25	0.17
Cu^{++}	0.37	0.20	0.05	0.17	0.10	0.11	0.15	0.08	0.02
Cd^{++}	0.37	0.22	0.07	0.19	0.10	0.11	0.15	0.08	0.03
Hg^{++}	0.7T	0.45	0.58T	T				T	0.64T
As^{+++}		0.45			0.45	0.45	0.48		
AsO_4^{---}	0.85		0.79					0.61	0.44
Sb^{+++}	0.01T		0					0.02	0.01
Sn^{++}			T		0.74	0.76	0.84		
Sn^{++++}	T				0.65	0.68	0.75	T	T
Al^{+++}	0.41		0.05	0.11	0.05	0.05	0.12	0.05	0.01
Cr^{+++}	0.43	0.19	0.05	0.15	0.06	0.06	0.15	0.06	0.01
Fe^{+++}	0.39	0.35	0.06	0.18	0.06	0.06	0.15	0.06	0.01
Zn^{++}	0.32		0.05	0.15	0.05	0.06	0.15	0.06	0.01
Mn^{++}	0.38		0.06	0.16	0.08	0.09	0.16	0.07	0.01
Co^{++}	0.32	0.20	0.03	0.17	0.08	0.09	0.16	0.05	0.02
Ni^{++}	0.26	0.18	0.02	0.17	0.05	0.06	0.15	0.05	0.01
Ca^{++}					0.06	0.06	0.14		
Sr^{++}		0.09		0.08	0.05	0.05	0.14		
Ba^{++}		0.04		0.08	0.05	0.06	0.13		
Mg^{++}					0.07	0.09	0.18		
Na^+					0.06	0.07	0.15		
K^+					0.06	0.07	0.15		
Rb^+				0.13					
Cs^+				0.13					
UO_2^{++}		0.60		0.40					
Tl^+		0.09		0.16					
Ge^{++++}	0.54		0.25					0.24	0.11
MoO_4^{--}	0.45	0.23	0.21	0.32				0.33	0.22
Ti^{++++}	T		0.15T					T	0.03
V^{5+}	0.49	0.28	0.27					0.20	0.17T
Zr^{++++}		0.02							
Th^{++++}				0.10					
PO_4^{---}	0.85		0.74					0.50	0.34
CrO_4^{--}	T		0.03					0.04	0.01
SeO_3^{--}				0.54 (1937a)			0.69 (1939)		
TeO_3^{--}				0.15 (1937a)			0.18 (1939)		
TcO_4^-				1.0 (1937a)					
ReO_4^-				1.0 (1937a)					

	Methyl n-propyl ketone 98 10 N HCl 2	Methyl n-propyl ketone 95 10 N HCl 5	Methyl n-propyl ketone 85 8 N HCl 15	Methyl n-propyl ketone 85 10 N HCl 15	Methyl n-propyl ketone 75 8 N HCl 25	Methyl isopropyl ketone 85 10 N HCl 15	Methyl isobutyl ketone 85 10 N HCl 15
Ag^+		0–0.49	0.16–0.55	0.26–0.57	0.26–0.98	0.33–0.63	0.05–0.34
Hg^+		0; 0.71–0.92	0.73–0.88	0.84–0.91	0; 0.82–0.97	0; 0.75–0.92	0.58–0.75
Hg^{++}		0.12–0.86	0.73–0.87		0.78–0.94	0.73–0.95	0.63–0.76
Tl^+		tail	0–0.19	0–0.19	0–0.38	0–0.19	0–0.17
Tl^{+++}				0–0.19			0–0.16
Pb^{++}		0.01–0.15	0.13–0.27	0.12–0.23	0.10–0.32	0.10–0.25	0.11–0.20
Bi^{+++}	0–0.38	0.31–0.54	0.35–0.43	0.41–0.60	0.47–0.61	0.32–0.51	0.30–0.36
As^{+++}			0.56–0.60				0.58–0.61
Sb^{+++}		0.83–0.97	0.78–0.96	0.91–0.98	0.86–0.97	0.85–0.94	0.82–0.87
Sn^{++}		0; 0.79–0.97	0.76–0.97	0.86–0.99	0.81–0.98	0.80–0.92	0.84–0.89
As^{+++++}		0.57–0.74	0.58–0.60	0.83–0.88	0.80–0.90	0.72–0.80	0.53–0.60
Sb^{+++++}			0; 0.80–0.93	0; 0.88–0.95	0.89–0.94		0; 0.81–0.98
Sn^{++++}		0	0	0	0; 0.90–0.94	0	0; 0.88–0.94
Cu^{++}	0.10–0.45	0.41–0.51	0.42–0.51	0.65–0.76	0.55–0.67	0.48–0.66	0.41–0.49
Cd^{++}	0.36–0.74	0.76–0.89	0.65–0.78	0.82–0.93	0.80–0.95	0.82–0.92	0.69–0.77
Mn^{++}	0	0.02–0.06	0.06–0.11	0.15–0.22	0.21–0.34	0.17–0.30	0.10–0.16
Co^{++}	0–0.05	0.17–0.31	0.25–0.34	0.65–0.82	0.66–0.85	0.48–0.76	0.48–0.56
Ni^{++}	0	0	0.03–0.08	0.01–0.06	0.02–0.07	0.01–0.05	0.02–0.07
Zn^{++}	0; 0.09–0.77	0.51–0.75	0.57–0.61	0.80–0.93	0.74–0.92	0.63–0.87	0.57–0.60
Fe^{++}	0–0.83	0.31–0.97 tail	0.62–0.95	0; 0.88–0.98	0.85–1.0 tail	0; 0.82–0.96	0; 0.72–0.94
Fe^{+++}		0; 0.82–0.91	0.71–0.76	0; 0.89–0.98	0.90–1.0 tail	0; 0.83–0.95	0; 0.90–0.98
Cr^{+++}		0	0–0.06	0–0.03	0–0.09	0–0.05	0–0.05
Al^{+++}		0	0–0.06	0–0.03	0.01–0.08	0–0.04	0–0.06
Ga^{+++}		0; 0.88–0.95	0; 0.84–0.91	0; 0.88–0.97	0.91–0.97	0; 0.88–0.93	0; 0.84–0.91
Ca^{++}		0	0.01–0.04	0.01–0.04	0.02–0.07	0–0.03	0.02–0.06
Ba^{++}			0	0		0	0
Sr^{++}			0–0.01	0	0–0.02	0	0–0.04

(Continued)

TABLE 147 (continued)

	Methyl n-propyl ketone 98 / 10 N HCl 2	Methyl n-propyl ketone 95 / 10 N HCl 5	Methyl n-propyl ketone 85 / 8 N HCl 15	Methyl n-propyl ketone 85 / 10 N HCl 15	Methyl n-propyl ketone 75 / 8 N HCl 25	Methyl isopropyl ketone 85 / 10 N HCl 15	Methyl isobutyl ketone 85 / 10 N HCl 15
Mg^{++}	0–0.02		0.07–0.11	0.01–0.05	0.05–0.12	0.01–0.05	0.03–0.09
Li^+	0–0.02			0.12–0.15	0.24–0.29	0.10–0.12	0.11–0.15
Na^+				0–0.06			0–0.08
K^+	0		0.05–0.06	0.03–0.05	0.08–0.12	0.02–0.05	0.05–0.07
Rb^+	0		0.05–0.06		0.10–0.12	0.03–0.06	0.09–0.11
Cs^+			0.07–0.08	0.05–0.07	0.16–0.20	0.06–0.09	0.06–0.10
Au^{+++}	0.88–0.97		0.82–0.89	0.84–0.93	0.90–0.97	0.87–0.94	0.90–0.95
Pd^{++}	0.16–0.20 tail		0.40–0.49	0.47–0.90	0.68–0.96	0; 0.33–0.52	0.28–0.36
Pt^{++++}	0.17–0.30, 0.40–0.51		0.59–0.68	0.45–0.83	0.70–0.83	0.43–0.81	0.49–0.65
Ir^{++++}			0.41–0.49	0.58–0.68	0.13–0.15	0.07–0.13	0.04–0.08, 0.39–0.45
Be^{++}	0–0.04		0.09–0.30	0–0.17	0.22–0.34	0–0.22	0.08–0.16
Ti^{+++}	0–0.02		0–0.07	0–0.11	0.08–0.15	0–0.08	0.04–0.09
VO^{++}	0–0.03		0.04–0.11	0.04–0.10	0.07–0.18	0–0.11	0.05–0.11
VO_4^{-3}			0.04–0.11 tail	0.03–0.10 tail	0.07–0.18		0.05–0.10 tail
MoO_4^{--}	0.89–0.97		0.78–0.92	0.88–0.94	0.87–0.93	0.88–0.94	0.87–0.94
WO_4^{--}	0; tail		0	0; tail	0; tail	0; tail	0
UO_2^{++}	0.34–0.61		0.58–0.64	0.74–0.89	0.81–0.94	0.81–0.96	0.62–0.74
Th^{++++}	0		0	0	0–0.03	0	0–0.03
Zr^{++++}	0–0.02		0–0.03	0; tail	0–0.22	0–0.12	
MnO_4^-	0.03–0.08		0.06–0.12	0.14–0.22	0.20–0.32	0.12–0.22	0.08–0.15
CrO_4^{--}	0–0.03		0–0.05	0–0.06	0–0.13	0–0.06	0–0.12
BO_3^{---}	0.18–0.25		0.34–0.44	0.32–0.43	0.45–0.67	0.41–0.52	0.28–0.36
Ce^{+++}	0		0.02–0.05	0–0.03	0–0.04	0–0.02	0–0.03
Y^{+++}	0		0–0.05	0–0.02	0–0.06	0–0.03	0.01–0.04
Pr^{+++}			0–0.04	0–0.03			0–0.03
La^{+++}			0–0.03	0–0.03			0–0.03
Acid front	0.54		0.59	0.85	0.67	0.67	0.60

TABLE 148. R_F values in butanol-HBr mixtures (Kertes and Lederer, *1698a*)

Solvents: I. Butanol 100 ml + hydrobromic acid 10 ml + water 90 ml (two phases)
II. Butanol fraction of solvent I + hydrobromic acid 10 ml (one phase)
III. Butanol fraction of solvent I + hydrobromic acid 20 ml (one phase)
IV. Butanol fraction of solvent I + hydrobromic acid 40 ml (one phase)
V. Butanol fraction of solvent I + hydrobromic acid 60 ml (one phase)

			Solvent mixtures		
Ion	I	II	III	IV	V
K	0.04	0.11			
Rb	0.04	0.11			
Cs	0.04	0.11			
Be	0.33	0.46	0.52	0.61	0.66
Mg	0.08	0.16	0.24	0.37	0.45
Ca	0.04	0.08	0.12	0.19	0.25
Sr	0.01	0.04	0.07	0.13	0.15
Ba	0.01	0.02	0.04	0.06	0.08
Zn	0.74	1.00	1.00	1.00	1.00
Cd	0.80	1.00	1.00	1.00	1.00
Hg(II)	1.00	1.00	1.00	1.00	1.00
Cu(II)	0.10	1.00	1.00	1.00	1.00
Ag	0.00	0.10 T	0.1–0.86 T	0.55–1.00 T	0.55–1.00 T
Sn(II)	0.83	1.00	1.00	1.00	1.00
Pb	0.40	0.60	0.60	0.64	0.65
As(III)	0.54	0.78	1.00		
Sb	0.90	1.00	1.00	1.00	1.00
Bi	0.80	1.00	0.92	0.85	0.78
Cr	0.03	0.06	0.09	0.23	0.30
Al	0.06	0.16	0.23	0.36	0.40
Fe(III)	0.11	0.58	1.00	1.00	1.00
Ni	0.07	0.13	0.20	0.32	0.33
Co	0.06	0.14	0.24	0.38	0.55
Rh	0.59	0.94	1.00	1.00	1.00
Pd	0.58	0.93	1.00	1.00	1.00
Pt	0.20 T	0.95	1.00	1.00 T	1.00 T
Au	0.67	0.97	1.00	1.00	1.00
Mn(II)	0.09	0.19	0.25	0.32	0.33
Sc	0.02	0.08	0.15	0.30	0.38
Y	0.02	0.07	0.13	0.24	0.30
La	0.00	0.03	0.08	0.17	0.24
Ti	0.07	0.19	0.26	0.43	0.49
V	0.12	0.20	0.27	0.43	0.47
Zr	0.00	0.01	0.03	0.11	0.14
Th	0.00	0.03	0.09	0.15	0.18
U(VI)	0.11	0.21	0.30	0.43	0.47
U(IV)	0.00	0.02	0.06	0.17	0.25
Ga	0.04	0.10	0.26	0.65	0.62
In	0.78	1.00	1.00	1.00	1.00
Tl(III)	0.93	1.00	1.00	1.00	1.00
Ce(III)	0.02	0.04	0.08	0.15	0.23

(Continued)

TABLE 148 *(Continued)*

Ion	Solvent mixtures				
	I	*II*	*III*	*IV*	*V*
$B_4O_7^{--}$	0.60	0.74	0.77	0.80	0.83
$Cr_2O_7^{--}$	0.11 T	0.17	0.22	0.36	0.40
SeO_3^{--}	0.63	0.78	1.00	1.00	1.00
TeO_3^{--}	0.18	0.91	1.00	1.00	1.00
MoO_4^{--}	0.21	0.38	0.48	0.53	0.55
PO_4^{---}	0.53	0.82	0.90	0.96	0.96
ReO_4^{-}	0.64	0.94	1.00	1.00	1.00
WO_4^{--}	0.00	0.00	0.00	0.00	0.00
$[Fe(CN)_6]^{----}$	0.64	1.00	1.00	1.00	1.00
$[Fe(CN)_6]^{---}$	0.52 T	0.91			
I^-	0.68	0.99	1.00	1.00	1.00
ClO_3^{-}	0.62	0.98	1.00	1.00	1.00

TABLE 148a. R_F values of cations in miscellaneous solvents

	Phenol shaken with 2 N HCl (1939)	*Butanol shaken with 20% aq. acetic acid (unpublished)*	*s-Collidine shaken with water (2540)*	*s-Collidine shaken with 0.4 N HNO₃ (878)*
Ag^+	0–0.68	0.3	0.90	0.78
Hg^+				0.0
Pb^{++}	0.0		0.10	0.0
Hg^{++}	0.045			0.0
Bi^{+++}	0.015	0.36	0.0	0.0
Cu^{++}	0.025	0.48	0.68	0.76
Cd^{++}	0.0	0.42	0.76	0.76
As^{+++}	0.30	0.42	0.68	0.65
Sb^{+++}	0.26		0	0.38
Sn^{++}	0.28		0	0
Sn^{++++}			0	0
Al^{+++}	0.0		0	0
Cr^{+++}	0.0		0	0
Fe^{+++}	0.01		0	0
Zn^{++}	0.038			0.75
Mn^{++}	0.0		0.28	0.71
Co^{++}	0.018		0.50	0.74
Ni^{++}	0.0		0.58	0.76
Ca^{++}	0.0			0.52
Sr^{++}	0.0			0.40
Ba^{++}	0.0			0.26
Mg^{++}	0.0			0.65
Na^+				0.32
K^+	0.19			0.42
Rb^+	0.26			
Cs^+	0.52			
UO_2^{++}		0.56		
Tl^+	0.56T	0.35		
SeO_3^{--}	0.23			
Au^{+++}	0.24			
Rare earths	0.0			

5. THE SEPARATION OF METALS IN THE DIFFERENT GROUPS

Paper chromatography is used at its best advantage if the groups of the cations are first separated by the usual means and each group then examined by paper chromatography.

(a) *Group 1*

Several methods are available for the separation of mixtures of Ag^+, Hg^+ and Pb^{++}. Frierson and Ammons (*1046*) recommend the use of butanol saturated with aqueous acetic acid (the exact concentration is not stated but appears to be 1 N aqueous acetic acid). Mercurous mercury moves fastest followed by silver, and lead moves slowest. Elbeih *et al.* (*2541*) separated the same mixture by the use of butanol saturated with 1 N HNO_3 containing 0.5% benzoylacetone. In this solvent, the following R_F values are obtained: Ag 0.1, Pb 0.03, Hg^+ 0.25; silver can also be removed from Hg^+ and lead with collidine saturated with 0.4 N HNO_3 (R_F values: Ag^+ 0.78, Hg^+ 0, Pb^{++} 0).

Harasawa (*1300*) separated Ag^+ and Pb^{++} using a butanol-pyridine-water mixture (100:20:20); R_F of Pb^{++} is 0 and of Ag 0.5–0.9; he also separated Ag^+, Pb^{++} and Hg^+ as nitrates. See also Martin (*2158*), Table 135.

Suzuki (*3160a*) separates group 1 with M NH_4CNS as solvent (R_F values Ag^+ 0.9, Pb^{++} 0.2 and Hg_2^{++} 0.8).

It is not mentioned by any of the above authors how the mixture of the insoluble halides obtained by precipitation is redissolved and placed on paper.

A satisfactory way of separating this mixture is the use of a "capillary separation" inside filter paper (*1918*). A small quantity (about 0.1 mg) is placed in the centre of a disc of No. 1 Whatman filter paper. From a pipette first two drops of water and then two drops of 5 N NH_4OH are added slowly to the precipitate, taking care that the liquid flows solely inside the paper. On holding the paper over H_2S, an outer ring of PbS, an inner ring of Ag_2S, and in the centre, black specks of Hg and HgS will be formed.

Thallous thallium, a rarer member of group 1, can be separated in solution from silver and mercurous mercury but not from lead, by using *iso*propyl alcohol containing 5% water and 10% concentrated HNO_3 (R_F values: Tl^+ 0.12, Ag^+ 0.24, Hg^+ 0.30, Pb^{++} 0.07, M. Lederer, unpublished results). See also Tewari (*3222, 3225*) and Weiss and Fallab (*3489*).

(b) *Group 2. Hg^{++}, Pb^{++}, Bi^{+++}, Cu^{++}, Cd^{++}*

For the separation of this group butanol saturated with 1 N HCl is employed (*1918*). Lead only travels a short distance and is not completely separated from copper. Cadmium travels further and forms an adjacent spot to bismuth while mercury is extracted into the butanol phase and travels ahead of the water front. After development, the paper is held over NH_3 and then over H_2S, precipitating all the cations as coloured sulphides. Various improvements have been made either by decreasing or increasing the amount of HCl. Frierson and Ammons (*1046*) use less acid (the amount is not stated) and obtain greater R_F differences between bismuth and cadmium. Burstall *et al.* (*467*) employ butanol saturated with 2 N or 3 N hydrochloric acid. They observe the variation of R_F values as given in Table 149. The reagent employed is dithizone in chloroform. Mercury is detected as a pink band

TABLE 149.   Variations in R_F values with change in acid concentration
(Burstall *et al.*, *467*)

Butanol	Cu	Pb	Bi	Cd	Water front	Hg
Satd. with H₂O	0.04, 0.04	0.05, 0.05	0.50, 0.49	0.60, 0.62	0.65, 0.66	0.73, 0.74
Satd. with 1 N HCl	0.08, 0.08	0.12, 0.11	0.55, 0.52	0.66, 0.63	0.70, 0.68	0.76, 0.78
Satd. with 2 N HCl	0.13, 0.11	0.17, 0.12	0.57, 0.54	0.69, 0.66	0.72, 0.69	0.76, 0.76
Satd. with 2.5 N HCl	0.19, 0.19	—	0.62, 0.62	0.73, 0.72	—	—
Satd. with 3 N HCl	0.20, 0.20	0.27, 0.27	0.60, 0.58	0.77, 0.77	0.79, 0.79	0.81, 0.81

behind the solvent front and is followed by cadmium as a purple band. Bismuth appeared purple and copper as a purplish-brown band. Lead, which gives only a weak colour with dithizone, was detected at the top portion of the strip by spraying with an aqueous solution of rhodizonic acid; a bright blue band due to lead appeared just below the copper. Amyl alcohol saturated with 3 N HCl is also a good solvent for group 2.

TABLE 150. R_F values using various alcohols saturated with HCl (*1925*)

	Cu	Pb	Bi	Cd	Hg
Ethyl alcohol 10% 5 N HCl	0.47	0.16	0.94	1.0	1.0
isoPropyl alcohol 10% 5 N HCl	0.28	0.08	0.84	0.84	1.0
Amyl alcohol satd. with 2 N HCl	0.02	0.0	0.22	0.23	0.55
Amyl alcohol satd. with 3 N HCl (*467*)	0.07, 0.07	—	0.36, 0.36	0.63, 0.64	0.70, 0.72

In Table 150 all separations were carried out by ascending development while the last one (amyl alcohol-3 N HCl) was developed by the descending method. Ethyl alcohol is a good solvent in the absence of mercury but no separation of cadmium and mercury can be achieved. In all the preceding solvents the separation of copper and lead is usually incomplete. In butanol saturated with 10 % hydrobromic acid good separations of lead and copper are obtained, but cadmium and bismuth do not separate at all (Table 151).

Shibata and Uemura (*2965*) employ butanol-3–4*N* HCl. Also see Harasawa (*1300*). Other solvents recommended (*467*) for group 2 are ethyl *iso*propyl ketone containing 10% v/v hydrochloric acid ($d = 1.18$) and pyridine. Ethyl *iso*propyl ketone is not completely miscible with 10% conc. HCl, but a homogenous solution is obtained on addition of small amounts of *iso*propyl alcohol or larger amounts of methyl alcohol. Lead, copper, cadmium, bismuth and mercury travel in that order. In pyridine, bismuth, copper, cadmium and mercury separate in this order. Lead and bismuth remained in the original spot, while mercury was just in front of, but overlapping, the cadmium. The original solution was spotted on the paper and made alkaline before development with pyridine by exposure to NH_3 fumes. Butanol containing either acetic acid, or acetic acid and acetoacetic ester can also be used. The first separates the mixture bismuth, lead, copper, cadmium and mercury in that order; for the second, see Table 151.

TABLE 151.   Further R_F values of group 2 metals

Solvent	Cu	Pb	Bi	Cd	Hg
Butanol sat. with 10% HBr (*1921*)	0.15	0.41	0.95	0.95	1.25
Butanol 50%, water 35%, acetoacetic ester 5%, acetic acid 10% (*2540*)	0.63	0.18	0.34	0.28	0.83

(c) *Group 3. As, Sb, Sn*

A large number of solvents give good separations of arsenic and antimony but only relatively few good separations of antimony and tin have been reported. Butanol with various concentrations of HCl separates arsenic from two adjacent bands of antimony and tin (*1919*). With low concentrations of hydrochloric acid, Pb^{++}, Cu^{++}, Cd^{++}, Bi^{+++}, As^{+++}, Sb^{+++}, Sn^{++++} travel in that order (*1046*). By the use of lower alcohols the separation of As^{+++} and Sb^{+++} can be carried out satisfactorily (see Table 152).

TABLE 152. R_F values of group 3 metals (*1925*)

Solvent	As^{+++}	Sb^{+++}	Sn^{++}
Amyl alcohol-2 *N* HCl	0.41	—	0.52
Butyl alcohol-1 *N* HCl	0.46	0.50	0.55
*iso*Propyl alcohol + 10% 5 *N* HCl	0.66	0.77	0.88
Ethyl alcohol 10% + 5 *N* HCl	0.50	0.85	0.97

Good separations of all three metals can be obtained by the use of collidine saturated with 0.4 N nitric acid, the R_F values being: As^{+++} 0.65, Sb^{+++} 0.38, Sn^{++} and Sn^{++++} 0 (*878*).

Burstall *et al.* (*467*) recommend acetylacetone saturated with water to which 0.5 % of conc. HCl and 25 % of acetone are added. A solution of As^{+++}, Sb^{+++} and Sn^{++} chlorides in dilute hydrochloric acid (2–4 N) is spotted on paper and allowed to dry for 15 minutes. The solvent was allowed to run over the strip for one hour, moving 15 cm in that time. The complexes formed with acetylacetone are extremely stable, particularly the one with tin. Before the strip is completely dry it is sprayed with a solution of dithizone in chloroform and then dried thoroughly. The tin is found in the solvent front and has a diffuse trailing. The antimony formed a wider band with a sharp front and a diffuse rear edge, $R_F = 0.5$. The arsenic forms a narrow band, $R_F = 0.2$.

Ethyl acetate saturated with water was also employed by Burstall *et al.* (*466, 467*) but later rejected owing to incomplete separation. Ethyl ether saturated with water also gives incomplete separations of Sb^{+++} and Sn^{++}. Other solvents tried were acetone, 2,2'-dichlorodiethyl ether, methyl *iso*butyl ketone, methyl ethyl ketone, cyclohexanone, *iso*propyl ether, Cellosolve, pyridine, dioxan, formic acid, acetic acid, triacetin and ethylene dichloride, all without success.

The rather difficult separation of Sb^{+++}–Sn^{++} is overcome by Harasawa (*1300*) by oxidising both elements to Sb^{+++++} and Sn^{++++} with hydrogen peroxide. In a butanol-HCl-water mixture (100:4:20) Sb^{+++++} has an R_F of 0, while Sn^{++++} has R_F 0.7. This could not be confirmed by the junior author. The separation advocated by Stefanović *et al.* (*3088a*) could not be confirmed either.

(d) *Groups 4 and 5. Fe, Al, Cr, Co, Ni, Mn, Zn, V, Ti and Mo*

(i) *Separation of Fe, Al, Cr*

Burstall *et al.* (*467*) recommend glacial acetic acid, containing 25% of dry methyl alcohol. The test solution of iron, aluminium and chromium chlorides was prepared in 5 N HCl, spotted onto a paper strip and allowed to evaporate to dryness in air. A solution containing insufficient free hydrochloric acid gave incomplete extraction and double band formation of iron. The experiment was allowed to run for twelve hours in an atmosphere saturated with respect to the mixed solvents. In order to detect the position of the bands, the solvent was allowed to evaporate and the strip was cut lengthwise into portions. One portion was sprayed with an alcoholic solution of alizarin, made alkaline by exposure to ammonia vapours and then warmed. Aluminium appeared as a red band, well separated from a purple band due to iron, which was some way behind the solvent front. The other portion of the strip was first sprayed with aqueous sodium peroxide solution and then with a solution of benzidine in acetic acid. Chromium was indicated as a bright blue band just behind the aluminium. See also Stefanović *et al.* (*3088a*).

(ii) *Separation of Fe, Co, Ni, Mn*

No satisfactory separation of this mixture can be obtained with aliphatic alcohols containing HCl. The low molecular weight ketones containing water and HCl are most satisfactory and a number of mixtures are recommended by Linstead *et al.* (*468*). A convenient solvent is 80% methyl *n*-propyl ketone, 10% acetone and 10% conc. HCl. Iron moves the greatest distance, followed by cobalt and manganese; nickel moves only slightly. The relative humidity of the atmosphere should be 65%. Nickel and cobalt are detected with rubeanic acid, manganese with ammoniacal $AgNO_3$, and iron with potassium ferrocyanide. Shibata and Uemura (*2965*) use acetone and 6–8 vol % of HCl. See also Airan (*24*), Airan and Barnabas (*26*) and Tewari (*3223*). The last mentioned author published some R_F values in butanol-2 N HCl which neither Burstall *et al.* (*467*) nor the junior author can confirm and which appear erroneous.

(iii) *Separation of Co, Ni, Mn, Zn*

A solution of these metals in dilute HCl is spotted on paper and allowed to dry in air. The solvent used consists of acetone containing 5% water and 8% conc. HCl. The solvent is allowed to travel 25 cm, the strip dried, held over ammonia and then sprayed with a mixture of alizarin, rubeanic acid and salicylaldoxime in alcohol. Nickel forms a blue band, manganese a brownish one, cobalt a brown one and zinc is at the solvent front. Approximate R_F values are: Ni = 0.07, Mn = 0.3, Co = 0.6, Zn = 0.9. Several other ketonic solvent mixtures were reported for the separation of Ni, Mn, Co, Cu, Fe, Ti, V and U, and are given in Table 153 (*467*).

TABLE 153. R_F values of group 4 metals (Burstall *et al.*, *467*)

Metal	Acetone 5% H_2O 5% HCl	Acetone 10% H_2O 5% HCl	Ethyl methyl ketone 6% HCl	Ethyl methyl ketone 8% HCl	Ethyl methyl ketone 5% H_2O 8% HCl
Ni	0.09	0.12	0.01	0.01	0.01
Mn	0.21	0.20	0.13	0.18	0.13
Co	0.43	0.27	0.42	0.54	0.30
Cu	0.61, 0.80	0.63, 0.92	0.65	0.71	0.72
Fe	0.97	0.97	0.90	0.93	0.91
Ti	—	0.17	—	—	—
V	0.18	0.17	0.07	0.14	0.09
U	0.64	0.81	0.76	0.87	0.73
Water front	0.64	0.64	0.50	0.82	0.72

Frierson and Ammons (*1046*) also claim the separation of Al from adjacent zones of Fe, Co and Ni, without giving the exact HCl concentration in butanol, the solvent employed. For the separation of zinc see also Hermanowicz and Sikorowska (*1372*) and Lamm (*1873*). A scheme for alloy steels was worked out by Venturello and Ghe (*3397*). For traces of Ni in Co see Harasawa and Takasu (*1303b*).

(iv) *Separation of Cr, Al, Zn*

These metals may be separated by the use of a mixture of butanol (50 %), acetic acid (10 %), acetoacetic ester (5 %), water (35 %). The R_F values are as follows: Zn = 0.30, Cr = 0.65, Al = 0.15. Several other solvents, for example alcohols containing HCl, separate zinc from aluminium but chromium and aluminium are not separated (*2540*). For the separation of *Be, Al and Fe* see Rao and Shankar (*2620*), also p. 506.

(v) *Separation of Fe, Ti, Al, and V*

Lacourt et al. (*1851*) studied the separation of this mixture as well as of Co and Ni. Employing solutions of these metals in 50 % HCl they used various solvents, recording the front and rear end of a band rather than R_F values thus indicating simultaneously the R_F and the efficiency of the operation. Using acetone, separation of iron and vanadium from Ti is obtained. Fe = 171/193, V = 34/56, Ti = 0/30, while Al interferes (Al = 10/50). In 6.5 N formic acid, Al separates from Fe and Ti. This separation will be described on p. 520.

Using different pure and mixed solvents, various separations of this mixture were also obtained. Alcohols give the values shown in Table 154.

TABLE 154. Separation of Fe, Ti, Al, V, Ni and Co (Lacourt et al., *1851*)

	Fe	Ti	Al	V	Ni	Co
MeOH	120/148		0/130	59/68	67/80	79/85
EtOH	70/91	5/35	0/56	—	58/77	52/82
Propanol	57/79	0/35	0/30	—	0/23	1/23
*iso*Propanol	45/74	0/10	0/15	—	0/20	0/20
*iso*Amyl alcohol	53/62	0/10	0/15	—	—	—

In dioxan, iron travels with the liquid front whereas the other metals do not move. The effects of small additions of pyridine to various solvents are tabulated in Table 155. For further work see Lacourt et al. (*1850, 1853, 1856*). For the analysis of aluminium alloys see Venturello and Ghe (*3394*).

(vi) *Separation of Cu and Co*

Von Hahn et al. (*1262*) described the separation of Cu and Co with butanol-acetic acid-acetoacetic acid-water (100:50:20:150). Co gives three adjacent spots (R_F 0.51, 0.46, 0.40) and Cu one spot (R_F of 0.64); 0.3 γ of both can be detected with violuric acid.

TABLE 155.   Separation of Fe, Ti, Al and V (Lacourt *et al.*, *1851*)

Solvent	Fe	Ti	Al	V
3% Pyridine in acetone	20/100	0/15	0/10	0/10
3% Pyridine in MeOH	52/90	0/45	70/97	0/90
2% Pyridine in water	21/33	24/32	98/105	56/95

(vii) *Separation of K, Ti, V, Mo and Fe*

Pollard *et al.* (*2542*) carried out the separation of K, Ti, V, Mo, and Fe using butanol-1 N HNO_3 containing 1% acetylacetone. R_F values: K = 0.07, Ti = 0.55, V = 0.17, Mo = 0.63 and Fe = 0.40.

For the separation of Ti, V and Mo, see also (*1929*).

(e) *Groups 6 and 7*

(i) *Alkaline earths*

For the separation of *Ba*, *Sr* and *Ca*, Arden *et al.* (*76*) used various solvents containing KCNS; calcium always travels fastest and is followed by Sr and Ba. Burstall *et al.* in a later paper (*467*), recommend pyridine containing 20% water and 1% KCNS. The chlorides are spotted on the paper, dried and developed for 6 hours in an atmosphere containing 65–80% water. After drying, Ca is shown up with alizarin, and Sr and Ba with rhodizonic acid. Walker (unpublished experiments) also separated Ca, Sr and Ba using 200 ml *iso*propyl alcohol with 19 grams NH_4CNS and 25 ml of water (R_F values: Ca 1.0, Sr 0.78, and Ba 0.47).

Miller and Magee (*2217*, *2218*) employ butanol-conc. HCl (5:95 v/v) saturated with *n*-butyl chloride for the separation of Sr and Ba. For the separation of radium and barium see page 528.

Separations of Be, Mg, Ca, Sr and Ba acetates using ethanol containing 20% of 2 N acetic acid were described by Erlenmeyer *et al.* (*915*), see R_F value Table 158. Elbeih *et al.* (*878*) also separate alkaline earths with collidine-0.4 N HNO_3 (see Table 148a).

(ii) *Alkali metals*

Linstead *et al.* (*467*) separated Li, Na and K chlorides with methanol as solvent. The spots are shown up by spraying with $AgNO_3$ and fluorescein to detect the accompanying chloride ions. Lithium is just below the liquid front and K just above the point of application. Further solvents for the separation of alkali chlorides are given by Chakrabarti and Burma (*539*), see Table 156.

TABLE 156. R_F values of alkali chlorides (Chakrabarti and Burma, *539*)

	Absolute methanol	Ethanol 10% water	Ethanol 20% water	Acetone 10% water	Acetone 20% water
K	0.22	0.13	0.25	0.09	0.19
Na	0.44	0.25	0.34	0.11	0.21
Li	0.72	0.53	0.60	0.21	0.40

Barnabas *et al.* (*143*) studied mixtures of methanol with other alcohols or ketones and obtained the R_F values shown in Table 157.

TABLE 157. R_F values of alkali chlorides (Barnabas *et al.*, *143*)

	Methyl ethyl ketone: methanol (2:3)		Acetone: methanol (2:3)		Ethanol: methanol (1:1)		Butanol: methanol (3:7)		Pentanol: methanol (3:7)	
	i	ii	i	ii	i	ii	i	ii	i	ii
K	0.14	0.40	0.15	0.42	0.08	0.27	0.11	0.31	0.09	0.30
Na	0.30	0.55	0.25	0.58	0.23	0.47	0.26	0.48	0.25	0.46
NH₄	0.33	0.60	0.28	0.63	0.32	0.59	0.34	0.57	0.33	0.53
Li	0.62	0.78	0.50	0.78	0.65	0.82	0.62	0.78	0.53	0.74

i ascending method
ii circular method

TABLE 158. R_F values of alkali and alkaline earth acetates with ethanol-20% 2 N acetic acid (Erlenmeyer *et al.*, *915*)

Metal	R_F value	Colour with violuric acid
Li	0.76 ± 0.02	red violet
Na	0.56 ± 0.06	violet
K	0.45 ± 0.06	violet
Be	0.86 ± 0.03	greenish yellow
Mg.	0.76 ± 0.03	orange
Ca	0.68 ± 0.06	orange
Sr	0.55 ± 0.07	red violet
Ba	0.43 ± 0.06	light red

Na and Li chlorides were separated by Miller and Magee *(2217, 2218)* with methanol-*n*-butanol (4:1). Erlenmeyer *et al.* *(915)* employ ethanol-20 % 2 N acetic acid for alkali and alkaline earth acetates and detect the spots by spraying with violuric acid and heating to 60°. Table 158 gives the R_F values and colour of the spots.

Quantitative determinations of alkalis and alkaline earths using the above technique and spot area measurement were described in a later paper by Seiler *et al.* *(2935)*. For the determination of Na, K, Mg and Ca in biological samples see *(2936)*. See also Gordon and Hewel *(1175)*, Pollard *et al.* *(2542b)* and Sommer *(3048a)*.

(iii) *Miscellaneous*

Separation of K, Na, NH_4, Mg *and* Li chlorides using propanol-10 % methanol as solvent was described by Sakaguchi and Yasuda *(2783)*. See also Burma *(456)* for the separation of alkali hydroxides, citrates and sulphates *(459)*, and Harasawa and Sakamoto *(1302)*. For a quantitative estimation of Na, K, Mg and Ca, see Tristram and Phillips *(3302)*.

The separation of tracer quantities of [22]Na from several grams of Mg was achieved by removing the bulk of Mg by heating to the oxide and chromatographing the soluble portion with ethanol-20 % water. Mg travels ahead of Na and a complete separation is obtained *(331)*.

K, Rb *and* Cs can be separated using a mixture of conc. HCl, methanol, *n*-butanol and *iso*butyl methyl ketone (55:35:5:5) (Miller and Magee, *2217*).

A more promising solvent for this separation was described by Steel *(3087)*, namely phenol saturated with 2 N HCl (R_F values K 0.19, Rb 0.27, Cs 0.43 and NH_4 0.11). This solvent proved useful for the purification of radioelements such as [86]Rb (Laberrigue and Lederer, *1841*) and [131]Cs (Lederer, *1935*; Fouarge and Duyckaerts *(1009a)*.

K in urine was determined by Beerstecher *(173)* by spot area techniques. Radioactive [42]K and [24]Na were separated by Frierson and Jones *(1048)* using a mixture of conc. HCl, water and 0.05 % tartaric acid (1:10:4) as solvent. Kumé *et al.* *(1830)* suggested a method for the electrolytic detection of alkali metals by placing the paper sprayed with phenolphthalein between two graphite electrodes (6 volts for 5 secs).

6. SPECIFIC TESTS FOR METALS

A number of solvents are available which specifically extract one metal from a mixture and thus permit its detection with an otherwise non-specific reagent. Such specific extractions are a most valuable addition to the field of spot tests and are in most cases as sensitive as the ordinary spot tests.

(a) *Mercury*

Mercury can be extracted as the chloride with methyl acetate containing 3 % methyl alcohol and 10 % water from mixtures with Pb^{++}, Cu^{++}, Bi^{+++}, Cd^{++}, As^{++}, Sb^{+++}, Fe^{+++}, Al^{+++}, Cr^{+++}, Ni^{++}, Co^{++}, Mn^{++} and Zn^{++}.

The test solution should not contain excess acid since that renders the mercuric chloride less soluble in the lipid solvent. The development is carried out at a temperature of 22° for 20–30 minutes. The paper is then dried, exposed to ammonia and sprayed with an alcoholic solution of diphenyl carbazide, giving a blue band at the solvent front when mercury is present. 1γ of mercury can be detected in the presence of a large amount of other metals. Some interference in the extraction of mercury occurs with high concentrations of copper (10 % in the test solution) but the extraction is largely independent of the ratio of mercury to copper. Similar separations of mercury can be achieved with tetrahydrosylvan, ethyl acetate and methyl propyl ketone (467).

(b) *Arsenic*

This element can be separated from all common metals by using butanol saturated with water and containing 1 % ammonium tartrate, 1 % ammonium borate and 0.5 % mannitol. The R_F of As $= 0.45$ while all other metals have an R_F of 0.22 or less. After development, the paper is acidified by spraying with a mixture of alcohol with 1 % nitric acid and 5 % glycerol and, after drying, it is sprayed with ammoniacal silver nitrate. The arsenic spot is converted presumably to silver arsenite (yellow) which, on irradiation with ultra-violet light whilst still wet, changes to silver arsenate (brown) and metallic arsenic (black). The spot is then easily visible against the purplish background from the silver nitrate spray. It was possible to detect 0.3γ of As in 0.01 ml of test solution applied to the paper (878).

For further work, see Elbeih (877a).

(c) *Antimony*

Trivalent antimony is readily hydrolysed and if a mixture containing its chloride is developed with water it separates from Hg^{++}, Pb^{++}, Bi^{+++}, Cu^{++}, Cd^{++}, As^{+++}, Sn^{++}, Fe^{++}, Co^{++}, Ni^{++}, Zn^{++}, Al^{+++}, Cr^{++++}. The antimony remains behind whereas all other metals are washed to the solvent front. By treating the paper with H_2S the antimony is detected as an orange spot (1914).

(d) *Gold*

A rapid method is available for the detection of small amounts of gold in chloride solutions of the platinum metals using ethyl ether containing 1–2 % dry HCl and 7.5 % dry methyl alcohol. This mixture extracts the gold as a narrow band in the solvent front while the platinum metals remain on the original spot. 1 γ of gold in several hundred times the amount of each platinum metal can be detected (467). The same separation can be achieved by the use of butanol, shaken with an aqueous solution of 10 % HCl and 10 % HNO_3. Gold travels as a narrow band at the solvent front and the platinum metals have R_F values between 0.1 and 0.7. In both separations the gold solution must contain an oxidising agent otherwise metallic gold deposits on application to the paper (*1921*). See also quantitative methods on pp. 511, 520, 522 and 526.

(e) *Vanadium*

This element can be separated from a large number of other metals by using diethyl ether or tetrahydrosylvan containing nitric acid and a small quantity of hydrogen peroxide. The vanadium moves as a pink band due to the formation of a soluble peroxy-compound. Colour development is achieved with 8-hydroxyquinoline (*76*).

(f) *Copper*

By the use of butanol saturated with 1.5 N NH_4OH to which 2 % of solid dimethylglyoxime is added, copper can be separated from Ag, Cd, Ni, Co, Fe, Mn, Bi, Hg^+, Hg^{++}, Pd, Pt and Au. Copper moves as a brown dimethyl-glyoxime complex ($R_F = 0.5$), while all other metals have $R_F = 0$–0.05. No reagent is required to show up the copper, but at lower concentrations the brown colour is weak (*58*). Almassy and Dezsö (*44a*) use ethanol-HCl-ether (17 : 3 : 80) for a specific separation of copper.

(g) *Thallium*

In amyl alcohol saturated with 2 N aqueous hydrochloride acid, thallic thallium can be removed from all other metals, even from the very soluble mercuric and auric chlorides. The R_F of thallium is 0.75, the highest R_F value of any other metal being that of Hg^{++} (0.55). The thallium can be detected by spraying with potassium iodide solution, which produces a dark brown spot (*1919*). For the detection of Tl in toxicological analysis see also Diller and Rex (*777*).

(h) *Uranium*

In the absence of phosphates, solutions in 25 % nitric acid are used without further treatment. Minerals and ores are treated with nitric acid with the addition of hydrofluoric acid and perchloric acid, or fused with NaOH or Na_2O_2, and then treated with nitric acid. Large amounts of sulphates are undesirable. It is necessary to remove phosphoric acid as a complex by ferric nitrate if large amounts are present. The best solvents are 2-methyl-tetra-hydrofuran and tetrahydropyran containing 2.5 % and 10 % of concentrated nitric acid respectively. Only 10 cm development are required and the uranium is extracted ahead of all other metals. By spraying with ferrocyanide a brown spot is obtained. Thorium is the only element likely to be extracted with the uranium but does not interfere since it gives no colour with ferrocyanide *(77)*. This test can be made quantitative; the method is given on page 521. See also Almassy and Vigvari *(44d)*.

(i) *Molybdenum*

Almassy and Straub *(44c)* worked out a specific separation for Mo. See also Ghe and Fiorentini *(1102a)*.

7. THE SEPARATION OF RARER METALS (PLATINUM METALS)

(a) *Gold, platinum, palladium, copper and silver*

A mixture or alloy of these metals is dissolved in aqua regia. The solution is then diluted with an equal amount of water and one drop placed on the paper strip. After development for twenty hours with butanol saturated with N HCl in an atmosphere saturated with the solvent, the strip is removed and held over NH_3 and then over H_2S. Silver is deposited at the starting point ($R_F = 0.0$), copper travels slowly ($R_F = 0.1$), palladium and platinum separate completely ($R_F = 0.6$ and 0.75 respectively) and gold is extracted with the butanol phase, which travels ahead of the water front. Silver and copper give black spots which fade in an hour. The spots of gold and platinum are yellow-brown and that of palladium orange-brown. A very small amount of gold is reduced and leaves a trail of colloidal gold of purple colour. The spots of platinum, palladium and gold deepen on storage and can be kept as permanent records of the analyses (*1915*) (see Fig. 135).

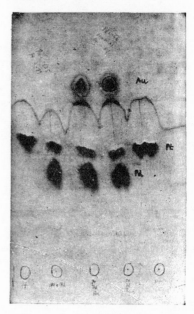

Fig. 135. The separation of gold, platinum and palladium with butanol-1 N HCl (Bouissières and Lederer, *331*)

(b) *Gold, platinum, palladium, rhodium, iridium, ruthenium and osmium*

In order to maintain standard conditions, Burstall *et al.* *(467)* recommend the preparation of a chloride solution by fusing together gold, palladium, platinum, rhodium and iridium chlorides with sodium peroxide, another solution by fusing ruthenium and osmium together with the other metals and Na_2O_2. This solution is spotted on the paper strip and dried in air. For the detection of the metals, stannous chloride was found to be a suitable reagent for Pt, Pd, Au and Rh. A more sensitive reagent is a mixture of stannous chloride and potassium iodide. The strip has to be warmed gently for the development of the full rhodium colour. Iridium, which is reduced to the colourless, trivalent state can be detected by re-oxidising with chlorine water. Treatment with thiourea in $5 N$ HCl and subsequent warming gives coloured spots of ruthenium and osmium. The solvents recommended by these authors are methyl propyl ketone containing 30 % of conc. HCl and methyl ethyl ketone containing 30 % of conc. HCl. Either of these solvents rapidly reduce tetravalent iridium to the trivalent state which moves with the rhodium. In general, four bands are formed: gold, platinum, palladium, whilst rhodium and iridium travel together as shown in Table 159. For the

TABLE 159. R_F values for platinum metals in ketonic solvents (Burstall *et al.*, *467*)

Solvent	Au	Os	Pt	Pd	Rh Ir	Ru
Methyl ethyl ketone, 30% HCl after 8 hours	0.95–0.97	0.92–0.93	0.76–0.84	0.61–0.65	0.06–0.16	0.07
Methyl propyl ketone, 30% HCl after 8 hours	0.91–0.92	0.89–0.91	0.75	0.53	0.11–0.17	0.07

separation of rhodium and iridium, after removal of gold, palladium and platinum with either solvent, the solvent is allowed to evaporate and the strip is sprayed with chlorine water to re-oxidise the Ir^{+++} to Ir^{++++}. Ir^{++++} is then extracted from the spot by developing with acetone containing 5% HCl. Other solvents tried by Burstall *et al.* are alcohols containing HCl, carbitol and cellosolve of which all give some separation. For further work see (*1682a*).

In these investigations Burstall *et al.* did not try to avoid the reduction of Ir^{++++} to Ir^{+++} on the paper. This can be done almost completely by including an oxidising agent in the solvent, for instance butanol shaken with an aqueous solution containing 10% HCl and 10% HNO_3. See also Fournier (*1009b*). The R_F values for this solvent and for butanol shaken with $1 N$ HCl are given in Table 160.

TABLE 160. R_F values of platinum metals (Lederer, *1915*)

Solvent	Au	Ir	Pt	Pd	Rh	Ru
Butanol sat. aq. 10% HCl, 10% HNO_3	1.0	0.64–0.77	0.67–0.75	0.53–0.58	0.15–0.20	0.10–0.17
Butanol sat. $1 N$ HCl	1.1	0.76 tails	0.76	0.6	0.07	—

8. OTHER RARE METAL MIXTURES

(a) Beryllium, aluminium and zirconium

Osborn and Jewsbury (2421) first separated Al and Be by using butanol containing HCl. Alcoholic 8-hydroxyquinoline and ammonia sprayed on the paper produces spots which are fluorescent in sunlight or ultra-violet light. In other solvents (see Table 161) beryllium can also be separated from other metals, such as the rare earths and zirconium. Another reagent which gives good results is quinalizarin in aqueous ammonia, giving purple spots for Be and Zr, and a red spot for Al. For a separation of Fe – Ti – Zr see Harasawa and Sakamoto (1303a). Ti – Zr – Th were separated by Almassy and Nagy (44b) using ethanol-HCl mixtures.

TABLE 161. R_F values of Be, Al, rare earths and Zr (Lederer, 1925)

Solvent	Be	Al	Rare earths	Zr
Butanol-1 N HCl	0.30	0.07	0.03–0.05	0.0
isoPropyl alcohol, 10% 5 N HCl .	0.62	0.35	0.06–0.07	0.02
Ethyl alcohol with 10% 5 N HCl.	0.70	0.37	0.07–0.11	0.0

(b) Aluminium, gallium, indium and zinc

This group has been studied by Arden et al. (76) using solutions of the chlorides and n-butyl alcohol containing HCl as the solvent. Aluminium scarcely moves at all, gallium travels with the solvent front, while indium appears between the two. Zinc appears between indium and gallium. Aluminium and gallium are developed with "aluminon", while dithizone is used for indium and zinc.

(c) Germanium

Germanium has been separated independently in three different laboratories (720, 1862, 1933, 2172).

The following R_F values were found

$$\begin{array}{ll}
\text{butanol- 1 } N \text{ HCl} & 0.26 \\
\text{butanol-10\% HBr} & 0.25\text{–}0.29 \\
\text{butanol-1.5 } N \text{ HNO}_3 & 0.235
\end{array}$$

This constancy of R_F values in various acids suggests that the same ion exists in all three solvents presumably a hydroxy-cation. As shown by Carvalho (720) the R_F values increase considerably in higher HCl concentration thus

$$\begin{array}{ll}
\text{butanol- 3 } N \text{ HCl} & 0.54 \\
\text{butanol- 6 } N \text{ HCl} & 0.93 \\
\text{butanol-12 } N \text{ HCl} & 1.00
\end{array}$$

For separations from other metals the general R_F tables should be consulted. As reagents, Ladenbauer and Hecht (*1862*) suggest *phenylfluorone*, Matsura (*2172*) *haematoxylin* which he found suitable for quantitative photometry. Lederer (*1933*) found visible reactions with *hydroxyquinoline*, also by passing H_2S over the paper, drying and then dipping into silver nitrate solution. For a spot area estimation method see (*1861*). See also Ladenbauer and Slama (*1862b*), Nagy and Polyik (*2334a*) and Bertorelle and Fanfani (*234a*).

(d) *Rare earths, scandium and thorium*

Both scandium and thorium may be separated from mixed rare earths. A solution of the mixture in dilute nitric acid is placed on the paper and dried. The solvent consisting of tetrahydrosylvan containing 5 % water and 10 % nitric acid was allowed to diffuse about 15 cm down the paper strip. By controlling the relative humidity of the atmosphere at about 80 % with a saturated solution of ammonium chloride at the bottom of the development chamber, a narrow and concentrated band of scandium is obtained. The solvent is allowed to evaporate and the strip is placed in an atmosphere of ammonia for about 10 min. The strip is then sprayed with an alcoholic solution of alizarin and finally with N acetic acid. Thorium is detected as a violet-blue band with a sharp front edge about 0.6 cm behind the solvent front. Scandium forms a more diffuse band about 2.5 cm in front of the rare earths which move only slightly. 1γ of scandium in 1000γ of rare earth can be detected (*467*).

Scandium can also be separated from the rare earths by using methyl acetate containing 10 % water and 5 % nitric acid as solvent. The strip is developed for 25 cm and scandium is again found about 2.5 cm ahead of the rare earths. Thorium forms a more diffuse band between the scandium and the solvent front. On a larger scale scandium can be separated from a large number of metals using a cellulose column. By working with the nitrates and by using ether containing 12.5 % HNO_3, scandium is washed through the column. Zirconium is also washed through, but may be held back by adding tartaric acid to the original solution.

The separation of rare earths by partition chromatography has presented difficulties as in almost all solvents there is considerable tailing. Pollard *et al.* (*2544, 2545*) worked with solvents containing 8-hydroxyquinoline and obtained some separations but no complete separation of adjacent elements. Lederer (*1930*) employed solvents containing acetylacetone and reduced comet formation by working at elevated temperatures. Under these conditions the rare earths are separated into five fractions and a few adjacent metals are resolved. However the system resolves only macro amounts and with carrier-free tracers only elongated trails are formed owing to adsorption on the paper. Table 162 shows the R_F values and Table 163 the reagents employed by Pollard *et al.* (*2545*). Lederer dips the paper after drying into

TABLE 162. R_F values of rare earths

Cation	Butanol sat. with water 100 ml 8-Hydroxyquinoline 5 g Acetic acid 20 ml	Butanol 100 ml Acetylacetone 30 ml Acetic acid 5 ml at 60° Water 65 ml
	Pollard et al. (2545)	Lederer (1930)
Y	0.47	0.59
La	0.34	0.31
Ce	0.40	0.38
Pr	0.40	0.38
Nd	0.43	—
Sm	0.44	0.47
Eu	—	0.49
Gd	0.44	0.43
Tb	0.47	—
Dy	0.50	0.62
Ho	—	0.56
Er	0.50	0.60
Tm	0.66	—
Yb	0.70	0.59
Lu	0.59–0.76	—
Sc	—	0.97
Ac	—	0.08

TABLE 163.   Reagents for the detection of rare earths (Pollard et al., 2545)
ff = brightly fluorescent, f = fluorescent, d = dark spot in ultra violet light.

Cation	2% Pyrogallol- 4-carboxylic acid in 50% alcohol	Aqueous ammonium purpurate ("murexide")	8-Hydroxy- quinoline 1% in 50% alcohol	Morin in 50% alcohol
Y	brown	pinkish orange	ff (green yellow)	ff (green)
La	bright purple	bright violet	ff (green)	ff (green)
Ce	purple turning brown	bright violet	d	d
Pr	bright purple	bright violet	d	d
Nd	bright purple	bright violet	d	d
Sm	dirty purple	purplish pink	d	d
Gd	purple brown	purplish pink	f (deep brown)	f (green)
Tb	brown	pink	d	d
Dy	brown	orange yellow	d	d
Er	brown	orange yellow	d	d
Tm	brown	orange yellow	d	d
Yb	brown	orange yellow	d	d
Lu	brown	orange yellow	ff (yellow green)	ff (green)

an aqueous solution of alizarin S which yields purple spots on a yellow background.

With the acetylacetone solvent at 60° the following mixtures could be resolved: La–Gd–Dy, Ac–La–Y–Sc, Pr–Nd–Dy, La–Sm–Er. The possibilities of the 8-hydroxyquinoline solvent are shown in Fig. 136. Column experiments were also made by Pollard *et al.* (*2544, 2545*) and yielded essentially the same results as paper strip separations.

Good separations of Ce–Pm–Eu and other mixtures were recently recorded with *non-complexed* rare earths on the tracer scale using ethanol containing 10 % 2 *N* HCl as solvent (*1940*).

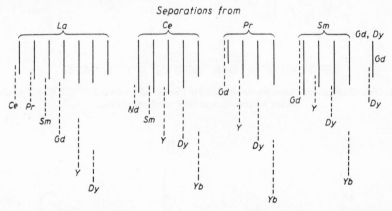

Fig. 136. Resolution obtained by running aqueous nitrate spots of the binary lanthanon mixtures upon acid-washed Whatman No. 1 paper strips in the eluant (*n*-butanol, 100 ml, acetic acid, 13 ml and water, 30 ml, mixed; to 110 ml of the mixture 5 g of oxine added, with warming to render the solution homogeneous, for approximately 14 h (Pollard *et al.*, *2545*)

This work was later extended and the sequence of all rare earths established as shown in Fig. 136a (*1941a*). Good separations of Sc from all rare earths are also possible with ethanol-10 % 2 *N* HCl as solvent. A very pure preparation of [147]Pm was prepared using this method in conjunction with paper electrophoresis (*1879a*).

Additions of NH_4CNS to the above solvent yielded complete separations of Nd–Sm–Eu as shown in Fig. 136b (*1941b*).

(e) *Uranium and thorium*

As shown in Tables 143 and 146 this pair of metals can be successfully separated with butanol-1.5 *N* HNO_3 as well as with other solvents. Pollard *et al.* (*2541*) also recommend butanol-1 *N* HNO_3 with 0.5 % benzoylacetone. Sarma (*2821*) employs butanol-3 *N* HCl or butanol-4 *N* HCl The R_F values for UO_2^{++} are 0.19 and 0.32 respectively, and for Th 0.05 and 0.13.

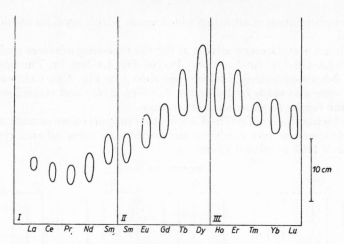

Fig. 136a. Rare earth chlorides developed for 1 week with ethanol-10% 2N HCl placed in order of atomic numbers (Lederer, *1941a*)

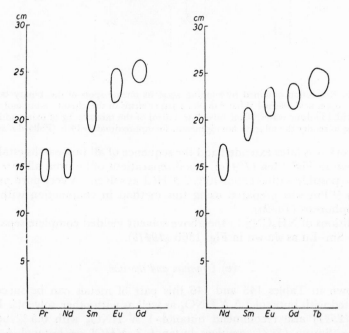

Fig. 136b. The movement of Pr, Nd, Sm, Eu, Gd and Tb when developed for 48 hours with ethanol 90 ml, 2N HCl 10 ml, NH$_4$CNS 1 g. The spots are revealed with ammoniacal alcoholic hydroxyquinoline (Lederer, *1941b*)

Further work by Sarma (*2822, 2823*) deals with the behaviour of various uranyl, thorium and rare earth salts with numerous solvents such as butanol-HCl, amyl alcohol-formic acid, ethyl acetate-acetic acid, cyclohexanol-HCl and ethyl ether-acid mixtures. His work confirms and extends numerous other separations already published. Suchy (*3154*) also published several solvents for the separation UO_2–Th and records the R_F values as follows:

Solvent	R_F values		temperature
	uranyl	thorium	
methanol:pyridine:toluene (7:2:1)	0.86	0.16	13°
methanol:pyridine:dioxane (7:2:1)	0.93	0.13	22°
butanol:pyridine:dioxane (7:2:1)	0.69	0.07	15.5°
methanol:chloroform:toluene (7:2:1)	0.72	0.13	20°
pyridine:chloroform:toluene (7:2:1)	0.45	0.13	19°

The separation of uranyl ions *from vanadium and chromium* was reported by Harasawa and Sakamoto (*1303*) with acetone-acetic acid-water and acetone-HCl-water mixtures. A spot area determination of microquantities of uranium and indications for a scheme of uranium estimation in minerals are given by Seiler *et al.* (*2934*). The solvent used is a methyl ethyl ketone-acetylacetone-acetic acid mixture and the spots are shown up with ferrocyanide.

A scheme for the determination of U in deep sea sediments using amongst other techniques paper chromatography with butanol-1.5 N HNO_3 is proposed by Hahofer and Hecht (*1264*). For quantitative methods see also p. 522, for specific tests, p. 503.

(f) Hafnium and zirconium

The separation of Hf from Zr was first proposed with ether containing 12.5 % HNO_3 as solvent (*468*). Kember and Wells (*1681*) later worked out a better separation using a solvent consisting of 70 ml of dichlorotriethyleneglycol to which 30 ml of concentrated nitric acid (*d* 1.42) were carefully added.

The sample to be separated had to be digested with concentrated nitric acid at 80° to avoid the presence of immobile basic nitrates. Downward development was employed; optimum separations were obtained with about 150 micrograms of the mixed oxides placed on a paper strip. The solution to be analysed is placed on the paper and developed without drying the spot to avoid hydrolysis. After 18 h development and allowing the solvent to run off the end of the paper the spots are shown up with saturated solution of alizarin in ethyl alcohol containing 5 % of 2 N HCl. The strip is then heated gently and resprayed if necessary. Under the conditions described the R_F values are 0.1 for Hf and 0.2 for Zr. None of the common metals give lakes with alizarin. The rare earths Fe, Ti, Cu, Co, Ni and Mn have R_F values between 0 and 0.2 while Th and U move in or near the solvent front.

(g) *The separation of different valencies of the same element*

Usually different valency states separate well if both are stable under the conditions of the development, e.g. Tl(I)–Tl(III),U(IV)–U(VI), Fe(II)–Fe(III) as indicated in the R_F value tables. Studies on the separation of valency states of Hg, Pt and V were published by Pollard *et al.* (*2539a*), of V and Co by Stevens (*3103a*), of Mo by Nicholas and Stevens (*2363a*), of Cr(II), Cr(III) and CrO_4^{--} by Bighi (*245*) and of Fe, Cu, Hg and Cr also by Bighi (*245a*).

9. THE SEPARATION OF ACIDS

(a) Cl^-, Br^-, I^- and CNS^-

This mixture is separated by using butanol saturated with 1.5 N NH_4OH as solvent. Thiocyanate ($R_F = 0.45$) and iodide ($R_F = 0.30$) are separated completely and can be detected by spraying, after drying the paper, with a mixture of H_2O_2 and ferric nitrate. A red $Fe(CNS)_3$ spot is given by CNS^- and a blue spot by iodine. This blue colour is due to the reaction of the liberated iodine with starch or hemicellulose in the paper and is not given by all papers, some only producing a brown spot. Chloride ($R_F = 0.10$) and bromine ($R_F = 0.16$) form two adjacent spots, which do not separate completely. Both can be shown up by spraying with $AgNO_3$. If the paper is exposed to sunlight, distinct dark black spots are produced on a grey background. However, faster results are obtained by washing the paper with dilute HNO_3 and holding over H_2S, Ag_2S being formed where the insoluble silver halides are situated (*1920*). A mixture of silver nitrate and fluorescein sprayed on the paper gives, on inspection under ultra-violet light, dark areas on a fluorescent background (*467*). Fluorides may be detected by spraying with zirconium-alizarin reagent. Indicators incorporated in the solvent (e.g. butanol-0.5 $N$$NH_4OH$ and 0.2 g bromocresol purple per 100 ml) also produce good spots with halides. Another solvent which gives the same separation is a mixture of 90 ml of *iso*propyl alcohol with 10 ml of conc. ammonia. The R_F values are more differentiated (Cl = 0.37, Br = 0.46, I = 0.6, CNS = 0.8). The spots spread more in this solvent and again CNS^- and I^- are completely separated and adjacent spots of Cl^- and Br^- are obtained (M. Lederer, unpublished experiments, see Table 164).

For the separation of Cl^-, Br^-, I^-, and fluoride in absence of CNS^-, pyridine containing 10 % of water, or acetone containing 20 % of water may be employed; in these solvents complete separation of Cl⁻ from Br⁻ can be obtained (467). See also Ando and Ishii (61) for the detection of halogens in organic compounds.

TABLE 164. R_F values of halides

	Butanol-1.5 $N\,NH_4OH$ (1920)	isoPropanol + 10% NH_4OH*	Acetone + 20% H_2O (467)	Pyridine + 10% H_2O (467)	Butanol-20% CH_3COOH*
Fluoride	0.0	—	0.25	0.0	—
Chloride	0.10	0.37	0.49–0.53	0.22–0.25	0.21
Bromide	0.16	0.46	0.60–0.62	0.47	0.16
Iodide	0.30	0.60	0.76–0.77	0.70–0.72	0.32
Thiocyanate	0.45	0.80	—	—	0.44

* M. Lederer, unpublished results

(b) *Selenite and tellurite*

A solution of selenite and tellurite in dilute nitric acid is spotted on a paper strip and allowed to dry. The solvent is prepared by mixing 60 volumes of dry butanol with 40 volumes of dry methanol. The atmosphere in the development chamber is saturated with the solvent and has to contain 50 % humidity. After 2 hours (8–10 cm) the strip is dried and sprayed with stannous chloride, tellurium forming a black band not far from the original spot and selenium an orange band, R_F approximately 0.5. Even 1 γ of Se can be detected in 1000 γ of Te (467).

The separation of selenite and tellurite was further examined by Lederer (1939) who noted the R_F values given in Table 165.

TABLE 165. R_F values of selenite and tellurite (1939)

Solvent	Selenite	Tellurite	Remarks
Butanol – water	0.57	0.15	salts diss. in HCl
Butanol – N HCl	0.67	0.18	salts diss. in HCl
Butanol – 1.5 N HCl	0.64	0.31	salts diss. in HCl
Butanol – 2 N HCl	0.71	0.47	salts diss. in HCl
Butanol – 3 N HCl	0.78	0.67	salts diss. in HCl
Butanol – water	0.54	0.004	salts diss. in HNO_3
Butanol – 1.5 N HNO_3	0.59	0.15	salts diss. in HNO_3
Butanol – 3 N HNO_3	0.69	0.18	salts diss. in HNO_3
Butanol – 10% HBr	0.64	0.46	salts diss. in HBr
Butanol – 20% HBr	0.73	0.92	salts diss. in HBr

This behaviour was discussed with reference to the general chemistry of selenite and tellurite (*1939*). Weatherly (*3480a*) worked out a specific separation and determination of Se.

(c) *Separation of tellurate and tellurite*

Ghosh Mazumdar and Lederer (*1102b*) list the following R_F values for tellurate and tellurite: butanol - 2 N HCl 0.08, 0.45; butanol - 3 N HCl 0.16, 0.67; acetone-water-HCl (90:5:5) 0.07, 0.95 (respectively). With butanol-HBr mixtures and with solvents containing methanol partial reduction of Te(VI) was observed.

(d) *Arsenate, arsenite, chromate and ferricyanide*

These four acids can be separated by using butanol, shaken with 20 % aqueous acetic acid as solvent. Arsenite travels fastest (R_F 0.41), followed by chromate (R_F 0.24), arsenate (R_F 0.15) and ferricyanide (R_F 0.03). Chromate, arsenate and ferricyanide can be detected by spraying with silver nitrate, and arsenite by H_2S to produce the sulphide. Chromate tails a little, probably due to slight reduction and to the formation of dichromate.

Separation of ferrocyanide, ferricyanide, sulphide, arsenate, phosphate and iodide was described by DeLoach and Drinkard (*732*) using *n*-butanol, 95 % ethanol and water (2:2:1); the approximate R_F values at 30° are given in Table 166.

TABLE 166. R_F values of some anions (De Loach and Drinkard, *732*)

Ferrocyanide	0.12	Arsenate	0.26
Ferricyanide	0.15	Phosphate	0.37
Sulphide	0.16	Iodide	0.92

Yamaguchi (*3643*) separates ferrocyanide and ferricyanide with 60 % aqueous methanol, also thiocyanate-ferrocyanide and arsenite with the same solvent.

(e) *Halogen-oxy-acids*

Separation and identification of IO_4^-, IO_3^-, BrO_3^-, ClO_4^- and ClO_4^- can be carried out with ethanol-water-15 N NH_4OH (30:10:5). The ethanol must be purified by distillation over Ag_2O to remove reducing materials. (Cohen and Lederer, unpublished experiments).

The R_F values are given in Table 167.

TABLE 167. R_F values of halogen (oxy-)acids

IO_4^-	0.0	ClO_3^-	0.73–0.79
IO_3^-	0.31–0.37	Cl^-	0.70
BrO_3^-	0.68	Br^-	0.71
ClO_4^-	0.80–0.85	I^-	0.71

Yamaguchi (*3643*) separates Cl^-, Br^-, I^-, CNS^-, ClO_3^-, BrO_3^- and IO_3^- with acetone-$\mathcal{N}$ NH_4OH (4:1) and chloride-chlorate with butanol-acetone-water (5:2:3). See also Servigne (*2949*).

(f) *Silicate, borate and molybdate*

Lacourt *et al.* (*1849*) separate this mixture by using 5% conc. HCl in acetone as solvent in an atmosphere saturated with 5% HCl in 2-butanone; for mixtures of 10 γ of each of the three ions the R_F values shown in Table 168 for the front and the rear edge of each band were obtained.

TABLE 168. R_F values of silicate, borate and molybdate (Lacourt *et al.*, *1849*)

	R_F	
	rear	front
Silicate	0.0	0.10
Borate	0.56	0.72
Molybdate	0.96	1.00

A colorimetric method for the quantitative determination of borates is described in the same paper. Muto (*2327a*) separates B from Ag, Hg and Pb.

A paper chromatographic detection of boron in mineral water was described by Quentin (*2607*). Baumann (*164a*) succeeded to separate several oligosilicic acids with *iso*propanol-water-acetic acid (20:5:2).

(g) *Acids of phosphorus*

Ebel and Volmar (*856*) separated numerous anions of pentavalent phosphorus, using filter paper washed with HCl or treated with 8-hydroxyquinoline. The R_F values in various solvents are given in Table 169.

Separation of *phosphates, phosphites and hypophosphites* was described by Bonnin and Sue (*311*) on paper washed with HCl. Using equal volumes of *n*-butanol, dioxan and 1 $\mathcal{N}$ NH_4OH, they obtain the R_F values shown in Table 170 for the rear and the front of the bands.

TABLE 169. R_F values of phosphoric acids (Ebel and Volmar, *856*)

	Phosphoric acid					
	Ortho-	Pyro-	Tripoly-	Trimeta-	Tetrameta-	Graham's
*iso*Propanol (70), water (30), trichloroacetic acid (5), ammonia (0.3)	0.76	0.56	0.44	0.34	0.22	0
tert. Butanol (40), *iso*-propanol (30), water (30), trichloroacetic acid (5), ammonia (0.3)	0.72	0.55	0.49	0.44	0.38	0
tert. Butanol (80), water (20), picric acid (4), ammonia (0.2)	0.68	0.49	0.38	0.34	0.25	0
*iso*Propanol (40), *iso*-butanol (20), water (39), conc. ammonia (1)	0.43	0.33	0.30	0.54	0.48	0
n-Propanol (30), ethanol (30), water (39), conc. ammonia (1)	0.45	0.36	0.30	0.74	0.61	0

TABLE 170. R_F values on HCl-washed Whatman No. 1 (Bonnin and Sue, *311*)

Quantity	Orthophosphate		Phosphite		Hypophosphite	
	front	rear	front	rear	front	rear
3.5γ	0.15	0.13	0.24	0.20	0.52–0.62	0.47–0.60
300γ	0.15	0.00	0.19	0.04	—	—

The work on phosphates using paper chromatography may be divided into two groups. (i) The elucidation of the structure and analysis of various phosphates which previously had been considered to be pure compounds and the study of reactions of various polyphosphates. The main work in this field was carried out by Ebel *et al.* (*852, 853, 854, 855, 3430, 3431*); see also Viscontini *et al.* (*2747*). (ii) The elaboration of technical analyses of commercial phosphates. Here the work of Westman *et al.* (*664, 665, 3516, 3517*) should be consulted. In one paper (*664*) most of the variables are discussed in detail and procedures suggested for various mixtures involving the analysis of lower or higher polymers as shown in Table 171. For further work see also Ando *et al.* (*62*), Yamaguchi (*3643*), Gauthier (*1092*), Meissner (*2182a*), Karl-Kroupa (*1632a*) and Shinagawa *et al.* (*2967a*).

TABLE 171.   Suggested procedures for various phosphate mixture (Crowther, *664*)

Mixture	Paper	Solvent	Time	Remarks
ortho, pyro, tri and higher	SS 589 orange	*tert.*-butanol 80 water 20 HCOOH 5	30h	very clear bands
ortho, pyro, tri, trimeta, tetra and higher	SS 589 black	*tert.*-butanol 80 water 20 HCOOH 5	30–40h	very clear bands but overloading possible
ortho, pyro, tri, trimeta, tetra, tetra-meta, penta, hexa and higher	SS 589 orange	*iso*propanol 80 water 20 HCOOH 5	48h	bands joined by streaks
pyro, tri, tetra, penta, hexa, septa, octa, nona and higher	SS 589 black	a) *tert.*-butanol 80 water 20 HCOOH 5	16h	
		b) *n*-propanol 60 water 20 NH₄OH 20	48h	bands joined by streaks

TABLE 172. R_F values of phosphor-organic compounds (Weil, *3487*)

Substance	Collidine-water 20°	Butanol-2N NH₄OH $R_{Ph-P(OH)_2}$* at 27°
Phenyl-P(OH)₂	0.62	1.00
Phenyl-PO(OH)₂	0.44	very small
(Phenyl)₂-POOH	0.72	1.60
p-Tolyl-P(OH)₂	0.66	1.26
p-Tolyl-PO(OH)₂	0.49	1.14
(*p*-Tolyl)₂-POOH	0.76	1.90
p-Ethylphenyl-P(OH)₂	0.71	1.52
p-Bromophenyl-P(OH)₂	0.69	1.52
p-Chlorophenyl-PO(OH)₂	0.59	0.17
m-Nitrophenyl-PO(OH)₂	0.48	very small
α-Naphthyl-P(OH)₂	0.66	1.42
β-Naphthyl-P(OH)₂	0.67	1.44
α-Naphthyl-PO(OH)₂ Na salt.	0.33	0.15
β-Naphthyl-PO(OH)₂	0.29	0.13

* $R_{Ph-P(OH)_2}$ is the speed of the spot with reference to the speed of a spot of Phenyl-P(OH)₂.

TABLE 173. R_F values and tests for acids

Anion	Butanol sat. with 1.5N NH₄OH (1916)	isoPropyl alcohol 10% 15N NH₄OH*	Butanol sat. 20% acetic acid*	Butanol (2) Pyridine (1) 1.5N NH₄OH(2) (2542)	Ethanol (80) NH₄OH(4) H₂O(6) (2045)	Tests
Thiocyanate	0.45	0.8	0.44	0.56		Ferric nitrate
Iodide	0.30	0.6	0.32	0.47	0.53	H₂O₂ or AgNO₃
Nitrate	0.24		0.25	0.40	0.48	Diphenylamine in conc. H₂SO₄
Arsenite	0.21		0.41	0.19		H₂S
Nitrite	0.20			0.25	0.47	KI and HCl
Bromide	0.16	0.46	0.16	0.36	0.48	AgNO₃
Bromate	0.13			0.25		KI and HCl
Chloride	0.10	0.37	0.21	0.24	0.43	AgNO₃
Iodate	0.03			0.09		KI and HCl
Fluoride	0.0				0.27	Fe(CNS)₃ or zirconium alizarin reagent
Cobaltinitrite	0.0					
Sulphide	0.0					AgNO₃
Thiosulphate	0.0					Dilute I₂
Periodate	0.0					KI and HCl
Chromate	0.0	0.01	0.24	0.0	0.09	—
Phosphate	0.0			0.04	0.02	Ammoniacal AgNO₃
Arsenate	0.0		0.15	0.05		Ammoniacal AgNO₃
Ferricyanide	0.0	0.01	0.03			Ferrous sulphate
Ferrocyanide	0.0	0.01				Ferric nitrate
Chlorate				0.42		
Carbonate				0.06		
Sulphate				0.07	0.09	
Sulphite					0.04	

* M. Lederer, unpublished experiments

(h) *Phosphonous, phosphinic and phosphinous acids*

Weil (*3487*) separated organic phosphonous, phosphinic and phosphinous acids by paper chromatography as shown in Table 172. Phenol red or silver nitrate and exposure to sunlight were used to reveal the spots.

(i) *Pertechnetate and perrhenate*

These two anions may be separated from other acids with butanol-1.5 N NH₄OH where they travel faster than any other anion on Table 173. Both travel with the acid front in butanol-N HCl and may thus be separated from molybdate (R_F 0.5) and Mn(II) (R_F 0.09) (*1939*). See also Table 144.

(j) *Polythionates*

Preliminary work on the separation of polythionates was published in a review by Pollard (*2536*) indicating that excellent separations can be effected. This was further elaborated by Pollard *et al.* (*2542a*). See also Bighi *et al.* (*245b, 245c, 245d*) and Scoffone and Carini (*2921a*).

(k) *General behaviour of acids*

When a mixture of acids is developed in a solvent containing ammonia, only monovalent acids are displaced. Hanes and Isherwood (*1297*) found that the polyvalent acids are precipitated by the calcium and magnesium ions, which are the usual impurities in filter paper; after washing the paper with hydrochloric acid, phosphate ions were found to travel a considerable distance.

Thus commercial filter paper containing the usual amounts of calcium and magnesium can be used as a means of determining the valency of an unknown acid. Relatively few solvents have been extensively examined and in Table 173 we can only list five solvents for anions. The reagents employed to detect the anion are also included in this table.

See also Nakano for further separations (*2338*). Butanol-N HCl has also been found useful mainly for polyvalent acids. For the separation of phosphate and sulphate see (*1937*), for selenite and tellurite p. 513.

(l) *Separation of cations and anions*

Separations of cations and anions have been successfully performed with ion exchangers (see Chapter 42, section 19); in many cases such separations are equally feasible by paper chromatography as many anions have high R_F values in acid solvents where cations have low ones. For example it is possible to separate tracer amounts of $H_3{}^{32}PO_4$ from Fe(III) or other carriers by simply chromatographing the mixture with butanol-N HCl as shown in Fig. 137. Some complexing appears to occur and retards the phosphate band a little but the separation is complete (*1934*).

10. QUANTITATIVE METHODS

The techniques of spot area, length or intensity measurements can all be applied to inorganic analysis. Erlenmeyer *et al.* (*2935*) applied the spot area technique to the determination of trace metals and Beerstecher (*173*) for potassium in urine; Arden (*77*) applied the visual comparison method to the estimation of uranium.

However, to obtain the usual accuracy of quantitative techniques, subsequent determination by other methods is usually employed.

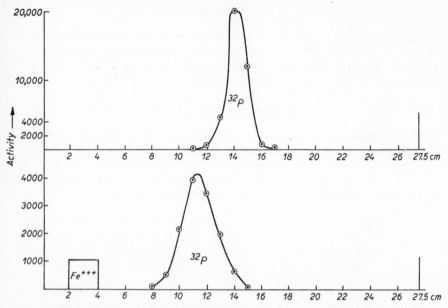

Fig. 137. Displacement of the pure tracer and the mixture $Fe^{+3} - {}^{32}PO_4{}^{-3}$ (Lederer, *1934*)

Paper chromatography with subsequent determination by other methods

(i) *Colorimetry*

Lacourt *et al.* (*1852*) separated 10 γ Al^{+++} from 10 γ of Fe^{+++} and 10 γ of Ti^{++++} by using 6.5 N formic acid as solvent. After drying, the strip is cut up and the Al^{+++} spot is eluted with water at 80°, reacted with Aluminon, the Fe and Ti spot with Tiron (after elution with 1 % HCl).

Co and Ni were also determined colorimetrically by Lacourt *et al.* (*1845*) after separation with acetone containing 1.4% alcohol and 1% dry HCl.

Borates were determined as the carmine complex, see p. 515.

1–10 γ of *gold* were separated from practically all other metals by Kember and Wells (*1680*) using ethyl acetate containing 5 % water and 10 % HNO_3. The gold travels with the liquid front and is allowed to drip off the paper for 80 minutes. The gold in the filtrate is determined colorimetrically with *p*-dimethylbenzylidene-rhodanine after evaporation and treatment with bromine and addition of NaF (to complex traces of Fe).

Further quantitative methods using colorimetry either after elution or on the paper were elaborated for the following elements: chromium (*665*), tungsten (*1848*), cobalt (*1846*), zinc (*1846*), copper (*1846, 2546*), molybdenum (*2546*), vanadium in presence of large amounts of dichromate (*2546a*) and iron (*2546*). For a general review, see Lacourt (*1843*).

(ii) *Polarography*
Preliminary work on Fe, Co and Ni was described in the report of the Chemistry Research Board (*745*). Uranium was determined polarographically by Arden (*77*) in micro-amounts, also by Lacourt *et al.* (*1857*). Lewis and Griffiths (*1996*) use paper chromatographic separation and subsequent polarographic determination in three problems.

a. *The separation of one element* from others. The example chosen was UO_2^{++}, which was separated with tetrahydrosylvan (5 % HNO_3) or tetrahydrofuran (7 % HNO_3) from 52 other metals. Subsequent polarography yielded good results in the range 20 γ–200 γ.

β. *Separation of a group of metals*: Ni, Mn, Co, Cu and Fe. The solvent employed was 50 % acetone, 8 % conc. HCl, 42 % methyl *n*-propyl ketone. The spots are ashed in separate crucibles, dissolved in a fuming agent, evaporated and redissolved in the supporting electrolyte and polarographed; an accuracy of ± 3 % was obtained in all cases. The method was applied to the analysis of alloy steels and iron pyrites.

γ. *Paper strip separation of metals which interfere in polarography*. By the use of butanol – 5 % conc. HCl one may divide metal mixtures into a band of low R_F (V, Cu, UO_2, Pb and Ti) and high R_F (Fe, Mo, Bi, Sb, and Co) which then do not interfere with each other in polarographic determinations.

Estimation of 0.1–1 mg of Cd^{++} in copper was carried out by Davies (*715*) by separating the two elements with methyl ethyl ketone – 30 % conc. HCl, eluting Cd with hot 1 % HCl and determining polarographically Cd in Cu containing 0.2–1.3 % of Cd. An accuracy of ± 1 % was obtained.

Bishop and Liebmann (*257*) elute Zn from a mixture of Sn, Al, Ni, Pb, Mn, Bi, Co, Zn, Cu and Fe on a cellulose column with butanol-conc. HCl (92/8) and estimate micro-quantities polarographically with an accuracy of ± 3 %.

(iii) *Titration methods*
Lacourt *et al.* (*1855*) titrated Mo(VI) with $Pb(NO_3)$ without removing the paper from the solution; 8–12 γ of Mo could be determined in a pyridine solution using diphenyl carbazide as indicator.

See also Lacourt *et al.* (*1854, 1858*) for the determination of Mo and V.

(iv) *Direct optical methods*
Vaeck (*3356, 3357*) uses reflectance measurements directly on the paper for quantitative estimations of inorganic ions on the paper.

11. MACRO METHODS

i. Cellulose columns

(a) *Cobalt, nickel and copper in steel*

Burstall *et al.* (*468*) employ chromatographic columns for separations on a large scale. The cellulose used is wood pulp, cotton linters or filter paper, which is boiled with dilute nitric acid for 12–20 minutes before use. Best form of cellulose are Whatman's ashless tablets, boiled to a pulp with 5 % HNO_3. The acid is washed out with water, alcohol and ether. The glass tube used to hold the column is rinsed out with dichlorodimethylsilane to prevent surface creep of the solvent.

By the use of such a column, steels can be analysed for Co, Ni and Cu. 0.2 g dissolved in 2.5 ml of dilute HCl are placed on the top of the column. Separate bands of Fe and Cu can be obtained if the column is washed with a solvent consisting of 100 parts methyl propyl ketone, 30 parts acetone, 4 parts water and 1 part HCl. Fe is first washed down as a yellow band followed by Cu as an amber band. Cobalt can then be removed from the Ni by washing with methyl propyl ketone containing 2 % HCl and the Ni is finally washed down with 2 N HCl. The combined washings are then evaporated, fumed with HNO_3 and evaporated twice with HCl. Copper is determined colorimetrically as the rubeanate, cobalt as the thiocyanate complex in presence of acetone, nickel with dimethylglyoxime. The results are within 0.1 % of the standard methods.

By essentially the same technique, $HgCl_2$ was separated from Cu, Bi, Pb and Cd, using dry methyl acetate as solvent. After extraction the mercury was precipitated with H_2S and weighed as the sulphide.

(b) *Quantitative estimation of gold*

Kember and Wells (*1680*) worked out an exact procedure for the quanti. tative estimation of gold, using a column of cellulose and ethyl acetate containing 2 % HNO_3 and 3 % water. Only Fe^{+++}, UO_2^{++} and Zn^{++} interfere. Details for the electrolytic determination of 0.0005–0.2 g of gold are given. Maximum error was 2 % for 5 mg Au and less, and 0.2 % for larger quantities.

(c) *Uranyl ion*

Burstall and Wells (*470*) discuss the macro-separation of UO_2^{++} with ether containing 3 % HNO_3. Th^{++++} is partially extracted with ether containing 5 % HNO_3, but not with ether containing 3 % HNO_3; Cl^-, Sn, V, PO_4 and Mo must be removed, also Pt, Pd and Ru.

A method is suggested for the estimation of U in monazite sands and other refractory ores; after elution, the U is determined with the usual methods with very good results. Szonntagh *et al.* (*3189b*) separate U from coal ash.

(d) *Extraction of thorium*

A method for the extraction of Th^{++++} with ether containing 12.5 % HNO_3 was described by Kember (*1678*). UO_2^{++} is first extracted with ether – 3 % HNO_3, which leaves the Th^{++++} on the top of the column. Yttrium will pass through ordinary paper pulp but not through activated pulp. Sc and Zr can be held on the column by addition of tartaric acid.

Further methods worked out in the laboratory of the DSIR (*2766, 3573*) use a compound column of alumina and cellulose for the determination of U and Th in minerals containing arsenic and molybdenum. A general method for uranium and thorium proposed in another paper is given below. For micro amounts of Th see (*3573*).

Detailed method for the estimation of U and Th in minerals

Reagents

All reagents should conform to recognized analytical standards:
Potassium hydroxide – Pellets.
Ammonium hydroxide – Sp. gr. 0.880.
Nitric acid, concentrated – Sp. gr. 1.42.
Nitric acid, diluted (1 + 3).
Hydrochloric acid, concentrated – Sp. gr. 1.18.
Hydrofluoric acid, diluted – Add 1 ml of 40% w/v hydrofluoric acid to 100 ml of water.
Oxalic acid – Crystals.
Ferric nitrate – Solid $Fe(NO_3)_3 \cdot 9H_2O$.
Di-sodium hydrogen phosphate – Solid Na_2HPO_4.
Hydrogen peroxyde – A 20-volume solution.
Ether – Redistilled; the water content must be less than 0.1 %.

Solvents

Ether containing 1% v/v of nitric acid – Add 1 ml of concentrated nitric acid, sp. gr. 1.42, to each 100 ml of ether.
Ether containing 12.5% v/v of nitric acid – Add 12.5 ml of concentrated nitric acid, sp. gr. 1.421, to each 87.5 ml of ether.

Adsorbents

Activated alumina – Type H, 100 to 200 mesh, as supplied by Peter Spence and Co., Ltd.
Activated cellulose – Whatman cellulose powder for chromatography.

Purification of ferric nitrate for micro-determination

Dissolve 250 g of ferric nitrate $Fe(NO_3)_3 \cdot 9H_2O$, in a final volume of 250 ml containing 25 ml of concentrated nitric acid. Transfer this solution to a 500 ml separating funnel and extract three times with 100 ml portions of ether. Warm the solution to remove ether, add 180 ml of concentrated nitric acid and dilute to 750 ml. Use 20 ml of this solution which contains approximately 25% of nitric acid and about 7 g of ferric nitrate, at the appropriate stage of the chromatographic extractions when very small amounts of uranium are to be determined.

Purification of alumina for micro-determination

Gradually pack about 1 kg of alumina into a glass column, 4.5 cm in diameter and 50 cm long, by pouring small quantities into the column together with ether containing 12.5% v/v of nitric acid, until a homogeneous column of the alumina is formed. Wash the column with a litre of ether containing 12.5% of nitric acid and then with 1 litre of ether alone. Transfer the alumina to a large dish, dry it under infra-red lamps and store it in a dry bottle.

Procedure for preparing the sample for chromatography

Weigh 10 g of potassium hydroxide pellets into a nickel crucible, 5 cm in diameter and 5 cm deep, and heat gently for about 10 minutes to remove water. Cool the melt and gently brush the accurately weighed sample, which should be about 1 g (or less if the thorium concentration exceeds 25 % of ThO_2), on to the surface of the cold melt. Cover the crucible and slowly heat the mixture to redness; continue heating for about 1 hour (monazites only need about 20 minutes, but more refractory ores such as pyrochlore or euxenite need longer). Allow the crucible to cool and leach the melt by immersing the crucible and lid in about 150 ml of dilute nitric acid in a 400 ml beaker covered with a clock glass. Wash the crucible with distilled water and then boil the contents of the beaker for 5 to 10 minutes. At this stage, the addition of 1 drop of dilute hydrofluoric acid sometimes helps to clear the solution, but some ores, such as those rich in zirconia, do not respond to this treatment. Allow the contents of the beaker to cool slightly and cautiously add ammonium hydroxide until the solution smells distinctly ammoniacal. A heavy hydroxide precipitate is usually formed and this contains uranium and thorium together with some phosphate ions. Filter the precipitate on Whatman No. 541 filter-paper and wash it twice with hot water containing a few drops of ammonium hydroxide. Only slight washing is required since most of the neutral salts will be in the original filtrate. Open out the paper over the original beaker and wash the precipitate into the beaker with a stream of 50 to 100 ml of hot water. Add about 40 ml of concentrated nitric acid to the beaker and then evaporate the solution until the residue is just moist. Add 20 ml of dilute nitric acid to the beaker, cover it with a clock glass, and heat the beaker on the edge of a hot-plate or below an infra-red lamp for 5 minutes. Then add 2 ml of hydrogen peroxide, replace the cover and heat the beaker for 10 minutes to reduce cerium to cerous state. Add 8 g of ferric nitrate and heat the beaker again for 15 minutes. Finally add 1.4 g of di-sodium hydrogen phosphate and heat for a further 15 minutes. The ferric nitrate must be added before the sodium phosphate, as otherwise the results will be low. With materials such as monazites, the addition of phosphate can be omitted or reduced to about 0.3 g, as such samples are usually low in zircon content; the quantity of ferric nitrate to be added is accordingly reduced to about 4 g. After thorough cooling the sample is ready for chromatography.

Procedure for the chromatographic extraction of uranium and thorium together and determination of thorium

Preparation of the column. The column consists of a tube of diameter 2.7 cm and about 30 cm long, with a funnel at its top. The bottom of the tube terminates in a short length of narrow-bore tubing to facilitate collection of the eluate. It is closed by a short length of polyvinyl chloride tubing with a screw clip. A number of indentations should be made near the base of the tube to support the packing. The whole column should preferably be water-jacketed in a manner similar to a condenser.

Make the column water repellent by treating with siliconing fluid. Place a small piece of Whatman's ashless block at the bottom of the column, add activated cellulose and then ether containing 12.5 % v/v of concentrated nitric acid. Add more pulp, if necessary, so that the cellulose reaches to a height of about 5 cm after "beating" with a glass plunger. Next pour activated alumina into the column so that after "beating" and allowing to settle, the alumina reaches to a height of about 6 cm. Cover the completed column with solvent and it is then ready for use.

Transfer of sample to column and extraction of uranium and thorium

Add approximately 50 g of alumina to the sample solution and thoroughly stir to give a homogeneous powder. Transfer the mixture to the prepared column with the aid of a stirring rod, when transference is complete, immerse the alumina mixture in ether 12.5 % nitric acid solvent, the final level of which should be about 1 cm above the top of the sample wad. Beat this wad gently with a glass plunger to ensure that the column is homogeneous; during this operation care should be taken not to penetrate far into the main column. Rinse the beaker with successive small amounts of ether solvent and pour the rinsings on the column. Continue to rinse until 500 ml of eluate have collected in a 1-litre conical flask; the level of the solvent in the column must not be allowed to fall below the top of the wad. Remove the conical flask and add 20 ml of water followed by 50 ml of ammonium hydroxide to neutralise most of the nitric acid; cautiously swirl the flask during the additions. Add two glass beads and remove the ether by heating the flask on a steam-bath. Transfer the aqueous residue to a 250 ml beaker, cover it and boil for 5 minutes.

Determination of thorium

Precipitate the hydroxides of thorium and uranium with ammonium hydroxide, filter, dissolve the precipitate in hot diluted nitric acid (1 + 3) containing 35 to 40 ml of concentrated nitric acid and collect the solution in the original beaker. Dilute the contents of the beaker to 250 ml, add ammonium hydroxide until the solution is just alkaline and then acidify with 10 ml of concentrated hydrochloric acid. Heat the solution to boiling, add 10 g of oxalic acid and continue boiling for 2 to 3 minutes. Set the beaker aside for at least 4 hours and then filter the precipitated oxalates, wash and ignite them to ThO_2 in the usual way and weigh the oxide.

For amounts of thorium giving a final precipitate of less than about 0.07 g of thoria it is advisable to precipitate the oxalate in a smaller volume, in which event the quantities of all reagents should be proportionally decreased. For less than 0.01 g of thoria the precipitation should be made in a volume of about 12.5 ml instead of 250 ml.

Procedure for chromatographic extraction of uranium and thorium separately and their determination in certain samples

Prepare the chromatographic column as described above, except that ether containing 1% of concentrated nitric acid is used as solvent.

Transfer the sample to the column as described above and elute the uranium and thorium in two extractions, first with ether containing 1% v/v of nitric acid, 400 ml of eluate being collected, to extract uranium, then with ether containing 12.5% v/v of nitric acid (the wad should be beaten up once on changing the solvent), the 400 ml of eluate collected containing the thorium.

Determine the uranium by a standard procedure, either colorimetrically or volumetrically, depending on the amount of uranium present.

Determine the thorium as described above.

(e) *Separation of tantalum and niobium*

Separation of tantalum and niobium was carried out by Burstall *et al.* (*469*). Ta is first eluted with methyl ethyl ketone saturated with water and the Nb with methyl ethyl ketone containing 7.5% hydrofluoric acid (40% aqueous solution). Good separations were achieved when the amount of Ta did not exceed that of Nb.

Sn, Zr, Ti, W when present contaminate the Nb fraction.

In a later paper (*3574*) the following method was suggested to eliminate Ti, Sn and Zr: A solution of the sample in hydrofluoric acid containing ammonium fluoride being used, the tantalum is extracted first in methyl ethyl ketone saturated with water. The column is then washed with methyl ethyl ketone containing 1% of hydrofluoric acid (40% solution) which arrests the movement of Ti, Sn and Zr. Niobium is then extracted with methyl ethyl ketone containing 12.5% of hydrofluoric acid. After removal of solvent the tantalum and niobium are determined in the appropriate fractions by ignition to the pentoxides. Chemical pretreatment (*2191*) is however necessary for low grade phosphatic and siliceous ores since the phosphate is partly extracted with the niobium. Separation of the earth acids from phosphate ions was achieved by prior precipitation with ammonia solution. In a later paper a shortened method (*2190*) is proposed to separate Nb together with Ta from all metals (except tungsten) by eluting from a short column with 400 ml of methyl ethyl ketone containing 15 % of hydrofluoric acid (40% aqueous solution). The separation Nb–Ta may then be carried out if desired. A rapid chromatographic method for the determination of Nb in low-grade samples has been worked out by ascending development on paper sheets with visual comparison of the spots revealed with tannic acid (*1487*). See also Burstall and Williams (*471*) and Bruninx *et al.* (*429*).

(f) *Separation of scandium from the rare earths*

Scandium nitrate can be separated from the rare earths with ether containing 12.5 % nitric acid as solvent. In presence of zirconium, tartaric acid is added to the solution to prevent Zr from contaminating the scandia (*466*).

(g) *Aluminium*

Bishop (*256a*) employs a cellulose column in the determination of Al in iron and steels.

ii. Strips of thick filter paper

(a) *Separation of thallium from Co, Ni, Cu and Fe*

Instead of using columns, Anderson and Lederer(*56*) employed strips of filter paper pulp, about 4–5 mm thick, which are developed by descending development; the solvent running over the strip is collected in a 50 ml beaker. 1 ml of the solution to be analysed is absorbed by the pulp as a small spot, and by using butanol shaken with 1 N HCl as solvent, 32.4 mg of trivalent thallium were quantitatively separated from Co, Ni, Cu and Fe. The thallium was reduced with SO_2 and precipitated as the chromate giving results within ± 1 %.

(b) *Separation of gold from platinum and palladium*

Gold was separated from platinum and palladium by the same method as for Tl, using ethyl ether containing HCl as the solvent and moistening the strip of paper pulp with aqueous hydrochloric acid (*57*).

(c) *Separation of uranium*

Another separation on thick filter paper strips was described by Frierson *et al.* (*1049*), who elute uranium from many other metals followed by polarographic determination. Recoveries of U appear satisfactory. By the use of radioisotopes the contamination of the eluate by other ions was investigated showing that as much as 0.1 % of Ru, Zr, Nb and the rare earth were eluted into the U.

12. COMPLEX IONS

As pointed out in the introductory section inorganic paper chromatography is mainly based on separations effected by the differential complexation of ions. In this section we shall review the work which concerns rather the study of complexes than their separation.

The phenomenon of *multiple spotting* was noted by Hahn *et al.* with Co(II) in a solvent containing acetoacetic ester, by Burstall *(465)* with vanadium and by Carvalho and Lederer *(720)* with Rh, Fe and V in several butanol-HCl mixtures.

This phenomenon was studied in detail by Erdem and Erlenmeyer *(910)* for the cation Cd in propanol-NH$_3$ as solvent where Cd yields three well separated spots of different ammines. Similar work on Cu complexes with amino acids was recorded by Erdem *et al.* *(911)*. For further work see Pollard *et al.* *(2542c)* and Erdem *(909a)*. See also Tanaka and Matsuo *(3200)* on the constancy of R_F values when different salts are placed on the paper.

Cobaltammines were studied by Yamamoto *et al.* *(3645)* for their possible separation with methanol-acetone-NH$_4$OH (7:2.5:1) as solvent. In the nitroammino series the R_F values were found to increase with the number of nitro groups in the complex. The nitro-ammino cobaltic series was also studied by Lederer *(1938)* who considered primarily the differences in R_F between this series and Co(II) in the conventional solvents such as acetone-HCl. *Cis-trans* isomers were separated by Stefanovic and Janjic *(3088)*.

Studies with 2-thenoyltrifluoroacetone complexes as well as with acetylacetonates were carried out by Berg and McIntyre *(194)* and Berg and Strassner *(196)*. The separations of the metals studied (Fe, Co, Ni, Mn and Cu) are no improvement on those previously known but these authors obtain some correlation between stability of the complexes and R_F values. This work was further described by Strassner *(3147a)*.

See also Venturello and Ghe *(3395, 3396, 3397a)* for separations of dithizone complexes and Fernando and Phillips *(944)* for oxine complexes.

13. PAPER CHROMATOGRAPHY OF RADIOELEMENTS

(a) *General*

We have already mentioned in the section on solvents that tracers behave like macro quantities in solvents containing strong acids. This was observed with RaE by Bouissières and Lederer (*331*), with numerous artificial tracers by Lederer(*1928*) and for RaD and RaE by Lederer (*1932*) and Lima (*2011*).

The divergence between the behaviour of tracers and macro quantities in solvents containing weak acids (acetic acid) and complexing agents was observed by Frierson *et al.* (*1047*) with ^{65}Zn and by Lederer (*1932*) with rare earths. Frierson and Jones (*1048*) also noted a difference between macro and tracer R_F values for Pb and Bi in the mixture butanol-HCl-H_2SO_4 (60:12:1).

(b) *Natural radioelements*

RaD–RaE–RaF(Po) were separated by Frierson and Jones (*1048*), Lederer (*1932*) and Dickey (*771*) with butanol-1 N HCl and by Lima (*2011*) with butanol - 3 N HCl. Fig. 138 shows the activity on a paper strip developed with butanol - 1 N HCl.

Francium and other radioelements of the actinium series were separated by Perey and Adloff (*2492*) using aqueous ammonium carbonate as solvents.

Ra–Ba were first separated by Bouissières and Lederer (*331*) with ethanol-20% water. This was successful with a mixture of AcX-Ba(Cl_2) precipitated with conc. HCl from an HCl solution of the actinium elements. In artificial mixtures trailing of the Ba was observed which prevented a complete

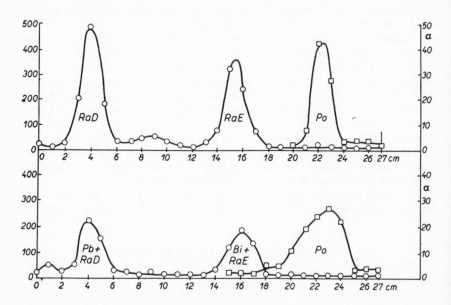

Fig. 138. Top: Separation RaD – RaE – Po. Bottom: Separation Pb – Bi – Po (Pb and Bi in macro quantities) (Lederer, *1932*). Solvent: butanol saturated with 1 N HCl

separation. Lederer (*1932*) proposed later ethanol-water-NH₄CNS-HCl (90:10:2:1) as solvent using descending development and allowing the solvent to run off the paper. Figure 139 shows a separation of Ba and AcX so obtained.

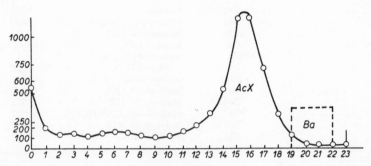

Fig. 139. Separation AcX – Ba (Lederer, *1932*). Descending development, 16 h. Solvent: C₂H₅OH 90 ml, H₂O 10 ml, NH₄CNS 2 g, conc. HCl 1 ml. – – – – Ba detected with rhodizonic acid

UX–UZ (*Th–Pa*) were separated with a mixture of acetone–HF–water the only solvent which ran fast enough to provide good yields of UZ (*1932*). *Pa* and *Nb* were separated by Fudge and Woodhead (*1056a*).

The purification of uranium was described by Arden using a column of paper pulp and ether containing 3 % HNO₃ as solvent. It was possible to prepare spectrographically pure uranium. For details see (*75*).

(c) *Artificial radioelements*

Identification. Paper chromatography was used to identify several contaminants in pile produced isotopes by Frierson and Jones (*1048*), for the identification of ⁶⁰Co in ⁵⁵Fe by Michalowics and Lederer (*2202*), for the study of fission products which do not precipitate with H₂S by Götte and Pätze (*1181*) and for the study of the spallation products of Cu by Carleson (*507*). Fission products were also separated by Matsuura (*2172a*).

Preparation. Since tracers have the same R_F values as macro amounts in numerous solvents containing strong acids, the preparation of carrier-free isotopes is relatively simple by paper chromatography provided the bulk of the target material is removed by suitable chemical reactions before placing the tracer on the paper. Table 174 lists the tracers prepared, the target materials and the solvents used. Usually the zone of activity is measured, cut out, eluted or ashed and contains then only the mineral constituents of the paper as impurities.

TABLE 174.   Tracers prepared by paper chromatography

Tracer prepared	Target material or carrier	Solvent used	Reference
^{102}Rh	Fe(OH)$_3$	butanol-HCl	1928
^{22}Na	Mg	ethanol-20% water	1928
^{103}Pd	Rh	butanol-1N HCl	1928
^{65}Zn	Cu	butanol-HCl	1928, 3315
^{233}U	^{233}Pa	butanol-HNO$_3$ and ether-HNO$_3$	1928
^{55}Fe	Mn	acetone-HCl	1932
^{74}As	Ge	butanol-1N HCl	1933
^{131}Cs	BaCO$_3$	phenol-2N HCl	1935
^{90}Y	^{90}Sr	ethanol-NH$_4$CNS-HCl	1932
^{113}Sn	In	butanol-1N HCl	1932
^{58}Co	Mn	acetone-HCl	1932

14. MISCELLANEOUS SEPARATIONS

(a) Hydrogen peroxide

Bruschweiler *et al.* (*433*) used paper chromatography (solvents: ether or isopropanol) to separate hydrogen peroxide from organic peroxides.

(b) Hydroxylamine and hydrazine

Bremner (*375*) examined numerous solvents (see Table 175) and obtained good separations of hydroxylamine and hydrazine. Picryl chloride and several other reagents are employed for the detection of micro amounts. For hydrazine derivatives see Hinman (*1394a*).

TABLE 175. R_F values of hydroxylamine and hydrazine on Whatman No. 4 paper (Bremner, *375*)

Solvent	Hydroxylamine	Hydrazine
Methanol (70), 6N hydrochloric acid (30)	0.53	0.34
Methanol (80), 90% formic acid (15), water (5) . . .	0.67	streak
Ethanol (70), 6N hydrochloric acid (30)	0.41	0.22
Ethanol (70), 90% formic acid (20), water (10). . . .	0.54	streak
n-Propanol (70), 6N hydrochloric acid (30).	0.36	0.21
n-Butanol	0.12	0.00
n-Butanol (40), acetic acid (50), water (10).	0.23	0.07
n-Butanol (50) 2N hydrochloric acid (50)	0.18	0.10
n-Butanol (50) 6N hydrochloric acid (50)	0.51	0.42
n-Butanol (70) 6N hydrochloric acid (30)	0.31	0.19
tert.-Butanol (70), 6N hydrochloric acid (30)	0.40	0.20
Diethyl ether (50), 90% formic acid (50)	0.36	streak

CHAPTER 44

Chromatographic separation of isotopes

1. ISOTOPE ENRICHMENT

The early work on chromatographic isotope separation by Taylor and Urey (*3212, 3213*) gave evidence only of isotope enrichment. These workers employed, what was probably the longest chromatographic columns ever used, 35 m long, and packed with artificial zeolites. The elements Li, K and N all showed a change in the isotope ratio after development, but no pure isotopes could be obtained. Still less successful were the attempts of Groth and Harteck (*1233*) who tried the separation of isotopes of Xe by fractional desorption from activated carbon and silica gel. No separation could be detected in these experiments.

2. SEPARATION OF LITHIUM ISOTOPES

The first successful separations were reported by Glueckauf *et al.* (*1148*).

Using a Zeo-Karb H.I. column containing 45 g of the exchanger and running N lithium acetate through it, milligram quantities of Li containing only as little as 5% of the natural ^{6}Li content were obtained. The outflowing solution was analysed in 3 drop portions for its isotope ratio, using the reaction of ^{6}Li with slow neutrons

$$\text{}^6_3\text{Li} + {}^1_0\text{n} \rightarrow {}^7_3\text{Li}^* \rightarrow {}^4_2\text{He} + {}^3_1\text{H} + 4.6 \text{ MeV}$$

and subsequent observation in an ionisation chamber as the technique for this determination. However, the isotope enrichment obtained is only a fraction of that expected from theory.

The relatively poor enrichment in Taylor and Urey's experiments could be due to large resin particles and fast flow-rates. No adequate explanation could, however, be given for the low yields obtained in the work of Glueckauf *et al.* (*1148*).

3. SEPARATION OF NEON ISOTOPES

This was carried out in a tube filled with charcoal, into one end of which was fed neon at atmospheric pressure. The other end led to a sampling device. The charcoal column was filled with N_2 and left in communication with a container of atmospheric neon. A Dewar vessel filled with liquid N_2 was then slowly raised, thus progressively cooling the column from the rear end. This caused a constant flow of neon into the charcoal up to the end of the low temperature region and the formation of an adsorption boundary which can be compared with a chromatographic boundary. Again, milligram amounts of almost pure isotopes were obtained both by the above technique and by the displacement of helium by neon in a charcoal column immersed from the start in liquid N_2 (*1148*).

This method is capable of producing larger quantities of pure isotopes; however, the separation factor must be sufficiently large and sharp boundaries must be maintained.

Glueckauf *et al.* (*1148*) suggest that the separation of ^3He is a case where these conditions might apply.

4. SEPARATION OF NITROGEN ISOTOPES

Spedding *et al.* (*3066b*) describe an ion exchange method for a complete separation of the natural N isotopes. They employ the NH_3-NH_4^+ equilibrium as well as the ion exchange equilibrium of NH_4^+ by displacing a band of ammonium ions with NaOH on a column. For the efficiency at various column lengths and theoretical considerations see the original paper.

5. MISCELLANEOUS

Very small separation effects were noted by Betts *et al.* (*235a*) in chromatographing a mixture of ^{22}Na and ^{24}Na on a column of Dowex-50 with 0.7 M HCl as eluant.

For the separation of Cl isotopes, see Langrad (*1880*); for theoretical considerations of isotope separation, see Mathur (*2171a*).

REFERENCES

References

1 M. Abdel Akher and F. Smith, *J. Am. Chem. Soc.*, 1951, *73*, 5859.
2 E. P. Abraham and E. Chain, *Nature*, 1942, *149*, 328.
3 E. Abrahamczik, *Mikrochim. Acta*, in press [presented to the Microchemical Congress (1955)].
4 A. M. Abu-Nasr and R. T. Holman, *J. Am. Oil Chemists' Soc.*, 1954, *31*, 41: *C.A.* 1954, *48*, 4229–i.
5 R. Acher, J. Chauvet, C. Crocker, U.-R. Laurila, J. Thaureaux and C. Fromageot, *Bull. soc. chim. biol.*, 1954, *36*, 167.
6 R. Acher and C. Crocker, *Biochim. Biophys. Acta*, 1952, *9*, 704.
7 R. Acher, C. Fromageot and M. Jutisz, *Biochim. Biophys. Acta*, 1950, *5*, 81.
8 R. Acher, M. Jutisz and C. Fromageot, *Biochim. Biophys. Acta*, 1952, *8*, 442.
9 R. Acher, M. Jutisz, C Fromageot and J. Chauvet, *Biochim. Biophys. Acta*, 1952, *8*, 442.
10 R. Acher, U.-R. Laurila, J. Thaureaux and C. Fromageot, *Biochim. Biophys. Acta*, 1954, *14*, 151.
11 R. Acher, J. Thaureaux, C. Crocker, M. Jutisz and C. Fromageot, *Biochim. Biophys. Acta*, 1952, *9*, 339.
12 B.-J. Ackerman and H. G. Cassidy, *Anal. Chem.*, 1954, *26*, 1874.
13 W. W. Ackermann and H. Kirby, *J. Biol. Chem.*, 1948, *175*, 483.
14 B. A. Adams and E. L. Holmes, *J. Soc. Chem. Ind.*, 1935, *54*, 1.
15 H. W. Adams and H. G. Stuart, *Analyst*, 1951, *76*, 553.
16 R. Adams and T. R. Govindachari, *J. Am. Chem. Soc.*, 1949, *71*, 1956.
17 R. S. Adams, *J. Am. Leather Chemists' Assoc.*, 1946, *41*, 552.
18 H. Adkins and G. Krsek, *J. Am. Chem. Soc.*, 1949, *71*, 3051.
19 A. Aebi and T. Reichstein, *Helv. Chim. Acta*, 1950, *33*, 1013.
20 O. T. Aepli, P. A. Munter and J. F. Gall, *Anal. Chem.*, 1948, *20*, 610.
21 G. Ågren and C. H. De Verdier, *Acta Chem. Scand.*, 1950, *4*, 1498: *C.A.*, 1951, *45*, 4297-f.
22 G. Ågren, C. H. De Verdier and J. Glomset, *Acta Chem. Scand.*, 1951, *5*, 324.
23 G. Ågren and J. Glomset, *Acta Chem. Scand.*, 1953, *7*, 1071.
24 J. W. Airan, *Sciecne and Culture*, 1952, *18*, 89: *C.A.*, 1953, *47*, 2625-h.
25 J. W. Airan and J. Barnabas, *Naturwissenschaften*, 1953, *40*, 510.
26 J. W. Airan and J. Barnabas, *Science and Culture*, 1953, *18*, 438: *C.A.*, 1953, *47*, 8580-b.
27 J. W. Airan, G. V. Joshi, J. Barnabas and R. W. P. Master, *Anal. Chem.*, 1953, *25*, 659.
28 B. V. Aivazov and D. A. Vyakhirev, *Zhur. Priklad. Khim.*, 1953, *26*, 505: *C.A.*, 1954, *48*, 6371-h.
29 S. Åkerfeldt, *Acta Chem. Scand.*, 1954, *8*, 521.

30 J. W. Albans and P. B. Baker, *Analyst*, 1950, *75*, 657.
31 A. Albert and D. J. Brown, *J. Chem. Soc.*, *1954*, 2060.
32 A. Albert, D. J. Brown and G. Cheeseman, *J. Chem. Soc.*, *1951*, 474.
32a P. A. Albertsson, *Nature*, 1956, *177*, 771
33 E. C. Albright, F. C. Larson and W. P. Deiss, *Proc. Soc. Exptl. Biol. Med.*, 1953, *84*, 240: *C.A.*, 1954, *48*, 2158-e.
34 M. Alcock and J. S. Cannell, *Analyst*, 1954, *79*, 389; *Nature*, 1956, *177*, 327.
35 A. Alessandro and C. A. Caldarera, *Boll. chim. farm.*, 1954, *93*, 404: *C.A.*, 1955, *49*, 4235-d.
36 G. B. Alexander, *J. Am. Chem. Soc.*, 1949, *71*, 3043.
37 M. Alfthan and A. I. Virtanen, *Acta Chem. Scand.*, 1955, *9*, 186.
37a I. P. Alimarin and A. M. Medvedeva, *Trudy Komissii Anal. Khim.*, *Akad. Nauk S.S.S.R.*, *Inst. Geokhim. i Anal. Khim.*, 1955, *6*, 351: *C.A.*, 1956, *50*, 12741.
38 L.-G. Allgén, *Acta Chem. Scand.*, 1954, *8*, 1101.
39 R. Allouf and M. Macheboeuf, *Bull. soc. chim. biol.*, 1952, *34*, 215.
40 R. Allouf and R. Munier, *Bull. soc. chim. biol.*, 1952, *34*, 196.
41 R. S. Alm, *Acta Chem. Scand.*, 1952, *6*, 1186.
42 R. S. Alm, R. J. P. Williams and A. Tiselius, *Acta Chem. Scand.*, 1952, *6*, 826.
43 A. K. Al-Mahdi and C. L. Wilson, *Mikrochem. ver. Mikrochim. Acta*, 1951, *36/37*, 218.
44 A. K. Al-Mahdi and C. L. Wilson, *Mikrochem. ver. Mikrochim. Acta*, 1952/53, *40*, 138.
44a G. Almassy and I. Dezsö, *Magyar Kém. Folyóirat*, 1955, *61*, 158.
44b G. Almassy and Z. Nagy, *Acta Chim. Acad. Sci. Hung.*, 1955, *7*, 325.
44c G. Almassy and J. Straub, *Acta Chim. Acad. Sci. Hung.*, 1955, *7*, 253.
44d G. Almassy and M. Vigvari, *Magyar Kém. Folyóirat*, 1955, *61*, 109.
45 A. M. Altschul, A. E. Sidwell and T. R. Hogness, *J. Biol. Chem.*, 1939, *127*, 123.
46 K. S. Ambe, L. Kulkarni and K. Sohonie, *J. Sci. Ind. Research*, 1954, *13B*, 380: *C.A.*, 1954, *48*, 13519-g.
46a D. Ambrose and R. R. Collerson, *Nature*, 1956, *177*, 84.
47 D. Amelung and P. Böhm, *Z. physiol. Chem.*, 1954, *298*, 199.
48 S. R. Ames and H. A. Risley, *Proc. Soc. Exptl. Biol. Med.*, 1948, *69*, 267.
49 D. Aminoff and W. T. J. Morgan, *Nature*, 1948, *162*, 579.
50 D. Aminoff, W. T. J. Morgan and W. M. Watkins, *Biochem. J.*, 1950, *46*, 426.
51 W. Andersen, *Acta Chem. Scand.*, 1954, *8*, 359 and 1723.
52 W. Andersen, *C. R. Trav. Lab. Carlsberg, Sér. Chim.*, 1953/55, *29*, 49: *C.A.*, 1955, *49*, 4068-g.
53 W. Andersen, C. A. Dekker and A. R. Todd, *J. Chem. Soc.*, *1952*, 2721.
54 A. S. Anderson, G. R. Barker, J. M. Gulland and M. V. Lock, *J. Chem. Soc.*, *1952*, 369.
55 G. Anderson, *Nature*, 1955, *175*, 863.
56 J. R. A. Anderson and M. Lederer, *Anal. Chim. Acta*, 1950, *4*, 513.
57 J. R. A. Anderson and M. Lederer, *Anal. Chim. Acta*, 1951, *5*, 321.
58 J. R. A. Anderson and M. Lederer, *Anal. Chim. Acta*, 1951, *5*, 396.
59 J. R. A. Anderson and E. C. Martin, *Anal. Chim. Acta*, 1953, *8*, 530.
60 J. R. A. Anderson and A. Whitley, *Anal. Chim. Acta*, 1952, *6*, 517.
61 T. Ando and S. Ishii, *Bull. Chem. Soc. Japan*, 1952, *25*, 106: *C.A.* 1953, *47*, 1538-c.
62 T. Ando, J. Ito, S. Ishii and T. Soda, *Bull. Chem. Soc. Japan*, 1952, *25*, 78: *C.A.*, 1953, *47*, 9860-g.

63 T. Ando and T. Yonemoto, *Rept. Radiation Chem. Research Inst. Tokyo Univ.*, 1948, *3*, 15: *C.A.*, 1950, *44*, 5414-h.
64 H. C. Andrews and F. Hagemann, *Report ANL 4176 July 15* (1948).
65 P. Andrews and J. K. N. Jones, *J. Chem. Soc.*, *1954*, 1724.
66 E. F. L. J. Anet, B. Lythgoe, M. H. Silk and S. Trippett, *J. Chem. Soc.*, *1953*, 309.
67 C. B. Anfinsen, R. R. Redfield, W. L. Choate, J. Page and W. R. Carroll, *J. Biol. Chem.*, 1954, *207*, 201.
68 E. F. Annison, *Biochem. J.*, 1954, *58*, 670.
69 E. F. Annison, A. T. James and W. T. J. Morgan, *Biochem. J.*, 1951, *48*, 477.
70 G. B. Ansell and R. M. C. Dawson, *Biochem. J.*, 1951, *50*, 241.
70a F. Aoki, *Bull. Chem. Soc. Japan*, 1953, *26*, 480.
71 N. Applezweig, *Ind. Eng. Chem. Anal. Ed.*, 1946, *18*, 82; *Ann. N.Y. Acad. Sci.*, 1948, *49*, 295.
72 B. A. Arbuzov and Z. G. Isaeva, *Izvest. Akad. Nauk S.S.S.R. Otdel. Khim. Nauk*, *1953*, 843.
73 R. M. Archibald, *J. Biol. Chem.*, 1944, *156*, 121.
74 R. M. Archibald, *J. Biol. Chem.*, 1945, *159*, 693. •
75 T. V. Arden, *J. Appl. Chem.*, 1954, *4*, 539.
76 T. V. Arden, F. H. Burstall, G. R. Davies, J. A. Lewis and R. P. Linstead, *Nature*, 1948, *162*, 691.
77 T. V. Arden, F. H. Burstall and R. P. Linstead, *J. Chem. Soc.*, *1949*, S311.
78 T. Arima, *J. Biochem. Japan*, 1955, *42*, 169.
79 J. B. Armitage, J. R. Cannon, A. W. Johnson, L. F. J. Parker, E. L. Smith, W. H. Stafford and A. R. Todd, *J. Chem. Soc.*, *1953*, 3849.
80 C. M. Armstrong, *Brit. Pat.* 612,077: *C.A.* 1949, *43*, 3029-i.
81 S. Aronoff, *Science*, 1949, *110*, 590.
82 S. Aronoff, *Arch. Biochem. Biophys.*, 1951, *32*, 237.
83 W. L. Arons and A. K. Solomon, *J. Clin. Invest.*, 1954, *33*, 995: *C.A.*, 1954, *48*, 11533-g.
84 G. Arroyave and L. R. Axelrod, *J. Biol. Chem.*, 1954, *208*, 579.
85 M. Asami, *Botan. Mag. (Tokyo)*, 1952, *65*, 217: *C.A.*, 1954, *48*, 7677-b.
86 T. Asaoka and K. Higashi, *J. Pharm. Soc. Japan*, 1954, *74*, 788: *C. A.*, 1955, *49*, 3472-b.
87 B. D. Ashley and U. Westphal, *Arch. Biochem. Biophys.*, 1955, *56*, 1.
88 B. A. Askonas, P. N. Campbell, C. Godin and T. S. Work, *Biochem. J.*, 1955, *61*, 105.
89 J. Asselineau, *Bull. soc. chim. France*, *1952*, 884.
90 J. Asselineau, N. Choucroun and E. Lederer, *Biochim. Biophys. Acta*, 1950, *5*, 197.
91 J. Asselineau and E. Lederer, *Biochim. Biophys. Acta*, 1951, *7*, 126.
92 J. Asselineau and E. Lederer, *Progr. Chem. Org. Nat. Prod.*, 1953, *10*, 170.
93 T. Astrup, A. Stage and E. Olsen, *Acta Chem. Scand.*, 1951, *5*, 1343.
94 E. B. Astwood, M. S. Raben, R. W. Payne and A. B. Grady, *J. Am. Chem. Soc.*, 1951, *73*, 2969.
95 H. F. Atkinson, *Nature*, 1948, *162*, 858.
96 H. F. Atkinson, *Nature*, 1949, *164*, 541.
97 R. O. Atkinson, R. G. Stuart and R. E. Stuckey, *Analyst*, 1950, *75*, 447.
98 R. W. Atteberry and G. E. Boyd, *J. Am. Chem. Soc.*, 1950, *72*, 4805.
99 R. W. Atteberry, Q. V. Larson and G. E. Boyd, *Abstract Am. Chem. Soc.*, 118th Meeting 8G (Chicago Sept. 1950).

100 J. L. Auclair and J. B. Maltais, *Nature*, 1952, *170*, 1114.
101 J. L. Auclair and R. L. Patton, *Rev. Can. Biol.*, 1950, *9*, 3.
102 I. Augestad and E. Berner, *Acta Chem. Scand.*, 1954, *8*, 251.
103 K.-B. Augustinsson and M. Grahn, *Acta Chem. Scand.*, 1953, *7*, 906.
103a G. V. Austerweil, *L'échange d'ions et les échangeurs*, Gauthier-Villars, Paris, 1955.
104 G. V. Austerweil and R. Pallaud, *J. Appl. Chem.*, 1955, *5*, 213.
105 J. Awapara, A. J. Landua and R. Fuerst, *J. Biol. Chem.*, 1950, *183*, 545.
106 J. Awapara, A. J. Landua, R. Fuerst and B. Seale, *J. Biol. Chem.*, 1950, *187*, 35.
107 L. R. Axelrod, *J. Am. Chem. Soc.*, 1953, *75*, 4074.
108 L. R. Axelrod, *J. Biol. Chem.*, 1953, *201*, 59.
109 L. R. Axelrod, *J. Biol. Chem.*, 1953, *205*, 173.
110 L. R. Axelrod, *Recent Progr. Hormone Research*, 1954, *9*, 69.
111 J. A. Ayres, *J. Am. Chem. Soc.*, 1947, *69*, 2879.
112 S. Baar, *Biochem. J.*, 1954, *58*, 175.
113 S. Baar and J. P. Bull, *Nature*, 1953, *172*, 414.
114 R. O. Bach, *Anales de la Asociación Quím. Argentina*, 1949, *37*, 55.
115 R. O. Bach, *Anales de la Asociación Quím. Argentina*, 1949, *37*, 69.
116 R. O. Bach, *Anales de la Asociación Quím. Argentina*, 1949, *37*, 274.
117 R. O. Bach, *Industria y quimica*, 1950, *12*, 283.
118 J. S. D. Bacon, *Biochem. J.*, 1954, *57*, 320.
119 J. S. D. Bacon and D. J. Bell, *J. Chem. Soc.*, *1953*, 2528.
120 J. Baddiley and E. M. Thain, *J. Chem. Soc.*, *1951*, 2253, 3421.
121 J. Baddiley and E. M. Thain, *J. Chem. Soc.*, *1952*, 3783.
122 L. C. W. Baker, B. Loev and T. P. McCutcheon, *J. Am. Chem. Soc.*, 1950, *72*, 2374.
123 P. B. Baker, F. Dobson and A. J. P. Martin, *Analyst*, 1950, *75*, 651.
124 P. B. Baker, F. Dobson and S. W. Stroud, *Nature*, 1951, *168*, 114.
125 W. Baker and C. B. Collis, *J. Chem. Soc.*, *1949*, S12.
126 W. Baker, J. B. Harborne and W. D. Ollis, *J. Chem. Soc.*, *1952*, 3215.
127 B. R. Baliga, K. Krishnamurthy, R. Rajagopalan and K. V. Giri, *J. Indian Inst. of Science*, 1955, *37*, 18.
128 W. Ballczo and W. Schenk, *Mikrochim. Acta*, *1954*, 163.
129 C. E. Ballou and A. B. Anderson, *J. Am. Chem. Soc.*, 1953, *75*, 648.
130 P. R. O. Bally, O. Schindler and T. Reichstein, *Helv. Chim. Acta*, 1952, *35*, 138.
131 J. N. Balston and B. E. Talbot, *A Guide to Filter Paper and Cellulose Powder Chromatography*, 145 pp., Reeve Angel & Co. Ltd., London, 1952.
132 S. L. Bandemer, *Univ. Microfilms*, Pub. No. 1289: *C.A.*, 1950, *44*, 5840.
133 R. S. Bandurski and B. Axelrod, *J. Biol. Chem.*, 1951, *193*, 405.
134 D. W. Bannister, C. S. G. Philips and R. J. P. Williams, *Anal. Chem.*, 1954, *26*, 1451.
135 M. Barbier and E. Lederer, *Biochim. Biophys. Acta*, 1952, *8*, 590.
136 C. J. Barker and R. H. Perry, *Chem. and Ind.*, *1955*, 588.
137 G. R. Barker and D. C. C. Smith, *Chem. and Ind.*, *1954*, 19.
138 H. Barker and A. H. Ennor, *Biochim. Biophys. Acta*, 1951, *7*, 272.
139 S. A. Barker, E. J. Bourne and D. M. O'Mant, *Chem. and Ind.*, *1955*, 425.
140 S. A. Barker, E. J. Bourne, S. Peat and I. A. Wilkinson, *J. Chem. Soc.*, *1950*, 3022.
141 J. Barnabas, *Naturwissenschaften*, 1954, *41*, 453.
142 J. Barnabas and G. V. Joshi, *Anal. Chem.*, 1955, *27*, 443.

143 T. Barnabas, M. G. Badve and J. Barnabas, *Naturwissenschaften*, 1954, *41*, 478.
144 D. L. Barney and J. W. Gryder, *J. Am. Chem. Soc.*, 1955, *77*, 3195.
145 J. Barrollier, *Naturwissenschaften*, 1955, *42*, 126.
146 J. Barrollier, *Naturwissenschaften*, 1955, *42*, 416.
147 G. T. Barry, J. D. Gregory and L. C. Craig, *J. Biol. Chem.*, 1948, *175*, 485.
148 V. C. Barry and P. W. D. Mitchell, *J. Chem. Soc.*, *1954*, 4020.
149 G. R. Bartlett and A. A. Marlow, *J. Lab. Clin. Med.*, 1953, *42*, 178: *C.A.*, 1953, *47*, 12470-f.
150 D. H. R. Barton and E. Miller, *J. Chem. Soc.*, *1949*, 337.
151 G. M. Barton, R. S. Evans and J. A. F. Gardner, *Nature*, 1952, *170*, 249.
152 E. C. Barton-Wright and G. Harris, *Nature*, 1951, *167*, 560.
153 R. K. Barua and R. A. Morton, *Biochem. J.*, 1949, *45*, 308.
154 R. E. Basford and F. M. Huennekens, *J. Am. Chem. Soc.*, 1955, *77*, 3878.
155 O. Bassir, *Chem. and Ind.*, *1954*, 709.
156 E. C. Bate-Smith, *Nature*, 1948, *161*, 835; *Biochem. J.*, 1948, *43*, lix.
157 E. C. Bate-Smith, *Partition Chromatography*, Biochemical Society Symposia 1950, No. 3, p. 62.
158 E. C. Bate-Smith and R. G. Westall, *Biochim. Biophys. Acta*, 1950, *4*, 427.
159 J. Baudet, *Compt. rend.*, 1952, *234*, 2454.
160 L. Bauer, *Naturwissenschaften*, 1952, *39*, 88.
161 W. S. Bauld, *Biochem. J.*, 1955, *59*, 294.
162 W. C. Bauman, *Ind. Eng. Chem.*, 1946, *38*, 46.
163 W. C. Bauman, R. E. Anderson and R. M. Wheaton, *Ann. Rev. Phys. Chem.*, 1952, *3*, 109.
164 W. C. Bauman and J. Eichhorn, *J. Am. Chem. Soc.*, 1947, *69*, 2830.
164a H. Baumann, *Naturwissenschaften*, 1956, *43*, 300.
165 J. G. Baxter and C. D. Robeson, *Brit. Pat.* 616,939 (1949): *C.A.*, 1949, *43*, 4818-g.
166 R. J. Bayly and E. J. Bourne, *Nature*, 1953, *171*, 385.
167 R. J. Bayly, E. J. Bourne and M. Stacey, *Nature*, 1951, *168*, 510.
168 R. J. Bayly, E. J. Bourne and M. Stacey, *Nature*, 1952, *169*, 876.
169 A. W. Bazemore, J. W. Richter, D. E. Ayer, J. Finnerty, N. G. Brink and K. Folkers, *J. Am. Chem. Soc.*, 1953, *75*, 1949.
170 R. N. Beale, R. J. C. Harris and E. M. F. Roe, *J. Chem. Soc.*, *1950*, 1397.
171 M. T. Beck and P. Ébrey, *Acta Chim. Acad. Sci. Hungar.*, 1954, *4*, 231: *C.A.*, 1955, *49*, 4458-g; *Biochim. Biophys. Acta*, 1956, *20*, 393.
172 H. F. Beckman, *Anal. Chem.*, 1954, *26*, 922.
173 E. Beerstecher Jr., *Anal. Chem.*, 1950, *22*, 1200.
174 P. H. Begemann, J. G. Keppler and H. A. Boekenoogen, *Rec. trav. chim.*, 1950, *69*, 439.
176 O. K. Behrens, J. Corse, J. P. Edwards, L. Garrison, R. G. Jones, Q. F. Soper, F. R. Van Abeele and C. W. Whitehead, *J. Biol. Chem.*, 1948, *175*, 793.
177 D. J. Bell, *J. Chem. Soc.*, *1944*, 473.
178 D. J. Bell, F. A. Isherwood and N. E. Hardwick, *J. Chem. Soc.*, *1954*, 3702.
179 D. J. Bell and A. Palmer, *J. Chem. Soc.*, *1949*, 2522.
180 P. H. Bell, F. J. Bone, J. P. English, C. E. Fellows, K. S. Howard, M. M. Rogers, R. G. Shepard, R. Winterbottom, A. C. Dornbush, S. Kushner and Y. Subbarow, *Ann. N.Y. Acad. Sci.*, 1949, *51*, 897.
181 L. J. Bellamy, J. H. Lawrie and E. W. S. Press, *Trans. Inst. Rubber Ind.*, 1947, *22*, 308; *23*, 15.

182   J. R. Bendall, S. M. Partridge and R. G. Westall, *Nature*, 1947, *160*, 374.
183   A. E. Bender, *Biochem. J.*, 1951, *48*, xv.
184   A. Bendich, J. R. Fresco, H. S. Rosenkranz and S. M. Beiser, *J. Am. Chem. Soc.*, 1955, *77*, 3671.
185   J. T. Benedict, W. C. Schumb and C. D. Coryell, *J. Am. Chem. Soc.*, 1954, *76*, 2036.
186   T. A. Bennet-Clark and N. P. Kefford, *Nature*, 1953, *171*, 645.
187   T. A. Bennet-Clark, M. S. Tambiah and N. P. Kefford, *Nature*, 1952, *169*, 452.
188   A. A. Benson, J. A. Bassham, M. Calvin, T. C. Goodale, V. A. Haas and W. Stepka, *J. Am. Chem. Soc.*, 1950, *72*, 1710.
189   A. A. Benson, J. A. Bassham, M. Calvin, A. G. Hall, H. E. Hirsch, S. Kawaguchi, V. Lynch and N. E. Tolbert, *J. Biol. Chem.*, 1952, *196*, 703.
190   H. R. Bentley and J. K. Whitehead, *Biochem. J.*, 1950, *46*, 341.
191   I. Berenblum, *Nature*, 1945, *156*, 601.
192   G. S. Berenson, S. Roseman and A. Dorfman, *Biochim. Biophys. Acta*, 1955, *17*, 75.
193   M. Berenstein, A. Georg and E. Briner, *Helv. Chim. Acta*, 1946, *29*, 258.
194   E. W. Berg and R. T. McIntyre, *Anal. Chem.*, 1954, *26*, 813.
195   E. W. Berg and W. L. Senn Jr., *Anal. Chem.*, 1955, *27*, 1255.
196   E. W. Berg and J. E. Strassner, *Anal. Chem.*, 1955, *27*, 127.
197   C. H. O. Berg, *U.S. Pat.* 2,519,342–3–4 (1950): *C.A.*, 1950, *44*, 9666-b.
197a C. Bergamini and A. Rovai, *Anal. Chim. Acta*, 1956, *15*, 43.
198   M. S. Bergdoll and D. M. Doty, *Ind. Eng. Chem. Anal. Ed.*, 1946, *18*, 600.
199   M. S. Bergdoll, B. Lavin, M. J. Surgalla and G. M. Dack, *Science*, 1952, *116*, 633.
200   B. Bergeret and F. Chatagner, *Biochim. Biophys. Acta*, 1954, *14*, 543.
201   R. Bergkvist and A. Deutsch, *Acta Chem. Scand.*, 1953, *7*, 1307.
202   R. Bergkvist and A. Deutsch, *Acta Chem. Scand.*, 1954, *8*, 1877.
203   R. Bergkvist and A. Deutsch, *Acta Chem. Scand.*, 1954, *8*, 1880.
204   R. Bergkvist and A. Deutsch, *Acta Chem. Scand.*, 1954, *8*, 1889.
205   R. Bergkvist and A. Deutsch, *Acta Chem. Scand.*, 1955, *9*, 1398.
206   K. G. Bergner and H. Sperlich, *Z. Lebensm.-Untersuch.-Forsch.*, 1953, *97*, 253: *C.A.*, 1954, *48*, 2273-g.
207   S. Bergström, *Nature*, 1945, *156*, 717.
208   S. Bergström and G. Hansson, *Acta Physiol. Scand.*, 1951, *22*, 87.
209   S. Bergström and K. Pääbo, *Acta Chem. Scand.*, 1949, *3*, 202.
210   S. Bergström and K. Pääbo, *Acta Chem. Scand.*, 1954, *8*, 1486.
211   S. Bergström and J. Sjövall, *Acta Chem. Scand.*, 1951, *5*, 1267.
212   S. Bergström and J. Sjövall, *Acta Chem. Scand.*, 1954, *8*, 611.
213   S. Bergström and J. Sjövall, *Acta Physiol. Scand.*, 1951, *23*, 91.
214   H. W. Berkhout and G. H. Jongen, *Chem. Weekblad*, 1953, *49*, 506: *C.A.*, 1953, *47*, 12107-d.
215   S. Berlingozzi, *Sperimentale, Sez. chim. biol.*, 1954, *5*, 33.
216   S. Berlingozzi, G. Adembri and G. Bucci, *Gazz. chim. ital.*, 1954, *84*, 383: *C.A.*, 1955, *49*, 5344-c.
217   S. Berlingozzi and L. Fabbrini, *Sperimentale, Sez. chim. biol.*, 1954, *5*, 1: *C.A.*, 1955, *49*, 5216-h.
218   S. Berlingozzi, G. Rapi and M. Dettori, *Sperimentale, Sez. chim. biol.*, 1953, *4*, 69: *C.A.*, 1954, *48*, 10822-h.
219   S. Berlingozzi, G. Rapi and M. Dettori, *Sperimentale, Sez. chim. biol.*, 1953, *4*, 75: *C.A.*, 1954, *48*, 10822-i.

220 S. Berlingozzi and G. Serchi, *Sperimentale, Sez. chim. biol.*, 1952, *3*, 1.
221 S. Berlingozzi, G. Serchi and G. Adembri, *Sperimentale, Sez. chim. biol.*, 1951, *2*, 89: *C.A.*, 1952, *46*, 9070-g.
222 S. Berlingozzi, G. Serchi, P. Michi and G. Rapi, *Sperimentale, Sez. chim. biol.*, 1952, *3*, 45: *C.A.*, 1953, *47*, 9916-h.
223 S. Berlingozzi, G. Serchi, F. Rapi and P. Michi, *Sperimentale, Sez. chim. biol.*, 1952, *3*, 125: *C.A.*, 1953, *47*, 9917-b.
224 D. J. Bernardi, *U.S. Pat.* 2,457,117: *C.A.*, 1949, *43*, 2268-e.
225 D. J. Bernardi and F. A. Dostal, *U.S. Pat.* 2,533,215 (1950): *C.A.*, 1951, *45*, 2529-i.
225a R. Bernasconi, H. P. Sigg and T. Reichstein, *Helv. Chim. Acta*, 1955, *38*, 1767.
226 S. A. Bernhard, *J. Am. Chem. Soc.*, 1952, *74*, 4946.
227 S. A. Bernhard and L. P. Hammett, *J. Am. Chem. Soc.*, 1953, *75*, 1798.
228 S. Berntsson and O. Samuelson, *Acta Chem. Scand.*, 1955, *9*, 277.
229 M. Beroza, *Anal. Chem.*, 1950, *22*, 1507.
230 H. K. Berry and L. Cain, *Arch. Biochem.*, 1949, *24*, 179.
231 T. Bersin and A. Müller, *Helv. Chim. Acta*, 1952, *35*, 475.
232 J. Bertetti, *Ann. chim. (Roma)*, 1953, *43*, 361: *C.A.*, 1954, *48*, 11299-g.
233 J. Bertetti, *Ann. chim. (Roma)*, 1954, *44*, 495: *C.A.*, 1955, *49*, 6783-i.
234 J. Berthet, *Biochim. Biophys. Acta*, 1954, *15*, 1.
234a E. Bertorelle and G. Fanfani, *Chimica e industria (Milano)*, 1955, *37*, 777.
235 S. P. Bessman, J. Magnes, P. Schwerin and H. Waelsch, *J. Biol. Chem.*, 1948, *175*, 817.
235a R. H. Betts, W. E. Harris and M. D. Stevenson, *Can. J. Chem.*, 1956, *34*, 65.
236 J. Beukenkamp and W. Rieman III, *Anal. Chem.*, 1950, *22*, 582.
237 J. Beukenkamp, W. Rieman III and S. Lindenbaum, *Anal. Chem.*, 1954, *26*, 505.
238 T. H. Bevan, G. I. Gregory, T. Malkin and A. G. Poole, *J. Chem. Soc.*, *1951*, 841.
239 J. Beyreder and H. Rettenbacher-Däubner, *Monatsh. Chem.*, 1953, *84*, 99; *C.A.*, 1953, *47*, 5475-i.
240 P. M. Bhargava and C. Heidelberger, *J. Am. Chem. Soc.*, 1955, *77*, 166.
241 E. M. Bickoff,·*Anal. Chem.*, 1948, *20*, 51.
242 R. G. S. Bidwell, G. Krotkov and G. B. Reed, *Arch. Biochem. Biophys.*, 1954, *48*, 72.
243 W. Bielenberg and L. Fischer, *Brennstoff-Chem.*, 1941, *22*, 278 and 1942, *23*, 283.
244 W. Bielenberg and H. Goldhahn, *Brennstoff-Chem.*, 1940, *21*, 236.
245 C. Bighi, *Ann. chim. (Roma)*, 1955, *45*, 1087.
245a C. Bighi, *Chimica e industria (Milano)*, 1955, *37*, 1066.
245b C. Bighi and I. Mantovani, *Boll. sci. fac. chim. ind. Bologna*, 1955, *13*, No. 4, 102.
245c C. Bighi and G. Trabanelli, *Ann. chim. (Roma)*, 1955, *45*, 1186.
245d C. Bighi and G. Trabanelli, *Boll. sci. fac. chim. ind. Bologna*, 1955, *13*, No. 4, 100.
246 S. B. Binkley, D. W. MacCorquodale, S. A. Thayer and E. A. Doisy, *J. Biol. Chem.*, 1939, *130*, 219.
246a W. W. Binkley, *Advances in Carbohydrate Chem.*, 1955, *10*, 55–94.
247 W. W. Binkley, M. G. Blair and M. L. Wolfrom, *J. Am. Chem. Soc.*, 1945, *67*, 1789.
248 W. W. Binkley and M. L. Wolfrom, *J. Am. Chem. Soc.*, 1946, *68*, 1720.
249 W. W. Binkley and M. L. Wolfrom, *J. Am. Chem. Soc.*, 1947, *69*, 664.
250 W. W. Binkley and M. L. Wolfrom, *Chromatography of Sugars and Related Substances* in *Repts. Sugar Research Foundation*, No. 10, New York, 1948.

251 W. W. Binkley and M. L. Wolfrom, *J. Am. Chem. Soc.*, 1950, *72*, 4778.
252 H. L. Bird Jr. and C. T. Pugh, *Antibiotics and Chemotherapy*, 1954, *4*, 750: *C.A.*, 1955, *49*, 3471-a.
253 R. Bird and R. H. Hopkins, *Biochem. J.*, 1954, *56*, 86 and 140.
254 G. Biserte, *Biochim. Biophys. Acta*, 1950, *4*, 416.
255 G. Biserte and P. Boulanger, *Bull. soc. chim. biol.*, 1950, *32*, 601.
256 G. Biserte and R. Osteux, *Bull. soc. chim. biol.*, 1951, *33*, 50.
256a J. R. Bishop, *Analyst*, 1956, *81*, 291.
257 J. R. Bishop and H. Liebmann, *Nature*, 1951, *167*, 524.
258 J. Bitman and J. F. Sykes, *Science*, 1953, *117*, 356.
259 C. O. Björling and A. Berggren, *Acta. Chem. Scand.*, 1955, *9*, 567.
260 S. Blackburn, *Biochem. J.*, 1949, *45*, 579.
261 S. Blackburn, *Biochem. J.*, 1951, *47*, 443.
262 S. Blackburn, *Chem. and. Ind.*, *1951*, 294.
263 S. Blackburn, R. Consden and H. Philips, *Biochem. J.*, 1944, *38*, 25.
264 S. Blackburn and G. R. Lee, *Chem. and Ind.*, *1954*, 1252.
265 S. Blackburn and A. G. Lowther, *Biochem. J.*, 1951, *48*, 126.
266 S. Blackburn and A. Robson, *Biochem. J.*, 1953, *54*, 295.
266a J. D. Blainey and H. J. Yardley, *Nature*, 1956, *177*, 83.
266b A. J. Blair and D. A. Pantony, *Anal. Chim. Acta*, 1956, *14*, 545.
267 D. E. Bland, *Nature*, 1949, *164*, 1093.
268 D. E. Bland and F. M. Gatley, *Nature*, 1954, *173*, 32.
269 E. Blasius and M. Negwer, *Naturwissenschaften*, 1952, *39*, 257.
270 E. Blasius and U. Wachtel, *Z. anal. Chem.*, 1954, *142*, 341.
271 J. Blass, O. Lecomte and M. Macheboeuf, *Bull. soc. chim. biol.*, 1951, *33*, 1522.
272 J. Blass, M. Macheboeuf and G. Núñez, *Bull. soc. chim. biol.*, 1950, *32*, 130.
273 K. Blau and S. G. Waley, *Biochem. J.*, 1954, *57*, 538.
274 H. S. Bloch, D. Zimmermann and S. L. Cohen, *J. Clin. Endocrinol. Metab.*, 1953, *13*, 1206: *C.A.*, 1954, *48*, 2165-d.
275 R. J. Block, *Proc. Soc. Exptl. Biol. Med.*, 1942, *51*, 252.
276 R. J. Block, *U.S. Pat.* 2,386,926 (1945): *C.A.*, 1946, *40*, 1172-2.
277 R. J. Block, in F. C. Nachod, *Ion Exchange; Theory and Application*, Academic Press Inc., New York, 1949.
278 R. J. Block, *U.S. Pat.* 2,462,597: *C.A.*, 1949, *43*, 3868-b.
279 R. J. Block, *Anal. Chem.*, 1950, *22*, 1327.
280 R. J. Block, *Anal. Chem.*, 1951, *23*, 298.
281 R. J. Block and D. Bolling, *The Amino Acid Composition of Proteins and Foods*, Charles C. Thomas, Springfield, Ill., 1951.
281a R. J. Block, E. L. Durrum and G. Zweig, *A Manual of Paper Chromatography and Paper Electrophoresis*, 484 pp., Academic Press, New York, 1955.
282 R. J. Block, R. Le Strange and G. Zweig, *Paper Chromatography, a Laboratory Manual*, 206 pp., Academic Press, New York, 1952.
283 A. Blumenthal, C. H. Eugster and P. Karrer, *Helv. Chim. Acta*, 1954, *37*, 787.
284 N. K. Boardman and S. M. Partridge, *J. Polymer. Sci.*, 1954, *12*, 281: *C.A.*, 1954, *48*, 5900-e.
285 N. K. Boardman and S. M. Partridge, *Biochem. J.*, 1955, *59*, 543.
286 R. M. Bock and N.-S. Ling, *Anal. Chem.*, 1954, *26*, 1543.
287 F. Bode, *Biochem. Z.*, 1955, *326*, 433.
288 H. A. Boekenoogen, *Chem. Weekblad*, 1942, *39*, 289: *C.A.*, 1945, *39*, 1769-1.
289 L. A. Boggs, *Anal. Chem.*, 1952, *24*, 1673.

290 L. A. Boggs, L. S. Cuendet, M. Dubois and F. Smith, *Anal. Chem.*, 1952, *24*, 1148.
291 L. A. Boggs, L. S. Cuendet, I. Ehrenthal, R. Koch and F. Smith, *Nature*, 1950, *166*, 520.
292 P. Böhm and G. Richarz, *Z. physiol. Chem.*, 1954, *298*, 110.
293 H. Böhme and H. Lampe, *Arch. Pharm.*, 1951, *284*, 227: *C.A.*, 1952, *46*, 9260-f.
294 R. A. Boissonnas, *Helv. Chim. Acta*, 1947, *30*, 1689.
295 R. A. Boissonnas, *Helv. Chim. Acta*, 1947, *30*, 1703.
296 R. A. Boissonnas, *Helv. Chim. Acta*, 1950, *33*, 1966.
297 R. A. Boissonnas, *Helv. Chim. Acta*, 1950, *33*, 1972.
298 R. A. Boissonnas, *Helv. Chim. Acta*, 1950, *33*, 1975.
299 R. A. Boissonnas and S. Lo Bianco, *Experientia*, 1952, *8*, 425.
300 J. Boldingh, *Experientia*, 1948, *4*, 270; *Rec. trav. chim.*, 1950, *69*, 247.
301 J. Boldingh, *International conference on biochemical problems of lipids*, Brussels, 1953, p. 64.
302 H. R. Bolliger and D. A. Prins, *Helv. Chim. Acta*, 1946, *29*, 1116.
303 D. Bolling, H. A. Sober and R. J. Block, *Federation Proc.*, 1949, *8*, 185.
304 H. G. Boman, *Nature*, 1952, *170*, 703.
305 H. G. Boman, *Biochim. Biophys. Acta*, 1955, *16*, 245.
306 H. G. Boman, *Nature*, 1955, *175*, 898.
307 P. Bonet-Maury, *Bull. soc. chim. France*, *1953*, 1066.
308 E. Bonetti and C. E. Dent, *Biochem. J.*, 1954, *57*, 77.
309 G. B. Bonino and V. Carassiti, *Rend. acad. naz. Lincei.*, *Ser. VIII*, 1951, *9*, 229.
310 G. B. Bonino and V. Carassiti, *Nature*, 1951, *167*, 569.
311 A. Bonnin and P. Sue, *Compt. rend.*, 1952, *234*, 960.
312 W. R. Boon, C. T. Calam, H. Gudgeon and A. A. Levi, *Biochem. J.*, 1948, *43*, 262.
313 I. O. Boone, D. F. Turney and K. T. Woodward, *Science*, 1954, *120*, 312.
314 V. H. Booth, *Analyst*, 1950, *75*, 109.
315 B. Borgström, *Acta Physiol. Scand.*, 1952, *25*, 101.
316 B. Borgström, *Acta Physiol. Scand.*, 1952, *25*, 111.
317 B. Borgström, *Acta Physiol. Scand.*, 1954, *30*, 231.
318 H. Borsook, C. L. Deasy, A. J. Haagen-Smit, G. Keighley and P. H. Lowy, *J. Biol. Chem.*, 1948, *173*, 423.
319 H. Borsook, C. L. Deasy, A. J. Haagen-Smit, G. Keighley and P. H. Lowy, *J. Biol. Chem.*, 1948, *174*, 1041.
320 H. Borsook, C. L. Deasy, A. J. Haagen-Smit, G. Keighley and P. H. Lowy, *J. Biol. Chem.*, 1948, *176*, 1383.
321 H. Borsook, C. L. Deasy, A. J. Haagen-Smit, G. Keighley and P. H. Lowy, *J. Biol. Chem.*, 1949, *179*, 705.
322 F. de A. Bosch, *Análisis cromatográfico*, Real. Acad. Med. Cirujia, Valencia, 1950: *C.A.*, 1952, *46*, 4957-a.
323 L. Bosch, *Biochim. Biophys. Acta*, 1953, *11*, 301.
324 R. J. Boscott, *Biochem. J.*, 1951, *48*, xlvii.
325 R. J. Boscott, *Chem. and Ind.*, *1952*, 472.
325a R. J. Boscott, *The technique and significance of oestrogen determinations*, in *Memoirs of the Soc. for Endocrinology*, No. 3, p. 23, Cambridge Univ. Press, London, 1955.
326 R. J. Boscott and H. Bickel, *Scand. J. Clin. & Lab. Invest.*, 1953, *5*, 380: *C.A.*, 1954, *48*, 5256-c.

327   H. Boser, *Z. physiol. Chem.*, 1954, *296*, 10.
328   H. Boser, *Z. physiol. Chem.*, 1954, *298*, 145.
329   W. Bottomley and P. I. Mortimer, *Australian J. Chem.*, 1954, *7*, 189: *C.A.*, 1955, *49*, 6975-c.
330   R. Bouchez and G. Kayas, *Compt. rend.*, 1949, *228*, 1222.
331   G. Bouissières and M. Lederer, *Bull. soc. chim. France*, 1952, 904–910.
332   P. Boulanger and G. Biserte, *Bull. soc. chim. biol.*, 1949, *31*, 696.
333   P. Boulanger and G. Biserte, *La chromatographie de partage*, 69 pp. in *Exposés annuels de Biochimie Médicale*, Vol. XI, Masson et Cie., Paris, 1950.
334   P. Boulanger and G. Biserte, *Compt. rend.*, 1951, *232*, 1451.
335   P. Boulanger and G. Biserte, *Compt. rend.*, 1951, *233*, 1498.
336   P. Boulanger and G. Biserte, *Bull. soc. chim. biol.*, 1951, *33*, 1930.
337   P. Boulanger and G. Biserte, *Bull. soc. chim. France*, 1952, 830–844.
338   P. Boulanger, G. Biserte and F. Courtot, *Bull. soc. chim. biol.*, 1952, *34*, 366.
339   P. Boulanger, G. Biserte and R. Scriban, *Ann. nutrition et aliment.*, 1951, *5*, 149: *C.A.*, 1951, *45*, 9423-b.
340   P. Boulanger and J. Montreuil, *Bull. soc. chim. biol.*, 1951, *33*, 784.
341   P. Boulanger and J. Montreuil, *Bull. soc. chim. biol.*, 1951, *33*, 791.
342   P. Boulanger and J. Montreuil, *Bull. soc. chim. France*, 1952, 844–852.
343   P. Boulanger, J. Montreuil and L. Masse, *Compt. rend.*, 1951, *232*, 1256.
344   P. Boulanger and R. Osteux, *Biochim. Biophys. Acta*, 1950, *5*, 416.
345   G. H. Bourne, *Nature*, 1949, *163*, 923.
346   J. C. Boursnell, *Nature*, 1950, *165*, 399.
347   G. Boutillon and M. Prettre, *Compt. rend.*, 1955, *240*, 1216.
348   J. P. Bowden and W. H. Peterson, *J. Biol. Chem.*, 1949, *178*, 533.
349   G. E. Boxer and V. C. Jelinek, *J. Biol. Chem.*, 1947, *170*, 491.
350   G. E. Boyd, *Ann. Rev. Phys. Chem.*, 1951, *2*, 309.
351   G. E. Boyd, A. W. Adamson and L. S. Myers Jr., *J. Am. Chem. Soc.*, 1947, *69*, 2836.
352   G. E. Boyd, L. S. Myers Jr. and A. W. Adamson, *J. Am. Chem. Soc.*, 1947, *69*, 2849.
353   G. E. Boyd, J. Schubert and A. W. Adamson, *J. Am. Chem. Soc.*, 1947, *69*, 2818.
354   J. W. Brackett Jr. and L. W. Bradford, *J. Criminal Law, Criminol. Police Sci.*, 1952, *43*, 530: *C.A.*, 1954, *48*, 6325-e.
355   A. E. Bradfield and E. C. Bate-Smith, *Biochim. Biophys. Acta*, 1950, *4*, 441.
356   A. E. Bradfield and A. E. Flood, *J. Chem. Soc.*, 1952, 4740.
357   J. E. S. Bradley, *Biochem. J.*, 1954, *56*, xlviii.
358   W. Bradley and R. A. Brindley, *Nature*, 1954, *173*, 312.
359   W. Bradley and G. C. Easty, *J. Chem. Soc.*, 1951, 499.
360   W. Bradley and G. C. Easty, *J. Chem. Soc.*, 1953, 1519.
361   H. Brandenberger, *Helv. Chim. Acta*, 1954, *37*, 97.
362   G. Brante, *Nature*, 1949, *163*, 651.
363   I. Brattsten and A. Nilsson, *Arkiv Kemi*, 1951, *3*, 337.
364   F. E. Brauns, J. B. Hlava and H. Seiler, *Anal. Chem.*, 1954, *26*, 607.
365   H. Braunsberg, *Nature*, 1952, *169*, 967.
366   H. Braunsberg, S. B. Osborn and M. I. Stern, *J. Endocrinol.*, 1954, *11*, 177: *C.A.*, 1954, *48*, 13783-i.
367   H. G. Bray, R. C. Clowes and W. V. Thorpe, *Biochem. J.*, 1952, *51*, 70.
368   H. G. Bray, H. J. Lake and W. V. Thorpe, *Biochem. J.*, 1951, *48*, 400.
369   H. G. Bray and W. V. Thorpe, *Meth. Biochem. Anal.*, 1954, *1*, 27.

370 H. G. Bray, W. V. Thorpe and K. White, *Biochem. J.*, 1950, *46*, 271.
371 H. G. Bray, W. V. Thorpe and P. B. Wood, *Biochem. J.*, 1951, *48*, 394.
372 H. Bredereck, H. Dürr and K. Ruck, *Chem. Ber.*, 1954, *87*, 526.
373 H. M. Bregoff, E. Roberts and C. C. Delwiche, *J. Biol. Chem.*, 1953, *205*, 565.
374 J. M. Bremner, *Biochem. J.*, 1950, *47*, 538.
375 J. M. Bremner, *Analyst*, 1954, *79*, 198.
376 J. M. Bremner and R. H. Kenten, *Biochem. J.*, 1951, *49*, 651.
377 M. Brenner and C. H. Burckhardt, *Helv. Chim. Acta*, 1951, *34*, 1070.
378 M. Brenner and R. Frey, *Helv. Chim. Acta*, 1951, *34*, 1701.
379 M. Brenner, H. R. Müller and R. W. Pfister, *Helv. Chim. Acta*, 1950, *33*, 568.
380 R. Bretschneider, *Monatsh. Chem.*, 1941, *74*, 53.
381 H. Breuer, *Naturwissenschaften*, 1955, *42*, 16.
382 R. C. Brimley, *Nature*, 1949, *163*, 215.
383 R. C. Brimley and A. Snow, *J. Sci. Instrum.*, 1949, *26*, 73.
384 H. Brindle, J. E. Carless and H. B. Woodhead, *J. Pharm. and Pharmacol.*, 1951, *3*, 793, 813: *C.A.*, 1952, *46*, 1710-d.
385 N. G. Brink, F. A. Kuehl Jr., E. H. Flynn and K. Folkers, *J. Am. Chem. Soc.*, 1946, *68*, 2557.
386 N. G. Brink, F. A. Kuehl Jr., J. W. Richter, A. W. Bazemore, M. A. P. Meisinger, D. E. Ayer and K. Folkers, *J. Am. Chem. Soc.*, 1952, *74*, 2120.
387 H. Brockmann, *Discuss. Faraday Soc.*, 1949, *7*, 58.
388 H. Brockmann and H. Gröne, *Naturwissenschaften*, 1953, *40*, 222.
389 H. Brockmann and H. Gröne, *Chem. Ber.*, 1954, *87*, 1036.
390 H. Brockmann, N. Grubhofer, W. Kass and H. Kalbe, *Chem. Ber.*, 1951, *84*, 260.
391 H. Brockmann and K. Müller, *Ann.*, 1939, *540*, 51.
392 H. Brockmann and H. Musso, *Naturwissenschaften*, 1951, *38*, 11.
393 H. Brockmann and H. Musso, *Chem. Ber.*, 1954, *87*, 681.
394 H. Brockmann and H. Musso, *Chem. Ber.*, 1954, *87*, 1779.
395 H. Brockmann, F. Pohl, K. Maier and M. N. Haschad, *Ann.*, 1942, *553*, 1.
396 H. Brockmann and H. Schodder, *Ber.*, 1941, *74*, 73.
397 H. Brockmann and F. Volpers, *Chem. Ber.*, 1947, *80*, 77.
398 H. Brockmann and F. Volpers, *Chem. Ber.*, 1949, *82*, 95.
399 M. C. Brooks and R. M. Badger, *J. Am. Chem. Soc.*, 1950, *72*, 4384.
400 R. V. Brooks, W. Klyne and E. Miller, *Biochem. J.*, 1951, *49*, xl, lxxii.
401 R. V. Brooks, W. Klyne and E. Miller, *Biochem. J.*, 1953, *54*, 212.
403 W. A. Brooksbank and G. W. Leddicotte, *J. Phys. Chem.*, 1953, *57*, 819.
404 E. Brouwer and H. J. Nijkamp, *Chem. Weekblad*, 1950, *46*, 37: *C.A.*, 1950, *44*, 5495-d.
405 E. Brouwer and H. J. Nijkamp, *Biochem. J.*, 1953, *55*, 444.
406 C. Brown and P. L. Kirk, *J. Criminal Law, Criminol. Police Sci.*, 1954, *45*, 335.
407 D. M. Brown, C. A. Dekker and A. R. Todd, *J. Chem. Soc.*, *1952*, 2715.
408 D. M. Brown, D. I. Magrath and A. R. Todd, *J. Chem. Soc.*, *1952*, 2708.
409 D. M. Brown and A. R. Todd, *J. Chem. Soc.*, *1952*, 44.
410 F. Brown, *Biochem. J.*, 1950, *47*, 598.
411 F. Brown, *Biochem. J.*, 1952, *51*, 237.
412 F. Brown, *Biochem. J.*, 1952, *52*, 523.
413 F. Brown and L. P. Hall, *Nature*, 1950, *166*, 66.
414 F. Brown, E. L. Hirst, L. Hough, J. K. N. Jones and H. Wadman, *Nature*, 1948, *161*, 720.

415   F. Brown and H. Jackson, *Biochem. J.*, 1954, *56*, 399.
416   F. B. Brown, J. C. Cain, D. E. Gant, L. F. J. Parker and E. L. Smith, *Biochem. J.*, 1955, *59*, 82.
416a G. L. Brown and A. V. Martin, *Nature*, 1955, *176*, 971.
417   G. L. Brown and M. Watson, *Nature*, 1953, *172*, 339.
418   G. M. Brown and E. E. Snell, *J. Biol. Chem.*, 1952, *198*, 375.
419   H. Brown, F. Sanger and R. Kitai, *Biochem. J.*, 1955, *60*, 556.
420   J. A. Brown, *Anal. Chem.*, 1953, *25*, 774.
421   J. A. Brown and M. M. Marsh, *Anal. Chem.*, 1952, *24*, 1952.
422   J. A. Brown and M. M. Marsh, *Anal. Chem.*, 1953, *25*, 1865.
423   R. Brown, A. D. Greenwood, A. W. Johnson, A. G. Long and G. J. Tyler, *Biochem. J.*, 1951, *48*, 564.
424   W. E. Brown and W. Rieman III, *J. Am. Chem. Soc.*, 1952, *74*, 1278.
425   W. G. Brown, *Nature*, 1939, *143*, 377.
426   J. Brüggemann and K. Drepper, *Naturwissenschaften*, 1952, *39*, 301.
427   J. Brüggemann, W. Krauss and J. Tiews, *Naturwissenschaften*, 1951, *38*, 562.
428   E. M. Brumberg, *Uspekhi Fiz. Nauk*, 1951, *43*, 600.
429   E. Bruninx, J. Eeckhout and J. Gillis, *Mikrochim. Acta*, in press [presented to the Microchemical Congress (1955)].
430   G. Brunisholz, *Helv. Chim. Acta*, 1952, *35*, 1003.
432   G. Brunisholz and A. Germano, *Helv. Chim. Acta*, 1954, *37*, 242.
433   H. Bruschweiler, G. J. Minkoff and K. C. Salooja, *Nature*, 1953, *172*, 909.
434   M. K. Brush, R. K. Boutwell, A. D. Barton and C. Heidelberger, *Science*, 1951, *113*, 4.
435   F. Bryant, *Australian J. Sci.*, 1950, *13*, 83: *C.A.*, 1951, *45*, 3447-c.
436   F. Bryant and B. T. Overell, *Nature*, 1951, *167*, 361.
437   F. Bryant and B. T. Overell, *Nature*, 1951, *168*, 167.
438   F. Bryant and B. T. Overell, *Biochim. Biophys. Acta*, 1953, *10*, 471.
439   L. H. Bryant, *Nature*, 1955, *175*, 556.
440   J. L. Bryson and T. J. Mitchell, *Nature*, 1951, *167*, 864.
441   C. Bublitz and E. P. Kennedy, *J. Biol. Chem.*, 1954, *211*, 963.
442   M. L. Buch, R. Montgomery and W. L. Porter, *Anal. Chem.*, 1952, *24*, 489.
443   J. G. Buchanan, *Nature*, 1951, *168*, 1091.
444   J. G. Buchanan, C. A. Dekker and A. G. Long, *J. Chem. Soc.*, *1950*, 3162.
445   G. Büchi, O. Jeger and L. Ruzicka, *Helv. Chim. Acta*, 1946, *29*, 442.
446   E. Bueding and H. W. Yale, *J. Biol. Chem.*, 1951, *193*, 411.
447   D. R. Buhler, R. C. Thomas, B. E. Christensen and C. H. Wang, *J. Am. Chem. Soc.*, 1955, *77*, 481.
448   W. A. Bulen, J. E. Varner and R. C. Burrell, *Anal. Chem.*, 1952, *24*, 187.
449   H. B. Bull, J. W. Hahn and V. H. Baptist, *J. Am. Chem. Soc.*, 1949, *71*, 550.
450   F. M. Bumpus, A. A. Green and I. H. Page, *J. Biol. Chem.*, 1954, *210*, 287.
451   F. M. Bumpus and I. H. Page, *J. Biol. Chem.*, 1955, *212*, 111.
452   F. M. Bumpus, W. R. Taylor and F. M. Strong, *J. Am. Chem. Soc.*, 1950, *72*, 2116.
453   C. A. Bunton, G. J. Minkoff and R. I. Reed, *J. Chem. Soc*, *1947*, 1416.
454   E. M. Buras Jr., and S. R. Hobart, *Anal. Chem.*, 1955, *27*, 1507.
455   D. P. Burma, *Nature*, 1951, *168*, 565.
456   D. P. Burma, *Analyst*, 1952, *77*, 382.
457   D. P. Burma, *J. Indian Chem. Soc.*, 1952, *29*, 567: *C.A.* 1953, *47*, 3915-d.
458   D. P. Burma, *Anal. Chem.*, 1953, *25*, 549.

459 D. P. Burma, *Anal. Chim. Acta*, 1953, *9*, 513.
460 D. P. Burma and B. Banerjee, *Science and Culture*, 1950, *15*, 363: *C.A.*, 1951, *45*, 3753-b.
461 S. I. Burmistrov, *Zhur. Anal. Khim.*, 1950, *5*, 39: *C.A.*, 1950, *44*, 4373-h.
462 J. B. Burnett, *Univ. Microfilms* (Ann. Arbor. Mich.), Publ. No. 7476: *C.A.*, 1954, *48*, 7104-b.
463 F. Burriel-Marti and F. Pino-Perez, *Anal. Chim. Acta*, 1949, *3*, 468.
464 S. Burrows, F. S. M. Grylls and J. S. Harrison, *Nature*, 1952, *170*, 800.
465 F. H. Burstall, private communication.
466 F. H. Burstall, G. R. Davies, R. P. Linstead and R. A. Wells, *Nature*, 1949, *163*, 64.
467 F. H. Burstall, G. R. Davies, R. P. Linstead and R. A. Wells, *J. Chem. Soc.*, *1950*, 516.
468 F. H. Burstall, G. R. Davies and R. A. Wells, *Discuss. Faraday Soc.*, 1949, *7*, 179.
469 F. H. Burstall, P. Swain, A. F. Williams and G. A. Wood, *J. Chem. Soc.*, *1952*, 1497.
470 F. H. Burstall and R. A. Wells, *Analyst*, 1951, *76*, 396.
471 F. H. Burstall and A. F. Williams, *Analyst*, 1952, *77*, 983.
472 D. Burton and G. Lee, *J. Intern. Soc. Leather Trades' Chemists*, 1945, *29*, 204: *C.A.*, 1946, *40*, 2332-g.
473 H. S. Burton, *Chem. and Ind.*, 1953, 1229.
474 H. S. Burton, *Nature*, 1954, *173*, 127.
475 H. S. Burton and E. P. Abraham, *Biochem. J.*, 1951, *50*, 168.
476 R. B. Burton, A. Zaffaroni and E. H. Keutmann, *J. Biol. Chem.*, 1951, *188*, 763.
477 R. B. Burton, A. Zaffaroni and E. H. Keutmann, *J. Biol. Chem.*, 1951, *193*, 769.
478 R. M. Burton and A. San Pietro, *Arch. Biochem. Biophys.*, 1954, *48*, 184.
479 H. Busch, R. B. Hurlbert and V. R. Potter, *J. Biol. Chem.*, 1952, *196*, 717.
480 I. E. Bush, *Brit. Med. Bull.*, 1954, *10*, 229.
481 I. E. Bush, *Recent Progr. Hormone Research*, 1954, *9*, 321.
482 I. E.Bush, *Biochem. J.*, 1955, *59*, xiv.
483 I. E. Bush and A. A. Sandberg, *J. Biol. Chem.*, 1953, *205*, 783, (790).
484 I. E.Bush and D. A. H. Taylor, *Biochem. J.*, 1952, *52*, 643.
485 W. Buser, *Helv. Chim. Acta*, 1951, *34*, 1635.
486 I. E. Bush, *Biochem. J.*, 1952, *50*, 370.
487 A. Butenandt and L. Poschmann, *Ber.*, 1944, *77*, 392.
488 J. A. V. Butler and J. M. L. Stephens, *Nature*, 1947, *160*, 469.
489 W. R. Butt, P. Morris and C. J. O. R. Morris, *First Intern. Congr. Biochem.*, Cambridge, *Abstrs. of Communs.*, 1949, 405: *C.A.*, 1953, *47*, 164-i.
490 Ng. Ph. Buu-Hoï and P. Cagniant, *Bull. soc. chim. France*, (5) 1944, *11*, 410.
491 W. L. Byrne and H. A. Lardy, *Biochim. Biophys. Acta*, 1954, *14*, 495.
491a M. J. Cabell and I. Milner, *Anal. Chim. Acta*, 1955, *13*, 258.
491b M. J. Cabell and A. Thomas, *AERE C/R, 1955*, 1725 (10 pp.); *Anal. Abstr.*, 1956, *3*, 925.
492 E. Cabib, *Biochim. Biophys. Acta*, 1951, *7*, 604.
493 E. Cabib, *Biochim. Biophys. Acta*, 1952, *8*, 607.
494 E. Cabib and L. F. Leloir, *J. Biol. Chem.*, 1954, *206*, 779.
495 R. S. Cahn, R. F. Phipers and J. J. Boam, *J. Soc. Chem. Ind.*, 1938, *57*, T200.
496 G. Cajelli, *Kautschuk*, 1940, *16*, 129: cf. *C.A.*, 1939, *33*, 9043-b.

497   P. C. Caldwell, *Biochem. J.*, 1955, *60*, xii.
498   D. N. Campbell and C. T. Kenner, *Anal. Chem.*, 1954, *26*, 560.
499   H. Campbell and J. A. Simpson, *Chem. and Ind.*, 1953, 342.
500   P. N. Campbell, S. Jacobs, T. S. Work and T. R. E. Kressman, *Chem. and Ind.*, 1955, 117.
501   P. N. Campbell, D. H. Simmonds and T. S. Work, *Biochem. J.*, 1951, *49*, xvi.
502   P. N. Campbell and T. S. Work, *Brit. Med. Bull.*, 1954, *10*, 196.
503   P. N. Campbell, T. S. Work and E. Mellanby, *Biochem. J.*, 1951, *48*, 106.
504   R. Caputto and R. E. Trucco, *Nature*, 1952, *169*, 1061.
505   J. Carles, *Bull. soc. chim. biol.*, 1955, *37*, 521.
506   G. Carleson, *Acta Chem. Scand.*, 1954, *8*, 1673.
507   G. Carleson, *Acta Chem. Scand.*, 1954, *8*, 1697.
508   J. E. Carless, *J. Pharm. and Pharmacol.*, 1953, *5*, 883: *C.A.*, 1954, *48*, 1628-g.
509   J. E. Carless and H. B. Woodhead, *Nature*, 1951, *168*, 203.
510   J. K. Carlton and W. C. Bradbury, *Anal. Chem.*, 1955, *27*, 67.
511   J. Carol, E. O. Haenni and D. Banes, *J. Biol. Chem.*, 1950, *185*, 267.
512   G. Caronna, *Chimica e industria (Milano)*, 1955, *37*, 113: *C.A.*, 1955, *49*, 7299-d.
513   K. K. Carroll, *Nature*, 1955, *176*, 398.
514   J. F. Carson, *J. Am. Chem. Soc.*, 1951, *73*, 4652.
515   M. E. Carsten, *J. Am. Chem. Soc.*, 1952, *74*, 5954.
516   M. E. Carsten and R. K. Cannan, *J. Am. Chem. Soc.*, 1952, *74*, 5950.
517   C. E. Carter, *J. Am. Chem. Soc.*, 1950, *72*, 1466.
518   C. E. Carter, *J. Am. Chem. Soc.*, 1951, *73*, 1508.
519   C. E. Carter and W. E. Cohn, *J. Am. Chem. Soc.*, 1950, *72*, 2604.
520   H. E. Carter, R. K. Clark Jr., S. R. Dickman, Y. H. Loo, P. S. Skell and W. A. Strong, *J. Biol. Chem.*, 1945, *160*, 337.
521   H. E. Carter, R. K. Clark Jr., P. Kohn, J. W. Rothrock, W. R. Taylor, C. A. West, G. B. Whitfield and W. G. Jackson, *J. Am. Chem. Soc.*, 1954, *76*, 566.
522   Y. Carteret, *J. chim. phys.*, 1954, *51*, 625: *C.A.*, 1955, 7324-e.
523   R. A. Cartwright and E. A. H. Roberts, *Chem. and Ind.*, 1954, 1389.
524   J. Cason and G. A. Gillies, *J. Org. Chem.*, 1955, *20*, 419.
525   J. Cason, G. Sumrell, C. F. Allen, G. A. Gillies and S. Elberg, *J. Biol. Chem.*, 1953, *205*, 435.
526   W. G. B. Casselman, *Biochim. Biophys. Acta*, 1954, *14*, 450.
527   H. G. Cassidy, *J. Am. Chem. Soc.*, 1940, *62*, 3073 and 3076.
528   H. G. Cassidy, *J. Am. Chem. Soc.*, 1941, *63*, 2735.
529   H. G. Cassidy, *J. Am. Chem. Soc.*, 1949, *71*, 402.
530   H. G. Cassidy, *Adsorption and Chromatography*, 360 pp., Interscience Publishers, New York, 1951.
531   H. G. Cassidy, *Anal. Chem.*, 1952, *24*, 1415.
532   H. G. Cassidy and S. E. Wood, *J. Am. Chem. Soc.*, 1941, *63*, 2628.
533   J. R. Catch, A. H. Cook and I. M. Heilbron, *Nature*, 1942, *150*, 633; *Brit. Pat.* 569,844.
534   D. Cavallini and N. Frontali, *Biochim. Biophys. Acta*, 1954, *13*, 439.
535   D. Cavallini, N. Frontali and G. Toschi, *Nature*, 1949, *163*, 568.
536   D. Cavallini, N. Frontali and G. Toschi, *Nature*, 1949, *164*, 792.
537   D. Cavallini, N. Frontali and G. Toschi, *Bull. soc. ital. biol. sper.*, 1949, *25*, 286.
538   G. Ceriotti, *Nature*, 1955, *175*, 897.
538a G. Cetini and F. Ricca, *Gazz. chim. ital.*, 1955, *85*, 419.

539 S. Chakrabarti and D. P. Burma, *Science and Culture*, 1951, *16*, 485: *C.A.*, 1952, *46*, 380-b.

540 H. C. Chakrabortty and D. P. Burma, *Current Sci. (India)*, 1953, *22*, 238: *C.A.*, 1954, *48*, 3438-b.

541 H. C. Chakrabortty and D. P. Burma, *Science and Culture*, 1954, *19*, 467: *C.A.*, 1954, *48*, 9869-f.

542 D. E. Chalkley and R. J. P. Williams, *J. Chem. Soc.*, *1954*, 1718.

543 M. A. Chambers, L. P. Zill and G. R. Noggle, *J. Am. Pharm. Assoc.*, 1952, *41*, 461.

544 S. K. Chanda, E. L. Hirst, J. K. N. Jones and E. G. V. Percival, *J. Chem. Soc.*, *1950*, 1289.

545 W.-H. Chang, R. L. Hossfeld and W. M. Sandstrom, *J. Am. Chem. Soc.*, 1952, *74*, 5766.

546 W.-S. Chang and W. H. Peterson, *J. Biol. Chem.*, 1951, *193*, 587.

547 E. Chargaff and J. Kream, *J. Am. Chem. Soc.*, 1952, *74*, 4274 and 5157.

548 E. Chargaff, C. Levine and C. Green, *J. Biol. Chem.*, 1948, *175*, 67.

549 E. Chargaff, C. Levine, C. Green and J. Kream, *Experientia*, 1950, *6*, 229.

550 E. Chargaff, R. Lipshitz, C. Green and M. E. Hodes, *J. Biol. Chem.*, 1951, *192*, 223.

551 E. Chargaff, B. Magasanik, R. Doniger and E. Vischer, *J. Am. Chem. Soc.*, 1949, *71*, 1513.

552 E. Chargaff and H. F. Saidel, *J. Biol. Chem.*, 1949, *177*, 417.

553 E. Chargaff, E. Vischer, R. Doniger, C. Green and F. Misani, *J. Biol. Chem.*, 1949, *177*, 405.

554 A. Charro Arias, *Anales bromatol. (Madrid)*, 1953, *5*, 359: *C.A.*, 1954, *48*, 5387-g.

555 F. W. Chattaway, D. E. Dolby, D. A. Hall and F. C. Happold, *Biochem. J.*, 1949, *45*, 592.

556 R. I. Cheftel, R. Munier and M. Machebeouf, *Bull. soc. chim. biol.*, 1951, *33*, 840.

557 R.-I. Cheftel, R. Munier and M. Macheboeuf, *Bull. soc. chim. biol.*, 1952, *34*, 380.

558 R.-I. Cheftel, R. Munier and M. Macheboeuf, *Bull. soc. chim. biol.*, 1953, *35*, 1085.

559 R.-I. Cheftel, R. Munier and M. Macheboeuf, *Bull. soc. chim. biol.*, 1953, *35*, 1091.

560 R.-I. Cheftel, R. Munier and M. Macheboeuf, *Bull. soc. chim. biol.*, 1953, *35*, 1095.

561 C. Chen and H. E. Tewell Jr., *Federation Proc.*, 1951, *10*, 377.

562 Y.-T. Chen, F. A. Isherwood and L. W. Mapson, *Biochem. J.*, 1953, *55*, 821.

563 A. Cherkin, F. E. Martinez and M. S. Dunn, *J. Am. Chem. Soc.*, 1953, *75*, 1244.

564 Y. Chiba, *Arch. Biochem. Biophys.*, 1955, *54*, 83.

565 Y. Chiba and I. Naguchi, *Cytologia*, 1954, *19*, 41: *C.A.*, 1954, *48*, 13792-d.

566 A. C. Chibnall, *Brit. Med. Bull.*, 1954, *10*, 183.

566a G. R. Choppin and R. J. Silva, *J. Inorg. Nuclear Chem.*, 1956, *3*, 153.

567 J. F. Christman and R. H. Trubey, *Stain Technol.*, 1952, *27*, 53: *C.A.*, 1952, *46*, 5650-f.

568 A. A. Christomanos, *Enzymologia*, 1953, *16*, 87.

569 T. C. Chu and E. J.-H. Chu, *J. Biol. Chem.*, 1954, *208*, 537.

570 T. C. Chu and E. J.-H. Chu, *J. Biol. Chem.*, 1955, *212*, 1.

571 T. C. Chu, A. A. Green and E. J.-H. Chu, *J. Biol. Chem.*, 1951, *190*, 643.
572 J. A. Cifonelli and F. Smith, *Anal. Chem.*, 1954, *26*, 1132.
573 I. Claesson and S. Claesson, *Arkiv Kemi, Mineral., Geol.*, 1944, *19A*, No. 5.
574 S. Claesson, *Arkiv Kemi, Mineral., Geol.*, 1941, *15*, A, No. 9.
575 S. Claesson, *The Svedberg Memorial Vol.*, 1944, 82: *C.A.*, 1945, *39*, 842-g.
576 S. Claesson, *Arkiv Kemi, Mineral., Geol.*, 1945, *20*, A, No. 3.
577 S. Claesson, *Arkiv Kemi, Mineral., Geol.*, 1946, *23*, A, No. 1.
578 S. Claesson, *Arkiv Kemi, Mineral., Geol.*, 1946, *24*, A, No. 7.
579 S. Claesson, *Rec. trav. chim.*, 1946, *65*, 571.
580 S. Claesson, *Arkiv Kemi, Mineral., Geol.*, 1947, *24*, *A*, No. 16.
581 S. Claesson, *Nature*, 1947, *159*, 708.
582 S. Claesson, *Ann. N.Y. Acad. Sci.*, 1948, *49*, 183.
583 S. Claesson, *Discuss. Faraday Soc.*, 1949, *7*, 34.
584 S. Claesson, *Discuss. Faraday Soc.*, 1949, *7*, 321.
585 H. Clauser and C. H. Li, *J. Am. Chem. Soc.*, 1954, *76*, 4337.
586 R. A. Clayton and F. M. Strong, *Anal. Chem.*, 1954, *26*, 579.
587 R. A. Clayton and F. M. Strong, *Anal. Chem.*, 1954, *26*, 1362.
588 C. S. Cleaver and H. G. Cassidy, *J. Am. Chem. Soc.*, 1950, *72*, 1147.
589 C. S. Cleaver, R. A. Hardy Jr. and H. G. Cassidy, *J. Am. Chem. Soc.*, 1945, *67*, 1343.
590 R. J. Clerc, C. B. Kincannon and T. P. Wier Jr., *Anal. Chem.*, 1950, *22*, 864.
591 K. Closs and C. M. Haug, *Chem. and Ind.*, *1953*, 103.
592 J. I. Coates and E. Glueckauf, *J. Chem. Soc.*, *1947*, 1302.
593 J. W. Cobble and A. W. Adamson, *J. Am. Chem. Soc.*, 1950, *72*, 2276.
594 J. R. Coffmann, *J. Biol. Chem.*, 1941, *140*, xxviii.
595 S. S. Cohen and D. B. M. Scott, *Science*, 1950, *111*, 543.
596 W. E. Cohn, *Science*, 1949, *109*, 377; *J. Am. Chem. Soc.*, 1949, *71*, 2275.
597 W. E. Cohn, *J. Am. Chem. Soc.*, 1950, *72*, 1471.
598 W. E. Cohn, *J. Am. Chem. Soc.*, 1950, *72*, 2811.
599 W. E. Cohn, *J. Am. Chem. Soc.*, 1951, *73*, 1539.
600 W. E. Cohn, in *The Nucleic Acids*, Vol. I, 211–242, Acad. Press, New York, 1955.
601 W. E. Cohn and C. E. Carter, *J. Am. Chem. Soc.*, 1950, *72*, 2606.
602 W. E. Cohn and C. E. Carter, *J. Am. Chem. Soc.*, 1950, *72*, 4273.
603 W. E. Cohn, and H. W. Kohn, *J. Am. Chem. Soc.*, 1948, *70*, 1986.
604 W. E. Cohn and E. Volkin, *J. Biol. Chem.*, 1953, *203*, 319.
605 P. G. Cole, G. H. Lathe and B. H. Billing, *Biochem. J.*, 1954, *57*, 514.
606 G. H. Coleman, A. G. Farnham and A. Miller, *J. Am. Chem. Soc.*, 1942, *64*, 1501.
607 G. H. Coleman and C. M. McCloskey, *J. Am. Chem. Soc.*, 1943, *65*, 1588.
608 G. H. Coleman, D. E. Rees, R. L. Sundberg and C. M. McCloskey, *J. Am. Chem. Soc.*, 1945, *67*, 381.
609 A. Comfort, *Biochem. J.*, 1955, *59*, x.
610 A. Comfort and M. Weatherall, *Biochem. J.*, 1953, *54*, 247.
611 J. J. Connell, R. M. Hainsworth, E. L. Hirst and J. K. N. Jones, *J. Chem. Soc.*, *1950*, 1696.
612 R. T. Conner and G. J. Straub, *Ind. Eng. Chem. Anal. Ed.*, 1941, *13*, 385.
613 R. E. Connick and S. W. Mayer, *J. Am. Chem. Soc.*, 1951, *73*, 1176.
614 A. L. Conrad, *Anal. Chem.*, 1948, *20*, 725.
615 R. Consden, *Brit. Med. Bull.*, 1954, *10*, 177.
616 R. Consden and A. H. Gordon, *Nature*, 1948, *162*, 180.

617 R. Consden and A. H. Gordon, *Biochem. J.*, 1950, *46*, 8.
618 R. Consden, A. H. Gordon and A. J. P. Martin, *Biochem. J.*, 1944, *38*, 224.
619 R. Consden, A. H. Gordon and A. J. P. Martin, *Biochem. J.*, 1946, *40*, 580.
620 R. Consden, A. H. Gordon and A. J. P. Martin, *Biochem. J.*, 1947, *41*, 590.
621 R. Consden, A. H. Gordon and A. J. P. Martin, *Biochem. J.*, 1948, *42*, 443.
622 R. Consden, A. H. Gordon and A. J. P. Martin, *Biochem. J.*, 1949, *44*, 548.
623 R. Consden, A. H. Gordon and A. J. P. Martin, *Transactions Intern. Congress of Pure and Applied Chemistry*, London, 1947.
624 R. Consden, A. H. Gordon, A. J. P. Martin, O. Rosenheim and R. L. M. Synge, *Biochem. J.*, 1945, *39*, 251.
625 R. Consden, A. H. Gordon, A. J. P. Martin and R. L. M. Synge, *Biochem. J.*, 1947, *41*, 596.
626 R. Consden and W. M. Stanier, *Nature*, 1952, *169*, 783.
627 A. H. Cook, *Biochem. J.*, 1942, *36*, xxiii.
628 A. H. Cook, S. F. Cox and T. H. Farmer, *Nature*, 1948, *162*, 61.
629 A. H. Cook and D. G. Jones, *J. Chem. Soc.*, *1939*, 1309.
630 A. H. Cook and A. L. Levy, *J. Chem. Soc.*, *1950*, 646.
631 A. H. Cook and A. L. Levy, *J. Chem. Soc.*, *1950*, 651.
632 E. R. Cook, S. R. Stitch, A. E. Hall and M. P. Feldman, *Analyst*, 1954, *79*, 24.
633 G. Cooley, B. Ellis and V. Petrow, *J. Pharm. and Pharmacol.*, 1950, *2*, 128.
634 M. L. Cooley, J. B. Christiansen and C. H. Schroeder, *Ind. Eng. Chem. Anal. Ed.*, 1945, *17*, 689.
635 J. W. Cornforth, C. E. Dalgliesh and A. Neuberger, *Biochem. J.*, 1951, *48*, 598.
636 P. Correale and I. Cortese, *Naturwissenschaften*, 1953, *40*, 57.
637 J. N. Counsell, L. Hough and W. H. Wadman, *Research*, 1951, *4*, 143.
638 L. Coutier, H. Andre and J. Prat, *Chim. anal.*, 1949, *31*, 201.
639 R. W. Cowgill, *Biochim. Biophys. Acta*, 1955, *16*, 614.
640 R. I. Cox and G. F. Marrian, *Biochem. J.*, 1953, *54*, 353.
641 L. C. Craig, *Anal. Chem.*, 1950, *22*, 1346.
642 P. N. Craig, *Ann. N.Y. Acad. Sci.*, 1953, *57*, 67.
643 F. Cramer, *Papierchromatographie*. Monographien zu "Angewandte Chemie" und "Chemie-Ingenieur Technik" No. 64. *3rd Ed.*, 136 pp., Verlag Chemie, Weinheim, 1954.
644 J. L. Crammer, *Nature*, 1948, *161*, 349.
645 C. F. Crampton, S. Moore and W. H. Stein, *J. Biol. Chem.*, 1955, *215*, 787.
646 R. K. Crane and F. Lipmann, *J. Biol. Chem.*, 1953, *201*, 235.
647 H. A. Cranston ans J. B. Thompson, *Ind. Eng. Chem. Anal. Ed.*, 1946, *18*, 323.
648 M. R. Craw and M. D. Sutherland, *Univ. Queensland Papers Dept. Chem.*, 1948, 1, No. 31: *C.A.*, 1950, *44*, 3945-a.
649 T. B. B. Crawford, *Biochem. J.*, 1951, *48*, 203.
650 E. Cremer and R. Müller, *Mikrochemie ver. Mikrochim. Acta*, 1951, *36/37*, 553.
651 E. Cremer and R. Müller, *Z. Elektrochem.*, 1951, *55*, 217.
652 H. D. Cremer and H. Berger, *Klin. Wochschr.*, 1947, *24–25*, 222: *C.A.*, 1948, *42*, 7814-i.
653 H. D. Cremer and A. Tiselius, *Biochem. Z.*, 1950, *320*, 273.
654 R. J. W. Cremlyn and C. W. Shoppee, *J. Chem. Soc.*, *1954*, 3515.
655 D. W. Criddle and R. L. LeTourneau, *Anal. Chem.*, 1951, *23*, 1620.
656 S. J. Cristol, S. B. Soloway, and H. L. Haller, *J. Am. Chem. Soc.*, 1947, *69*, 510.
657 U. Croatto, *Atti reale I. Veneto di Sci.*, 1942–43, *12*, 2, p. 103.

658   W. M. L. Crombie, R. Comber and S. G. Boatman, *Biochem. J.*, 1955, *59*, 309.
659   E. M. Crook and S. P. Datta, *Chem. and Ind.*, *1951*, 718.
660   E. M. Crook, H. Harris, F. Hassan and F. L. Warren, *Biochem. J.*, 1954, *56*, 434.
661   F. R. Cropper, *Analyst*, 1946, *71*, 263.
662   F. R. Cropper and A. Heywood, *Nature*, 1954, *174*, 1063.
663   M. O'L. Crowe, *Ind. Eng. Chem. Anal. Ed.*, 1941, *13*, 845.
664   J. P. Crowther, *Anal. Chem.*, 1954, *26*, 1383.
665   J. P. Crowther, *Nature*, 1954, *173*, 486.
666   H. R. Crumpler and C. E. Dent, *Nature*, 1949, *164*, 441.
667   H. R. Crumpler, C. E. Dent, H. Harris and R. G. Westall, *Nature*, 1951, *167*, 307.
668   K. Cruse and R. Mittag, *Z. anal. Chem.*, 1950, *131*, 273.
669   K. Cruse and R. Mittag, *Z. Elektrochem.*, 1950, *54*, 418.
670   E. Cruz-Coke, F. Gonzales and W. Hulsen, *Science*, 1945, *101*, 340.
671   G. Csobán, *Magyar Kém. Folyóirat*, 1950, *56*, 449: *C.A.*, 1952, *46*, 1384-g.
672   F. W. Cuckow, R. J. C. Harris and F. E. Speed, *J. Soc. Chem. Ind.*, 1949, *68*, 208.
673   L. S. Cuendet, R. Montgomery and F. Smith, *J. Am. Chem. Soc.*, 1953, *75*, 2764.
674   C. C. J. Culvenor, L. J. Drummond and J. R. Price, *Austral. J. Chem.*, 1954, *7*, 277: *C.A.*, 1955, *49*, 8998-h.
675   J. G. Cuninghame, M. L. Sizeland, H. H. Willis, J. Eakins and E. R. Mercer, *J. of Inorg. & Nuclear Chem.*, 1955, *1*, 163.
676   A. S. Curry, *Nature*, 1953, *171*, 1026.
677   A. S. Curry, *J. Criminal Law, Criminol. Police Sci.*, 1954, *44*, 787: *C.A.*, 1955, *49*, 12175-f.
678   A. S. Curry and H. Powell, *Nature*, 1954, *173*, 1143.
679   R. Curti and U. Colombo, *J. Am. Chem. Soc.*, 1952, *74*, 3961.
680   G. Curzon and J. Giltrow, *Nature*, 1954, *173*, 314.
681   W. F. J. Cuthbertson and D. M. Ireland, *Biochem. J.*, 1952, *52*, xxxiv.
682   K. Dakshinamurti, *Current Sci. (India)*, 1954, *23*, 89: *C.A.*, 1954, *48*, 10820-h.
683   G. F. D'Alelio, *U.S. Pat.* 2,366,007 (1945): *C.A.*, 1945, *39*, 4418-3.
684   C. E. Dalgliesh, *Nature*, 1950, *166*, 1076.
685   C. E. Dalgliesh, *J. Chem. Soc.*, *1952*, 3940.
686   C. E. Dalgliesh, *J. Chem. Soc.*, *1952*, 3943.
686a C. E. Dalgliesh, *J. Clin. Pathol.*, 1955, *8*, 73.
687   C. E. Dalgliesh, A. W. Johnson, A. R. Todd and L. V. Vining, *J. Chem. Soc.*, *1950*, 2946.
688   S. Dal Nogare, *Anal. Chem.*, 1953, *25*, 1874.
689   M. M. Daly, V. G. Allfrey and A. E. Mirsky, *J. Gen. Physiol.*, 1950, *33*, 497.
690   M. M. Daly and A. E. Mirsky, *J. Biol. Chem.*, 1949, *179*, 981.
691   C. P. H. Dam and J. P. J. Glavind, *Danish Pat.* 57,897: *C.A.*, 1946, *40*, 3235-9.
692   H. Dam and L. Lewis, *Biochem. J.*, 1937, *31*, 17.
693   D. Daniel, E. Lederer and L. Velluz, *Bull. soc. chim. biol.*, 1945, *27*, 218.
694   C. E. Danielsson, *Arkiv Kemi*, 1952, *5*, 173.
695   M. Danilovič, *Acta Pharm. Jugoslav.*, 1953, *3*, 219: *C.A.*, 1954, *48*, 10300-b.
696   J. D'Ans, E. Blasius, H. Guzatis and U. Wachtel, *Chem.-Ztg.*, 1952, *76*, 811, 841: *C.A.*, 1953, *47*, 3169-i and 4781-a.
697   J. D'Ans, E. Blasius, H. Guzatis and U. Wachtel, *Chem.-Ztg.*, 1952, *76*, 841: *C.A.*, 1953, *47*, 4781-a.

698   J. D'Ans, G. Heinrich and D. Jänchen, *Chem.-Ztg.*, 1953, *77*, 240: *C.A.*, 1953, *47*, 8460-e.

699   P. Darmon and D. Faucquembergue, *Ann. pharm. franç.*, 1954, *12*, 766.

700   W. Dasler and C. D. Bauer, *Ind. Eng. Chem. Anal. Ed.*, 1946, *18*, 52.

701   S. P. Datta, C. E. Dent and H. Harris, *Biochem. J.*, 1950, *46*, xlii.

702   S. P. Datta, C. E. Dent and H. Harris, *Science*, 1950, *112*, 621.

703   S. P. Datta and H. Harris, *Nature*, 1951, *168*, 296.

704   S. P. Datta and H. Harris, *Ann. Eugenics*, 1953, *18*, 107.

705   S. P. Datta, H. Harris and K. R. Rees, *Biochem. J.*, 1950, *46*, xxxvi.

706   S. P. Datta and B. G. Overell, *Biochem. J.*, 1949, *44*, xliii.

707   S. P. Datta, B. G. Overell and M. Stack-Dunne, *Nature*, 1949, *164*, 673.

708   W. G. Dauben, H. L. Bradlow, N. K. Freeman, D. Kritchevsky and M. R. Kirk, *J. Am. Chem. Soc.*, 1952, *74*, 4321.

709   M. Dautrevaux and G. Biserte, *Compt. rend.*, 1955, *240*, 1153.

710   J. B. Davenport and M. D. Sutherland, *Univ. Queensland Papers, Dept. Chem.*, 1950, *1*, No. 39: *C.A.*, 1951, *45*, 2910.

711   C. W. Davies, *Biochem. J.*, 1949, *45*, 38.

712   C. W. Davies, *Research*, 1950, *3*, 447.

713   C. W. Davies, R. B. Hughes and S. M. Partridge, *J. Chem. Soc.*, 1950, 2285.

714   C. W. Davies and G. G. Thomas, *J. Chem. Soc.*, *1951*, 2624.

715   R. L. Davies, *Nature*, 1951, *168*, 834.

716   R. B. Davis, J. M. McMahon and G. Kalnitsky, *J. Am. Chem. Soc.*, 1952, *74*, 4483.

717   H. O. Day Jr., J. S. Gill, E. V. Jones and W. L. Marshall, *Anal. Chem.*, 1954, *26*, 611.

718   J. A. Dean, *Anal. Chem.*, 1951, *23*, 202.

719   J. A. Dean, *Anal. Chem.*, 1951, *23*, 1096.

719a R. G. De Carvalho, paper read to the International Congress of Analytical Chemistry at Lisbon (1956).

720   R. G. De Carvalho and M. Lederer, *Anal. Chim. Acta*, 1955, *13*, 437.

721   P. Decker, *Z. physiol. Chem.*, 1955, *300*, 245.

722   P. Decker, W. Riffart and G. Oberneder, *Naturwissenschaften*, 1951, *38*, 288.

723   R. Dedonder, *Bull. soc. chim. France*, *1952*, 874.

724   R. Dedonder, *Bull. soc. chim. biol.*, 1952, *34*, 144, 157, 171.

725   R. C. DeGeiso, W. Rieman III and S. Lindenbaum, *Anal. Chem.*, 1954, *26*, 1840.

726   V. R. Deitz, *Ann. N.Y. Acad. Sci.*, 1948, *49*, 315.

727   C. A. Dekker and D. T. Elmore, *J. Chem. Soc.*, *1951*, 2864.

728   C. A. Dekker and A. R. Todd, *Nature*, 1950, *166*, 557.

729   J. De Ley, *Enzymologia*, 1954, *17*, 55.

730   J. De Ley, *Naturwissenschaften*, 1955, *42*, 96.

731   P. J. Delmon, *Nature*, 1954, *174*, 755.

732   W. S. DeLoach and C. Drinkard, *J. Chem. Educ.*, 1951, *28*, 461.

733   H. Demarteau, *Compt. rend.*, 1951, *232*, 2494.

734   P. De Moerloose, *Mededel. Vlaam. Chem. Ver.*, 1953, *15*, 13; *C.A.*, 1953, *47*, 6604-i.

735   D. von Denffer, M. Behrens and A. Fischer, *Naturwissenschaften*, 1952, *39*, 258.

736   D. von Denffer, M. Behrens and A. Fischer, *Naturwissenschaften*, 1952, *39*, 550.

737   D. von Denffer and A. Fischer, *Naturwissenschaften*, 1952, *39*, 549.

738   F. W. Denison Jr. and E. F. Phares, *Anal. Chem.*, 1952, *24*, 1628.

739 A. Denoël, F. Jaminet, E. Philippot and M. J. Dallemagne, *Arch. Intern. Physiol.*, 1951, *59*, 341 : *C.A.*, 1953, *47*, 5627-f.
740 C. E. Dent, *Biochem. J.*, 1947, *41*, 240.
741 C. E. Dent, *Biochem. J.*, 1948, *43*, 169.
742 C. E. Dent and J. A. Schilling, *Biochem. J.*, 1949, *44*, 318.
743 C. E. Dent, W. Stepka and F. C. Steward, *Nature*, 1947, *160*, 682.
744 C. E. Dent and J. M. Walshe, *Brit. Med. Bull.*, 1954, *10*, 247.
745 Department of Scientific and Industrial Research, *Report of the Chemistry Research Board, 1947*, p. 32 (1949).
746 P. L. De Reeder, *Anal. Chim. Acta*, 1953, *8*, 325.
747 J. De Repentigny and A. T. James, *Nature*, 1954, *174*, 927.
748 R. Derungs and H. Deuel, *Helv. Chim. Acta*, 1954, *37*, 657.
749 J. Deshusses and P. Desbaumes, *Mitt. Gebiete Lebensm. u. Hyg.*, 1953, *44*, 500 : *C.A.*, 1954, *48*, 6079-f.
750 P. Desuelle and C. Fabre, *Biochim. Biophys. Acta*, 1955, *18*, 49.
751 P. Desnuelle, M. Naudet and J. Rouzier, *Biochim. Biophys. Acta*, 1948, *2*, 561.
752 P. Desnuelle, M. Rovery and G. Bonjour, *Biochim. Biophys. Acta*, 1950, *5*, 116.
753 V.Desreux, *Rec. trav. chim.*, 1949, *68*, 789.
754 H. Deuel and F. Hostettler, *Experientia*, 1950, *6*, 445.
755 H. Deuel, J. Solms and L. Anyas-Weisz, *Helv. Chim. Acta*, 1950, *33*, 2171.
756 H. Deuel, J. Solms, L. Anyas-Weisz and G. Huber, *Helv. Chim. Acta*, 1951, *34*, 1849
757 A. Deutsch and R. Nilsson, *Acta Chem. Scand.*, 1953, *7*, 858.
758 A. Deutsch and R. Nilsson, *Acta Chem. Scand.*, 1953, *7*, 1288.
759 D. DeVault, *J. Am. Chem. Soc.*, 1943, *65*, 532.
760 J. E. DeVay, W. H. Chang and R. L. Hossfeld, *J. Am. Chem. Soc.*, 1951, *73*, 4977.
761 C.-H. De Verdier, *Nature*, 1952, *170*, 804.
762 C.-H. De Verdier and G. Ågren, *Acta Chem. Scand.*, 1948, *2*, 783.
763 C.-H. De Verdier and C. I. Sjöberg, *Acta Chem. Scand.*, 1954, *8*, 1161.
764 G. De Vries, *Nature*, 1954, *173*, 735.
765 G. De Vries and E. Van Dalen, *Rec. trav. chim.*, 1954, *73*, 1028.
766 J. De Wael and R. Diaz Cadavieco, *Rec. trav. chim.*, 1954, *73*, 333.
767 J. B. DeWitt and M. X. Sullivan, *Ind. Eng. Chem. Anal. Ed.*, 1946, *18*, 117.
768 C. Dhéré, *Candollea (Genève)*, 1943, *10*, 23–73.
769 R. M. Diamond, K. Street Jr. and G. T. Seaborg, *Report UCRL 1034 Dec.* (1950).
770 R. M. Diamond, K. Street Jr. and G. T. Seaborg, *J. Am. Chem. Soc.*, 1954, *76*, 1461.
771 E. E. Dickey, *J. Chem. Educ.*, 1953, *30*, 525 : *C.A.*, 1954, *48*, 3158-h.
772 E. E. Dickey and M. L. Wolfrom, *J. Am. Chem. Soc.*, 1949, *71*, 825.
773 F. H. Dickey, *Proc. Nat. Acad. Sci.*, 1949, *35*, 229.
773a F. H. Dickey, *J. Phys. Chem.*, 1955, *59*, 695.
774 J. W. Dieckert and R. Reiser, *Science*, 1954, *120*, 678.
775 W. Dihlmann, *Naturwissenschaften*, 1953, *40*, 510.
776 W. Dihlmann, *Biochem. Z.*, 1954, *325*, 295.
776a S. Dikstein, F. Bergmann and M. Chaimovitz, *J. Biol. Chem.*, 1956, *221*, 239.
777 H. Diller and O. Rex, *Z. anal. Chem.*, 1952/53, *137*, 241.
778 E. Dimant, D. R. Sanadi and F. M. Huennekens, *J. Am. Chem. Soc.*, 1952, *74*, 5440.
779 R. J. Dimler, W. C. Schaefer, C. S. Wise and C. E. Rist, *Anal. Chem.*, 1952, *24*, 1411.

780 R. J. Dimler, J. W. Van Cleve, E. M. Montgomery, L. R. Bair, F. J. Castle and J. A. Whitehead, *Anal. Chem.*, 1953, *25*, 1428.
781 G. Di Modica and E. Angeletti, *La Ricerca scientifica*, 1952, *22*, 715: *C.A.*, 1953, *47*, 6918-d.
782 G. Di Modica and P. F. Rossi, *Chim. anal.*, 1952, *34*, 271: *C.A.*, 1953, *47*, 1543-i.
783 G. Di Modica and C. Spriano, *Ann. chim. (Roma)*, 1951, *41*, 64: *C.A.*, 1952, *46*, 426-d.
784 E. Dingemanse, L. G. Huis in 't Veld and S. L. Hartogh-Katz, *J. Clin. Endocrinol. and Metabolism*, 1952, *12*, 66: *C.A.*, 1953, *47*, 3381-e.
785 E. Dingemanse, L. G. Huis in 't Veld and B. M. De Laat, *J. Clin. Endocrin.*, 1946, *6*, 535.
786 G. U. Dinneen, C. J. Thompson, J. R. Smith and J. S. Ball, *Anal. Chem.*, 1950, *22*, 871.
787 Distillation Products Inc., *Brit. Pat.* 621,412 (1949): *C.A.*, 1949, *43*, 6657-a.
788 A. S. J. Dixon, *Biochem. J.*, 1955, *60*, 165.
789 H. B. F. Dixon, S. Moore, M. P. Stack-Dunne and F. G. Young, *Nature*, 1951, *168*, 1044.
790 Z. Dizdar, *Rec. trav. inst. recherches structure matière*, 1953, *2*, 77: *C.A.*, 1953, *47*, 6817.
791 K. M. Dzhemukhadze and G. A. Shal'neva, *Biokhimia*, 1955, *20*, 336: *C.A.*, 1955, *49*, 16084-h.
792 R. Djurfelt, J. Hansen and O. Samuelson, *Svensk Kem. Tid.*, 1947, *59*, 13: *C.A.*, 1947, *41*, 3009-b.
793 R. Djurfelt and O. Samuelson, *Acta Chem. Scand.*, 1950, *4*, 165.
794 K. Dobriner, S. Lieberman and C. P. Rhoads, *J. Biol. Chem.*, 1948, *172*, 241.
795 M. S. Dobro and S. Kusafuka, *J. Criminal Law, Criminol. Police Sci.*, 1953, *44*, 247: *C.A.*, 1954, *48*, 3637-h.
796 B. M. Dobyns and S. R. Barry, *J. Biol. Chem.*, 1953, *204*, 517.
797 K. S. Dodgson, J. Pryde and A. L. Sims, *Biochem. J.*, 1950, *47*, xxiii.
798 G. F. Doebbler, *Texas J. Sci.*, 1953, *5*, 443: *C.A.*, 1954, *48*, 5920-h.
799 M. Dohnal and J. Bialá, *Chem. Listy*, 1954, *48*, 1261: *C.A.*, 1954, *48*, 14116-f.
800 D. Dolar and Z. Draganic, *Rec. trav. inst. recherches structure matière*, 1953, *2*, 77: *C.A.*, 1953, *47*, 6816.
801 N. G. Doman and Z. S. Kagan, *Biokhimiya*, 1952, *17*, 719: *C.A.*, 1953, *47*, 4795-f.
802 K. O. Donaldson, V. J. Tulane and L. M. Marshall, *Anal. Chem.*, 1952, *24*, 185.
803 G. D. Dorough and D. L. Seaton, *J. Am. Chem. Soc.*, 1954, *76*, 2873.
804 C. D. Douglass, W. L. Howard and S. H. Wender, *J. Am. Chem. Soc.*, 1949, *71*, 2658.
805 R. Douin, *Rev. gen. botanique*, 1953, *60*, 777.
806 L. Douste-Blazy, J. Polonovski and P. Valdiguié, *Compt. rend.*, 1952, *235*, 1643: *C.A.*, 1953, *47*, 5981-a.
807 Dow Chemical Company, Midland, Mich., U.S.A., *Dowex Fine Mesh Resins*, Publication No. 2, 1954.
808 Dow Chemical Company, Midland, Mich., U.S.A., *Ion Exclusion*, Technical Service and Development Pamphlet, 1952.
809 M. W. Dowler and D. H. Laughland, *Anal. Chem.*, 1952, *24*, 1047.
810 Y. P. Dowmont and J. S. Fruton, *J. Biol. Chem.*, 1952, *197*, 271.
811 B. Drake, *Nature*, 1947, *160*, 602.

812 B. Drake, *Anal. Chim. Acta*, 1949, *3*, 452.
813 B. Drake, *Acta Chem. Scand.*, 1950, *4*, 554.
814 B. Drake and S. Gardell, *Arkiv Kemi*, 1952, *4*, 469.
814a B. Drake, *Arkiv Kemi*, 1955, *8*, 1.
815 N. A. Drake, *J. Am. Chem. Soc.*, 1950, *72*, 3803.
816 O. J. Draper and A. L. Pollard, *Science*, 1949, *109*, 448.
817 W. Drell, *J. Am. Chem. Soc.*, 1953, *75*, 2506.
818 F. D. Drew, L. M. Marshall and F. Friedberg, *J. Am. Chem. Soc.*, 1952, *74*, 1852.
819 R. E. A. Drey and G. E. Foster, *J. Pharm. and Pharmacol.*, 1953, *5*, 839: *C.A.*, 1954, *48*, 1627-h.
820 A. Drèze and A. De Boeck, *Arch. Intern. Physiol.*, 1952, *60*, 201.
821 A. Drèze, S. Moore and E. J. Bigwood, *Anal. Chim. Acta*, 1954, *11*, 554.
822 J. W. Dubnoff, *J. Biol. Chem.*, 1941, *141*, 711.
823 M. Dubois, K. Gilles, J. K. Hamilton, P. A. Rebbers and F. Smith, *Nature*, 1951, *168*, 167.
824 R. Dubrisay, *Compt. rend.*, 1947, *225*, 300.
825 R. B. Duff and D. J. Eastwood, *Nature*, 1950, *165*, 848.
826 L. R. Dugan Jr., B. W. Beadle and A. S. Henick, *J. Am. Oil Chemists' Soc.*, 1948, *25*, 153: *C.A.* 1948, *42*, 6142-d.
827 J. Dulberg, W. G. Roessler, T. H. Sanders and C. R. Brewer, *J. Biol. Chem.*, 1952, *194*, 199.
828 C. Dumazert and M. Bozzi-Tichadou, *Bull. soc. chim. biol.*, 1955, *37*, 169.
829 J. E. Dunabin, H. Mason, A. P. Seyfang and F. J. Woodman, *Nature*, 1949, *164*, 916.
830 J. F. Duncan and B. A. J. Lister, *J. Chem. Soc.*, *1949*, 3285.
831 J. F. Duncan and B. A. J. Lister, *Discuss. Faraday Soc.*, 1949, *7*, 104.
832 G. Dupont, R. Dulou and M. Vilkas, *Bull. soc. chim. France*, *1948*, 785.
833 P. Dupuy and J. Puisais, *Compt. rend.*, 1955, *240*, 1802.
834 J. A. Durant, *Nature*, 1952, *169*, 1062.
835 S. V. Durmishidze and N. N. Nutsubidze, *Doklady Akad. Nauk S.S.S.R.*, 1954, *96*, 1197: *C.A.*, 1954, *48*, 13831-g.
836 G. Durr, *Compt. rend.*, 1952, *235*, 1314.
837 E. L. Durrum, *J. Am. Chem. Soc.*, 1950, *72*, 2943.
838 D. F. Durso, E. D. Schall and R. L. Whistler, *Anal. Chem.*, 1951, *23*, 425.
839 R. Duschinsky and E. Lederer, *Bull. soc. chim. biol.*, 1935, *17*, 1534.
840 J. P. Dustin, *Ind. chim. belge*, 1952, *17*, 257: *C.A.*, 1952, *46*, 6867-a.
841 N. L. Dutta, *Nature*, 1955, *175*, 85.
842 H. J. Dutton, *J. Phys. Chem.*, 1944, *48*, 179.
843 H. J. Dutton and C. L. Reinbold, *J. Am. Oil Chemists' Soc.*, 1948, *25*, 120: *C.A.*, 1948, *42*, 3973-f.
844 F. P. Dwyer, N. S. Gill, E. C. Gyarfas and F. Lyons, *Australian J. Sci.*, 1950, *13*, 52.
845 F. Dybing, O. Dybing and K. B. Jensen, *Acta Pharmacol. et Toxicol.*, 1951, *7*, 337.
846 F. Dybing, O. Dybing and K. B. Jensen, *Acta Pharmacol. et Toxicol.*, 1954, *10*, 93.
847 J. Dykyj and J. Černý, *Chem. Listy*, 1945, *39*, 84: *C.A.*, 1950, *44*, 9868-h.
848 D. Dyrssen, *Svensk Kem. Tid.*, 1950, *62*, 153: *C.A.*, 1951, *45*, 960-b.
849 M. Eastham, *Biochem. J.*, 1949, *45*, xiii.
850 J. E. Eastoe, *Nature*, 1954, *173*, 540.
851 J. E. Eastoe and A. W. Kenchington, *Nature*, 1954, *174*, 966.
852 J. P. Ebel, *Thèse de doctorat*, Strasbourg (1951).

853   J. P. Ebel, *Bull. soc. chim. biol.*, 1952, *34*, 321.
854   J. P. Ebel, *Bull. soc. chim. France*, 1953, 991, 998, 1089 and 1096.
855   J. P. Ebel, *Mikrochim. Acta*, 1954, 679.
856   J. P. Ebel and Y. Volmar, *Compt. rend.*, 1951, *233*, 415.
857   H. Eberle, *Naturwissenschaften*, 1954, *41*, 479.
858   J. Edelman and J. S. D. Bacon, *Biochem. J.*, 1951, *49*, 529.
859   J. Edelman and R. V. Martin, *Biochem. J.*, 1952, *50*, xxi.
860   E. Eden, *Biochem. J.*, 1950, *46*, 259.
861   D. G. Edgar, *Biochem. J.*, 1953, *54*, 50.
862   D. G. Edgar, *J. Endocrinol.*, 1953, *10*, 54: *C.A.*, 1954, *48*, 5976-b.
863   P. Edman, *Arkiv Kemi, Mineral., Geol.*, 1945, *22*A, No. 3.
864   P. Edman, *Acta Chem. Scand.*, 1948, *2*, 592.
865   P. Edman, *Acta Chem. Scand.*, 1950, *4*, 283.
866   P. Edman, E. Hammarsten, B. Löw and P. Reichard, *J. Biol. Chem.*, 1949, *178*, 395.
867   J.-E. Edström, *Nature*, 1951, *168*, 876.
868   J.-E. Edström, *Biochim. Biophys. Acta*, 1952, *9*, 528.
869   J. T. Edward and S. Nielsen, *Chem. and Ind.*, 1953, 197.
870   J. T. Edward and D. M. Waldron, *J. Chem. Soc.*, 1952, 3631.
871   P. W. R. Eggitt and L. D. Ward, *J. Sci. Food. Agr.*, 1953, *4*, 177: *C.A.*, 1953, *47*, 5982-b.
872   L. V. Eggleston and R. Hems, *Biochem. J.*, 1952, *52*, 156.
873   H. Egner, E. Erikson and A. Emanuelsson, *K. Lantbruks Högskol. Ann.*, 1949, *16*, 593: *C.A.*, 1950, *44*, 385-c.
874   H. C. Ehrmantraut and A. Weinstock, *Biochim. Biophys. Acta*, 1954, *15*, 589.
875   E. Ekedahl, E. Högfeldt and L. G. Sillén, *Nature*, 1950, *166*, 723.
876   B. Ekman, *Acta Chem. Scand.*, 1948, *2*, 383.
877   T. Ekstrand, *Acta Chem. Scand.*, 1954, *8*, 1099.
877a I. I. M. Elbeih, *Chemist Analyst*, 1955, *44*, 20.
878   I. I. M. Elbeih, J. F. W. McOmie and F. H. Pollard, *Discuss. Faraday Soc.*, 1949, *7*, 183.
879   D. Elenkov, *Bull. Inst. Chim. Acad. Bulgare Sci.*, 1953, *2*, 177: *C.A.*, 1955, *49*, 5379-f.
880   M. F. S. El Hawary and R. H. S. Thompson, *Biochem. J.*, 1953, *53*, 340.
881   M. F. S. El Hawary and R. H. S. Thompson, *Biochem. J.*, 1954, *58*, 518.
882   J. Elks, R. M. Evans, A. G. Long and G. H. Thomas, *J. Chem. Soc.*, 1954, 451.
883   N. Ellfolk and R. L. M. Synge, *Biochem. J.*, 1955, *59*, 523.
884   P. Ellinger and R. A. Coulson, *Nature*, 1943, *152*, 383.
885   B. Ellis, V. Petrow and G. F. Snook, *J. Pharm. and Pharmacol.*, 1949, *1*, 950: *C.A.*, 1950, *44*, 1972-i.
886   W. H. Ellis and R. L. LeTourneau, *Anal. Chem.*, 1953, *25*, 1269.
887   D. T. Elmore, *Nature*, 1948, *161*, 931.
888   D. T. Elmore, *J. Chem. Soc.*, 1950, 2094.
889   S. R. Elsden, *Biochem. J.*, 1946, *40*, 252.
890   K. El-Shazly, *Biochem. J.*, 1952, *51*, 640.
891   J. A. Elvidge and M. Whalley, *Chem. and Ind.*, 1955, 589.
892   N. D. Embree and O. Hawks, *U.S. Pat. Appl.* 653,948: *C.A.*, 1951, *45*, 7311-d.
893   N. D. Embree and E. M. Shantz, *J. Biol. Chem.*, 1940, *132*, 619.
894   A. J. Emery Jr. and E. Stotz, *Stain Technol.*, 1952, *27*, 21: *C.A.*, 1952, *46*, 5650-c.

895 A. Emmerie, *Ann. N.T. Acad. Sci.*, 1949/50, *52*, 309.
896 A. Emmerie, *Rec. trav. chim.*, 1953, *72*, 893.
897 A. Emmerie and C. Engel, *Rec. trav. chim.*, 1939, *58*, 283.
898 P. H. Emmett, *Chem. Revs.*, 1948, *43*, 69.
899 D. T. Englis and H. A. Fiess, *Ind. Eng. Chem.*, 1944, *36*, 604.
900 J. English Jr., *J. Am. Chem. Soc.*, 1941, *63*, 941.
901 J. Enselme, R. Creyssel and A. Rapatel, *Bull. soc. chim. biol.*, 1947, *29*, 939.
902 J. Entel, C. H. Ruof and H. C. Howard, *Anal. Chem.*, 1953, *25*, 616.
903 Ya. A. Epshteïn, *Akad. Nauk. S.S.S.R. Otdel. Khim. Nauk, 1950*, 211: *C.A.*, 1954, *48*, 2443.
904 Ya. A. Ephsteïn and M. P. Fomina, *Biokhimiya*, 1950, *15*, 321: *C.A.*, 1950, *44*, 206-i.
905 O. Erämetsä, *Ann. Acad. Sci. Fennicae*, 1940, *57*, 3.
906 O. Erämetsä, *Bull. comm. géol. Finlande*, 1941, *14*, 36: *C.A.*, 1943, *37*, 3316-5.
907 O. Erämetsä, *Suomen Kemistilehti*, 1943, *16B*, 13: *C.A.*, 1946, *40*, 4620-8.
908 O. Erämetsä, Th. G. Sahama and V. Kanula, *Ann. Acad. Sci. Fennicae*, 1941, *A57*, Nos. 3, 5–20: *C.A.*, 1944, *38*, 4490-5.
909 H. Erbring and P. Patt, *Naturwissenschaften*, 1954, *41*, 216.
909a B. Erdem, *Rev. fac. sci. univ. Istanbul*, 1955, *20*, 332 and 346.
910 B. Erdem and H. Erlenmeyer, *Helv. Chim. Acta*, 1954, *37*, 2220.
911 B. Erdem, B. Prijs and H. Erlenmeyer, *Helv. Chim. Acta*, 1955, *38*, 267.
912 L. Eriksen, *Scand. J. Clin. & Lab. Invest.*, 1953, *5*, 155: *C.A.*, 1953, *47*, 10603-f.
913 S. Eriksson and J. Sjövall, *Acta Chem. Scand.*, 1954, *8*, 1099.
914 H. Erlenmeyer and H. Dahn, *Helv. Chim. Acta*, 1939, *22*, 1369.
915 H. Erlenmeyer, H. von Hahn and E. Sorkin, *Helv. Chim. Acta*, 1951, *34*, 1419.
916 H. Erlenmeyer and J. Schmidlin, *Helv. Chim. Acta*, 1941, *24*, 1213.
917 H. Erlenmeyer and W. Schoenauer, *Helv. Chim. Acta*, 1941, *24*, 878.
918 K. Erler, *Z. anal. Chem.*, 1949, *129*, 209.
919 K. Erler, *Z. anal. Chem.*, 1950, *131*, 106.
920 V. Erspamer and G. Boretti, *Arch. intern. pharmacodynamie*, 1951, *88*, 296: *C.A.*, 1952, *46*, 2700-h.
921 H. von Euler and A. Fonó, *Arkiv Kemi, Mineral., Geol.*, 1947, *25A*, No. 15: *C.A.*, 1948, *42*, 6870-d.
922 U. S. von Euler and R. Eliasson, *Nature*, 1952, *170*, 664.
923 U. S. von Euler and U. Hamberg, *Nature*, 1949, *163*, 642.
924 U. S. von Euler, U. Hamberg and S. Hellner, *Biochem. J.*, 1951, *49*, 655.
925 J. von Euler, A. Lardon and T. Reichstein, *Helv. Chim. Acta*, 1944, *27*, 1287.
925a D. E. M. Evans and J. C. Tatlow, *J..Chem. Soc., 1955*, 1184.
926 E. E. Evans and J. W. Mehl, *Science*, 1951, *114*, 10.
927 E. E. Evans and K. W. Walls, *J. Bact.*, 1952, *63*, 422.
928 G. G. Evans and W. S. Reith, *Biochem. J.*, 1954, *56*, 111.
929 R. A. Evans, W. H. Parr and W. C. Evans, *Nature*, 1949, *164*, 674.
930 D. T. Ewing, T. D. Schlabach, M. J. Powell, J. W. Vaitkus and O. D. Bird, *Anal. Chem.*, 1954, *26*, 1406.
931 M. Ezrin and H. G. Cassidy, *Ann. N.T. Acad. Sci.*, 1953, *57*, 79.
932 D. Fairbairn and R. P. Harpur, *Nature*, 1950, *166*, 789.
933 J. E. Falk, *Brit. Med. Bull.*, 1954, *10*, 211.
934 J. E. Falk and A. Benson, *Biochem. J.*, 1953, *55*, 101.
935 J. E. Falk and A. Benson, *Arch. Biochem. Biophys.*, 1954, *51*, 528.

936 K. H. Fantes, J. E. Page, L. F. J. Parker and E. Lester Smith, *Proc. Roy. Soc. London*, 1950, *B 136*, 592.

937 J. Farradane, *Nature*, 1951, *167*, 120.

938 G. V. Fedorova, *Izvest. V.T.I.*, 1946, *15*, No. 2, 28; *C.A.*, 1946, *40*, 5509-6.

939 P. Feigelson, J. N. Williams Jr. and C. A. Elvehjem, *J. Biol. Chem.*, 1950, *185*, 741.

940 F. Feigl, *Spot Tests*, 2 vols., 5th Ed., Elsevier, Amsterdam, 1956.

941 L. I. Feldman and I. G. Gunsalus, *J. Biol. Chem.*, 1950, *187*, 821.

942 K. Felix and A. Krekels, *Z. physiol. Chem.*, 1952, *290*, 78.

943 I. G. Fels and A. Tiselius, *Arkiv Kemi*, 1951, *3*, 369: *C.A.*, 1952, *46*, 2692-i.

944 Q. Fernando and J. P. Philips, *Anal. Chem.*, 1953, *25*, 819.

945 M. E. Fewster ans D. H. Hall, *Nature*, 1951, *168*, 78.

946 L. F. Fieser, *J. Am. Chem. Soc.*, 1951, *73*, 5007.

947 L. F. Fieser, *J. Am. Chem. Soc.*, 1953, *75*, 4377.

949 L. F. Fieser and R. Stevenson, *J. Am. Chem. Soc.*, 1954, *76*, 1728.

950 D. L. Fillerup and J. F. Mead, *Proc. Soc. Exptl. Biol. Med.*, 1953, *83*, 574: *C.A.*, 1953, *47*, 12480-a.

951 H. G. Fillinger, *J. Chem. Educ.*, 1947, *24*, 444.

952 D. F. Fink, R. W. Lewis and F. T. Weiss, *Anal. Chem.*, 1950, *22*, 850.

953 D. F. Fink, R. W. Lewis and F. T. Weiss, *Anal. Chem.*, 1950, *22*, 858.

954 K. Fink and R. M. Fink, *Proc. Soc. Exptl. Biol. Med.*, 1949, *70*, 654.

955 K. Fink, R. B. Henderson and R. M. Fink, *Proc. Soc. Exptl. Biol. Med.*, 1951, *78*, 135.

956 K. Fink, R. B. Henderson and R. M. Fink, *J. Biol. Chem.*, 1952, *197*, 441.

957 R. M. Fink, C. F. Dent and K. Fink, *Nature*, 1947, *160*, 801.

958 H. Fischbach, T. E. Eble and M. Mundell, *J. Am. Pharm. Ass.*, 1947, *36*, 220.

959 H. Fischbach and J. Levine, *Science*, 1955, *121*, 602.

960 H. Fischbach, M. Mundell and T. E. Eble, *Science*, 1946, *104*, 84.

961 A. Fischer, *Planta*, 1953/54, *43*, 288: *C.A.* 1954, *48*, 7097-f.

962 A. Fischer and M. Behrens, *Z. physiol. Chem.*, 1952, *291*, 14.

963 A. Fischer and M. Behrens, *Z. physiol. Chem.*, 1952, *291*, 242.

964 E. H. Fischer and W. Settele, *Helv. Chim. Acta*, 1953, *36*, 811.

965 F. G. Fischer and H. Dörfel, *Biochem. Z.*, 1953, *324*, 544.

966 F. G. Fischer and H. Dörfel, *Z. physiol. Chem.*, 1954, *297*, 164.

967 H. Fischer and M. Conrad, *Ann.*, 1939, *538*, 143.

968 H. Fischer and H. Gibian, *Ann.*, 1942, *550*, 208, *552*, 153.

969 H. Fischer, A. Krekels and O. Hövels, *Klin. Wochschr.*, 1952, *30*, 137: *C.A.*, 1952, *46*, 5114-h.

970 P. Fischer Jørgensen, *Dansk Tids. Farm.*, 1950, *24*, 1: *C.A.*, 1950, *44*, 2893-e.

970a W. Fischer and K. E. Niemann, *Z. anorg. u. allgem. Chem.*, 1956, *283*, 96.

971 R. B. Fisher and R. Holmes, *Biochem. J.*, 1949, *44*, liv.

972 R. B. Fisher, D. S. Parsons and R. Holmes, *Nature*, 1949, *164*, 183.

973 R. B. Fisher, D. S. Parsons and G. A. Morrison, *Nature*, 1948, *161*, 764.

974 N. R. Fisk, *Paint Technol.*, 1945, *10*, 85.

975 F. T. Fitch and D. S. Russell, *Can. J. Chem.*, 1951, *29*, 363.

976 F. T. Fitch and D. S. Russell, *Anal. Chem.*, 1951, *23*, 1469.

977 J. F. Flagg and W. E. Bleidner, *J. Chem. Phys.*, 1945, *13*, 269.

978 M. Flavin, *J. Biol. Chem.*, 1954, *210*, 771.

979 M. Flavin, *Nature*, 1954, *173*, 214.

980 M. Flavin and C. B. Anfinsen, *J. Biol. Chem.*, 1954, *211*, 375.

981 C. M. Fletcher, A. G. Lowther and W. S. Reith, *Biochem. J.*, 1954, *56*, 106.
982 E. Fletcher and F. H. Malpress, *Nature*, 1953, *171*, 838.
983 A. E. Flood, E. L. Hirst and J. K. N. Jones, *Nature*, 1947, *160*, 86; *J. Chem. Soc.*, *1948*, 1679.
984 H. Flood, *Z. anal. Chem.*, 1940, *120*, 237.
985 H. Flood, *Tids. Kjemi og Bergvesen*, 1940, *20*, 111.
986 H. Flood, *Tids. Kjemi Bergvesen Met.*, 1943, *3*, 9: *C.A.*, 1944, *38*, 2895-8.
987 H. Flood, *Discuss. Faraday Soc.*, 1949, *7*, 180.
988 H. Flood and E. Risberg, *Tids. Kjemi Bergvesen Met.*, 1942, *2*, 36: *C.A.*, 1944, *38*, 2586-2.
989 H. Flood and A. Smedsaas, *Tids. Kjemi Bergvesen Met.*, 1941, *1*, 150: *C.A.*, 1943, *37*, 4319-2.
990 H. Flood and A. Smedsaas, *Tids. Kjemi Bergvesen Met.*, 1942, *2*, 1: *C.A.*, 1943, *37*, 4319-4.
991 H. Flood and A. Smedsaas, *Tids. Kjemi Bergvesen Met.*, 1942, *2*, 17: *C.A.*, 1943, *37*, 5334-8.
992 D. Florentin and M. Heros, *Bull. soc. chim. France*, *1947*, 210.
993 H. M. Flowers and W. S. Reith, *Biochem. J.*, 1953, *53*, 657.
994 E. H. Flynn, T. J. Bond, T. J. Bardos and W. Shive, *J. Am. Chem. Soc.*, 1951, *73*, 1979.
995 O. Folin and V. Ciocalteu, *J. Biol. Chem.*, 1927, *73*, 627.
996 K. Folkers and J. Shavel Jr., *J. Am. Chem. Soc.*, 1942, *64*, 1892.
997 K. Folkers and J. Shavel Jr., *U.S. Pat.* 2,573,702 (1951): *C.A.*, 1952, *46*, 693-g.
998 B. F. Folkes, R. A. Grant and J. K. N. Jones, *J. Chem. Soc.*, *1950*, 2136.
999 J. E. Ford, E. S. Holdsworth and S. K. Kon, *Biochem. J.*, 1955, *59*, 86.
1000 G. Foreman, G. W. H. Thompson and J. D. Tolliday, *J. Am. Leather Chemists' Assoc.*, 1950, *45*, 378: *C.A.*, 1950, *44*, 11144-i.
1001 H. S. Forrest and H. K. Mitchell, *J. Am. Chem. Soc.*, 1954, *76*, 5656.
1002 H. S. Forrest and A. R. Todd, *J. Chem. Soc.*, *1950*, 3295.
1003 W. Forsling, *Arkiv Kemi*, 1953, *5*, 489, 503.
1004 D. A. Forss, E. A. Dunstone and W. Stark, *Chem. and Ind.*, *1954*, 1292.
1005 W. G. C. Forsyth, *Nature*, 1948, *161*, 239.
1006 W. G. C. Forsyth, *Biochem. J.*, 1955, *60*, 108.
1007 W. G. C. Forsyth and N. W. Simmonds, *Proc. Roy. Soc. London*, 1954, *B 142*, 549.
1008 L. S. Fosdick and R. Q. Blackwell, *Science*, 1949, *109*, 314.
1009 G. E. Foster, J. Macdonald and T. S. G. Jones, *J. Pharm. and Pharmacol.*, 1949, *1*, 802: *C.A.*, 1950, *44*, 1229-h.
1009a J. Fouarge and G. Duyckaerts, *Anal. Chim. Acta*, 1956, *14*, 527.
1009b R. Fournier, *Rev. mét.*, 1955, *52*, 596.
1010 L. Fowden, *Biochem. J.*, 1951, *48*, 327.
1011 L. Fowden and J. R. Penney, *Nature*, 1950, *165*, 846.
1012 L. Fowden and J. A. Webb, *Biochem. J.*, 1955, *59*, 228.
1013 H. D. Fowler, *Nature*, 1951, *168*, 1123.
1014 J. Franc, *Chem. Listy*, 1954, *48*, 1526: *C.A.*, 1955, *49*, 1484-b.
1015 J. Franc, *Collection Czechoslov. Chem. Communs.*, 1955, *20*, 298.
1016 J. Franc and J. Latinák, *Chem. Listy*, 1955, *49*, 317: *C.A.*, 1955, *49*, 9352-e.
1017 J. Franc and J. Latinák, *Chem. Listy*, 1955, *49*, 325: *C.A.*, 1955, *49*, 9352-e.
1018 J. Franc and J. Latinák, *Chem. Listy*, 1955, *49*, 328: *C.A.*, 1955, *49*, 9352-e.
1019 J. Francis, H. M. Macturk, J. Madinaveitia and G. S. Snow, *Biochem. J.*, 1953, *55*, 596.

1020 G. T. Franglen, *Nature*, 1955, *175*, 134.
1021 H. Frank and H. Petersen, *Z. physiol. Chem.*, 1955, *299*, 1.
1022 M. Frankel and R. Wolovsky, *Experientia*, 1954, *10*, 367.
1023 A. E. Franklin and J. H. Quastel, *Science*, 1949, *110*, 447.
1024 A. E. Franklin and J. H. Quastel, *Proc. Soc. Exptl. Biol. Med.*, 1950, *74*, 803.
1025 A. E. Franklin, J. H. Quastel and S. F. Van Straten, *Nature*, 1951, *168*, 687; *Proc. Soc. Exptl. Biol. Med.*, 1951, *77*, 783.
1026 G. S. Fraps, A. R. Kemmerer and S. M. Greenberg, *Ind. Eng. Chem. Anal. Ed.*, 1940, *12*, 16.
1027 D. Fraser, *Anal. Chem.*, 1954, *26*, 1858.
1028 S. Freed, K. M. Sancier and A. H. Sporer, *J. Am. Chem. Soc.*, 1954, *76*, 6006.
1029 F. Freeman, K. Gardner and D. W. Pound, *J. Appl. Chem.*, 1953, *3*, 160.
1030 J. H. Freeman, *Anal. Chem.*, 1952, *24*, 955.
1031 L. Kh. Freïdlin, L. F. Vereshchagin, I. E. Neïmark, I. U. Numanov and R. Yu. Sheïnfaïn, *Izvest. Akad. Nauk. S.S.S.R. Otdel Khim. Nauk, 1953*, 945: *C.A.*, 1954, *48*, 4929-f.
1032 E. C. Freiling, *J. Am. Chem. Soc.*, 1955, *77*, 2067.
1033 E. C. Freiling and L. R. Bunney, *J. Am. Chem. Soc.*, 1954, *76*, 1021.
1034 D. French and D. W. Knapp, *J. Biol. Chem.*, 1950, *187*, 463.
1035 D. French and G. M. Wild, *J. Am. Chem. Soc.*, 1953, *75*, 2612.
1036 D. French and G. M. Wild, *J. Am. Chem. Soc.*, 1953, *75*, 4490.
1037 K. Freudenberg and L. Hartmann, *Ann.*, 1954, *587*, 207.
1038 K. Freudenberg, H. Walch and H. Molter, *Naturwissenschaften*, 1942, *30*, 87.
1039 R. Fricke and W. Neugebauer, *Naturwissenschaften*, 1950, *37*, 427.
1040 R. Fricke and H. Schmäh, *Z. anorg. Chem.*, 1948, *255*, 253.
1041 F. Friedberg, *Naturwissenschaften*, 1954, *41*, 141.
1042 J. Fried and E. Titus, *J. Am. Chem. Soc.*, 1948, *70*, 3615.
1043 J. Fried, H. L. White and O. Wintersteiner, *J. Am. Chem. Soc.*, 1950, *72*, 4621.
1044 J. Fried and O. Wintersteiner, *Science,* 1945, *101*, 613.
1045 M. Friedkin and D. Roberts, *J. Biol. Chem.*, 1954, *207*, 257.
1046 W. J. Frierson and M. J. Ammons, *J. Chem. Educ.*, 1950, *27*, 37.
1047 W. J. Frierson, S. L. Hood, I. B. Whitney and C. L. Comar, *Arch. Biochem. Biophys.*, 1952, *38*, 397.
1048 W. J. Frierson and J. W. Jones, *Anal. Chem.*, 1951, *23*, 1447.
1049 W. J. Frierson, P. F. Thomason and H. P. Raaen, *Anal. Chem.*, 1954, *26*, 1210.
1050 H. Fritz and A. Bauer, *Chem.-Ing.-Tech.*, 1954, *26*, 609.
1051 L. D. Frizzell, *Ind. Eng. Chem. Anal. Ed.*, 1944, *16*, 615.
1052 C. E. Frohman and J. M. Orten, *J. Biol. Chem.*, 1953, *205*, 717.
1053 C. E. Frohman, J. M. Orten and A. H. Smith, *J. Biol. Chem.*, 1951, *193*, 277.
1054 C. Fromageot, M. Jutisz and E. Lederer, *Biochim. Biophys. Acta*, 1948, *2*, 487.
1055 C. Fromageot, M. Jutisz, D. Meyer and L. Pénasse, *Biochim. Biophys. Acta*, 1950, *6*, 283.
1056 C. Fromageot, M. Jutisz and P. Tessier, *Bull. soc. chim. biol.*, 1949, *31*, 689.
1056a A. J. Fudge and J. L. Woodhead, *Analyst*, 1956, *81*, 417.
1057 R. Fuerst, A. J. Landua and J. Awapara, *Science*, 1950, *111*, 635.
1058 Y. Fujisawa, *J. Osaka City Med. Center*, 1951, *1*, 7: *C.A.*, 1954, *48*, 13550-d.
1059 M. Fujiwara and H. Shimizu, *Anal. Chem.*, 1949, *21*, 1009.
1060 N. A. Fuks, *Zavodskaya Lab.*, 1950, *16*, 878: *C.A.*, 1951, *45*, 971-c.
1061 N. A. Fuks and L. S. Chetverikova, *Zhur. Anal. Khim.*, 1948, *3*, 220: *C.A.*, 1949, *43*, 8978-g.

1062 N. A. Fuks and M. A. Rappoport, *Doklady Akad. Nauk S.S.S.R.*, 1948, *60*, 1219: *C.A.*, 1949, *43*, 121-i.
1063 D. K. Fukushima, S. Lieberman and B. Praetz, *J. Am. Chem. Soc.*, 1950, *72*, 5205.
1064 N. W. Furby, *Anal. Chem.*, 1950, *22*, 876.
1065 S. A. Fusari, R. P. Frohardt, A. Ryder, T. H. Haskell, D. W. Johannessen, C. C. Elder and Q. R. Bartz, *J. Am. Chem. Soc.*, 1954, *76*, 2878.
1066 G. Gabrielson and O. Samuelson, *Svensk Kem. Tids.*, 1950, *62*, 214: *C.A.*, 1951, *45*, 4168-i.
1067 G. Gabrielson and O. Samuelson, *Svensk Kem. Tids.*, 1952, *64*, 150: *C.A.*, 1952, *46*, 9018-i.
1068 S. Gaddis, *J. Chem. Educ.*, 1942, *19*, 327.
1069 T. B. Gage, C. D. Douglass and S. H. Wender, *Anal. Chem.*, 1951, *23*, 1582.
1070 T. B. Gage, Q. L. Morris, W. E. Detty and S. H. Wender, *Science*, 1951, *113*, 522.
1071 T. B. Gage and S. H. Wender, *Anal. Chem.*, 1950, *22*, 708.
1072 B. D. E. Gaillard, *Nature*, 1953, *171*, 1160.
1073 E. M. Gal, *Science*, 1950, *111*, 677.
1074 E. M. Gal and D. M. Greenberg, *Proc. Soc. Exptl. Biol. Med.*, 1949, *71*, 88: *C.A.*, 1949, *43*, 6269-a.
1075 F. Galinovsky and O. Vogl, *Monatsh. Chem.*, 1948, *79*, 325.
1076 U. Gallo, *Boll. chim. farm.*, 1953, *92*, 332: *C.A.*, 1954, *48*, 1712-a.
1077 O. L. Galmarini and V. Deulofeu, *Ciencia e invest. (Buenos Aires)*, 1947, *3*, 479.
1078 N. C. Ganguli, *Naturwissenschaften*, 1953, *40*, 624.
1079 N. C. Ganguli, *Nature*, 1954, *174*, 189.
1080 N. C. Ganguli, *Naturwissenschaften*, 1954, *41*, 282.
1081 N. C. Ganguli, *Anal. Chim. Acta*, 1955, *12*, 335.
1081a E. N. Gapon and I. M. Belenkaya, *Akad. Nauk S.S.S.R.*, *1950*, 35.
1082 E. N. Gapon and T. N. Chernikova, *Doklady Vsesoyuz. Akad. Sel'sko-Khoz. Nauk im. V. I. Lenina S.S.S.R.*, 1948, No. 7, 26: *C.A.*, 1949, *43*, 3549-h.
1083 E. N. Gapon and T. B. Gapon, *Zhur. Fiz. Khim.*, 1948, *22*, 979: *C.A.*, 1949, *43*, 465-a.
1084 T. B. Gapon and E. N. Gapon, *Zhur. Anal. Khim.*, 1948, *3*, 203: *C.A.*, 1949, *43*, 8971-g.
1085 T. B. Gapon and E. N. Gapon, *Doklady Akad. Nauk S.S.S.R.*, 1948, *60*, 401: *C.A.*, 1948, *42*, 6603-e.
1086 S. Gardell, *Acta Chem. Scand.*, 1951, *5*, 1011.
1087 S. Gardell, *Acta Chem. Scand.*, 1953, *7*, 201.
1088 S. Gardell, *Acta Chem. Scand.*, 1953, *7*, 207.
1089 J. E. Gardiner and V. P. Whittaker, *Biochem. J.*, 1954, *58*, 24.
1090 B. Gastambide, *Ann. chim. (Paris)*, 1954, (12) *9*, 257: *C.A.*, 1955, *49*, 9568-f.
1091 H. Gault and C. Ronez, *Bull. soc. chim. France*, *1950*, 597.
1092 P. Gauthier, *Bull. soc. chim. France*, *1955*, 981.
1093 J. D. Geerdes, B. A. Lewis, R. Montgomery and F. Smith, *Anal. Chem.*, 1954, *26*, 264.
1094 T. A. Geissman, P. Deuel, E. K. Bonde and F. A. Addicott, *J. Am. Chem. Soc.*, 1954, *76*, 685.
1095 T. A. Geissman, E. C. Jorgensen and B. Lennart Johnson, *Arch. Biochem. Biophys.*, 1954, *49*, 368.
1096 T. Gendre and E. Lederer, *Biochim. Biophys. Acta*, 1952, *8*, 49.

1097 T. Gendre and E. Lederer, *Ann. Acad. Sci. Fenn.*, Ser. A II, *60*, 313.
1098 J. A. R. Genge, A. Holroyd, J. E. Salmon and J. G. L. Wall, *Chem. and Ind.*, *1955*, 357.
1099 L. W. Georges, R. S. Bower and M. L. Wolfrom, *J. Am. Chem. Soc.*, 1946, *68*, 2169.
1100 E. Gerlach and H.-J. Döring, *Naturwissenschaften*, 1955, *42*, 344.
1101 W. Gerok, *Z. physiol. Chem.*, 1955, *299*, 112.
1102 I. I. Geschwind, J. O. Porath and C. H. Li, *J. Am. Chem. Soc.*, 1952, *74*, 2121.
1102a A. M. Ghe and A. R. Fiorentini, *Ann. chim. (Roma)*, 1955, *45*, 400.
1102b A. S. Ghosh Mazumdar and M. Lederer, *J. Inorg. Nuclear Chem.*, 1956, in press.
1103 J.-M. Ghuysen, *Compt. rend. soc. biol.*, 1952, *146*, 1812.
1104 C. Giddey, *Schweiz. med. Wochschr.*, 1953, *83*, 331: *C.A.*, 1954, *48*, 3434-f.
1105 G. A. Gilbert and A. J. Swallow, *Biochem. J.*, 1950, *47*, 502.
1106 V. E. Gilbert and M. Stacey, *J. Chem. Soc.*, *1948*, 1560.
1107 C. H. Giles, S. K. Jain and A. S. A. Hassan, *Chem. and Ind.*, *1955*, 629.
1108 J. M. Gillespie, M. A. Jermyn and E. F. Woods, *Nature*, 1952, *169*, 487.
1109 A. R. Gilson, *Chem. and Ind.*, *1951*, 185.
1110 M. E. Gilwood, Chapter 4 in *Ion Exchange* by Nachod, Acad. Press, New York, 1949.
1111 K. V. Giri, *Nature*, 1953, *171*, 1159.
1112 K. V. Giri, *Nature*, 1954, *173*, 1194.
1113 K. V. Giri, *Experientia*, 1955, *11*, 165.
1114 K. V. Giri, *J. Indian Inst. of Science*, 1955, *37*, 1.
1115 K. V. Giri, G. D. Kalyankar and C. S. Vaidyanathan, *Naturwissenschaften*, 1954, *41*, 14.
1116 K. V. Giri, K. Krishnamurthy and T. A. Venkatasubramanian, *Current Sci. (India)*, 1952, *21*, 44: *C.A.*, 1952, *46*, 1051-d.
1117 K. V. Giri, K. Krishnamurthy and T. A. Venkatasubramanian, *J. Indian Inst. Sci., Sect. A*, 1952, *34*, 209: *C.A.*, 1953, *47*, 2249-i.
1118 K. V. Giri, P. R. Krishnaswamy, G. D. Kalyankar and P. L. Narasimha Rao, *Experientia*, 1953, *9*, 296.
1119 K. V. Giri, D. V. Krishna Murthy and P. L. Narasimha Rao, *J. Indian Inst. Sci., Sect. A*, 1953, *35*, 93: *C.A.*, 1953, *47*, 4248-i.
1120 K. V. Giri and A. Nagabhushanam, *Naturwissenschaften*, 1952, *39*, 548.
1121 K. V. Giri and V. N. Nigam, *Naturwissenschaften*, 1953, *40*, 343.
1122 K. V. Giri and V. N. Nigam, *J. Indian Inst. Sci., Sect. A*, 1954, *36*, 49: *C.A.*, 1954, *48*, 7497-g.
1123 K. V. Giri, V. N. Nigam and K. Saroja, *Naturwissenschaften*, 1953, *40*, 484.
1124 K. V. Giri and D. B. Parihar, *Nature*, 1955, *175*, 304.
1125 K. V. Giri and A. L. N. Prasad, *Nature*, 1951, *167*, 859.
1126 K. V. Giri and A. L. N. Prasad, *Nature*, 1951, *168*, 786.
1127 K. V. Giri, A. L. N. Prasad, S. Gowri Devi and J. Sri Ram, *Biochem. J.*, 1952. *51*, 123.
1128 K. V. Giri, A. N. Radhakrishnan and C. S. Vaidyanathan, *Anal. Chem.*, 1952, *24*, 1677.
1129 K. V. Giri, A. N. Radhakrishnan and C. S. Vaidyanathan, *J. Indian Inst. Sci., Sect. A*, 1953, *35*, 145: *C.A.*, 1953, *47*, 10044-a.
1130 K. V. Giri, A. N. Radhakrishnan and C. S. Vaidyanathan, *Nature*, 1952, *170*, 1025.

1131 K. V. Giri and N. A. N. Rao, *Nature*, 1952, *169*, 923; *J. Indian Inst. Science*, *Sect. A*, 1952, *34*, 95: *C.A.*, 1952, *46*, 9019-f.
1132 K. V. Giri and N. A. N. Rao, *J. Indian Inst. Sci.*, *Sect. A*, 1953, *35*, 343; *C.A.*, 1954, *48*, 3203-f.
1133 K. V. Giri, P. L. Narasimha Rao, K. Saroja and R. Venkataraman, *Naturwissenschaften*, 1953, *40*, 484.
1134 J. A. Gladner and H. Neurath, *J. Biol. Chem.*, 1953, *205*, 345.
1135 A. J. Glazko and W. A. Dill, *Anal. Chem.*, 1953, *25*, 1782.
1136 A. J. Glazko, W. A. Dill and M. C. Rebstock, *J. Biol. Chem.*, 1950, *183*, 679.
1137 G. I. Gleason, *J. Biol. Chem.*, 1955, *213*, 837.
1138 R. A. Glenn, J. S. Wolfarth and C. W. DeWalt Jr., *Anal. Chem.*, 1952, *24*, 1138.
1139 J. Glover, T. W. Goodwin and R. A. Morton, *Biochem. J.*, 1947, *41*, 94.
1140 E. Glueckauf, *Nature*, 1945, *155*, 205; *156*, 571.
1141 E. Glueckauf, *Nature*, 1945, *156*, 748.
1142 E. Glueckauf, *Nature*, 1947, *160*, 301.
1143 E. Glueckauf, *J. Chem. Soc.*, *1947*, 1321.
1144 E. Glueckauf, *J. Chem. Soc.*, *1949*, 3280.
1145 E. Glueckauf, *Discuss. Faraday Soc.*, 1949, *7*, 12.
1146 E. Glueckauf, *J. Am. Chem. Soc.*, 1951, *73*, 849.
1147 E. Glueckauf, *Trans. Faraday Soc.*, 1955, *51*, 34.
1148 E. Glueckauf, K. H. Barker and G. P. Kitt, *Discuss. Faraday Soc.*, 1949, *7*, 199.
1149 P. Godin, *Chem. and Ind.*, *1954*, 1424.
1150 P. Godin, *Nature*, 1954, *174*, 134.
1151 M. Godlewicz, *Nature*, 1949, *164*, 1132.
1152 M. Goehring and I. Darge, *Z. anal. Chem.*, 1943, *125*, 180.
1153 M. Goehring and I. Darge, *Z. anal. Chem.*, 1943, *125*, 373.
1154 J. P. Goeller and S. Sherry, *Proc. Soc. Exptl. Biol. Med.*, 1950, *74*, 381.
1155 M. Goldenberg, M. Faber, E. J. Alston and E. Chargaff, *Science*, 1949, *109*, 534.
1156 S. Goldschmidt and H. Burkert, *Z. physiol. Chem.*, 1955, *300*, 188.
1157 P. M. Good and A. W. Johnson, *Nature*, 1949, *163*, 31.
1158 R. R. Goodall and A. A. Levi, *Nature*, 1946, *158*, 675; *Analyst*, 1947, *72*, 277.
1159 M. Goodman, *U.S. Atomic Energy Comm. Natl. Sci. Foundation*, *Wash. D.C. UCRL*, 1952, *1961*, 80: *C.A.*, 1953, *47*, 12539-h.
1160 T. W. Goodwin, *Biochem. J.*, 1952, *50*, 550.
1161 T. W. Goodwin, *Biochem. J.*, 1952, *51*, 458.
1162 J. Gootjes and W. T. Nauta, *Rec. trav. chim.*, 1954, *73*, 886.
1163 F. Goppelsroeder, *Mittlgen. des k.k. Technol. Gewerbe-Museums in Wien, II*, 1888, Nos. 3–4 and *III*, 1889, Nos. 1–4.
1164 F. Goppelsroeder, *Verhandl. Naturforsch. Ges. Basel*, (1901) Vol, *14*, (1904) *17*, (1907) *19* Heft 2.
1165 G. Gorbach, *Mikrochem. ver. Mikrochim. Acta*, 1952, *39*, 204.
1166 A. H. Gordon, *Biochem. J.*, 1949, *45*, 99.
1167 A. H. Gordon, B. Keil, K. Sebesta, O. Knessl and F. Šorm, *Collection Czechoslov. Chem. Communs.*, 1950, *15*, 1: *C.A.*, 1950, *44*, 8992-b.
1168 A. H. Gordon, A. J. P. Martin and R. L. M. Synge, *Biochem. J.*, 1943, *37*, 79.
1169 A. H. Gordon, A. J. P. Martin and R. L. M. Synge, *Biochem. J.*, 1943, *37*, 86.
1170 A. H. Gordon, A. J. P. Martin and R. L. M. Synge, *Biochem. J.*, 1943, *37*, 92.
1171 A. H. Gordon, A. J. P. Martin and R. L. M. Synge, *Biochem. J.*, 1943, *37*, 312.
1172 A. H. Gordon, A. J. P. Martin and R. L. M. Synge, *Biochem. J.*, 1944, *38*, 65.

1173 A. H. Gordon and P. Reichard, *Biochem. J.*, 1951, *48*, 569.
1174 B. E. Gordon, F. Wopat Jr., H. D. Burnham and L. C. Jones Jr., *Anal. Chem.*, 1951, *23*, 1754.
1175 H. T. Gordon and C. A. Hewel, *Anal. Chem.*, 1955, *27*, 1471.
1176 N. Gordon and M. Beroza, *Anal. Chem.*, 1952, *24*, 1968.
1177 S. Gordon and G. L. Nardi, *J. Lab. Clin. Med.*, 1954, *43*, 827: *C.A.*, 1954, *48*, 9441-g.
1178 D. N. Gore, *Chem. and Ind.*, *1951*, 479.
1179 T. S. Gore and K. Venkataraman, *Proc. Indian Acad. Sci.*, 1951, *34A*, 368: *C.A.*, 1953, *47*, 1635-h.
1180 H. Goto and Y. Kakita, *J. Chem. Soc. Japan*, 1942, *63*, 120: *C.A.*, 1942, *36*, 3010-c.
1181 S. Gottlieb, *J. Am. Chem. Soc.*, 1948, *70*, 423.
1182 H. Götte and D. Pätze, *Z. Elektrochem.*, 1954, *58*, 636.
1183 J. D. S. Goulden, *Nature*, 1954, *173*, 646.
1184 M. M. Graff and E. L. Skau, *Ind. Eng. Chem. Anal. Ed.*, 1943, *15*, 340.
1185 C. Graichen, *J. Assoc. Offic. Agr. Chemists*, 1951, *34*, 795: *C.A.*, 1952, *46*, 2875-g.
1185a J. A. Grande and J. Beukenkamp, *Anal. Chem.*, 1956, *28*, 1497.
1186 R. Grangaud and I. Garcia, *Bull. soc. chim. biol.*, 1952, *34*, 754.
1187 C. Granger and J. P. Zwilling, *Bull. soc. chim. France*, *1950*, 873.
1188 R. A. Grant and S. R. Stitch, *Chem. and Ind.*, *1951*, 230.
1189 H. Grasshof, *Angew. Chem.*, 1951, *63*, 96.
1190 W. Grassmann and G. Deffner, *Chem.-Zeitung*, 1952, *76*, 623: *C.A.*, 1953, *47*, 4689-a.
1191 W. Grassmann and G. Deffner, *Z. physiol. Chem.*, 1953, *293*, 89.
1192 W. Grassmann and K. Hannig, *Naturwissenschaften*, 1950, *37*, 397.
1193 W. Grassmann and K. Hannig, *Z. physiol. Chem.*, 1952, *290*, 1.
1194 W. Grassmann and K. Hannig, *Klin. Wochschr.*, 1954, *32*, 838: *C.A.*, 1954, *48*, 13780-b.
1195 W. Grassmann, K. Hannig and M. Knedel, *Deut. med. Wochschr.*, 1951, *76*, 333: *C.A.*, 1951, *45*, 6680-g.
1196 W. Grassmann, K. Hannig and M. Plöckl, *Z. physiol. Chem.*, 1955, *299*, 258.
1197 W. Grassmann, H. Hörmann and H. Endres, *Z. physiol. Chem.*, 1954, *296*, 208.
1198 W. Grassmann, H. Hörmann and H. Endres, *Chem. Ber.*, 1955, *88*, 102.
1199 F. V. Gray, A. F. Pilgrim, H. J. Rodda and R. A. Weller, *Nature*, 1951, *167*, 954.
1200 I. Gray, S. Ikeda, A. A. Benson and D. Kritchevsky, *Rev. Sci. Instruments*, 1950, *21*, 1022: *C.A.*, 1951, *45*, 4100-i.
1201 R. A. Gray, *Arch. Biochem. Biophys.*, 1952, *38*, 305.
1202 R. A. Gray, *Science*, 1952, *115*, 129.
1203 F. C. Green and L. M. Kay, *Anal. Chem.*, 1952, *24*, 726.
1204 J. Green, *Biochem. J.*, 1951, *49*, 45.
1205 J. Green, *Biochem. J.*, 1951, *49*, 232 and 243.
1206 J. Green, *Biochem. J.*, 1952, *51*, 144.
1206a J. Green, S. Marcinkiewicz and P. R. Watt, *J. Sci. Food Agr.*, *1955*, 274.
1207 J. Green, N. Mower, C. W. Picard and F. S. Spring, *J. Chem. Soc.*, *1944*, 527.
1208 J. Green and D. O. Singleton, *Analyst*, 1954, *79*, 431.
1209 J. P. Green, *Nature*, 1954, *174*, 369.
1210 J. P. Green and H. Dam, *Acta Chem. Scand.*, 1954, *8*, 1093.

1211   J. P. Green and H. Dam, *Acta Chem. Scand.*, 1954, *8*, 1341.
1212   J. W. Green, *J. Am. Chem. Soc.*, 1954, *76*, 5791.
1213   T. Green, F. O. Howitt and R. Preston, *Chem. and Ind.*, *1955*, 591.
1214   G. R. Greenberg, *J. Biol. Chem.*, 1951, *190*, 611.
1215   R. M. Greenway, P. W. Kent and M. W. Whitehouse, *Research (London)*, 1953, *6*, Suppl. No. 1, p. 65: *C.A.*, 1954, *48*, 1203-i.
1216   R. I. Gregerman and G. Wald, *J. Gen. Physiol.*, 1952, *35*, 489: *C.A.*, 1952, *46*, 3668-a.
1217   H. P. Gregor, *J. Am. Chem. Soc.*, 1948, *70*, 1293.
1218   H. P. Gregor, *J. Am. Chem. Soc.*, 1951, *73*, 642.
1219   G. F. Gregory, *Science*, 1955, *121*, 169.
1220   G. I. Gregory and T. Malkin, *J. Chem. Soc.*, *1951*, 2453.
1221   J. D. Gregory and F. Lipmann, *J. Am. Chem. Soc.*, 1952, *74*, 4017.
1222   A. Grieg, *Nature*, 1952, *170*, 845.
1223   G. Gries, P. Gedigk and J. Georgi, *Z. physiol. Chem.*, 1954, *298*, 132.
1224   J. H. Griffiths and C. S. G. Phillips, *J. Chem. Soc.*, *1954*, 3446.
1225   M. Grinstein, S. Schwartz and C. J. Watson, *J. Biol. Chem.*, 1945, *157*, 323.
1226   M. Grinstein and C. J. Watson, *J. Biol. Chem.*, 1943, *147*, 667.
1227   J. Gripenberg, *Acta Chem. Scand.*, 1952, *6*, 1152.
1228   N. Grobbelaar, R. M. Zacharius and F. C. Steward, *J. Am. Chem. Soc.*, 1954, *76*, 2912.
1229   J. Gross, *Brit. Med. Bull.*, 1954, *10*, 218.
1230   J. Gross, C. P. Leblond, A. E. Franklin and J. H. Quastel, *Science*, 1950, *111*, 605.
1231   J. Gross and R. Pitt-Rivers, *Biochem. J.*, 1953, *53*, 645.
1232   D. P. Groth, G. C. Mueller and G. A. LePage, *J. Biol. Chem.*, 1952, *199*, 389.
1233   W. Groth and P. Harteck, *Naturwissenschaften*, 1941, *29*, 535.
1234   N. Grubhofer and L. Schleith, *Naturwissenschaften*, 1953, *40*, 508.
1235   N. Grubhofer and L. Schleith, *Z. physiol. Chem.*, 1954, *296*, 262.
1236   W. Gruch, *Naturwissenschaften*, 1954, *41*, 39.
1237   A. Grüne, *Arzneimittel-Forsch.*, 1954, *4*, 347.
1238   R. Grüttner, *Klin. Wochschr.*, 1954, *32*, 263: *C.A.*, 1954, *48*, 6900-i.
1240   J. Guerillot, A. Guerillot-Vinet and L. Delmas, *Compt. rend.*, 1952, *235*, 1295.
1241   J. Guérin, *Bull. soc. chim. biol.*, 1954, *36*, 1453.
1242   I. Günder, *Naturwissenschaften*, 1953, *40*, 20.
1243   D. R. Gupta and A. K. Bhattacharya, *J. Ind. Chem. Soc.*, 1953, *30*, 661: *C.A.*, 1954, *48*, 6778-b.
       N. C. S. Gupta, see N. C. Sen Gupta.
1244   C. Gustafsson, J. Sundman and T. Lindh, *Paper and Timber (Finland)*, 1951, *33*, 1: *C.A.*, 1951, *45*, 4175-e.
1245   K. H. Gustavson, *Svensk Kem. Tid.*, 1944, *56*, 14: *C.A.*, 1945, *39*, 4557-8.
1246   K. H. Gustavson, *J. Intern. Soc. Leather Trades' Chemists* 1945, *29*, 204.
1247   K. H. Gustavson, *Svensk Kem. Tid.*, 1946, *58*, 274: *C.A.*, 1947, *41*, 7790-i.
1248   K. H. Gustavson, *J. Am. Leather Chemists' Assoc.*, 1950, *45*, 536: *C.A.*, 1950, *44*, 11146-e.
1249   K. H. Gustavson, *J. Intern. Soc. Leather Trades' Chemists*, 1950, *34*, 259: *C.A.*, 1950, *44*, 11148-f.
1249a K. H. Gustavson, *The Chemistry of Tanning Processes*, Acad. Press, New York, 1956.
'250   G. L. Haberland, F. Bruns and K. I. Altman, *Biochim. Biophys. Acta*, 1954, *15*, 578.

1251 M. H. Hack, *Biochem. J.*, 1953, *54*, 602.
1252 R. H. Hackman and V. M. Trikojus, *Biochem. J.*, 1952, *51*, 653.
1253 E. O. Haenni, J. Carol and D. Banes, *J. Am. Pharm. Assoc.*, 1953, *42*, 167.
1254 L. Hagdahl, *Acta Chem. Scand.*, 1948, *2*, 574.
1255 L. Hagdahl, *Science Tools*, 1954, *1*, 21: *C.A.*, 1955, *49*, 5896-b.
1256 L. Hagdahl and C. E. Danielson, *Nature*, 1954, *174*, 1062.
1257 L. Hagdahl and R. T. Holman, *J. Am. Chem. Soc.*, 1950, *72*, 701.
1258 L. Hagdahl and A. Tiselius, *Nature*, 1952, *170*, 799.
1259 L. Hagdahl, R. J. P. Williams and A. Tiselius, *Arkiv Kemi*, 1952, *4*, 193.
1260 F. Hagemann and H. C. Andrews, *Report ANL* 4215, Oct. 18 (1948).
1261 H. Haglund and A. Tiselius, *Acta Chem. Scand.*, 1950, *4*, 957.
1262 H. von Hahn, E. Sorkin and H. Erlenmeyer, *Experientia*, 1951, *7*, 358.
1263 R. B. Hahn, C. Backer and R. Backer, *Anal. Chim. Acta*, 1953, *9*, 223.
1264 E. Hahofer and F. Hecht, *Mikrochim. Acta*, 1954, 417.
1265 W. J. Haines, *Recent Progr. Hormone Research*, 1952, *7*, 255.
1266 W. J. Haines and J. N. Karnemaat, *Meth. Biochem. Anal.*, 1954, *1*, 171.
1267 I. M. Hais, *Chem. Listy*, 1951, *45*, 76: *C.A.*, 1951, *45*, 6680-c.
1268 I. M. Hais and O. Horešovský, *Chem. Listy*, 1954, *48*, 549: *C.A.*, 1954, *48*, 8302-d.
1269 I. M. Hais and K. Macek, *Papirová chromatografie*, Nákladatelství Československe Akademie Věd, Praha, 1954: *C.A.*, 1955, *49*, 12104-i.
1270 I. M. Hais, K. Macek and V. Francová-Klimova, *Čs. Farmacie*, 1955, *4*, 127.
1271 I. M. Hais and L. Pecáková, *Nature*, 1949, *163*, 768.
1272 M. Haïssinsky, *J. chim. phys.*, 1952, *49*, C 133.
1273 D. K. Hale, *Chem. and Ind.*, 1955, 1147.
1274 D. K. Hale, A. R. Hawdon, J. I. Jones and D. I. Packham, *J. Chem. Soc.*, 1952, 3503.
1275 D. K. Hale and D. Reichenberg, *Discuss. Faraday Soc.*, 1949, *7*, 79.
1276 Halfter, *Z. ges. Schiess. u. Sprengstoffw.*, 1943, *38*, 173: *C.A.*, 1945, *39*, 3671-i.
1277 D. A. Hall, *Nature*, 1951, *168*, 124.
1278 D. A. Hall and A. Tiselius, *Acta Chem. Scand.*, 1951, *5*, 854.
1279 D. A. Hall and F. Wewalka, *Nature*, 1951, *168*, 685.
1280 N. F. Hall and D. H. Johns, *J. Am. Chem. Soc.*, 1953, *75*, 5787.
1281 H. L. Haller, P. D. Bartlett, N. L. Drake, M. S. Newman, S. J. Cristol, C. M. Eaker, R. A. Hayes, G. W. Kilmer, B. Magerlein, G. P. Mueller, A. Schneider and W. Wheatley, *J. Am. Chem. Soc.*, 1945, *67*, 1591.
1282 G. Halliwell, *Nature*, 1952, *169*, 1063.
1283 J. G. Hamilton, *Univ. Microfilms Publ.*, 1954, No. 7223: *C.A.*, 1954, *48*, 7094-f.
1284 J. G. Hamilton Jr. and R. T. Holman, *Arch. Biochem. Biophys.*, 1952, *36*, 456.
1285 J. G. Hamilton and R. T. Holman, *J. Am. Chem. Soc.*, 1954, *76*, 4107.
1286 P. B. Hamilton, *Anal. Chem.*, 1954, *26*, 1857.
1287 P. B. Hamilton and R. A. Anderson, *J. Biol. Chem.*, 1954, *211*, 95.
1288 P. B. Hamilton and R. A. Anderson, *J. Biol. Chem.*, 1955, *213*, 249.
1289 P. B. Hamilton and P. J. Ortiz, *Anal. Chem.*, 1950, *22*, 948.
1290 G. C. M. Hamoir, *Biochem. J.*, 1945, *39*, 485.
1291 D. J. Hanahan, *J. Biol. Chem.*, 1954, *207*, 879.
1292 D. J. Hanahan, *J. Biol. Chem.*, 1954, *211*, 313.
1293 D. J. Hanahan and M. E. Jayko, *J. Am. Chem. Soc.*, 1952, *74*, 5070.
1294 D. J. Hanahan, M. B. Turner and M. E. Jayko, *J. Biol. Chem.*, 1951, *192*, 623.
1295 W. E. Hanby and H. N. Rydon, *Biochem. J.*, 1946, *40*, 297.

1296   C. S. Hanes, F. J. R. Hird and F. A. Isherwood, *Biochem. J.*, 1952, *51*, 25.
1297   C. S. Hanes and F. A. Isherwood, *Nature*, 1949, *164*, 1107.
1298   R. S. Hansen and K. Gunnar, *J. Am. Chem. Soc.*, 1949, *71*, 4958.
1299   R. S. Hansen, K. Gunnar, A. Jacobs and C. R. Simmons, *J. Am. Chem. Soc.*, 1950, *72*, 5043.
1300   S. Harasawa, *J. Chem. Soc. Japan, Pure Chem. Sect.*, 1951, *72*, 107, 236, 423: *C.A.*, 1952, *46*, 850-c, 1917-d, 3449-d.
1301   S. Harasawa, *Kagaku no Ryoiki*, 1951, *5*, 461: *C.A.*, 1954, *48*, 11245.
1302   S. Harasawa and T. Sakamoto, *J. Chem. Soc. Japan, Pure Chem. Sect.*, 1952, 73, 614: *C.A.*, 1953, *47*, 3752-f.
1303   S. Harasawa and T. Sakamoto, *J. Chem. Soc. Japan, Pure Chem. Sect.*, 1953, *74*, 285: *C.A.*, 1953, *47*, 9855-g.
1303a   S. Harasawa and T. Sakamoto, *J. Chem. Soc. Japan, Pure Chem. Sect.*, 1956, *77*, 165.
1303b   S. Harasawa and K. Takasu, *J. Chem. Soc. Japan, Pure Chem. Sect.*, 1955, *76*, 173.
1304   E. Hardegger, *Helv. Chim. Acta*, 1946, *29*, 1195.
1305   E. Hardegger, H. Heusser and F. Blank, *Helv. Chim. Acta*, 1946, *29*, 477.
1306   E. Hardegger, L. Ruzicka and E. Tagmann, *Helv. Chim. Acta*, 1943, *26*, 2205.
1307   T. L. Hardy and D. O. Holland, *Chem. and Ind.*, *1954*, 517.
1308   T. L. Hardy, D. O. Holland and J. H. C. Nayler, *Anal. Chem.*, 1955, *27*, 971.
1309   A. Harjanna, *Suomen Kemistilehti*, 1955, *28 B*, 37: *C.A.*, 1955, *49*, 8743-f.
1310   R. P. Harker, *Chem. and Ind.*, *1955*, 592.
1311   D. H. Harris and E. R. Tompkins, *J. Am. Chem. Soc.*, 1947, *69*, 2792.
1312   G. Harris and I. C. MacWilliam, *Chem. and Ind.*, *1954*, 249.
1313   J. I. Harris and C. H. Li, *J. Am. Chem. Soc.*, 1954, *76*, 3607.
1314   J. O. Harris, *Chem. and Ind.*, *1951*, 255.
1315   R. Harris and A. N. Wick, *Ind. Eng. Chem. Anal. Ed.*, 1946, *18*, 276.
1316   R. J. C. Harris and J. F. Thomas, *Nature*, 1948, *161*, 931.
1317   R. J. C. Harris and J. F. Thomas, *J. Chem. Soc.*, *1948*, 1936.
1318   J. S. Harrison, *Analyst*, 1951, *76*, 77.
1318a   H. Hartkamp and H. Specker, *Naturwissenschaften*, 1955, *42*, 534.
1319   N. Hartler and O. Samuelson, *Anal. Chim. Acta*, 1953, *8*, 130.
1320   Z. V. Harvalik, *Anal. Chem.*, 1950, *22*, 1149.
1321   D. Harvey and D. E. Chalkley, *Fuel*, 1955, *34*, 191: *C.A.*, 1955, *49*, 7436-i.
1322   Y. Hashimoto, *Pharm. Bulletin*, 1953, *1*, 176: *C.A.*, 1954, *48*, 7433-h.
1323   Y. Hashimoto and I. Mori, *J. Pharm. Soc. Japan*, 1952, *72*, 1552: *C.A.*, 1953, *47*, 1434-e.
1324   Y. Hashimoto and I. Mori, *Nature*, 1952, *170*, 1024.
1325   Y. Hashimoto and I. Mori, *Nature*, 1953, *172*, 542.
1326   T. Hashizume, *Nature*, 1954, *173*, 645.
1327   T. H. Haskell, S. A. Fusari, R. P. Frohardt and Q. R. Bartz, *J. Am. Chem. Soc.*, 1952, *74*, 599.
1328   A. L. Haskins Jr., A. I. Sherman and W. M. Allen, *J. Biol. Chem.*, 1950, *182*, 429.
1329   G. A. D. Haslewood, *Biochem. J.*, 1954, *56*, 581.
1330   G. A. D. Haslewood and J. Sjövall, *Biochem. J.*, 1954, *57*, 126.
1331   H. B. Hass, T. De Vries and H. H. Jaffé, *J. Am. Chem. Soc.*, 1943, *65*, 1486.
1332   C. H. Hassall and K. E. Magnus, *Experientia*, 1954, *10*, 425.
1333   C. H. Hassall and S. L. Martin, *J. Chem. Soc.*, *1951*, 2766.

1334 C. H. Hassall and K. Reyle, *Biochem. J.*, 1955, *60*, 334.
1335 G. Haugaard and T. D. Kroner, *J. Am. Chem. Soc.*, 1948, *70*, 2135.
1336 E. G. E. Hawkins and R. F. Hunter, *J. Chem. Soc.*, *1944*, 411.
1337 R. D. Haworth, R. MacGillivray and D. H. Peacock, *Nature*, 1951, *167*, 1068.
1338 J. R. Hawthorne, *Nature*, 1947, *160*, 714.
1339 F. Haxo, C. O'hEocha and P. Norris, *Arch. Biochem. Biophys.*, 1955, *54*, 162.
1340 O. Hayaishi and A. Kornberg, *J. Am. Chem. Soc.*, 1951, *73*, 2975.
1341 E. Hayek and F. Lorenz, *Monatsh. Chem.*, 1953, *84*, 647: *C.A.*, 1954, *48*, 1879-f.
1342 E. E. Hays, I. C. Wells, P. A. Katzman, C. K. Cain, F. A. Jacobs, S. A. Thayer, E. A. Doisy, W. L. Gaby, E. C. Roberts, R. D. Muir, C. J. Carroll, L. R. Jones and N. J. Wade, *J. Biol. Chem.*, 1945, *159*, 725.
1343 E. Hecht and C. Mink, *Biochim. Biophys. Acta*, 1952, *8*, 641.
1344 C. G. Hedén, *Nature*, 1950, *166*, 999.
1345 E. Heftmann, *Science*, 1950, *111*, 571.
1346 E. Heftmann, *J. Am. Chem. Soc.*, 1951, *73*, 851.
1346a E. Heftmann, *Chem. Revs.*, 1955, *55*, 679.
1347 E. Heftmann, P. Berner, A. L. Hayden, H. K. Miller and E. Mosettig, *Arch. Biochem. Biophys.*, 1954, *51*, 329.
1348 E. Heftmann and A. L. Hayden, *J. Biol. Chem.*, 1952, *197*, 47.
1349 E. Heftmann and A. J. Levant, *J. Biol. Chem.*, 1952, *194*, 703.
1350 H. Hegedüs, C. Tamm and T. Reichstein, *Helv. Chim. Acta*, 1953, *36*, 357.
1351 M. L. Heideman Jr., *Endocrinology*, 1953, *53*, 640: *C.A.*, 1954, *48*, 13794-e.
1352 T. Heikel, L. Bäckström and M. Hakala, *Scand. J. Clin. & Lab. Invest.*, 1954, *6*, 79: *C.A.*, 1954, *48*, 13783-d.
1353 D. Heikens, P. H. Hermans and P. F. van Velden, *Nature*, 1954, *174*, 1187.
1354 R. Heilmann et R. Glénat, *Compt. rend.*, 1955, *240*, 2317.
1355 W. Heimann, R. Strohecker and F. Matt, *Z. Lebensm.-Untersuch. u.-Forsch.*, 1953, *97*, 263: *C.A.*, 1954, *48*, 2812-e.
1356 D. W. Hein, *J. Am. Chem. Soc.*, 1955, *77*, 2797.
1357 P. Heinänen, S. Nyyssönen and L. Tuderman, *Farm. Aikakauslehti*, 1951, *5*, 84: *C.A.*, 1952, *46*, 7286-g.
1358 G. Heinrich, *Naturwissenschaften*, 1952, *39*, 257.
1359 N. Hellström, *Acta Chem. Scand.*, 1953, *7*, 329.
1360 N. Hellström and H. Borgiel, *Acta Chem. Scand.*, 1949, *3*, 401.
1361 O. M. Helmer, *Proc. Soc. Exptl. Biol. Med.*, 1950, *74*, 642.
1362 H. Helrich and W. Rieman III, *Anal. Chem.*, 1947, *19*, 651.
1362a H. B. Henbest, E. R. H. Jones, T. C. Owen and V. Thaller, *J. Chem. Soc.*, *1955*, 2763.
1363 H. B. Henbest, E. R. H. Jones and G. F. Smith, *J. Chem. Soc.*, *1953*, 3796.
1364 G. M. Henderson and H. G. Rule, *J. Chem. Soc.*, *1939*, 1568.
1365 R. Henry and M. Thevenet, *Bull. soc. chim. biol.*, 1952, *34*, 839.
1366 L. A. Heppel and P. R. Whitfeld, *Biochem. J.*, 1955, *60*, 1.
1367 L. A. Heppel, P. R. Whitfeld and R. Markham, *Biochem. J.*, 1955, *60*, 8.
1368 S. F. Herb, L. P. Witnauer and R. W. Riemenschneider, *J. Am. Oil Chemists' Soc.*, 1951, *28*, 505: *C.A.*, 1952, *46*, 1271-d.
1369 R. H. Herber and J. W. Irvine Jr., *J. Am. Chem. Soc.*, 1954, *76*, 987.
1370 D. Herbert and A. J. Pinsent, *Biochem. J.*, 1948, *43*, 193.
1371 E. Herbert, V. R. Potter and Y. Takagi, *J. Biol. Chem.* 1955, *213*, 923.
1372 W. Hermanowicz and C. Sikorowska, *Przemysł Chem.*, 1952, *31* (8), 238: *C.A.*, 1953, *47*, 9856-b.

1373 P. H. Hermans, *Physics and Chemistry of Cellulose Fibres*, 543 pp., Elsevier Publishing Co., Amsterdam, 1949.

1374 D. S. Herr, *Ind. Eng. Chem.*, 1945, *37*, 631.

1375 D. F. Hersey and S. J. Ajl, *J. Biol. Chem.*, 1951, *191*, 113.

1376 G. P. Hess and F. H. Carpenter, *J. Am. Chem. Soc.*, 1952, *74*, 4971.

1377 H. Hess, P. Speiser, O. Schindler and T. Reichstein, *Helv. Chim. Acta*, 1951, *34*, 1854.

1378 W. C. Hess, *J. Lab. Clin. Med.*, 1947, *32*, 1163: *C.A.*, 1948, *42*, 241-g.

1379 G. Hesse, I. Daniel and G. Wohlleben, *Angew. Chem.*, 1952, *64*, 103.

1380 G. Hesse and B. Tschachotin, *Naturwissenschaften*, 1942, *30*, 387.

1381 C. Heusghem, *Nature*, 1953, *171*, 42.

1382 H. Heusser, E. V. Jensen, N. Frick and P. A. Plattner, *Helv. Chim. Acta*, 1949, *32*, 1326.

1383 H. Heymann, F. L. Chan and C. W. Clancy, *J. Am. Chem. Soc.*, 1950, *72*, 1112.

1384 K. Heyns and G. Anders, *Z. physiol. Chem.*, 1951, *287*, 8.

1385 K. Heyns and G. Anders, *Z. physiol. Chem.*, 1951, *287*, 109.

1386 K. Heyns, G. Anders and E. Becker, *Z. physiol. Chem.*, 1951, *287*, 120.

1387 K. Heyns and W. Walter, *Z. physiol. Chem.*, 1951, *287*, 15.

1388 K. Heyns and W. Walter, *Z. physiol. Chem.*, 1953, *294*, 111.

1389 H. J. Hibshman, *Ind. Eng. Chem.*, 1950, *42*, 1310.

1390 R. J. Hickey and W. F. Phillips, *Anal. Chem.*, 1954, *26*, 1640.

1391 J. L. Hickson and R. L. Whistler, *Anal. Chem.*, 1953, *25*, 1425.

1392 G. H. Higgins and K. Street Jr., *J. Am. Chem. Soc.*, 1950, *72*, 5321.

1393 T. Higuchi, N. C. Hill and G. B. Corcoran, *Anal. Chem.*, 1952, *24*, 491.

1394 E. Hiller, F. Zinnert and G. Frese, *Biochem. Z.*, 1952/53, *323*, 245.

1394a R. L. Hinman, *Anal. Chim. Acta*, 1956, *15*, 125.

1395 K. Hinrichs, H. Poppe, H. Steyer and H. Weidemann, *Arch. exptl. Pathol. Pharmakol.*, 1954, *223*, 493: *C.A.*, 1955, *49*, 3298-f.

1396 S. Hirase, C. Araki and S. Nakanishi, *Bull. Chem. Soc. Japan*, 1953, *26*, 183: *C.A.*, 1954, *48*, 5737-i.

1397 F. J. R. Hird, *Australian J. Sci.*, 1949, *11*, 170: *C.A.*, 1949, *43*, 7092-a.

1398 F. J. R. Hird and E. V. Rowsell, *Nature*, 1950, *166*, 517.

1399 F. J. R. Hird and P. H. Springell, *Biochem. J.* 1954, *56*, 417.

1400 F. J. R. Hird and V. M. Trikojus, *Australian J. Sci.*, 1948, *10*, 185: *C.A.*, 1948, *42*, 8852-e.

1401 C. H. W. Hirs, *J. Biol. Chem.*, 1953, *205*, 93.

1402 C. H. W. Hirs, S. Moore and W. H. Stein, *J. Biol. Chem.*, 1952, *195*, 669.

1403 C. H. W. Hirs, S. Moore and W. H. Stein, *J. Am. Chem. Soc.*, 1954, *76*, 6063.

1404 G. H. W. Hirs, W. H. Stein and S. Moore, *J. Am. Chem. Soc.*, 1951, *73*, 1893.

1405 C. H. W. Hirs, W. H. Stein and S. Moore, *J. Biol. Chem.*, 1954, *211*, 941.

1406 A. E. Hirschler, *U.S. Pat.* 2,472,250 (1949): *C.A.*, 1949, *43*, 6011-f.

1407 A. E. Hirschler, *U.S. Pat.* 2,518,236 (1950): *C.A.*, 1950, *44*, 9982-d.

1408 A. E. Hirschler, *U.S. Pat.* 2,559,157 (1951): *C.A.*, 1952, *46*, 519-h.

1409 A. E. Hirschler and S. Amon, *Ind. Eng. Chem.*, 1947, *39*, 1585.

1410 A. E. Hirschler and M. R. Lipkin, *U.S. Pat.* 2,441,572 (1948): *C.A.*, 1948, *42*, 8457-i.

1411 E. L. Hirst, L. Hough and J. K. N. Jones, *Nature*, 1949, *163*, 177.

1412 E. L. Hirst, L. Hough and J. K. N. Jones, *J. Chem. Soc.*, 1949, 928.

1413 E. L. Hirst, L. Hough and J. K. N. Jones, *J. Chem. Soc.*, 1949, 3145.

1414　E. L. Hirst, L. Hough and J. K. N. Jones, *J. Chem. Soc.*, *1951*, 323.
1415　E. L. Hirst, D. I. McGilvray and E. G. V. Percival, *J. Chem. Soc.*, *1950*, 1297.
1416　E. L. Hirst and J. K. N. Jones, *J. Chem. Soc.*, *1949*, 1659.
1417　E. L. Hirst and J. K. N. Jones, *Discuss. Faraday Soc.*, 1949, 7, 268.
1418　E. R. Hiscox and N. J. Berridge, *Nature*, 1950, *166*, 522.
1419　D. J. D. Hockenhull, *Nature*, 1953, *171*, 982.
1420　D. J. D. Hockenhull and G. D. Floodgate, *Biochem. J.*, 1952, *52*, 38.
1421　D. J. D. Hockenhull and D. Herbert, *Biochem. J.*, 1945, *39*, 102.
1422　D. J. D. Hockenhull, G. D. Hunter and M. W. Herbert, *Chem. and Ind.*, *1953*, 127.
1423　F. Hoffmann–La Roche, *Brit. Pat.* 599,626: *C.A.*, 1948, *42*, 6063-f.
1424　C. L. Hoffpauir and J. D. Guthrie, *J. Biol. Chem.*, 1949, *178*, 207.
1425　H. Hofmann and Hj. Staudinger. *Biochem. Z.*, 1951, *322*, 230.
1426　H. Hofmann and Hj. Staudinger, *Naturwissenschaften*, 1951, *38*, 213; *Biochem. Z.*, 1952, *322*, 230.
1427　B. Högberg, *Acta Chem. Scand.*, 1954, *8*, 1098.
1428　A. Holasek and K. Winsauer, *Monatsh. Chem.*, 1954, *85*, 796: *C.A.*, 1955, *49*, 2757-a.
1429　E. R. Holiday and E. A. Johnson, *Nature*, 1949, *163*, 216.
1430　R. T. Holman, *Anal. Chem.*, 1950, *22*, 832.
1431　R. T. Holman, *J. Am. Chem. Soc.*, 1951, *73*, 1261.
1432　R. T. Holman, *J. Am. Chem. Soc.*, 1951, *73*, 3337.
1433　R. T. Holman, *J. Am. Chem. Soc.*, 1951, *73*, 5289.
1434　R. T. Holman, *Chromatography of Fatty Acids and Related Substances*; p. 104–126 in *Progress in the Chemistry of Fats and Other Lipids*, Vol. I, Pergamon Press Ltd., London, 1952.
1435　R. T. Holman and L. Hagdahl, *Arch. Biochem.*, 1948, *17*, 301.
1436　R. T. Holman and L. Hagdahl, *J. Biol. Chem.*, 1950, *182*, 421.
1437　R. T. Holman and L. Hagdahl, *Anal. Chem.*, 1951, *23*, 794.
1438　R. T. Holman and W. T. Williams, *J. Am. Chem. Soc.*, 1951, *73*, 5285.
1439　P. Holton, *Nature*, 1949, *163*, 217.
1440　L. Holzapfel and W. Engel, *Naturwissenschaften*, 1949, *36*, 375.
1441　M. Honda, *J. Chem. Soc. Japan*, 1949, *70*, 52, 55, 103, 163, 165; 1950, *71*, 59; 1951, *72*, 361: *C.A.*, 1951, *45*, 4169-g, 5053-e; 1952, *46*, 847-h.
1442　M. Honda, *Japan Analyst*, 1954, *3*, 132: *C.A.*, 1954, *48*, 9868-b.
1442a　M. Honda, Y. Sasaki and H. Natsume, *Japan Analyst*, 1955, *4*, 240.
1443　P. P. Hopf, *J. Chem. Soc.*, *1946*, 785.
1444　P. P. Hopf, *Ind. Eng. Chem.*, 1947, *39*, 983.
1445　P. P. Hopf, C. G. Lynam and H. Weil, *Brit. Pat.* 585,224 (1947): *C.A.*, 1950, *44*, 3750-c.
1446　B. L. Horecker, P. Z. Smyrniotis and J. E. Seegmiller, *J. Biol. Chem.*, 1951, *193*, 383.
1447　R. E. Horne and A. L. Pollard, *J. Bact.*, 1948, *55*, 231.
1448　L. Horner, W. Emrich and A. Kirschner, *Z. Elektrochem.*, 1952, *56*, 987.
1449　R. H. Horrocks, *Nature*, 1949, *164*, 444.
1450　R. H. Horrocks and G. B. Manning, *Lancet*, 1949, *256*, 1042.
1451　R. L. Hossfeld, *J. Am. Chem. Soc.*, 1951, *73*, 852.
1452　R. D. Hotchkiss, *J. Biol. Chem.*, 1948, *175*, 315.
1453　F. W. Hougen, *Biochem. J.*, 1955, *59*, 302.
1454　L. Hough, *Nature*, 1950, *165*, 400.

1455   L. Hough, *Meth. Biochem. Anal.*, 1954, *1*, 205.
1456   L. Hough and J. K. N. Jones, *J. Chem. Soc.*, *1950*, 1199.
1457   L. Hough, J. K. N. Jones and E. L. Hirst, *Nature*, 1950, *165*, 34.
1458   L. Hough, J. K. N. Jones and W. H. Wadman, *Nature*, 1948, *162*, 448; *J. Chem. Soc.*, *1949*, 2511.
1459   L. Hough, J. K. N. Jones and W. H. Wadman, *J. Chem. Soc.*, *1950*, 1702.
1460   G. A. Howard and A. J. P. Martin, *Biochem. J.*, 1950, *46*, 532.
1461   G. A. Howard and A. R. Tatchell, *Chem. and Ind.*, *1954*, 219.
1462   E. E. Howe and M. Tishler, *U.S. Pat.* 2,480,654 (1949): *C.A.*, 1950, *44*, 6086-i.
1463   D. R. Howton, *Science*, 1955, *121*, 704.
1464   H. Hoyer, *Z. Elektrochem.*, 1950, *54*, 413.
1465   H. Hoyer, *Kolloid-Z.*, 1950, *116*, 121.
1466   H. Hoyer, *Kolloid-Z.*, 1951, *121*, 121; 1951, *122*, 142: *C.A.*, 1951, *45*, 6528-e, 8848-b.
1467   H. Hoyer, *Chem. Ber.*, 1953, *86*, 1016.
1468   P. T. Huang, R. T. Holman and W. M. Potts, *J. Am. Oil Chemists' Soc.*, 1949, *26*, 405: *C.A.*, 1949, *43*, 7721-b.
1469   H. J. Hübener, F. Bode, H. J. Mollat and M. Wehner, *Z. physiol. Chem.*, 1952, *290*, 136.
1470   H. J. Hübener, E. Hoffmann and F. Bode, *Z. physiol. Chem.*, 1952, *289*, 102.
1471   S. Hudeček, *Chem. Listy*, 1955, *49*, 60: *C.A.*, 1955, *49*, 5216-a.
1472   C. F. Huebner, *Nature*, 1951, *167*, 119.
1473   F. M. Huennekens, D. J. Hanahan and M. Uziel, *J. Biol. Chem.*, 1954, *206*, 443.
1474   J. W. Huff and W. A. Perlzweig, *J. Biol. Chem.*, 1943, *150*, 395.
1475   E. H. Huffman and G. M. Iddings, *J. Am. Chem. Soc.*, 1952, *74*, 4714.
1476   E. H. Huffman and R. C. Lilly, *J. Am. Chem. Soc.*, 1949, *71*, 4147.
1477   E. H. Huffman and R. L. Oswalt, *J. Am. Chem. Soc.*, 1950, *72*, 3323.
1478   A. C. Hulme, *Nature*, 1953, *171*, 610.
1479   A. C. Hulme, *Biochim. Biophys. Acta*, 1954, *14*, 36.
1480   A. C. Hulme, *Biochim. Biophys. Acta*, 1954, *14*, 44.
1481   A. C. Hulme, *Nature*, 1954, *174*, 1055.
1482   A. C. Hulme and W. Arthington, *Nature*, 1950, *165*, 716.
1483   A. C. Hulme and W. Arthington, *Nature*, 1952, *170*, 659.
1484   A. C. Hulme and A. Richardson, *J. Sci. Food Agr.*, 1954, *5*, 221: *C.A.*, 1954, *48*, 8984-h.
1485   A. C. Hulme and T. Swain, *Nature*, 1951, *168*, 254.
1486   J. P. Hummel and O. Lindberg, *J. Biol. Chem.*, 1949, *180*, 1.
1487   E. C. Hunt and R. A. Wells, *Analyst*, 1954, *79*, 351.
1488   C. D. Hurd, G. R. Thomas and A. A. Frost, *J. Am. Chem. Soc.*, 1950, *72*, 3733.
1489   C. D. Hurd and R. P. Zelinski, *J. Am. Chem. Soc.*, 1947, *69*, 243.
1490   R. B. Hurlbert, H. Schmitz, A. F. Brumm and V. R. Potter, *J. Biol. Chem.*, 1954, *209*, 23.
1491   R. O. Hurst and G. C. Butler, *J. Biol. Chem.*, 1951, *193*, 91.
1492   R. O. Hurst, J. A. Little and G. C. Butler, *J. Biol. Chem.*, 1951, *188*, 705.
1493   R. O. Hurst, A. M. Marko and G. C. Butler, *J. Biol. Chem.*, 1953, *204*, 847.
1494   E. K. Hyde, *Natl. Nuclear Energ. Ser.*, Div. IV, *Plutonium Project Record 14A, Actinide elements*, 542 (1954): *C.A.*, 1954, *48*, 13455-i.
1495   C. H. Ice and S. H. Wender, *Anal. Chem.*, 1952, *24*, 1616.

1496   D. R. Idler, *J. Am. Chem. Soc.*, 1949, *71*, 3854.
1497   D. R. Idler and C. A. Baumann, *J. Biol. Chem.*, 1952, *195*, 623.
1498   S. Iijima, T. Sato and T. Kamoshita, *Bull. Inst. Phys. Chem. Research Tokyo*, 1944, *23*, 181, 233, 284: *C.A.*, 1948, *42*, 7197-i.
1499   R. M. Ikeda, A. D. Webb and R. E. Kepner, *Anal. Chem.*, 1954, *26*, 1228.
1500   B. Imelik, M. V. Mathieu, M. Prettre and S. Teichner, *J. chim. phys.*, 1954, *51*, 651: *C.A.*, 1955, *49*, 7324-f.
1501   V. M. Ingram, *Nature*, 1950, *166*, 1038.
1501a   Y. Inouye, A. Kawamura, K. Wada and H. Okamura, *Japan Analyst*, 1955, *4*, 277.
1502   Y. Inouye and M. Noda, *J. Agr. Chem. Soc. Japan*, 1950, *23*, 294: *C.A.*, 1951, *45*, 8449-f.
1503   Y. Inouye and M. Noda, *J. Agr. Chem. Soc. Japan*, 1950, *23*, 291, 295, 368: *C.A.*, 1952, *46*, 6408-a-g.
1504   Y. Inouye and M. Noda, *J. Agr. Chem. Soc. Japan*, 1952, *26*, 634: *C.A.*, 1953, *47*, 9635-d.
1505   Y. Inouye and M. Noda, *J. Agr. Chem. Soc. Japan*, 1953, *27*, 50: *C.A.*, 1953 *47*, 9635-g.
1506   Y. Inouye, M. Noda and C. Hirayama, *J. Am. Oil Chemists' Soc.*, 1955, *32*, 132: *C.A.*, 1955, *49*, 6783-f.
1507   Y. Inouye, K. Onodera, S. Kitaoka and J. Shishiyama, *J. Agr. Chem. Soc. Japan*, 1953, *27*, 1: *C.A.*, 1955, *49*, 871-i.
1508   Interchemical Corp., *Brit. Pat.* 671.042: *C.A.* 1952, *46*, 8171-b.
1509   G. E. Irish and A. C. Karbum, *Anal. Chem.*, 1954, *26*, 1445.
1510   F. Irreverre and W. Martin, *Anal. Chem.*, 1954, *26*, 257.
1511   H. M. Irving and R. J. P. Williams, *Science Progr.*, 1953, *41*, 418: *C.A.*, 1953, *47*, 8460-c.
1512   F. A. Isherwood, *Biochem. J.*, 1946, *40*, 688.
1513   F. H. Isherwood, *Brit. Med. Bull.*, 1954, *10*, 202.
1514   F. A. Isherwood, Y. T. Chen and L. W. Mapson, *Biochem. J.*, 1954, *56*, 17.
1515   F. A. Isherwood and D. H. Cruickshank, *Nature*, 1954, *173*, 121.
1516   F. A. Isherwood and D. H. Cruickshank, *Nature*, 1954, *174*, 123.
1517   F. A. Isherwood and C. S. Hanes, *Biochem. J.*, 1953, *55*, 824.
1518   F. A. Isherwood and M. A. Jermyn, *Biochem. J.*, 1951, *48*, 515.
1519   F. A. Isherwood and R. L. Jones, *Nature*, 1955, *175*, 419.
1520   Y. Ishida, N. Inagaki, A. Shiota and R. Watanabe, *J. Pharm. Soc. Japan*, 1953, *73*, 736: *C.A.*, 1953, *47*, 10141-e.
1521   S. Ishii and T. Andô, *Science (Japan)*, 1950, *20*, 24: *C.A.*, 1951, *45*, 10, 141-c.
1522   H. C. Isliker, *Ann. N.Y. Acad. Sci.*, 1953, *57*, 225.
1523   T. Itai, T. Oba and S. Kamiya, *Bull. Natl. Hyg. Lab.*, 1954; *72*, 87: *C.A.*, 1955, *49*, 6350-h.
1524   M. Ito, S. Wakamatsu and H. Kawahara, *J. Chem. Soc. Japan, Pure Chem. Sect.*, 1954, *75*, 413: *C.A.*, 1954, *48*, 13172-d.
1525   H. Iwainsky, *Z. physiol. Chem.*, 1954, *297*, 194.
1526   V. K. Iya and J. Loriers, *Compt. rend.*, 1953, *257*, 1413.
1527   M. Jaarma, *Acta Chem. Scand.*, 1954, *8*, 860.
1528   S. S. Jackel, E. H. Mosbach, and C. G. King, *Arch. Biochem. Biophys.*, 1951, *31*, 442.
1529   P. W. M. Jacobs and F. C. Tompkins, *Trans. Faraday Soc.*, 1945, *41*, 388.
1530   P. W. M. Jacobs and F. C. Tompkins, *Trans. Faraday Soc.*, 1945, *41*, 395, 400.

1531   W. A. Jacobs and L. C. Craig, *J. Biol. Chem.*, 1941, *141*, 67.
1532   J. Jacques and J. P. Mathieu, *Bull. soc. chim. France*, *1946*, 94.
1533   L. Jaenicke and K. von Dahl, *Naturwissenschaften*, 1952, *39*, 87.
1534   G. Jakovliv and G. Colpé, *Ann. Fals. et Fraudes*, 1953, *46*, No. 533.
1535   A. T. James, *Biochem. J.*, 1952, *52*, 242.
1536   A. T. James and A. J. P. Martin, *Analyst*, 1952, *77*, 915.
1537   A. T. James and A. J. P. Martin, *Biochem. J.*, 1952, *50*, 679.
1538   A. T. James and A. J. P. Martin, *Biochem. J.*, 1954, *57*, v.
1539   A. T. James and A. J. P. Martin, *Brit. Med. Bull.*, 1954, *10*, 170.
1539a A. T. James and A. J. P. Martin, *Biochem. J.*, 1956, *63*, 144.
1540   A. T. James, A. J. P. Martin and G. Howard Smith, *Biochem. J.*, 1952, *52*, 238.
1541   A. T. James, A. J. P. Martin and S. S. Randall, *Biochem. J.*, 1951, *49*, 293.
1542   D. H. James and C. S. G. Phillips, *J. Chem. Soc.*, *1953*, 1600.
1543   D. H. James and C. S. G. Phillips, *J. Chem. Soc.*, *1954*, 1066.
1544   R. A. James and W. P. Bryan, *J. Am. Chem. Soc.*, 1954, *76*, 1982.
1545   W. O. James, *Nature*, 1948, *161*, 851.
1546   W. O. James and N. Kilbey, *Nature*, 1950, *166*, 67.
1547   F. Jaminet, *J. pharm. Belg.*, 1950, *5*, 297: *C.A.*, 1951, *45*, 4404-d.
1548   F. Jaminet, *J. pharm. Belg.*, 1951, *6*, 81: *C.A.*, 1951, *45*, 9801-c.
1549   F. Jaminet, *J. pharm. Belg.*, 1953, *8*, 339 and 449: *C.A.*, 1954, *48*, 8482-c.
1550   M. M. Jamison and E. E. Turner, *J. Chem. Soc.*, *1942*, 611.
1551   J. Janák, *Chem. Listy*, 1953, *47*, 464; *Collection Czechoslov. Chem. Communs.*, 1953, *18*, 798 (in Russian): *C.A.*, 1954, *48*, 3196-h.
1552   J. Janák, *Collection Czechoslov. Chem. Communs.*, 1954, *19*, 684.
1553   J. Janák, *Collection Czechoslov. Chem. Communs.*, 1954, *19*, 700.
1554   J. Janák, *Collection Czechoslov. Chem. Communs.*, 1954, *19*, 917.
1555   J. Janák, *Mikrochim. Acta*, in press [presented to the Microchemical Congress (1955)].
1556   J. Janàk and M. Rusek, *Chem. Listy*, 1954, *48*, 207: *C.A.*, 1954, *48*, 6321-f.
1557   J. Janák and M. Rusek, *Collection Czechoslov. Chem. Communs.*, 1955, *20*, 343.
1558   J. Janák and K. Tesařík, *Collection Czechoslov. Chem. Communs.*, 1955, *20*, 348.
1559   M.-M. Janot, E. Saïas and M. Foucher, *Bull. soc. chim. biol.*, 1953, *35*, 1101.
1560   F. G. Jarvis and M. J. Johnson, *J. Am. Chem. Soc.*, 1949, *71*, 4124.
1561   H. Jatzkewitz, *Z. physiol. Chem.*, 1953, *292*, 94.
1562   H. Jatzkewitz and N.-D. Tam, *Z. physiol. Chem.*, 1954, *296*, 188.
1563   P. Jax and H. Aust, *Milchwiss. Ber.*, *1953*, 145: *C.A.*, 1954, *48*, 7218-d.
1564   A. Jeanes, C. A. Wilham, R. W. Jones, H. M. Tsuchiya and C. E. Rist, *J. Am. Chem. Soc.*, 1953, *75*, 5911.
1565   A. Jeanes, C. S. Wise and R. J. Dimler, *Anal. Chem.*, 1951, *23*, 415.
1566   R. Jeanloz, D. A. Prins and T. Reichstein, *Helv. Chim. Acta*, 1946, *29*, 371.
1567   R. N. Jeffrey, *Anal. Chem.*, 1951, *23*, 936.
1568   O. Jeger, M. Montavon and L. Ruzicka, *Helv. Chim. Acta*, 1946, *29*, 1124.
1569   O. Jeger, C. Nisoli and L. Ruzicka, *Helv. Chim. Acta*, 1946, *29*, 1183.
1570   O. Jeger, J. Redel and R. Nowak, *Helv. Chim. Acta*, 1946, *29*, 1241.
1571   O. Jeger, R. Rüegg and L. Ruzicka, *Helv. Chim. Acta*, 1947, *30*, 1294.
1572   P. H. Jellinek, *Nature*, 1953, *171*, 750.
1573   H. Jenny, *Kolloid-Beihefte*, 1927, *23*, 428.
1574   J. L. Jensen, E. M. Shantz, N. D. Embree, J. D. Cawley and P. L. Harris, *J. Biol. Chem.*, 1943, *149*, 473.

1575 K. B. Jensen, *Acta Pharmacol. et Toxicol.*, 1952, *8*, 110: *C.A.*, 1952, *46*, 10538-c.
1576 K. B. Jensen, *Acta Pharmacol. et Toxicol.*, 1953, *9*, 99, 275: *C.A.*, 1954, *48*, 2322-c-e.
1577 K. B. Jensen, *Acta Pharmacol. et Toxicol.*, 1953, *9*, 99; 1954, *10*, 69: *C.A.*, 1954, *48*, 2322-c.
1578 K. B. Jensen and A. B. Svendsen, *Pharm. Acta Helv.*, 1950, *25*, 31: *C.A.*, 1950, *44*, 5532-g.
1579 K. B. Jensen and K. Tennöe, *J. Pharm. and Pharmacol.*, 1955, *7*, 334.
1580 J. B. Jepson and I. Smith, *Nature*, 1953, *172*, 1100.
1581 D. Jerchel and W. Jacobs, *Angew. Chem.*, 1953, *65*, 342.
1582 D. Jerchel and W. Jacobs, *Angew. Chem.*, 1954, *66*, 298.
1583 D. Jerchel and R. Müller, *Naturwissenschaften*, 1951, *38*, 561.
1584 M. A. Jermyn, *Nature*, 1952, *169*, 488.
1585 M. A. Jermyn, *Australian J. Biol. Sci.*, 1953, *6*, 77: *C.A.*, 1953, *47*, 6479-f.
1586 M. A. Jermyn and F. A. Isherwood, *Biochem. J.*, 1949, *44*, 402.
1587 M. A. Jermyn and R. Thomas, *Nature*, 1953, *172*, 728.
1588 M. A. Jermyn and R. G. Tomkins, *Biochem. J.*, 1950, *47*, 437.
1589 G. Johansson, *Svensk Kem. Tids.*, 1953, *65*, 79: *C.A.*, 1953, *47*, 9855-e.
1590 E. A. Johnson, *Biochem. J.*, 1952, *51*, 133.
1591 G. R. A. Johnson, G. Stein and J. Weiss, *J. Chem. Soc.*, *1951*, 3275.
1592 C. D. Johnston, *Science*, 1947, *106*, 91.
1593 G. Jollès and C. Fromageot, *Biochim. Biophys. Acta*, 1953, *11*, 95.
1594 P. Jollès and C. Fromageot, *Biochim. Biophys. Acta*, 1954, *14*, 228.
1595 P. Jollès and J. de Repentigny, *Biochim. Biophys. Acta*, 1954, *15*, 161.
1596 A. R. Jones, *Anal. Chem.*, 1952, *24*, 1055.
1597 A. R. Jones, E. J. Dowling and W. J. Skraba, *Anal. Chem.*, 1953, *25*, 394.
1598 J. I. M. Jones and S. E. Michael, *Nature*, 1950, *165*, 685.
1599 J. K. N. Jones, *J. Chem. Soc.*, *1944*, 333.
1600 J. K. N. Jones, *J. Chem. Soc.*, *1949*, 3141.
1601 J. K. N. Jones and S. R. Stitch, *Biochem. J.*, 1953, *53*, 679.
1602 T. S. G. Jones, *Biochem. J.*, 1948, *42*, lix.
1603 T. S. G. Jones, *Ann. N.Y. Acad. Sci.*, 1949, *51*, 909.
1604 T. S. G. Jones, *Brit. Med. Bull.*, 1954, *10*, 224.
1605 C. Jørgensen and H. Kofod, *Acta Chem. Scand.*, 1954, *8*, 941.
P. F. Jørgensen, see P. Fischer Jørgensen.
1606 N. R. Joseph, *J. Biol. Chem.*, 1938, *126*, 403.
1607 W. Juda, J. A. Marinsky and N. W. Rosenberg, *Ann. Rev. Phys. Chem.*, 1953, *4*, 373.
1608 M. Jutisz, *Thèse Doctorat ès Sciences*, Paris, 1949.
1609 M. Jutisz, *Chromatographie d'acides amines, peptides, protéines et constituants azotés des nucléoprotéines*, Centre de Documentation, Paris, 1950.
1610 M. Jutisz, *Bull. soc. chim. France*, *1952*, 821.
1611 M. Jutisz and E. Lederer, *Médecine et Biologie*, *1947*, No. 5, 229.
1612 M. Jutisz and E. Lederer, *Nature*, 1947, *159*, 445.
1613 M. Jutisz, M. Privat de Garilhe, M. Suquet and C. Fromageot, *Bull. soc. chim. biol.*, 1954, *36*, 117.
1614 M. Jutisz and S. Teichner, *Bull. soc. chim. France*, *1947*, 389.
1614a S. Kahn and D. E. Hawkinson, *J. Inorg. Nuclear Chem.*, 1956, *3*, 155.
1615 F. Kaiser, *Chem. Ber.*, 1955, *88*, 556.

1616   H. Kakihana, J. Chem. Soc. Japan, Pure Chem. Sect., 1950, 71, 481; 1951, 72,
       200: C.A., 1951, 45, 6125-c; 1952, 46, 3449-b.
1617   H. Kakihana, J. Chem. Soc. Japan, Pure Chem. Sect., 1951, 72, 255: C.A., 1952,
       46, 847-g.
1618   K. Kakihana and S. Kojima, Japan Analyst, 1953, 2, 421: C.A., 1954, 48,
       6307-i.
1619   H. Kalbe, Z. physiol. Chem., 1954, 297, 19.
1620   D. R. Kalkwarf and A. A. Frost, Anal. Chem., 1954, 26, 191.
1621   J. P. Kaltenbach and G. Kalnitsky, J. Biol. Chem., 1951, 192, 629.
1622   F. Kalz, P. Telner, J. H. Quastel and S. F. van Straten, Arch. Dermatol.
       Syphilol., 1953, 68, 167.
1623   B. Kamieński, Polska Akad. Umiejetności Rozprawy Wydziału Matemat-Przyrod.,
       1949, 74 A, 41 (No. 3): C.A., 1952, 46, 11013-c.
1624   B. Kamieński, Bull. intern. acad. polon. sci., Classe sci. math. nat., A, 1950,
       73.
1625   J. V. Karabinos, Euclides, 1954, 14, 263: C.A., 1955, 49, 4533-a.
1626   J. V. Karabinos and P. M. Hyde, J. Am. Chem. Soc., 1948, 70, 428.
1627   G. Karagunis, Helv. Chim. Acta, 1949, 32, 1840.
1628   G. Karagunis and G. Coumoulos, Nature, 1938, 142, 162.
1629   T. Kariyone, Y. Hashimoto and M. Kimura, Nature, 1951, 168, 511.
1630   T. Kariyone and Y. Hashimoto, Nature, 1951, 168, 739.
1631   T. Kariyone, Y. Hashimoto and M. Kimura, J. Pharm. Soc. Japan, 1953, 73,
       1093: C.A., 1954, 48, 1631-b.
1632   T. Kariyone and T. Inoue, J. Pharm. Soc. Japan, 1954, 74, 301: C.A., 1954,
       48, 7710-g.
1632a E. Karl-Kroupa, Anal. Chem., 1956, 28, 1091.
1633   M. L. Karnovsky and M. J. Johnson, Anal. Chem., 1949, 21, 1125.
1634   C. Karr Jr., W. D. Weatherford Jr. and R. G. Capell, Anal. Chem., 1954,
       26, 252.
1635   C. Karr Jr., W. D. Weatherford Jr., T. R. Kendrick III and R. G. Capell,
       Anal. Chem., 1954, 26, 1841.
1636   P. Karrer, A. Geiger, R. Legler, A. Rüegger and H. Salomon, Helv. Chim.
       Acta, 1939, 22, 1464.
1637   P. Karrer and C. H. Eugster, Helv. Chim. Acta, 1950, 33, 1172.
1638   P. Karrer and E. Jucker, Helv. Chim. Acta, 1945, 28, 300, 471.
1639   P. Karrer and E. Jucker, Helv. Chim. Acta, 1945, 28, 717.
1640   P. Karrer and E. Jucker, Carotenoids, Elsevier, New York-Amsterdam, 1950.
1641   P. Karrer, E. Jucker, J. Rutschmann and K. Steinlin, Helv. Chim. Acta, 1945,
       28, 1146.
1642   P. Karrer, R. Keller and G. Szönyi, Helv. Chim. Acta, 1943, 26, 38.
1643   P. Karrer and H. Schmidt, Helv. Chim. Acta, 1946, 29, 1853.
1644   M. Karschulin and Z. Svarc, Kem. Vjestnik (Archiv Kem. i Tehnol.), 1943, 17,
       99: C.A., 1946, 40, 5352-9.
1645   A. Katz and T. Reichstein, Helv. Chim. Acta, 1945, 28, 476.
1646   J. Katz and I. L. Chaikoff, J. Biol. Chem., 1954, 206, 887.
1647   J. Katz and I. L. Chaikoff, J. Am. Chem. Soc., 1955, 77, 2659.
1648   E. R. Katzenellenbogen, K. Dobriner and T. H. Kritchevsky, J. Biol. Chem.,
       1954, 207, 315.
1649   P. A. Katzman, M. Godfrid, C. K. Cain and E. A. Doisy, J. Biol. Chem.,
       1943, 148, 501.

1650 P. A. Katzman, E. E. Hays, C. K. Cain, J. J. Van Wyck, F. J. Reithel, S. A. Thayer, E. A. Doisy, W. L. Gaby, C. J. Carroll, R. D. Muir, L. R. Jones and N. J. Wade, *J. Biol. Chem.*, 1944, *154*, 475.

1651 H. P. Kaufmann, *Fette u. Seifen*, 1939, *46*, 268; 1940, *47*, 460.

1652 H. P. Kaufmann, *Angew. Chem.*, 1940, *53*, 98.

1653 H. P. Kaufmann, *French Pat.* 835,065 (1940): *C.A.*, 1942, *36*, 2436-3, 4728-8.

1654 H. P. Kaufmann, *Fette u. Seifen*, 1950, *52*, 713.

1655 H. P. Kaufmann, *Olearia*, 1950, *4*, 101: *C.A.*, 1950, *44*, 7073-g.

1656 H. P. Kaufmann and J. Budwig, *Fette u. Seifen*, 1951, *53*, 69, 253, 390, 408.

1657 H. P. Kaufmann and J. Budwig, *Fette u. Seifen*, 1952, *54*, 156: *C.A.*, 1952, *46*, 8703-b.

1658 H. P. Kaufmann and P. Kirsch, *Fette u. Seifen*, 1942, *49*, 841.

1659 H. P. Kaufmann and O. Schmidt, *Fette u. Seifen*, 1940, *47*, 294.

1660 H. P. Kaufmann and W. Wolf, *Fette u. Seifen*, 1943, *50*, 519.

1661 O. Kaufmann, *Österr. Chem.-Ztg.*, 1953, *54*, 110: *C.A.*, 1953, *47*, 7364-c.

1662 V. V. Kaval'skiĭ, *Biokhimiya*, 1948, *13*, 131: *C.A.*, 1948, *42*, 7817-a.

1662a A. Kawamura and H. Okamura, *Japan Analyst*, 1955, *4*, 163.

1662b A. Kawamura, H. Okamura and N. Kaneko, *Japan Analyst*, 1955, *4*, 157.

1663 E. Kawerau, *Biochem. J.*, 1951, *48*, 281.

1664 E. Kawerau and T. Wieland, *Nature*, 1951, *168*, 77.

1665 L. M. Kay and K. N. Trueblood, *Anal. Chem.*, 1954, *26*, 1566.

1666 G. Kayas, *Compt. rend.*, 1949, *228*, 1002.

1667 G. Kayas, *J. chim. phys.*, 1950, *47*, 408.

1668 M. A. G. Kaye and M. Stacey, *Biochem. J.*, 1951, *48*, 249.

1669 F. Kazmeier and A. Gassen, *Klin. Wochschr.*, 1954, *32*, 81: *C.A.*, 1954, *48*, 7674-e.

1670 J. Kebrle, H. Schmid, P. Waser and P. Karrer, *Helv. Chim. Acta*, 1953, *36*, 345.

1671 G. Kegeles and H. A. Sober, *Anal. Chem.*, 1952, *24*, 654.

1672 R. Kehl and B. Günther, *Naturwissenschaften*, 1954, *41*, 118.

1673 R. Kehl and B. Günther, *Z. physiol. Chem.*, 1954, *297*, 254.

1674 R. Kehl and W. Stich, *Z. physiol. Chem.*, 1951, *289*, 6.

1675 B. Keil, *Collection Czechoslov. Chem. Communs.*, 1954, *19*, 1006 (in Russian).

1676 A. E. Kellie, E. R. Smith and A. P. Wade, *Biochem. J.*, 1953, *53*, 578.

1677 A. E. Kellie and A. P. Wade, *Biochem. J.*, 1953, *53*, 582.

1678 N. F. Kember, *Analyst*, 1952, *77*, 78.

1679 N. F. Kember, P. J. Macdonald and R. A. Wells, *J. Chem. Soc.*, *1955*, 2273.

1680 N. F. Kember and R. A. Wells, *Analyst*, 1951, *76*, 579.

1681 N. F. Kember and R. A. Wells, *Chem. and Ind.*, *1952*, 1129.

1682 N. F. Kember and R. A. Wells, *Nature*, 1955, *175*, 512.

1682a N. F. Kember and R. A. Wells, *Analyst*, 1955, *80*, 735.

1683 A. R. Kemble and H. T. Macpherson, *Nature*, 1952, *170*, 664.

1684 A. R. Kemble and H. T. Macpherson, *Biochem. J.*, 1954, *56*, 548.

1685 A. R. Kemmerer and G. S. Fraps, *Ind. Eng. Chem. Anal. Ed.*, 1943, *15*, 714.

1686 W. Kemula, *Roczniki Chem.*, 1952, *26*, 694, 696.

1687 J. E. Kench, C. Gardikas and J. F. Wilkinson, *Biochem. J.*, 1950, *47*, 129.

1688 E. P. Kennedy, *J. Biol. Chem.*, 1953, *201*, 399.

1689 E. P. Kennedy and H. A. Barker, *Anal. Chem.*, 1951, *23*, 1033.

1690 E. P. Kennedy and S. W. Smith, *J. Biol. Chem.*, 1954, *207*, 153.

1691 G. Y. Kennedy, *Scand. J. Clin. & Lab. Invest.*, 1953, *5*, 281: *C.A.*, 1954, *48*, 1460-b.

1692 G. W. Kenner, A. R. Todd and R. F. Webb, *J. Chem. Soc.*, *1954*, 2843.
1693 G. W. Kenner, A. R. Todd and F. J. Weymouth, *J. Chem. Soc.*, *1952*, 3675.
1694 P. W. Kent, *Research (London)*, 1950, *3*, 427: *C.A.*, 1951, *45*, 1522-d.
1695 P. W. Kent, *J. Chem. Soc.*, *1951*, 364.
1696 P. W. Kent, G. Lawson and A. Senior, *Science*, 1951, *113*, 354.
1697 P. W. Kent and M. W. Whitehouse, *Biochemistry of the Aminosugars*, Butterworth, London, 1955.
1698 G. P. Kerby, *Proc. Soc. Exptl. Biol. Med.*, 1953, *83*, ?63; *J. Clin. Invest.*, 1954, *33*, 1168: *C.A.*, 1953, *47*, 10803-e; 1954, *48*, 13888-f.
1698a S. Kertes and M. Lederer, *Anal. Chim. Acta*, 1956, *15*, 543.
1699 A. S. Keston and S. Udenfriend, *Cold Spring Harbor Symposia Quant. Biol.*, 1950, *14*, 92: *C.A.*, 1950, *44*, 6471-g.
1700 A. S. Keston, S. Udenfriend and M. Levy, *J. Am. Chem. Soc.*, 1947, *69*, 3151.
1701 A. S. Keston, S. Udenfriend and M. Levy, *J. Am. Chem. Soc.*, 1950, *72*, 748.
1702 B. H. Ketelle and C. E. Boyd, *J. Am. Chem. Soc.*, 1947, *69*, 2800.
1703 B. H. Ketelle and G. E. Boyd, *J. Am. Chem. Soc.*, 1951, *73*, 1862.
1704 W. Keup, *Z. physiol. Chem.*, 1952, *291*, 223.
1705 J. A. Keverling Buisman, W. Stevens, and J. Van der Vliet, *Rec. trav. chim.*, 1947, *66*, 83.
1706 N. A. Khan, W. O. Lundberg and R. F. Holman, *J. Am. Chem. Soc.*, 1954, *76*, 1779.
1707 O. I. Khokhlova, *Aptechnoe Delo*, 1953, *2*, 22: *C.A.*, 1953, *47*, 10174-h.
1708 A. I. Khomlev and L. I. Tsimbalista, *Zhur. Anal. Khim.*, 1953, *8*, 217: *C.A.*, 1953, *47*, 11068.
1709 J. X. Khym and W. E. Cohn, *J. Am. Chem. Soc.*, 1953, *75*, 1153.
1710 J. X. Khym and W. E. Cohn, *Biochim. Biophys. Acta*, 1954, *15*, 139.
1711 J. X. Khym and D. G. Doherty, *J. Am. Chem. Soc.*, 1952, *74*, 3199.
1712 J. X. Khym, D. G. Doherty and W. E. Cohn, *J. Am. Chem. Soc.*, 1954, *76*, 5523.
1713 J. X. Khym and L. P. Zill, *J. Am. Chem. Soc.*, 1951, *73*, 2399; 1952, *74*, 2090.
714 H. Kiessling and J. Porath, *Acta Chem. Scand.*, 1954, *8*, 859.
1715 H. Kihara, W. G. McCullough and E. E. Snell, *J. Biol. Chem.*, 1952, *197*, 783.
1716 K. Kimura, N. Saito, H. Kakihana and T. Ishimori, *J. Chem. Soc. Japan, Pure Chem. Sect.*, 1953, *74*, 305: *C.A.*, 1953, *47*, 9850.
1717 B. H. Kindt, E. W. Balis and H. A. Liebhafsky, *Anal. Chem.*, 1952, *24*, 1501.
1718 E. J. King, M. Gilchrist and A. L. Tarnoky, *Biochem. J.*, 1946, *40*, 706.
1719 E. L. King and E. B. Dismukes, *J. Am. Chem. Soc.*, 1952, *74*, 1674.
1720 E. L. King and R. R. Walters, *J. Am. Chem. Soc.*, 1952, *74*, 4471.
1721 R. P. King, D. W. Bolin, W. E. Dinusson and M. L. Buchanan, *J. Animal Sci.*, 1953, *12*, 628: *C.A.*, 1954, *48*, 2946-h.
1722 D. S. Kinnory, Y. Takeda and D. M. Greenberg, *J. Biol. Chem.*, 1955, *212*, 379, 385.
1723 J. J. Kipling, *J. Chem. Soc.*, *1948*, 1487.
1724 J. G. Kirchner and G. J. Keiler, *J. Am. Chem. Soc.*, 1950, *72*, 1867.
1725 J. G. Kirchner and J. M. Miller, *Ind. Eng. Chem.*, 1952, *44*, 318.
1726 J. G. Kirchner and J. M. Miller, *J. Agr. Food Chem.*, 1953, *1*, 512: *C.A.*, 1953, *47*, 8924-g.
1727 J. G. Kirchner, J. M. Miller and G. J. Keller, *Anal. Chem.*, 1951, *23*, 420.
1728 J. G. Kirchner, A. N. Prater and A. J. Haagen-Smit, *Ind. Eng. Chem. Anal. Ed.*, 1946, *18*, 31.

1729 A. V. Kiselev, I. A. Vorms, V. V. Kiseleva and N. A. Shtokvish, *Zhur. Fiz. Khim.*, 1945, *19*, 83: *C.A.*, 1945, *39*, 3715-b.
1730 A. Kjaer and K. Rubinstein, *Acta Chem. Scand.*, 1953, *7*, 528.
1731 A. Kjaer and K. Rubinstein, *Nature*, 1953, *171*, 840.
1732 C. Klatzkin, *Nature*, 1952, *169*, 422.
1733 R. Klement, *Z. anal. Chem.*, 1944, *127*, 2.
1734 R. Klement, *Z. anorg. Chem.*, 1949, *260*, 267.
1735 R. Klement and R. Dmytruk, *Z. anal. Chem.*, 1948, *128*, 106.
1736 W. Klementschitz and P. Mathes, *Scientia Pharm.*, 1952, *20*, 65: *C.A.*, 1952, *46*, 9781-g.
1737 E. Klenk, *Z. physiol. Chem.*, 1951, *288*, 216.
1738 E. Klenk and W. Bongard, *Z. physiol. Chem.*, 1952, *290*, 181.
1739 E. Klenk and W. Bongard, *Z. physiol. Chem.*, 1953, *292*, 51.
1740 R. Klevstrand and A. Nordal, *Acta Chem. Scand.*, 1950, *4*, 1320.
1741 M. W. Klohs, R. Arons ,M. D. Draper, F. Keller, S. Koster, W. Malesh and F. J. Petracek, *J. Am. Chem. Soc.*, 1952, *74*, 5107.
1742 R. G. Kluener, *J. Bact.*, 1949, *57*, 101: *C.A.*, 1949, *43*, 4322-e.
1743 M. Klungsøyr, R. J. Sirny and C. A. Elvehjem, *J. Biol. Chem.*, 1951, *189*, 557.
1744 O. Knessl, B. Keil, A. Malý and F. Šorm, *Collection Czechoslov. Chem. Communs.*, 1950, *15*, 918.
1745 O. Knessl and A. Vlastiborová, *Collection Czechoslov. Chem. Communs.*, 1954, *19*, 782.
1746 C. A. Knight, *J. Biol. Chem.*, 1951, *190*, 753.
1747 H. S. Knight and S. Groennings, *Anal. Chem.*, 1954, *26*, 1549.
1748 V. Kobrle and R. Zahradník, *Chem. Listy*, 1954, *48*, 1189: *C.A.*, 1954, *48*, 13546-h.
1749 J. M. Koch, *Ind. Eng. Chem. Anal. Ed.*, 1944, *16*, 25.
1750 R. Koch and H. Hanson, *Z. physiol. Chem.*, 1953, *292*, 180.
1751 C. D. Kochakian and G. Stidworthy, *J. Biol. Chem.*, 1952, *199*, 607.
1752 E. Kodicek and D. R. Ashby, *Biochem. J.*, 1954, *57*, xii.
1753 E. Kodicek and D. R. Ashby, *Biochem. J.*, 1954, *57*, xiii.
1754 E. Kodicek and K. K. Reddi, *Nature*, 1951, *168*, 475.
1755 B. A. Koechlin and H. D. Parish, *J. Biol. Chem.*, 1953, *205*, 597.
1756 H. J. Koepsell, H. M. Tsuchiya, N. N. Hellman, A. Kazenko, C. A. Hoffman, E. S. Sharpe and R. W. Jackson, *J. Biol. Chem.*, 1953, *200*, 793.
1757 M. Kofler, *Helv. Chim. Acta*, 1942, *25*, 1469; 1943, *26*, 2166; 1945, *28*, 26.
1758 M. Kofler, *Helv. Chim. Acta*, 1945, *28*, 702.
1759 M. Kofler, *Helv. Chim. Acta*, 1947, *30*, 1053.
1760 W. Kofler, *Monatsh. Chem.*, 1949, *80*, 694.
1761 E. Kofrányi, *Z. physiol. Chem.*, 1955, *299*, 129.
1762 R. R. Kohn, *Nature*, 1953, *172*, 1185.
1763 S. Kojima and H. Kakihana, *Japan Analyst*, 1954, *3*, 42: *C.A.*, 1954, *48*, 6899-h.
1764 A. J. Kolka, H. D. Orloff and M. E. Griffing, *J. Am. Chem. Soc.*, 1954, *76*, 3940.
1765 J. Komenda, *Chem. Listy*, 1953, *47*, 1877: *C.A.*, 1954, *48*, 3850-g.
1765a S. K. Kon, *Biochem. Soc. Symposium No. 13* ("*The Biochemistry of Vitamin B₁₂*"(, p. 13 (Cambridge University Press, 1955).
1766 I. M. Korenman and Z. V. Kraïnova, *Zhur. Priklad. Khim.*, 1946, *19*, 604: *C.A.*, 1947, *41*, 2347-h.
1767 S. Kořístek and Ž. Procházka, *Chem. Listy*, 1951, *45*, 272.

1768   A. Kornberg and W. E. Pricer Jr., *J. Am. Chem. Soc.*, 1952, *74*, 1617.

1769   A. Kornberg and W. E. Pricer Jr., *J. Biol. Chem.*, 1953, *204*, 345.

1770   W. Koschara, *Z. physiol. Chem.*, 1943, *277*, 159.

1771   J. V. Koštíř and K. Slavík, *Collection Czechoslov. Chem. Communs.*, 1950, *15*, 17: *C.A.*, 1950, *44*, 8817-e.

1772   M. Kotake, T. Sakan, N. Nakamura and S. Senoh, *J. Am. Chem. Soc.*, 1951, *73*, 2973.

1773   G. N. Kowkabany, *Adv. Carbohydrate Chem.*, 1954, *9*, 304.

1774   G. N. Kowkabany and H. G. Cassidy, *Anal. Chem.*, 1950, *22*, 817.

1775   G. N. Kowkabany and H. G. Cassidy, *Anal. Chem.*, 1952, *24*, 643.

1776   R. Kozak and H. F. Walton, *J. Phys. Chem.*, 1945, *49*, 471.

1777   P. J. G. Kramer and H. van Duin, *Rec. trav. chim.*, 1954, *73*, 63.

1778   V. S. Krasnova, *Zhur. Priklad. Khim.*, 1945, *18*, 86: *C.A.*, 1945, *39*, 5399-6.

1778a   K. A. Kraus, T. A. Carlson and J. S. Johnson, *Nature*, 1956, *177*, 1128.

1779   K. A. Kraus and G. E. Moore, *J. Am. Chem. Soc.*, 1949, *71*, 3263.

1780   K. A. Kraus and G. E. Moore, *J. Am. Chem. Soc.*, 1949, *71*, 3855.

1781   K. A. Kraus and G. E. Moore, *J. Am. Chem. Soc.*, 1950, *72*, 4293.

1782   K. A. Kraus and G. E. Moore, *J. Am. Chem. Soc.*, 1950, *72*, 5792.

1783   K. A. Kraus and G. E. Moore, *J. Am. Chem. Soc.*, 1951, *73*, 9.

1784   K. A. Kraus and G. E. Moore, *J. Am. Chem. Soc.*, 1951, *73*, 13.

1785   K. A. Kraus and G. E. Moore, *J. Am. Chem. Soc.*, 1951, *73*, 2900.

1786   K. A. Kraus and G. E. Moore, *J. Am. Chem. Soc.*, 1953, *75*, 1460.

1787   K. A. Kraus and F. Nelson, *J. Am. Chem. Soc.*, 1953, *75*, 3273.

1788   K. A. Kraus and F. Nelson, *J. Am. Chem. Soc.*, 1954, *76*, 984.

1789   K. A. Kraus, F. Nelson and J. F. Baxter, *J. Am. Chem. Soc.*, 1953, *75*, 2768.

1790   K. A. Kraus, F. Nelson, F. B. Clough and R. C. Carlston, *J. Am. Chem. Soc.*, 1955, *77*, 1391.

1791   K. A. Kraus, F. Nelson and G. W. Smith, *J. Phys. Chem.*, 1954, *58*, 11.

1792   K. A. Kraus and G. W. Smith, *J. Am. Chem. Soc.*, 1950, *72*, 4329.

1793   H. R. Kraybill, M. H. Thornton and K. E. Eldridge, *Ind. Eng. Chem.*, 1940, *32*, 1138.

1794   F. Krczil, *Aktive Tonerde, ihre Herstellung und Anwendung*, Stuttgart, 1938.

1795   H. Krebs and R. Rasche, *Naturwissenschaften*, 1954, *41*, 63.

1796   H. A. Krebs and R. Hems, *Biochim. Biophys. Acta*, 1953, *12*, 172.

1797   H. Krebs and A. Wankmüller, *Naturwissenschaften*, 1953, *40*, 623.

1797a   T. R. E. Kressman, *J. Phys. Chem.*, 1952, *56*, 118.

1798   T. R. E. Kressman and J. A. Kitchener, *J. Chem. Soc.*, 1949, 1190.

1799   T. R. E. Kressman and J. A. Kitchener, *Discuss. Faraday Soc.*, 1949, *7*, 90.

1800   V. L. Kretovich and A. A. Bundel, *Doklady Akad. Nauk S.S.S.R.*, 1950, *73*, 137: *C.A.*, 1950, *44*, 10603-a.

1801   V. L. Kretovich, T. V. Drozdova and I. S. Petrova, *Doklady Akad. Nauk S.S.S.R.*, 1951, *80*, 409: *C.A.*, 1952, *46*, 1923-c.

1802   K. A. Krieger, *J. Am. Chem. Soc.*, 1941, *63*, 2712.

1803   D. Kritchevsky and M. Calvin, *J. Am. Chem. Soc.*, 1950, *72*, 4330.

1804   D. Kritchevsky and M. R. Kirk, *Arch. Biochem. Biophys.*, 1952, *35*, 346.

1805   D. Kritchevsky and M. R. Kirk, *J. Am. Chem. Soc.*, 1952, *74*, 4484.

1806   D. Kritchevsky and M. R. Kirk, *J. Am. Chem. Soc.*, 1952, *74*, 4713.

1807   T. H. Kritchevsky and A. Tiselius, *Science*, 1951, *114*, 299.

1808   M. G. Kritsman and M. B. Lebedeva, *Ukrain. Biokhim. Zhurn.*, 1950, *22*, 430: *C.A.*, 1954, *48*, 2159-g.

1809 T. D. Kroner, W. Tabroff and J. J. McGarr, *J. Am. Chem. Soc.*, 1953, *75*, 4084.
1810 W. Kruckenberg, *Z. physiol. Chem.*, 1949, *284*, 19, 40.
1811 J. Kruh, *Bull. soc. chim. biol.*, 1952, *34*, 778.
1812 J. Kruh, J. C. Dreyfus and G. Schapira, *Bull. soc. chim. biol.*, 1952, *34*, 773.
1813 H. Kubli, *Helv. Chim. Acta*, 1947, *30*, 453.
1814 F. A. Kuehl Jr., M. N. Bishop, L. Chaiet and K. Folkers, *J. Am. Chem. Soc.*, 1951, *73*, 1770.
1815 F. A. Kuehl Jr., R. L. Peck, C. E. Hoffhine Jr., R. P. Graber and K. Folkers, *J. Am. Chem. Soc.*, 1946, *68*, 1460.
1816 C. A. Kuether and M. E. Haney Jr., *Science*, 1955, *121*, 65.
1817 R. Kuhn and R. Brossmer, *Chem. Ber.*, 1954, *87*, 123.
1818 R. Kuhn, A. Gauhe and H. H. Baer, *Chem. Ber.*, 1953, *86*, 827.
1819 R. Kuhn, A. Gauhe and H. H. Baer, *Chem. Ber.*, 1954, *87*, 289.
1820 R. Kuhn and W. Kirschenlohr, *Chem. Ber.*, 1954, *87*, 1547.
1821 R. Kuhn and E. Lederer, *Naturwissenschaften*, 1931, *19*, 306; *Ber.*, 1931, *64*, 1349.
1822 R. Kuhn and H. W. Ruelius, *Chem. Ber.*, 1950, *83*, 420.
1823 R. Kuhn and K. Wallenfels, *Ber.*, 1939, *72*, 1407.
1824 R. Kuhn and T. Wieland, *Ber.*, 1940, *73*, 962.
1825 R. Kuhn, A. Winterstein and E. Lederer, *Z. physiol. Chem.*, 1931, *197*, 141.
1826 R. Kuhn, F. Zilliken and A. Gauhe, *Chem. Ber.*, 1953, *86*, 466.
1827 E. Kulonen, *Scand. J. Clin. & Lab. Invest.*, 1953, *5*, 72: *C.A.*, 1953, *47*, 7573-e.
1828 E. Kulonen, *Suomen Kemistilehti*, 1955, *28 B*, 105: *C.A.*, 1955, *49*, 8744-b.
1829 E. Kulonen, E. Carpén and T. Ruokolainen, *Scand. J. Clin. & Lab. Invest.*, 1952, *4*, 189: *C.A.*, 1953, *47*, 1766-c.
1830 S. Kumé, K. Otozai and H. Watanabé, *Nature*, 1950, *166*, 1076.
1831 S. Kumé, T. Yamamoto, K. Otozai and S. Fukushima, *Bull. Chem. Soc. Japan*, 1953, *26*, 93: *C.A.*, 1953, *47*, 12130-f.
1832 R. Kunin, *Anal. Chem.*, 1949, *21*, 87.
1833 R. Kunin, *Anal. Chem.*, 1950, *22*, 64.
1834 R. Kunin, *Anal. Chem.*, 1951, *23*, 45.
1835 R. Kunin, *Ind. Eng. Chem.*, 1951, *43*, 102.
1836 R. Kunin, *U.S. Pat.* 2,549,378 (1951): *C.A.*, 1951, *45*, 8033-h.
1837 R. Kunin and R. J. Myers, *J. Am. Chem. Soc.*, 1947, *69*, 2874.
1838 R. Kunin and R. J. Myers, *Ion Exchange Resins*, John Wiley and Sons, New York, 1950.
1839 F. E. Kurtz, *J. Am. Chem. Soc.*, 1952, *74*, 1902.
1840 M. Kutàček and J. Koloušek, *Sborník Českoslov. Akad. Zeměděl. Ved.*, 1953, *26 A*, 575: *C.A.*, 1954, *48*, 5259-c.
1841 J. Laberrigue-Frolow and M. Lederer, *J. phys. radium*, 1955, *16*, 346.
1842 G. Labruto and G. Stagno d'Alcontres, *Ann. chim. applicata*, 1948, *38*, 320.
1843 A. Lacourt, *Mikrochim. Acta*, 1955, 824.
1844 A. Lacourt, H. Gillard and M. Van der Walle, *Mikrochem. ver. Mikrochim. Acta*, 1950, *35*, 262.
1845 A. Lacourt, J. Gillard and M. Van der Walle, *Nature*, 1950, *166*, 225.
1846 A. Lacourt and P. Heyndryckx, *Mikrochim. Acta*, *1954*, 630.
1847 A. Lacourt and G. Sommereyns, *Mikrochim. Acta*, *1954*, 550.
1848 A. Lacourt and G. Sommereyns, *Mikrochim. Acta*, *1954*, 604.
1849 A. Lacourt, G. Sommereyns and M. Claret, *Mikrochem. ver. Mikrochim. Acta*, 1951, *38*, 444.

1850   A. Lacourt, G. Sommereyns and E. DeGeyndt, *Mededel. Vlaam. Chem. Ver.*, 1950, *12*, 91: *C.A.*, 1950, *44*, 10572-g.

1851   A. Lacourt, G. Sommereyns, E. DeGeyndt, J. Baruh and J. Gillard, *Mikrochem. ver. Mikrochim. Acta*, 1949, *34*, 215.

1852   A. Lacourt, G. Sommereyns, E. DeGeyndt, J. Baruh and J. Gillard, *Nature*, 1949, *163*, 999.

1853   A. Lacourt, G. Sommereyns, E. DeGeyndt and O. Jacquet, *Mikrochem. ver. Mikrochim. Acta*, 1951, *36/37*, 117.

1854   A. Lacourt, G. Sommereyns, J. Hoffmann, S. Frank-Frederic and G. Wantier, *Anal. Chim. Acta*, 1953, *8*, 444.

1855   A. Lacourt, G. Sommereyns, J. Hoffmann, A. Stadler and G. Wantier, *Compt. rend.*, 1952, *234*, 2365.

1856   A. Lacourt, G. Sommereyns, O. Jacquet and G. Wantier, *Bull. soc. chim. France, 1951*, 873.

1857   A. Lacourt, G. Sommereyns and J. Soete, *Mikrochem. ver. Mikrochim. Acta*, 1951, *38*, 348.

1858   A. Lacourt, G. Sommereyns, A. Stadler-Denis and Wantier, *Mikrochemie ver. Mikrochim. Acta*, 1952/53, *40*, 268.

1859   A. Lacourt, G. Sommereyns and G. Wantier, *Compt. rend.*, 1951, *232*, 2426.

1860   A. Lacourt, G. Sommereyns and G. Wantier, *Mikrochim. Acta, 1954*, 240.

1861   I. M. Ladenbauer, *Mikrochim. Acta, 1955*, 139.

1862   I. M. Ladenbauer, L. K. Bradacs and F. Hecht, *Mikrochim. Acta, 1954*, 388.

1862a   I. M. Ladenbauer and F. Hecht, *Mikrochim. Acta, 1954*, 397.

1862b   I. M. Ladenbauer and O. Slama, *Mikrochim. Acta, 1955*, 903.

1863   O. Lagerstrom, O. Samuelson and A. Scholander, *Svensk Papperstidn.*, 1948, *108*, 439.

1864   O. Lagerstrom, O. Samuelson and A. Scholander, *Svensk Papperstidn.*, 1949, *52*, 113: *C.A.*, 1949, *43*, 4851-d.

1865   A. Lahiri and E. Mikolajewski, *Nature*, 1945, *155*, 77.

1866   R. A. Laidlaw and S. G. Reid, *Nature*, 1950, *166*, 476.

1867   T. K. Lakshmanan and S. Lieberman, *Arch. Biochem. Biophys.*, 1954, *53*, 258.

1868   K. Lakshminarayanan, *Arch. Biochem. Biophys.*, 1954, *49*, 396.

1869   K. Lakshminarayanan, *Proc. Indian Acad. Sci.*, 1954, *40 B*, 167: *C.A.*, 1955, *49*, 7635-f.

1870   S. G. Laland and W. G. Overend, *Acta Chem. Scand.*, 1954, *8*, 192.

1871   W. A. LaLande Jr., *Ind. Eng. Chem.*, 1941, *33*, 108.

1872   J. P. Lambooy, *J. Am. Chem. Soc.*, 1954, *76*, 133.

1873   C.-G. Lamm, *Acta Chem. Scand.*, 1953, *7*, 1420.

1874   J. O. Lampen, *J. Biol. Chem.*, 1953, *204*, 999.

1875   I. Landler, *Compt. rend.*, 1947, *225*, 234.

1876   W. A. Landmann, M. P. Drake and J. Dillaha, *J. Am. Chem. Soc.*, 1953, *75*, 3638.

1877   A. J. Landua and J. Awapara, *Science*, 1949, *109*, 385.

1878   A. J. Landua, R. Fuerst and J. Awapara, *Anal. Chem.*, 1951, *23*, 162.

1879   J.-M. Landucci and M. Pimont, *Bull. soc. chim. biol.*, 1953, *35*, 1041.

1879a   H. Langevin-Joliot and M. Lederer, *J. phys. radium*, 1956, *17*, 497.

1880   T. Langvad, *Acta Chem. Scand.*, 1954, *8*, 526.

1881   M. C. Lanning and S. S. Cohen, *J. Biol. Chem.*, 1951, *189*, 109.

1882   J. E. Larson and S. H. Harvey, *Chem. and. Ind.*, *1954*, 45.

1883   D. E. Laskowski and W. C. McCrone, *Anal. Chem.*, 1951, *23*, 1579.

1884 D. E. Laskowski and R. E. Putscher, *Anal. Chem.*, 1952, *24*, 965.
1885 J. Latinák, *Chem. Listy*, 1954, *48*, 843: *C.A.*, 1954, *48*, 14213-f.
1886 W. Lautsch, G. Manecke and W. Broser, Z. *Naturforsch.*, 1953, *8 B*, 232.
1887 D. Lawday, *Nature*, 1952, *170*, 415.
1888 A. S. C. Lawrence and D. Barby, *Discuss. Faraday Soc.*, 1949, *7*, 255.
1889 A. Lawson, H. V. Morley and L. I. Woolf, *Biochem. J.*, 1950, *47*, 513.
1890 C. H. Lea and D. N. Rhodes, *Biochem. J.*, 1953, *54*, 467.
1891 C. H. Lea and D. N. Rhodes, *Biochem. J.*, 1954, *57*, xxiii.
1892 C. H. Lea and D. N. Rhodes, *Biochem. J.*, 1955, *59*, v.
1893 B. E. Leach, W. H. DeVries, H. A. Nelson, W. G. Jackson and J. S. Evans, *J. Am. Chem. Soc.*, 1951, *73*, 2797.
1894 B. E. Leach and C. M. Teeters, *J. Am. Chem. Soc.*, 1951, *73*, 2794.
1895 H. Lecoq, *Bull. Soc. Roy. Sci. Liège*, 1943, *12*, 316: *C.A.*, 1928, *42*, 7490-a.
1896 H. Lecoq, *Bull. Soc. Roy. Sci. Liège*, 1943, *12*, 323: *C.A.*, 1948, *42*, 7464-g.
1897 H. Lecoq, *Bull. Soc. Roy. Sci. Liège*, 1944, *13*, 20: *C.A.*, 1948, *42*, 6703-h.
1898 E. Lederer, *Bull. soc. chim. biol.*, 1938, *20*, 554.
1899 E. Lederer, *Bull. soc. chim. France*, (5), 1939, *6*, 897.
1900 E. Lederer, *Compt. rend.*, 1939, *209*, 528.
1901 E. Lederer, *Bull. soc. chim. biol.*, 1943, *25*, 1073.
1902 E. Lederer, *Biochim. Biophys. Acta*, 1952, *9*, 92.
1903 E. Lederer, *Bull. soc. chim. France, 1952*, 815.
1904 E. Lederer, *Mises au point de chimie analytique*, 3ème série, Masson, Paris, 1955.
1905 E. Lederer and R. Glaser, *Compt. rend.*, 1938, *207*, 454.
1906 E. Lederer, and C. Huttrer, *Bull. soc. chim. biol.*, 1942, *24*, 1055.
1907 E. Lederer, F. Marx, D. Mercier and G. Pérot, *Helv. Chim. Acta*, 1946, *29*, 1354.
1908 E. Lederer, D. Mercier and G. Pérot, *Bull. soc. chim. France, 1947*, 345.
1909 E. Lederer and J. Polonsky, *Bull. soc. chim. biol.*, 1942, *24*, 1386.
1910 E. Lederer and J. Pudles, *Bull. soc. chim. biol.*, 1951, *33*, 1003.
1911 E. Lederer and P. K. Tchen, *Biochim. Biophys. Acta*, 1947, *1*, 35.
1912 E. Lederer, P. K. Tchen, H. Pénau and G. Hagemann, *Bull. soc. chim. biol.*, 1944, *26*, 1032.
1913 E. Lederer and R. Tixier, *Compt. rend.*, 1947, *225*, 531.
1914 M. Lederer, *Anal. Chim. Acta*, 1948, *2*, 261.
1915 M. Lederer, *Nature*, 1948, *162*, 776.
1916 M. Lederer, *Australian J. Sci.*, 1949, *11*, 174.
1917 M. Lederer, *Australian J. Sci.*, 1949, *11*, 208.
1918 M. Lederer, *Nature*, 1949, *163*, 598.
1919 M. Lederer, *Anal. Chim. Acta*, 1949, *3*, 476.
1920 M. Lederer, *Science*, 1949, *110*, 115.
1921 M. Lederer, *Anal. Chim. Acta*, 1950, *4*, 629.
1922 M. Lederer, *J. Proc. Roy. Australian Chem. Inst.*, 1950, *17*, 308: *C.A.*, 1951, *45*, 3222-c.
1923 M. Lederer, *Nature*, 1950, *165*, 529.
1924 M. Lederer, *Science*, 1950, *112*, 504.
1925 M. Lederer, *Anal. Chim. Acta*, 1951, *5*, 185.
1926 M. Lederer, *Anal. Chim. Acta*, 1952, *6*, 267.
1927 M. Lederer, *Anal. Chim. Acta*, 1952, *7*, 458.
1928 M. Lederer, *Anal. Chim. Acta*, 1953, *8*, 134.
1929 M. Lederer, *Anal. Chim. Acta*, 1953, *8*, 259.

1930 M. Lederer, *Compt. rend.*, 1953, *236*, 1557.
1931 M. Lederer, *Nature*, 1953, *172*, 727.
1932 M. Lederer, *Thèse de doctorat*, Paris (1954).
1933 M. Lederer, *Anal. Chim. Acta*, 1954, *11*, 132.
1934 M. Lederer, *Anal. Chim. Acta*, 1954, *11*, 524.
1935 M. Lederer, *Anal. Chim. Acta*, 1954, *11*, 528.
1936 M. Lederer, *Introduction to Paper Electrophoresis and related Methods*, Elsevier, Amsterdam, 1955.
1937 M. Lederer, *Anal. Chim. Acta*, 1955, *12*, 142.
1938 M. Lederer, *Anal. Chim. Acta*, 1955, *13*, 350.
1939 M. Lederer, *Anal. Chim. Acta*, 1955, *12*, 146.
1940 M. Lederer, *Nature*, 1955, *176*, 462.
1941 M. Lederer, *Research (London)*, 1955, *8*, 357.
1941a M. Lederer, *Anal. Chim. Acta*, 1956, *15*, 46.
1941b M. Lederer, *Anal. Chim. Acta*, 1956, *15*, 122.
1942 M. Lederer and I. Cook, *Australian J. Sci.*, 1951, *14*, 56: *C.A.*, 1952, *46*, 372-g.
1942a M. Lederer and S. Kertes, *Anal. Chim. Acta*, 1956, *15*, 226.
1943 M. Lederer and H. Silberman, *Anal. Chim. Acta*, 1952, *6*, 133.
1944 M. Lederer and F. L. Ward, *Australian J. Sci.*, 1951, *13*, 114: *C.A.*, 1951, *45*, 4603-c.
1945 M. Lederer and F. L. Ward, *Anal. Chim. Acta*, 1952, *6*, 355.
1946 H. Lee, *J. Soc. Leather Trades' Chemists*, 1950, *34*, 150: *C.A.*, 1950, *44*, 9714-d.
1947 N. S. Leeds, D. K. Fukushima and T. F. Gallagher, *J. Am. Chem. Soc.*, 1954, *76*, 2943.
1948 J. Legge, private communication.
1949 A. L. Lehninger and G. D. Greville, *Biochim. Biophys. Acta*, 1953, *12*, 188.
1950 E. Leifer, W. H. Langham, J. F. Nyc and H. K. Mitchell, *J. Biol. Chem.*, 1950, *184*, 589.
1951 E. Leifer, L. J. Roth, D. S. Hogness and M. H. Corson, *J. Biol. Chem.*, 1951, *190*, 595.
1952 T. Leigh, *Discuss. Faraday Soc.*, 1949, *7*, 311.
1953 J. G. Leitner and G. P. Kerby, *Stain Technol.*, 1954, *29*, 257: *C.A.*, 1954, *48*, 13778-f.
1954 R. Lemberg, J. P. Callaghan, D. E. Tandy and N. E. Goldsworthy, *Australian J. Exptl. Biol. Med. Sci.*, 1948, *26*, 9: *C.A.*, 1950, *44*, 3079-b.
1955 R. Lemberg, D. E. Tandy and N. E. Goldsworthy, *Nature*, 1946, *157*, 103.
1956 R. U. Lemieux and H. F. Bauer, *Anal. Chem.*, 1954, *26*, 920.
1957 R. M. Lemmon, W. Tarpey and K. G. Scott, *J. Am. Chem. Soc.*, 1950, *72*, 758.
1958 J. Lenoir, *Bull. soc. chim. France*, (5), 1942, *9*, 475.
1959 J. Lens and A. Evertzen, *Rec. trav. chim.*, 1952, *71*, 43.
1960 J. Lens, H. G. Wijmenga, R. Wolff, R. Karlin, K. C. Winkler and P. G. De Haan, *Biochim. Biophys. Acta*, 1952, *8*, 56.
1961 N. J. Leonard and W. J. Middleton, *J. Am. Chem. Soc.*, 1952, *74*, 5114.
1962 G. Leonhardi, *Naturwissenschaften*, 1954, *41*, 141.
1963 G. A. LePage and G. C. Mueller, *J. Biol. Chem.*, 1949, *180*, 975.
1964 S. Lepkovsky and E. Nielsen, *J. Biol. Chem.*, 1942, *144*, 135.
1965 S. Lepkovsky, E. Roboz and A. J. Haagen-Smit, *J. Biol. Chem.*, 1943, *149*, 195.
1966 P. Lerch and S. Neukomm, *Schweiz. med. Wochschr.*, 1954, *84*, 515: *C.A.*, 1954, *48*, 11121-c.

1967　L. S. Lerman, *Nature*, 1953, *172*, 635.
1968　L. S. Lerman, *Proc. Natl. Acad. Sci. U.S.*, 1953, *39*, 232: *C.A.*, 1953, *47*, 8167-h.
1969　L. S. Lerman, *Biochim. Biophys. Acta*, 1955, *18*, 132.
1970　M. Lerner and W. Rieman III, *Anal. Chem.*, 1954, *26*, 610.
1971　A. L. LeRosen, *J. Am. Chem. Soc.*, 1942, *64*, 1905.
1972　A. L. LeRosen, *J. Am. Chem. Soc.*, 1945, *67*, 1683.
1973　A. L. LeRosen, *J. Am. Chem. Soc.*, 1947, *69*, 87.
1974　A. L. LeRosen, J. K. Carlton and P. B. Moseley, *Anal. Chem.*, 1953, *25*, 666.
1975　A. L. LeRosen and A. May, *Anal. Chem.*, 1948, *20*, 1090.
1976　A. L. LeRosen, P. H. Monaghan, C. A. Rivet and E. D. Smith, *Anal. Chem.*, 1951, *23*, 730.
1977　A. L. LeRosen, P. H. Monaghan, C. A. Rivet and E. D. Smith, *Proc. Louisiana Acad. Sci.*, 1951, *12*, 99.
1978　A. L. LeRosen, P. H. Monaghan, C. A. Rivet, E. D. Smith and H. A. Suter, *Anal. Chem.*, 1950, *22*, 809.
1979　A. L. LeRosen, R. T. Moravek and J. K. Carlton, *Anal. Chem.*, 1952, *24*, 1335.
1980　A. L. LeRosen and G. A. Rivet, *Anal. Chem.*, 1948, *20*, 1093.
1981　A. L. LeRosen and L. Zechmeister, *J. Am. Chem. Soc.*, 1942, *64*, 1075.
1982　S. Leskowitz and E. A. Kabat, *J. Am. Chem. Soc.*, 1954, *76*, 4887.
1983　R. J. LeStrange and R. H. Müller, *Anal. Chem.*, 1954, *26*, 953.
1984　F. Leuschner, *Arch. exptl. Pathol. Pharmakol.*, 1954, *221*, 323: *C.A.*, 1954, *48*, 8853-h.
1985　F. Leuthardt and E. Testa, *Helv. Chim. Acta*, 1951, *34*, 931.
1986　A. A. Levi, *Biochem. J.*, 1948, *43*, 257.
1987　A. A. Levi, *Discuss. Faraday Soc.*, 1949, *7*, 124.
1988　A. A. Levi, S. G. Terjesen and I. C. I., *Brit. Pat.* 569,844 (1945): *C.A.*, 1947, *41*, 6026-h.
1989　C. Levine and E. Chargaff, *J. Biol. Chem.*, 1951, *192*, 465.
1990　C. Levine and E. Chargaff, *J. Biol. Chem.*, 1951, *192*, 481.
1991　A. L. Levy, *Nature*, 1954, *174*, 126.
1992　A. L. Levy and D. Chung, *Anal. Chem.*, 1953, *25*, 396.
1993　M. F. Levy, *Anal. Chem.*, 1954, *26*, 1849.
1994　B. W. Lew, M. L. Wolfrom and R. M. Goepp Jr., *J. Am. Chem. Soc.*, 1945, *67*, 1865.
1995　B. W. Lew, M. L. Wolfrom and R. M. Goepp Jr., *J. Am. Chem. Soc.*, 1946, *68*, 1449.
1996　J. A. Lewis and J. M. Griffiths, *Analyst*, 1951, *76*, 388.
1997　J. C. Lewis and N. S. Snell, *J. Am. Chem. Soc.*, 1951, *73*, 4812.
1998　H. Leyon, *Arkiv Kemi*, 1949, *1*, 313.
1999　C. H. Li and L. Ash, *J. Biol. Chem.*, 1953, *203*, 419.
2000　C. H. Li, L. Ash and H. Papkoff, *J. Am. Chem. Soc.*, 1952, *74*, 1923.
2001　C. H. Li, I. I. Geschwind, J. S. Dixon, A. L. Levy and J. I. Harris, *J. Biol. Chem.*, 1955, *213*, 171.
2002　C. H. Li and K. O. Pedersen, *Arkiv Kemi*, 1950, *1*, 533.
2003　C. H. Li, A. Tiselius, K. O. Pedersen, L. Hagdahl and H. Carstensen, *J. Biol. Chem.*, 1951, *190*, 317.
2004　L. A. Liberman, A. Zaffaroni and E. Stotz, *J. Am. Chem. Soc.*, 1951, *73*, 1387.
2004a　D. H. Lichtenfels, S. A. Fleck and F. H. Burow, *Anal. Chem.*, 1955, *27*, 1510.
2005　H. F. Liddell and H. N. Rydon, *Biochem. J.*, 1944, *38*, 68.

2006　S. Lieberman, K. Dobriner, B. R. Hill, L. F. Fieser and C. P. Rhoads, *J. Biol. Chem.*, 1948, *172*, 263.

2007　S. Lieberman and D. K. Fukushima, *J. Am. Chem. Soc.*, 1950, *72*, 5211.

2008　S. Lieberman, D. K. Fukushima and K. Dobriner, *J. Biol. Chem.*, 1950, *182*, 299.

2009　O. Liebknecht, *U.S. Pat.* 2,191,060, 2,206,007: *C.A.*, 1940, *34*, 4501–1; 7503–6.

2010　H. Lieck and H. Willstaedt, *Svensk Kem. Tid.*, 1945, *57*, 134: *C.A.*, 1946, *40*, 4759-3.

2011　F. W. Lima, *J. Chem. Educ.*, 1954, *31*, 153: *C.A.*, 1954, *48*, 6174-b.

2012　E. F. Lind, H. C. Lane and L. S. Gleason, *Plant Physiol.*, 1953, *28*, 325: *C.A.*, 1953, *47*, 9437-d.

2013　O. Lindan and E. Work, *Biochem. J.*, 1951, *48*, 337.

2014　B. Lindberg and B. Wickberg, *Acta Chem. Scand.*, 1954, *8*, 569.

2015　O. Lindberg and L. Ernster, *Science Tools*, 1955, *2*, 7.

2016　S. Lindenbaum, T. V. Peters Jr. and W. Rieman III, *Anal. Chim. Acta*, 1954, *11*, 530.

2017　G. Lindhard Christensen and B. K. Jensen, *Dansk Tids. Farm.*, 1947, *21*, 68.

2018　R. Lindner, *Z. physik. Chem.*, 1944, *194*, 51.

2019　R. Lindner, *Z. Naturforsch.*, 1947, *2a*, 329.

2020　R. Lindner, *Z. Naturforsch.*, 1947, *2a*, 333.

2021　R. Lindner, *Z. Naturforsch.*, 1948, *3b*, 219.

2022　R. Lindner, *Z. Elektrochem.*, 1950, *54*, 421.

2023　R. Lindner and O. Peter, *Z. Naturforsch.*, 1946, *1*, 67.

2024　B. Lindqvist and T. Storgårds, *Acta Chem. Scand.*, 1953, *7*, 87.

2025　B. Lindqvist and T. Storgårds, *Nature*, 1955, *175*, 511.

2026　I. Lindqvist, *Acta Chem. Scand.*, 1948, *2*, 88.

2027　G. Lindstedt, *Acta Chem. Scand.*, 1950, *4*, 448.

2028　G. Lindstedt and A. Misiorny, *Acta Chem. Scand.*, 1951, *5*, 121; 1952, *6*, 744.

2029　G. Lindstedt and B. Zacharias, *Acta Chem. Scand.*, 1955, *9*, 781.

2030　J. Links and M. S. De Groot, *Rec. trav. chim.*, 1953, *72*, 57.

2031　H. F. Linskens, *Papierchromatographie in der Botanik*, Springer Verlag, Berlin, 1955.

2032　S. Lissitzky, I. Garcia and J. Roche, *Experientia*, 1954, *10*, 379.

2033　S. Lissitzky and R. Michel, *Bull. soc. chim. France*, 1952, 891.

2034　P. H. List, *Naturwissenschaften*, 1954, *41*, 454.

2035　B. A. J. Lister, *Discuss. Faraday Soc.*, 1949, *7*, 237.

2036　B. A. J. Lister and M. L. Smith, *J. Chem. Soc.*, 1948, 1272.

2037　W. C. Lister, *Chem. and Ind.*, 1955, 583.

2038　A. B. Littlewood, C. S. G. Phillips and D. T. Price, *J. Chem. Soc.*, 1955, 1480.

2039　C. Loeb and M. J. Lichtenberger, *Bull. soc. chim. France*, 1950, 362.

2040　H. Loebl, G. Stein and J. Weiss, *J. Chem. Soc.*, 1951, 405.

2041　J. E. Löffler and E. R. Reichl, *Mikrochim. Acta*, 1953, 79.

2042　N. Löfgren, *Acta Chem. Scand.*, 1952, *6*, 1030.

2043　K. Löhr, *Biochem. Z.*, 1950, *320*, 115.

2044　M. E. Lombardo, T. A. Viscelli, A. Mittelman and P. B. Hudson, *J. Biol. Chem.*, 1955, *212*, 353.

2045　A. G. Long, J. R. Quayle and R. J. Stedman, *J. Chem. Soc.*, 1951, 2197.

2046　W. H. Longenecker, *Science*, 1948, *107*, 23.

2047　W. H. Longenecker, *Anal. Chem.*, 1949, *21*, 1502.

2048 J. Loriers and D. Carminati, *Compt. rend.*, 1953, *237*, 1328.
2049 J. Loriers and J. Quesney, *Compt. rend.*, 1954, *239*, 1643.
2050 H. S. Loring, H. W. Bortner, L. W. Levy and M. L. Hammell, *J. Biol. Chem.*, 1952, *196*, 807.
2051 M. Loury, *Bull. mat. grasses inst. colonial Marseille*, 1943, *27*, 151: *C.A.*, 1945, *39*, 3446-g.
2052 A. Lowman, *Science*, 1942, *96*, 211.
2053 J. Lucas and J. M. Orten, *J. Biol. Chem.*, 1951, *191*, 287.
2054 L. C. Luckwill, *Nature*, 1952, *169*, 375.
2055 O. Lüderitz and O. Westphal, *Z. Naturforsch.*, 1952, *7b*, 136.
2056 J. W. H. Lugg, *Brit. Med. Bull.*, 1954, *10*, 192.
2057 J. W. H. Lugg and B. T. Overell, *Nature*, 1947, *160*, 87.
2058 J. W. H. Lugg and B. T. Overell, *Australian J. Sci. Res.*, 1948, *1*, 98.
2059 J. W. H. Lugg and R. A. Weller, *Biochem. J.*, 1948, *42*, 408.
2060 N. A. Lund, F. S. M. Grylls and J. S. Harrison, *Nature*, 1954, *173*, 544.
2061 Yu. Yu. Lure and N. A. Filippova, *Zavodskaya Lab.*, 1949, *15*, 771: *C.A.*, 1950, *44*, 476-c.
2062 D. J. Lussman, E. R. Kirch and G. L. Webster, *J. Am. Pharm. Assoc.*, 1951, *40*, 368: *C.A.*, 1951, *45*, 9219.
2063 C. G. Lynam and H. Weil, *Manuf. Chemist*, 1950, *21*, 195-9, 205: *C.A.*, 1950, *44*, 7594-c.
2064 C. G. Lynam and H. Weil, *Manuf. Chemist*, 1950, *21*, 228: *C.A.*, 1950, *44*, 9590-e.
2065 C. G. Lynam and H. Weil, *Ind. Chemist*, 1950, *26*, 109: *C.A.*, 1950, *44*, 5158-i.
2066 R. Ma and T. D. Fontaine, *Science*, 1949, *110*, 232.
2067 V. Mačák, I. Bartosová and F. Šantavý, *Ann. pharm. franc.*, 1954, *12*, 555: *C.A.*, 1955, *49*, 4942-b.
2068 A. K. Macbeth, J. A. Mills and R. Pettit, *J. Chem. Soc.*, *1950*, 3538.
2069 H. McCormick, *Analyst*, 1953, *78*, 562.
2070 R. M. McCready and W. Z. Hassid, *J. Am. Chem. Soc.*, 1944, *66*, 560.
2071 R. M. McCready and E. A. McComb, *Anal. Chem.*, 1954, *26*, 1645.
2072 H. J. McDonald, *Ionography, Electrophoresis in stabilised media*, The Year Book Publishers, Chicago, 1955.
2073 H. J. McDonald, M. C. Urbin and M. B. Williamson, *Science*, 1950, *112*, 227.
2074 I. W. McDonald, *Biochem. J.*, 1954, *57*, 566.
2075 S. McDonough, *Nature*, 1954, *173*, 645.
2076 K. Macek, *Chem. Listy*, 1954, *48*, 1181: *C.A.*, 1954, *48*, 13550-b.
2077 K. Macek and M. Tadra, *Chem. Listy*, 1952, *46*, 450: *C.A.*, 1952, *46*, 11049-g.
2078 K. Macek, S. Vaněček and Z. J. Vejdělek, *Chem. Listy*, 1955, *49*, 539.
2078a K. Macek and Z. J. Vejdelek, *Nature*, 1955, *176*, 1173.
2079 E. F. McFarren, *Anal. Chem.*, 1951, *23*, 168.
2080 E. F. McFarren, K. Brand and H. R. Rutkowski, *Anal. Chem.*, 1951, *23*, 1146.
2081 E. F. McFarren and J. A. Mills, *Anal. Chem.*, 1952, *24*, 650.
2082 M. G. McGeown and F. H. Malpress, *Nature*, 1954, *173*, 212.
2083 M. Macheboeuf, M. Cachin and J. Blass, *Ann. biol. clin.*, 1952, *10*, 22.
2084 M. Macheboeuf, P. Rebeyrotte and M. Brunerie, *Bull. soc. chim. biol.*, 1951, *33*, 1543.
2085 F. C. McIntire, L. W. Roth and J. L. Shaw, *J. Biol. Chem.*, 1947, *170*, 537.
2086 F. C. McIntire and J. R. Schenck, *J. Am. Chem. Soc.*, 1948, *70*, 1193.
2087 K. W. McKerns, *Can. J. Med. Sci.*, 1951, *29*, 59: *C.A.*, 1951, *45*, 8584-g.

2088 J. M. McKibbin and W. E. Taylor, *J. Biol. Chem.*, 1952, *196*, 427.

2089 G. Mackinney, S. Aronoff and B. T. Bornstein, *Ind. Eng. Chem. Anal. Ed.*, 1942, *14*, 391.

2090 N. F. Maclagan, C. H. Bowden and J. H. Wilkinson, *Rec. trav. chim.*, 1955, *74*, 633.

2091 C. K. McLane and S. Peterson, *Paper 19.6 of the "Transuranium elements" National Nuclear Energy Series*, Div. IV, Volume 14B, Part II, p. 1385, McGraw Hill, New York, 1949.

2092 J. M. McMahon, R. B. Davies and G. Kalnitsky, *Proc. Soc. Exptl. Biol. Med.*, 1950, *75*, 799: *C.A.*, 1951, *45*, 3011-d.

2093 W. H. McNeely, W. W. Binkley and M. L. Wolfrom, *J. Am. Chem. Soc.*, 1945, *67*, 527.

2094 W. M. MacNevin and W. B. Crummett, *Anal. Chem.*, 1953, *25*, 1628.

2095 W. M. MacNevin and W. B. Crummett, *Anal. Chim. Acta*, 1954, *10*, 323.

2096 L. B. Macpherson, *Nature*, 1954, *173*, 1195.

2097 E. B. McQuarrie, A. J. Liebman, R. G. Kluener and A. T. Venosa, *Arch. Biochem.*, 1944, *5*, 307.

2098 R. R. McSwiney, R. E. H. Nicholas and F. T. G. Prunty, *Biochem. J.*, 1950, *46*, 147.

2099 C. Mader, *Anal. Chem.*, 1954, *26*, 566.

2100 C. Mader and G. Mader Jr., *Anal. Chem.*, 1953, *25*, 1423.

2101 A. C. Maehly and T. Reichstein, *Helv. Chim. Acta*, 1947, *30*, 496.

2102 B. Magasanik and E. Chargaff, *Biochim. Biophys. Acta*, 1951, *7*, 396.

2103 B. Magasanik and H. E. Umbarger, *J. Am. Chem. Soc.*, 1950, *72*, 2308.

2104 B. Magasanik, E. Vischer, R. Doniger, D. Elson and E. Chargaff, *J. Biol. Chem.*, 1950, *186*, 37.

2105 L. B. Magnusson, S. G. Thompson and G. T. Seaborg, *Phys. Rev.*, 1950, *78*, 363.

2106 B. J. Mair, *Ann. N.Y. Acad. Sci.*, 1948, *49*, 218.

2107 B. J. Mair, A. L. Gaboriault and F. D. Rossini, *Ind. Eng. Chem.*, 1947, *39*, 1072.

2108 P. H. Mars, *Pharm. Weekblad*, 1953, *88*, 319: *C.A.*, 1953, *47*, 8585-f.

2109 K. Makino, K. Satoh, T. Fujiki and K. Kawaguchi, *Nature*, 1952, *170*, 977.

2110 C. Malatesta, *Boll. oculist.*, 1949, *28*, 727.

2111 E. W. Malmberg, *J. Am. Chem. Soc.*, 1954, *76*, 980.

2112 E. W. Malmberg, *Anal. Chem.*, 1955, *27*, 840.

2113 E. W. Malmberg, K. N. Trueblood and T. D. Waugh, *Anal. Chem.*, 1953, *25*, 901.

2114 F. H. Malpress and A. B. Morrison, *Nature*, 1949, *164*, 963.

2115 F. H. Malpress and A. B. Morrison, *Nature*, 1952, *169*, 1103.

2116 E. Malyoth, *Naturwissenschaften*, 1951, *38*, 478.

2117 G. Malyoth and H. W. Stein, *Biochem. Z.*, 1952, *322*, 165.

2118 G. Malyoth and H. W. Stein, *Biochem. Z.*, 1952/53, *323*, 265.

2119 J. M. Mancuro and A. Zygmuntowicz, *Endocrinology*, 1951, *48*, 114: *C.A.*, 1953, *47*, 1212-a.

2120 G. Manecke and K.-E. Gillert, *Naturwissenschaften*, 1955, *42*, 212.

2121 T. B. Mann, *Analyst*, 1944, *69*, 34.

2122 G. J. Mannering, A. C. Dixon, N. V. Carroll and O. B. Cope, *J. Lab. Clin. Med.*, 1954, *44*, 292: *C.A.*, 1954, *48*, 13782-a.

2123 W. M. Manning and H. H. Strain, *J. Biol. Chem.*, 1943, *151*, 1.

2124 L. A. Manson and J. O. Lampen, *J. Biol. Chem.*, 1951, *193*, 539.

2125 C. L. Mantell, *Adsorption*, McGraw-Hill Book Company, New York and London, 1945.
2126 L. W. Mapson and S. M. Partridge, *Nature*, 1949, *164*, 479.
2127 J. G. Marchal and T. Mittwer, *Compt. rend. soc. biol.*, 1951, *145*, 417.
2128 J. G. Marchal and T. Mittwer, *Koninkl. Ned. Akad. Wetenschap. Proc.*, 1951, *C 54*, 391: *C.A.*, 1952, *46*, 2442-e.
2129 E. Margoliash, *Nature*, 1952, *170*, 1014.
2130 E. Margoliash, *Biochem. J.*, 1954, *56*, 529 and 535.
2131 G. Marinetti and E. Stotz, *J. Am. Chem. Soc.*, 1954, *76*, 1347.
2132 G. B. Marini-Bettòlo, *Chimia e Industria (Milano)*, 1952, *34*, 269.
2133 G. B. Marini-Bettòlo-Marconi and S. Guarino, *Experientia*, 1950, *6*, 309.
2134 J. A. Marinsky, L. E. Glendenin and C. D. Coryell, *J. Am. Chem. Soc.*, 1947, *69*, 2781.
2135 S. Markees, *Biochem. J.*, 1954, *56*, 703.
2136 R. Markham, *Brit. Med. Bull.*, 1954, *10*, 214.
2137 R. Markham and J. D. Smith, *Nature*, 1949, *163*, 250; *Biochem. J.*, 1949, *45*, 294.
2138 R. Markham and J. D. Smith, *Biochem. J.*, 1950, *46*, 513.
2139 R. Markham and J. D. Smith, *Biochem. J.*, 1951, *49*, 401.
2140 F. Markwardt, *Naturwissenschaften*, 1954, *41*, 139.
2141 D. H. Marrian, *Biochim. Biophys. Acta*, 1954, *13*, 278.
2142 G. F. Marrian and W. S. Bauld, *Biochem. J.*, 1955, *59*, 136.
2143 A. Marshak, *J. Biol. Chem.*, 1951, *189*, 607.
2144 A. Marshak and H. J. Vogel, *J. Biol. Chem.*, 1951, *189*, 597.
2145 L. M. Marshall, K. O. Donaldson and F. Friedberg, *Anal. Chem.*, 1952, *24*, 773.
2146 A. E. Martin and J. Smart, *Nature*, 1955, *175*, 422.
2147 A. J. P. Martin, *Endeavour*, 1947, *6*, 21.
2148 A. J. P. Martin, *Ann. N.Y. Acad. Sci.*, 1948, *49*, 249.
2149 A. J. P. Martin, *Ann. Repts. Progress Chem.*, 1949, *45*, 267.
2150 A. J. P. Martin, *Biochem. Soc. Symposia*, 1949, No. 3, p. 4, Cambridge University Press.
2151 A. J. P. Martin, *Ann. Rev. Biochem.*, 1950, *19*, 517.
2152 A. J. P. Martin, *Brit. Med. Bull.*, 1954, *10*, 161.
2152a A. J. P. Martin and A. T. James, *Biochem. J.*, 1956, *63*, 138.
2153 A. J. P. Martin and R. Mittelmann, *Biochem. J.*, 1948, *43*, 353.
2154 A. J. P. Martin and R. R. Porter, *Biochem. J.*, 1951, *49*, 215.
2155 A. J. P. Martin and R. L. M. Synge, *Biochem. J.*, 1941, *35*, 91.
2156 A. J. P. Martin and R. L. M. Synge, *Biochem. J.*, 1941, *35*, 1358.
2157 A. J. P. Martin and R. L. M. Synge, *Advances in Protein Chemistry*, 1945, *2*, 1–83.
2158 E. C. Martin, *Anal. Chim. Acta*, 1951, *5*, 511.
2159 G. J. Martin, *Ion Exchange and Adsorption Agents in Medicine*, Little, Brown & Co., Boston, 1955.
2160 J. R. Martin and J. R. Hayes, *Anal. Chem.*, 1952, *24*, 182.
2161 S. M. Martin, *Chem. and Ind.*, *1955*, 427.
2162 C. S. Marvel and R. E. Light Jr., *J. Am. Chem. Soc.*, 1950, *72*, 3887.
2163 C. S. Marvel and R. D. Rands Jr., *J. Am. Chem. Soc.*, 1950, *72*, 2642.
2164 H. Masamune and M. Maki, *Tôhoku J. Exptl. Med.*, 1952, *55*, 299: *C.A.*, 1953, *47*, 11303-a.

2165   H. S. Mason and E. F. Davis, *J. Biol. Chem.*, 1952, *197*, 41.
2166   M. Mason and C. P. Berg, *J. Biol. Chem.*, 1951, *188*, 783.
2167   P. Mastagli et G. V. Durr, *Bull. soc. chim. France, 1955*, 268.
2168   S. Masuyama, *J. Agr. Chem. Soc. Japan*, 1949, *23*, 45: *C.A.*, 1950, *44*, 3440-h.
2169   S. Masuyama, *J. Chem. Soc. Japan*, 1949, *70*, 232: *C.A.*, 1951, *45*, 5606-c.
2170   W. Matthias, *Der Züchter*, 1954, *24*, 313.
2171   M. Mathieu, *Bull. soc. chim. France, 1947*, 14.
2171a P. B. Mathur, *J. Sci. Ind. Research (India)*, 1955, *14B*, No. 2, 667.
2171b T. Matsuo and A. Iwase, *Japan Analyst*, 1955, *4*, 148.
2172   J. Matsuura, *Japan Analyst*, 1953, *2*, 135: *C.A.*, 1953, *47*, 7937-i.
2172a J. Matsuura, *Japan Analyst*, 1955, *4*, 242.
2173   W. Matthias, *Naturwissenschaften*, 1954, *41*, 17.
2174   G. A. Maw, *Nature*, 1947, *160*, 261; *Biochem. J.*, 1948, *43*, 139.
2175   S. W. Mayer and E. C. Freiling, *J. Am. Chem. Soc.*, 1953, *75*, 5647.
2176   S. W. Mayer and E. R. Tompkins, *J. Am. Chem. Soc.*, 1947, *69*, 2866.
2177   S. W. Mayer and S. D. Schwartz, *Naval Radiol. Defense Lab. San Francisco Calif., Unclassified Report* AD 128c (June 1949) and N.R.D.C. 624 (October 1949).
2178   J. F. Mead, G. Steinberg and D. R. Howton, *J. Biol. Chem.*, 1953, *205*, 683, 687.
2179   D. F. Meigh, *Nature*, 1952, *169*, 706.
2180   D. F. Meigh, *Nature*, 1952, *170*, 579.
2181   J. E. Meinhard and N. F. Hall, *Anal. Chem.*, 1949, *21*, 185.
2182   J. E. Meinhard and N. F. Hall, *Anal. Chem.*, 1950, *22*, 344.
2182a J. Meissner, *Z. anorg. u. allgem. Chem.*, 1955, *281*, 293.
2183   P. D. Meister, D. H. Peterson, S. H. Eppstein, H. C. Murray, L. M. Reineke, A. Weintraub and H. M. L. Osborn, *J. Am. Chem. Soc.*, 1954, *76*, 5679.
2184   D. P. Mellor, *Australian J. Sci.*, 1950, *12*, 183: *C.A.*, 1950, *44*, 10445-b.
2185   D. B. Melville, W. H. Horner and R. Lubschez, *J. Biol. Chem.*, 1954, *206*, 221.
2186   R. Mendenhall and C. H. Li, *Proc. Soc. Exptl. Biol. Med.*, 1951, *78*, 668: *C.A.*, 1952, *46*, 2591-e.
2187   B. Mendioroz, A. Charbonnier and R. Bernard, *Compt. rend. soc. biol.*, 1951, *145*, 1483.
2188   K. H. Meng, *Brit. Pat.* 621,620 (1949): *C.A.*, 1949, *43*, 6013-c.
2189   C. Mentzer, D. Molho et L. Molho-Lacroix, *Bull. soc. chim. France, 1953*, 636.
2190   R. A. Mercer and R. A. Wells, *Analyst*, 1954, *79*, 339.
2191   R. A. Mercer and A. F. Williams, *J. Chem. Soc., 1952*, 3399.
2192   R. B. Merrifield and D. W. Woolley, *J. Biol. Chem.*, 1952, *197*, 521.
2193   J. K. Mertzweiller, D. M. Carney and F. F. Farley, *J. Am. Chem. Soc.*, 1943, *65*, 2367.
2194   R. L. Metzenberg and H. K. Mitchell, *J. Am. Chem. Soc.*, 1954, *76*, 4187.
2195   P. Meunier, *Bull. soc. chim. France, 1946*, 73, 77.
2196   P. Meunier, J. Jouanneteau and G. Zwingelstein, *Compt. rend.*, 1950, *231*, 1170, 1570.
2197   P. Meunier and A. Vinet, *Bull. soc. chim. biol.*, 1942, *24*, 365.
2198   P. Meunier and A. Vinet, *Bull. soc. chim. biol.*, 1945, *27*, 186.
2199   P. Meunier and A. Vinet, *Chromatographie et Mésomérie*, pp. 126, Masson et Cie., Paris, 1947.
2200   K. Meyer, *Helv. Chim. Acta*, 1946, *29*, 718.
2201   G. Meyerheim and H. J. Hübener, *Naturwissenschaften*, 1952, *39*, 482.

2201a J. Michal, *Collection Czechoslov. Chem. Communs.*, 1956, *21*, 576.
2202 A. Michalowicz and M. Lederer, *J. phys. radium*, 1952, [8], *13*, 669.
2203 F. Micheel and P. Albers, *Mikrochim. Acta, 1954*, 489.
2204 F. Micheel and H. Schweppe, *Angew. Chem.*, 1954, *66*, 136.
2205 F. Micheel and H. Schweppe, *Mikrochim. Acta, 1954*, 53.
2206 F. Micheel and F.-P. van de Kamp, *Angew. Chem.,* 1952, *64*, 607.
2207 G. Michel and E. Lederer, *Compt. rend.*, 1955, *240*, 2454.
2208 H. Michl, *Naturwissenschaften*, 1953, *40*, 390.
2209 W. R. Middlebrook, *Nature*, 1949, *164*, 501.
2210 .J. K. Miettinen, *Ann. Acad. Sci. Fenn.*, 1955, Ser. A II, *60*, 520.
2211 J. K. Miettinen, S. Kari, T. Moisio, M. Alfthan, and A. I. Virtanen, *Suomen Kemistilehti*, 1953, *26 B*, 26: *C.A.*, 1953, *47*, 7608-b.
2212 J. K. Miettinen and T. Moisio, *Acta Chem. Scand.*, 1953, *7*, 1225.
2213 J. K. Miettinen and A. I. Virtanen, *Acta Chem. Scand.*, 1949, *3*, 459.
2214 H. T. Miles, E. R. Stadtman and W. W. Kielley, *J. Am. Chem. Soc.*, 1954, *76*, 4041.
2215 B. Milićević, *Bull. soc. chim. Belgrade*, 1951, *16*, 101: *C.A.*, 1952, *46*, 4319-g.
2216 C. C. Miller and J. A. Hunter, *Analyst*, 1954, *79*, 483.
2217 C. C. Miller and R. J. Magee, *J. Chem. Soc.*, *1951*, 3183.
2218 C. C. Miller and R. J. Magee, *J. Chem. Soc.*, *1951*, 3188.
2219 H. Miller and D. M. Kraemer, *Anal. Chem.*, 1952, *24*, 1371.
2220 J. M. Miller and J. G. Kirchner, *Anal. Chem.*, 1951, *23*, 428.
2221 J. M. Miller and J. G. Kirchner, *Anal. Chem.*, 1952, *24*, 1480.
2222 J. M. Miller and J. G. Kirchner, *Anal. Chem.*, 1953, *25*, 1107.
2223 J. M. Miller and J. G. Kirchner, *Anal. Chem.*, 1954, *26*, 2002.
2224 J. M. Miller and L. B. Rockland, *Arch. Biochem. Biophys.*, 1952, *40*, 416.
2225 G. L. Mills, *Nature*, 1950, *165*, 403.
2226 J. S. Mills and A. E. A. Werner, *Nature*, 1952, *169*, 1064; *J. Chem. Soc.*, *1955*, 3132.
2227 M. R. Mills, *Paint Technol.*, 1945, *10*, 107: *C.A.*, 1945, *39*, 5091-3.
2228 M. Milone and G. Cetini, *Ann. chim. (Roma)*, 1953, *43*, 648: *C.A.*, 1954, *48*, 5720-g.
2229 M. Milone and G. Cetini, *Chimica e Industria (Milano)*, 1953, *35*, 346.
2229a M. Milone and G. Cetini, *Atti accad. sci. Torino*, 1955/56, *90*, 1. •
2230 M. Milone, G. Cetini and F. Ricca, *Ann. chim. (Roma)*, 1953, *43*, 659: *C.A.*, 1954, *47*, 5721-a.
2231 D. W. Minty, R. McNeil, M. Ross, E. A. Swinton and D. E. Weiss, *Australian J. Appl. Sci.*, 1953, *4*, 530.
2232 D. W. Minty, M. Ross and D. E. Weiss, *Australian J. Appl. Sci.*, 1953, *4*, 519.
2233 F. L. Mitchell, *Nature*, 1952, *170*, 621.
2234 F. L. Mitchell and R. E. Davies, *Biochem. J.*, 1954, *56*, 690.
2235 H. K. Mitchell, M. Gordon and F. A. Haskins, *J. Biol. Chem.*, 1949, *180*, 1071.
2236 H. K. Mitchell and F. A. Haskins, *Science*, 1949, *110*, 278.
2237 L. C. Mitchell, *J. Assoc. Offic. Agr. Chemists*, 1952, *35*, 920: *C.A.*, 1953, *47*, 5061-b.
2238 L. C. Mitchell, *J. Assoc. Offic. Agr. Chemists*, 1953, *36*, 943: *C.A.*, 1954, *48*, 9693-h.
2239 L. C. Mitchell, *J. Assoc. Offic. Agr. Chem.*, 1953, *36*, 1123: *C.A.*, 1954, *48*, 14007-e.

2240   L. C. Mitchell, *J. Assoc. Offic. Agr. Chemists*, 1953, *36*, 1183: *C.A.*, 1954, *48*, 14088-e.

2241   L. C. Mitchell, *J. Assoc. Offic. Agr. Chemists*, 1954, *37*, 216: *C.A.*, 1954, *48*, 14088-f.

2242   L. C. Mitchell, *J. Assoc. Offic. Agr. Chemists*, 1954, *37*, 530, 996: *C.A.*, 1955, *49*, 559-b.

2243   L. C. Mitchell, *J. Assoc. Offic. Agr. Chemists*, 1954, *37*, 1021: *C.A.*, 1955, *49*, 2248-f.

2244   L. C. Mitchell and W. I. Patterson, *J. Assoc. Offic. Agr. Chemists*, 1953, *36*, 553: *C.A.*, 1953, *47*, 12724-i.

2245   L. C. Mitchell and W. I. Patterson, *J. Assoc. Offic. Agr. Chemists*, 1953, *36*, 1127: *C.A.*, 1954, *48*, 13543-a.

2246   V. K. Mohan Rao, *Experientia*, 1953, *9*, 151.

2247   M. Moilanen and H. Richtzenhain, *Acta Chem. Scand.*, 1954, *8*, 704.

2248   G. Moir, M. Ross and D. E. Weiss, *Australian J. Appl. Sci.*, 1953, *4*, 543.

2249   S. G. Mokrushin, *Soobshcheniya Nauch. Rabot Vsesoyuz. Khim. Obshchestva im. Mendeleeva*, 1953, No. 2, 26: *C.A.*, 1955, *49*, 2149-h.

2250   D. Molho and L. Molho-Lacroix, *Compt. rend.*, 1952, *235*, 522.

2251   T. Momose and A. Yamada, *J. Pharm. Soc. Japan*, 1951, *71*, 980: *C.A.*, 1952, *46*, 1921-e.

2252   P. H. Monaghan, P. B. Moseley, T. S. Burkhalter and O. A. Nance, *Anal. Chem.*, 1952, *24*, 193.

2253   P. H. Monaghan, H. A. Suter and A. L. LeRosen, *Anal. Chem.*, 1950, *22*, 811.

2254   R. Monier and M. Jutisz, *Biochim. Biophys. Acta*, 1954, *15*, 62.

2255   R. Monier and L. Pénasse, *Compt. rend.*, 1950, *230*, 1176.

2256   R. Montgomery, *J. Am. Chem. Soc.*, 1952, *74*, 1466.

2257   J. Montreuil, *Compt. rend.*, 1954, *239*, 510.

2258   J. Montreuil and P. Boulanger, *Compt. rend.*, 1950, *231*, 247.

2259   J. Montreuil and P. Boulanger, *Compt. rend.*, 1953, *236*, 337.

2260   J. Montreuil and P. Boulanger, *Compt. rend.*, 1954, *239*, 367.

2261   J. Montreuil and R. Scriban, *Bull. soc. chim. biol.*, 1952, *34*, 674.

2262   A. M. Moore and J. B. Boylen, *Science*, 1953, *118*, 19.

2263   G. E. Moore and K. A. Kraus, *J. Am. Chem. Soc.*, 1950, *72*, 5792.

2264   L. A. Moore, *Ind. Eng. Chem. Anal. Ed.*, 1942, *14*, 707.

2265   S. Moore and W. H. Stein, *Ann. N.Y. Acad. Sci.*, 1948, *49*, 265.

2266   S. Moore and W. H. Stein, *J. Biol. Chem.*, 1948, *176*, 367.

2267   S. Moore and W. H. Stein, *J. Biol. Chem.*, 1949, *178*, 53.

2268   S. Moore and W. H. Stein, *J. Biol. Chem.*, 1951, *192*, 663.

2269   S. Moore and W. H. Stein, *Ann. Rev. Biochem.*, 1952, *21*, 521.

2270   S. Moore and W. H. Stein, *J. Biol. Chem.*, 1954, *211*, 893.

2271   S. Moore and W. H. Stein, *J. Biol. Chem.*, 1954, *211*, 907.

2272   M. C. Moreira de Almeida, *Rev. Fac. Sci. Univ. Lisboa*, 1952–53, 2a, Ser. B 2, 59: *C.A.*, 1955, *49*, 1133-c.

2273   I. Mori, *J. Pharm. Soc. Japan*, 1953, *73*, 958: *C.A.*, 1954, *48*, 792-a.

2274   I. Mori, *J. Pharm. Soc. Japan*, 1954, *74*, 213: *C.A.*, 1954, *48*, 5709-d.

2275   I. Mori, *J. Pharm. Soc. Japan*, 1954, *74*, 525: *C.A.*, 1954, *48*, 9857-h.

2276   I. Mori, *Science*, 1954, *119*, 653.

2277   I. M. Morice and J. C. E. Simpson, *J. Chem. Soc.*, 1941, 181.

2278   G. A. Morin, A. Cheutin, O. Costerousse and M. Fouquet, *Bull. soc. chim. biol.*, 1952, *34*, 193.

2279 I. Moring-Claesson, *Biochim. Biophys. Acta*, 1948, *2*, 389.

2280 C. J. O. R. Morris, *Biochem. J.*, 1944, *38*, 203.

2281 C. J. O. R. Morris and D. C. Williams, *Biochem. J.*, 1953, *54*, 470.

2282 D. L. Morris, *Science*, 1948, *107*, 254.

2283 D. L. Morris, *U.S. Pat.*, 2,700,672 (1955): *C.A.*, 1955, *49*, 4776-f.

2284 M. Morrison, R. W. Estabrook and E. Stotz, *J. Am. Chem. Soc.*, 1954, *76*, 6409.

2284a M. Morrison and E. Stotz, *J. Biol. Chem.*, 1955, *213*, 373.

2285 R. I. Morrison, *Biochem. J.*, 1952, *50*, xiv.

2286 D. C. Mortimer, *Can. J. Chem.*, 1952, *30*, 653.

2287 E. H. Mosbach, M. Nierenberg and F. E. Kendall, *J. Am. Chem. Soc.*, 1953, *75*, 2358.

2288 E. H. Mosbach, C. Zomzely and F. E. Kendall, *Arch. Biochem. Biophys.*, 1954, *48*, 95.

2289 P. B. Moseley, A. L. LeRosen and J. K. Carlton, *Anal. Chem.*, 1954, *26*, 1563.

2290 M. Mottier and M. Potterat, *Mitt. Gebiete Lebensm. u. Hyg.*, 1952, *43*, 118: *C.A.*, 1952, *46*, 7942-f.

2291 M. Mottier and M. Potterat, *Mitt. Gebiete Lebensm. u. Hyg.*, 1953, *44*, 293: *C.A.*, 1953, *47*, 10752-i.

2292 M. Mottier and M. Potterat, *Anal. Chim. Acta*, 1955, *13*, 46.

2293 N. Mower, J. Green and F. S. Spring, *J. Chem. Soc.*, 1944, 256.

2294 D. F. Mowery Jr., *J. Am. Chem. Soc.*, 1951, *73*, 5047.

2295 D. F. Mowery Jr., *J. Am. Chem. Soc.*, 1951, *73*, 5049.

2296 D. F. Mowery Jr., *J. Am. Chem. Soc.*, 1955, *77*, 1667.

2297 D. F. Mowery Jr. and G. R. Ferrante, *J. Am. Chem. Soc.*, 1954, *76*, 4103.

2298 V. Moyle, E. Baldwin and R. Scarisbrick, *Biochem. J.*, 1948, *43*, 308.

2299 P. Moynihan and P. O'Colla, *Chem. and Ind.*, 1951, 407.

2300 V. Mráz, *Chem. Listy*, 1950, *44*, 259: *C.A.*, 1951, *45*, 5654-f.

2301 S. Mukherjee and M. L. Sen Gupta, *J. Proc. Inst. Chemists (India)*, 1949, *21*, 83: *C.A.*, 1950, *44*, 3676-b.

2302 S. Mukherjee and H. C. Srivastava, *Nature*, 1952, *169*, 330.

2303 R. P. Mull, R. W. Townley and C. R. Scholz, *J. Am. Chem. Soc.*, 1945, *67* 1626.

2304 A. F. Müller and F. Leuthardt, *Helv. Chim. Acta*, 1949, *32*, 2289.

2305 P. B. Müller, *Helv. Chim. Acta*, 1943, *26*, 1945.

2306 P. B. Müller, *Helv. Chim. Acta*, 1944, *27*, 443.

2307 P. B. Müller, *Helv. Chim. Acta*, 1947, *30*, 1172.

2308 R. Müller, *Beitr. Biol. Pflanz.*, 1953, *30*, 1: *C.A.*, 1955, *49*, 4065-i.

2309 R. H. Müller, *Anal. Chem.*, 1950, *22*, 72.

2310 R. H. Müller and D. L. Clegg, *Anal. Chem.*, 1949, *21*, 1123.

2311 R. H. Müller and D. L. Clegg, *Anal. Chem.*, 1949, *21*, 1429.

2312 R. H. Müller and D. L. Clegg, *Anal. Chem.*, 1951, *23*, 396.

2313 R. H. Müller and E. N. Wise, *Anal. Chem.*, 1951, *23*, 207.

2314 P. K. Mulvany, H. D. Agar, Q. P. Peniston and J. L. McCarthy, *J. Am. Chem. Soc.*, 1951, *73*, 1255.

2315 A. Munch-Petersen, H. M. Kalckar, E. Cutolo and E. E. B. Smith, *Nature*, 1953, *172*, 1036.

2316 R. Munier, *Bull. soc. chim. biol.*, 1951, *33*, 857.

2317 R. Munier, *Bull. soc. chim. biol.*, 1951, *33*, 862.

2318   R. Munier, *Bull. soc. chim. France, 1952*, 852.
2319   R. Munier, *Bull. soc. chim. biol.*, 1953, *35*, 1225.
2320   R. Munier and M. Macheboeuf, *Bull. soc. chim. biol.*, 1949, *31*, 1144.
2321   R. Munier and M. Macheboeuf, *Bull. soc. chim. biol.*, 1950, *32*,192.
2322   R. Munier and M. Macheboeuf, *Bull. soc. chim. biol.*, 1950, *32*, 904.
2323   R. Munier and M. Macheboeuf, *Bull. soc. chim. biol.*, 1951, *33*, 846.
2324   R. Munier, M. Macheboeuf and N. Cherrier, *Bull. soc. chim. biol.*, 1951, *33*, 1919.
2325   R. Munier, M. Macheboeuf and N. Cherrier, *Bull. soc. chim. biol.*, 1952, *34*, 204.
2326   T. Münz, *Naturwissenschaften*, 1954, *41*, 553.
2326a   A. Murata, *J. Chem. Soc. Japan, Pure Chem. Sect.*, 1955, *76*, 517.
2327   D. Müting, *Naturwissenschaften*, 1952, *39*, 303.
2327a   S. Muto, *J. Chem. Soc. Japan, Pure Chem. Sect.*, 1955, *76*, 294.
2328   R. J. Myers, *Advances in Colloid Science*, 1942, *1*, 317.
2329   R. J. Myers, *Ind. Eng. Chem.*, 1943, *35*, 858.
2330   R. J. Myers, J. W. Eastes and D. Urquhart, *Ind. Eng. Chem.*, 1941, *33*, 1270.
2331   K. Myrbäck and E. Willstaedt, *Arkiv Kemi*, 1953, *6*, 417.
2332   F. C. Nachod, *Ion Exchange*, 411 pp., Acad. Press, New York, 1949.
2332a   F. C. Nachod and J. Schubert, *Ion Exchange Technology*, 660 pp., Acad. Press, New York, 1956.
2333   L. Naftalin, *Nature*, 1948, *161*, 763.
2334   J. Naghski, C. S. Fenske Jr. and J. F. Couch, *J. Am. Pharm. Assoc.*, 1951, *40*, 613.
2334a   Z. Nagy and E. N. Polyik, *Magyar Kém. Folyóirat*, 1955, *61*, 248.
2335   T. Naito and N. Takahashi, *Japan Analyst*, 1954, *3*, 125: *C.A.*, 1954, *48*, 9859-e.
2336   V. A. Najjar and K. C. Ketron, *J. Biol. Chem.*, 1944, *152*, 579.
2337   K. Nakanishi and L. F. Fieser, *J. Am. Chem. Soc.*, 1952, *74*, 3910.
2338   S. Nakano, *J. Chem. Soc. Japan, Pure Chem. Sect.*, 1954, *75*,71,150: *C.A.*,1954, *48*, 8126-c, 11237-f.
2339   H. A. Nash and A. R. Smashey, *Arch. Biochem.*, 1951, *30*, 237.
2340   W. T. Nauta, H. K. Oosterhuis, A. C. Van der Linden, P. Van Duyn and J. W. Dienske, *Rec. trav. chim.*, 1946, *65*, 865.
2341   J. Navarro Sagristá, *Afinidad*, 1954, *31*, 325: *C.A.*, 1954, *48*, 12608-e.
2342   Y. R. Naves and E. Perrottet, *Helv. Chim. Acta*, 1941, *24*, 3.
2343   G. W. Nederbragt and J. J. De Jong, *Rec. trav. chim.*, 1946, *65*, 831.
2344   R. Neher and A. Wettstein, *Helv. Chim. Acta*, 1951, *34*, 2278.
2345   R. Neher and A. Wettstein, *Helv. Chim. Acta*, 1952, *35*, 276.
2346   J. B. Neilands, *J. Biol. Chem.*, 1952, *197*, 701.
2347   J. B. Neilands and Å. Åkeson, *J. Biol. Chem.*, 1951, *188*, 307.
2348   M. B. Neïman, V. N. Levkovskii and A. F. Lukovnikov, *Doklady Akad. Nauk. S.S.S.R.*, 1951, *81*, 841: *C.A.*, 1955, *49*, 7489-c.
2349   I. E. Neïmark, I. B. Slinyakova and F. I. Khatset, *Akad. Nauk S.S.S.R., Otdel. Khim. Nauk, 1950*, 98: *C.A.*, 1954, *46*, 2442-g.
2350   A. C. Neish, *Can. J. Res.*, 1949, *27 B*, 6.
2351   A. C. Neish, *Can. J. Res.*, 1950, *28 B*, 535.
2352   A. C. Neish, *Can. J. Chem.*, 1951, *29*, 552.
2353   D. H. Nelson and L. T. Samuels, *J. Clin. Endocrinol. and Metabolism*, 1952, *12*, 519: *C.A.*, 1953, *47*, 4405-i.
2353a   F. Nelson, *J. Am. Chem. Soc.*, 1955, *77*, 813.

2354   F. Nelson and K. A. Kraus, *J. Am. Chem. Soc.*, 1955, *77*, 329.
2355   R. Nelson and H. J. Walton, *J. Phys. Chem.*, 1944, *48*, 406.
2355a W. E. Nervik, *J. Phys. Chem.*, 1955, *59*, 690.
2356   F. H. M. Nestler and H. G. Cassidy, *J. Am. Chem. Soc.*, 1950, *72*, 680.
2357   R. M. Nettleton Jr. and R. B. Mefferd Jr., *Anal. Chem.*, 1952, *24*, 1687.
2358   W. P. Neumann, *Naturwissenschaften*, 1955, *42*, 370.
2359   M. B. Neuworth, *J. Am. Chem. Soc.*, 1947, *69*, 1653.
2360   G. T. Newbold and F. S. Spring, *J. Chem. Soc.*, *1944*, 249.
2361   A. G. Newcombe and S. G. Reid, *Nature*, 1953, *172*, 455.
2362   I. E. Newnham, *J. Am. Chem. Soc.*, 1951, *73*, 5899.
2363   G. G. F. Newton, E. P. Abraham and N. J. Berridge, *Nature*, 1953, *171*, 606.
2363a D. J. D. Nicholas and H. M. Stevens, *Nature*, 1955, *176*, 1066.
2364   R. E. H. Nicholas, *Biochem. J.*, 1951, *48*, 309.
2365   R. E. H. Nicholas and A. Comfort, *Biochem. J.*, 1949, *45*, 208.
2366   R. E. H. Nicholas and C. Rimington, *Scand. J. Clin. Lab. Invest.*, 1949, *1*, 12:
       C.A., 1950, *44*, 2065-c.
       Nguyen-Van Thoai, see Thoai.
2367   R. E. H. Nicholas and C. Rimington, *Biochem. J.*, 1951, *48*, 306.
2368   R. E. H. Nicholas and C. Rimington, *Biochem. J.*, 1951, *50*, 194.
2369   D. E. Nicholson, *Nature*, 1949, *163*, 954.
2370   H. J. Nijkamp, *Chem. Weekblad*, 1949, *45*, 480: *C.A.*, 1950, *44*, 977-c.
2371   H. J. Nijkamp, *Anal. Chim. Acta*, 1951, *5*, 325.
2372   H. J. Nijkamp, *Nature*, 1953, *172*, 1102.
2373   G. R. Noggle, *Arch. Biochem. Biophys.*, 1953, *43*, 238.
2374   G. R. Noggle and L. P. Zill, *Arch. Biochem. Biophys.*, 1952, *41*, 21.
2375   G. R. Noggle and L. P. Zill, *Plant Physiol.*, 1953, *28*, 731: *C.A.*, 1954, *48*,
       2166-b.
2376   H. Noll, H. Bloch, J. Asselineau and E. Lederer, *Biochim. Biophys. Acta*, 1956,
       20, 299.
2377   E. J. Norberg, I. Auerbach and R. M. Hixon, *J. Am. Chem. Soc.*, 1945, *67*, 342.
2378   A. Nordal and R. Klevstrand, *Acta Chem. Scand.*, 1951, *5*, 85.
2379   R. Nordmann, O. Gauchery, J.-P. du Ruisseau, Y. Thomas and J. Nordmann,
       Bull. soc. chim. biol., 1954, *36*, 1461.
2380   R. Nordmann, O. Gauchery, J.-P. du Ruisseau, Y. Thomas et J. Nordmann,
       Bull. soc. chim. biol., 1954, *36*, 1641.
2381   C. G. Nordström and T. Swain, *J. Chem. Soc.*, *1953*, 2764.
2382   A. Norman, *Acta Chem. Scand.*, 1953, *7*, 1413.
2383   F. C. Norris and J. J. R. Campbell, *Can. J. Res.*, 1949, *27 C*, 253.
2384   L. B. Norton and R. Hansberry, *J. Am. Chem. Soc.*, 1945, *67*, 1609.
2385   L. Novellie, *Nature*, 1950, *166*, 745.
2386   L. Novellie, *Nature*, 1950, *166*, 1000.
2387   L. Novellie, *Nature*, 1952, *169*, 672.
2388   L. Novellie and H. M. Schwartz, *Nature*, 1954, *173*, 450.
2389   G. Nunez and J. Spiteri, *Bull. soc. chim. biol.*, 1953, *35*, 851.
2390   J. F. Nyc, J. E. Garst, H. B. Friedgood and D. M. Maron, *Arch. Biochem.*,
       1950, *29*, 219.
2391   J. F. Nyc, D. M. Maron, J. B. Garst and H. B. Friedgood, *Proc. Soc. Exptl.
       Biol. Med.*, 1951, *77*, 466.
2392   F. Nydahl, *Anal. Chem.*, 1954, *26*, 580.
2393   D. J. O'Connor and F. Bryant, *Nature*, 1952, *170*, 84.

2394 J. T. Odencrantz and W. Rieman III, *Anal. Chem.*, 1950, *22*, 1066.

2395 L. Odin and I. Werner, *Nord. Med.*, 1950, *43*, 470: *C.A.*, 1950, *44*, 7913-g.

2396 G. O'Donnell, *Anal. Chem.*, 1951, *23*, 894.

2397 G. Oertel, *Acta Endocrinol.*, 1954, *16*, 263: *C.A.*, 1954, *48*, 13786-b.

2398 G. Oertel, *Acta Endocrinol.*, 1954, *16*, 267: *C.A.*, 1954, *48*, 13786-d.

2399 A. C. Offord and J. Weiss, *Discuss. Faraday Soc.*, 1949, *7*, 26.

2400 S. Ogawa, *J. Pharm. Soc. Japan*, 1953, *73*, 59 and 94: *C.A.*, 1953, *47*, 4042-c, d.

2401 T. Ogawa and M. Ohno, *J. Chem. Soc. Japan, Ind. Chem. Sect.*, 1950, *53*, 170: *C.A.*, 1952, *46*, 10257-d.

2402 M. Ohara and Y. Suzuki, *Science (Japan)*, 1951, *21*, 362: *C.A.*, 1951, *45*, 10286-h.

2403 V. Öhman, *The Svedberg Memorial Vol.*, 1944, 413: *C.A.*, 1945, *39*, 1113-7.

2404 Y. Oka and A. Murata, *J. Chem. Soc. Japan, Pure Chem. Sect.*, 1952, *73*, 494: *C.A.*, 1953, *47*, 2631-f.

2405 Y. Oka and A. Murata, *J. Chem. Soc. Japan, Pure Chem. Sect.*, 1951, *73*, 496: *C.A.*, 1953, *47*, 2631-f.

2406 Y. Oka and A. Murata, *Sci. Repts. Research Insts. Tôhoku Univ. Ser. A*, 1951, *3*, 82: *C.A.*, 1953, *47*, 66-g.

2407 Y. Oka and A. Murata, *Sci. Repts. Research Insts. Tôhoku Univ. Ser. A*, 1953, *5*, 343: *C.A.*, 1954, *48*, 7480-f.

2408 A. Okáč and P. Černý, *Chem. Listy*, 1952, *46*, 181.

2409 A. Okáč and P. Černý, *Collection Czechoslov. Chem. Communs.*, 1953, *18*, 73 (in Russian): *C.A.*, 1953, *47*, 7365-c.

2410 M. Okada, A. Yamada and K. Kometani, *J. Pharm. Soc. Japan*, 1952, *72*, 930: *C.A.*, 1953, *47*, 3324-c.

2411 H. Okuno, M. Honda and T. Ishimori, *Japan Analyst*, 1953, *2*, 428: *C.A.*, 1954, *48*, 6320-c.

2412 J. L. Olsen, *U.S. Pat.* 2,564,717: *C.A.*, 1951, *45*, 9312-g.

2413 K. M. Ol'shanova and K. V. Chmutov, *Zhur. Anal. Khim.*, 1954, *9*, 67: *C.A.*, 1954, *48*, 6899-i.

2414 P. A. Ongley, *J. Chem. Soc.*, *1954*, 3634.

2415 K. Onoe, *J. Chem. Soc. Japan, Pure Chem. Sect.*, 1952, *73*, 337: *C.A.*, 1953, *47*, 3757-e.

2416 J. Opieńska-Blauth, *Biokhimiya*, 1953, *18*, 748.

2417 J. Opieńska-Blauth, I. Madecka-Borkowska and T. Borkowski, *Nature*, 1952, *169*, 798.

2418 J. Opieńska-Blauth, O. Sakławska-Szymonowa and M. Kanski, *Nature*, 1951, *168*, 511.

2419 I. Oreskes and A. Saifer, *Anal. Chem.*, 1955, *27*, 854.

2420 W. Oroshnik, A. D. Mebane and G. Karmas, *J. Am. Chem. Soc.*, 1953, *75*, 1050.

2420a G. H. Osborn, *Synthetic Ion-Exchangers*; *Recent Developments in Theory and Application*, Chapman & Hall, London, 1955.

2421 G. H. Osborn and A. Jewsbury, *Nature*, 1949, *164*, 443.

2422 Y. Oshima and T. Nakabayashi, *J. Agr. Chem. Soc. Japan*, 1952, *26*, 754: *C.A.*, 1954, *48*, 5942-h, i.

2423 Y. Oshima, T. Nakabayashi, N. Hada and S. Matsuyama, *J. Agr. Chem. Soc. Japan*, 1954, *28*, 618 and 621: *C.A.*, 1955, *49*, 7065-d, f.

2424 R. Osteux and J. Laturaze, *Compt. rend.*, 1954, *239*, 512.

2425 H. Otsuka and T. Kimura, *J. Biochem. Japan*, 1955, *42*, 81.

2426 G. H. Ott and T. Reichstein, *Helv. Chim. Acta*, 1943, *26*, 1799.
2427 M. Ottesen and C. Villee, *Compt. rend. trav. lab. Carlsberg, Sér. chim.*, 1951, *27*, 421: *C.A.*, 1952, *46*, 7132-d.
2428 T. C. J. Ovenston, *Nature*, 1952, *169*, 924.
2429 B. T. Overell, *Australian J. Sci.*, 1952/53, *15*, 28.
2430 H. S. Owens, A. E. Goodban and J. B. Stark, *Anal. Chem.*, 1953, *25*, 1507.
2431 H.Pachéco, *Bull. soc. chim. biol.*, 1951, *33*, 1915.
2432 E. Pacsu, T. P. Mora and P. W. Kent, *Science*, 1949, *110*, 446.
2433 E. Pacsu and J. W. Mullen, *J. Am. Chem. Soc.*, 1941, *63*, 1168.
2434 M. Paecht and A. Katchalsky, *J. Am. Chem. Soc.*, 1954, *76*, 6197.
2435 A. C. Paladini and L. C. Craig, *J. Am. Chem. Soc.*, 1954, *76*, 688.
2436 A. C. Paladini and L. F. Leloir, *Anal. Chem.*, 1952, *24*, 1024.
2437 A. C. Paladini and L. F. Leloir, *Biochem. J.*, 1952, *51*, 426.
2438 S. Paléus and J. B. Neilands, *Acta Chem. Scand.*, 1950, *4*, 1024.
2439 S. C. Papastamatis and J. E. Kench, *Nature*, 1952, *170*, 33.
2440 S. C. Papastamatis and J. F. Wilkinson, *Nature*, 1950, *167*, 724.
2441 A. B. Pardee, *J. Biol. Chem.*, 1951, *190*, 757.
2442 D. B. Parihar, *Naturwissenschaften*, 1954, *41*, 427.
2443 R. Paris, *Bull. soc. chim. biol.*, 1952, *34*, 767.
2444 D. V. Parke and R. T. Williams, *Biochem. J.*, 1951, *48*, 621.
2445 M. W. Partridge, *J. Pharm. Pharmacol.*, 1952, *4*, 217: *C.A.*, 1952, *46*, 7283-i.
2446 M. W. Partridge and J. Chilton, *Nature*, 1951, *167*, 79.
2447 S. M. Partridge, *Nature*, 1946, *158*, 270.
2448 S. M. Partridge, *Biochem. J.*, 1948, *42*, 238.
2449 S. M. Partridge, *Biochem. J.*, 1948, *42*, 251.
2450 S. M. Partridge, *Nature*, 1949, *164*, 443.
2451 S. M. Partridge, *Discuss. Faraday Soc.*, 1949, *7*, 296.
2452 S. M. Partridge, *Biochem. J.*, 1949, *44*, 521.
2453 S. M. Partridge, *Biochem. J.*, 1949, *45*, 459.
2454 S. M. Partridge, *Chem. and Ind.*, *1950*, 383.
2455 S. M. Partridge, *Nature*, 1952, *169*, 496.
2456 S. M. Partridge, *Brit. Med. Bull.*, 1954, *10*, 241.
2457 S. M. Partridge and R. C. Brimley, *Biochem. J.*, 1949, *44*, 513.
2458 S. M. Partridge and R. C. Brimley, *Biochem. J.*, 1951, *48*, 313.
2459 S. M. Partridge and R. C. Brimley, *Biochem. J.*, 1951, *49*, 153.
2460 S. M. Partridge and R. C. Brimley, *Biochem. J.*, 1952, *51*, 628.
2461 S. M. Partridge, R. C. Brimley and K. W. Pepper, *Biochem. J.*, 1950, *46*, 334.
2462 S. M. Partridge and T. Swain, *Nature*, 1950, *166*, 272.
2463 S. M. Partridge and R. G. Westall, *Biochem. J.*, 1949, *44*, 418.
2464 S. M. Partridge, R. G. Westall and J. R. Bendall, *Brit. Pat.* 644,382: *C.A.*, 1951, *45*, 2530-d.
2465 A. Patschky, *Angew. Chem.*, 1950, *62*, 50: *C.A.*, 1952, *46*, 9790-e.
2466 E. L. Patterson, J. V. Pierce, E. L. R. Stokstad, C. E. Hoffmann, J. A. Brockman Jr., F. P. Day, M. E. Macchi and T. H. Jukes, *J. Am. Chem. Soc.*, 1954, *76*, 1823.
2467 A. R. Patton and P. Chism, *Nature*, 1951, *167*, 406.
2468 A. R. Patton and E. M. Foreman, *Science*, 1949, *109*, 339.
2469 A. R. Patton, E. M. Foreman and P. C. Wilson, *Science*, 1949, *110*, 593.
2470 H. W. Patton, J. S. Lewis and W. I. Kaye, *Anal. Chem.*, 1955, *27*, 170.
2471 R. Paulais, *Ann. pharm. franç.*, 1946, *4*, 101: *C.A.*, 1947, *41*, 3500-d.

2472 J. C. Paulson, F. E. Deatherage and E. F. Almy, *J. Am. Chem. Soc.*, 1953, *75*, 2039.
2473 R. W. Payne, M. S. Raben and E. B. Astwood, *J. Biol. Chem.*, 1950, *187*, 719.
2474 W. J. Payne and R. J. Kieber, *Arch. Biochem. Biophys.*, 1954, *52*, 1.
2475 J. H. Pazur, *J. Biol. Chem.*, 1952, *199*, 217.
2476 J. H. Pazur, *J. Biol. Chem.*, 1953, *205*, 75.
2477 J. H. Pazur, *Science*, 1953, *117*, 355.
2478 J. H. Pazur and D. French, *J. Biol. Chem.*, 1952, *196*, 265.
2479 J. H. Pazur and A. L. Gordon, *J. Am. Chem. Soc.*, 1953, *75*, 3458.
2480 I. A. Pearl and D. L. Beyer, *J. Am. Chem. Soc.*, 1954, *76*, 6106.
2481 I. A. Pearl and E. E. Dickey, *J. Am. Chem. Soc.*, 1951, *73*, 863.
2482 I. A. Pearl and E. E. Dickey, *J. Am. Chem. Soc.*, 1952, *74*, 614.
2483 M. M. Pechet, *J. Clin. Endocrinol. Metab.*, 1953, *13*, 1542: *C.A.*, 1954, *48*, 4038-f.
2484 M. M. Pechet, *Science*, 1955, *121*, 39.
2485 R. L. Peck, *Ann. N.Y. Acad. Sci.*, 1948, *49*, 235: *C.A.*, 1948, *42*, 5619-c.
2486 R. L. Peck, A. Walti, R. P. Graber, E. Flynn, C. E. Hoffhine Jr., V. Allfrey and K. Folkers, *J. Am. Chem. Soc.*, 1946, *68*, 772.
2487 H. Pénau, E. Saïas and C. Andreetti, *Ann. pharm. franç.*, 1952, *10*, 514: *C.A.*, 1953, *47*, 2813-h.
2488 K. W. Pepper, *School Science Review*, 1950, *31*, 164: *C.A.*, 1950, *44*, 10393-f.
2489 Q. P. Peniston, H. D. Agar and J. L. McCarthy, *Anal. Chem.*, 1951, *23*, 994.
2490 K. W. Pepper, *J. Appl. Chem.*, 1951, *1*, 124.
2491 A. Pereira and J. A. Serra, *Science*, 1951, *113*, 387.
2492 M. Perey and J.-P. Adloff, *Compt. rend.*, 1953, *236*, 1664.
2493 B. Pernis and C. Wunderly, *Biochim. Biophys. Acta*, 1953, *11*, 209.
2494 J. C. Perrone, *Nature*, 1951, *167*, 513.
2495 D. H. Peterson, A. H. Nathan, P. D. Meister, S. H. Eppstein, H. C. Murray, A. Weintraub, L. M. Reineke and H. M. Leigh, *J. Am. Chem. Soc.*, 1953, *75*, 419.
2496 D. H. Peterson and L. M. Reineke, *J. Am. Chem. Soc.*, 1950, *72*, 3598.
2497 E. A. Peterson and H. A. Sober, *J. Am. Chem. Soc.*, 1954, *76*, 169.
2498 M. H. Peterson and M. J. Johnson, *J. Biol. Chem.*, 1948, *174*, 775.
2499 M. H. Peterson, M. J. Johnson and W. V. Price, *J. Dairy Sci.*, 1949, *32*, 862: *C.A.*, 1950, *44*, 3168-e.
2500 S. Peterson, *J. Chem. Educ.*, 1951, *28*, 22.
2501 C. W. Pettinga, W. M. Stark and F. R. Van Abeele, *J. Am. Chem. Soc.*, 1954, *76*, 569.
2502 E. Pfau and S. Bergt, *Pharm. Zentralhalle*, 1950, *89*, 303.
2503 N. Pfennig, *Naturwissenschaften*, 1954, *41*, 62.
2504 G. Pfleiderer and W. Schulz, *Ann.*, 1953, *580*, 237.
2505 E. F. Phares, E. H. Mosbach and F. W. Denison Jr., *Anal. Chem.*, 1952, *24*, 660.
2505a C. Phillips, *Gas Chromatography*, 105 pp., Butterworth, London, 1956.
2506 C. S. G. Phillips, *Discuss. Faraday Soc.*, 1949, *7*, 241.
2507 D. M. P. Phillips, *Nature*, 1948, *161*, 53.
2508 D. M. P. Phillips, *Nature*, 1948, *162*, 29.
2509 D. M. P. Phillips, *Nature*, 1949, *164*, 545.
2510 D. M. P. Phillips, *Biochim. Biophys. Acta*, 1949, *3*, 341.
2511 D. M. P. Phillips, *Biochim. Biophys. Acta*, 1954, *13*, 560.

2512 D. M. P. Phillips and J. M. L. Stephen, *Nature*, 1948, *162*, 152.
2513 J. D. Phillips and A. Pollard, *Nature*, 1953, *171*, 41.
2514 N. Piantanida and N. Muiç, *Arch. Biochem. Biophys.*, 1953, *46*, 110.
2515 W. F. Pickering, *Anal. Chim. Acta*, 1953, *8*, 344.
2516 J. G. Pierce and V. Du Vigneaud, *J. Biol. Chem.*, 1950, *186*, 77.
2517 K. A. Piez, *J. Biol. Chem.*, 1954, *207*, 77.
2518 K. A. Piez, E. B. Tooper and L. S. Fosdick, *J. Biol. Chem.*, 1952, *194*, 669.
2519 L. O. Pilgeram, E. M. Gal, E. N. Sassenrath and D. M. Greenberg, *J. Biol. Chem.*, 1953, *204*, 367.
2520. J. H. Pinckard, A. Chatterjee and L. Zechmeister, *J. Am. Chem. Soc.*, 1952, *74*, 1603.
2521 J. H. Pinckard, B. Wille and L. Zechmeister, *J. Am. Chem. Soc.*, 1948, *70*, 1938.
Ping Shu, see Shu.
2522 Z. Pinterović, *Kem. Vjestnik Zagreb*, 1941/42, *15/16*, 45: *C.A.*, 1946, *40*, 4617-6.
2523 Z. Pinterović, *Bull. soc. chim. belges*, 1949, *58*, 522.
2524 P. A. Plattner, A. Fürst and W. Keller, *Helv. Chim. Acta*, 1949, *32*, 2464.
2525 P. A. Plattner and U. Nager, *Helv. Chim. Acta*, 1948, *31*, 2203.
2526 P. A. Plattner and A. S. Pfau, *Helv. Chim. Acta*, 1937, *20*, 224.
2527 P. A. Plattner, L. Ruzicka, H. Heusser and K. Meier, *Helv. Chim. Acta*, 1946, *29*, 2023.
2528 G. W. E. Plaut and K. A. Plaut, *Arch. Biochem. Biophys.*, 1954, *48*, 189.
2529 M. Pöhm, *Mitt. chem. Forsch.-Inst. Wirtsch. Österr.*, 1953, *7*, 121: *C.A.*, 1954, *48*, 5439-e.
2530 M. Pöhm and L. Fuchs, *Naturwissenschaften*, 1954, *41*, 63.
2531 M. Pöhm and F. Galinovsky, *Monatsh. Chem.*, 1953, *84*, 1197: *C.A.*, 1954, *48*, 6509-a.
2532 M. Pöhm and M. Wichtl, *Scientia Pharm.*, 1955, *23*, 31.
2533 B. D. Polis and J. G. Reinhold, *J. Biol. Chem.*, 1944, *156*, 231.
2534 B. D. Polis and H. W. Shmukler, *J. Biol. Chem.*, 1953, *201*, 475.
2535 F. H. Pollard, *Advancement of Sci.*, 1950, *7*, 322.
2536 F. H. Pollard, *Brit. Med. Bull.*, 1954, *10*, 187.
2537 F. H. Pollard and C. J. Hardy, *Chem. and Ind.*, *1955*, 1145.
2538 F. H. Pollard and J. F. W. McOmie, *Endeavour*, 1951, *10*, 213: *C.A.*, 1952, *46*, 374-b.
2539 F. H. Pollard and J. F. W. McOmie, *Chromatographic Methods of Inorganic Analysis*, Butterworth, London, 1953.
2539a F. H. Pollard, J. F. W. McOmie and A. J. Banister, *Chem. and Ind.*, *1955*, 1598.
2540 F. H. Pollard, J. F. W. McOmie and I. I. M. Elbeih, *Nature*, 1949, *163*, 292.
2541 F. H. Pollard, J. F. W. McOmie and I. I. M. Elbeih, *J. Chem. Soc.*, *1951*, 466.
2542 F. H. Pollard, J. F. W. McOmie and I. I. M. Elbeih, *J. Chem. Soc.*, *1951*, 470.
2542a F. H. Pollard, J. F. W. McOmie and D. J. Jones, *J. Chem. Soc.*, *1955*, 4337.
2542b F. H. Pollard, J. F. W. McOmie and J. V. Martin, *Analyst*, 1956, *81*, 353.
2542c F. H. Pollard, J. F. W. McOmie, J. V. Martin and C. J. Hardy, *J. Chem. Soc.*, *1955*, 4332.
2543 F. H. Pollard, J. F. W. McOmie and H. M. Stevens, *J. Chem. Soc.*, *1951*, 771.
2544 F. H. Pollard, J. F. W. McOmie and H. M. Stevens, *J. Chem. Soc.*, *1952*, 4730.
2545 F. H. Pollard, J. F. W. McOmie and H. M. Stevens, *J. Chem. Soc.*, *1954*, 3435.
2546 F. H. Pollard, J. F. W. McOmie, H. M. Stevens and J. G. Maddock, *J. Chem. Soc.*, *1953*, 1338.

2546a F. H. Pollard, G. Nickless and A. J. Banister, *Analyst*, 1956, *81*, 577.
2547   M. Polonovski, R. G. Busnel and M. C. Pesson, *Compt. rend.*, 1943, *217*, 163.
2548   M. Polonovski, F. Penaranda and L. Robert, *Bull. soc. chim. biol.*, 1953, *35*, 801.
2549   J. Polonsky, *Compt. rend.*, 1951, *232*, 1878.
2550   J. Polonsky and E. Lederer, *Bull. soc. chim. France*, 1954, 504.
2551   A. Polson, *Nature*, 1948, *161*, 351.
2552   A. Polson, *Biochim. Biophys. Acta*, 1949, *3*, 205.
2553   A. Polson, V. M. Mosley and R. W. G. Wyckoff, *Science*, 1947, *105*, 603.
2554   A. Polson and R. W. G. Wyckoff, *Science*, 1948, *108*, 501.
2555   M. H. Pond, *Lancet*, 1952, *263*, 906.
2556   C. G. Pope and M. F. Stevens, *Biochem. J.*, 1939, *33*, 1070.
2557   E. A. Popenoe and V. du Vigneaud, *J. Biol. Chem.*, 1954, *206*, 353.
2558   J. Porath, *Acta Chem. Scand.*, 1952, *6*, 1237.
2559   J. Porath, *Acta Chem. Scand.*, 1954, *8*, 1813.
2560   J. Porath, *Arkiv Kemi*, 1954, *7*, 535.
2560a J. Porath, *Nature*, 1955, *175*, 478.
2561   J. Porath and P. Flodin, *Nature*, 1951, *168*, 202.
2562   J. Porath and C. H. Li, *Biochim. Biophys. Acta*, 1954, *13*, 268.
2563   J. W. Porter and R. E. Lincoln, *Arch. Biochem.*, 1950, *27*, 390.
2564   J. W. Porter and F. P. Zscheile, *Arch. Biochem.*, 1946, *10*, 537.
2565 R. R. Porter, *Biochem. J.*, 1953, *53*, 320.
2566   R. R. Porter, in *The Chemical Structure of Proteins*, Ciba Foundation Symposium, 1953, p. 31.
2567   R. R. Porter, *Brit. Med. Bull.*, 1954, *10*, 237.
2568   R. R. Porter, *Methods of Enzymology*, Vol. 1, Acad. Press, New York, 1954.
2569   R. R. Porter, *Biochem. J.*, 1955, *59*, 405.
2570   R. R. Porter and F. Sanger, *Biochem. J.*, 1948, *42*, 287.
2571   W. L. Porter, *Anal. Chem.*, 1951, *23*, 412.
2572   W. L. Porter, *Anal. Chem.*, 1954, *26*, 439.
2573   W. L. Porter and N. Hoban, *Anal. Chem.*, 1954, *26*, 1846.
2574   W. L. Porter, J. Naghski and A. Eisner, *Arch. Biochem.*, 1949, *24*, 461.
2575   A. Post, *J. Soc. Cosmet. Chem.*, 1954, *5*, 23.
2576   T. Posternak, D. Reymond et W. Haerdi, *Helv. Chim. Acta*, 1955, *38*, 191.
2577   J. L. Potter, K. D. Brown and M. Laskowski, *Biochim. Biophys. Acta*, 1952, *9*, 150.
2578   A. M. Potts and T. F. Gallagher, *J. Biol. Chem.*, 1944, *154*, 349.
2579   J. J. Pratt and J. L. Auclair, *Science*, 1948, *108*, 213.
2580   V. Prelog and H. G. Beyerman, *Helv. Chim. Acta*, 1945, *28*, 350.
2581   V. Prelog, L. Frenkiel and S. Szpilfogel, *Helv. Chim. Acta*, 1946, *29*, 484.
2582   V. Prelog and J. Führer, *Helv. Chim. Acta*, 1945, *28*, 583.
2583   V. Prelog and U. Geyer, *Helv. Chim. Acta*, 1945, *28*, 576.
2584   V. Prelog and M. Osgan, *Helv. Chim. Acta*, 1952, *35*, 981.
2585   V. Prelog and L. Ruzicka, *Helv. Chim. Acta*, 1944, *27*, 61.
2586   V. Prelog, L. Ruzicka and P. Stein, *Helv. Chim. Acta*, 1943, *26*, 2222.
2587   V. Prelog, L. Ruzicka and F. Steinmann, *Helv. Chim. Acta*, 1944, *27*, 674.
2588   V. Prelog and S. Szpilfogel, *Helv. Chim. Acta*, 1944, *27*, 390.
2589   V. Prelog, S. Szpilfogel and J. Battegay, *Helv. Chim. Acta*, 1947, *30*, 366.
2590   V. Prelog and P. Wieland, *Helv. Chim. Acta*, 1944, *27*, 1127.
2591   B. A. Prescott and H. Waelsch, *J. Biol. Chem.*, 1946, *164*, 331.
2592   T. D. Price, P. B. Hudson and D. F. Ashman, *Nature*, 1955, *175*, 45.
2593   V. E. Price and R. E. Greenfield, *J. Biol. Chem.*, 1954, *209*, 363.

2594 D. A. Prins, *Helv. Chim. Acta*, 1946, *29*, 1.
2595 H. K. Prins and T. H. J. Huisman, *Nature*, 1955, *175*, 903.
2596 G. W. Probst and M. O. Schultze, *J. Biol. Chem.*, 1950, *187*, 453.
2597 Ž. Procházka, *Chem. Listy*, 1950, *44*, 43: *C.A.*, 1951, *45*, 5561-c.
2598 Ž. Procházka, *Collection Czechoslov. Chem. Communs.*, 1954, *19*, 98; *Chem. Listy*, 1953, *47*, 1637; *Českoslov. farm.*, 1953, *2*, 17: *C.A.*, 1954, *48*, 2805-c, 4033-f.
2599 Ž. Procházka, L. Lábler and Z. Kotásek, *Collection Czechoslov. Chem. Communs.*, 1954, *19*, 1258.
2600 A. A. Prokrovskii, *Byull. Eksptl. Biol. i Med.*, 1954, *38*, No. 12, 69: *C.A.*, 1955, *49*, 6358-i.
2601 H. Proom and A. J. Woiwod, *J. Gen. Microbiol.*, 1949, *3*, 319: *C.A.*, 1950, *44*, 3560-h.
2602 B. Puri and T. R. Seshadri, *J. Sci. Ind. Res.*, 1954, *13B*, 321: *C.A.*, 1955, *49*, 8856-b.
2603 J. H. Purnell and M. S. Spencer, *Nature*, 1955, *175*, 988.
2604 E. W. Putman and W. Z. Hassid, *J. Biol. Chem.*, 1952, *196*, 749.
2605 M. L. Quaife, *J. Biol. Chem.*, 1948, *175*, 605.
2606 J. H. Quastel and S. F. Van Straten, *Proc. Soc. Exptl. Biol. Med.*, 1952, *81*, 6: *C.A.*, 1953, *47*, 662-a.
2607 K.-E. Quentin, *Z. Lebensm.-Untersuch. u. -Forsch.*, 1952, *95*, 305: *C.A.*, 1953, *47*, 2082-f.
2608 I. D. Raacke-Fels, *Arch. Biochem. Biophys.*, 1953, *43*, 289.
2609 P. Radhakrishna, *Anal. Chim. Acta*, 1952, *6*, 351.
2610 P. Radhakrishna, *Anal. Chim. Acta*, 1953, *8*, 140.
2611 R. Radhakrishnamurty and P. S. Sarma, *Current Sci. (India)*, 1953, *22*, 209: *C.A.*, 1954, *48*, 3431-a.
2612 C. M. Rafique and F. Smith, *J. Am. Chem. Soc.*, 1950, *72*, 4634.
2613 H. W. J. Ragetli and J. P. H. van der Want, *Koninkl. Ned. Akad. Wetenschap. Proc.*, 1954, *C 57*, 621: *C.A.*, 1955, *49*, 6388-h.
2613a K. S. Rajan and J. Gupta, *J. Sci. Ind. Research (India)*, 1955, *14B*, 453; *Anal. Abstr.*, 1956, *4*, 961.
2614 L. L. Ramsey, *J. Assoc. Offic. Agr. Chem.*, 1948, *31*, 164.
2615 L. L. Ramsey and W. I. Patterson, *J. Assoc. Offic. Agr. Chem.*, 1945, *28*, 644.
2616 L. L. Ramsey and W. I. Patterson, *J. Assoc. Offic. Agr. Chem.*, 1946, *29*, 337.
2617 L. L. Ramsey and W. I. Patterson, *J. Assoc. Offic. Agr. Chem.*, 1948, *31*, 139.
2618 L. L. Ramsey and W. I. Patterson, *J. Assoc. Offic. Agr. Chem.*, 1948, *31*, 441.
2619 S. S. Randall and A. J. P. Martin, *Biochem. J.*, 1949, *44*, ii.
2620 C. L. Rao and J. Shankar, *Anal. Chim. Acta*, 1953, *8*, 491.
2621 N. A. N. Rao and P. K. Wadhwani, *Current Sci. (India)*, 1954, *23*, 359: *C.A.*, 1955, *49*, 5214-d.
2622 N. R. Rao, K. H. Shah and K. Venkataraman, *Current Sci. (India)*, 1950, *19*, 149: *C.A.*, 1950, *44*, 9154-g.
2623 N. R. Rao, K. H. Shah and K. Venkataraman, *Current Sci. (India)*, 1951, *20*, 66: *C.A.*, 1951, *45*, 8775-h.
2624 P. S. Rao, R. M. Beri and P. R. Rao, *Proc. Indian Acad. Sci.*, 1951, *34 A*, 236: *C.A.*, 1952, *46*, 9020-g.
2625 T. Rao and K. V. Giri, *J. Indian Inst. Sci., Sect. A*, 1953, *35*, 137: *C.A.*, 1953, *47*, 10043-i.
2626 D. A. Rappoport, C. R. Calvert, R. K. Loeffler and J. H. Gast, *Anal. Chem.*, 1955, *27*, 820.

2627 M. M. Rapport, K. Meyer and A. Linker, *J. Am. Chem. Soc.*, 1951, *73*, 2416.
2628 P. S. Rasmussen, *Biochim. Biophys. Acta*, 1954, *14*, 567.
2629 P. S. Rasmussen, *Biochim. Biophys. Acta*, 1955, *16*, 157.
2630 H. M. Rauen, *Angew. Chem.*, 1948, *60*, A 250.
2631 H. M. Rauen and K. Felix, *Z. physiol. Chem.*, 1948, *283*, 139.
2632 H. M. Rauen, G. Leonhardi and M. Buchka, *Z. physiol. Chem.*, 1949, *284*, 178.
2633 H. M. Rauen, W. Stamm and K.-H. Kimbel, *Z. physiol. Chem.*, 1952, *289*, 80.
2634 H. M. Rauen and L. Wolf, *Z. physiol. Chem.*, 1948, *283*, 233.
2635 G. Ray, N. C. Ganguli and S. C. Roy, *Nature*, 1953, *172*, 809.
2636 N. H. Ray, *J. Appl. Chem.*, 1954, *4*, 21.
2637 C. H. Rayburn, W. R. Harlan and H. R. Hanmer, *Anal. Chem.*, 1953, *25*, 1419.
2638 L. Rebenfeld and E. Pacsu, *J. Am. Chem. Soc.*, 1953, *75*, 4370.
2639 F. Reber and T. Reichstein, *Helv. Chim. Acta*, 1945, *28*, 1164.
2640 F. Reber and T. Reichstein, *Helv. Chim. Acta*, 1946, *29*, 343.
2641 R. R. Redfield, *Biochim. Biophys. Acta*, 1953, *10*, 344.
2642 R. R. Redfield and E. S. Guzman Barron, *Arch. Biochem. Biophys.*, 1952, *35*, 443.
2643 L. J. Reed, *J. Biol. Chem.*, 1950, *183*, 451.
2644 L. J. Reed and B. G. DeBusk, *J. Biol. Chem.*, 1952, *199*, 873.
2645 E. T. Reese and W. Gilligan, *Arch. Biochem. Biophys.*, 1953, *45*, 74.
2646 J. Reese, *Angew. Chem.*, 1954, *66*, 170.
2647 W. A. Reeves and T. B. Crumpler, *Anal. Chem.*, 1951, *23*, 1576.
2648 R. M. Reguera and I. Asimov, *J. Am. Chem. Soc.*, 1950, *72*, 5781.
2649 H. Reich, D. H. Nelson and A. Zaffaroni, *J. Biol. Chem.*, 1950, *187*, 411.
2650 H. Reich and T. Reichstein, *Helv. Chim. Acta*, 1943, *26*, 562.
2651 H. Reich, S. J. Sanfilippo and K. F. Crane, *J. Biol. Chem.*, 1952, *198*, 713.
2652 W. S. Reich, *Biochem. J.*, 1939, *33*, 1000.
2653 P. Reichard, *Nature*, 1948, *162*, 662; *J. Biol. Chem.*, 1949, *179*, 763.
2654 D. Reichenberg, K. W. Pepper and D. J. McCauley, *J. Chem. Soc.*, *1951*, 493.
2655 R. Reichert, *Helv. Chim. Acta*, 1945, *28*, 484.
2655a E. R. Reichl, *Monatsh. Chem.*, 1955, *86*, 69.
2655b E. R. Reichl, *Mikrochim. Acta*, *1956*, 958.
2656 T. Reichstein and J. van Euw, *Helv. Chim. Acta*, 1938, *21*, 1197.
2657 T. Reichstein and C. Montigel, *Helv. Chim. Acta*, 1939, *22*, 1212.
2658 T. Reichstein and C. W. Shoppee, *Discuss. Faraday Soc.*, 1949, *7*, 305.
2659 A. F. Reid, *Brit. Pat.* 682,768: *C.A.*, 1953, *47*, 4559-d.
2660 A. F. Reid, *U.S. Pat.* 2,669,559: *C.A.*, 1954, *48*, 7094-i.
2661 A. F. Reid and F. Jones, *Ind. Eng. Chem.*, 1951, *43*, 1074.
2662 R. L. Reid and M. Lederer, *Biochem. J.*, 1951, *50*, 60.
2663 W. W. Reid, *Nature*, 1950, *166*, 569.
2664 W. W. Reid, *Nature*, 1951, *168*, 739.
2665 F. Reimers and K. R. Gottlieb, *Quart. J. Pharm. Pharmacol.*, 1947, *20*, 99.
2666 F. Reindel and W. Hoppe, *Naturwissenschaften*, 1953, *40*, 245.
2667 F. Reindel and W. Hoppe, *Chem. Ber.*, 1954, *87*, 1103.
2668 F. Reinhardt, *Mikrochim. Acta*, *1954*, 219.
2669 R. H. Reitsema, *Anal. Chem.*, 1954, *26*, 960.
2670 F. E. Resnik, L. A. Lee and W. A. Powell, *Anal. Chem.*, 1955, *27*, 928.
2671 A. Resplandy, *Compt. rend.*, 1954, *239*, 496.
2672 T. M. Reynolds, *Nature*, 1955, *175*, 46.

2673 L. E. Rhuland, E. Work, R. F. Denman and D. S. Hoare, *J. Am. Chem. Soc.*, 1955, *77*, 4844.
2674 J. Ribéreau-Gayon and P. Ribéreau-Gayon, *Compt. rend.*, 1954, *238*, 2114.
2675 F. A. H. Rice and A. G. Osler, *J. Biol. Chem.*, 1951, *189*, 115.
2676 R. G. Rice, G. J. Keller and J. G. Kirchner, *Anal. Chem.*, 1951, *23*, 194.
2677 A. Richardson and A. C. Hulme, *Nature*, 1955, *175*, 43.
2678 R. W. Richardson, *J. Chem. Soc.*, *1951*, 910.
2679 J. P. R. Riches, *Nature*, 1946, *158*, 96.
2680 J. W. Richter, D. E. Ayer, A. W. Bazemore, N. G. Brink and K. Folkers, *J. Am. Chem. Soc.*, 1953, *75*, 1952.
2681 C. Riebeling and H. Burmeister, *Klin. Wochschr.*, 1954, *32*, 1057: *C.A.*, 1955, *49*, 1855-f.
2682 B. Riegel, C. E. Schweitzer and P. G. Smith, *J. Biol. Chem.*, 1939, *129*, 495.
2683 W. Rieman III and S. Lindenbaum, *Anal. Chem.*, 1952, *24*, 1199.
2683a R. W. Riemenschneider, S. F. Herb and P. L. Nichols Jr., *J. Am. Oil Chemists' Soc.*, 1949, *26*, 371: *C.A.*, 1949, *43*, 6839-i.
2684 R. F. Riley, *J. Am. Chem. Soc.*, 1950, *72*, 5782.
2685 V. T. Riley, *J. Natl. Cancer Inst.*, 1950, *11*, 199: *C.A.*, 1951, *45*, 3019-i.
2686 V. T. Riley, *J. Natl. Cancer Inst.*, 1950, *11*, 215: *C.A.*, 1951, *45*, 3019-d.
2687 V. T. Riley and M. W. Woods, *Proc. Soc. Exptl. Biol. Med.*, 1950, *73*, 92.
2688 V. T. Riley, *Science*, 1948, *107*, 573.
2689 V. T. Riley, M. L. Hesselbach, S. Fiala, M. W. Woods and D. Burk, *Science*, 1949, *109*, 361.
2690 C. Rimington, *Biochem. J.*, 1943, *37*, 443.
2691 C. Rimington, *Biochem. J.*, 1946, *40*, 669.
2692 E. A. H. Roberts, *Chem. and Ind.*, *1955*, 631.
2693 E. A. H. Roberts and D. J. Wood, *Biochem. J.*, 1951, *49*, 414.
2694 E. A. H. Roberts and D. J. Wood, *Biochem. J.*, 1953, *53*, 332.
2695 E. J. Roberts, *U.S. Pat.* 2,590,209: *C.A.*, 1952, *46*, 5111-d.
2696 H. R. Roberts and E. F. McFarren, *J. Dairy Sci.*, 1953, *36*, 620: *C.A.*, 1953, *47*, 8112-b.
2697 J. C. Roberts and K. Selby, *J. Chem. Soc.*, *1949*, 2785.
2698 J. D. Roberts and C. Green, *Ind. Eng. Chem. Anal. Ed.*, 1946, *18*, 335.
2699 G. K. Robertson, *Med. J. Australia*, 1954, *1*, 698: *C.A.*, 1954, *48*, 11614-g.
2700 C. D. Robeson, *U.S. Pat.* 2,552,908 (1951): *C.A.*, 1951, *45*, 8211-g.
2701 A. M. Robinson, *Recent Progr. Hormone Research*, 1954, *9*, 163.
2702 D. Robinson, J. N. Smith and R. T. Williams, *Biochem. J.*, 1951, *50*, 221.
2703 D. A. Robinson and G. F. Mills, *Ind. Eng. Chem.*, 1949, *41*, 2221.
2704 F. A. Robinson and K. L. A. Fehr, *Biochem. J.*, 1952, *51*, 298.
2705 G. Robinson, *Metallurgia*, 1947, *37*, 45, 107: *C.A.*, 1948, *42*, 2204-c.
2706 G. Robinson, *Discuss. Faraday Soc.*, 1949, *7*, 195.
2707 R. Robinson, *Nature*, 1951, *168*, 512.
2708 W. Robson and A. S. M. Selim, *Biochem. J.*, 1953, *53*, 431.
2709 J. Roche and M. Eysseric-Lafon, *Bull. soc. chim. biol.*, 1951, *33*, 1437.
2710 J Roche, M. Jutisz, S. Lissitzky and R. Michel, *Biochim. Biophys. Acta*, 1951, *7*, 257.
2711 J. Roche, S. Lissitzky, O. Michel and R. Michel, *Biochim. Biophys. Acta*, 1951, *7*, 439.
2712 J. Roche, S. Lissitzky and R. Michel, *Biochim. Biophys. Acta*, 1952, *8*, 339.

2713 J. Roche, S. Lissitzky and R. Michel, *Biochim. Biophys. Acta*, 1953, *11*, 215, 220.

2714 J. Roche, S. Lissitzky and R. Michel, *Meth. of Biochem. Anal.*, 1954, *1*, 243,

2715 J. Roche, R. Michel, S. Lissitzky and S. Mayer, *Biochim. Biophys. Acta*, 1951, *7*, 446.

2716 J. Roche, R. Michel, S. Lissitzky and Y. Yagi, *Bull. soc. chim. biol.*, 1954, *36*, 143.

2717 J. Roche, R. Michel, J. Nunez and G. Lacombe, *Compt. rend.*, 1955, *240*, 464.

2718 J. Roche, R. Michel and J. Tata, *Biochim. Biophys. Acta*, 1953, *11*, 543.

2719 J. Roche, R. Michel and E. Volpert, *Compt. rend. soc. biol.*, 1954, *148*, 21.

2720 J. Roche, R. Michel and W. Wolf, *Compt. rend.*, 1955, *240*, 251.

2721 J. Roche, R. Michel, W. Wolf and J. Nunez, *Compt. rend.*, 1955, *240*, 921.

2722 J. Roche, Ng.-V. Thoai and J. L. Hatt, *Biochim. Biophys. Acta*, 1954, *14*, 71.

2723 J. Roche, Y. Yagi, R. Michel, S. Lissitzky and M. Eysseric-Lafon, *Bull. soc. chim. biol.*, 1951, *33*, 526.

2724 L. B. Rockland, J. L. Blatt and M. S. Dunn, *Anal. Chem.*, 1951, *23*, 1142.

2725 L. B. Rockland and M. S. Dunn, *Science*, 1949, *109*, 539.

2726 L. B. Rockland and M. S. Dunn, *Science*, 1950, *111*, 332.

2727 L. B. Rockland and M. S. Dunn, *J. Am. Chem. Soc.*, 1949, *71*, 4121.

2728 L. B. Rockland, J. Lieberman and M. S. Dunn, *Anal. Chem.*, 1952, *24*, 778.

2729 L. B. Rockland and J. C. Underwood, *Anal. Chem.*, 1954, *26*, 1557.

2730 C. J. Rogers, J. R. Kimmel, M. E. Hutchin and H. A. Harper, *J. Biol. Chem.*, 1954, *206*, 553.

2731 M. Rohdewald and L. Zechmeister, *Enzymologia*, 1951, *15*, 109: *C.A.*, 1953, *47*, 4387-h.

2732 J. F. Roland Jr. and A. M. Gross, *Anal. Chem.*, 1954, *26*, 502.

2733 L. P. Romanoff and R. S. Wolf, *Recent Progr. Hormone Research*, 1954, *9*, 337.

2734 I. A. Rose and B. S. Schweigert, *J. Am. Chem. Soc.*, 1951, *73*, 5903.

2735 S. Rosebeek, *Chem. Weekblad*, 1950, *46*, 813: *C.A.*, 1951, *45*, 4298-d.

2736 S. Roseman, R. H. Abeles and A. Dorfman, *Arch. Biochem. Biophys.*, 1952, *36*, 232.

2737 S. Roseman and J. Ludowieg, *J. Am. Chem. Soc.*, 1954, *76*, 301.

2738 C. A. Rossi, *Boll. soc. ital. biol. sper.*, 1950, *26*, 1563.

2739 F. D. Rossini, *Frontiers in Chemistry*, Vol. 7, *Recent Advances in Analytical Chemistry*, p. 157–182, Interscience Publishers, New York, 1949.

2740 L. J. Roth, E. Leifer, J. R. Hogness and W. H. Langham, *J. Biol. Chem.*, 1949, *178*, 963.

2741 W. A. Roth and A. L. LeRosen, *Anal. Chem.*, 1948, *20*, 1092.

2742 H. Röttger, *Experientia*, 1953, *9*, 150.

2743 A. Roudier, *Mém. services chim. état (Paris)*, 1951, *36*, 155; *Chim. anal.*, 1951, *33*, No. spécial 11 bis, p. 391.

2744 A. Roudier, *Compt. rend.*, 1953, *237*, 662.

2745 A. Roudier and L. Eberhard, *Compt. rend.*, 1955, *240*, 2012.

2746 D. G. Roux, *Nature*, 1951, *168*, 1041.

2747 H. Roux, E. Thilo, H. Grunze and M. Viscontini, *Helv. Chim. Acta*, 1955, *38*, 15.

2748 M. Rovery and P. Desnuelle, *Bull. soc. chim. biol.*, 1954, *36*, 95.

2749 M. Rovery and C. Fabre, *Bull. soc. chim. biol.*, 1953, *35*, 541.

2750 E. V. Rowsell, *Nature*, 1951, *168*, 104.

2751 B. L. Rubin, H. Rosenkrantz, R. I. Dorfman and G. Pincus, *J. Clin. Endocrin. Metab.*, 1953, *13*, 568: *C.A.*, 1953, *47*, 7580-e.

2752 H. Rückert and O. Samuelson, *Svensk Kem. Tid.*, 1954, *66*, 337: *C.A.*, 1955, *49*, 8658-h.
2753 G. Runneberg, *Svensk Kem. Tid.*, 1945, *57*, 114: *C.A.*, 1946, *40*, 2416-g.
2754 G. Runneberg, and O. Samuelson, *Svensk Kem. Tid.*, 1945, *57*, 91: *C.A.*, 1946, *40*, 2416-8.
2755 G. Runneberg and O. Samuelson, *Svensk Kem. Tid.*, 1945, *57*, 250: *C.A.*, 1946, *40*, 2417-5.
2756 A. S. Russell and C. N. Cochran, *Ind. Eng. Chem.*, 1950, *42*, 1336.
2757 R. G. Russell and D. W. Pearce, *J. Am. Chem. Soc.*, 1943, *65*, 595.
2758 L. Rutter, *Nature*, 1949, *163*, 487; *Analyst*, 1950, *75*, 37.
2759 L. Ruzicka, O. Jeger and J. Norymberski, *Helv. Chim. Acta*, 1944, *27*, 1185.
2760 L. Ruzicka, P. A. Plattner and H. Heusser, *Helv. Chim. Acta*, 1946, *29*, 473.
2761 L. Ruzicka, P. A. Plattner and J. Pataki, *Helv. Chim. Acta*, 1945, *28*, 1360.
2762 L. Ruzicka and V. Prelog, *Helv. Chim. Acta*, 1943, *26*, 975.
2763 L. Ruzicka, V. Prelog and E. Tagmann, *Helv. Chim. Acta*, 1944, *27*, 1149.
2764 L. Ruzicka, E. Rey and A. C. Muhr, *Helv. Chim. Acta*, 1944, *27*, 472.
2765 D. I. Ryabchikov and A. I. Lazarev, *Doklady Akad. Nauk S.S.S.R.*, 1953, *92*, 777: *C.A.*, 1954, *48*, 5722-d.
2765a D. I. Ryabchikov and V. F. Osipova, *Doklady Akad. Nauk S.S.S.R.*, 1954, *96*, 761; *Anal. Abstr.*, 1956, *3*, 100.
2766 W. Ryan and A. F. Williams, *Analyst*, 1952, *77*, 293.
2767 D. Rybář, B. Toušek and I. M. Hais, *Collection Czechoslov. Chem. Communs.*, 1955, *20*, 724.
2768 S. Rydel and M. Macheboeuf, *Bull. soc. chim. biol.*, 1949, *31*, 1265.
2769 H. N. Rydon and P. W. G. Smith, *Nature*, 1952, *169*, 922.
2770 A. P. Ryle, F. Sanger, L. F. Smith and R. Kitai, *Biochem. J.*, 1955, *60*, 541.
2771 L. Sacconi, *Gazz. chim. ital.*, 1948, *78*, 583.
2772 L. Sacconi, *Gazz. chim. ital.*, 1949, *79*, 141.
2773 L. Sacconi, *Gazz. chim. ital.*, 1949, *79*, 152.
2774 L. Sacconi, *Nature*, 1949, *164*, 70.
2775 L. Sacconi, *Discuss. Faraday Soc.*, 1949, *7*, 173.
2776 J. Sacks, L. Lutwak and P. D. Hurley, *J. Am. Chem. Soc.*, 1954, *76*, 424.
2777 I. Saenz-Lascaño-Ruiz, *Ind. Parfum.*, 1946, *1*, 187: *C.A.*, 1948, *42*, 5675-c.
2778 I. Saenz-Lascaño-Ruiz, P. Chovin and H. Moureu, *Bull. soc. chim. France*, *1946*, 592.
2779 I. Saenz-Lascaño-Ruiz, P. Chovin and H. Moureu, *La séparation chromatographique des colorants alimentaires et son application à la détection des fraudes*, 59 pages. Actualités scientifiques et industrielles, No. 1046, Hermann et Cie, Paris, 1948.
2780 E. L. Saier, A. Pozefsky and N. D. Coggeshall, *Anal. Chem.*, 1954, *26*, 1258.
2781 A. Saifer and I. Oreskes, *Anal. Chem.*, 1953, *25*, 1539.
2782 A. Saifer and I. Oreskes, *Science*, 1954, *119*, 124.
2783 T. Sakaguchi and H. Yasuda, *J. Pharm. Soc. Japan*, 1951, *71*, 1469: *C.A.*, 1952, *46*, 3452-i.
2784 E. H. Sakal and E. J. Merrill, *Science*, 1953, *117*, 451.
2785 J. E. Salmon and H. R. Tietze, *J. Chem. Soc.*, *1952*, 2324.
2786 L. T. Samuels, *J. Biol. Chem.*, 1947, *168*, 471.
2787 O. Samuelson, *Z. anal. Chem.*, 1939, *116*, 328.
2788 O. Samuelson, *Svensk Kem. Tid.*, 1939, *51*, 195: *C.A.*, 1940, *34*, 1271-2.
2789 O. Samuelson, *Svensk Kem. Tid.*, 1940, *52*, 115: *C.A.*, 1940, *34*, 6539-7.

2790   O. Samuelson, *Svensk Kem. Tid.*, 1940, *52*, 241: *C.A.*, 1941, *35*, 1340-6.
2791   O. Samuelson, *Svensk Kem. Tid.*, 1941, *53*, 422: *C.A.*, 1944, *38*, 3889-7.
2792   O. Samuelson, *Svensk Kem. Tid.*, 1942, *54*, 124: *C.A.* 1944, *38*, 2896-6.
2793   O. Samuelson, *Svensk Papperstid.*, 1945, *48*, 55: *C.A.*, 1945, *39*, 3157-7.
2794   O. Samuelson, *Dissertation Tekniska Högskola*, Stockholm, 1944.
2795   O. Samuelson, *Svensk Kem. Tid.*, 1945, *57*, 158: *C.A.*, 1946, *40*, 2417-2.
2796   O. Samuelson, *Iva*, 1946, *17*, 5: *C.A.*, 1946, *40*, 5657-9.
2797   O. Samuelson, *Svensk Kem. Tid.*, 1946, *58*, 247: *C.A.*, 1947, *41*, 1571-h.
2798   O. Samuelson, *Tekn. Tid.*, 1946, *76*, 561: *C.A.*, 1946, *40*, 5657-9.
2799   O. Samuelson, *Ion Exchangers in Analytical Chemistry*, Almquist & Wiksell, Stockholm, 1952.
2800   O. Samuelson, *Svensk Kem. Tid.*, 1953, *65*, 275: *C.A.*, 1954, *48*, 6309-c.
2801   O. Samuelson, *Z. Elektrochem.*, 1953, *57*, 207.
2802   O. Samuelson, R. Djurfeldt and A. Scholander, *Elementa*, 1947, *30*, 107.
2803   O. Samuelson and F. Gärtner, *Acta Chem. Scand.*, 1951, *5*, 596.
2804   O. Samuelson, L. Lundén and K. Schramm, *Z. anal. Chem.*, 1953, *140*, 330.
2805   G. L. San and A. J. Ultée Jr., *Nature*, 1952, *169*, 586.
2806   A. Sandoval and L. Zechmeister, *J. Am. Chem. Soc.*, 1947, *69*, 553.
2807   F. Sanger, *Biochem. J.*, 1945, *39*, 507.
2808   F. Sanger, *Biochem. J.*, 1946, *40*, 261.
2809   F. Sanger, *Biochem. J.*, 1949, *44*, 126.
2810   F. Sanger, *Biochem. J.*, 1949, *45*, 563.
2811   F. Sanger, *Advances in Protein Chemistry*, 1952, *7*, 1–67.
2812   F. Sanger and H. Tuppy, *Biochem. J.*, 1951, *49*, 463.
2813   F. Sanger and H. Tuppy, *Biochem. J.*, 1951, *49*, 481.
2814   C. Sannié and H. Lapin, *Bull. soc. chim. France*, *1952*, 1080.
2815   G. Sansone and F. Cusmano, *Boll. soc. ital. biol. sper.*, 1950, *26*, 1343, 1680: *C.A.*, 1951, *45*, 8628-e.
2816   G. Sansone and F. Cusmano, *Boll. soc. ital. biol. sper.*, 1951, *27*, 1369, 1370: *C.A.*, 1952, *46*, 6176-a.
2817   G. Sansone, F. Cusmano and C. Ravazzoni, *Boll. soc. ital. biol. sper.*, 1951, *27*, 1371: *C.A.*, 1952, *46*, 6222-d.
2818   B. Sansoni, *Naturwissenschaften*, 1954, *41*, 213.
2819   D. I. Sapozhnikov, I. A. Bronshteïn and T. A. Krasovskaya, *Biokhimiya*, 1955, *20*, 286: *C.A.*, 1955, *49*, 16084-f.
2820   L. H. Sarett, *J. Am. Chem. Soc.*, 1948, *70*, 1454.
2821   B. Sarma, *Science and Culture*, 1950, *16*, 165: *C.A.*, 1951, *45*, 3683-a.
2822   B. Sarma, *Science and Culture*, 1951, *17*, 139: *C.A.*, 1953, *47*, 6217-i.
2823   B. Sarma, *Trans. Bose Research Inst.*, Calcutta, 1949/51, *18*, 105: *C.A.*, 1953, *47*, 11067-e.
2824   K. Satake and T. Seki, *J. Japan. Chemistry*, 1950, *4*, 557: *C.A.*, 1951, *45*, 4604-f.
2825   L. Sattler and F. W. Zerban, *Anal. Chem.*, 1952, *24*, 1862.
2826   H. E. Sauberlich, *J. Biol. Chem.*, 1952, *195*, 337.
2827   L. Saunders and R. Srivastava, *J. Chem. Soc.*, *1950*, 2915.
2828   K. Savard, *J. Biol. Chem.*, 1953, *202*, 457.
2829   K. Savard, *Recent Progr. Hormone Research*, 1954, *9*, 185.
2830   K. Savard, H. W. Wotiz, P. Marcus and H. M. Lemon, *J. Am. Chem. Soc.*, 1953, *75*, 6327.
2831   P. Savary, *Bull. soc. chim. biol.*, 1954, *36*, 927.
2832   P. Savary and P. Desnuelle, *Bull. soc. chim. France*, *1953*, 939.

2833   P. Savary and P. Desnuelle, *Bull. soc. chim. France*, *1954*, 936.
2834   F. Savoia, *Chimica (Milano)*, 1954, *9*, 223.
2835   R. Scarisbrick, E. Baldwin and V. Moyle, *Biochem. J.*, 1948, *42*, xiv.
2836   H. Schäfer and W. Neugebauer, *Z. anorg. u. allgem. Chem.*, 1953, *274*, 114.
2836a H. K. Schauer and R. Burlisch, *Z. Naturforsch.*, 1955, *10b*, 683.
2837   R. W. Schayer, *J. Biol. Chem.*, 1951, *189*, 301.
2838   R. W. Schayer, *J. Biol. Chem.*, 1952, *196*, 469.
2839   H. W. Scheeline, *U.S. Pat.* 2,523,149: *C.A.*, 1950, *44*, 10388-h.
2840   E. Schenker, A. Hunger and T. Reichstein, *Helv. Chim. Acta*, 1954, *37*, 680, 1004.
2841   H. H. Schenker and W. Rieman III, *Anal. Chem.*, 1953, *25*, 1637.
2842   O. Schindler and T. Reichstein, *Helv. Chim. Acta*, 1951, *34*, 18.
2843   O. Schindler and T. Reichstein, *Helv. Chim. Acta*, 1951, *34*, 108, 608.
2844   O. Schindler and T. Reichstein, *Helv. Chim. Acta*, 1951, *34*, 1732.
2845   O. Schindler and T. Reichstein, *Helv. Chim. Acta*, 1952, *35*, 307.
2846   O. Schindler and T. Reichstein, *Helv. Chim. Acta*, 1952, *35*, 673.
2847   T. D. Schlabach Jr., *Univ. Microfilms* (Ann. Arbor, Mich.), Publ. No. 7479: *C.A.*, 1954, *48*, 7104-c.
2848   D. Schleede, *Brennstoff-Chem.*, 1955, *36*, 78: *C.A.*, 1955, *49*, 8742-g.
2849   K. Schlögl and A. Siegel, *Mikrochem. ver. Mikrochim. Acta*, 1952/53, *40*, 202.
2850   K. Schlögl and A. Siegel, *Z. physiol. Chem.*, 1953, *292*, 263.
2851   K. Schlögl, F. Wessely and E. Wawersich, *Monatsh. Chem.*, 1954, *85*, 957.
2852   H. H. Schlubach and P. Hauschildt, *Ann.*, 1952, *577*, 54.
2853   H. H. Schlubach and A. Heesch, *Ann.*, 1951, *572*, 114: *C.A.*, 1951, *45*, 7918-f.
2854   H. Schlüssel, W. Maurer, A. Hock and O. Hummel, *Biochem. Z.*, 1951, *322*, 226.
2855   K. Schmeiser and D. Jerchel, *Angew. Chem.*, 1953, *65*, 366.
2856   H. Schmid, A. Ebnöther and M. Burger, *Helv. Chim. Acta*, 1950, *33*, 609.
2857   H. Schmid, A. Ebnöther and P. Karrer, *Helv. Chim. Acta*, 1950, *33*, 1486.
2858   H. Schmid and P. Karrer, *Helv. Chim. Acta*, 1950, *33*, 512.
2859   H. Schmid, J. Kebrle and P. Karrer, *Helv. Chim. Acta*, 1952, *35*, 1864.
2860   M. D. Schmid and H. R. Bolliger, *Helv. Chim. Acta*, 1954, *37*, 884.
2861   B. Schmidli, *Helv. Chim. Acta*, 1955, *38*, 1078.
2862   H. Schmidt and H. Staudinger, *Angew. Chem.*, 1954, *66*, 711.
2863   H. Schmidt and H. Staudinger, *Biochem. Z.*, 1955, *326*, 343.
2864   O. T. Schmidt and R. Lademann, *Ann.*, 1951, *571*, 41.
2865   H. Schmitz, *Biochim. Biophys. Acta*, 1954, *14*, 160.
2866   H. Schmitz, R. B. Hurlbert and V. R. Potter, *J. Biol. Chem.*, 1954, *209*, 41.
2867   F. Schneider and G. A. Erlemann, *Naturwissenschaften*, 1952, *39*, 160.
2868   K. W. Schneider and R. Oberkobusch, *Brennstoff-Chem.*, 1951, *32*, 110: *C.A.*, 1951, *45*, 5573-b.
2869   A. Schöberl and P. Rambacher, *Biochem. Z.*, 1940, *305*, 223.
2870   G. Schoen, *Acta Vitaminol.*, 1953, *7*, 151: *C.A.*, 1954, *48*, 13851-h.
2871   G. Schoen, *Analisi cromatografica su carta*, Soc. Ed. Farm., Milano, 1954: *C.A.*, 1954, *48*, 9278-h.
2872   A. Schönberg, A. Mustafa and W. Asker, *J. Am. Chem. Soc.*, 1952, *74*, 5640.
2873   T. Schönfeld and E. Broda, *Mikrochem. ver. Mikrochim. Acta*, 1951, *36/37*, 537.
2874   F. Schoofs and H. Lecoq, *Bull. acad. roy. med. Belg.*, 1944, *9*, 122.
2875   J. Schormüller, *Z. Lebensm.-Untersuch. u. -Forsch.*, 1948, *88*, 576: *C.A.*, 1949, *43*, 3864-c.

2876 E. Schram and E. J. Bigwood, *Anal. Chem.*, 1953, *25*, 1424.

2877 E. Schram, J. P. Dustin, S. Moore and E. J. Bigwood, *Anal. Chim. Acta*, 1953, *9*, 149.

2878 E. Schram, S. Moore and E. J. Bigwood, *Biochem. J.*, 1954, *57*, 33.

2879 G. Schramm and G. Braunitzer, *Z. Naturforsch.*, 1950, *5b*, 297: *C.A.*, 1951, *45*, 10295-g.

2880 G. Schramm and J. Primosigh, *Ber.*, 1943, *76*, 373.

2881 G. Schramm and J. Primosigh, *Ber.*, 1944, *77*, 417.

2882 G. Schramm and J. Primosigh, *Z. physiol. Chem.*, 1947, *282*, 271.

2883 W. A. Schroeder, *Ann. N.Y. Acad. Sci.*, 1948, *49*, 204.

2884 W. A. Schroeder, *J. Am. Chem. Soc.*, 1952, *74*, 281.

2885 W. A. Schroeder, *Progr. Chem. Org. Nat. Prod.*, 1954, *11*, 240.

2886 W. A. Schroeder and L. R. Honnen, *J. Am. Chem. Soc.*, 1953, *75*, 4615.

2887 W. A. Schroeder, L. R. Honnen and F. G. Green, *Proc. Natl. Acad. Sci. U.S.*, 1953, *39*, 23: *C.A.*, 1953, *47*, 8101-c.

2888 W. A. Schroeder, L. M. Kay, J. LeGette, L. R. Honnen and F. C. Green, *J. Am. Chem. Soc.*, 1954, *76*, 3556.

2889 W. A. Schroeder, L. M. Kay and I. C. Wells, *J. Biol. Chem.*, 1950, *187*, 221.

2890 W. A. Schroeder, E. W. Malmberg, L. L. Fong, K. N. Trueblood, J. D. Landerl and E. Hoerger, *Ind. Eng. Chem.*, 1949, *41*, 2818.

2891 J. Schubert, *J. Phys. Chem.*, 1948, *52*, 340.

2892 J. Schubert, *Anal. Chem.*, 1950, *22*, 1359.

2893 J. Schubert, *J. Colloid Sci.*, 1950, *5*, 376.

2894 J. Schubert and A. Lindenbaum, *Nature*, 1950, *166*, 913; 118th Meeting A.C.S. Chicago (1950).

2895 J. Schubert and J. W. Richter, *J. Phys. Chem.*, 1948, *52*, 350.

2896 J. Schubert and J. W. Richter, *J. Am. Chem. Soc.*, 1948, *70*, 4259.

2897 J. Schubert, E. R. Russell and L. B. Farabee, *Science*, 1949, *109*, 316.

2898 J. Schubert, E. R. Russell and L. S. Myers Jr., *J. Biol. Chem.*, 1950, *185*, 387.

2899 J. Schubert, E. R. Russell and L. S. Myers Jr., in F. C. Nachod, *Ion Exchange, Theory and Application*, Acad. Press, New York, 1949, pp. 167–221.

2900 H. A. Schuette, M. H. Khan and S. W. Nicksic, *J. Biol. Chem.*, 1953, *200*, 319.

2901 J. A. Schufle and H. M. Eiland, *J. Am. Chem. Soc.*, 1954, *76*, 960.

2902 R. H. Schuler, A. C. Boyd Jr. and D. J. Kay, *J. Chem. Educ.*, 1951, *28*, 192: *C.A.*, 1951, *45*, 4977-a.

2903 J. Schulman Jr. and R. P. Keating, *J. Biol. Chem.*, 1950, *183*, 215.

2904 M. P. Schulman and J. M. Buchanan, *J. Biol. Chem.*, 1952, *196*, 513.

2905 K. E. Schulte and H. Krause, *Biochem. Z.*, 1951, *322*, 168.

2906 H. J. Schümann, *Arch. exptl. Pathol. Pharmakol.*, 1955, *225*, 110: *C.A.*, 1955, *49*, 9059-d.

2907 J. B. Schute, *Nature*, 1953, *171*, 839.

2908 G. M. Schwab, *Discuss. Faraday Soc.*, 1949, *7*, 170.

2909 G. M. Schwab and G. Dattler, *Angew. Chem.*, 1937, *50*, 691.

2910 G. M. Schwab and A. N. Ghosh, *Angew. Chem.*, 1939, *52*, 389.

2911 G. M. Schwab and A. N. Ghosh, *Angew. Chem.*, 1940, *53*, 39.

2912 G. M. Schwab and A. N. Ghosh, *Z. anorg. Chem.*, 1949, *258*, 323.

2913 G. M. Schwab and A. Issidoridis, *Z. physik. Chem. B*, 1942, *53*, 1.

2914 G. M. Schwab and K. Jockers, *Naturwissenschaften*, 1937, *25*, 44.

2915 G. M. Schwab and K. Jockers, *Angew. Chem.*, 1937, *50*, 646.

2916 S. Schwartz and C. J. Watson, *Proc. Soc. Exptl. Biol. Med.*, 1941, *47*, 390.

2917 V. Schwarz, *Biochem. J.*, 1953, *53*, 148.
2918 E. Schwerdtfeger, *Naturwissenschaften*, 1953, *40*, 201.
2919 E. Schwerdtfeger, *Chem. Tech.* 1954, *6*, 192 : *C.A.*, 1954, *48*, 11246-i.
2920 E. Schwerdtfeger, *Naturwissenschaften*, 1954, *41*, 18.
2921 S. Schwimmer, *Nature*, 1953, *171*, 442.
2921a E. Scoffone and E. Carini, *Ricerca sci.*, 1955, *25*, 2109.
2922 J. E. Scott and L. Golberg, *Chem. and Ind.*, *1954*, 48.:
2923 R. W. Scott, *Anal. Chem.*, 1955, *27*, 367.
2924 J. M. Searle and R. G. Westall, *Biochem. J.*, 1951, *48*, 1.
2925 J. W. Sease, *J. Am. Chem. Soc.*, 1947, *69*, 2242.
2926 J. W. Sease, *J. Am. Chem. Soc.*, 1948, *70*, 3630.
2927 J. W. Sease, *Anal. Chem.*, 1949, *21*, 1430.
2928 J. W. Sease and L. Zechmeister, *J. Am. Chem. Soc.*, 1947, *69*, 270.
2929 E. Seebeck and T. Reichstein, *Helv. Chim. Acta*, 1943, *26*, 536.
2930 E. Seebeck and O. Schindler, *Helv. Chim. Acta*, 1946, *29*, 317.
2931 J. E. Seegmiller and B. L. Horecker, *J. Biol. Chem.*, 1952, *194*, 261.
2932 R. A. Seibert, C. E. Williams and R. A. Huggins, *Science*, 1954, *120*, 222.
2933 C. F. Seidel, P. H. Müller and H. Schinz, *Helv. Chim. Acta*, 1944, *27*, 738.
2934 H. Seiler, M. Schuster and H. Erlenmeyer, *Helv. Chim. Acta*, 1954, *37*, 1252.
2935 H. Seiler, E. Sorkin and H. Erlenmeyer, *Helv. Chim. Acta*, 1952, *35*, 120.
2936 H. Seiler, E. Sorkin and H. Erlenmeyer, *Helv. Chim. Acta*, 1952, *35*, 2483.
2937 H. Selbach and W. Trappe, *Arch. Psychiat. Nervenheilk.*, 1944, *117*, 541 : *C.A.*, 1949, *43*, 9139-d.
2938 R. B. Seligman, M. D. Edmonds, A. E. O'Keeffe and L. A. Lee, *Chem. and Ind.*, *1954*, 1195.
2939 D. Seligson and B. Shapiro, *Anal. Chem.*, 1952, *24*, 754.
2940 K. Semm and R. Fried, *Naturwissenschaften*, 1952, *39*, 326.
2940a B. N. Sen, *Australian J. Science*, 1950, *13*, 49.
2941 B. N. Sen, *Australian J. Science*, 1952/53, *15*, 133.
2942 B. N. Sen, *Z. anorg. u. allgem. Chem.*, 1953, *273*, 183.
2943 S. P. Sen and A. C. Leopold, *Physiologia Plantarum*, 1954, *7*, 98 : *C.A.*, 1955, *49*, 13574-g.
2944 N. C. Sen Gupta and A. Gupta, *Science and Culture*, 1951, *17*, 265 : *C.A.*, 1952, *46*, 10049-f.
2945 G. Serchi, *Ann. chim. (Roma)*, 1953, *43*, 253 : *C.A.*, 1953, *47*, 12129-c.
2946 G. Serchi and G. Rapi, *Sperimentale, Sez. chim. biol.*, 1952, *3*, 107 : *C.A.*, 1953, *47*, 2814-b.
2947 R. F. Serro and R. J. Brown, *Anal. Chem.*, 1954, *26*, 890.
2948 M. Servigne, P. Guérin de Montgareuil and M. Pinta, *Fractionnement chromatographique et dosage de la vitamine A*, Centre National de la Recherche Scientifique, Paris, 1951.
2949 Y. Servigne, *Compt. rend.*, 1954, *239*, 272.
2950 S. M. Shea, *Nature*, 1950, *165*, 729.
2951 W. H. Shearon and O. F. Gee, *Ind. Eng. Chem.*, 1949, *41*, 218.
2952 J. C. Sheehan and W. A. Bolhofer, *J. Am. Chem. Soc.*, 1950, *72*, 2466.
2953 F. M. Shemyakin, P. P. Kharlamov and E. S. Mitselovskii, *Zavodskaya Lab.*, 1950, *16*, 1126 : *C.A.*, 1951, *45*, 1908-b.
2954 F. M. Shemyakin and E. S. Mitselovskii, *Zhur. Anal. Khim.*, 1948, *3*, 349 : *C.A.*, 1949, *43*, 8973-b.

2955 F. M. Shemyakin and E. S. Mitselovskiĭ, *Zavodskaya Lab.*, 1950, *16*, 748: *C.A.*, 1951, *45*, 974-e.
2956 F. M. Shemyakin, E. S. Mitselovskiĭ and D. V. Romanov, *Izvest. Sektora Fiz-Khim. Anal. Akad. Nauk S.S.S.R.*, 1953, *23*, 334: *C.A.*, 1954, *48*, 9151.
2957 C. C. Shepard, *J. Immunol.*, 1952, *68*, 179.
2958 C. C. Shepard and A. Tiselius, *Discuss. Faraday Soc.*, 1949, *7*, 275.
2959 C. C. Shepard and W. G. Woodend, *J. Immunol.*, 1951, *66*, 385: *C.A.*, 1951, *45*, 4779-e.
2960 D. M. Shepherd and G. B. West, *Nature*, 1952, *169*, 797.
2961 D. M. Shepherd, G. B. West and V. Erspamer, *Nature*, 1953, *172*, 357.
2962 C. W. Sheppard, W. E. Cohn and P. J. Mathias, *Arch. Biochem. Biophys.*, 1953, *47*, 475.
2963 J. M. Shewan, L. I. Fletcher, S. M. Partridge and R. C. Brimley, *J. Sci. Food Agric.*, 1952, *3*, 394: *C.A.*, 1953, *47*, 4517-d.
2964 M. Shibata, *Science (Japan)*, 1949, *19*, 570.
2965 M. Shibata and T. Uemura, *J. Chem. Soc. Japan, Pure Chem. Sect.*, 1951, *72*, 541: *C.A.*, 1952, *46*, 3458-d.
2966 S. Shibata, M. Takito and O. Tanaka, *J. Am. Chem. Soc.*, 1950, *72*, 2789.
2967 Z. V. Shidkova, *Uspekhi Fiz. Nauk*, 1951, *43*, 369.
2967a M. Shinagawa, J. Takanaka, Y. Kiso, A. Tsukiji and Y. Matama, *Bull. Chem. Soc. Japan*, 1955, *28*, 565, 568.
2968 R. Ya. Shkol'nik, *Doklady Akad. Nauk S.S.S.R.*, 1954, *98*, 443: *C.A.*, 1955, *49*, 3321-h.
2969 C. W. Shoppee and D. A. Prins, *Helv. Chim. Acta*, 1943, *26*, 201.
2970 C. W. Shoppee and G. H. R. Summers, *J. Chem. Soc.*, *1950*, 687.
2971 P. Shu, *Can. J. Research*, 1950, *28 B*, 527: *C.A.*, 1951, *45*, 4175.
2972 A. T. Shulgin, O. G. Lien Jr., E. M. Gal and D. M. Greenberg, *J. Am. Chem. Soc.*, 1952, *74*, 2427.
2973 G. M. Shull, J. L. Sardinas and R. C. Nubel, *Arch. Biochem. Biophys.*, 1952, *37*, 186.
2974 O. Sibliková-Zbudovská and I. M. Hais, *Chem. Listy*, 1954, *48*, 1263: *C.A.* 1954, *48*, 14116-g.
2975 W. Siedel and W. Fröwis, *Z. physiol. Chem.*, 1941, *267*, 37.
2976 W. Siedel and E. Grams, *Z. physiol. Chem.*, 1941, *267*, 49.
2977 W. Siedel and H. Möller, *Z. physiol. Chem.*, 1940, *264*, 64.
2978 A. Siegel and K. Schlögl, *Mikrochem. ver. Mikrochim. Acta*, 1953, *40*, 383.
2979 B. Siegel, G. A. Candela and R. M. Howard, *J. Am. Chem. Soc.*, 1954, *76*, 1311.
2980 J. M. Siegel, *J. Biol. Chem.*, 1954, *208*, 205.
2981 H. W. Siegelman, *J. Biol. Chem.*, 1955, *213*, 647.
2982 G. Siewert and H. Jungnickel, *Ber.*, 1943, *76*, 210.
2983 H. Silberman and S. Silberman-Martyncewa, *J. Biol. Chem.*, 1946, *165*, 359.
2984 H. Silberman and R. H. Thorp, *J. Pharm. and Pharmacol.*, 1953, *5*, 438: *C.A.*, 1953, *47*, 11659-c.
2985 N. Siliprandi and P. Bianchi, *Biochim. Biophys. Acta*, 1955, *16*, 424.
2986 D. Siliprandi and N. Siliprandi, *Biochim. Biophys. Acta*, 1954, *14*, 52.
2987 M. H. Silk and H. H. Hahn, *Biochem. J.*, 1954, *56*, 406.
2988 M. H. Silk and H. H. Hahn, *Biochem. J.*, 1954, *57*, 577 and 582.
2989 M. Silverman and J. C. Keresztesy, *J. Am. Chem. Soc.*, 1951, *73*, 1897.
2989a V. Simanek and J. Janak, *Chem. Listy*, 1954, *48*, 1623.
2990 J. Simek, *Chem. Przmysl*, 1954, *4*, (29), 56: *C.A.*, 1955, *49*, 4439-h.

2991 R. O. Simmons and F. W. Quackenbush, *J. Am. Oil Chem. Soc.*, 1953, *30*, 614: *C.A.*, 1954, *48*, 1705-i.
2992 P. Simonart and K.-Y. Chow, *Bull. soc. chim. Belg.*, 1950, *59*, 417.
2993 P. Simonart and K.-Y. Chow, *Enzymologia*, 1951, *14*, 356: *C.A.*, 1953, *47*, 3358-f.
2994 P. Simonart and K.-Y. Chow, *Netherlands Milk Dairy J.*, 1952, *6*, 206: *C.A.*, 1953, *47*, 2811.
2995 S. A. Simpson and J. F. Tait, *Mem. Soc. Endocrinol.*, 1953, *2*, 9: *C.A.*, 1955, *49*, 12576-g.
2996 S. A. Simpson, J. F. Tait, A. Wettstein, R. Neher, J. von Euw, O. Schindler and T. Reichstein, *Helv. Chim. Acta*, 1955, *37*, 1163.
2997 A. J. Singer and L. Kenner, *Anal. Chem.*, 1951, *23*, 387.
2998 R. L. Sinsheimer, *J. Biol. Chem.*, 1954, *208*, 445.
2999 R. L. Sinsheimer and J. F. Koerner, *Science*, 1951, *114*, 42.
3000 M. D. Siperstein, F. M. Harold, I. L. Chaikoff and W. G. Dauben, *J. Biol. Chem.*, 1954, *210*, 181.
3001 C. Sironval, *Arch. Intern. Physiol.*, 1953, *61*, 563.
3002 N. M. Sisakyan, E. N. Bezinger, P. G. Garkavi and G. Ya. Kivman, *Doklady Akad. Nauk S.S.S.R.*, 1954, *96*, 343: *C.A.*, 1954, *48*, 10821-i.
3003 J. Sjöquist, *Acta Chem. Scand.*, 1953, *7*, 447.
3004 J. Sjöquist, *Biochim. Biophys. Acta*, 1955, *16*, 283.
3005 E. Sjöström, *Svensk Kem. Tid.*, 1952, *64*, 301: *C.A.*, 1953, *47*, 4248-c.
3006 E. Sjöström, *Acta Polytech.*, 1954, *144*, 7: *C.A.*, 1955, *49*, 1483-f.
3007 J. Sjövall, *Acta Chem. Scand.*, 1952, *6*, 1552.
3008 J. Sjövall, *Acta Chem. Scand.*, 1954, *8*, 339.
3009 A. Skogseid, *Thesis*, University of Oslo (1948).
3010 L. Æ. Sluyterman and H. J. Veenendaal, *Rec. trav. chim.*, 1949, *68*, 717.
3011 L. Æ. Sluyterman and H. J. Veenendaal, *Rec. trav. chim.*, 1951, *70*, 1049.
3012 J. K. Small, *U.S. Pat.* 2,548,502: *C.A.*, 1951, *45*, 5467-d.
3013 P. Smit, *U.S. Pat.* 2,191,063, 2,205,635: *C.A.*, 1940, *34*, 4500-7, 7504-3.
3014 W. M. Smit, *Anal. Chim. Acta*, 1948, *2*, 671: *C.A.*, 1949, *43*, 7868-d.
3015 W. M. Smit, *Discuss. Faraday Soc.*, 1949, *7*, 248.
3016 D. H. Smith and F. E. Clark, *Soil Science*, 1951, *72*, 353: *C.A.*, 1953, *47*, 3502-b.
3017 D. H. Smith and F. E. Clark, *Soil Sci. Soc. Am., Proc.*, 1952, *16*, 170: *C.A.*, 1952, *46*, 8574-g.
3018 E. D. Smith and A. L. LeRosen, *Anal. Chem.*, 1951, *23*, 732.
3019 E. D. Smith and A. L. LeRosen, *Anal. Chem.*, 1954, *26*, 928.
3020 E. D. Smith, W. A. Mueller and L. N. Rogers, *Anal. Chem.*, 1952, *24*, 1117.
3021 E. Lester Smith, *Biochem. J.*, 1942, *36*, xxii.
3022 E. Lester Smith, *Biochem. Soc. Symposia*, 1949, No. 3, 89.
3023 E. Lester Smith, *Discuss. Faraday Soc.*, 1949, *7*, 317.
3024 E. Lester Smith and D. Allison, *Analyst*, 1952, *77*, 29.
3025 E. Lester Smith, *Nature*, 1952, *169*, 60.
3026 E. Lester Smith and W. F. J. Cuthbertson, *Biochem. J.*, 1949, *45*, xii.
3027 Eliane L. Smith and E. F. Tuller, *Arch. Biochem. Biophys.*, 1955, *54*, 114.
3028 F. Smith and D. Spriestersbach, *Nature*, 1954, *174*, 466.
3029 G. H. Smith, *J. Chem. Soc.*, *1952*, 1530.
3030 G. N. Smith and C. S. Worrel, *Arch. Biochem.*, 1950, *28*, 1.
3031 I. Smith, *Nature*, 1953, *171*, 43.
3032 J. D. Smith and R. Markham, *Biochem. J.*, 1950, *46*, 509.
3033 J. D. Smith and G. R. Wyatt, *Biochem. J.*, 1951, *49*, 144.

3034   J. N. Smith and R. T. Williams, *Biochem. J.*, 1955, *60*, 284.
3035   J. R. Smith, C. R. Smith Jr. and G. U. Dinneen, *Anal. Chem.*, 1950, *22*, 867.
3036   K. C. Smith and F. W. Allen, *J. Am. Chem. Soc.*, 1953, *75*, 2131.
3037   M. A. Smith and B. R. Willeford Jr., *Anal. Chem.*, 1954, *26*, 751.
3038   O. C. Smith, *Inorganic Chromatography*, D. Van Nostrand Company Inc., New York, 1953.
3039   S. C. Smith and S. H. Wender, *J. Am. Chem. Soc.*, 1948, *70*, 3719.
3040   O. Snellman and B. Gelotte, *Nature*, 1951, *168*, 461.
3041   J. Q. Snyder and S. H. Wender, *Arch. Biochem. Biophys.*, 1953, *46*, 465.
3042   H. A. Sober, G. Kegeles and F. J. Gutter, *J. Am. Chem. Soc.*, 1952, *74*, 2734.
3043   H. A. Sober and E. A. Peterson, *J. Am. Chem. Soc.*, 1954, *76*, 1711.
3044   W. T. Sokolski, H. Koffler and P. A. Tetrault, *Arch. Biochem. Biophys.*, 1953, *43*, 236.
3045   J. Solms, *Helv. Chim. Acta*, 1955, *38*, 1127.
3046   I. A. Solomons and P. P. Regna, *J. Am. Chem. Soc.*, 1950, *72*, 2974.
3047   A. H. Soloway, W. J. Considine, D. K. Fukushima and T. F. Gallagher, *J. Am. Chem. Soc.*, 1954, *76*, 2941.
3048   A. W. Somerford, *J. Criminal Law, Criminol. Police Sci.*, 1952, *43*, 124.
3048a   G. Sommer, *Z. anal. Chem.*, 1956, *151*, 336.
3049   M. Soodak, A. Pircio and L. R. Cerecedo, *J. Biol. Chem.*, 1949, *181*, 713.
3050   F. Šorm and B. Keil, *Collection Czechoslov. Chem. Communs.*, 1951, *16*, 366.
3051   F. Šorm and Z. Šormová, *Collection Czechoslov. Chem. Communs.*, 1951, *16*, 207.
3052   F. Šorm, M. Suchý and V. Herout, *Chem. Listy*, 1952, *46*, 55: *C.A.*, 1952, *46*, 11134-b.
3053   T. H. Soutar and E. Hampton, *Nature*, 1954, *174*, 801.
3054   J. C. Sowden, *J. Am. Chem. Soc.*, 1954, *76*, 4487.
3055   E. C. Spaeth and D. H. Rosenblatt, *Science*, 1949, *110*, 258.
3056   D. H. Spalding and T. G. Metcalf, *J. Bact.*, 1954, *68*, 160: *C.A.*, 1954, *48*, 13038-h.
3057   E. Späth, F. Galinovsky and M. Mayer, *Ber.*, 1942, *75*, 805.
3058   H. Specker and H. Hartkamp, *Naturwissenschaften*, 1953, *40*, 271.
3059   F. H. Spedding, *Discuss. Faraday Soc.*, 1949, *7*, 214.
3060   F. H. Spedding and J. L. Dye, *J. Am. Chem. Soc.*, 1950, *72*, 5350.
3061   F. H. Spedding, E. I. Fulmer, B. Ayers, T. A. Butler, J. E. Powell, A. D. Tevebaugh and R. Thompson, *J. Am. Chem. Soc.*, 1948, *70*, 1671.
3062   F. H. Spedding, E. I. Fulmer, T. A. Butler, E. M. Gladrow, M. Gobush, P. E. Porter, J. E. Powell and J. M. Wright, *J. Am. Chem. Soc.*, 1947, *69*, 2812.
3063   F. H. Spedding, E. I. Fulmer, T. A. Butler and J. E. Powell, *J. Am. Chem. Soc.*, 1950, *72*, 2349.
3064   F. H. Spedding, E. I. Fulmer, J. E. Powell and T. A. Butler, *J. Am. Chem. Soc.*, 1950, *72*, 2354.
3065   F. H. Spedding and J. E. Powell, *J. Am. Chem. Soc.*, 1954, *76*, 2545.
3066   F. H. Spedding and J. E. Powell, *J. Am. Chem. Soc.*, 1954, *76*, 2550.
3066a   F. H. Spedding and J. E. Powell, in *Ion Exchange Technology* (Edited by Nachod and Schubert), Acad. Press, New York, 1956.
3066b   F. H. Spedding, J. E. Powell and H. J. Svec, *J. Am. Chem. Soc.*, 1955, *77*, 6125.
3067   F. H. Spedding, J. E. Powell and E. J. Wheelwright, *J. Am. Chem. Soc.*, 1954, *76*, 612.
3068   F. H. Spedding, J. E. Powell and E. J. Wheelwright, *J. Am. Chem. Soc.*, 1954, *76*, 2557.

3069 F. H. Spedding, A. F. Voigt, E. M. Gladrow and N. R. Sleight, *J. Am. Chem. Soc.*, 1947, *69*, 2777.

3070 F. H. Spedding, A. F. Voigt, E. M. Gladrow, N. R. Sleight, J. E. Powell, J. M. Wright, T. A. Butler and P. Figard, *J. Am. Chem. Soc.*, 1947, *69*, 2786.

3071 J. R. Spies, E. J. Coulson, H. S. Bernton and H. Stevens, *J. Am. Chem. Soc.*, 1940, *62*, 1420.

3072 J. Spiteri, *Bull. soc. chim. biol.*, 1954, *36*, 1355.

3073 A. H. Sporer, S. Freed and K. M. Sancier, *Science*, 1954, *119*, 68.

3074 T. Sproston Jr. and E. G. Bassett, *Anal. Chem.*, 1954, *26*, 552.

3075 B. S. Srikantan and V. Krishnan, *J. Indian Chem. Soc.*, 1949, *26*, 415.

3076 B. S. Srikantan and V. Krishnan, *J. Indian Chem. Soc.*, 1950, *27*, 34.

3077 B. S. Srikantan and C. N. Venkatachallum, *J. Indian Chem. Soc.*, 1953, *30*, 167.

3078 M. Stacey and G. Swift, *J. Chem. Soc.*, *1948*, 1555.

3079 E. R. Stadtman, *J. Biol. Chem.*, 1952, *196*, 535.

3080 E. R. Stadtman and H. A. Barker, *J. Biol. Chem.*, 1950, *184*, 769.

3081 E. R. Stadtman and A. Kornberg, *J. Biol. Chem.*, 1953, *203*, 47.

3082 F. H. Stadtman, *J. Am. Chem. Soc.*, 1948, *70*, 3583.

3083 Standard Oil Development Co., *Brit. Pat.* 610,187 (1948): *C.A.*, 1949, *43*, 2215-h.

3084 J. B. Stark, A. E. Goodban and H. S. Owens, *Anal. Chem.*, 1951, *23*, 413.

3085 H. E. Stavely, *J. Am. Chem. Soc.*, 1941, *63*, 3127.

3086 A. E. Steel, *Nature*, 1951, *168*, 877.

3087 A. E. Steel, *Nature*, 1954, *173*, 315.

3088 G. Stefanović and T. Janjić, *Anal. Chim. Acta*, 1954, *11*, 550.

3088a G. Stefanović, T. Janjić and R. Crnojević, *Bull. soc. chim. Belgrade*, 1955, *20*, 343.

3089 M. Steiger and T. Reichstein, *Helv. Chim. Acta*, 1938, *21*, 546.

3090 W. H. Stein, in *The Chemical Structure of Proteins*, Ciba Foundation Symposium, 1953, p. 17.

3091 W. H. Stein, *J. Biol. Chem.*, 1953, *201*, 45.

3092 W. H. Stein, A. C. Paladini, C. H. W. Hirs and S. Moore, *J. Am. Chem. Soc.*, 1954, *76*, 2848.

3093 W. H. Stein and S. Moore, *J. Biol. Chem.*, 1948, *176*, 337.

3094 W. H. Stein and S. Moore, *J. Biol. Chem.*, 1949, *178*, 79.

3095 W. H. Stein and S. Moore, *Cold Spring Harbor Symposia Quant. Biol.*, 1949, *14*, 179: *C.A.*, 1950, *44*, 6910-d.

3096 W. H. Stein and S. Moore, *J. Biol. Chem.*, 1951, *190*, 103.

3097 W. H. Stein and S. Moore, *J. Biol. Chem.*, 1954, *211*, 915.

3098 M. Steiner and E. Stein von Kamienski, *Naturwissenschaften*, 1953, *40*, 483.

3099 E. P. Stepanyan, *Klin. Med.*, 1954, *32*, 42: *C.A.*, 1955, *49*, 5562-f.

3100 M. H. Stern, C. D. Robeson, L. Weisler and J. G. Baxter, *J. Am. Chem. Soc.*, 1947, *69*, 869.

3101 M. I. Stern and G. I. M. Swyer, *Nature*, 1952, *169*, 796.

3102 J. Sternberg, *Mikrochim. Acta*, in press [presented to the Microchemical Congress (1955)].

3103 C. M. Stevens, P. E. Halpern and R. P. Gigger, *J. Biol. Chem.*, 1951, *190*, 705.

3103a H. M. Stevens, *Anal. Chim. Acta*, 1956, *15*, 51.

3104 P. C. Stevenson, A. A. Franke, R. Borg and W. Nervik, *J. Am. Chem. Soc.*, 1953, *75*, 4876.

3105 F. C. Steward, W. Stepka and J. F. Thompson, *Science*, 1948, *107*, 451.

3106 F. C. Steward, J. F. Thompson and C. E. Dent, *Science*, 1949, *110*, 439.
3107 F. C. Steward, R. H. Wetmore, J. F. Thompson and J. P. Nitsch, *Am. J. Botany*, 1954, *41*, 123: *C.A.*, 1954, *48*, 8342-b.
3108 F. C. Steward, R. M. Zacharius and J. K. Pollard, *Ann. Acad. Sci. Fenn.*, 1955, Ser. A II, *60*, 321.
3109 A. Stewart, *Discuss. Faraday Soc.*, 1949, *7*, 65.
3110 A. Stewart and Imp. Chem. Ind., *Brit. Pat.* 565,405 (1944): *C.A.*, 1946, *40*, 4490-2.
3111 D. C. Stewart, *Anal. Chem.*, 1955, *27*, 1279.
3112 H. B. Stewart, *Biochem. J.*, 1953, *55*, xxvi.
3113 B. F. Stimmel, *J. Biol. Chem.*, 1944, *153*, 327; 1946, *162*, 99.
3114 B. F. Stimmel, *J. Biol. Chem.*, 1946, *162*, 99.
3115 B. F. Stimmel, J. D. Randolph and W. M. Conn, *J. Clin. Endocrinol. and Metabolism*, 1952, *12*, 371: *C.A.*, 1953, *47*, 3912-d.
3116 W. Stich, R. Kehl and H. R. Walter, *Z. physiol. Chem.*, 1953, *292*, 178.
3117 W. Stich and G. Stärk, *Naturwissenschaften*, 1953, *40*, 56.
3118 A. Stöckli, *Helv. Chim. Acta*, 1954, *37*, 1581.
3119 F. H. Stodola, O. L. Shotwell, A. M. Borud, R. G. Benedict and A. C. Riley Jr., *J. Am. Chem. Soc.*, 1951, *73*, 2290.
3120 P. J. Stoffyn and R. W. Jeanloz, *Arch. Biochem. Biophys.*, 1954, *52*, 373.
3121 W. M. Stokes, F. C. Hickey, P. O. Fish and W. A. Fish, *J. Am. Chem. Soc.*, 1954, *76*, 5174.
3122 E. L. R. Stokstad, B. L. Hutchings and Y. Subbarow, *Ann. N.Y. Acad. Sci.*, 1946, *48*, 261: *C.A.*, 1947, *41*, 6259-f.
3123 J. Stolkowski, *Compt. rend.*, 1947, *225*, 312.
3124 A. Stoll, E. Angliker, F. Barfuss, W. Kussmaul and J. Renz, *Helv. Chim. Acta*, 1951, *34*, 1460.
3125 A. Stoll and W. Kreis, *Helv. Chim. Acta*, 1951, *34*, 1431.
3126 A. Stoll and A. Rüegger, *Helv. Physiol. Pharmacol. Acta*, 1952, *10*, 385: *C.A.*, 1953, *47*, 448-g.
3127 A. Stoll and A. Rüegger, *Helv. Chim. Acta*, 1954, *37*, 1725.
3128 M. Stoll, *Helv. Chim. Acta*, 1947, *30*, 991.
3129 S. Stoll et Y. Bouteville, *Ann. fals. et fraudes*, 1954, *47*, 183: *C.A.*, 1954, *48*, 13171-c.
3130 S. Stoll and Y. Bouteville, *Chim. anal.*, 1954, *36*, 33: *C.A.*, 1954, *48*, 4716-c.
3131 A. Stolman and C. P. Stewart, *Analyst*, 1949, *74*, 536, 543.
3132 I. D. E. Storey and G. J. Dutton, *Biochem. J.*, 1955, *59*, 279.
3133 B. B. Stowe and K. V. Thimann, *Arch. Biochem. Biophys.*, 1954, *51*, 499.
3134 E. Strack and I. Lorenz, *Z. physiol. Chem.*, 1954, *298*, 27.
3135 H. H. Strain, *J. Am. Chem. Soc.*, 1939, *61*, 1292.
3136 H. H. Strain, *Chromatographic Adsorption Analysis*, Interscience Publ., New York, 1942.
3137 H. H. Strain, *J. Phys. Chem.*, 1942, *46*, 1151.
3138 H. H. Strain, *Ind. Eng. Chem. Anal. Ed.*, 1946, *18*, 605.
3139 H. H. Strain, *J. Am. Chem. Soc.*, 1948, *70*, 588.
3140 H. H. Strain, *Ind. Eng. Chem.*, 1950, *42*, 1307.
3141 H. H. Strain, *Frontiers in Chromatographic Analysis*, in *Frontiers in Colloid Chemistry*, Vol. VIII, p. 29–63, Interscience Publ., New York, 1950.
3142 H. H. Strain, *J. Phys. Chem.*, 1953, *57*, 638.
3143 H. H. Strain and W. M. Manning, *J. Biol. Chem.*, 1942, *144*, 625.

3144 H. H. Strain and W. M. Manning, *J. Biol. Chem.*, 1942, *146*, 275.

3145 H. H. Strain, W. M. Manning and G. Hardin, *J. Biol. Chem.*, 1943, *148*, 655.

3146 H. H. Strain and G. W. Murphy, *Anal. Chem.*, 1952, *24*, 50.

3147 H. H. Strain and J. C. Sullivan, *Anal. Chem.*, 1951, *23*, 816.

3147a J. E. Strassner, *Dissertation Abstr.*, 1955, *15*, (8), 1309.

3148 K. Street Jr., *U.S. Pat.* 2,546,953: *C.A.*, 1951, *45*, 6128.

3149 K. Street Jr. and G. T. Seaborg, *J. Am. Chem. Soc.*, 1948, *70*, 4268.

3150 K. Street Jr. and G. T. Seaborg, *J. Am. Chem. Soc.*, 1950, *72*, 2790.

3151 K. Street Jr., S. G. Thompson and G. T. Seaborg, *J. Am. Chem. Soc.*, 1950, *72*, 4832.

3152 A. J. Streiff, B. J. Mair and F. D. Rossini, *Ind. Eng. Chem.*, 1949, *41*, 2037.

3153 U. Ströle, *Z. anal. Chem.*, 1955, *144*, 256.

3154 K. Suchý, *Chem. Listy*, 1954, *48*, 1084: *C.A.*, 1954, *48*, 13530-h.

3155 J. B. Summer, A. L. Dounce and V. L. Frampton, *J. Biol. Chem.*, 1940, *136*, 343.

3156 J. G. Surak, N. Leffler and R. Martinovich, *J. Chem. Educ.*, 1953, *30*, 20: *C.A.*, 1953, *47*, 4781-c.

3157 J. G. Surak and D. P. Schlueter, *J. Chem. Educ.*, 1952, *29*, 144.

3159 D. A. Sutton and J. Dutta, *J. Chem. Soc.*, *1949*, 939.

3160 W. J. L. Sutton and E. F. Almy, *J. Dairy Sci.*, 1953, *36*, 1248: *C.A.*, 1954, *48*, 1594-d.

3160a K. Suzuki, *J. Chem. Soc. Japan, Pure Chem. Sect.*, 1955, *76*, 184.

3161 A. B. Svendsen and K. B. Jensen, *Pharm. Acta Helv.*, 1950, *25*, 241: *C.A.*, 1951, *45*, 1726-a.

3162 L. Svennerholm, *Acta Chem. Scand.*, 1954, *8*, 1108.

3163 H. Svensson, *Acta Chem. Scand.*, 1950, *4*, 399, 1329.

3164 H. Svensson, *Acta Chem. Scand.*, 1951, *5*, 72, 1301, 1410.

3165 H. Svensson, *Swed. Pat.* 133,951: *C.A.*, 1952, *46*, 4863-g.

3166 H. Svensson, C.-E. Agrell, S.-O. Dehlén and L. Hagdahl, *Science Tools*, 1955, *2*, 17: *C.A.*, 1955, *49*, 14389-e.

3167 H. Svensson and I. Brattsten, *Arkiv Kemi*, 1949, *1*, 401.

3168 T. Swain, *Biochem. J.*, 1953, *53*, 200.

3169 R. L. Swank, A. E. Franklin and J. H. Quastel, *Proc. Soc. Exptl. Biol. Med.*, 1950, *75*, 850: *C.A.*, 1951, *45*, 3471-a.

3170 R. L. Swank, A. E. Franklin and J. H. Quastel, *Proc. Soc. Exptl. Biol. Med.*, 1951, *76*, 183: *C.A.*, 1951, *45*, 5291-i.

3171 M. L. Sweat, *Anal. Chem.*, 1954, *26*, 1964.

3172 M. L. Sweat and G. Farrell, *Federation Proc.*, 1953, *12*, 141.

3173 T. R. Sweeney and J. D. Bultman, *Anal. Chem.*, 1953, *25*, 1358.

3174 R. C. Sweet, W. Rieman III and J. Beukenkamp, *Anal. Chem.*, 1952, *24*, 952.

3175 C. E. Swift, W. G. Rose and G. S. Jamieson, *Oil and Soap*, 1943, *20*, 249.

3176 S. M. Swingle and A. Tiselius, *Biochem. J.*, 1951, *48*, 171.

3177 E. A. Swinton and D. E. Weiss, *Australian J. Appl. Sci.*, 1953, *4*, 316.

3178 G. I. M. Swyer and H. Braunsberg, *J. Endocrinol.*, 1950/51, *7*, lx.

3179 V. Sýkora and Ž. Procházka, *Chem. Listy*, 1953, *47*, 1674: *C.A.*, 1954, *48*, 3852-e.

3180 N. D. Sylvester, A. N. Ainsworth and E. B. Hughes, *Analyst*, 1945, *70*, 295.

3181 R. L. M. Synge, *Biochem. J.*, 1944, *38*, 285.

3182 R. L. M. Synge, *Biochem. J.*, 1945, *39*, 363.

3183 R. L. M. Synge, *Analyst*, 1946, *71*, 256.

3184   R. L. M. Synge, *Biochem. J.*, 1949, *44*, 542.
3185   R. L. M. Synge, *Biochem. J.*, 1951, *48*, 429.
3186   R. L. M. Synge, *Biochem. J.*, 1951, *49*, 642.
3187   R. L. M. Synge and A. Tiselius, *Acta Chem. Scand.*, 1947, *1*, 749.
3188   R. L. M. Synge and A. Tiselius, *Acta Chem. Scand.*, 1949, *3*, 231.
3189   D. Szafarz and C. Paternotte, *Bull. soc. chim. biol.*, 1951, *33*, 1518.
3189a P. Szarvas, T. Balogh and B. Toth, *Magyar Kém. Folyóirat*, 1956, *62*, 68.
3189b J. Szonntagh, L. Farady and A. Janosi, *Magyar Kém. Folyóirat*, 1955, *61*, 312.
3190   J. Tabone, N. Mamounas and S. Thomassey, *Bull. soc. chim. biol.*, 1951, *33*, 1557.
3191   J. Tabone, D. Robert, S. Thomassey and N. Mamounas, *Bull. soc. chim. biol.*, 1950, *32*, 529.
3192   J. Tabone, D. Robert and J. Troestler, *Bull. soc. chim. biol.*, 1948, *30*, 547.
3193   E. Tagmann, V. Prelog and L. Ruzicka, *Helv. Chim. Acta*, 1946, *29*, 440.
3194   P. W. Talboys, *Nature*, 1950, *166*, 1077.
3195   H. H. Tallan, S. Moore and W. H. Stein, *J. Biol. Chem.*, 1954, *211*, 927.
3196   H. H. Tallan and W. H. Stein, *J. Am. Chem. Soc.*, 1951, *73*, 2976.
3197   H. H. Tallan, W. H. Stein and S. Moore, *J. Biol. Chem.*, 1954, *206*, 825.
3198   E. A. Talley, D. R. Reynolds and W. L. Evans, *J. Am. Chem. Soc.*, 1943, *65*, 575.
3199   Z. Tamura, *Japan Analyst*, 1952, *1*, 117: *C.A.*, 1953, *47*, 4792-g.
3200   K. Tanaka and T. Matsuo, *Kagaku no Ryoiki*, 1951, *5*, 533: *C.A.*, 1954, *48*, 12609-a.
3201   K. Tanaka and T. Sugawa, *J. Pharm. Soc. Japan*, 1952, *72*, 616: *C.A.*, 1952, *46*, 9782-b.
3202   M. Tanaka, T. Ashizawa and M. Shibata, *Chem. Researches (Japan) 5, Inorg. and Anal. Chem.*, 1949, 35: *C.A.*, 1949, *43*, 8945-h.
3203   M. Tanaka and M. Shibata, *J. Chem. Soc. Japan*, 1950, *71*, 254, 312: *C.A.*, 1951, *45*, 4525-a, 4997-c.
3204   M. Tanaka and M. Shibata, *J. Chem. Soc. Japan, Pure Chem. Sect.*, 1951, *72*, 221: *C.A.*, 1953, *46*, 8506-b.
3205   D. S. Tarbell, E. G. Brooker, A. Vanterpool, W. Conway, C. J. Claus and T. J. Hall, *J. Am. Chem. Soc.*, 1955, *77*, 767.
3206   H. Tauber and E. L. Petit, *J. Am. Chem. Soc.*, 1952, *74*, 2865.
3207   A. Taurog, F. N. Briggs and I. L. Chaikoff, *J. Biol. Chem.*, 1951, *191*, 29 and 1952, *194*, 655.
3208   A. Taurog, I. L. Chaikoff and W. Tong, *J. Biol. Chem.*, 1949, *178*, 997.
3209   A. Taurog, C. Entenman, B. A. Fries and I. L. Chaikoff, *J. Biol. Chem.*, 1944, *155*, 19.
3210   A. Taurog, W. Tong and I. L. Chaikoff, *J. Biol. Chem.*, 1950, *184*, 83.
3211   S. P. Taylor Jr., *Proc. Soc. Exptl. Biol. Med.*, 1954, *85*, 226: *C.A.*, 1954, *48*, 7097-c.
3212   T. I. Taylor and H. C. Urey, *J. Chem. Phys.*, 1937, *5*, 597.
3213   T. I. Taylor and H. C. Urey, *J. Chem. Phys.*, 1938, *6*, 429.
3214   A. F. Teague, W. A. Gey and R. W. Van Dolah, *Anal. Chem.*, 1955, *27*, 785.
3215   B. Tegethoff, *Z. Naturforsch.*, 1953, *8b*, 374.
3216   H. Teicher and L. Gordon, *Anal. Chem.*, 1951, *23*, 930.
3217   H. Teicher and L. Gordon, *Anal. Chim. Acta*, 1953, *9*, 507.
3218   C. Teige, *Mem. soc. roy. lettres et sci. Classe sci.*, 1949, No. 5: *C.A.*, 1954, *48*, 5022.
3219   D. M. Tennent, J. B. Whitla and K. Florey, *Anal. Chem.*, 1951, *23*, 1748.

3220 H. M. Tenney and F. E. Sturgis, *Anal. Chem.*, 1954, *26*, 946.

3221 R. Ter Heide and J. F. Lemmens, *Perfumery Essent. Oil Record*, 1954, *45*, 21: *C.A.*, 1954, *48*, 7852-i.

3222 S. N. Tewari, *Z. anal. Chem.*, 1954, *141*, 401; *Kolloid-Z.*, 1954, *135*, 159: *C.A.*, 1954, *48*, 8117-b.

3223 S. N. Tewari, *Kolloid-Z.*, 1954, *138*, 178.

3224 S. N. Tewari, *Naturwissenschaften*, 1954, *41*, 217.

3225 S. N. Tewari, *Z. anal. Chem.*, 1954, *141*, 401.

3226 H. Thaler and R. Scheler, *Z. Lebensm.-Untersuch. u. -Forsch.*, 1952, *95*, 1: *C.A.*, 1952, *46*, 10479-e.

3227 S. J. Thannhauser, N. F. Boncoddo and G. Schmidt, *J. Biol. Chem.*, 1951, *188*, 417.

3228 G. Thibaudet, *Compt. rend.*, 1945, *220*, 751.

3229 H. Thies and F. W. Reuther, *Naturwissenschaften*, 1954, *41*, 230.

3230 Ng.-V. Thoai, J. Roche, Y. Robin and Ng.-V. Thiem, *Biochim. Biophys. Acta*, 1953, *11*, 593.

3231 G. Thomas, J. Ransy, P. Roland and A. Vanden Bulcke, *J. pharm. Belg.*, (2), 1950, *5*, 263: *C.A.*, 1951, *45*, 4406-f.

3232 A. Thompson, K. Anno, M. L. Wolfrom and M. Inatome, *J. Am. Chem. Soc.*, 1954, *76*, 1309.

3233 A. Thompson and M. L. Wolfrom, *J. Am. Chem. Soc.*, 1951, *73*, 5849.

3234 A. R. Thompson, *Australian J. Sci. Res.*, 1951, *B4*, 180: *C.A.*, 1952, *4* .27- .

3235 A. R. Thompson, *Nature*, 1951, *168*, 390.

3236 A. R. Thompson, *Nature*, 1952, *169*, 495.

3236a A. R. Thompson, *Biochim. Biophys. Acta*, 1954, *14*, 581.

3237 D. F. Thompson, P. L. Bayless and C. R. Hauser, *J. Org. Chem.*, 1954, *19*, 1490.

3238 E. O. P. Thompson and A. R. Thompson, *Progr. Chem. Org. Nat. Prod.*, 1955, *12*, 270.

3239 J. F. Thompson, J. K. Pollard and F. C. Steward, *Plant Physiol.*, 1953, *28*, 401: *C.A.*, 1955, *49*, 9743-f.

3240 J. F. Thompson and F. C. Steward, *J. Exptl. Botany*, 1952, *3*, 170: *C.A.*, 1954, *48*, 3489-i.

3241 J. F. Thompson, R. M. Zacharius and F. C. Steward, *Plant Physiol.*, 1951, *26*, 375, 421–40: *C.A.*, 1951, *45*, 9104-i, 9105-e.

3242 S. G. Thompson, B. B. Cunningham and G. T. Seaborg, *J. Am. Chem. Soc.*, 1950, *72*, 2798.

3243 S. G. Thompson, B. G. Harvey, G. R. Choppin and G. T. Seaborg, *UCRL*-2591 Rev.

3244 S. G. Thompson, R. A. James and L. O. Morgan, *The Tracer Chemistry of Am and Cm*, U.S. Atomic Energy Commission AECD No. 1907: cf. *C.A.*, 1950, *44*, 3376-b.

3245 R. H. K. Thomson, *Biochem. J.*, 1952, *51*, 118.

3246 M. H. Thornton, H. R. Kraybill and F. K. Broome, *J. Am. Chem. Soc.*, 1941, *63*, 2079.

3247 N. Tikhomiroff, *Compt. rend.*, 1953, *236*, 1263.

3248 D. H. Tilden, *J. Assoc. Offic. Agr. Chem.*, 1952, *35*, 423: *C.A.*, 1952, *46*, 11473-i.

3249 D. H. Tilden, *J. Assoc. Offic. Agr. Chem.*, 1953, *36*, 802: *C.A.*, 1954, *48*, 9694-i.

3250   J. Tischer and E. Illner, *Fette u. Seifen*, 1940, *47*, 578: *C.A.*, 1942, *36*, 674-8.
3251   J. Tischer and E. Tögel, *Z. physiol. Chem.*, 1947, *282*, 103.
3252   A. Tiselius, *Arkiv Kemi, Mineral. Geol.*, 1940, *14 B*, No. 22.
3253   A. Tiselius, *Arkiv Kemi, Mineral. Geol.*, 1941, *14 B*, No. 32 and *15 B*, No. 6: *C.A.*, 1941, *35*, 5407-1; 1942, *36*, 369-2.
3254   A. Tiselius, *Science*, 1941, *94*, 145.
3255   A. Tiselius, *Advances in Colloid Sci.*, 1941, *1*, 81: *C.A.*, 1942, *36*, 3413-8.
3256   A. Tiselius, *Kolloid-Z.*, 1943, *105*, 101.
3257   A. Tiselius, *Arkiv Kemi, Mineral. Geol.*, 1943, *16 A*, No. 18: *C.A.*, 1944, *38*, 2895-7.
3258   A. Tiselius, *The Svedberg Memorial Volume*, 1945, 370: *C.A.*, 1945, *39*, 1117-7.
3259   A. Tiselius, *Advances in Protein Chem.*, 1947, *3*, 67.
3260   A. Tiselius, *Arkiv Kemi, Mineral. Geol.*, 1948, *26 B*, No. 1: *C.A.*, 1949, *43*, 1624-e.
3261   A. Tiselius, *Naturwissenschaften*, 1950, *37*, 25.
3262   A. Tiselius and S. Claesson, *Arkiv Kemi, Mineral. Geol.*, 1942, *15 B*, No. 18: *C.A.*, 1944, *38*, 35-b.
3263   A. Tiselius, *Endeavour*, 1952, *11*, 5.
3264   A. Tiselius, *Arkiv Kemi*, 1954, *7*, 443.
3265   A. Tiselius, *Gazz. chim. ital.*, 1954, *84*, 1177: *C.A.*, 1955, *49*, 9052-f.
3266   A. Tiselius, *Angew. Chem.*, 1955, *67*, 245.
3267   A. Tiselius, *Ann. Acad. Sci. Fenn.*, 1955, Ser. A II, *60*, 257.
3268   A. Tiselius, B. Drake and L. Hagdahl, *Experientia*, 1947, *3*, 21.
3269   A. Tiselius and L. Hagdahl, *Acta Chem. Scand.*, 1950, *4*, 394.
3270   A. Tiselius and L. Hahn, *Kolloid-Z.*, 1943, *105*, 177.
3271   A. Tiselius and F. Sanger, *Nature*, 1947, *160*, 433.
3272   G. H. Tishkoff, R. Bennett, V. Bennett and L. L. Miller, *Science*, 1949, *110*, 452.
3273   E. Titus and J. Fried, *J. Biol. Chem.*, 1948, *174*, 57.
3274   R. Tixier, *Bull. soc. chim. biol.*, 1945, *27*, 621, 627.
3275   R. Tixier, *Ann. Inst. Océanographique de Monaco*, 1945, *22*, 343: *C.A.*, 1952, *46*, 10463-f.
3276   R. Tixier, *Mém. Mus. Nat. Hist. Nat.*, 1952, Série Zool. *5*, No. 2.
3277   G. Toennies and J. J. Kolb, *Anal. Chem.*, 1951, *23*, 823.
3278   J. D. Tolliday, G. W. H. Thompson and G. Foreman, *J. Soc. Leather Trades' Chemists*, 1950, *34*, 221: *C.A.*, 1950, *44*, 10362-i.
3279   R. M. Tomarelli and K. Florey, *Science*, 1948, *107*, 630.
3280   E. R. Tompkins, *J. Am. Chem. Soc.*, 1948, *70*, 3520.
3281   E. R. Tompkins, *J. Chem. Educ.*, 1949, *26*, 32, 92.
3282   E. R. Tompkins, *Anal. Chem.*, 1950, *22*, 1352.
3283   E. R. Tompkins, *U.S. Pat.* 2,554,648 (1951): *C.A.*, 1951, *45*, 7472-d.
3284   E. R. Tompkins, J. X. Khym and W. E. Cohn, *J. Am. Chem. Soc.*, 1947, *69*, 2769.
3285   E. R. Tompkins and S. W. Mayer, *J. Am. Chem. Soc.*, 1947, *69*, 2859.
3286   W. Tong, A. Taurog and I. L. Chaikoff, *J. Biol. Chem.*, 1954, *207*, 59.
3287   W. Tong, A. Taurog and I. L. Chaikoff, *J. Biol. Chem.*, 1951, *191*, 665.
3288   N. E. Topp and K. W. Pepper, *J. Chem. Soc.*, 1949, 3299.
3289   T. Tošić and T. Moore, *Biochem. J.*, 1945, *39*, 498.
3290   G. Tòth and J. Bársony, *Enzymologia*, 1943, *11*, 19: *C.A.*, 1945, *39*, 2772-1.
3291   G. H. N. Towers and D. C. Mortimer, *Nature*, 1954, *174*, 1189.

3292 G. H. N. Towers and F. C. Steward, *J. Am. Chem. Soc.*, 1954, *76*, 1959.
3293 G. H. N. Towers, J. F. Thompson and F. C. Steward, *J. Am. Chem. Soc.*, 1954, *76*, 2392.
3294 W. Trappe, *Biochem. Z.*, 1940, *305*, 150.
3295 W. Trappe, *Biochem. Z.*, 1940, *306*, 316.
3296 W. Trappe, *Biochem. Z.*, 1941, *307*, 97; *Z. Krebsforsch.*, 1942, *53*, 47: *C.A.*, 1944, *38*, 4676-4.
3297 E. M. Trautner and M. Roberts, *Analyst*, 1948, *73*, 140.
3298 E. Treiber and H. Koren, *Monatsh. Chem.*, 1953, *84*, 478: *C.A.*, 1954, *48*, 1078-c.
3299 N. R. Trenner, C. V. Warren and S. L. Jones, *Anal. Chem.*, 1953, *25*, 1685.
3300 W. E. Trevelyan, D. P. Procter and J. S. Harrison, *Nature*, 1950, *166*, 444.
3301 E. B. Trickey, *J. Am. Chem. Soc.*, 1950, *72*, 3474.
3302 D. R. Tristram and C. S. G. Phillips, *J. Chem. Soc.*, 1955, 580.
3303 G. R. Tristram, *Biochem. J.*, 1946, *40*, 721.
3304 F. Trombe and J. Loriers, *Compt. rend.*, 1953, *236*, 1567.
3305 R. H. Trubey and J. F. Christman, *Stain Technol.*, 1952, *27*, 87: *C.A.*, 1952, *46*, 5650-g, h.
3306 K. N. Trueblood and E. W. Malmberg, *Anal. Chem.*, 1949, *21*, 1055.
3307 K. N. Trueblood and E. W. Malmberg, *J. Am. Chem. Soc.*, 1950, *72*, 4112.
3308 R. Tschesche, G. Grimmer and F. Seehofer, *Chem. Ber.*, 1953, *86*, 1235.
3309 R. Tschesche and F. Korte, *Chem. Ber.*, 1951, *84*, 641.
3310 R. Tschesche and F. Korte, *Chem. Ber.*, 1951, *84*, 801.
3311 T. C. Tso and R. N. Jeffrey, *Arch. Biochem. Biophys.*, 1953, *43*, 269.
3312 K. Tsuda and T. Matsumoto, *J. Pharm. Soc. Japan*, 1947, *67*, 238: *C.A.*, 1951, *45*, 9468-e.
3313 K. Tsukida and M. Yoneshige, *J. Pharm. Soc. Japan*, 1954, *74*, 379: *C.A.*, 1955, *49*, 5410-c.
3314 M. Tswett, *Trav. soc. nat. Varsovie*, 1903, *14*; *Ber. dtsch. botan. Ges.*, 1906, *24*, 316, 384; *Biochem. Z.*, 1907, *5*, 6; *Chromofilli w rastitelnom i schivotnom mirje*, Warsaw, 1910.
3315 R. Tupper and R. W. E. Watts, *Nature*, 1954, *173*, 349.
3316 H. Tuppy, *Monatsh. Chem.*, 1953, *84*, 342: *C.A.*, 1954, *48*, 4476-h.
3317 H. Tuppy and G. Bodo, *Monatsh. Chem.*, 1954, *85*, 1024.
3318 H. Tuppy and S. Paléus, *Acta Chem. Scand.*, 1955, *9*, 353.
3319 F. Turba, *Ber.*, 1941, *74*, 1829.
3320 F. Turba, *Chromatographische Methoden in der Protein-Chemie, einschliesslich verwandter Methoden wie Gegenstromverteilung, Papier-ionophorese*, 358 pp., Springer-Verlag, Berlin, 1954.
3321 F. Turba and H. J. Enenkel, *Naturwissenschaften*, 1950, *37*, 93.
3322 F. Turba and G. Gundlach, *Biochem. Z.*, 1955, *326*, 322.
3323 F. Turba, H. Pelzer and G. Schuster, *Z. physiol. Chem.*, 1954, *296*, 97.
3324 F. Turba and M. Richter, *Ber.*, 1942, *75*, 340.
3325 F. Turba, M. Richter and F. Kuchar, *Naturwissenschaften*, 1943, *31*, 508.
3326 F. Turba and E. von Schrader-Beielstein, *Naturwissenschaften*, 1947, *34*, 57.
3327 F. Turba and E. von Schrader-Beielstein, *Naturwissenschaften*, 1948, *35*, 123.
3328 F. Turba and M. Turba, *Naturwissenschaften*, 1951, *38*, 188.
3329 N. M. Turkel'taub, *Zavodskaya Lab.*, 1949, *15*, 653: *C.A.*, 1950, *44*, 9747-e.
3330 N. M. Turkel'taub, *Zhur. Anal. Khim.*, 1950, *5*, 200: *C.A.*, 1950, *44*, 9856-b.

3331   N. M. Turkel'taub, V. P. Shvartsman, T. V. Georgievskaya, O. V. Zolotareva and A. I. Karymova, *Zhur. Fiz. Khim.*, 1953, *27*, 1827: *C.A.*, 1954, *48*, 10476-f.
3332   J. Turkevich, *Ind. Eng. Chem. Anal. Ed.*, 1942, *14*, 792.
3333   N. C. Turner, *Petroleum Refiner*, 1943, *22*, 140: *C.A.*, 1943, *37*, 3914-4.
3334   R. A. Turner, J. G. Pierce and V. Du Vigneaud, *J. Biol. Chem.*, 1951, *191*, 21.
3335   D. Turnock, *Nature*, 1953, *172*, 355.
3336   C. N. Turton and E. Pacsu, *J. Am. Chem. Soc.*, 1955, *77*, 1059.
3337   V. E. Tyler Jr. and A. E. Schwarting, *J. Am. Pharm. Assoc.*, 1952, *41*, 354.
3338   S. Udenfriend, *J. Biol. Chem.*, 1950, *187*, 65.
3339   S. Udenfriend, C. T. Clark and E. Titus, *Experientia*, 1952, *8*, 379.
3340   S. Udenfriend and M. Gibbs, *Science*, 1949, *110*, 708.
3341   S. Udenfriend and S. F. Velick, *J. Biol. Chem.*, 1951, *190*, 733.
3342   M. Ulmann, *Biochem. Z.*, 1950/51, *321*, 377.
3343   M. Ulmann, *Pharmazie*, 1954, *9*, 523: *C.A.*, 1955, *49*, 9723-e.
3344   A. J. Ultée Jr. and J. Hartel, *Anal. Chem.*, 1955, *27*, 557.
3345   H. E. Umbarger and E. A. Adelberg, *J. Biol. Chem.*, 1951, *192*, 883.
3346   H. E. Umbarger and B. Magasanik, *J. Am. Chem. Soc.*, 1952, *74*, 4253.
3347   E. J. Umberger and J. M. Curtis, *J. Biol. Chem.*, 1949, *178*, 265.
3348   F. Umland and W. Fischer, *Naturwissenschaften*, 1953, *40*, 439.
3349   J. C. Underwood and L. B. Rockland, *Anal. Chem.*, 1954, *26*, 1553.
3350   H. Uno and A. Koyama, *J. Ferment. Technol. (Japan)*, 1951, *29*, 219: *C.A.* 1953, *47*, 1008-c.
3351   I. H. Updegraff and H. G. Cassidy, *J. Am. Chem. Soc.*, 1949, *71*, 407.
3352   K. F. Urbach, *Proc. Soc. Exptl. Biol. Med.*, 1949, *70*, 146: *C.A.*, 1949, *43*, 3478-d.
3353   K. F. Urbach, *Science*, 1949, *109*, 259.
3354   Yu. I. Usatenko and O. V. Datsenko, *Zavodskaya Lab.*, 1949, *15*, 145: *C.A.*, 1949, *43*, 5336-g.
3355   Yu. I. Usatenko and O. V. Datsenko, *Zavodskaya Lab.*, 1949, *15*, 779: *C.A.*, 1950, *44*, 478-a.
3356   S. V. Vaeck, *Nature*, 1953, *172*, 213.
3357   S. V. Vaeck, *Anal. Chim. Acta*, 1954, *10*, 48.
3358   M. Vahrman, *Nature*, 1950, *165*, 404.
3359   J. Valentin and G. Kirchübel, *Arch. Pharm.*, 1951, *284*, 114: *C.A.*, 1951, *45*, 10488-e.
3360   L. Vámos, *Magyar Kém. Folyóirat*, 1953, *59*, 253: *C.A.*, 1954, *48*, 506-d.
3361   J. C. Vanatta and C. C. Cox, *J. Biol. Chem.*, 1954, *210*, 719.
3362   J. C. Vanatta and C. C. Cox, *J. Biol. Chem.*, 1955, *212*, 599.
3363   J. H. Van de Kamer, K. W. Gerritsma and E. J. Wansink, *Biochem. J.*, 1955, *61*, 174.
3364   F. A. Vandenheuvel and E. R. Hayes, *Anal. Chem.*, 1952, *24*, 960.
3365   M. J. Vander Brook, A. N. Wick, W. H. DeVries, R. Harris and G. F. Cartland, *J. Biol. Chem.*, 1946, *165*, 463.
3366   P. C. Van der Schaaf and T. H. J. Huisman, *Biochim. Biophys. Acta*, 1955, *17*, 81.
3367   H. Van Duin, *Biochim. Biophys. Acta*, 1953, *10*, 198.
3368   H. Van Duin, *Biochim. Biophys. Acta*, 1953, *10*, 343.
3369   H. Van Duin, *Biochim. Biophys. Acta*, 1953, *12*, 489.
3370   H. Van Duin, *Biochim. Biophys. Acta*, 1953, *12*, 490.
3372   H. Van Duin, *Rec. trav. chim.*, 1954, *73*, 68.

3373 P. C. Van Erkelens, *Nature*, 1953, *172*, 357.
3374 J. Van Espen, *J. pharm. Belg.*, 1950, (2), *5*, 130: *C.A.*, 1950, *44*, 8054-b.
3375 J. Van Espen, *Mededel. Vlaam. Chem. Ver.*, 1953, *15*, 66: *C.A.*, 1953, *47*, 11661-d.
3376 M. B. Van Halteren, *Nature*, 1951, *168*, 1090.
3377 A. P. Vanselow, *Soil Science*, 1932, *33*, 95.
3378 A. P. Vanselow, *J. Am. Chem. Soc.*, 1932, *54*, 1307.
3379 L. G. Vanyarkho and V. A. Garanina, *Aptechnoe Delo*, 1952, No. 3, 22: *C.A.*, 1953, *47*, 439-h.
3380 K. C. Varma, J. B. Burt and A. E. Schwarting, *J. Am. Pharm. Assoc.*, 1952, *41*, 318.
3381 G. Vavon and B. Gastambide, *Compt. rend.*, 1948, *226*, 1201.
3382 G. Vavon and B. Gastambide, *Compt. rend.*, 1949, *228*, 1236.
3383 G. Vavon and B. Gastambide, *Compt. rend.*, 1950, *231*, 1151.
3384 C. Vavon and G. Medynski, *Compt. rend.*, 1949, *229*, 655.
3385 M. Večeřa and J. Gasparič, *Collection Czechoslov. Chem. Communs.*, 1954, *19*, 1175.
3386 F. P. Veitch Jr. and H. S. Milone, *J. Biol. Chem.*, 1945, *157*, 417.
3387 S. F. Velick and S. Udenfriend, *J. Biol. Chem.*, 1951, *190*, 721.
3388 S. F. Velick and S. Udenfriend, *J. Biol. Chem.*, 1951, *191*, 233.
3389 S. F. Velick and L. F. Wicks, *J. Biol. Chem.*, 1951, *190*, 741.
3390 H. Venner, *Naturwissenschaften*, 1955, *42*, 179.
3391 G. Venturello, *Chimica e Industria (Milano)*, 1944, *26*, 72: *C.A.*, 1946, *40*, 3035-g.
3392 G. Venturello and N. Agliardi, *Ann. chim. appl.*, 1940, *30*, 220, 224: *C.A.*, 1942, *36*, 5439-b.
3393 G. Venturello and A. Burdese, *Ann. chim. appl.*, 1951, *41*, 155.
3394 G. Venturello and A. M. Ghe, *Anal. Chim. Acta*, 1952, *7*, 261 and 268.
3395 G. Venturello and A. M. Ghe, *Ann. chim. (Roma)*, 1953, *43*, 267: *C.A.*, 1954, *48*, 1194-h.
3396 G. Venturello and A. M. Ghe, *Anal. Chim. Acta*, 1954, *10*, 335.
3397 G. Venturello and A. M. Ghe, *Ann. chim. (Roma)*, 1954, *44*, 960.
3397a G. Venturello and A. M. Ghe, *Ann. chim. (Roma)*, 1955, *45*, 1054.
3397b G. Venturello and C. Gualandi, *Ann. chim. (Roma)*, 1956, *46*, 229.
3398 G. Venturello and G. Saini, *Ann. chim. appl.*, 1949, *39*, 375.
3399 G. Verhaar, *Cinchona (Arch. Kinacultuur)*, 1950, *11*, No. 1: *C.A.*, 1951, *45*, 9800-b.
3400 M. Verzele, *Bull. soc. chim. Belges*, 1955, *64*, 70.
3401 M. Verzele, private communication.
3402 R. C. Vickery, *J. Chem. Soc.*, *1952*, 4357.
3403 R. C. Vickery, *Nature*, 1952, *170*, 665.
3404 P. Vignes, M. Robey and H. Simonnet, *Bull. soc. chim. biol.*, 1954, *36*, 1163.
3405 E. Vilkas and E. Lederer, *Bull. soc. chim. biol.*, 1956, *38*, 111.
3406 L. C. Vining and S. A. Waksman, *Science*, 1954, *120*, 388.
3407 A. I. Virtanen and M. Alfthan, *Acta Chem. Scand.*, 1954, *8*, 1720.
3408 A. I. Virtanen and M. Alfthan, *Acta Chem. Scand.*, 1955, *9*, 188.
3409 A. I. Virtanen and A.-M. Berg, *Acta Chem. Scand.*, 1955, *9*, 553.
3410 A. I. Virtanen and P. K. Hietala, *Acta Chem. Scand.*, 1955, *9*, 175.
3411 A. I. Virtanen and S. Kari, *Acta Chem. Scand.*, 1954, *8*, 1290.
3412 A. I. Virtanen and S. Kari, *Acta Chem. Scand.*, 1955, *9*, 170.
3413 A. I. Virtanen and P. Linko, *Acta Chem. Scand.*, 1955, *9*, 531.
3414 A. I. Virtanen and P. Linko, *Acta Chem. Scand.*, 1955, *9*, 551.

3415 A. I. Virtanen and J. K. Miettinen, *Biochim. Biophys. Acta*, 1953, *12*, 181.
3416 A. I. Virtanen, J. K. Miettinen and H. Kunttu, *Acta Chem. Scand.*, 1954, *7*, 38.
3417 E. Vischer and E. Chargaff, *J. Biol. Chem.*, 1947, *168*, 781.
3418 E. Vischer and E. Chargaff, *J. Biol. Chem.*, 1948, *176*, 703.
3419 E. Vischer and E. Chargaff, *J. Biol. Chem.*, 1948, *176*, 715.
3420 E. Vischer and T. Reichstein, *Helv. Chim. Acta*, 1944, *27*, 1332.
3421 E. Vischer, S. Zamenhof and E. Chargaff, *J. Biol. Chem.*, 1949, *177*, 429.
3422 M. Viscontini, G. Bonetti, C. Ebnöther and P. Karrer, *Helv. Chim. Acta*, 1951, *34*, 1384.
3423 M. Viscontini, D. Hoch und P. Karrer, *Helv. Chim. Acta*, 1955, *38*, 642.
3424 G. Vitte and E. Boussemart, *Bull. trav. soc. pharm. Bordeaux*, 1951, *88*, 181: *C.A.*, 1951, *45*, 7299-f.
3425 J. C. Vitucci, N. Bohonos, O. P. Wieland, D. V. Lefemine and B. L. Hutchings, *Arch. Biochem. Biophys.*, 1951, *34*, 409.
3426 W. Vogt, *Chem. Ing. Tech.*, 1951, *23*, 580.
3427 E. Volkin and C. E. Carter, *J. Am. Chem. Soc.*, 1951, *73*, 1516.
3428 E. Volkin and W. E. Cohn, *J. Biol. Chem.*, 1953, *205*, 767.
3429 E. Volkin, J. X. Khym and W. E. Cohn, *J. Am. Chem. Soc.*, 1951, *73*, 1533.
3430 Y. Volmar, J.-P. Ebel and Y. Fawzi-Bassili, *Compt. rend.*, 1952, *235*, 372.
3431 Y. Volmar, J. P. Ebel and Y. Fawzi-Bassili, *Bull. soc. chim. France, 1953*, 1085.
3432 J. L. Wachtel and H. G. Cassidy, *Science*, 1942, *95*, 233.
3433 J. L. Wachtel and H. G. Cassidy, *J. Am. Chem. Soc.*, 1943, *65*, 665.
3434 C. A. Wachtmeister, *Acta Chem. Scand.*, 1951, *5*, 976.
3435 C. A. Wachtmeister, *Acta Chem. Scand.*, 1952, *6*, 818.
3436 H. E. Wade and D. M. Morgan, *Nature*, 1953, *171*, 529.
3437 H. E. Wade and D. M. Morgan, *Biochem. J.*, 1955, *60*, 264.
3438 W. H. Wadman, G. J. Thomas and A. B. Pardee, *Anal. Chem.*, 1954, *26*, 1192.
3439 H. G. Wager, *Analyst*, 1954, *79*, 34.
3440 G. Wagner, *Arch. Pharm.*, 1952, *285*, 409: *C.A.*, 1954, *48*, 4416-i.
3441 G. Wagner, *Arch. Pharm.*, 1953, *286*, 232: *C.A.*, 1954, *48*, 9626-b.
3442 G. Wagner, *Arch. Pharm.*, 1953, *286*, 269: *C.A.*, 1954, *48*, 7495-d.
3443 G. Wagner, *Pharmazie*, 1954, *9*, 123, 631: *C.A.*, 1955, *49*, 9225-c.
3444 A. Wahhab, *J. Am. Chem. Soc.*, 1948, *70*, 3580.
3445 P. N. Wahi and R. G. S. Nigam, *Indian J. Med. Res.*, 1954, *42*, 261: *C.A.*, 1954, *48*, 13550-c.
3446 A. Waksman and E. J. Bigwood, *Ann. Acad. Sci. Fenn.*, 1955, Ser. A II, *60*, 49.
3447 G. Wald, *J. Gen. Physiol.*, 1948, *31*, 489: *C.A.*, 1949, *43*, 157-i.
3448 G. Wald and G. Allen, *J. Gen. Physiol.*, 1946, *30*, 41: *C.A.*, 1946, *40*, 7419-1.
3449 D. M. Waldron-Edward, *Chem. and Ind., 1954*, 104.
3450 E. Waldschmidt-Leitz, J. Ratzer and F. Turba, *J. prakt. Chem.*, 1941, *158*, 72.
3451 E. Waldschmidt-Leitz and F. Turba, *J. prakt. Chem.*, 1940, *156*, 55.
3452 S. G. Waley and J. Watson, *Biochem. J.*, 1954, *57*, 529.
3453 B. Waligóra and Z. Bylo, *Bull. intern. acad. polon. sci., Classe III*, 1953, *1*, 143: *C.A.*, 1954, *48*, 4930-c.
3454 D. G. Walker and F. L. Warren, *Biochem. J.*, 1951, *49*, xxi.
3455 D. M. Walker, *Biochem. J.*, 1952, *52*, 679.
3456 F. T. Walker, *J. Oil Colour Chem. Assoc.*, 1945, *28*, 119: *C.A.*, 1945, *39*, 5090-4.
3457 F. T. Walker and M. R. Mills, *J. Soc. Chem. Ind.*, 1942, *61*, 125.

3458 T. K. Walker, A. N. Hall and J. W. Hopton, *Nature*, 1951, *168*, 1042.
3459 W. R. Walker, *Australian J. Sci.*, 1950, *13*, 26: *C.A.*, 1951, *45*, 1912-g.
3460 W. R. Walker, *Australian J. Sci.*, 1950, *13*, 84: *C.A.*, 1951, *45*, 3146-d.
3461 W. R. Walker and M. Lederer, *Anal. Chim. Acta*, 1951, *5*, 191.
3462 J. S. Wall, *Anal. Chem.*, 1953, *25*, 950.
3464 K. Wallenfels, *Naturwissenschaften*, 1950, *37*, 491.
3465 K. Wallenfels, *Naturwissenschaften*, 1951, *38*, 238.
3466 K. Wallenfels, *Naturwissenschaften*, 1951, *38*, 306.
3467 K. Wallenfels, E. Bernt and G. Limberg, *Angew. Chem.*, 1953, *65*, 581.
3468 K. Wallenfels, E. Bernt and G. Limberg, *Ann.*, 1953, *579*, 113.
3469 J. Walravens, *Arch. Intern. Physiol.*, 1952, *60*, 191.
3470 H. F. Walton, *J. Franklin Inst.*, 1941, *232*, 305.
3471 H. F. Walton, *Ion Exchange Equilibria*, in F. C. Nachod, *Ion Exchange - Theory and Application*, Academic Press, New York, 1949, p. 3–28.
3472 H. F. Walton, *Scientific American*, 1950, *183*, 48.
3473 T. P. Wang and J. O. Lampen, *J. Biol. Chem.*, 1952, *194*, 785.
3474 A. Wankmüller, *Naturwissenschaften*, 1952, *39*, 133.
3475 A. Wankmüller, *Naturwissenschaften*, 1952, *39*, 302.
3476 H. Wanner, *Helv. Chim. Acta*, 1952, *35*, 460.
3477 I. W. Wark, *Proc. Roy. Soc. N.S.W.*, 1929, *63*, 47: *C.A.*, 1930, *24*, 1568.
3478 C. J. Watson and M. Berg, *J. Biol. Chem.*, 1955, *214*, 537.
3479 C. J. Watson, M. Berg and V. Hawkinson, *J. Biol. Chem.*, 1955, *214*, 547.
3480 C. J. Watson, S. Schwartz and V. Hawkinson, *J. Biol. Chem.*, 1945, *157*, 345.
3480a E. G. Weatherly, *Analyst*, 1956, *81*, 404.
3481 E. Wegner, *Naturwissenschaften*, 1953, *40*, 580.
3482 R. Weichert, *Acta Chem. Scand.*, 1955, *9*, 547.
3483 H. Weil, *Paint Technol.*, 1949, *14*, 391, 439: *C.A.*, 1950, *44*, 1267-h.
3484 H. Weil, *Chimie et Industrie*, 1950, *64*, 432.
3485 H. Weil and T. I. Williams, *Nature*, 1950, *166*, 1000.
3486 H. Weil and T. I. Williams, *Nature*, 1951, *167*, 906.
3487 T. Weil, *Helv. Chim. Acta*, 1955, *38*, 1274.
3488 H. Weil-Malherbe, *J. Chem. Soc.*, *1943*, 303.
3489 A. Weiss and S. Fallab, *Helv. Chim. Acta*, 1954, *37*, 1253.
3490 D. E. Weiss, *Nature*, 1948, *162*, 372.
3491 D. E. Weiss, *Discuss. Faraday Soc.*, 1949, *7*, 142.
3492 D. E. Weiss, *Nature*, 1950, *166*, 66.
3493 D. E. Weiss, *Roy. Australian Chem. Inst. & Proc.*, 1950, *17*, 141.
3494 D. E. Weiss, *Australian J. Appl. Sci.*, 1953, *4*, 510.
3495 J. Weiss, *J. Chem. Soc.*, *1943*, 297.
3496 B. Weissmann, P. A. Bromberg and A. B. Gutman, *Proc. Soc. Exptl. Biol. Med.*, 1954, *87*, 257: *C.A.*, 1955, *49*, 2552-h.
3497 B. Weissmann, K. Meyer, P. Sampson and A. Linker, *J. Biol. Chem.*, 1954, *208*, 417.
3498 E. Weitz and F. Schmidt, *Ber.*, 1939, *72*, 1740, 2099.
3499 E. Weitz, F. Schmidt and J. Singer, *Z. Elektrochem.*, 1940, *46*, 222.
3500 L. E. Weller, S. H. Wittwer and H. M. Sell, *J. Am. Chem. Soc.*, 1954, *76*, 629.
3501 E. F. Wellington, *Can. J. Chem.*, 1952, *50*, 581.
3502 I. C. Wells, *J. Biol. Chem.*, 1952, *196*, 331.
3503 R. A. Wells, *Quart. Rev.*, 1953, *7*, 307.

3504 K. Wendel, *Planta*, 1949/50, *37*, 604: *C.A.*, 1950, *44*, 4306-c.
3505 S. H. Wender and T. B. Gage, *Science*, 1949, *109*, 287.
3506 W. Werle and J. Koch, *Naturwissenschaften*, 1951, *38*, 333.
3507 E. Werle and D. Palm, *Biochem. Z.*, 1952/53, *323*, 255.
3508 L. B. Werner and I. Perlman, *The Preparation and Isolation of Cm*, U.S. Atomic Energy Comm. AECD, No. 1898: *C.A.*, 1950, *44*, 7671-g.
3509 G. B. West and J. F. Riley, *Nature*, 1954, *174*, 882.
3510 R. G. Westall, *Biochem. J.*, 1948, *42*, 249.
3511 R. G. Westall, *J. Sci. Food Agr.*, 1950, *1*, 191: *C.A.*, 1951, *45*, 223-i.
3512 R. G. Westall, *Nature*, 1950, *165*, 717.
3513 R. G. Westall, *Biochem. J.*, 1952, *52*, 638.
3514 R. G. Westall, *Biochem. J.*, 1953, *55*, 244.
3515 R. G. Westall, *Biochem. J.*, 1955, *60*, 247.
3516 A. E. R. Westman and A. E. Scott, *Nature*, 1951, *168*, 740.
3517 A. E. R. Westman, A. E. Scott and J. T. Pedley, *Chemistry in Can.*, 1952, *4*, 189.
3518 U. Westphal, P. Gedigk and F. Meyer, *Z. physiol. Chem.*, 1950, *285*, 36.
3519 A. Wetterholm, *Harald Nordenson Anniv. Vol.*, 1946, 460: *C.A.*, 1949, *43*, 5325-d.
3520 F. Weygand, *Arkiv Kemi*, 1950, *3*, 11.
3521 F. Weygand and L. Birkhofer, *Z. physiol. Chem.*, 1939, *261*, 172.
3522 F. Weygand and E. Csendes, *Chem. Ber.*, 1952, *85*, 45.
3523 F. Weygand and H. Hofmann, *Chem. Ber.*, 1950, *83*, 405.
3524 R. M. Wheaton and W. C. Bauman, *Ann. N.Y. Acad. Sci.*, 1953, *57*, 159.
3525 R. M. Wheaton and W. C. Bauman, *Ind. Eng. Chem.*, 1953, *45*, 228.
3526 W. J. Whelan and K. Morgan, *Chem. and Ind.*, *1954*, 78.
3527 R. L. Whistler, *Science*, 1954, *120*, 899.
3528 R. L. Whistler, H. E. Conrad and L. Hough, *J. Am. Chem. Soc.*, 1954, *76*, 1668.
3529 R. L. Whistler and J. H. Duffy, *J. Am. Chem. Soc.*, 1955, *77*, 1017.
3530 R. L. Whistler and D. F. Durso, *J. Am. Chem. Soc.*, 1950, *72*, 677.
3531 R. L. Whistler and J. L. Hickson, *J. Am. Chem. Soc.*, 1954, *76*, 1671.
3532 R. L. Whistler and C.-C. Tu, *J. Am. Chem. Soc.*, 1953, *75*, 645.
3533 L. G. Whitby, *Nature*, 1950, *166*, 479.
3534 D. White and D. W. Grant, *Nature*, 1955, *175*, 513.
3535 J. W. White Jr., *Anal. Chem.*, 1948, *20*, 726.
3536 J. W. White Jr. and E. C. Dryden, *Anal. Chem.*, 1948, *20*, 853.
3537 M. F. White and J. B. Brown, *J. Am. Chem. Soc.*, 1948, *70*, 4269.
3538 W. F. White and W. L. Fierce, *J. Am. Chem. Soc.*, 1953, *75*, 245.
3539 P. R. Whitfeld, L. A. Heppel and R. Markham, *Biochem. J.*, 1955, *60*, 15.
3540 V. P. Whittaker and S. Wijesundera, *Biochem. J.*, 1952, *51*, 348; *52*, 475.
3541 J. P. Wibaut, H. C. Beyerman and P. H. Enthoven, *Rec. trav. chim.*, 1954, *73*, 102.
3542 A. Wickström and B. Salvesen, *J. Pharm. and Pharmacol.*, 1952, *4*, 98: *C.A.*, 1952, *46*, 4742-d.
3543 A. Wickström and B. Salvesen, *J. Pharm. and Pharmacol.*, 1952, *4*, 631: *C.A.*, 1952, *46*, 10539-d.
3544 G. Wiegner, *J. Soc. Chem. Ind.*, 1931, *50*, 65T.
3545 G. Wiegner and K. W. Müller, *Z. Pflanz. Düngung u. Bodenk.*, 1929, *14 A*, 321: *C.A.*, 1929, *23*, 5400.
3546 H. Wieland, K. Bähr and B. Witkop, *Ann.*, 1941, *547*, 156.

3547 H. Wieland, and G. Coutelle, *Ann.*, 1941, *548*, 270.
3548 O. P. Wieland, B. L. Hutchings and J. H. Williams, *Arch. Biochem. Biophys.*, 1952, *40*, 205.
3549 T. Wieland, *Z. physiol. Chem.*, 1942, *273*, 24.
3550 T. Wieland, *Ber.*, 1942, *75*, 1001.
3551 T. Wieland, *Naturwissenschaften*, 1942, *30*, 374.
3552 T. Wieland, *Angew. Chem.*, 1943, *56*, 213.
3553 T. Wieland, *Ber.*, 1944, *77*, 539.
3554 T. Wieland, *Angew. Chem.*, 1948, *60*, 313.
3555 T. Wieland, *Fortschr. Chem. Forsch.*, 1949/50, *1*, 211: *C.A.*, 1950, *44*, 5949-c.
3556 T. Wieland, *Angew. Chem.*, 1951, *63*, 171.
3557 T. Wieland and L. Bauer, *Angew. Chem.*, 1951, *63*, 511.
3558 T. Wieland and A. Berg, *Angew. Chem.*, 1952, *64*, 418.
3559 T. Wieland and U. Feld, *Angew. Chem.*, 1951, *63*, 258.
3560 T. Wieland and E. Fischer, *Naturwissenschaften*, 1948, *35*, 29.
3561 T. Wieland and E. Fischer, *Naturwissenschaften*, 1949, *36*, 219.
3562 T. Wieland and H. Fremerey, *Ber.*, 1944, *77*, 234.
3563 T. Wieland and W. Paul, *Ber.*, 1944, *77*, 34.
3564 T. Wieland, K. Schmeiser, E. Fischer and G. Maier-Leibnitz, *Naturwissenschaften*, 1949, *36*, 280.
3565 T. Wieland and G. Schmidt, *Ann.*, 1952, *577*, 215.
3565a T. Wieland and W. Schön, *Ann.*, 1955, *593*, 157.
3566 T. Wieland and L. Wirth, *Ber.*, 1943, *76*, 823.
3567 T. Wieland and L. Wirth, *Angew. Chem.*, 1951, *63*, 171.
3568 L. F. Wiggins and J. H. Williams, *Nature*, 1952, *170*, 279.
3569 J. B. Wilkes, *Ind. Eng. Chem. Anal. Ed.*, 1946, *18*, 329.
3570 J. B. Wilkes, *Ind. Eng. Chem. Anal. Ed.*, 1946, *18*, 702.
3571 J. B. Wilkie and S. W. Jones, *Anal. Chem.*, 1952, *24*, 1409.
3572 J. B. Wilkie and S. W. Jones, *J. Assoc. Offic. Agr. Chem.*, 1954, *37*, 880: *C.A.*, 1954, *48*, 12212-d.
3573 A. F. Williams, *Analyst*, 1952, *77*, 297.
3574 A. F. Williams, *J. Chem. Soc.*, *1952*, 3155.
3575 A. H. Williams, *Chem. and Ind.*, *1955*, 120.
3576 B. L. Williams and S. H. Wender, *J. Am. Chem. Soc.*, 1952, *74*, 5919.
3577 K. A. Williams, *Analyst*, 1946, *71*, 259.
3578 K. T. Williams and A. Bevenue, *Science*, 1951, *113*, 582.
3579 K. T. Williams and A. Bevenue, *Anal. Chem.*, 1955, *27*, 331.
3580 R. Williams Jr. and J. V. Hightower, *Chem. Eng.*, 1948, *55*, 133.
3581 R. J. Williams and H. Kirby, *Science*, 1948, *107*, 481.
3582 R. J. P. Williams, *Analyst*, 1952, *77*, 905.
3583 R. J. P. Williams, L. Hagdahl and A. Tiselius, *Arkiv Kemi*, 1954, *7*, 1.
3584 R. R. Williams and R. E. Smith, *Proc. Soc. Exptl. Biol. Med.*, 1951, *77*, 169: *C.A.*, 1951, *45*, 7622-i.
3585 T. I. Williams, *An Introduction to Chromatography*, 100 pp., Blackie and Son, London, 1946.
3586 T. I. Williams, *Manuf. Chemist*, 1949, *20*, 16.
3587 T. I. Williams, *The Elements of Chromatography*, Blackie & Son, London, 1954.
3588 M. B. Williamson and J. M. Passmann, *J. Biol. Chem.*, 1952, *199*, 121.
3589 H. Willstaedt, *Enzymologia*, 1940/41, *9*, 260: *C.A.*, 1942, *36*, 842-8.
3590 H. Willstaedt, *Svensk Kem. Tid.*, 1946, *58*, 23, 81: *C.A.*, 1946, *40*, 4720-1.

3591   J. N. Wilson, *J. Am. Chem. Soc.*, 1940, *62*, 1583.
3592   E. Windsor, *J. Biol. Chem.*, 1951, *192*, 595.
3593   H. M. Winegard, G. Toennies and R. J. Block, *Science*, 1948, *108*, 506.
3594   W. J. Wingo, *Anal. Chem.*, 1953, *25*, 1939.
3595   W. J. Wingo, *Anal. Chem.*, 1954, *26*, 1527.
3596   W. J. Wingo and I. Browning, *Anal. Chem.*, 1953, *25*, 1426.
3597   W. A. Winsten, *Science*, 1948, *107*, 605.
3598   W. A. Winsten and E. Eigen, *J. Am. Chem. Soc.*, 1948, *70*, 3333.
3599   W. A. Winsten and E. Eigen, *Proc. Soc. Exptl. Biol. Med.*, 1948, *67*, 513: *C.A.*, 1948, *42*, 6874-i.
3600   W. A. Winsten and E. Eigen, *J. Biol. Chem.*, 1949, *181*, 109.
3601   W. A. Winsten and A. H. Spark, *Science*, 1947, *106*, 192.
3602   F. P. W. Winteringham, A. Harrison and R. G. Bridges, *Nature*, 1950, *166*, 999.
3603   F. P. W. Winteringham, A. Harrison and R. G. Bridges, *Analyst*, 1952, *77*, 19.
3604   H. Wislicenus, *Kolloid-Z.*, 1942, *100*, 66.
3605   T. K. With, *Z. physiol. Chem.*, 1942, *275*, 166.
3606   J. Wittenberg and D. Shemin, *J. Biol. Chem.*, 1950, *185*, 103.
3607   A. J. Woiwod, *Biochem. J.*, 1948, *42*, xxviii.
3608   A. J. Woiwod, *Nature*, 1948, *161*, 169.
3609   A. J. Woiwod, *Biochem. J.*, 1949, *45*, 412.
3610   A. J. Woiwod, *J. Gen. Microbiol.*, 1949, *3*, 312.
3611   A. J. Woiwod, *Nature*, 1950, *166*, 272.
3612   A. J. Woiwod and F. V. Linggood, *Nature*, 1948, *162*, 219.
3613   A. J. Woiwod and F. V. Linggood, *Nature*, 1949, *163*, 218.
3614   A. J. Woiwod and H. Proom, *J. Gen. Microbiol.*, 1950, *4*, 501: *C.A.*, 1951, *45*, 9111-i.
3615   M. L. Wolfrom, R. S. Bower and G. G. Maher, *J. Am. Chem. Soc.*, 1951, *73*, 875.
3616   M. L. Wolfrom and J. C. Dacons, *J. Am. Chem. Soc.*, 1952, *74*, 5331.
3617   M. L. Wolfrom and B. W. Lew, *U.S. Pat.* 2,524,414 (1950): *C.A.*, 1951, *45*, 2975-f.
3618   M. L. Wolfrom, A. Thompson, T. T. Galkowski and E. J. Quinn, *Anal. Chem.*, 1952, *24*, 1670.
3619   M. L. Wolfrom, A. Thompson, A. N. O'Neill and T. T. Galkowski, *J. Am. Chem. Soc.*, 1952, *74*, 1062.
3620   W. Q. Wolfson, C. Cohn and W. A. Devaney, *Science*, 1949, *109*, 541.
3621   E. G. Wollish, M. Schmall and E. G. E. Shafer, *Anal. Chem.*, 1951, *23*, 768.
3622   T. Wood, *Nature*, 1955, *176*, 175.
3623   H. B. Woodruff and J. C. Foster, *J. Biol. Chem.*, 1950, *183*, 569.
3624   C. C. Woodward and G. S. Rabideau, *Anal. Chem.*, 1954, *26*, 248.
3625   L. I. Woolf, *Nature*, 1953, *171*, 841.
3626   E. O. Woolfolk, F.-E. Beach and S. P. McPherson, *J. Org. Chem.*, 1955, *20*, 391.
3627   E. Work, *Lancet*, 1949, *256*, 652: *C.A.*, 1949, *43*, 5441-b.
3628   E. Work, *Biochim. Biophys. Acta*, 1950, *5*, 204.
3629   E. Work, *Biochem. J.*, 1951, *49*, 17.
3630   E. Work, S. M. Birnbaum, M. Winitz and J. P. Greenstein, *J. Am. Chem. Soc.*, 1955, *77*, 1916.
3631   L. D. Wright, E. L. Cresson and C. A. Driscoll, *Proc. Soc. Exptl. Biol. Med.*, 1954, *86*, 480: *C.A.*, 1954, *48*, 12874-b.

3632 L. D. Wright, E. L. Cresson, H. R. Skeggs, T. R. Wood, R. L. Peck, D. E. Wolf and K. Folkers, *J. Am. Chem. Soc.*, 1952, *74*, 1996.

3633 Ch. Wunderly, *Die Papierelektrophorese*, Sauerländer & Co., Aarau, 1954.

3634 Ch. Wunderly, *Nature*, 1954, *173*, 267.

3635 G. R. Wyatt, *Biochem. J.*, 1951, *48*, 581.

3636 G. R. Wyatt, *Biochem. J.*, 1951, *48*, 584.

3637 G. R. Wyatt, in *The Nucleic Acids*, Vol. I, 243–256, Acad. Press, New York, 1955.

3638 V. Wynn, *Nature*, 1949, *164*, 445.

3639 V. Wynn and G. Rogers, *Australian J. Sci. Research*, 1950, *B3*, 124: *C.A.*, 1950, *44*, 8975-h.

3640 E. C. Yackel and W. O. Kenyon, *J. Am. Chem. Soc.*, 1942, *64*, 121.

3641 K. Yagi, J. Okuda, Y. Matsuoka, *Nature*, 1955, *175*, 555.

3642 Y. Yagi, R. Michel and J. Roche, *Ann. pharm. franç.*, 1953, *11*, 30: *C.A.*, 1954, *48*, 2005-b.

3643 K. Yamaguchi, *J. Pharm. Soc. Japan*, 1953, *73*, 1285: *C.A.*, 1954, *48*, 3200-h.

3644 M. Yamaguchi and F. D. Howard, *Anal. Chem.*, 1955, *27*, 332.

3645 Y. Yamamoto, A. Nakahara and R. Tsuchida, *J. Chem. Soc. Japan, Pure Chem. Sect.*, 1954, *75*, 232: *C.A.*, 1954, *48*, 11243.

3646 J. T. Yang, *Anal. Chim. Acta*, 1950, *4*, 59.

3647 J. T. Yang and M. Haïssinsky, *Bull. soc. chim. France*, *1949*, 546.

3648 C. Yanovsky, E. Wassermann and D. M. Bonner, *Science*, 1950, *111*, 61.

3649 S. Yasunaga and O. Shimomura, *J. Pharm. Soc. Japan*, 1953, *73*, 1346 and 1350: *C.A.*, 1954, *48*, 3188-g.

3650 S. Yasunaga and O. Shimomura, *J. Pharm. Soc. Japan*, 1953, *73*, 1353: *C.A.*, 1953, *47*, 3188-i.

3651 A. Yoda, *J. Chem. Soc. Japan, Pure Chem. Sect.*, 1952, *73*, 18: *C.A.*, 1953, *47*, 3185-c.

3652 Y. Yoshino, *Bull. Chem. Soc. Japan*, 1953, *26*, 401: *C.A.*, 1954, *48*, 6907-g.

3653 Y. Yoshino, *Japan Analyst*, 1954, *3*, 121: *C.A.*, 1954, *48*, 9867-d.

3654 Y. Yoshino and M. Kojima, *Bull. Chem. Soc. Japan*, 1950, *23*, 46.

3654a Y. Yoshino and M. Kojima, *Japan Analyst*, 1955, *4*, 311.

3655 N. A. Yudaev, *Doklady Akad. Nauk S.S.S.R.*, 1949, *68*, 119: *C.A.*, 1950, *44*, 686-e.

3656 R. M. Zacharius, J. F. Thompson and F. C. Steward, *J. Am. Chem. Soc.*, 1952, *74*, 2949.

3657 A. Zaffaroni and R. B. Burton, *J. Biol. Chem.*, 1951, *193*, 749.

3658 A. Zaffaroni, R. B. Burton and E. H. Keutmann, *J. Biol. Chem.* 1949, *177*, 109; *Science*, 1950, *111*, 6.

3659 S. E. Zager and T. C. Doody, *Ind. Eng. Chem.*, 1951, *43*, 1070.

3660 H. Zahn, *Textil-Praxis*, 1951, *6*, 127: *C.A.*, 1951, *45*, 8251-c.

3661 R. K. Zahn, W. Stamm and H. M. Rauen, *Angew. Chem.*, 1951, *63*, 280.

3662 R. J. Zahner and W. B. Swann, *Anal. Chem.*, 1951, *23*, 1093.

3663 S. F. Zakrzewski and C. A. Nichol, *J. Biol. Chem.*, 1953, *205*, 361.

3664 M. Zalokar, *J. Am. Chem. Soc.*, 1952, *74*, 4213.

3665 S. Zamenhof and E. Chargaff, *Nature*, 1951, *168*, 604.

3666 M. Zaoral, *Chem. Listy*, 1953, *47*, 1872: *C.A.*, 1955, *49*, 966-i.

3667 M. N. Zaprometov, *Doklady Akad. Nauk S.S.S.R.*, 1952, *87*, 649: *C.A.*, 1954, *48*, 153-d.

3668 M. N. Zaprometov et G. A. Soboleva, *Doklady Akad. Nauk S.S.S.R.*, 1954, *96*, 1205: *C.A.*, 1954, *48*, 14034-i.

3669   K. Zarudnaya, *Univ. Microfilms Pub.*, No. 2054: *C.A.*, 1951, *45*, 7201-f.
3670   L. J. Zatman, N. O. Kaplan and S. P. Colowick, *J. Biol. Chem.*, 1953, *200*, 197.
3671   V. Zbinovsky, *Anal. Chem.*, 1955, *27*, 764.
3672   O. Zbudovská and I. M. Hais, *Chem. Listy*, 1952, *46*, 307: *C.A.*, 1953, *47*, 270-g.
3673   L. Zechmeister, *Progress in Chromatography, 1938–1947*, 368 pp., Chapman and Hall, London, 1950.
3674   L. Zechmeister, *Chem. Revs.*, 1944, *34*, 267.
3675   L. Zechmeister, *Isis*, 1946, *36*, No. 104, 108: *C.A.*, 1946, *40*, 2703-4.
3676   L. Zechmeister, *Ann. N.Y. Acad. Sci.*, 1948, *49*, 220.
3677   L. Zechmeister, *Nature,* 1951, *167*, 405.
3678   L. Zechmeister, *Science*, 1951, *113*, 35.
3679   L. Zechmeister and L. von Cholnoky, *Principles and Practice of Chromatography* Chapman and Hall, London; J. Wiley and Sons, New York, 2nd Ed., 1943.
3680   L. Zechmeister, L. von Cholnoky and E. Ujhelyi, *Bull. soc. chim. biol.*, 1936, *18*, 1885.
3681   L. Zechmeister and R. B. Escue, *J. Biol. Chem.*, 1942, *144*, 321.
3682   L. Zechmeister and A. L. LeRosen, *J. Am. Chem. Soc.*, 1942, *64*, 2755.
3683   L. Zechmeister and W. H. McNeely, *J. Am. Chem. Soc.*, 1942, *64*, 1919.
3684   L. Zechmeister, W. H. McNeely and G. Sólyom, *J. Am. Chem. Soc.*, 1942, *64*, 1922.
3685   L. Zechmeister and J. H. Pinckard, *Experientia*, 1953, *9*, 16.
3686   L. Zechmeister and A. Polgár, *Science*, 1944, *100*, 317.
3687   L. Zechmeister and M. Rohdewald, *Enzymologia*, 1949, *13*, 388: *C.A.*, 1950, *44*, 6465-d.
3688   L. Zechmeister and M. Rohdewald, *Fortschr. Chem. Org. Naturstoffe*, 1951, *8*, 341: *C.A.*, 1952, *46*, 1078-i.
3689   L. Zechmeister and A. Sandoval, *Science*, 1945, *101*, 585.
3690   L. Zechmeister and W. A. Schroeder, *J. Biol. Chem.*, 1942, *144*, 315.
3691   L. Zechmeister and W. A. Schroeder, *J. Am. Chem. Soc.*, 1942, *64*, 1173.
3692   L. Zechmeister and G. Tóth, *Enzymologia*, 1939, *7*, 165: *C.A.*, 1940, *34*, 1340-g.
3693   L. Zechmeister, G. Tóth and P. Fürth, *Enzymologia*, 1941, *9*, 155: *C.A.*, 1942, *36*, 790-1.
3694   L. Zechmeister, G. Tóth and E. Vajda, *Enzymologia*, 1939, *7*, 170: *C.A.*, 1940, *34*, 1340-8.
3695   K. Zeile and M. Oetzel, *Z. physiol. Chem.*, 1949, *284*, 1.
3696   A. A. Zhukhovitskiĭ, O. V. Zolotareva, V. A. Sokolov and N. M. Turkel'taub, *Doklady Akad. Nauk S.S.S.R.*, 1951, *77*, 435: *C.A.*, 1952, *46*, 11011-b.
3697   L. P. Zill, J. X. Khym and G. M. Cheniae, *J. Am. Chem. Soc.*, 1953, *75*, 1339.
3698   F. Zilliken, G. A. Braun, and P. György, *Arch. Biochem. Biophys.*, 1955, *54*, 564.
3699   G. Zimmermann, *Z. anal. Chem.*, 1953, *138*, 321.
3700   G. Zimmermann, *Naturwissenschaften*, 1955, *42*, 257.
3701   G. Zimmermann and K. H. Kludas, *Chem. Tech.*, 1953, *5*, 203: *C.A.*, 1953, *47*, 10048-e.
3702   G. Zimmermann and K. Nehring, *Angew. Chem.*, 1951, *63*, 556.
3703   V. L. Zolotavia, *Sbornik Statei Obshchei Khim. Nauk S.S.S.R.*, 1953, *1*, 34: *C.A.*, 1954, *48*, 12508-b.
3704   G. Zwingelstein, H. Pachéco and J. Jouanneteau, *Compt. rend.*, 1953, *236*, 1561.

APPENDIX AND INDEXES

APPENDIX

Newer references
on gas-liquid chromatography

GENERAL ARTICLES

Gas-liquid partition chromatography. D. H. Lichtenfels, S. A. Fleck and F. H. Burow, *Anal. Chem.*, 1955, *27*, 1510.
Gas-liquid partition chromatography. G. Dijkstra, J. G. Keppler and J. A. Scholz, *Rec. trav. chim.*, 1955, *74*, 805.
Symposium on gas chromatography. *Analyst*, 1956, *81*, 52.
Vapor-phase chromatography. W. J. Podbielniak and S. T. Preston, *Petroleum refiner*, 1956, *35*, 215: *C.A.*, 1956, *50*, 8368-f.

PHYSICO-CHEMICAL STUDIES

The selectivity of the stationary liquid in vapor-phase chromatography. A. I. M. Keulemans, A. Kwantes and P. Zaal, *Anal. Chim. Acta*, 1955, *13*, 357.
A consideration of factors governing the separation of substances by gas-liquid partition chromatography. D. W. Grant and G. A. Vaughan, *J. Appl. Chem.* (*London*), 1956, *6*, 145.
Temperature effects in gas-liquid partition chromatography. M. R. Hoare and J. H. Purnell, *Trans. Faraday Soc.*, 1956, *52*, 222.
Application of gas-liquid partition chromatography to problems in chemical kinetics. A. B. Callear and R. J. Cvetanović, *Can. J. Chem.*, 1955, *33*, 1256: *C.A.*, 1955, *49*, 14421-i.
New microcatalytic gas-chromatographic technique for studying catalytic reactions. R. J. Kokes, H. Tobin and P. H. Emmett, *J. Am. Chem. Soc.*, 1955, *77*, 5860.
Solvent effects in gas-liquid partition chromatography. G. J. Pierotti, C. H. Deal, E. L. Derr and P. E. Porter, *J. Am. Chem. Soc.*, 1956, *78*, 2989.
The determination of partition coefficients from gas-liquid partition chromatography. P. E. Porter, C. H. Deal and F. H. Stross, *J. Am. Chem. Soc.*, 1956, *78*, 2999.
Elution time and resolution in vapor chromatography. A. K. Wiebe, *J. Phys. Chem.*, 1956, *60*, 685.
Molecular weight determination of components by gas-phase chromatography. A. Liberti, L. Conti and V. Crescenzi, *Nature*, 1956, *178*, 1067.
Partition coefficients from gas-liquid partition chromatography. J. R. Anderson, *J. Am. Chem. Soc.*, 1956, *78*, 5692.

TECHNICAL DETAILS

The gas-chromatographic separation on a preparative scale. D. E. M. Evans and J. C. Tatlow, *J. Chem. Soc.*, *1955*, 1184.

The application of vapour-phase chromatography to the preparation of pure materials. F. H. Pollard and C. J. Hardy, *Chem. and Ind.*, *1956*, 527.

Use of gas-liquid partition chromatography as a preparative method. D. Ambrose and R. R. Collerson, *Nature*, 1956, *177*, 84.

Apparatus requirements for quantitative application of gas-liquid partition chromatography. M. Dimbat, P. E. Porter and F. H. Stross, *Anal. Chem.*, 1956, *28*, 290.

The use of helium as the mobile phase in gas chromatography. W. A. Wiseman, *Chem. and Ind.*, *1956*, 127.

DETECTORS

A new detector for vapour phase partition chromatography. R. P. W. Scott, *Nature*, 1955, *176*, 793.

A radiological detector for gas chromatography. C. H. Deal, J. W. Otvos, V. N. Smith and P. S. Zucco, *Anal. Chem.*, 1956, *28*, 1958.

The use of thermistor detectors in gas chromatography. A. D. Davis and G. A. Howard, *Chem. and Ind.*, *1956*, R 25.

A new detector for vapour phase chromatography. J. Harley and V. Pretorius, *Nature*, 1956, *178*, 1244.

Vapour phase chromatographic determination of benzene, naphthalene and other hydrocarbons in wash oil. Application of a simple and accurate detection method for vapour phase chromatography. L. Blom and L. Edelhausen, *Anal. Chim. Acta*, 1956, *15*, 559.

The micro-flame detector in gas-liquid partition chromatography: correlation of response with heats of combustion. J. I. Henderson and J. H. Knox, *J. Chem. Soc.*, *1956*, 2299.

Gas-liquid chromatography: the gas-density meter, a new apparatus for the detection of vapours in flowing gas streams. A. J. P. Martin and A. T. James, *Biochem. J.*, 1956, *63*, 138.

Discussion on katharometers as recorders in gas chromatography. *Analyst*, 1956, *81*, 57.

APPLICATIONS

Hydrocarbons

Analysis of gaseous hydrocarbons by gas-liquid partition chromatography. E. M. Fredericks and F. R. Brooks, *Anal. Chem.*, 1956, *28*, 297.

Gas-liquid chromatography-separation of hydrocarbons using various stationary phases. V. T. Brooks and G. A. Collins, *Chem. and Ind.*, *1956*, 921.

Use of liquid-modified solid adsorbent to resolve C_5 and C_6 saturates. F. T. Eggersten, H. S. Knight and S. Groennings, *Anal. Chem.*, 1956, *28*, 303.

Retention volumes of isomeric hexenes and hexanes in gas-liquid partition chromatography using phthalate esters as liquid phase. J. J. Sullivan, J. R. Lotz and C. B. Willingham, *Anal. Chem.*, 1956, *28*, 495.

The separation and identification of some volatile paraffinic, naphthenic, olefinic and aromatic hydrocarbons. A. T. James and A. J. P. Martin, *J. Appl. Chem. (London)*, 1956, *6*, 105.

Gas chromatographic determination of some hydrocarbons in cigarette smoke. H. W. Patton and G. P. Touey, *Anal. Chem.*, 1956, *28*, 1685.

Acids

Gas-liquid chromatography: the separation and identification of the methyl esters of saturated and unsaturated acids from formic acid to *n*-octadecanoic acid. A. T. James and A. J. P. Martin, *Biochem. J.*, 1956, *63*, 144.

Studies of sebum. 6. The determination of the component fatty acids of human forearm sebum by gas-liquid chromatography. A. T. James and V. R. Wheatley, *Biochem. J.*, 1956, *63*, 269.

The volatile acids of mutton fat. A. G. McInnes, R. P. Hansen and A. S. Jessop, *Biochem. J.*, 1956, *63*, 702.

Amines

Gas-liquid chromatography. Separation and microestimation of volatile aromatic amines. A. T. James, *Anal. Chem.*, 1956, *28*, 1564.

Separation of pyridine bases by vapour phase chromatography. V. T. Brooks and G. A. Collins, *Chem. and Ind.*, *1956*, 1021.

The determination of small amounts of γ-picoline in aqueous solutions of β-picoline by vapour phase chromatography. W. J. Murray and A. F. Williams, *Chem. and Ind.*, *1956*, 1020.

Miscellaneous

Application of gas phase partition chromatography to competitive chlorination reactions. J. H. Knox, *Chem. and Ind.*, *1955*, 1631.

Separation of mercaptans by gas-liquid partition chromatography. S. Sunner, K. J. Karrman and V. Sundén, *Mikrochim. Acta*, *1956*, 1144.

Determination of amino acids by ninhydrin oxidation and gas chromatography. Separation of leucine and iso-leucine. I. R. Hunter, K. P. Dimick and J. W. Corse, *Chem. and Ind.*, *1956*, 294.

Use of gas-phase chromatography for the separation of mixtures of carrier free radioactive substances: products of chemical reactions activated by nuclear processes. J. B. Evans and J. E. Willard, *J. Am. Chem. Soc.*, 1956, *78*, 2908.

Author Index

Löfgren, N., 363, 369, 586
Löhr, K., 169, 586
Lombardo, M. E., 277, 586
Long, A. G., 62, 138, 160, 161, 196, 198, 262, 368, 518, 546, 557, 586
Longenecker, W. H., 130, 131, 221, 586
Loo, Y. H., 414, 548
Lorenz, F., 430, 569
Lorenz, I., 401, 614
Loriers, J., 459, 460, 461, 573, 587, 619
Loring, H. S., 370, 587
Lotz, J. R., 632
Loury, M., 171, 271, 587
Löw, B., 361, 557
Lowman, A., 13, 587
Lowther, A. G., 331, 542, 560
Lowy, P. H., 306, 337, 543
Lubschez, R., 291, 590
Lucas, J., 386, 587
Luckwill, L. C., 409, 587
Lüderitz, O., 136, 587
Ludowieg, J. 263, 604
Lugg, J. W. H., 121, 191, 192, 322, 587
Lukovnikov, A. F., 170, 594
Lund, N. A., 99, 587
Lundberg, W. O., 174, 578
Lundén, L., 473, 606
Lure, Yu. Yu., 472, 587
Lussman, D. J., 212, 587
Lutwak, L., 370, 605
Lynam, C. G., 17, 18, 443, 571, 587
Lynch, V., 253, 540
Lyons, F., 80, 556
Lythgoe, B., 56, 537

Ma, R., 131, 587
Mačák, V., 212, 587
Macbeth, A. K., 421, 587
McCarthy, J. L., 11, 262, 593, 598
McCauley, D. J., 81, 602
Macchi, M. E., 400, 597
McCloskey, C. M., 238, 550
McComb, E. A., 258, 587
McCormick, H., 18, 587
MacCorquodale, D. W., 72, 395, 541
McCready, R. M., 225, 258, 587
McCrone, W. C., 442, 443, 582
McCullough, W. G., 337, 578
McCutcheon, T. P., 473, 538
McDonald, H. J., 11, 12, 358, 587

McDonald, I. W., 307, 363, 587
Macdonald, J., 211, 560
Macdonald, P. J., 88, 577
McDonough, S., 282, 587
Macek, K., v, 103, 212, 259, 357, 567, 587
McFarren, E. F., 121, 242, 256, 258, 318, 331, 587, 603
McGarr, J. J., 334, 348, 581
McGeown, M. G., 260, 587
MacGillivray, R., 295, 569
McGilvray, D. I., 261, 571
Macheboeuf, M., 11, 103, 115, 122, 127, 133, 163, 191, 192, 203, 210, 211, 212, 214, 259, 325, 367, 536, 542, 549, 587, 594, 605
McInnes, A. G., 633
McIntire, F. C., 201, 587
McIntyre, R. T., 527, 540
McKerns, K. W., 359, 587
McKibbin, J. M., 272, 588
Mackinney, G., 379, 588
Maclagan, N. F., 408, 588
McLane, C. K., 461, 588
McMahon, J. M., 393, 553, 588
McNeely, W. H., 55, 239, 425, 588, 628
McNeil, R., 18, 591
MacNevin, W. M., 471, 588
McOmie, J. F. W., 122, 429, 477, 478, 480, 481, 484, 485, 488, 492, 493, 495, 496, 498, 499, 501, 502, 507, 508, 509, 518, 519, 520, 527, 557, 599
Macpherson, H. T., 309, 318, 577
Macpherson, L. B., 271, 588
McPherson, S. P., 156, 626
McQuarrie, E. B., 352, 588
McSwiney, R. R., 386, 588
Macturk, H. M., 62, 560
MacWilliam, I. C., 253, 568
Maddock, J. G., 520, 599
Madecka-Borkowska, I., 262, 596
Mader, C., 17, 43, 588
Mader, G., Jr., 17, 588
Madinaveitia, J., 62, 560
Maehly, A. C., 241, 588
Magasanik, B., 194, 195, 363, 373, 549, 588, 620
Magee, R. J., 499, 501, 591
Magerlein, B., 216, 567
Magnes, J., 335, 541

Ruzicka, L., 62, 155, 159, 174, 176, 274, 277, 426, 546, 568, 574, 599, 600, 605, 616
Ryabchikov, D. I., 470, 474, 605
Ryan, W., 523, 605
Rybář, D., 221, 605
Rydel, S., 163, 605
Ryder, A., 99, 417, 562
Rydon, H. N., 108, 182, 265, 304, 309, 314, 322, 339, 567, 585, 605
Ryle, A. P., 345, 605

Sacconi, L., 429, 430, 431, 433, 605
Sacks, J., 370, 605
Saenz-Lascaño-Ruiz, I., 49, 232, 233, 605
Sahama, Th. G., 432, 558
Saïas, E., 206, 266, 574, 598
Saidel, H. F., 362, 549
Saier, E. L., 152, 605
Saifer, A., 136, 309, 311, 596, 605
Saini, G., 432, 621
Saito, N., 470, 578
Sakaguchi, T., 501, 605
Sakal, E. H., 282, 605
Sakamoto, T., 501, 506, 511, 568
Sakan, T., 422, 580
Saklawska-Szymonowa, O., 191, 192, 596
Salmon, J. E., 468, 470, 563, 605
Salomon, H., 395, 576
Salooja, K. C., 530, 546
Salvesen, B., 367, 624
Sampson, P., 243, 623
Samuels, L. T., 274, 277, 594, 605
Samuelson, O., 71, 97, 168, 176, 243, 470, 472, 473, 541, 555, 562, 568, 582, 605, 606
San, G. L., 221, 606
Sanadi, D. R., 372, 554
Sancier, K. M., 388, 561, 613
Sandberg, A. A., 282, 547
Sanders, T. H., 227, 556
Sandoval, A., 66, 382, 425, 606, 628
Sandstrom, W. M., 166, 549
Sanfilippo, S. J., 274, 602
Sanger, F., 289, 316, 330, 333, 340, 345, 348, 546, 600, 605, 606, 618
Sannié, C., 287, 606
San Pietro, A., 377, 547

Sansone, G., 358, 359, 606
Sansoni, B., 126, 606
Šantavý, F., 212, 587
Sapozhnikov, D. I., 388, 606
Sardinas, J. L., 282, 610
Sarett, L. H., 64, 606
Sarma, B., 509, 511, 606
Sarma, P. S., 395, 601
Saroja, K., 260, 563, 564
Sasaki, Y., 470, 571
Sassenrath, E. N., 207, 599
Satake, K., 119, 187, 188, 606
Sato, T., 435, 573
Satoh, K., 199, 588
Sattler, L., 257, 606
Sauberlich, H. E., 400, 606
Saunders, L., 209, 606
Savard, K., 279, 282, 284, 285, 286, 606
Savary, P., 88, 177, 181, 190, 270, 606, 607
Savoia, F., 138, 607
Scarisbrick, R., 178, 593, 607
Schaefer, W. C., 259, 554
Schäfer, H., 430, 607
Schall, E. D., 16, 556
Schapira, G., 359, 581
Schauer, H. K., 117, 118, 607
Schayer, R. W., 145, 206, 406, 607
Scheeline, H. W., 150, 607
Scheler, R., 233, 617
Schenck, J. R., 201, 587
Schenk, W., 435, 538
Schenker, E., 267, 607
Schenker, H. H., 176, 607
Schilling, J. A., 326, 554
Schindler, O., 21, 65, 265, 266, 267, 278, 279, 384, 400, 538, 570, 607, 609, 611
Schinz, H., 155, 609
Schlabach, T. D., 22, 392, 558, 607
Schleede, D., 164, 607
Schleith, L., 98, 421, 566
Schlögl, K., 129, 158, 166, 217, 341, 607, 610
Schlubach, H. H., 244, 352, 607
Schlueter, D. P., 477, 615
Schlüssel, H., 323, 607
Schmäh, H., 434, 561
Schmall, M., 399, 626
Schmeiser, K., 144, 217, 220, 607, 625
Schmid, H., 187, 209, 210, 577, 607

Subject Index